Commercial Flowers

— Volume 4 —

THE EDITORS

Professor T.K. Bose obtained his M.Sc. (Ag.) and Ph.D. degree from Calcutta University, West Bengal, India. Trained at the Royal Botanic Garden, Kew, UK. Former Secretary of the Agri-Horticultural Society of India; Project Coordinator, All India Coordinated Floriculture Improvement Project (ICAR); Professor and Head, Department of Horticulture; Dean, Faculty of Agriculture; Director of Research, Bidhan Chandra Krishi Viswavidyalaya, West Bengal; Member, State Agriculture Commission, West Bengal; Member of Scientific Advisory Committees of ICAR, CSIR (Govt. of India) and Government of West Bengal. Awarded (D.Sc.) Honoris Causa from OUAT, Bhubaneshwar and UBKV, Coochbehar, India. Guided 45 Ph.D. students, published more than 250 research papers and authored 30 books on horticultural crops.

Dr. Chenna Reddy Aswath obtained his B.Sc. (Hort), M.Sc. (Hort.) in Floriculture from University of Agricultural Sciences, Bengaluru and Ph.D. degree in Biotechnology from Guwhati University, Assam. Serving as Principal Scientist in ICAR-IIHR, Bengaluru and has made significant contributions in the field of Floriculture and Biotechnology. Recipient of "Dr BP Pal Best Scientist Award for Agriculture" from NABS, Chennai, became FELLOW of the NABS, Chennai, ISOH, New Delhi and SAB, Dharwad, CHAI, New Delhi. Received the 2012 SAB Award of Excellence in Horticultural Biotechnology. Recipient of Brain pool Post-Doctoral Fellowship from South Korea Science Federation for the outstanding research contributions in Horticultural Biotechnology. His short term fellowship in Japan under INSA has recognized him among the Orchid tissue culture group. Published more than 110 research papers besides several books. Guided more than 10 post-graduate students.

Dr. Reeta Bhatia is working as senior scientist at the Division of Floriculture and Landscaping, ICAR- Indian Agricultural Research Institute, New Delhi. She has more than 13 years of research and teaching experience in the field of floriculture. She has obtained her Master's degree from Punjab Agricultural University, Ludhiana and doctoral degree from Indian Agricultural Research Institute, New Delhi. She has been awarded with the prestigious Endeavor Research Fellowship by the Government of Australia and pursued her postdoctoral research at the University of Western Australia. She has an excellent academic and research career and awarded with several medals. She has published more than 65 research papers in the reputed high impact scientific journals and written more than 30 review/popular/scientific articles. She is also serving as editor for the Journal of Ornamental Horticulture.

Dr. Tarak Nath Saha graduated in Agriculture from Banaras Hindu University, Varanasi (Uttar Pradesh), obtained his Masters and Doctoral degree in Horticulture with specialization in Floriculture from India Agricultural Research Institute (IARI), New Delhi. Presently working as Senior Scientist at ICAR-Directorate of Floricultural Research, Pune. He has an experience of more than 12 year in research, teaching and extension, besides coordinating work of All India Coordinated Research Project (AICRP) on Floriculture. He has contributed 25 research papers, 41 technical articles, 5 book chapters, 20 technical bulletins, 6 folders, besides in the editorial team of 30 Annual Reports and 10 AICRP workshop proceedings. Developed one variety and registered one genetic stocks of tuberose with ICAR-NBPGR, New Delhi. He is actively involved in PG teaching; recognized as faculty of MPKV, Rahuri (also at IARI, New Delhi 2009-11) and guided more than 5 students in Floriculture as advisory.

Dr. Rajiv Kumar obtained his B.Sc. (Ag and AH) and M.Sc. (Ag) Horticulture from C.S. Azad University of Agriculture and Technology, Kanpur, Uttar Pradesh, India and completed his Ph.D. (Horticulture) from I.A.R.I., New Delhi. Presently working as Principal Scientist at ICAR-IIHR, Bengaluru. Recipient of Young Scientist Associate Award by HTHS, U.P.; Young Scientist Award by SSDAT, U.P.; Gold Medal by HTHS, Meerut, and BR and IC Pvt. Ltd.; Fellow (Horticulture) by HAHS, Muzaffarnagar, U.P., AABS, Karnakata, and ISOH, New Delhi. Published 110 research articles, 25 popular articles, 6 edited books, 20 book chapters, 8 technical bulletins and 6 extension folders. Developed two varieties of China aster, associated in five and three varieties of Gerbera and gladiolus, respectively; registered 17 genetic stocks of China aster, Gerbera, Chrysanthemum and Gladiolus with ICAR-NBPGR, New Delhi. Guided two M.Sc. (Hort.) and two Ph.D. (Hort.) students in Floriculture.

Dr. Koushik Dutta obtained his M.Sc. (Hort.) from CCS Haryana Agricultural University, Hisar, Haryana and Ph.D. from Uttar Banga Krishi Viswavidyalaya, West Bengal. Acting as Assistant Secretary of The Agri-Horticultural Society of India, Kolkata, West Bengal. Recipient of Inspire Fellowship from Department of Science and Technology, New Delhi. Published several scientific papers in national journals on different aspect of horticulture particularly in tissue culture and written book chapters on nursery management and jasmine. Associated with micropropagation of different of horticultural crops during last 7 years.

Commercial Flowers

— *Volume 4* —

— Senior Editor —

C. Aswath
Division of Flower and Medicinal Crops
ICAR-Indian Institute of Horticultural Research
Bengaluru 560089, Karnataka, India

— Editors —

T.K. Bose
Faculty of Horticulture
Bidhan Chandra Krishi Viswavidyalaya
West Bengal 741 252, India

Reeta Bhatia
Division of Floriculture and Landscaping,
ICAR- Indian Agricultural
Research Institute
New Delhi 110012

Tarak Nath Saha
ICAR-Directorate of
Floricultural Research,
Pune, Maharashtra, India
Bengaluru 560089, Karnataka, India

Rajiv Kumar
Division of Flower and Medicinal
Crops, ICAR-Indian Institute of
Horticultural Research
Bengaluru 560089, Karnataka, India

Koushik Dutta
The Agri-Horticultural Society of India
1, Alipore Road,
Kolkata 700027
West Bengal, India

2022

Daya Publishing House®
A Division of
Astral International Pvt. Ltd.
New Delhi – 110002

Third Revised & Illustrated edition 2021

Cataloging in Publication Data--DK

Courtesy: D.K. Agencies (P) Ltd. <docinfo@dkagencies.com>

Commercial flowers / senior editor, C. Aswath ; editors, T.K. Bose [and four others]. -- Third revised edition.

Volumes cm

Contributed articles.

Includes index.

ISBN 978-93-90435-45-6 (HB)

1. Floriculture. 2. Flowers--Pictorial works. I. Aswath, C. (Principal scientist at the Indian Institute of Horticulture Research), editor. II. Bose, T. K., editor.

LCC SB405.C66 2021 | DDC 635.9 23

Published by : **Daya Publishing House®**
A Division of
Astral International Pvt. Ltd.
– ISO 9001:2015 Certified Company –
4736/23, Ansari Road, Darya Ganj
New Delhi-110 002
Ph. 011-43549197, 23278134
E-mail: info@astralint.com
Website: www.astralint.com

Digitally Printed at : **Replika Press Pvt. Ltd.**

From the Desk of Prof. T.K. Bose

The undersigned had initiated the publications entitled 'Fruits: Tropical and Sub-tropical', 'Vegetable Crops' and 'Commercial Flowers', one volume each in the 1980's in collaboration with co-editors, humbly expresses from the core of his heart that the updated revised editions of the above mentioned publications with colour illustrations in thirteen volumes which was planned in June, 2020, are being printed from January this year, a rare event in Horticultural Science has been possible by the blessing of Almighty God, with the cooperation of dedicated horticulturists as editors and contributors.

T.K. Bose
Editor Emeritus

Preface to the Third Edition

The first edition of the book entitled **'Commercial Flowers'** published in 1989, reprinted in 1993 and 1998 was greatly appreciated and proved valuable scientific review of commercially important flowers, to the students, teachers, researchers and extension specialists in horticulture in general and floriculture in particular, in different parts of the world. The book was revised in 2002 with the same title in two volumes due to its increasing demand, which contained updated scientific information of commercially grown flower crops and considered as valuable contribution on the subject. After about 18 years from the last revised publication in 2002, voluminous information pertaining to crop improvement, production, protection, postharvest management, biotechnology has been generated necessitating further revision of the existing chapters and inclusion of more ornamentals plants of commercial importance in four volumes, with relevant colour illustrations.

Fourth volume in this edition includes chapters on Anthurium, Gerbera, Antirrhinum, Petunia, Ranunculus, Marigold, China aster, Zinnia, Lupine, Salvia, Stock, Celosia, Gypsophila, Rudbeckia, Calendula, Gaillardia, Clarkia and Schizanthus with relevant colour illustrations, covering more than 600 pages.

Apart from own collection of photographs, editors have also taken some photographs which are in public domain from different websites, as far as possible. Therefore, we acknowledge with enormous thanks to all State Agricultural Universities and Research Institutes under Indian Council of Agricultural Research, New Delhi, India; Public and Private International Agricultural Institutes across the world for using photographs from their websites to make this book a valuable source of knowledge for the readers. We think images can be powerful teaching tools, as illustrations to in-class lectures, or for studying concepts outside of the classroom. We also think that identification of flower crops, their importance and uses, species and cultivars, methods of propagation, cultivation practices including greenhouse cultivation, diseases and pests, harvesting and postharvest technology and achievement of biotechnology, with the help of suitable illustrations in a consolidated manner will make a dent for the students, researchers and teachers of floriculture and ornamental horticulture.

We are grateful to Dr. Arup Chattopadhayay, Professor, Bidhan Chandra Krishi Vishwavidyalaya, West Bengal for his enormous support in preparing the revised illustrated manuscripts. We sincerely thank the contributors for preparing updated revised manuscripts. Our thanks are due to Shri Prateek Mittal, Director, Astral International (P) Ltd., New Delhi, for his keen interest in bringing out this publication.

Editors

Kolkata

Contents

List of Contributors

Arora, J.S.

Department of Vegetable Crops and Landscaping, Punjab Agricultural University, Ludhiana 141004, Punjab

Aswath, C.

Division of Flower and Medicinal Crops, ICAR-Indian Institute of Horticultural Research, Bengaluru 560089, Karnataka

Ayesha

ICAR-Indian Institute of Horticultural Research Bengaluru 560089, Karnataka

Bal, S.

Department of Vegetable Science, Faculty of Horticulture, Bidhan Chandra Krishi Viswavidyalaya, Mohanpur Nadia 741252, West Bengal

Bennurmath, P.

Division of Flower and Medicinal Crops, ICAR-Indian Institute of Horticultural Research, Bengaluru 560089, Karnataka

Bhatt, N.R.

Department of Horticulure, Mahatma Phule Krishi Vidyapeeth, Rahuri 413722, Maharashtra

Bhatia, R.

Division of Floriculture and Landscaping, ICAR-Indian Agricultural Research Institute, New Delhi 110012

Bhatia, S.

Division of Floriculture and Landscape Architecture, Dr. Y.S.P. University of Horticulture and Forestry, Nauni, Solan 173230, Himachal Pradesh

Biswas, B.

Department of Horticulture, Bidhan Chandra Krishi Vishwavidyalaya, Mohanpur 741252, West Bengal

Bose, T.K.

Department of Horticulture, Bidhan Chandra Krishi Vishwavidyalaya, Mohanpur 741252, West Bengal

Chatterjee, S.

Department of Horticulture, Centurion University of Technology and Management, Paralakhemundi 761211, Odisha

Chawla, S.L.

Navsari Agricultural University, Navsari 396450, Gujarat

Das, P.

Regional Plant Resource Centre, Bhubaneswar 751015, Odisha

Desai, B.B.

Department of Horticulture, Mahatma Phule Krishi Vidyapeeth, Rahuri 413722, Maharashtra

Dhuna, R.S.

Bidhan Chandra Krishi Vishwavidyalaya, Mohanpur 741252, West Bengal

Dutta, K.

The Agri-Horticultural Society of India, 1 Alipore Road, Kolkata 700027, West Bengal

Ganga, M.

Department of Floriculture and Landscape Architecture, Horticultural College and Research Institute, Tamil Nadu Agricultural University, Coimbatore 641003, Tamil Nadu

Gurung, A.

Division of Floriculture and Medicinal Crops, ICAR-Indian Institute of Horticultural Research, Bengaluru 560089, Karnataka

Jana, B.K.

Department of Horticulture, Bidhan Chandra Krishi Vishwavidyalaya, Mohanpur 741252, West Bengal

Jyoti, C.

ICAR-Indian Institute of Horticultural Research, Bengaluru 560 089, Karnataka

Jyothi, R.

University of Agricultural Sciences, Raichur, Karnataka

Karjee, S.

ICAR-Indian Institute of Horticultural Research, Bengaluru 560089, Karnataka

Kumar, R.

Division of Flower and Medicinal Crops, ICAR-Indian Institute of Horticultural Research, Bengaluru 560089, Karnataka

Kumari, S.

Dr. Kalam Agricultural College, Bihar Agricultural University, Kishanganj 855107, Bihar

Madhubala

Department of Floriculture and Landscaping, Punjab Agricultural University, Ludhiana 141004, Punjab

Mitra, M.

Department of Horticulture, Bidhan Chandra Krishi Vishwavidyalaya, Mohanpur 741252, West Bengal

Mukherjee, D.

Department of Vegetable Science, Faculty of Horticulture, Bidhan Chandra Krishi Viswavidyalaya, Mohanpur 741252, West Bengal

Namita

Division of Floriculture and Landscaping, ICAR-Indian Agricultural Research Institute, New Delhi 110012

Pal, P.

Department of Horticulture, Bidhan Chandra Krishi Vishwavidyalaya, Mohanpur 741252, West Bengal

Panwar, S.

Division of Floriculture and Landscaping, ICAR-Indian Agricultural Research Institute, New Delhi 110012

Parthasarathy, V.A.

ICAR-Indian Institute of Spices Research, 32/482 C Narmada Nilayam, Bharathan Bazar Chelavoor, Calicut 673571, Kerala

Poulose, B.

Division of Floriculture and Landscaping, ICAR-Indian Agricultural Research Institute, New Delhi 110012

Rihne, T.

Division of Floriculture and Landscaping, ICAR-Indian Agricultural Research Institute, New Delhi 110012

Safeena

Division of Flower and Medicinal Crops, ICAR-Indian Institute of Horticultural Research, Bengaluru 560089, Karnataka

Saha, S.

ICAR-Indian Agricultural Research Institute, Regional Station Pune 411008, Maharashtra

Saha, T.N.

ICAR-Directorate of Floricultural Research, College of Agriculture Campus, Shivajinagar, Pune 411005, Maharashtra

Samanta, P.K.S.

Department of Horticulture, Orissa University of Agriculture and Technology, Bhubaneswar 751002, Odisha

Sangeetha, Priya S.

Division of Flower and Medicinal Crops, ICAR-Indian Institute of Horticultural Research, Bengaluru 560089, Karnataka

Sanyal, D.

Department of Crop Improvement Horticulture and Agricultural Botany, Institute of Agriculture, Viswa-Bharathi Sriniketan 731236, West Bengal

Shisa, Ullas P.

Division of Floriculture and Landscaping, ICAR-Indian Agricultural Research Institute, New Delhi 110012

Visalakshi, M.

Department of Floriculture and Landscape Architecture, Horticultural College and Research Institute, Tamil Nadu Agricultural University, Coimbatore 641003, Tamil Nadu

1

Anthurium

Parvathi Bennurmath, C. Aswath, Rajiv Kumar, K. Dutta, N.R. Bhatt, B.B. Desai and D. Sanyal

1.0 Introduction

Anthuriums are tropical plants of great beauty and grown either for the showy cut flowers or for their unusually attractive foliage. They are very popular cut flowers because of the bold effect and lasting qualities. These contribute to the elegance and attractiveness which are the pre-requisities for a quality design. Provided right greenhouse conditions they are easy enough to grow.

The name anthurium is derived from Greek *anthos,* flower, and *oura,* tail, referring to the spadix. These evergreen plants are native to Colombia, Peru, Central and South America, Brazil and Venezuela and have been grown in England since the early nineteenth century. Anthurium is one of the Hawaii's principal cut flower for export. The area under anthurium is fast increasing. It is also under commercial cultivation in Philippines, Poland, Mexico, Taiwan, Jamaica, Costa Rica, Trinidad and Dominican Republic. Germany is the highest consumer in European market. In 1994, imports stood at 1.5 million stems, up from 608,000 stems in 1988 (Singh, 1998). World trade in anthurium is second only to orchids among tropical flowers. In Asia in particular, its popularity is increasing. Though bacterial diseases have caused production in traditional producing countries to decline by 25 per cent since 1986, new production in other geographical locations is now contributing to market (Laws and Galinsky, 1996).

It is a tropical shade plant, does not thrive well under high light intensities and shade must be provided for its satisfactory growth and flowering (Hlatshwayo and Wahome, 2010). The important states cultivating anthuriums are Assam, Kerala, Tamil Nadu (Salem) and Karnataka (Coorg) where the favourable climate exists (Diaz *et al.,* 2010).

2.0 Importance and Uses

2.1.0 Aesthetic Uses

Anthuriums not only add an exotic look to arrangements but can also be used with all types of flowers and foliages. Hybrid anthuriums are popular for floral arrangements due to their wide range of colors (white to pale green to bright red and many shades of pink and purple, with spots, speckles and splashes). They can make spectacular house plants, as long as they are not exposed to direct sunlight and their soil is kept constantly moist. The extraordinary beauty of its flower, both on the plant and as cut, attracts the gaze and overcomes or even annihilates the fact that it is not fragrance. It is used as cut flower, specimen, pots and containers, indoor, outdoor besides in parks and gardens. The pot plants are useful for purifying the air from dangerous toxic substances, such as formaldehyde, ammonia, xylene and toluene, and this is another reason for its popularity as house plant. In addition, the flowers, of course, are suitable for decorating vases or potteries and as groundcover plant (www.kalliergeia.com).

Uses of Anthurium

In Garden

In Pot

In Greenhouse

In Flower Arrangement

In Bed

Around a Plant

In Pathway

In Flower Vase

In Front of a Residential Area

2.2.0 Therapeutic and other Uses

Anthurium steam can be used to ease joint pain during arthritis and rheumatism. As the person sweats and pores open, the medicinal properties enter the body. It is also useful as a poultice to aid muscle aches and cramps (www.ugaoo.com). The whole plant is toxic. If penetrated in mucous membrane causes painful irritations in mouth and throat and contact with skin causes erythema, blisters *etc.* (www.en.wikipedia.org).

3.0 Morphology and Classification

Anthuriums are perennial plants with creeping, climbing, assurgent or arborescent stems. Leaves variable, evergreen, net-veined, with a prominent mid-nerve and lateral nerves, and a well-defined nerve at or near the margin. The hermaphrodite flowers which are small and insignificant are densely packed on a cylindrical spadix subtended in large heart-shaped spathe. The spathe may either be flat or slightly undulated ending with a pointed tip. Spathe and in many cases spadix are brilliantly coloured ranging from scarlet, red, salmon, orange, pink to white. Ovary 2-celled with 1-2 ovules, fruits berry.

The genus *Anthurium* belongs to the family Araceae. It seems to fall naturally into two sections or groups, *viz.* foliage and flowering. Although most anthuriums flower, those of the foliage group have large handsome velvety leaves, may bear conspicuous or unattractive flowers, while those in the flowering group have remarkably showy spathes and spadices but less handsome foliage.

4.0 Species and Cultivars

4.1.0 Species

The genus *Anthurium* consists of some 500-600 known species, though there are probably not more than fifty in cultivation and perhaps not more than ten or fifteen known to the trade (Bailey, 1963). Some of the important cultivated species belonging to both flowering and foliage groups are mentioned below.

4.1.1 Flowering Group

Anthurium andraeanum (*A. Andreanum*), *A. bakeri, A. brownii, A. x ferrierense, A. ornatum, A. regale, A. regnellianum, A. robustum, A. scherzerianum.*

4.1.2 Foliage Group

A. clarinervium, A. corrugatum, A. crystallinum, A. holtonianum, A. leuconerum, A. magnificum, A. panduratum, A. papilionensis, A. splendidum, A. Veitchii and *A. warocqueanum.*

Among the various species in flower group, *A. andraeanum* and *A. scherzerianum* are cultivated extensively for the production of flowers. The characterisic features of these two species are given below:

***Anthurium andraeanum*: Oil Cloth Flower, Tail Flower, Painter's Palette**

An erect plant. Leaves oblong, heart-shaped, 20-35 cm long, 15-20 cm wide; spathe heart-shaped, lacquered reddish orange or scarlet, 10-15 cm long; spadix

Species of Anthurium

Anthurium andraeanum

Anthurium bakeri

Anthurium brownii

Anthurium ornatum

Anthurium regale

Anthurium scherzerianum

Anthurium clarinervium

Anthurium splendidum

Anthurium veitchii

yellow and white. It is suitable for greenhouse and is widely grown for its handsome foliage and coloured spathe.

A. *scherzerianum* : Flamingo Flower, Flame Plant

The plant is better known and more compact. Leaves narrow, 15-20 cm long, 4.6-6.6 cm wide; spathe ovate, brilliant scarlet; spadix spirally twisted, golden-yellow. Flowers chiefly from February to July and needs keeping moist. This is a popular house plant of the genus.

4.2.0 Cultivars

A good anthurium variety should have the qualities like compact plants, producing suckers profusely, bright clear coloured, showy, heart-shaped spathe, with plenty of blisters and symmetrical overlapping of basal lobes. Spadix reclining to the spathe, shorter in length than the spathe, oriented at an angle less than 30°, an erect, long flower stem, above five times the length of the spathe, resistance to common diseases.

The cultivars based on spathe colour and types are:

Colour/Type	*Cultivars*
Red	Altiplano, Avo-Claudia, Cancan, Kaumana, Ozaki, Quito an Tropical
Orange	Casino, Nitta, Sun Burst
White	Acropolis, Angel, Chameleon, Cotopaxi, Lambada, Manova Mist,Trinidad
Pink	Abe, Avo-Anneke, Fair Lady, Marian Seefurth, Rosa, Sonata
Green	Midori
Obake	Cultivars with bi-coloured spathe. Extremely variable in size and shape and contain some development of chlorophyll in the spathe. Important ones are Avenue, Chameleon, Cardinal, Paradiso
Novelties	Tulip-type, miniatures. Spathe cupped, tulip-type; spadix straight, erect and not reclining as in other varieties. Important varieties are Calypso and Trinidad
Exclusive	Cascade - White, green spadix Cheers - Light pink, green spadix Choco - Dark brown-green spadix Jumbo - Obake-cream Safari - Brown-red white veins Taquila - Cream-green red veins
Double	Double flowering anthurium plants produce one small and one large spathe on the same stem with a single spadix. Red, pink, and orange doubles are reported.

The present-day flowering anthuriums are mostly hybrids of different species, involving mainly *A. andraeanum* and *A. scherzerianum*. The popular cultivars grown throughout the world both for cut flower and in pots are Abe (bright pink), Aneunue (green and coral-pink), Avo-Anneke (pink), Avo-Jose (white), Avo-Claudia (red), Chameleon (white), Favoriet (orange), Haga White, Horning Orange, Horning Rubin, Jamaica (white), Kaumana (red), Kozohara (red), Kansako No. 1 (red), Mauna Kea (white with green border), Marian Seefurth (rose pink), Mirjam (red), Manova Mist (white), Nitta (orange), Nova Aurora (red), Ozaki (red), Red Elf Rico (rose), Sun Burst (bright orange), Sarina (white and rose) and Uniwai (white). Cultivars Calypso (dark pink on inner surface and light pink on outer side), Trinidad (off white), Blush (red veins on spathe) and Double (different colours) are the novelties and some of the important cultivars.

In order to select suitable cultivars, some investigations were made in different parts of the world. Of 113 clones evaluated by Kamemoto and Nakasone (1963), 13 were recommended for commercial cut flower production, including Haga White, Nitta, Kaumana, Ozaki, Kansako No. 1 and Hirose. In an evaluation trial conducted by Leeuwen (1984), cultivars Avo Nette, Avo Tineke, Favoriet, Germa, Avo-Claudia, Avo-Ingrid, Nova Aurora, Avo-Jose, Jamaica, Hoenette, Sarina and Avo-Anneke were considered the best. Promising cultivars including Favoriet, Mirjam, Rico, Sarina and Poolster were also selected by Gelder *et al.* (1984).

The German variety Arabella is an early flowering, medium-sized plant with short dark green leaves. The spathe is broad with free lobes, a shallow sinus and recurving spadix. Spathe and spadix are red. Crossing between *Anthurium antioquiense* x Rotolante evolved the variety Ruth Morat syn. Lady Ruth. Spathes are red larger than those of Lady Jane, with a mean width and length of 50.1 and 76.8 mm, respectively. *Anthurium* x Red Hot is a new interspecific hybrid pot plant, originating from a cross between *cv.* Southern Blush (an F_1 hybrid of *A. andraeanum* and *A. amnicola*) and *cv.* Lady Jane. The typical characteristics are : leaves dark green, lanceolate, 18-20 cm long, base 11-12 cm wide; peduncle grey-orange, 20-28 cm above soil surface when the spathe is fully open; spathes 6-7 cm long, 4-5 cm wide and medium red at anthesis, gradually changing to a lighter red prior to senescence; spadix orange-red apically, blending to red at base, 3-4 cm long and 5-6 mm wide (Henny, 1999).

A new method of developing *A. scherzerianum* varieties using seeds cultured *in vitro* has been described. It involves selecting genotypes with a high regenerative ability, carefully adjusting the NH_4 : NO_3 ratio of the liquid medium for regeneration and multiplication, maintaining clonal explants for up to about 2 years at 11-15°C and 300 lux light in twist-off glass vessels which allow for a degree of ventilation, and carefully selecting for horticulturally desireable traits (under greenhouse conditions), especially flower colour, flowering onset and cold requirement for flowering (Zimmer, 1990).

Cultivar 'Favorita' has medium size plant suitable for production in 15 cm to 20 cm containers, with vigorous growth, full and symmetrical growth habit with early and abundant branching and year-round flowering and resistance to *Xanthomonas campestris pv. dieffenbachiae*. Exceptionally dark green, relatively large,

Cultivars of Anthurium

Cancan　Flamingo Red　Ozaki

Acropolis　Casino　Angel

Rosa　Midori　Obake

Cheers　Choco　Safari

Cultivars of Anthurium

Avo Gabriella

Red King

Avo Summer

Tequila

Xavia

Tropic Night

glossy leaves create a sharp contrast with bright orange-red and glossy spathes held immediately above foliage on straight, strong, relatively thick peduncles (Osiecki *et al.*, 2002).

Cultivar 'Cherry Red' is particularly distinguished by its dark-red, heart-shaped spathes, which are durable and may retain their color for 8 weeks or more. The foliage of is thick, durable, dark-green and shiny with reddish brown petioles and midribs. Plants are large in stature and best suited for pot sizes 20 cm and larger (Lamb, 2002).

Cultivar 'Patriot' is upright and outwardly spreading plant habit having glossy dark green-colored leaves with reddish brown-colored venation and petioles, dark red-colored spathes with whitish-colored spadices and excellent inflorescence longevity with spathes maintaining dark red coloration for at least eight weeks (Lamb and Knauss, 2004).

Cultivar 'True Love' is characterized by its upright and outwardly spreading plant habit with durable dark green leaves, red-colored spathes with yellow orange-colored spadices that are positioned slightly above and beyond the foliage on strong and erect scapes with freely flowering habit and good inflorescence longevity (Van Rijn, 2004).

Cultivar 'Bardonta' is characterized by its upright with uniform plant habit, glossy medium green-colored leaves, large red-colored spathes (Van der Knaap., 2010).

Cultivar 'Orange hot' is a sport of tissue cultured plants of 'Red Hot'. It has orange spathe and spadix colour is stable (Henny *et al.*, 2003).

Twelve new anthurium varieties are presented; they are named for Brazilian Indian tribes (Tombolato *et al.*, 2002).

- ✰ Aikanã (IAC NL 79) – green spathe and spadix 'Apalai' (IAC NK 130) – light red spathe and white spadix.
- ✰ Aruak (IAC NK 142-143-144) – white spathe with rose nerves and white spadix.
- ✰ Ianomami (IAC NM 84-85-86-87) – deep orange bicolor spathe with green edges and white spadix.
- ✰ Kauê (IAC NK 151-152) – brown spathe and green spadix.
- ✰ Krenak (IAC NL 89-90) – white (slightly pink) bicolor spathe with green edges and white spadix.
- ✰ Kinã (IAC NM 70) – green spathe with brown shades on the nerves and green spadix.
- ✰ Krahô (IAC NK 10) – large red spathe and white spadix.
- ✰ Parakanã (IAC NK 50-51) – white (slightly pink) spathe and white spadix.
- ✰ Terena (IAC NN 155-156) – pink bicolor spathe with green edges and white spadix.

- ✰ Xavante (IAC NK 129-131) - salmon bicolor spathe with green edges and white pinkish spadix.
- ✰ Zoé (IAC NM 157-158-159) – pink spathe and spadix.

5.0 Genetics and Breeding

5.1.0 Genetics

The basic chromosome numbers of anthurium are n = 15, 16 and 22. The species like *A. andraeanum* (2n = 45), *A. hookeri* (2n = 30) and *A. magnificum* (2n = 32) are diploid, *A. scandens* (2n = 30) is triploid and *A. digitatum* and *A. wallisii* (4n = 60) are tetraploid (Gaiser, 1927). The chromosome number of *A. warocqueanum* was found to be 2n = 30+3 B chromosomes. These were classified as 2 pairs of large chromosomes, 1 pair of satellite chromosomes, 12 pairs of medium to small chromosomes, and 3 B chromosomes. Self-pollination resulted in offsprings with 2, 3 and 4 chromosomes, indicating the transmission of B chromosomes through both pollen and egg (Keneko and Kamemoto, 1979). Sheffer and Croat (1983) worked out the chromosome number in 86 anthurium species from North, Central and South America. Fiftyone species had 2n = 24 to 66, 30 being the most common. Paleoneuploidy, polyploidy and B chromosomes have been the basic features of the genus, but aneuploidy has not been found.

Cytological analysis of *Anthurium andraeanum* revealed 2n = 30 as the number of chromosomes with 4 fairly large, 22 medium and 4 small chromsomes (Sengupta *et al.*, 1989). Chromosome number of 4 species of *Anthurium* from Bahia, Brazil, were determined. Number for *A. longipes* and *A. affine* (2n = 30) were reported for the first time. The 2n = 90 for *A. bellum* differ from earlier reports, whereas 2n = 30 and 2n = 60 for *A. pentophyllum* var. *pentaphyllum* confirms previous counts. *A. affine* had 1- 4 B chromosomes. It is suggested that secondary base number x = 15 for *Anthurium* was produced by hybridization and duplications (Cotias de Oliveira *et al.*, 1999). Somatic chromosome number for *Anthurium andraeanum* was 2n = 30 + 2B, irrespective of the varietal type. A high percentage of meiotic abnormalities and karyotype differences within the species indicated a hybrid origin for the species. *A. andraeanum* is a secondary polyploid with a probable basic chromosome number of x = 6. All 5 varieties studied were grouped into karyotype asymmetry category 3 B which indicates that the species is highly evolved (Bindu and Mercy, 1997). Karyotypes and meiotic configurations of *A. andraeanum* and closely related taxa were analysed. Karyotypes of *A. andraeanum, A. caperatum, A. formosum, A. kamemotoanum, A. lindenianum, A. roseospadix, A.* cf.*sanctifidense, A. subsig-natum, A. garagaranum,* and an unidentified *Anthurium* sp. commonly consisted of 4 large metacentric or submetacentric chromosomes, 2 fairly large acrocentric chromosomes, 2 satellite chromsomes and 22 smaller chromosomes. Variation in the karyotypes of *A. nymphaeifolium* and *A. ochranthum* suggested chromosomal rearrangement in the genus. All taxa showed 15 pairs of chromosomes at prometa-phase I of meiosis in PMCs. Four large chromosomes appeared as ring bivalents, and the rest of the chromsomes appeared as either ring or rod bivalents. Regular bivalent formation at prometaphase I of meiosis in PMCs of species hybrids suggested close genomic relationship among parental taxa. On the other hand, reduction of pollen fertility estimated by pollen stainability in those hybrids suggested genetic divergence of species (Marutani *et al.*, 1993).

Floral fragrance compounds of *A. armeniense, A. fragrantissimum, A. lindenianum, A. orchranthum* and *A. roseospadix* and 3 hybrids of *A. armeniense* were studied and twenty-eight compounds were identified. Their concentrations (per cent) were assessed at the scent-emitting pistillate stage of the spadix. Variation of compounds in the head-space in response to the time of day were determined in the hybrid Tatsuta Pink Obake (UH 1299), which emitted strong fragrance all day. Chemical profiles from the headspace and from spadices soaked in CH_2Cl_2 were compared in 3 species. It was concluded that combination of these two data sets would provide more complete chemotaxonomic profiles in *Anthurium* (Kuanprasert *et al.*, 1988).

Few controlled hybridization indicated that neither white nor red flower colour was dominant and pink was an intermediate heterozygous condition (Kamemoto and Nakasone, 1955). Anthocyanins in the spathes of various *A. andraeanum* cultivars were identified as cyanidin 3-rhamnosylglucoside and pelargonidin 3-rhamnosyl-glucoside. Both pigments were presents in the red *cvs*. Ozaki, Kaumana, Kozohara, Kansako No. I and Nakazawa and in pink cultivar Marian Seefurth. The orange *cv.* Nitta and the coral coloured *cv.* Teteishi Coral contained only pelargonidin 3-rhamnosylglucoside (Iwata *et al.*, 1979). According to them (1985) spathe colour in these species was determined by the relative concentrations of the anthocyanins: a predominance of cyanidin 3-rhamnosylglucoside which results in pink to dark red colours whereas a predominance of pelargonidin 3-rhamnosylglucoside resulted in coral to orange. A flavone present in large and variable amount was characterized but not demonstrated in having a modifying effect on cyanic shades.

Maurer (1979) described the techniques of cross pollinating *A. scherzerianum* and discussed the presence of recessive characters (*A* = with anthocyanin, *a* = without anthocyanin, *B* = whole spathe coloured and *b* = spotted spathe). Wannakrairoj and Kamemoto, (1990) reported that when the parents were *Aa/Bb*, the decendents were 9 red (*AB*), 3 red spots on white (*Abb*) and 4 white (*aaB* and *aabb*). The deficit in white plants was provisionally attributed to their lack of vigour. The recessive allele *p* modifies the colour of anthocyanins controlled by the *M* and *O* loci. A spathe is purple when the genotype is *M-O-pp*. If the *P* locus is dominant, *M-O* is red, while *mmO* is orange (Table 1). The *p* allele has no effect on the *oo* (white) genotypes.

Taxonomic differentiation of the species is difficult, mainly due to their morphological plasticity. The most recent anatomical studies on *Anthurium* are based on the anatomy of the leaf and spathe (Coelho and Catharino, 2008).

The inheritance of spathe color in anthurium (*Anthurium andraeanum*) was studied in 77 crosses involving 59 parental genotypes. The progeny segregation into colored : white did not fit the genetic model proposed by Kamemoto *et al.* (1988), but fit a duplicate recessive epistasis model involving genes O and R, in which either O or R or both in the recessive form would result in white-spathed progeny. The progeny segregation into red or orange group was in agreement with the Kamemoto *et al.* (1988) model and is determined by the status of the M gene. The dosage effect of particularly the R gene accounted for the differentiation of reds from pinks and oranges from corals. Putative roles for the three genes (M, O, and R) in the flavonoid biosynthetic pathway are postulated (Elibox and Umaharan, 2008).

Table 1: Spathe Colours of Interspecific Hybrids and Multiway Crosses in *Anthurium*

Cross Numbers	*Parents*	*Spathe Colours of Offspring*
	F_1 hybrids	
552, 559	*A. lindenianum* (W) × *A. amnicola* (P)	R
568, 572	*A. amnicola* (P) × *A. formosum* (P)	P
569, 575	*A. amnicola*(P) × *A. lindenianum* (W)	R
717, 718	*A. andraeanum* (O) × *A. amincola* (P)	R
755	*A. andraeanum* 'Uniwai' (W) × *A. amnicola*(P)	R
756	*A. andraeanum* 'Marian Seefurth' (R) × *A. amnicola* (P)	R
	Multiway crosses	
573	*A. amnicola* × (*A. kamemotoanum*× *A. formosum*)	P,R
587	[(*A. andraeanum* 'Uniwai' × *A. kamemotoanum*) × *A. formosum*] × *A. amnicola*	P,R
634	Cross 552 (R) × *A. amnicola*	P,R
692, 696, 697 and 753	[(*A. andraeanum* 'Uniwai' × *A. lindenianum*) × *A. amnicola*)] × *A. formosum*	P,R,W
733	(*A. andraeanum* 'UH 712' (R) × *A. amnicola*) × *A. formosum*	P,R
739	Cross 587 (R) × *A. formosum*	P,R
747, 750	Cross 552 (R) × (*A. kamemotoanum* × *A. formosum*)	P,R,W
751	Cross 572 (P) × *A. formosum*	P,W
762	[(*A. andraeanum* × *A. kamemotoanum*) × *A. formosum*] × *A. formosum*	P,R,W

Spathe colours are Purple (P), Pink-red (R), Orange (O), and White (W)*.

* Wannakrairoj and Kamemoto (1990).

5.2.0 Breeding

Hybridization and selection are the common methods for improving anthuriums. Productivity, flower colour, shape and texture, short internodes and suckering ability are different characters to be considered in selection (Kamemoto and Nakasone, 1955). Two cultivars, Uniwai (an exceptionally high yielding white) and Marian Seefurth with a rose opal spathe were evolved by clonal selection (Kamemoto and Nakasone, 1963). Kamemoto and Sheffer (1978) made successful crosses between *A. scherzerianum* and *A. wendlingerii* to produce a hybrid with a greyish-orange spathe. Other characteristics, such as the length and coil of the spadix and the length and position of the leaf blade were intermediate between the highly contrasting characters of the parental species (Table 2). Fertility in this hybrid was very good, indicating the relatively close taxonomic relationship of the two species.

Gajekand-Schwarz (1980) described five *A. andraeanum* hybrids, valuable for all the year round cut flower which use a limited energy input. The best of the cultivars are the medium-sized Iga Gold with a shiny red spathe and a white spadix with yellow tip, and the compact Ellrina with a vermillion light salmon spathe and a sulphur-yellow spadix.

Table 2: Morphological Characteristics of *Anthurium scherzerianum*, *A. wendlingerii* and the Species Hybrid*

*Parent and Cross***	*Spathe Colour*	*Spathe Length (cm)*	*Spathe Width (cm)*	*Spadix Length (cm)*	*Spadix Coil (No.)*	*Peduncle Length (cm)*	*Leaf Blade Length (cm)*	*Leaf Blade Width (cm)*	*Petiole Length (cm)*
A. scherze-rianum	Scarlet (R.H.S.C.C. 34C)	20.00	13.5	15.0	1.5	31.5	28.5	10.0	13.0
A. wendlin-gerii	Purplish-brown	15.0	2.0	43.0	5.0	27.0	106.0	7.5	19.3
A. scherze-rianum x A. wendlin-gerii	(R.H.S.C.C. 173A)	15.9 ± 0.6	5.2 ± 0.4	23.2 ± 0.4	3.1 ± 0.2	37.6 ± 1.2	49.3 ± 1.2	10.0 ± 0.3	23.2 ± 0.0

* Kamemoto and Sheffer (1978).

** A single plant each of both parental species and 20 species hybrids are involved.

R.H.S.C.C: Royal Horticultural Society Colour Chart.

Techniques to cross anthurium varieties were studied by Chen *et al.* (1999). The best time for pollination was taken when mature stigmas on the spadix produced a sticky material. The cooler period from December to April was the time when most anthurium varieties shed their pollen in an environment-controlled greenhouse. Pollen-shedding habits varied greatly, dependent on the variety. Most varieties shed pollen after their stigmas were no longer receptive. However, some varieties occasionally shed pollen soon after the spathe unfurled. The transparent minigrip PE (polyethylene) bag proved a good envelope to encase the anthurium flower in order to prevent undesired pollination. Compared to other pollination methods, the pen brush yielded the highest fertility and seed-bearing on the spadix.

Current control measures for bacterial blight included strict sanitation and in some cases antibiotic treatments. Attempts to tranfer systemic resistance from *A. antioquiense* to the cultivated *A. andraeanum* produced resistant F_1 hybrids. Backcrossing to *A. andraeanum* to produce cultivated varieties with horticulturally desirable characteristics takes many years because it is a perennial crop, with a long juvenile stage (2 to 3 years) and slow seed maturation (6 months). As the genetics of the available resistance was not properly understood, resistant cultivars which were released soon became susceptible to blight. Two cultivars, Rudolph and UH 1060, were transformed with vectors containing antibacterial genes and synthetic derivatives from *Hyalophora cecropia* and bacteriophages. Regenerated plants showed a delay in the disease symptoms development compared with non-transformed controls (Kuehnle *et al.*, 1995).

Humming birds and passerine birds have been reported to visit some Colombian *Anthurium* that retain many typical aroid traits. In addition, several arthropod taxa were documented as *Anthurium* visitors for the first time consider the potential for a bird or a mixed pollination system (Bleiweiss *et al.*, 2019).

Relative flower size (RFS) is defined as the ratio of spadix diameter and transverse floral width in the same region of the spadix. Variation was examined within and between inflorescences, populations and species. Within a single spadix, RFS may vary significantly between basal and upper spadix zones but usually not between basal and middle zones. Within populations, RFS does not differ significantly between spadix zones (Pontes *et al.*, 2014). The Brazilian flora has a large number of species with ornamental potential. De Morais *et al.* (2017) conducted a study to evaluate the potential use of four accessions belonging to the species of *Anthurium plowmanii, A. raimundii, A. bonplandii* and *A. Affine* for cutting foliage. The characteristics such as fruit color, petiole, spadix and spathe length are highly variable within the species whereas some characters like petiole length and inflorescence position are particularly important as indicators for the use as cut foliage. The production of leaves per year ranged between 6.6 (*A. plowmanii*) to 10.5 (*A. bonplandii*), which might be increased with the improvement of cultivation aspects, with an emphasis on spacing, plant management and fertilization. The leaves of *Anthurium plowmanii, A. raimundii, A. bonplandii* and *A. affine* present high postharvest durability and great potential as cut foliage, possessing florist acceptance as well.

Anatomical studies clustered the genotypes *A. andraeanum* and distinguished the hybrid (*A. lindmanianum* × *A. guianesis*) and *A. bomplandii* from the other genotypes, with anatomical features specific to each. Differences in DNA content, assessed by flow cytometry, allowed the differentiation between the genotypes and the hybrid. Cluster analysis allowed grouping of the genotypes *A. raimundii* and *A. affine, A. bomplandii* and *A. gracile, A. jemanii* and *A. plowmanii*, and the cultivars Astral and Rubi (*A. andraeanum*), and isolated the hybrid (*A. lindmaniarum* × *A. guianesis*) (Dias *et al.*, 2017).

One variegated plantlet resulting from chlorophyll loss of partly parenchymal tissue was obtained from the adventitious shoots regenerated from callus of Anthurium and achieved three types of chimeras in the process of propagation by indirect shoot regeneration via callus or axillary bud emergence, namely variegated types-albino types and jade-green types, in which the chlorophyll content was lower than that of the normal *in vitro* plantlets and recommended that the chimeras, especially the variegated plantlets, might be potential breeding materials or new cultivars (Xu *et al.*, 2006).

5.2.1 Polyploidy Breeding

Tian and Ma (2008) reported polyploid *A. andraeanum* 'Arizona' produced from a callus clump and young seedlings by treatment with colchicine and found that the optimum condition for the induction of polyploidy was the use of adventitious roots that had been soaked in 0.3 per cent colchicine for 7 h, resulting in highest polyploid induction of 63.3 per cent. The characteristic tetraploids were similar to those obtained by Zhang *et al.* (2007). Tetraploids plants of *Anthurium andraeanum* "Arizona" were successfully induced after treating diploid tissue masses with 0.3 per cent colchicine with 5-h treatment and percentage of induced tetraploids ranged from 0.2 to 7.6 per cent and reported larger stomatal size, stronger petioles, thicker and deeper green leaves, and thicker and longer lived spathes and tetraploid plantlets could be regenerated via aerial roots (Chen *et al.,* 2011). However, Tian *et al.* (2013) reported a higher induction rate of tetraploidy (17.5 per cent) when *A. andraeanum* 'Alabama' callus was cultured in 0.2 g/l colchicine in MS liquid medium for 15 d under constant rotation using the method of Tian and Ma (2008).

5.2.2 Characteristics for Cut-flower Selection

The following are important characteristics of the hybrid seedlings: Spathe medium size (10-15 cm long), brilliant color, flat, opened, hard texture, prominent nerves, well developed lobes on heart shape, uniform coloration (except for bicolor), minimum 15 days of keeping quality after harvest. Spadix -2/3-3/4 as long as the spathe length, contrasting color to the spathe, except for the white, held at an acute angle to the spathe. Flowering stem - erect, hard, minimum 60 cm long and plant should have short internodes, compact growing habit, non sprouting, tolerant or resistant to the main diseases: anthracnose (*Colletotrichum gloeosporioides*), bacteriosis (*Xanthomonas axonopodis* pv. *dieffembacchiae*) and viral pathogens (Tombolato *et al.,* 2005).

6.0 Growth and Flowering

The growth and development of *A. andraeanum cv.* Kaumana flowers (spathe, spadix and peduncle) before and after emergence from the subtending leaf base was studied by Dai and Paull (1990). Eighty days before emergence, the flower was approximately 0.3 cm long, enclosed by 2 tightly rolled stipules at the base of the subtending leaf petiole. During the rapid elongation stage of the leaf petiole, the flower (0.8-1.0 cm long) entered a period of slow growth 40-60 days before flower emergence. After the subtending leaf blade unfurled and had a positive photosyn-thetic rate, flower growth resumed. Spathe colour development started approximately 28 days before emergence when the flower had reached 50 per cent of its final length. At flower emergence, approximately 80 per cent of the spathe was red. The lobes did not develop full redness until 7-10 days after emergence. Peduncle growth was sigmoidal with the maximum growth rate occurring 21 days after emergence. Spathe growth is characterized by a double sigmoid curve. The young, growing, subtending leaf blade had a negative net photosynthetic rate. Removal of this leaf blade advanced flower emergence by 18 days. The soft green leaf (25-30 days after leaf emergence) had a slightly positive measured net photosynthetic rate, and the removal of this leaf advanced flower emergence by 11 days. A mature subtending leaf had the highest measured net photosynthetic rate, and its removal had little effect on flower emergence. The subtending leaf acted as a source of nutrients required for the developing flower. One of the main factors that affect the production of cut anthurium flowers is the slow growth of the inflorescences, the speed of which is controlled by the source-sink ratio (Dufour and Gurein, 2003).

6.1.0 Environmental Factors

6.1.1 Light

In commercial practice anthuriums are given 80 per cent shade during the summer months. In an experiment, plants were placed under 27, 43, 57 or 73 per cent shade. Decreasing the shade did not affect flower production but reduced stem length and the leaves of plants under 27 per cent shade became chlorotic (Poole and McConnell, 1971). In another trial with 75, 50 or 25 per cent shade of full sunlight, the largest number of flowers was produced with the least shading, but flower quality was better under more shade. Plants from cuttings without the apical bud showed less vegetative growth and did not show a marked response to light intensity (Boula *et al.*, 1973). Similarly, Leffring (1975) reported that growth rate increased and average flower production rose from 5 to 12 flowers per plant per year when plants were shaded so as to receive at least 45 per cent of the available light by means of an automatic system outside the greenhouse. Overhead sprinklers were also used to prevent leaf scorch on sunny days. The range of light intensity for plant growth is between 2000-6000 lux. The unsuitable growing conditions stimulated the development of abnormal spathe and/or spadix and reduced the productivity of plant (Steen and Vijvosberg, 1973).

In an evaluation of response of anthurium *cv.* Lady Jane liners to different light and fertilizer levels, Henny and Fooschee (1988) observed that plants were taller, had larger leaves, were of better quality and had greater top and root fresh weight at lowest light level (172 $\mu mol/s/m^2$), while plants were smaller at higher light levels (344 or 487 $\mu mol/s/m^2$). An experiment was carried out on *A. andraeanum* grown at low altitude (Ehaliyagoda) and at high altitude site (Hatton) under 3 shade levels (approximately 40 per cent, 65 per cent and 85 per cent) in Sri Lanka. Shade level did not affect the number of leaves/plant or growth rate at either location but a faster growth rate was observed at the low altitude site. Flower production was highest under medium and low shade levels at Hatton and under medium and high shade levels at Ehaliyagoda, spathe size and flower stalk length increased with increasing shade at both locations. On average, high altitude produced larger spathes and low altitude produced longer flower stalks. Low shade level was required for sucker production at Ehaliyagoda. It was concluded that high shade levels at low altitude were best for flower production, while high altitude was more suitable for the production of suckers for use as planting material (Kuruppu and Yogaratnam, 1989).

In an experiment, Prakash *et al.* (1998) observed that as the shade intensity increased the plant spread in both the directions increased significantly. Maximum spread, petiole length, number of branches and suckers were produced under 80 per cent shade. The duration of successive inflorescence emergence and number of spikes per annum were enhanced and the colour of the flower spikes was brilliant at 80 per cent shade. Similar type of results was also obtained by Valsalakumari *et al.* (1998). Heavy shading is commonly applied during production of pot-plants in order to avoid damage caused by high light intensities. Experiments were carried out with two cultivars (Royal Champion and Pink Champion) in a conventional modern glasshouse compartment covered by clear glass with DLI (Daily light interval) limited to 7.5 mol/m$^{2/}$d and reported that the stimulating effect of diffuse glasscover on dry mass production per unit intercepted PPFD (RUE) in Anthurium pot-plants is cultivar specific. Increasing DLI underdiffuse glass cover not only stimulates biomass production, but also improves plant ornamental quality (Li *et al.*, 2014).

Roni *et al.* (2016) reported that anthurium petiole length elongation was recorded maximum in 60 per cent reduced light due to the increase of cell length and cell number. Khawlhring *et al.* (2012) similarly showed taller plant heights from those grown under a shade house with 75 per cent shade. Treatment with different light intensities (20, 100, and 400 µmol/s/m^2) for 15 d on seedlings of wild-type and etiolate mutant plants of *Anthurium andraeanum* cultivar 'Sonate' to analyze leaf plastid development and pigment content and the results suggested that light intensity is a crucial environmental factor influencing plastid development and leaf color formation in cultivar 'Sonate' (Wang *et al.*, 2018). According to Lima *et al.* (2010), the black screen with 75 per cent of shading results higher leaf area and best condition for growth. While, Ravinath (2007) reported that the optimum light intensity for growing anthurium is 15,000–20,000 Lux.

6.1.2 Temperature

Anthuriums undergo two distinct developmental stages, a juvenile phase (with only vegetative buds in the axil) and generative phase. The flower buds produced in the axils of leaves during the generative phase remain dormant after initiation and develop into flower, depending upon the dormancy breaking. Unfavourable environ-mental condition during this phase can damage the flower bud before normal development or remain vegetative, depending on the stage at which the stress situation is created (Christensen, 1971). For both *A. andraeanum* and *A. scherzerianum*, the optimum temperature for vegetative growth is 18.3°C. Lower growing temperature cause spotting on leaves. Higher temperature (21.1 to 23.9°C) is needed to force the plant to initiate flower buds. A temperature below 18.3°C, especially in *A. scherzerianum* markedly reduced the growth of the plants. Temperature requirements for satisfactory flowering of *A. scherzerianum* in pots were also investigated by Vonk-Noordegraaf (1969). Below 15°C leaf production was somewhat slower and the leaves were less stiff whereas above 18°C the flower number and size were generally reduced. Klapwijk and Spek (1984) studied the monthly patterns of leaf and inflorescence formation for 4 years in 6 anthurium cultivars. The average leaf number/m^2 of glasshouse rose from 1.5 in March to 5 in June, thereafter declining until the following March. The time required for buds to develop into blooms ready for harvest ranged from about 45 to 53 days during May to October and from about 65 to 75 days during December to March.

Investigations on the effects of air temperature (13, 16, 19 or 22°C) with a constant peat substrate temperature of 22°C on flowering of 5 cultivars of *A. andraeanum* made by Schenk and Brundert (1981) revealed that *cvs.* Horning Orange and Horning Rubin were normally temperature neutral but in Horning Rubin the total number of flowers/plant was reduced at 22°C. For the 3 light red Vogel cultivars, yield and quality of flowers were best at 19°C at which Vogel No. 202 produced 7 flowers/

plant, the highest yield. At 10°C Vogel plants died and at 13°C they showed some leaf necrosis. In another experiment, Maatsch and Bachthaler (1964) observed that plant vigour increased with rising temperature. A night temperature of 10°C delayed flowering considerably in *A. scherzerianum*. At 15, 20 and 25°C night temperature all the plants flowered at the same time. Plants grown at 15°C night temperature produced the maximum flowers, at 15 and 20°C flower quality was good, at 25 and 10°C flower stalks were short.

Ten genotypes, obtained by *in vitro* cloning of single seeds from 2 crosses and 2 selfings, were compared over a 16-month period (October-January) at 18 or 22°C for flowering behaviour. In December, the clones were transferred to lower temperatures (10/10, 14/14 or 20/14°C) for 6 weeks. The plants were 19 months old at the start. The juvenile phase ended in all cases with the development of the 6th leaf. At this point, a flower bud was initiated in each axil. Some genotypes did not require chilling for premature flower development. Others flowered in response to reduced temperature. Intermediate types were also found. Inflorescences remained dormant in the axils for up to 7 months (Schaper and Zimmer, 1991). Nitric oxide could increase chilling tolerance in anthurium by stimulating the antioxidant defense system, decreasing the accumulation of ROS, and elevating antioxidant contents. (Liang *et al.*, 2018).

6.1.3 CO_2

Enrichment of CO_2 improved and advanced plant growth (by about 3 weeks) and increased the market value (Schmidt and Lauterbach, 1984). A concentration of 900 ml/m^3 proved more beneficial than 600 ml/m^3.

6.1.4 Electrical Conductivity (EC)

The experiment on effect of electrical conductivity on the growth of *Anthurium andreanum* and its tolerance patterns, throughout special period of time on two varieties of Anthurium (Red queen-Elizabeth) in the greenhouse, supplemented by two kinds of water. The first is distilled water withsome added salts of compound fertilizer of N: P: K with 1:0.5:1.5 to create different levels of electrical conductivity (0-1-1.5-2) ds/m. The second is tap water where the electrical conductivity is 0.30 ds/m with added salts of compound fertilizer of N: P: K with 1:0.5:1.5 to it to create other levels of electrical conductivity (1-1.5-2) ds/m. One year old plants were used and many variables were studied. Results showed that when EC was increased above 1ds/m, the growth of the plant decreased including the number of flowers and leaves. It was clear that tap was better than distilled water (Eid *et al.*, 2016).

6.2.0 Growth Regulators

Certain chemicals were also found to be effective on plant growth. Treatment of plants with PBA, BA or ethephon induced adventitious buds. Maximum shoot formation was observed with BA at 100 mg/l (3.0 shoots/plant) followed by PBA 1500 mg/l (2.2) and ethephon at 1000 mg/l (1.8). Control plants exhibited no adventitious shoot formation (Higaki and Rasmussen, 1979). Nakasone and Kamemoto (1962) failed to demonstrate any significant effect of GA (10, 25, 50 and 100 ppm) application on stem length, flower production or flower size. In an experiment, juvenile plants were treated by topping and/or with foliar sprays of GA_3 and BA. It was noticed that with increasing concentration (0 to 500 ppm) of GA_3 topped plants showed an increase in lateral shoots (Table 3), while at rising concentration of BA (0 to 1000 ppm), the number of lateral shoots increased on both topped and intact plants (Imamura and Higaki, 1988). *A. scherzerianum cv.* Amazonas grown in the greenhouse were treated with 0, 500 or 1000 ppm GA_3 at potting or with 500 ppm GA_3 at potting + 500 ppm GA_3 30 days later. The split treatment resulted in the largest floral buds without reducing leaf number (Boschi*et al.*, 1998). Light intensity and GA_3 concentration are reported to be important factors in regulating anthurium bloom count. *A. scherzerianum* 'Renate' grown at high light and receiving high GA_3 levels (375 or 500 mg/l) had high bloom counts (Henny and Hamilton, 1992).

Table 3: Effect of GA_3 and Topping Treatments on 'Mauna Kea' Anthurium Shoot Production*

Topping Treatment	*GA_3 (ppm)*	*No. of Shoots*
Intact	0	1.6 cZ
	500	2.0 b
Topped	0	3.3 b
	500	5.8 a

Z Mean separation by Duncan's multiple range test, 5 per cent level.

* Imamura and Higaki (1988).

Zheng *et al.* (2004) found that GA_3 100 mg/l and GA_3+KT (50+50) mg/l treated plants bloomed earlier 29 and 33 days, respectively and had high flower quality. The ability to produce suckers is variety dependent and suckering capacity can be improved by exogenous application of growth regulators at frequent intervals (Srinivasa, 2006). *Anthurium andreanum* var. Tropical under naturally ventilated polyhouse treatedwith benzyl adenine at 750 ppm took least (88.67) days to sucker production, which is on par with BA at 500 ppm (90.33), BA at 250 and at the final stage of the experiment (at 240 days after spraying) number of suckers per plant was recorded in highest in BA at 750 ppm (3.33) (Seemanthini *et al.*, 2018).

Foliar spray of humic acid and gibberillic acid can evince maximum plant height, plant spread, number of flowers per plant, flower stalk length, spathe length, spathe breadth and early flower bud appearance (Ajish Muraleedharan and Karuppaiah, 2015). Foliar application of gibberellic acid at 500 ppm can be effectively used for the improvement on growth, flowering, yield, quality attributes and for extending the vase life of *Anthurium andreanum cv.* Tropical. The best performance might be due to appropriate combination of shade, growing media and growth regulator (Muraleedharan *et al.*, 2020).

Bordoloi and Talukdar, (2019) reported that treatment (comprising) RDF+*Azospirillum*+100 ppm GA_3 and RDF+100 ppm GA_3 were found to be the most efficient for both growth and flowering. Hence, these may be adopted by the growers for commercial cultivation of Anthurium.

7.0 Propagation

Anthuriums are generally propagated through seed, division and cutting. In recent years, micropropagation is also being employed for commercial production of planting material.

7.1.0 Seed Propagation

Seeds should be harvested at proper stage of berry ripeness for successful germination. In *A. scherzerianum*, the best germination resulted from picking berries at the orange-red stage and fermenting them for 4 days in water at 22°C to separate the seeds from the pulp. Storage was sometimes necessary because berries from the same inflorescence ripened at different times, but the maximum time for storage of fermented seed in water was 5 days. This reduced germination from 99 to 80 per cent, and 10 days in water reduced it to 53 per cent (Bell, 1971). Bachthaler (1978) reported that seeds extracted from unripe (green), half ripe (reddish) and ripe (red) berries showed 100 per cent germination as compared to only 42 per cent in those from overripe (reddish brown) berries. It was also suggested that seeds taken from half-ripe and ripe berries were the first to germinate and were the most suitable for commercial seed production. Of several methods tried for extracting seed from crushed mature fruits, soaking in 13 per cent crystalline sodium carbonate at 20°C for 2.5 hours or in 6 per cent pectinase solution at 26-30°C for 5 hours were the two methods resulting in seeds being clean enough to sow singly (Maurer and Brandes, 1979a).

The optimum storage temperature is very important for successful germination. Seeds of *A. scherzerianum* hybrids stored within the berries at 2°C suffered from cold injury, at 5°C they were affected by low temperature and fungal (*Botrytis* sp.) infection and at 10°C only with fungal infection. The best storage temperature was 10°C and after 6 weeks 60 per cent of the seeds germinated. When seeds from berries were treated with aatiram (thiram) dust before storage, 95 per cent germinated after 12 weeks at 10°C and 60 per cent after 16 weeks. Berry disinfection with euparen (dichlofluanid), benomyl, maneb, phytox 80 or orthocide 50 (captan) was as effective as with aatiram but with polyramcombi (metiram), seed germination was poor (Bachthaler, 1979). He (1980) also observed that seeds from berries stored at 10°C for up to 6 weeks germinated in 4 days whereas those stored at 5°C and 2°C, took 5 and 7-8 days, respectively. Germination was 100 per cent in seeds from berries stored at 5°C and 2°C, respectively, for the same period. Seed from berries treated with thiram germinated well (in 4 days) after 16 weeks' storage. After 12 weeks' storage at 10°C, seeds from the untreated berries failed to germinate, whereas germination was over 90 per cent with seeds from the treated berries. Maurer and Brandes (1979b) stored seeds in spadices, in berries or as cleaned seeds in dry condition, above water, in water or in pulp from the berries, at 20, 8 or 5°C. They observed that seeds left in spadices or berries became too dry or rotted. Germination percentage decreased in cleaned seed with drying. Some seeds stored at 20°C germinated, but those stored at 5°C usually became non-viable, although seeds were successfully stored at 8°C above water for 10 weeks. Seeds stored in water rotted. Storage at lower pressures (700 and 400 hPa) resulted in germination reductions of about 70 and 50 per cent,

Propagation of Anthurium

Matured Seeds

Harvested Seeds

Germination of Seeds

Stem Cuttings

Propagation by Division

Micropropagation

1-6: *In vitro* Regeneration of Plants in *Anthurium andraeanum.*

1: Induction of callus from leaf segment in dark condition. 2: Regeneration of multiple shoots from leaf derived callus. 3: Regeneration of multiple shoots from spadix derived callus. 4: Growth and elongation of multiple shoots. 5: Induction of roots from the regenerated shoots. 6: Plantlets established in soil.

Source: Jahan, *et al.* (2010).

respectively, though with little effect on the time-course of germination. Seeds lost their germinative capacity after storage for 20 weeks at 100 hPa (Bachthaler, 1993).

Optimum atmospheric temperature and suitable medium are also necessary for proper seed germination. Temperature ranging from 20-25°C was reported to be optimum for germination of *A. scherzerianum* hybrid seeds (Bachthaler, 1977). Seeds of *A. andraeanum* germinated better at 28°C in high peat substrate having pH between 4.0 and 5.0 and under continuous lighting whereas those of *A. scherzerianum* preferred high peat substrate with a pH 4.0 (Szendel *et al.*, 1982).

7.2.0 Vegetative Propagation

Anthuriums are propagated asexually to ensure uniformity of cultivars. Conventional methods of propagation involve sprouting of vegetative buds positioned opposite each leaf at alternate nodes (Higaki, 1976).

7.2.1 Division

Anthurium can easily be propagated by division of offshoots with aerial roots from the main stem. Plants obtained through this method flower early in comparison with those from cuttings or other methods. Cultivars belonging to *A. scherzerianum* generally produce more number of offshoots (on an average 17.2 per plant per year) as compared to those of *A. andraeanum*. In the latter species, maximum of 14.2 new plants per plant per year could be obtained by covering the stem with peat moss (Zafirova, 1982).

7.2.2 Cutting

Propagation by terminal cuttings rooted under intermittent mist is also possible. In a trial comparing the effects of intermittent mist and daily sprinkling for 30 minutes on the rooting of *A. andraeanum cv.* Marian Seefurth, the rooting percentage and average root length were much greater under mist. Moreover, mist accelerated rooting and enhanced survival. In another trials on the effect of intermittent mist on rooting of 1, 2- and 3-leaved terminal cuttings there was no difference in rooting percentage between the type of cuttings. However, after 4 weeks the average root length of 2- and 3-leaved cuttings was greater, and after 7 weeks more roots had emerged than on the 1-leaved cuttings (Kunisaki and Sagawa, 1971). Hetman and Pudelska (1984) studied the effect of plant age, rooting preparation and methods of transpiration reduction on the rooting of single node cuttings and found that plant age had no marked effect on rooting but more leaves were produced by rooted cuttings taken from the older plants. Seradix 1, a hormone preparation gave the best rooting while reduction of the transpiration area had no effect on root formation or top growth. In addition to enhancing rooting, hot water + IBA increased the number of shoots per cutting in *A. andraeanum cv.* Marian Seefurth (Hata *et al.*, 1994).

7.3.0 Micropropagation

With tissue culture technique, anthuriums are multiplied very rapidly and flower production has been reported to be higher than those from seedlings (Leeuwen, 1980). A modified MS medium was suggested by several workers. The growth of the callus tissue from adult *A. andraeanum* plants was found to be the best in such medium. The genotype A 42-3 reached a fresh weight multiplication rate of 30.7 when grown for 5 weeks at a rotation speed of 120 rev/min, at 25°C and in continuous darkness (Pierik, 1975). Both *A. andraeanum* and *A. scherzerianum* explants produced many shoots when cultivated *in vitro* in the presence of BA and 2,4-D. However, yeast extract stimulated shoot growth of *A. scherzerianum* but restricted it in *A. andraeanum* and root growth on leafy shoots, a spontaneous phenomenon in *A. andraeanum*, required stimulation in *A. scherzerianum*. Pot plants of *A. scherzerianum* could be obtained more quickly than those of *A. andraeanum* owing to the former's rapid callus formation and shoot growth (Fersing and Lutz, 1977). Yu (1995) conducted experiments on the effect of the macroelement strength of the media (0.3, 0.5, 0.8, 1.0, 1.5 or 2-strength of macroelements in MS medium) on the organogenesis of shoot-tip cultures of *A. andreanum* cultivars Hazrija and Ingrid and *A. scherzerianum cv.* Belinda. Shoot growth *in vitro* was enhanced on media containing 0.3-0.8 strength of microelements.

Kunisaki (1980) obtained high yield of viable *A. andraeanum* cultures using small explants of vegetative buds and a long disinfection soak in 0.26 per cent sodium hypochlorite. Explants were grown into plantlets in modified MS medium supplemented with 15 per cent coconut water. For clonal multiplication the best result was obtained when stem sections from aseptically grown plantlets were cultured in medium containing 0.2 mg/l BA. Formation of callus and adventitious shoots by shoot tip explants was increased significantly by changing the NH_4-N: NO_3-N ratio from 1:1 to 1:5. The increase in productivity differed between clones. In liquid media fresh weight production within 6 weeks was 3 times higher than on solid media with both NH_4-N : NO_3-N ratios (Zens and Zimmer, 1986).

Regeneration from leaf explant has been found better than other plant parts. Pierik *et al.* (1975) suggested a method of propagation that can be used in commercial practice. After callus has formed on pieces of leaf tissue it is transferred to a solid medium. When the callus starts growth it is transferred to a liquid medium in Erlenmeyer flask, which is placed on a shaking machine rotating at 120 revolutions per minute. All the 7 cultivars tested in this medium grew well but there were marked differences in growth and vigour among them. To obtain plantlets the callus clumps are transferred again to a solid medium in the dark, where shoots are formed; these shoots are later rooted in light. Multiplication by this method is far more rapid than with the use of only solid agar medium. The factors affecting callus formation from pieces of leaf tissue of *Anthurium* spp. were studied by Leffring *et al.* (1976). Disinfection of the tissue for 20 minutes resulted in less damage than disinfection for 3 minutes. Tissue from the leaf margins and that from the apical half of the leaf produced callus better than that from the basal half. Regeneration was also highly dependent on genotype and leaf age. It was found that once leaf hardening had started, regeneration decreased and was nil in fully hardened leaves (Geier, 1986). Leffring and Soede (1979a), working with *A. andraeanum*, reported that addition of 3 mg 2-isopentyladenine (2ip) to a modified MS medium containing 1 mg BA/l in which leaf sections were cultured, resulted in widespread shoot formation from 65 per cent of the stock plants tested. They (1979b) also observed that when shoots had already formed on leaf pieces kinetin was preferable to benzyladenine for inducing shoot growth and side shoot produciton, and 3 mg/l was usually the best concentration. Application of 2-isopentyladenine at this stage of growth was unsatisfactory. A liquid agar-based medium was better than a solid one. Roots formed readily with all concentrations of agar but 5 g was better than 6 or 7 g/l.

Successful regeneration of leaf explants of *A. andraeanum* cultured on a basic nutrient agar medium was also recorded by Pierik *et al.* (1979). Regeneration was optimum with 0.125-1.00 ml Tween-20 per litre during sterilization, by using growth substances (adenine 0.1 mg/l, zeatin 1 mg/l or 2, 4-D 0.08 ml/l), culturing for 16 weeks in darkness followed by 4 weeks light and by including 206 mg/l NH_4NO_3. Shoot regeneration in callus was promoted by NH_4^+ ion and not by NO_3^- ion.

Sreelatha *et al.* (1998) observed that callus initiation in leaf explants of 4 *Anthurium* sp. was best in MS medium with combination of 2,4-D and BA. Modified MS medium with a reduced salt concentration was ideal for callus initiation. Callus initiation was better from basal portions of leaves than from apical portions. Continuous darkness was necessary for callus initiation. MS medium with 1/4 strength nutrients was sufficient for callus multiplication. Shoot regeneration and growth were best in medium supplemented with 0.5 mg BA + 2.0 mg IAA/l. No rooting treatments were found necessary as the shoots rooted spontaneously in the shoot regulation medium itself.

Leaves of *A. andraeanum* (*A. andreanum*) cultivars Tropical Pink, Premium Red, White and Tulip were soaked in 32.5 per cent benlate (benomyl) solution for 24 h, sterilized by rinsing in 70 per cent alcohol for 45 s and soaking in 1.25 per cent sodium hypochlorite for 15 minutes, then rinsed 3 times in sterile distilled water for 15 minutes. Leaf explants (1-2 cm^2) were taken and inoculated onto callus induction plates with Nitsch's basal medium. The best callus formation was obtained with *cv.* Tulip at 0.5 mg 2,4-D/l; Tropical Pink gave good results with 0.05-0.5 mg 2,4-D/l. Good shoot multiplication (3-5 shoots/callus) was obtained at 0.2-0.8 mg BA/l in all cultivars. Root formation was not affected by varying concentrations of ammonium nitrate. However, larger leaves and more prolific leaf production occurred with increased ammonium nitrate concentrations (Lightbourn and Prasad, 1990).

Leaf explant of 7 cultivars of Hawaiian anthuriums (*A. andraeanum*) produced callus most successfully after 2 to3 months on a modified Pierik medium containing 0.36 µM 2,4-D and 4.4 µM BA. Petiole explants callused best on Pierik, modified Pierik, and Finnie and van Stadenmedia. Adventitious plantlets were recovered from callus planted on a Kunisakimedium containing 2.2 or 22 µM BA. Regeneration appeared to be organogenicrather than embryogenic and varied among the genotypes tested (Kuehnle and Sugii, 1991).

Spadix explants had a better capacity for regeneration than leaf segments on the modified Nitsch media used, and the plantlets derived from spadix segments were less variable (Singh *et al.*, 1991). Histological examination of *in vitro* cultured *A. andraeanum* plantlets showed that plantlets were formed from adventitious buds. Light was superior to darkness for the induction and growth of callus. Comparing the effects of different concentrations of sucrose and glucose, it was found that 3 per cent glucose had the greatest inductive effect on callus formation. Fullstrength MS medium + 1.0 ppm BA had the best inductive effect on bud formation, while half-strength MS medium + 0.1 ppm NAA was best for inducing root formation (Cen *et al.*, 1993). Callus culture was more promising than shoot culture for micropropagation (Yu and Paek, 1995).

Seeds collected from physiologically ripe inflorescences were disinfected by dipping in 70 per cent alcohol for 1 s, dipping in sodium hypochlorite for 5 min rinsing 3 times with distilled water, dipping in sodium hypochloride for 8 min and rinsing again 3 times. All explants from this treatment were healthy and seed germination was 100 per cent. Seedling emergence was enhanced by a photoperiod of 16 h light and a temperature of 27 ± 1°C. More than 15 adventitious sprout/explant was produced within 60 days on MS medium supplemented with BA at 0.22-0.88 µM. Percentage survival was high (90 per cent) when 3- to 4-leaved plantlets were transplanted to pots containing a mixture of soil and organic matter (1:1) (Montes *et al.*, 1999).

Farsi *et al.* (2012) were able to establish producible protocol for indirect *in vitro* regeneration of the Terra genotype and found callus with the highest fresh and dry weight was produced on modified Murashige and Skoog medium containing 0.1 mg/l 2,4-dichlorophenoxy acetic acid (2,4-D) and 1.5 mg/l 6-benzylaminopurine (BAP). For regeneration stage, induced calli were transferred to modified MS medium without plant growth regulators, and after six weeks, shoots formed. For rooting, regenerated shoot were transferred to MS medium without any plant growth regulators. Regenerated plantlets potted on mixture of coco pit and perlite were succesfully acclimated.

Maitra *et al.* (2012) observed that ½ MS medium supplemented with coconut water 15 per cent (v/v), NAA (1 mg/l) and BAP (3 mg/l) was effective for shoot regeneration of anthurium from adventitious bud explants. For early rooting ½ ms medium supplemented with NAA (1 mg/l) and IBA (2 mg/l) was found ideal. This medium (AR2) started rooting on 39.10 days after culture. Though culture on ½ ms medium supplemented with NAA (1 mg/l)+IBA (3 mg/l) started rooting 42.67 days after transfer but this medium (AR3) represented 84 per cent explant survivability hence, may be advocated for rooting of anthurium micro-cuttings.

MS and ½ MS culture medium at pH 5.7 showed the most beneficial effects on growth parameters of anthurium 'Tropical' plantlets (Kanani *et al.*, 2017). Plant regeneration of *A. andraeanum* varieties "Red Love" and "Pink Champion" have been achieved through adventitious shoot formation from calli and the combination of 0.1mg/l 2,4-D and 1.5mg/l BAP found found to be the best combination for calli biomass production (Aygün and Aki, 2016).

According to Toppo and Beura (2018), the leaf explants of *Anthurium cv.* Xavia, exposed for 5 minutes to 0.1 per cent $HgCl_2$ resulted in minimum fungal (0.00 per cent) and bacterial (0.43 per cent) contamination with maximum survival percentage (97.41 per cent) and 0.1 per cent $HgCl_2$ can assure the explants sterility and regeneration capacity.

Table 4: Organogenesis *In vitro* from *Anthurium andraeanum*

Cultivar	*Explant*	*Media*	*Hormones*	*Response*	*Sources*
Anthurium andraeanum *cv.* Rubrun	Seed	MS Basal medium with 1.2µM Thaimin+0.6µM myoinositol+0.3 per cent sucrose+0.2 per cent Gelrite at pH 5.8 before autoclaving.	Callus 8.9 µM BA +2.7 µM NAA and were incubated under continuous fluorescent Light (50 m mol/m²/s) at 25 °C.	80 per cent of plant Acclimatization was obtained.	Vargas *et al.* (2004)
Anthurium andraeanum *cv.* Nitta (red color)	Leaf mid rib	Callus and shoot Basal MS medium -Rooting MS Basal medium	-Callus and Shoot MS medium with hormone combination 0.5,1 and 1.5mg/l NAA+0.5,1,1.5and 2 mg/l BAP – Rooting MS medium with 0.5,1 and 1.5 IBA+0.5,1,1.5 and 2 BAP	1mg/l NAA+1mg/l BAP best for survivability 1mg/l IBA+1mg/l BAP best for shoot formation	Islam *et al.* (2010)
Anthurium (Dual light condition)	Spathe	Callus Nitsch and Nitsch medium -Shoot Nitsch and Nitsch	Callus 1mg/l 2,4-D + 1mg/l Kinetin + 1mg/l BA - Shoot 1 mg/l 2,4 D + (1, 2, and 3) mg/l BA	Better callus development in dark light after 60 days.	Budirto *et al.* (2007)

IBA: Indole-3-butyric acid; NAA: Naphthalene acetic acid; 2,4-D: 2,4-Dichloropenoxyacetic acid; BA-6: Benzy laminopurine.

8.0 Climate

Anthuriums require warm greenhouse with shading from direct sunshine and a humid condition. The optimum temperature for growths is 18-21°C and the minimum temperature should not be less than 10-13°C for a short period. The relative humidity which also plays an important role in the growth and development of anthurium should be around 80 per cent (Ignasse, 1984), higher humidity has, however, marginal effect on the plants (Papenhagen, 1986). Bright but filtered light is essential for abundant flowering. The ideal conditions for growing anthurium are 25 to 30°C temperature, 80 per cent relative humidity and shade up to 75 per cent (Mandhar *et al.*, 1998).

9.0 Cultivation

9.1.0 Greenhouse

Construction of suitable greenhouse to provide ideal growing environment including light, temperature, *etc.* is very essential for commercial production of anthurium. Three different systems of glasshouse shading were compared by Meij (1976). Flower productivity was lowest and the flowers were smallest in the house with whitewash on the roof, roof sprayers and a plastic screen. Flower yield was,

Growing of Anthurium

Seedlings of Anthurium

Planting of Anthurium

Planting of Greenhouse

Flowering in Anthurium

Cultivation in Pots

Cultivation in Greenhouse

however, better in a house with whitewash, roof sprayers and crop sprinklers but no plastic screen. A house with movable external screens and crop sprinklers was intermediate in productivity. Berg and Valentin (1977) suggested plastic film and cloth screening for saving energy. In a glasshouse planted with *A. andraeanum*, mobile horizontal screens of transparent polythene and 2 types of cloth material were compared for preventing heat loss. All three types of screens resulted in a fuel saving of about 30 per cent with a wind speed of 1-7 m/sec. As the wind speed increased the fuel saving also increased greatly. Similarly, Breedveld and Glass (1984) compared the yield of *A. andraeanum* grown under single glass with those under double ones and recorded slight differences in cut flower production/m^2, flower quality and size.

Vegetatively propagated lines of 6 hybrid anthurium cultivars (Pink Aristocrat, Patty Anne, Purple Viking, Royal Pink, Royal Orange and Royal Red) were planted in pots and grown under warm (highest temperature of 30°C) or hot (highest temperature of 35°C) conditions in greenhouses. Plants under warm and hot conditions, except for Pink Aristocrat, had similar degrees of foliar injury in April, but plants under warm conditions had better quality in July than those under hot conditions. Growing these cultivars at highest air temperature of 30°C is recommended for good quality and high flower count (Wang and Wang, 1999). Zicsi *et al.* (1999) reported that an introduced earthworm (*Amynthas corticis*) caused damage to potted anthurium plants in a greenhouse at Hohenberg, Austria.

The use of screen or plastic films of different nature or color can change the quality of spectral radiation, and as consequence the growth and production of a crop. The evaluation of *Anthurium andraeanum* 'Apalai' under 70 per cent shading (blue, red, black and thermo-reflective) screens revealed that the black screen gave the best results, providing the best conditions for growth and production of anthurium, and the larger dimensions flowers stem, that are considered attributes of quality to cut flower (Nomura *et al.*, 2009).

9.2.0 Soil and Media

The media for growing anthuriums, whether in pot or bed, should be light, well drained and rich in organic matter. For this, a number of materials have been tested and the most suitable ones are selected. In comparison of several growing media, leaf growth and flower production of *A. andraeanum* was better in gravel than in peat or in a mixture of sphagnum and coniferous forest soil. Growth differences between the media were less marked with *A. scherzerianum*, but flower production was again highest in gravel (Steiner and Van Zon, 1957). Nakasone and Kamemoto (1957) also investigated the effects of six different media and found that the 1:1 mixture of wood shavings and soil or 5 : 1 mixture of wood shavings and cow manure and tree fern fibre produced the best plant growth. Rodriguez *et al.* (1979) compared several other materials with peat moss for the production of cut flowers. After more than three years the results showed that materials such as sugarcane bagasse, coffee leaf mould, wood shavings, spent ground coffee, cured coffee pulp or coffee parchment, chicken manure, tree bark and top soil + filter press cake were as effective as peat moss for the production of flowers. Plants on bagasse produced the highest number of flowers (1421 dozens/ha/week) followed by those on spent ground coffee (1391 dozens/ha/week). Similarly, sphagnum (Kuhle, 1960), pumice (Bik *et al.*, 1960), peat (Jungbauer, 1973; Boertje, 1978; Straver, 1980; Turski *et al.*, 1983), bark (Tesi and Faro, 1985), *etc.* have been suggested to be used for growing anthuriums.

The current trend is towards the use of peat or bark-based compost. Presence of peat in the potting mixture is associated with higher yield of quality flowers. Out of the six substances compared by Nikolova and Zafirova (1980), a medium containing peat + pinebark + perlite (2:2:1) gave 98.8 per cent top grade flowers and 18 per cent higher flower yield as compared to other media used for growing *A. andraeanum*. Hetman *et al.* (1981) suggested a medium containing peat + superphosphate + perlite (equal quantity) for better growth and flowering in this species. Pot trials conducted by Turski *et al.* (1983) showed that the best substrates were generally those in which the basic component was high peat. The tallest plants with the highest number of leaves were found on a 2:1:1 mixture of high peat : perlite : sphagnum moss. With CO_2 enrichment (1200 µl/l) growth was greatest in rockwool + perlite medium (Matysiak and Nowak, 1996). The effects of organic media [peat + perlite (3:1) or Jiffy-7] and mineral media (perlite or rockwool) on the growth of rooted micro-cuttings of *A. cultorum* (*A. ferrierense*), at CO_2 concentrations of 350 or 1200 µl/l were investigated by Matysiak *et al.* (1996). Peat + perlite (3:1) and Jiffy-7 were the best substrates for *A.* x *cultorum* grown at low CO_2 concentrations. At high CO_2 concentrations, good shoot growth of *A.* x *cultorum* was observed on all growing media except perlite.

Table 5: Influence of Medium and Harvest Time on Flower Production of Ozaki Red Anthurium*

Medium	*Number of Flower per Plant*		
	First Year	*Second Year*	*Third Year*
Bagasse	4.3aZ	4.0 ab	3.8 c
Wood shaving	4.0 ab	4.1 ab	3.6 cd
Cinder/Bagasse	4.6 cd	3.7 cd	3.7 cd
Tree fern chips	3.3 c	3.0 f	2.8 g
Black cinder	3.1f	2.9 fg	2.7 g

Z: Mean separation by Duncan's multiple range test, 5 per cent level. Nos. are average of 48 units.

* Higaki and Poole (1978).

Based on a review on the use of powdered bark as a constituent of the growing medium for container grown plants, Tesi and Faro (1985) concluded that anthuriums could easily be grown in a bark based medium. In an experiment on *A. andraeanum cv.* Ozaki, Higaki and Poole (1978) concluded that sugarcane bagasse was the best media for flower production followed by wood shavings, black cinder (Table 5). Higaki and Imamura (1985) also studied the performance of wood products as media for culture of anthuriums. For flower production, stem length and flower size, the effectiveness decreased in the order: wood bark > wood chips > bagasse =

red wood. Thus, a damp but well drained spongy compost consisting of peat and sphagnum moss is ideal for most of the anthuriums. Voogt (1979) reported that the flower yield was inversely correlated with the salinity of the soil. Out of the salts present in soil sodium chloride was most detrimental to growth and flowering.

Growth of anthurium *cv.* Lady Jane was compared in container media in which the primary organic component was sphagnum peat, sedge peat or coir dust. Growth index and shoot DW were only marginally higher in the coir than in the sedge peat medium, and root DWs were comparable. It was concluded that high quality coir dust appears to be an acceptable substitute for sphagnum or sedge peat in soilless container media (Meerow, 1995). In recent years, the anthurium industry in Jamaica has been affected by several problems associated with the use of coconut husk as a growing medium. Trials were carried out on the use of washed river gravel as an economic alternative growing medium. This medium is not biodegradable, is easily managed, readily available and also has the required physical and chemical properties for anthurium growth. Preliminary results indicated that there was good growth and yield in this medium with minimal disease problems (Tai and Ricketts, 1991). Tree ferns (*Dicksonia squarrosa*) was evaluated as a growing medium for a range of plants. The growth of anthurium was better in tree fern medium (Prasad and Fietje, 1989).

Soluble salt concentrations in leachates from a growing substrate (peat : pine bark, 1 : 1) receiving various amounts of fertilizers, were examined and related to the growth rates of anthurium. The fertilizer applied was 19-2.6-10 Osmocote at 1-13 g per 10-or 15-cm pot. It was observed that anthurium plants tolerated a wide range of Osmocote levels (Poole and Chase, 1989). Four soilless media, 1 : 1 (v/v) pine bark compost : perlite, volcanic : cinder, 2 : 1 (v/v) cinder : peat, and a 1 : 2 : 1 (v/v/v) rockwool : cinder : peat, were evaluated by Wang *et al.* (1997) in the greenhouse for their effects on the growth of anthurium cultivars Alii and Midori and on the population increase of nematode (*Radopholus similis*). Plant growth expressed as shoot and root dry weight, root vigour, shoot damage, number of new leaves and flower number was less affected by nematode damage in cinder than in the other media. These effects were also more pronounced in Alii than in Midori. Orthogonal contrast analysis indicated that nematode damage on Alii was favoured by the presence of organic matter in the medium.

Anthurium solitarium Schott is a species that might be used both as potted plant and cut foliage. Plants grown in substrates with 100 per cent bamboo compost (BC) showed higher growth, ornamental potential, market quality and acceptance than those grown in coconut fiber (CF) (Maitan and Jasmim, 2020).

In India, anthurium cultivation under protected conditions influenced by supply of nutrients and varieties grown. The fertilizer dose of 36:36:72 g NPK/plant/year found to be significantly superior in quality and quantity parameters of anthurium *cv.* 'Flame' (Giridhar *et al.*, 2012).

To standardize growing medium in pots for anthurium *cv.* Flame under protected conditions of 75 per cent shade net house, Singh *et al.* (2011) noted that best medium comprised of saw dust + brick pieces + wooden charcoal + soil + sand

+ FYM (2:1:1: 1: 1:1) for leaf area, petiole length, minimum days to flowering, stalk length, spathe length and width and number of flowers per plant whereas, the maximum number of leaves per plant, highest number of suckers per plant and longest inflorescence longevity were recorded in treatment combination involving saw dust + wooden charcoal + soil + sand + FYM (2:1:1: 1: 1).

Paramveer *et al.* (2011) noted that the pot planted anthurium *cv.* Flame under protected conditions of 75 per cent shade net house with the treatment saw dust + brick pieces + wooden charcoal + soil + sand + FYM (2:1:1: 1: 1:1) influenced the vegetative growth, flowering pattern and flower production of anthurium. The sawdust mulching is an asset to anthurium cultivation by improving the soil's organic matter content, stimulating plants growth and yield when compared results with bare soil. The application of 40 l/m^2 of sawdust every three months is recommended (Lima *et al.*, 2016).

Tissue culture plantlets of *A. andreanum* cv. Evita planted in cocopeat, charcoal and vermicompost in the proportion of 6.5:1:1 filled plastic pot placed under shade house (75 per cent shade net) and fertilized with NPK (19:19:19) at 25 g/pot/year can enhance growth and flowering characters (Khawlhring *et al.*, 2019)

With respect to vegetative growth, Basheer and Thekkayam (2012) observed that sand + coir pith compost was found to produce the maximum plant height, leaf area and petiole length at the fourth week after emergence and shortest phyllachron, highest fresh and dry weight of leaves and their N and K content at 225 days after planting. Vegetative growth enhancement was also obtained with 4 g/l and 6 g/l fresh cow dung extract. Plants receiving 25 g organic manure mix recorded greater plant height and leaf area during the latter stages of growth. Plants grown in coconut husk + charcoal media (3:1) with foliar spray of 30: 10:10 NPK at 0.2 per cent on anthurium plants performed superior with respect to vegetative growth parameters such as plant height, number of leaves per plant, leaf area and plant spread was recorded.

Plants grown in coconut husk and charcoal as growing media with foliar application of 12: 61: 40 NPK at 0.2 per cent weekly produced significantly maximum flower stalk length, spathe length, spathe width and maximum number of flowers per plant and advanced flowering (Sumathi *et al.*, 2019).

9.3.0 Potting and Planting

Potting is an essential operation for anthuriums. As soon as the seedlings are large enough to handle, cuttings have made good roots, *in vitro* cultured plants bear fully developed roots and shoots or offshoots are ready for division, they should be moved into pots. Care should be taken that the pots are of suitable sizes. Potting may be done fairly firmly when using loam-based composts, firming with fingers, but peat-based composts should be only lightly firmed and watered to settle the compost. Peat compost should never be allowed to dry.

If planting is to be done in beds, plants should properly be spaced. Four plant densities (5.2, 7, 8.7 and 10.5 plants/m^2 of bed) were compared by Steen (1976, 1977) for growing *A. andraeanum*. It was recorded that the number of flower/m^2 increased with plant number, this effect being greater in the first year than in the second. However, the productivity per plant declined with increasing plant density. The number of flowers/m^2 increased slightly as more leaves were retained on the plants. Leaving side shoots on the plants had a positive effect on the number of flower/m^2, but this effect became less pronounced as plant density increased. When all side shoots were removed, plant density did not influence flower diameter but when they were retained increasing density had a slightly adverse effect on diameter.

Plants should be carefully examined to decide when repotting is due. The new pots should not be very much larger than the previous ones. Young plants need repotting every year, while the adult ones every 2-3 years (Chiusoli and Boriani, 1986).

9.4.0 Irrigation

Anthurium requires generous watering. However, the right amount of water and proper intervals between successive applications of water are determined by several factors. Microclimate in the greenhouse, season, size of the plant in relation to the diameter of pot, stage of plant growth, type of container and the compost in which the plant is being grown are important. Plentiful watering should be done during spring and summer, and in winter, during resting period the surface soil is allowed to dry out between successive watering.

The quality of water also has considerable influence on plant growth and development. The cut flower yield of *A. andraeanum* declined progressively as the salinity of the water used for glasshouse irrigation increased. Water containing sodium chloride was particularly detrimental. The threshold electrical conductivity (EC) value above which flower yield reduction occurs was found to be 0.6 m mhos/cm (Voogt, 1979; Sonneveld and Voogt, 1983). Similarly, Spek and Sonneveld (1985) reported that excess NaCl, Na_2SO_4 or $MgCl_2$ was highly detrimental for lowering flower production by 40 to 46 per cent. The EC value of irrigation water higher than about 1.5 m mhos/cm reduced the flower weight, the effect being most marked with NaCl alone. The availability of most of the minor elements is adversely affected at pH above 6.5 and the plant shows deficiency symptoms. Soluble salts or EC of good water should be less than 1.0 m mhos/m. High Na level causes substrate to tighten up, holding more water and less air, reducing root growth significantly (Aswath *et al.*, 1998).

9.5.0 Manuring and Fertilization

Anthurium need adequate amount of nutrients for their proper growth and flowering. Among the major elements, application of nitrogen, potassium and calcium markedly improve the yield and quality of flowers. Deficiencies of these nutrients, on the other hand, adversely affect the plant growth and developments. Insufficient levels of nitrogen and potassium are associated with lower flower yield, reduced stem length and smaller flowers. Leaf necrosis and dead root tips were observed with K deficiency (Bik and Straver, 1978; Imamura and Higaki, 1984). Application of 126 mg N/12.51 container per week was suggested for *A. andraeanum* by Bik (1976). An increase in K from 19 mg/container per week to 225 mg improved both flower yield and quality, but a further increase had a slightly negative effect. Higaki and Poole (1978) reported that low fertilizer rate gave as good or better results in flower production as the high rates (Table 6). In an experiment with 5 levels of N and 3 of K, the best results were obtained with an annual dressing of 29 g N + 30 g K_2O/m^2. At optimum nutrient levels, mature leaves contained 2 per cent N and 3 per cent K (Boertje, 1978). Nitrogen also had a beneficial effect on the quality of potted *A. scherzerianum*, the best result being obtained with a medium dose of 21.6 mg N/pot/week. Application of 22.5 mg K/pot/week gave much better-quality plants than 1.9 mg but raising the dose further did not result in any improvement (Bik and Straver, 1978).

Table 6: Influence of Fertilizer Levels on Flower Oroduction of Ozaki Red Anthurium for 3 Years*

Fertilizer levels (kg N-P-K/ha/year)	*Number of Flowers per Plant*		
	First Year	*Second Year*	*Third Year M*
448-196-370	3.7a	3.6 a	3.3 b
896-392-740	3.6 a	3.6 a	3.3 d
1344-588-1110	3.7 a	3.4 a	3.1 c

Mean separation by Duncan's multiple range test, 5 per cent level. Nos. are average of 80 units.

* Higaki and Poole (1978).

It was observed that growth of tissue cultured plants was best at the lowest light intensity (120 ft-candles) and with 20 : 20 : 20 NPK fertilizer at 200 and 400 g/l (Henny and Fooschee, 1988). *A. andraeanum* cultivars Tropical and Cuba were grown in granulated polyphenol foam for 2 years. The nutrient solution used was added to the irrigation water. Flower production and vegetative growth decreased linearly with increasing nutrient concentrations. In cultivar Tropical the total flower weight decreased by 8 per cent and the vegetative plant weight by 9 per cent per unit EC. In cultivar Cuba the corresponding decreases were 13 and 11 per cent, respectively. The results suggested that a nutrient solution EC of 0.7 dS/m can be recommended for optimal production (Sonneveld *et al.*, 1993).

Optimum flower production was achieved with annual rates of 312 kg N + 448 kg P + 375 kg K/ha. Increased N and K application resulted in a linear increase in flower size. Flower stem length also increased with increasing N, P and K rates. Maximum flower yield occurred when leaf-tissue levels were 1.87 per cent N, 0.17 per cent P and 2.07 per cent K. Flower stem length and flower size were at their maximum with leaf N at 1.59 per cent and 1.67 per cent and K at 2.20 per cent and 1.86 per cent, respectively. No relationship was observed between leaf P content, flower size or stem length (Higaki *et al.*, 1992). Paull *et al.* (1992) studied the effect of fertilizer on postharvest life of flower. High N rate reduced postharvest life but this was alleviated by high K, P had no effect. Spadix necrosis and spathe condition signalled the end of vase life at low N and P and at high K rate. Loss of spathe gloss was signifcant at high N and low K rates. Optimum annual fertilizer rates for maximum postharvest life were 0 kg N/ha and 224 or 448 kg K/ha.

Adequate level of calcium is also necessary for obtaining optimum yield and to stabilize spathe colour. Deficiency of calcium resulted in colour break down, instability of the middle lamella and cell separation and collapse in proximal section of the lobe and spathe. The problem is more serious at lower pH of the substrate (3.0 to 4.0). The optimum calcium content in the lobe and leaf tissues have been found to be 0.16 and 0.54 per cent, respectively (Higaki, 1977; Higaki *et al.*, 1980a,b). Penningsfeld (1977) reported the significance of soil reaction and boron application on growth and flowering of *A. scherzerianum* in a peat substrate. Plants having the highest fresh and dry weights, the largest number of flowers, the longest stems and the best quality flowers with no leaf chlorosis were observed with 4 g $CaCO_3$ and no boron. However, the longest spadix and flower fresh weight were obtained with 4 g $CaCO_3$ and 2.0 mg boron.

NPK at 30:10:10 at 0.2 per cent foliar spray and GA_3 at 100 ppm along with bio-fertilizers (*Azospirillum*, Phospho-bacteria and VAM each at 2 g per plant) significantly influenced the flower quality, vase life and sucker production of Anthurium in *cv.* Tropical Red (Handaragall *et al.*, 2013).

Experiment conducted by El-Naggar *et al.* (2016) reported that planting under greenhouse conditions with applying 10 g/pot of mineral NPK monthly during the growing seasons, combined with active dry yeast at 6 g/l improved the vegetative growth and flowering parameters (plant height, number of leaves, leaf area, leaf dry weight, number of flowers/plant, flower dry weight and leaf chemical composition parameters [total chlorophyll and mineral content (N, P and K per cent).

The growth response and sustainability of flamingo flower in different growth media were evaluated, results indicated that growth media significantly affected the percentage of surviving plants. Sawdust + poultry manure (3:1) (93.33 per cent) was similar to those in coconut coir + poultry manure (3:1) (80.00 per cent). Plants in the medium, coconut coir + poultry manure (3:1) (7.33) were the better in terms of visual quality at 85 days after planting (85 DAP) (AdeOluwa *et al.*, 2017).

Anion proportion in the nutrient solution impacts the growth and nutrient status of Anthurium increased growth of anthurium was attained at either high or low NO_3^{-1} proportion and it is able to cope with high SO_4^{-2} by avoiding the transport of S to the shoot, decreasing SO_4^{-2} intake, maintaining a favorable internal N/S and S/P proportion, and increasing P tissue concentration (Juárez-Maldonado, 2017).

Nitrogen concentration of 105 or 158 mg/l sufficiently met the nutrient requirement of Anthurium cultivated in coir during the growing and flowering stages. Regardless of plant growth, flower yield, and nutrient uptake, there was interactions between N treatments and seasonal fluctuations in a subtropical region during year-round cultivation of Anthurium (Chang *et al.*, 2012). While Dufour and Guérin, (2005) found that concentration of 8.9 mmol N/l of nutrient solution is sufficient for a good yield and flower.

The growth, development, yield, and cut flower quality of *A. andreanum* receiving the PRHC (pea and rice hull compost) treatment were the same as those receiving CNS (chemical nutrient solution) and CRF(controlled release fertilizer), indicating that PRHC can replace the chemical fertilization management system for the cut flower production of anthurium cultivated under soilless conditions. The nutritional status of plants that received the PRHC treatment also demonstrated that the supply of nutrients from PRHC was adequate and met the requirements of *A. andreanum* for C assimilation, growth, plant development, and nutrient uptake (Chang *et al.*, 2010).

9.5.1 Nutrient Deficiency

Nitrogen deficiency in anthurium can diminish yield and flower quality, besides it increases the juvenile period. Roots damage and secondary infection by fungi and bacteria are also associated with nitrogen deficiency.

Phosphorus deficiency in anthurium results in atrophic plants, with small, dark-green and narrow leaves, with short petioles; old leaves are chlorotic, with necrotic areas along the margins also affecting the root system.

The supply of potassium directly affects the plant development by controlling absorption and evaporation of water. It plays an essential role in yield and flower quality. Therefore, potassium deficiency diminishes the length of the floral stem-the red and orange spathes can show bluish points or areas (Tombolato *et al.*, 2002).

9.6.0 Weeding

In trials to evaluate herbicides for use in *A. andraeanum cv.* Ozaki Red, De Frank *et al.* (1989) observed that 2.2-9.2 kg oxyfluorfen/ha + 1.1 or 4.5 kg oryzalin, 3.3 or 6.6 kg oxadiazon and woven plastic ground cover provided excellent weed control. Spray applications of oxadiazon were phytotoxic to anthurium leaves and emerging flowers, but the granular material, used > 268 days after planting (DAP), was not. Oxyfluorfen + oryzalin (9.2 + 4.5 kg) and both rates of oxadiazon reduced marketable flower yield. At 733 DAP, all plants treated with oxadiazon and oxyfluorfen + oryzalin (4.5 + 2.2 and 9.2 + 4.5 kg) were extremely stunted.

10.0 Diseases and Pests

10.1.0 Diseases

Fungal Diseases

10.1.1 Anthracnose

This is the most damaging disease of anthuriums, caused by *Colletotrichum gleosporioides*. Also known as spadix rot or black nose and it is a problem in high rainfall areas. Spadices are damaged and flowers become unsuitable for commercial purpose. Circular or irregularly shaped lesions were observed on leaves of *A. scherzerianum* and *A. andraeanum* in Japan. *Colletotrichum gloeosporioides* (*Glomerella cingulata*) was isolated, and its pathogenicity was confirmed (Kagiwata, 1990).

Severe anthracnose was observed on *A. andraeanum* in vinyl-housed fields and greenhouses in Gyunggi province, Korea Republic (Kim *et al.*, 1993). The fungus (*C. gloeosporioides*) caused typical anthracnose symptoms on leaves, flower stalks, spathes and spikes (Deshmukh *et al.*, 1999).

In December 2016, Anthurium flowers with circular necrotic spot symptoms were collected in Brazil. Based on morphological and molecular characterisation, the fungus isolates were identified as *Colletotrichum theobromicola*, which were detected for the first-time causing anthracnose on Anthurium in the world (Chaves *et al.*, 2020).

Maneb at 2 lb/100 gal and dodine or dyrene at 1 lb/100 gal gave good control of the disease. It was suggested that application should be made at 2-week interval and a sticker should be used. Studies of maneb treated anthurium spadices inoculated with strain of *C. gleosporioides* indicated that no maneb resistant strains of the fungus had evolved (Aragaki, 1964; Aragaki and Ishii, 1960). Certain cultivars like Marian Seefurth, Uniwai and Manova Mist are reported to be resistant to the pathogen (Kamemoto *et al.*, 1971).

10.1.2 Black Rot

Black rot is reported to be caused by *Phytophthora citrophthora* and *P. parasitica* (Pitta *et al.*, 1991; Ann, 1992). Symptoms occur on the bright coloured spathe, spadix, leaves, stalks and petioles as black rot with wet tissue at the margin (Pitta *et al.*, 1991).

10.1.3 Root Rot

Root rot of *A. andraeanum*, incited by *Pythium splendens* often causes serious losses in Puerto Rico, especially during rainy season. *Calonectria crotalariae, P. splendens, P. spinosum, P. vexans* and/or binucleate *Rhizoctonia* spp. were isolated from 11 anthurium cultivars showing symptoms of root rot in Hawaii, USA (Guo and Ko, 1994). The herbicide diuron increased the disease incidence caused by all most all the pathogens (Guo and Ko, 1994, 1996). The disease can be controlled by a soil application of PCNB (quintozene) at 8 oz/50 gal water (Liu and Serapion, 1977). Root and Crown Rot of Anthurium Caused by *Calonectria Ilicicola* in Iran (Mirabolfathy *et al.*, 2010).

10.1.4 Leaf Spot

Leaf spot disease is caused by *Septoria anthurii* and controlled by sprays of 300 g carbileen (zineb) in 100 l water, repeated at interval of 2-3 weeks. A non-parasitic leaf spotting may occur under unfavourable cultural conditions. Among the several fungicides tested *in vitro* under different conditions of temperature and humidity, oxyquinoline sulphate gave best control of *S. minima* (Bosmans and Veldeman, 1964).

10.1.5 Powdery Mildew

This disease, caused by *Erysiphe communis*, was first reported on *Anthurium scherzerianum* grown in Germany (Schneider and Kiewniek, 1974). Treatment with 0.1 per cent benomyl easily controls the disease.

10.1.6 Blight

It is caused by *Xanthomonas campestris* pv. *dieffenbachiae*. In Hawaii, there had been two major outbreaks, in 1971 and 1980, and the disease now pose a considerable threat to the commercial anthurium growers because of systemic infection. Symptoms of this disease included necrotic lesions surrounded by a yellow halo, water-soaked spots and combinations of these symptoms. The causal organism was identified as *Xanthomonas axonopodis* pv. *dieffenbachiae*. Among 75 strains of the pathogen tested, 68 were resistant to streptomycin and 6 were tolerant of copper hydroxide (Hseu and Lin, 1998). The structure of the hydathodes in *A. andraeanum* is such that bacteria entering through the pores may be able to swim all the way to the xylem vessels in the guttation fluid (Sakai *et al.*, 1990b).

Natural and Alvarej (1990) noted differences in varietal reactions and *cv.* Ozaki being very susceptible and Nitta having some tolerance. Of 10 cultivars grown in new and old cinder media without the removal of diseased leaves, the most resistant to bacterial blight (*X. campestris* pv. *dieffenbachiae*) was Mauna Kea (91.1 per cent survival after 11 months) followed by Gervais Orange Obake (86.6 per cent) and Kaumana (74.1 per cent), compared with the most susceptible, Splash, with only 2.7 per cent survival (Sakai *et al.*, 1990a). Varying degrees of resistance were observed among the 15 cultivars tested. The pot cultivars Julia and Gemini were the most resistant, while the cut flower cultivar Hearts Desire was the most susceptible. Each cultivar displayed different degrees of resistance to individual isolates of the pathogen (Norman *et al.*, 1999). The cultivar susceptibility of anthuriums to bacterial blight may differ depending on the phase of disease progression, and thus evaluation of disease phases is essential for complete understanding of cultivar susceptibility (Fukui *et al.*, 1998).

Strict sanitation measures, the removal of affected leaves and spraying with streptomycin sulphate or oxytetracycline are recommended. The main practices include growing plants from clean tissue cultured material, proper bed preparation, preventing disease from entering fields by minimizing movement of personnel and tools, preventing the spread within the field by following sanitation procedures for blight removal and harvesting, and providing the best cultural conditions (Sewake and Alvarez, 1990). The bacterium appears to be resistant to copper-based preparatios, and those can also be phytotoxic to anthuriums (Sato, 1983; Nishijima and Fujiyama, 1985).

The contact biocide agribrom killed the pathogens of blight. Although the plants are fairly sensitive to agribrom, a concentration of 100 ppm caused only minor chlorosis to some cultivars. The bacterium was also killed *in vitro* by exposure to oxytetracycline at 32 ppm for 16 d and *in vitro* by drenching infected plants with oxytetracycline at 1000 or 1500 ppm 3 or 4 times at weekly intervals (Nishijma and Alvarez, 1990).

B. subtilis EPB14 demonstrated the best antimicrobial activity against the pathogen as a biological agent in controlling bacterial leaf blight of anthurium (Kumsingkaew and Akarapisan, 2014).

Bacterial blight of anthurium (*Anthurium andreanum*) can be managed by using ecofriendly materials such as five sprays (at weekly intervals) of both turmeric powder and streptocycline (Dhanya and Mary, 2007).

10.1.7 Spadix Rot

Spadix rot was reported in *Anthurium andraeanum* cultivation farms in southern Thailand and the pathogen was identified as *Lasiodiplodia theobromae*, based on morphology and the nucleotide sequences of the internal transcribed spacer (ITS) and translation elongation factor 1-α (*EF1*-α) genes (Daengsuwan *et al.*, 2020).

10.1.8 Bacterial Wilt

Piriformospora indica has strong potential as bio-fertilizer for utilization in ornamental plant cultivation. The study was conducted to support the disease resistance conferred by *P. indica* through the plant-fungal symbiosis and reported faster elongation of roots early in the interaction. *P. indica*-colonized plants absorb more phosphorus and exhibit higher photosynthesis rates. Moreover, higher activities of stress-related enzymes, of jasmonic acid levels and mRNA levels of jasmonic acid-responsive genes suggested that the fungus prepares the plant to respond more efficiently to potentially upcoming threats, including bacterial wilt (Lin *et al.*, 2019).

Viral Diseases

Mosaic and malformation of leaves and spathes were observed in different anthurium cultivars in commercial nurseries near Caracas, Venezuela, up to 20 per cent infection was present in pink and red cultivars and 94 per cent in white cultivars. The virus could be transmitted by *Bemisia tabaci* and possibly also by grafting a piece of infected leaf onto the stem (Herold, 1967).

10.1.9 Cucumber Mosaic Virus

Miura *et al.* (2013) suggested that it is possible to avoid or reduce CMV dissemination in anthurium commercial crops through elimination of weeds, mainly *Commelina* and *Tradescantia*, those are natural hosts of the virus; removal of symptomatic anthuriums with the destruction of these plants away from the cultivation area; and use of healthy propagative material. The control of aphid population not only in the production area but also around is important to reduce viral inoculum in medium and long term.

10.1.10 Bud Necrosis

Groundnut bud necrosis virus infecting anthurium (*Anthurium andreanum*) in India (Amruta *et al.*, 2020).

10.2.0 Pests

Though pests are not a problem in anthurium cultivation, insects like aphids, scales, mealy bugs, thrips, white flies, spider mites are found to attack the plants and cause considerable damage. Among these, aphids, thrips, scale insects and spider mites are the major pests.

Diseases and Pests of Anthurium

Bacterial Blight

Anthracnose

Leaf Spot

Root Rot

Aphids Infestations

Spider Mites Infestation

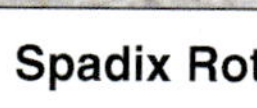

Spadix Rot

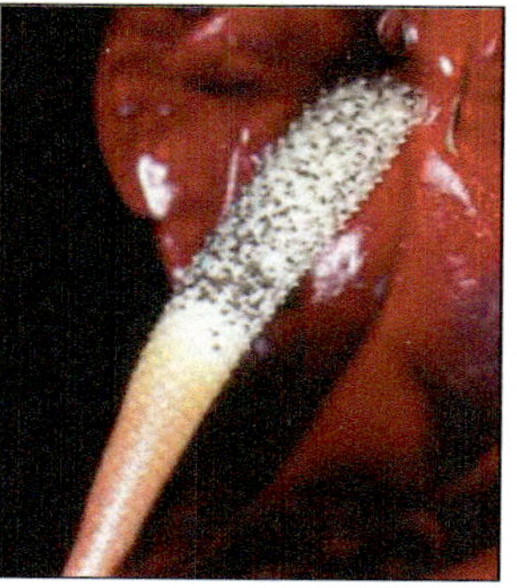

Scale Insects

10.2.1 Aphids

These insects suck the plant's sap, cause yellowish and distortion of leaves and poor growth. They secrete a sticky, sugary substance called 'honeydew' upon which a black mould often grows. The mould not only looks very unsightly but interferes with the leaves as well. Systemic insecticides containing dimethoate and malathion (0.2 per cent) effectively control the aphids. A formulation containing pyrethrum extract gave complete control of *Myzus circumflexus* on *A. scherzerianum* plants in the greenhouse, without phytotoxicity (Pignon, 1963). The biological control of aphids was successfully practised at various culture periods under glass in Germany using the predatory cecidomyiid *Aphidoletes aphidimyza*. Suppresion of aphid population was obtained with 5-6 gall midges/m^2 (Schmidt *et al.*, 1989).

10.2.2 Scale Insects

Scale insects infest stems and leaves, suck the sap and weaken the plants. Wiping the insects of the plants with a cotton-wool swab soaked in methylated spirit or spray with 0.2 per cent malathion work well against these insects.

10.2.3 Spider Mites

These minutes mites cause yellowish mottling of the leaves which may become brown and shrivel. A fine white web spun on the underside of leaves is another sign. Spraying of affected plants with 0.2 per cent malathion or kelthane (8 ml in 10 l of water) was found to be very effective to control the mites.

10.2.4 Thrips

Thrips suck the sap from leaves and cause a mottled effect on foliage and flower. Insecticide granules were very effective in decreasing injury to flowers by *Chaetanaphothrips orchidii* for 5 months. Aldicarb 10 g and disulfoton 15 g were most effective, followed by chlorpyrifos 2 g and oxamyl 10 g. This study suggested the use of granular insecticides at 5-month intervals or foliar sprays at 2-week intervals to reduce populations of *C. orchidii* to below economic injury levels in a pest management strategy (Hata and Hara, 1992a).

10.2.5 White Fly

The life cycle of *Aleurotulus anthuricola* was completed within the inner surface of anthurium sheaths and took 35.5 ± 1.2 days from egg to adult. Systemic insecticides were more effective than contact insecticides. Under greenhouse conditions, systemic insecticides like methomyl 1.81 and dimethoate 40 EC significantly reduced number of eggs, nymphs, pupae and adults of *A. anthuricola*. Malathion 50 EC was only effective against adults and pupae and bifenthrin 10 WP was effective only against adults. In field trails, methomyl 1.81 and acephate 75 SP significantly reduced numbers of nymphs and pupae. Numbers of eggs were also significantly reduced in acephate 75 SP-treated plots (Hata and Hara, 1992b).

An integrated pest management programme (IPM) for *A. andraeanum* in Hawaii successfully reduced pesticide applications without reducing production and marketability. Insecticides, acaricide and fungicide application at 3 farms studied were reduced by 45, 79 and 96 per cent, respectively, by economically justifying spray applications. No significant increase in thrips, mite or anthracnose injuries occurred, nor did total cost for treatment and pest damage increase. The study indicated that the IPM concept implemented on a floricultural crop can reduce pesticide applications and increase profitability (Hara *et al.*, 1990).

10.2.6 Nematode

A survey of plant parasitic nematodes revealed heavy infestations of *Meloidogyne incognita* in standard and *Radopholus similis* in hybrid anthuriums. Evidence suggests that management, in particular propagation techniques, is responsible for the widespread infestations of these ornamentals (Cadet *et al.*, 1993). Six anthurium species and three *A. andraeanum* cultivars (Midori, Ozaki and Mauna Kea) were screened *in vitro* for tolerance and resistance to the nematode *R. similis.* Based on a lower relative symptom index, *A. pittieri, A. ravenii, A. antioquiense* and *A. aripoense* had reduced nematode damage as compared to the reference standard, Midori. However, *A. Antioquiense* and *A. aripoense* had higher root damage than Midori. Lower nematode damage in *A. Pittieri* and *A. ravenii* is positively correlated with greater plant vigour or to fewer target roots for nematode infection. *A. ravenii* was among the most resistant species as measured by nematode reproduction. This is followed by *A. aripoense* and *A. pittieri* (Wang *et al.*, 1998). *Radopholus similis* occurred in all stem section, leaf petioles, and roots in both Alii and Midori. Nematode distribution differed between the two cultivars. Midori had higher numbers of nematode in the roots whereas Alii had higher numbers of nematode in the stem sections and first leaf petiole. Anthurium apical stem cuttings could be contaminated with *R. similis* and may not be a nematode free propagation material (Wang *et al.*, 1999). Fenamiphos was the most effective treatment in reducing anthurium decline in terms of enhanced growth while nematrol (ground sesame plant) decreased decline slightly (Sipes and Delate, 1996).

Pin Nematode Slow Decline of *Anthurium andraeanum*, a new disease caused by the pin nematode *Paratylenchus shenzhenensis* (Wang *et al.*, 2016).

Burrowing nematode (*Radopholus similes*) causes an estimated 50 per cent loss in flowering yield of Anthurium spp., although heat treatment (hot water at 50°C for 12 min or hot air at 50°C, 60 per cent RH for 35 min) may eliminate this nematode and bacterial blight from cut flowers (Tsang *et al.*, 2010).

11.0 Harvesting and Postharvest Management

11.1.0 Harvesting

Anthuriums are generally harvested when the spadix is almost fully developed. Flowers picked too early wilt quickly. Development of true flowers on spadix is also used as a criterion for determining the harvest stage. In Hawaii, growers normally harvest anthurium when one-third of the true flowers on spadix are fully developed (Akamine and Goo, 1981). Mechanical injury is the most common reason for rejection of flowers after harvest, followed by curved petiole (Hara *et al.*, 1989). The commercial yield of anthurium varies from 4 to 9 floral stems per plant per year (Singh *et al.*, 2011).

11.2.0 Yield

The flower, after harvest, is graded following USDA standards as mentioned below (Singh, 1998).

Grade	*Spadix Measure (Average length and width in cm)*
Miniature	< 8
Small	8-10
Medium	10-13
Large	13-15
Extra-large	> 15

11.3.0 Postharvest Management

11.3.1 Grading and Packaging

The postharvest life is determined by a number of senescence symptoms – loss of spathe gloss, blueing of spathe, spadix dehydration and necrosis and stem collapse (Paull, 1982). The flower stems are placed in lukewarm water (38°C) after the harvest and allowed to stand overnight prior to shipment. There are several methods for packing cut flowers. In one of the earliest methods, the stems are placed in rubber balloon filled with water and the balloon is tied to the flower stems while in the other, cotton pad saturated with water and enclosed in wax paper is placed at the basal end of the flower stem and tied. In both the methods the flowers are then placed in corrugated cartons and the stems are properly tied to prevent displacement and mechanical injuries. In recent past, other methods, such as packing the flowers in plastic bags before placing in cartons or dipping the spadix in melted paraffin to reduce moisture loss were also very popular in Hawaii and other North American States (Akamine, 1976). The growers in the Netherlands follow the recommendations made by Segers (1976) in which the flower stems placed in flask containing water are packed in moist boxes and soft protection material is put in between spathe and spadix. Foam plastic supports are provided in the box and flowers are secured carefully with tape. The most commonly used box sizes for packing anthurium flowers are 21.6 × 50.8 × 91.4 cm or 27.9 × 43.2 × 101.6 cm which can accommodate at least 10 dozen flowers. Lining of cartons with polythene sheet and moist paper insulation are necessary to maintain proper humidity as well as to prevent injuries to the flowers.

Mechanizing the processes like classification, grading and packing will improve the product quality and shorten the delivery time from field or greenhouse to the consumer. The time between picking to the consumer should be as short as possible. On the other hand, the postharvest operations in fields, such as bunching, packing, and grading, are an important determinant of quality and life of cut flowers and foliage. Therefore, developing automatic systems to implement the postharvest operations is a need for Anthurium, to speed up the processing operations. Among instrumental-oriented inspection approaches, computer vision is one of the methods

Harvesting and Postharvest Management

Harvesting Stage in Greenhouse

Harvesting in Open

Packing in Polythene

Packing of Flowers with Stems

Packing in Cardboard Box

Marketed as Pot Plants

Anthurium as Cut Flowers

Display of Anthurium

in which there is no contact between the instrument and object. Also, a lot of data is obtained via only one scanning procedure.

Performance and efficiency of these systems are extremely depended on the robustness of applied image processing algorithms and machine learning techniques (Pour *et al.*, 2018).

Soleimanipour and Chegini (2020) had developed a hybrid approach for highly accurate and effective identification of Anthurium flower cultivars in a computer vision-based sorting machine. In an automatic grading machine for flowers, cultivar classification of flowers is an important step for subsequent grading tasks and suggested that cultivar identification approach could be implemented in postharvest systems to classify the flower cultivar before grading and packing operations.

11.3.2 Pulsing

The postharvest longevity of flowers could be considerably increased (19 days) by a short-term pulsing treatment of the flowers for 8 hours with a solution of BA 150 ppm (Adbussamed *et al.*, 2000).

Sahare and Alka (2015) suggested pulsing with 10 mg/l BA and10 mg/l GA_3 + 4 per cent sugar in *cv* Xavia significantly enhance vase life and delaye flower abscission up to 19 days and improve physiological parameters like water uptake (ml), exhibite spike fresh weight (per cent) and diminishe electrolyte leakage (per cent) in flower spathe and improve flower quality as it decreases colour fading in spathe tissue.

Maitra and Roychowdhury (2005) found that pulsing solutions containing 8-Hydroxyquinoline 500 ppm + sucrose 5 per cent increased the postharvest life (20.17 day) of cut anthurium *cv*. Nitta flowers.

11.3.3 Storage

Anthurium can easily be stored at 13°C for 2 to 3 weeks and will last 2-4 weeks in an arrangement (Auman, 1980). Exposure of flowers to temperature below 13°C causes the red flowers to turn blue, dark red flowers being most susceptible. Individual flowers showed a great difference in response to temperature, some not turning blue even at 5°C (Sytsema and Barendse, 1975). Where refrigeration facilities are not available, storage in 2-10 per cent O_2 can be used advantageously at ambient temperatures of 24-25°C (Akamine and Goo, 1981). The optimum storage temperature, as recorded by Paull (1987) for flowers of anthurium (*A. andaearum*) *cvs*. Kaumana, Nitta, and Ozaki, was between 14 and 17°C.

Sugar content, respiration rate, and spathe quality of cut flowers of *A. andraeanum cv.* Scarlette were determined after storage at chilling temperatures. The content of reducing sugar (glucose + fructose) in the spathe generally decreased during storage with rapid decreases usually corresponding with sudden increase in respiration rate. Storage at a chilling temperature of 4°C tended to slow the loss of reducing sugars; this was probably due to slower respiration. The rate of spathe discoloration (blueing) was not directly related to the content of reducing sugars (Pritchard *et al.*, 1991).

In order to decrease the metabolic activity, one alternative is the use of low temperatures. However, these inflorescences may show chilling injury (CI) at temperatures below 12°C (Soleimani Aghdam *et al.*, 2016), and recommended storage temperature is 12.5–20°C (Promyou *et al.*, 2012). Exogenous melatonin can ameliorate chilling injury in cut anthurium flowers during low temperature storage (Aghdam *et al.*, 2019).

The spectral light composition during storage of Anthurium cut flowers is an important environmental factor determining postharvest flower performance. A high percentage of blue in the light spectrum was associated with increased electrolyte leakage of the spathe and a shorter vase life. Therefore, postharvest handling of Anthurium cut flowers should preferably be performed in an environment with limited blue spectrum when exposures to low temperatures are inevitable (Aliniaeifard *et al.*, 2020)

The application of GA_3 + SPM by spraying can be used to reduce the senescence in anthurium *cv.* Arizona cut flowers stored at 20°C, and improve the commercial quality of the inflorescences (Do Nascimento Simões *et al.*, 2018).

Postharvest spray of 6- benzylaminopurine (100 and 200 mg/l) in anthurium revealed that by lowering BAP spray concentration and cold storing stems for two days exert positive effect on spadix necrosis but not observed for spathe parameters (Favero *et al.*, 2012).

Foliage stems of anthurium (*Anthurium andraeanum*) *cv.* Apalai treated with gibberellic acid at 200 mg/l concentration maintained the higher scores with enabled efficient storage for 22 days (Marsala *et al.*, 2014).

11.3.4 Vase Life

Anthurium cut flowers are not sensitive to problems of water uptake, probably due to their low transpiration rate (Doorn *et al.*, 1999). However, Paull and Goo (1985) reported that senescence of anthurium flowers is caused by plugging of the stem vascular tissue. A circadian rhythm in CO_2 production was observed by Hew *et al.* (1994) in flower of 3 cultivars of *A. andraeanum* (Jaya, Fla-range and Leonette). The rhythmicity was not affected by light, darkness or detachment. Spadix respiration accounted for 90 per cent of the total flower respiration.

Kamemoto (1962) studied some factors affecting the vase life of flowers and observed that the keeping quality of flowers increased as they developed and was maximum when 3/4th of the length of the spadix had changed colour. Large and medium-sized flowers kept better than small and miniature ones. The flowers kept best at 13.0°C. Precooling or short refrigeration period did not extend the total life of the flowers. Similarly, the degree of shade under which the plants were grown did not affect flower keeping quality. According to Kalkman (1983), vase life is the longest in the flowers, cut when the spadix was almost completely white, the average for flowers cut at this stage in winter and summer being 21.5 and 25.2 days, respectively.

Cultivars also greatly differ in their postharvest life. Cut flowers placed directly in water lasted 22-25 days, regardless of shading treatment. When the flowers were stored in fives in plastic film for 1 or 1.5 days at 23 or 25°C, vase life (10-26 days) was similar for shaded and unshaded plants (Vierig, 1990).

Various commercial preservatives (floralife, everbloom, roselife) and chemicals and beverages (sodium benzoate, benzoic acid, glucose, sodium hypochlorite, hydrochloric acid, 7-Up, *etc.*) can be used to prolong the shelf-life of anthurium flowers (Akamine and Goo, 1975). Cytokinin also affected postharvest life, apparently *via* effects on water balance (Paull and Goo, 1985). A pre-shipping dip of flower stems in solutions of 2.25 per cent 7-Up (a carbonated beverage), 500 ppm benzoic acid or 7.3 ppm of sodium hypochlorite remarkably extended the vase life of flowers (Akamine and Goo, 1975). Treatment with benzyladenine reduced the respiration rate of flowers and imparted some tolerance to chilling and extended the saleable period (Shirakawa *et al.*, 1964). Dipping of flower stems for 10-60 min in 4 mM silver nitrate solution (within 12 hours of harvest) also extended the vase life by 40-60 per cent (Paull and Goo, 1982; Paull, 1987). The effect of silver nitrate on prolonging vase life might be due to its role as antiethylene agent and/or as a bactericide. The inhibitors Co^{2+} and amino-ethoxyvinyl-glycine significantly reduced the peak of ethylene produced, whereas inhibitors of the ethylene response (Ag^+, nitrogen and CO_2 bubbling) reduced anticatalytic ethylene production from cut anthurium flower stem sections (Paull and Goo, 1985). Paull (1983) suggested the use of wax for extending postharvest vase life of flowers. Out of eight products tested, FMC-819 carnanba-based wax was most effective, increasing the vase life from 18 days in untreated control to 36 and imparting a high gloss.

Time to spadix necrosis was the only senescence indicator that occurred in all cultivars so, it is suggested as a simplified criterion for vase life assessment and vascular occlusion is a mechanism for and as a result of water stress that induces senescence in Anthurium (Elibox and Umaharan, 2008).

Sankari *et al.* (2013) reported that pre-transit pulsing treatment with sodium hypochlororite 500 ppm took long days to develop spadix blackening (20 d), spathe blueing (21 d) and loss of glossiness of spathe (19 d) and recorded an increased vase-life period of 13 d over control. The cut spikes of Anthurium *cv.* Temptation kept in holding solution of sucrose 5 per cent + NaOCl, $16H_2O$ - 50 ppm + kinetin - 25 ppm took maximum days for anther bursting (18 d), discolouration of the spathe (29 d) and vaselife period (30 d), while, the untreated flowers had a reduced vase-life of 15 days only. However, Reddy *et al.* (2013) found that holding solution having 3 per cent sucrose + 250 ppm cobalt chloride resulted in increased water uptake, water loss, fresh weight, vase life, low microbial count and the benefits.

Improving the quality and vase life of cut flowers of Anthurium *cv.* Marshall through the application of thidiazuron (TDZ) was investigated and reported that TDZ 10 and 15 µM (21.5 and 21 d, respectively) found the most effective in prolonging the vase life of flowers (Thawiang *et al.*, 2007). However, treating stems with up to 2 per cent ethanol solution shortened the vase life of flower (16 d) (Thawiang *et al.*, 2012)

High light intensity increased the longevity of the long vase life cultivar but not the short. Exposing certain cut flowers to high light appears to extend the time that the pigments are maintained. Pigment composition in general and the extent of the reflectance dip in the red region in particular may provide a useful marker for vase life (Evelyn *et al.*, 2020).

12.0 Biotechnology

Micropropagation of *Anthurium andraeanum* and *A. scherzerianum* from vegetative parts and seed have been summarised in propagation.

12.1.0 Tissue Culture

12.1.1 Somatic Embryogenesis

Whole leaf blade explants of four *A. andraeanum* cultivars (UH 780, UH 965, UH 1060 and UH 1003) from plants grown *in vitro*, formed translucent embryogenic calluses at their basal ends within one month of culture in the dark. Secondary somatic embryos formed frequently and without an intervening callus on surfaces of primary embryos. Embryogenesis was induced with UH 965, UH 1060 and UH780 using a modified half strength MS medium supplemented with 1.0-4.0 mg 2,4-D and 0.33-1.0 mg kinetin/l. Rooted plantlets were recovered and transferred into pots with tree fern fibre medium and grown in the greenhouse (Kuehnle *el al.*, 1992). Morphological, anatomical and histochemical aspects of zygotic embryogenesis by *A. Andraeanum* cv. Kalapana were investigated from 4 to 24 weeks post-pollination. Anatomical features were correlated with the morphology of the spadix and the capacity of embryos to germinate *in vitro*. Development from a single-celled zygote to fully mature seed took 24 weeks. It was apparent by week 8, substantial until week 14, then diminished rapidly until its absence by week 22. Differentiation of the shoot apex, cotyledon and protoderm occurred at 14 weeks. The embryo started to derive nutrition from the endosperm at this time and germination of cultured ovules reached 56 per cent. By 20 weeks the shoot apex had visible leaf primordia and the root apex was clearly defined (Matsumoto *et al.*, 1998).

Xu *et al.* (2010) reported some physiological and biochemical characteristics of different developmental stages of somatic embryogenesis of A. andraeanum 'Amigo' and found that the embryogenic callus formation stage was important for the regulation of somatic embryogenesis. A recent study by Pinheiro *et al.* (2014) on *A. andraeanum* 'Eidibel' showed how the choice of explant, in this case nodal segments, being a superior explant, could influence the outcome of somatic embryogenesis.

Pinheiro *et al.* (2014) developed a highly efficient protocol for the induction of somatic embryogenesis in anthurium (*Anthurium andraeanum cv.* Eidibel) using nodal segments from *in vitro* established plantlets that were cultured in Pierik medium supplemented with 10 µM NAA. This auxin resulted in a well-defined procambium, a clear delimitation of the protoderm layer and signs of polarization, which are important features for further maturation of the somatic embryo and the effective conversion into plants. This protocol is useful not only for large-scale vegetative propagation but also for genetic transformation or synthetic seed production and can be combined with other biotechnological approaches.

Guadalupe and Mayanin (2017) suggested the protocol which induced adventitious shoot organogenesis were Murashige and Skoog medium at half ionic strength supplemented with 0.2 µM of thidiazuron (TDZ) and 2.2 µM of 6-benzylaminopurine (BAP), 0.2 µM of TDZ and 6.6 µM of zeatin (ZEA) and the treatment which induced somatic embryogenesis was 0.4 µM of TDZ and 0.5 µM of 2, 4-dichlorophenoxyacetic acid (2, 4-D).

12.1.2 Protoplast Culture

Protoplasts capable of first division were isolated from *A. andraeanum*using etiolated shoots derived from callus cultures of selections UH965, UH1060 and UH1003. Protoplasts cultured in modified MS medium, divided in 2-6 days after incubation in the dark at 22°C. However, sustained divisions were not seen (Kuehnle and Nan, 1991).

Protoplast fusion is an important *in vitro* breeding technique to produce somatic hybrids and may be a starting point for the complete regeneration of somatic intergeneric hybrids in the Araceae family, and the establishment of a model system for asexual asymmetric fusion in monocot ornamental crops (Duquenne *et al.*, 2007).

12.1.3 Somaclonal Variation

Although there is evidence for some genome instability in callus, only a small pro-portion of variants seems to be transmitted to regenerated plants. Chromosome counts in root-tip cells of 169 plants derived from a single spadix explant through three passages on callus medium revealed very little variation in ploidy level (Geier, 1985). The extent of morphological variability was clearly less than that usually observed in seed propagated selections of *A. scherzerianum* and showed no obvious relationship to the mode and number of cycles of multiplication.

Geier (1987) examined plants regenerated from spadix and leaf segments for variation in ploidy and morphological characters. Chromosome counts and/or measurements of stomatal guard cell length of 377 plants revealed all but 2 to be diploids; one was tetraploid and another periclinal chimera with diploid epidermis and tetraploid core. Morphological variation was reported in the form of malformed leaves and inflorescences, leaf variegation, dwarf plants and degree of branching.

Both microprojectile bombardment and *Agrobacterium* mediated gene transformation systems appeared promising for introducing resistance to blight (*X. campestris* pv. *dieffenbachiae*) into anthurium (Kuehnle *et al.*, 1990).

12.1.4 Anther Culture

To develop and establish an anther or half-anther culture system for *A. andraeanum*, 'Tropical' serves as an appropriate plant model. Half-anthers were the most responsive and suitable explant and responsive anthers were isolated from the transition area of the spadix with 50 per cent of the stigmas being receptive (Rachmawati, 2007). The best culture media for callus induction, formation and regeneration were either WT medium supplemented with 0.5 mg/l TDZ, 1.0 mg/l BA and 0.01 mg/l NAA or NTW medium containing 1.5 mg/l TDZ, 0.75 mg/l BA and 0.02 mg/l NAA (Winarto, 2009). Based on the results of 'Tropical' anther culture, anther cultures for *A. andraeanum* 'Carnaval', 'Casino', 'Laguna' and 'Safari' and a local Indonesian accession of anthurium were successfully developed (Winarto and Mattjik, 2009).

Plantlets derived from *A. andreanum* 'Tropical' half-anther culture showed mixed ploidy, namely haploids, diploids and triploids could be successfully acclimatized with as high as 100 per cent survival. Treatments started by removing plantlets carefully from culture bottles, washing plantlet roots gently under tap water, immersing in a biocide solution containing 0.5 per cent benomyl (w/v) and 0.2 per cent streptomycin sulphate (w/v) for 1 min, planting in a plastic pot containing burned-rice husk, raw rice husk and organic manure (2:2:1, v/v/v), covering the pot with transparent plastic for 7 d and placing rooted plantlets in a glasshouse under low light intensity (37–74 mol/m^2/s) (Winarto *et al.*, 2011).

12.2.0 Genetic Transformation

Genetic engineering of anthurium has become feasible after some cultivars could be infected by *Agrobacterium tumefaciens* (Kuhnle and Sugii, 1991). They devised a method for infecting *A. andraeanum* with *A. tumefaciens*. Tumours were obtained on stems 7-10 weeks after inoculation with oncogenic *A. tumefaciens* strains C 58 and A 281 cultured previously on an induction medium containing 200 µM acetosyringone (3,5-methoxy 4-hydroxyacetophenone) at pH 5.5. A higher percentage of tumours were formed *in vitro* on etiolated internodes (32 per cent) than on green leaf (2 per cent) or petiole explants (3 per cent) 2 weeks after inoculation with induced C 58.

Several procedures were tested in development of a gene transfer protocol for anthurium. Etiolated internode segments of cultivars Rudolph and UH1060 were co-cultured with *Agrobacterium tumefaciens* LBA 4404 carrying the chimaeric genes *neo*, for antibiotic resistance, and *att* encoding antibacterial attacin. Assays of genomic DNA and RNA from kanamycin-resistant Rudolph and DNA and UH1060 plantlets, recovered as soon as 1 year after culture on selection media, indicated the presence of introduced genes, including *neo* and *att*, and transcription of *att*. Western analysis confirmed the expression of attacin protein in calli induced from laminae of regenerated kanamycin-resistant Rudolph plantlets (Chen *et al.*, 1996).

Chi *et al.* (1998) coated different plasmids carrying different promoters (pUGCI-maize ubiquitin promoter; pEF-1 alpha-*Arabidopsis thaliana* promoter; and p70gus-CaMV promoter) and a bacterial gene encoding beta-glucuronidase (*gus*) with gold microcarriers and used for bombardment with the biolistic PSD-1000/He. The system operated at 900, 1100, 1300 or 1550 psi and the target tissues used were calli and whole leaves from *A. andraeanum*. Transient GUS expression was monitored 24 or 48 h post bombardment and the highest transient expression of GUS in *A. andraeanum* callus tissues using pUGCI-maize ubiquitin was found. This indicated that the promoter from maize, a monocot, is more efficient in expressing the *gus*gene in anthurium than dicot promoters.

Two 'Paradise Pink' lines out of 13 lines transformed with a derivative of the cecropin gene showed increased tolerance to bacterial blight over the most tolerant industry standard, 'Kalapana', based on bioassays with bioluminescent bacteria (Kuehnle *et al.*, 2004).

Based on the results of six genes transferred to two cultivars (Marian Seefurth and Midori) using different tissues, it was concluded that the best anthurium tissues to transform are embryogenic calli on Medium using LBA4404 or AGLØ cocultivation for 10 to 14 d. Selection is recommended using G418 concentrations based on kill curve results for individual cultivars. The improvement in selection time and efficiency using embryogenic calli allows for high-throughput transformation in which attributes like pest resistance and improvement in quality depend on the chance good fortune of inserted gene location (Fitch *et al.*, 2011)

In most studies, the embryogenic callus of *A. andreanum*, *A. limdeniarum* and *A. kamemotoanum* cultivars was used as the target tissue for the genetic transformation (Hosein *et al.*, 2012).

Cystatin, NPRI, attacin, T4 lysozyme, cowpea trypsin inhibitor genes were introduced by Fitch *et al.* (2011) to *A. andraeanum* 'Midori' and 'Marian Seefurth' hybrids for resistance to bacterial blight and nematodes, and even though transgene insertions were confirmed by PCR and Southern hybridization, no pest-resistance assays were performed.

After insect resistance genes (API-A, API-B; arrowhead proteinase inhibitor) were introduced to *A. andraeanum* 'Pink Champion', nine kanamycin-resistant plantlets (29 per cent) were obtained, seven of which were positive for transgenes as assayed by PCR, but no insect-resistance assays were performed (Zhang *et al.*, 2009)

Two transgenic lines of 'Paradise Pink' selected by Kuehnle *et al.* (2004) showed the production of the cecropin-like Shival lytic peptide and tolerance to Anthurium blight caused by *Xanthomonas campestris* pv. Dieffenbachiae.

12.2.1 Cryopreservation

Wang *et al.* (2010) reported a vitrification cryopreservation protocol for *A. andraeanum* embryonic suspension cells. Hanan and Sherif (2013) also reported conservation experiments with *A. andraeanum* plantlets that consisted of three treatments for chemical conservation and four treatments (excluding the control), for cryopreservation. In the former method, micronodes (3–4 cm long shoots with 4–5 well developed leaves) were cultured on MS medium (modified PVS2) containing different concentrations of sorbitol (0.5, 1.0 and 1.5 g/l) for 8 months. In the latter method, micronodes were treated with different protection solutions for 10 min in two treatments; while cultured on MS medium containing different protection solutions for 7 d in the other two treatments. The explants were immersed in liquid nitrogen for 12 d after the independent treatments then rapidly thawed in a water bath at 40°C.

12.2.2 Synthetic Seed Technology

Synthetic seed (synseed) technology allows for the conservation and storage of rare, endangered and desirable genotypes while providing easy handling and transportation. Nhut *et al.* (2004) encapsulated the leaf-derived embryogenic callus of 'Tropical' that had been derived in the presence of 1.0 mg/l BA and 0.08 mg/l 2,4- D on 1/2 MS medium.

13.0 References

Adbussamed, K.P., Valsalakumari, P.K., Rajeevan, P.K. and Geetha, C.K. (2000) *Proc. 12th Kerala Sci. Cong.*, Kumily, pp. 505-508.

AdeOluwa, O.O., Akinkunmi, O.Y. and Akintoye, H.A. (2017) *J. Appl. Hortic.*, **19** : 46-50.

Aghdam, M.S., Jannatizadeh, A., Nojadeh, M.S. and Ebrahimzadeh, A. (2019) *Postharvest Biol. Technol.*, **148** : 184-191.

Ajish Muraleedharan and Karuppaiah, P. (2015) *Int. J. Curr. Res.* **7** : 21627-21629.

Akamine, E.K. (1976) *HortScience,* **11** : 125-127.

Akamine, E.K. and Goo, T. (1975) *Flor. Rev.,* **155** : 14-15.

Akamine, E.K. and Goo, T. (1981) *HortScience,* **16** : 206-207.

Aliniaeifard, S., Falahi, Z., Dianati Daylami, S., Li, T. and Woltering, E. (2020) *Front. Plant Sci.,* **11** : 846. http://doi.org/10.3389/fpls.2020.06846.

Amruta, B.S., LaxmiDevi, V., Ramegowda, G.K., Seetharamu, G.K., Usharani, T. R. and Krishnareddy, M. (2020) *New Dis. Rep.*, **41** : 14-14.

Ann, P.J. (1992) *Plant Pathology Bull.,* **1** : 79-89.

Aragaki, M. (1964) *Hawaii Fm. Sci.,* **13** : 16.

Argaki, M. and Ishii, M. (1960) *Plant Dis. Reptr.,* **44** : 865-867.

Aswath, C., Choudhury, M.L., Prakesh, D. and Prasad, K.V. (1998) *Souvenir and Abstract, Nat. Sem.On Anthurium Production, 2-3 June, 1998, Viv. Ornamental Crops and Central Hort. Expt. Sta.,* IIHR, Bangalore, India, pp. 20.

Auman, C.W. (1980) In : *Introduction of Floriculture* (Ed. R. A. Larson), Academic Press, New York.

Aygün, G. and Aki, C. (2016) *Ann. Biol. Res.*, **7**: 30-33.

Bachthaler, E. (1977) *Gartenbauwiss,* **42** : 136-138.

Bachthaler, E. (1978) *Gartenerborse and Gartenwelt,* **78** : 1249-1250.

Bachthaler, E. (1979) *Gartenbauwiss,***44** : 251-255.

Bachthaler, E. (1993) *Gartenbauwiss,* **58** : 21-24.

Bailey, L.H. (1963) *The Standard Cyclopedia of Horticulture,* The Macmillan Co., New York.

Basheer, S.N. and Thekkayam, S.G. (2012) *Asian J. Horti.*, **7** : 354-358.

Bell, E. (1971) *B.V.O. Mededelingen,* No. 59, pp. 5.

Berg, G.A.V.D. and Valentin, J.C.M. (1977) *Vakbladvoor de Bloemisterij,* **32** : 17.

Bik, R.A. (1976) *Vakbladvoor de Bloemisterij,* **31** : 15.

Bik, R.A. and Straver, N.A. (1978) *Vakbladvoor de Bloemisterij,* **33** : 36.

Bik, R.A., Van Zon, A.G. and Steiner, R.A. (1960) *Jversl. Proefst Bloom. Aalsmeer,* pp. 40-45.

Bindu, M.R. and Mercy, S.T. (1997) *South Indian Hort.,* **45** : 134-138.

Bleiweiss, R., Molina, F.S., Freire, E. and Croat, T.B. (2019) *Flora,* **255** : 80-85.

Boertje, G.A. (1978) *Acta Hortic.,* **82** : 159-164.

Bordoloi, S. and Talukdar, M.C. (2019) *Int. J. Curr.Microbiol. App. Sci.,* **8** : 1157-1165.

Boschi, C., Benedetto, A. di and di Benedetto, A. (1998) *Revista de la Facultad de Agronomia Universidad de Buenos Aires,* **18** : 89-92.

Bosmans, P. and Veldeman, R. (1964) *Meded. Landb. Hogesch. Gent.,* **29** : 956-963.

Boula, R., Fougerouze, J., Bonhomme, K. and Schoch, P.G. (1973) *Pepinierstes Horticultures Maraichers,* No. 134, pp. 13-16.

Breedveld, L. and Glass, R. (1984) *Vakbladvoor de Bloemisterij,* **39** : 310.

Budiarto, K. and Handayati, W. Agrivita (2007) *Flora,* **39** : 261-268.

Cadet, P., Berg, E. van den, Nema, L. and Van den Berg, E. (1993) *PHM Revue Horticole,* **341** : 53-58.

Cardoso, J.C. (2019) *Sci. Hortic.,* **253**: 358-363.

Cen, Y.Q., Jiang, R.M., Deng, Z.L. and Ni, D.X. (1993) *Acta HorticulturaeSinica,* **20** : 187-192.

Chang, K.H., Wu, R.Y., Chang, G.P., Hsieh, T.F., and Chung, R.S. (2012) *HortScience,* **47** : 515-521.

Chang, K.H., Wu, R.Y., Chuang, K.C., Hsieh, T.F. and Chung, R.S. (2010) *Sci Hortic.,* **125** : 434-441.

Chaves, T.P., Miranda, A.R.G.S., da Paz, L.C., Netto, M.S.B., Lima, G.S.A. and Assunção, I. P. (2020) *Australasian Plant Disease Notes,* **15** : 1-3.

Chen, C., Hou, X., Zhang, H., Wang, G. and Tian, L. (2011) *Euphytica,* **181** : 137-145.

Chen, F.C., Kuehnle, A.R. and Chen, F.C. (1996) *J. Amer. Soc. Hort. Sci.,* **121** : 47-51.

Chen, J.B., Chen, W.H. and Wu, G.D. (1999) *Report of the Taiwan Sugar Research Inst.,* No. 164, pp. 41-57.

Chi, G.L., Goh, H.K.L., Hoo, K.Y. and Legavre, T. (1998) *Acta Hortic.,* **461** : 379-383.

Chiusoli, A. and Boriani, M.L. (1986) *The Macdonald Encyclopedia of House Plants,* Macdonald and Co. (Publishers) Ltd. Londan.

Christensen, O.V. (1971) *Tidsskr. Planteavl.,* **75** : 793-798.

Coelhom, M.A.N. and Catharino, E.L.M. (2008) *Rodriguésia,* **59** : 829-833.

Cotias de Oliveira, A.L.P., Guedes, M.L.S. and Barreto, E.C. (1999) *General Mol. Biol.,* **22** : 237-242.

Daengsuwan, W., Wonglom, P. and Sunpapao, A. (2020) *J. Phytopathol.,* **168** : 129-133.

Dai, J.W. and Paull, R.E. (1990) *J. Amer. Soc. Hort. Sci.,* **115** : 901-905.

De Morais, É.B., de Castro, A.C.R., de Souza Aragão, F.A., Silva, T.F., da Silva, J.P. and Soares, N.S. (2017) *Ornam. Hortic.,* **23** : 07-14.

Deshmukh, H.V., Pawar, D.R. and Joshi, M.S. (1999) *Orissa J. Hort.,* **27** : 26-28.

Dhanya, M.K. and Mary, C.A. (2007) *J. Trop. Agric.,* **44** : 74-75.

Dias, G.D.M.G., Soares, J.D.R., Castro, A.C.R., Pasqual, M., Ascari, I.J. and Costa, A.C. (2017) *Crop Breed Appl Biotechnol.,* **17 :** 341-349.

Diaz, L.P., Namur, J.J., Bollati, S.A. and Arce, O.E.A. (2010) *Rev. Colomb.Biotechnol.,* **12** : 27-40.

Do Nascimento Simões, A., Diniz, N.B., da Silva Vieira, M.R., Ferreira-Silva, S.L., da Silva, M.B., Minatel, I.O. and Lima, G.P.P. (2018) *Sci. Hortic.,* **241** : 178-186.

Doorn, W.G.van, van Doorn, W.G., Fischer, G. and Angarita, A. (1999) *Acta Hortic.,* **482** : 65-69.

Dufour, L. and Guérin, V. (2005) *Sci. Hortic.,* **105** : 269-282.

Dufour, L. and Guerin, V. (2003) *Sci. Hortic.,* **98** : 25-35.

Duquenne, B., Eeckhaut, T., Werbrouck, S. and Van Huylenbroeck, J. (2007) *Plant Cell Tissue Organ Cult.,* **91** : 165–173.

Eid, G.M., Albatal, N. and Haddad, S. (2016) *Int. J. Hortic.,* **6** : 128-131.

Elibox, W. and Umaharan, P. (2008) *HortScience,* **43** : 825-831.

Elibox, W. and Umaharan, P. (2008) *HortScience,* **43** : 787-791.

El-Naggar, A.H., Esmaiel, N.M. and El-Naggar, A.A. (2016) *Alex. Sci. Exchange J.,* **37** : 1123-1140.

Evelyn, S., Farrella, A.D., Eliboxa, W., Abreua, K.D. and Umaharana, P. (2020) *PBT,* **159** : 10-1016.

Farsi, M., Taghavizadeh, Y.M. and Qasemiomran, V. (2012) *African J. Biotech.,* **11** : 13162-13166.

Favero, B.T., Lima, G.P.P., Cavasini, R. and Laschi, D. (2012) In: *X International Symposium on Postharvest Quality of Ornamental Plants* **1060.** pp. 183-186.

Fersing, G. and Lutz, A. (1977) *Competes Rendus Hebdomadaires des Searces de l'Academic des Sciences,* D., **284** : 2231-2233.

Fitch, M.M.M., Leong, T.C.W., He, X., McCafferty, H.R.K., Zhu, Y.J., Moor, P.H., Gonsalves, D., Aldwinckle, H.S. and Atkinson, H.J. (2011) *HortScience,* **46** : 358–364.

Fitch, M.M., Leong, T.C., He, X., McCafferty, H.R., Zhu, Y.J., Moore, P.H., and Atkinson, H.J. (2011) *HortScience,* **46** : 358-364.

Fukui, H., Alvarez, A.M. and Fukui, R. (1998) *Plant Diseases,* **82** : 800-806.

Gaiser, L.W. (1927) *Trans. Roy. Soc. Can.,* **21** : 1.

Gajek, W. and Schwarz, K.H. (1980) *Gartenbau,* **27** : 343.

Geier, T. (1986) *Plant Cell Tissue and Organ Culture,* **6** : 115-125.

Geier, T. (1987) *Acta Hortic.,* **226** : 293-298.

Geier, T. (1985) *Vortragepflanzenzuch,* **8** : 169-181.

Gelder, A.De.Reijnders, M.A.C., Kalkaman, E.C. and Glas, R. (1984) *Vakbladvoor de Bloemisterij,* **39** : 42-43, 45.

Giridhar, J., Ambad, S.N., Shivanand, H. and Viresh, H. (2012) *Asian J. Hortic.,* **7** : 276-280.

Guadalupe, L.O.P. and Mayanin, R.I.B.I. (2017) *African J. Biotechnol.,* **16** : 2092-2099.

Guo, L.Y. and Ko, W.H. (1994) *Plant Pathology Bull,* **3** : 18-23.

Guo, L.Y. and Ko, W.H. (1996) *J. Phytopath.,* **144** : 7-11.

Hanan, M.A.Y. and Sherif, S.S. (2013) *Middle East J. Agric. Res.,* **2** : 36–43.

Handaragall, A.G., Jayanthi, R. and Hemalatha, R. (2013) *Asian J. Hortic.,* **8** : 686-689.

Hara, A.H., Nishijima, W.T., Bushe, B.C. and Sato, D.M. (1989) *HortSciences,* **24** : 846-847.

Hara, A.H., Nishijima, W.T., Hansen, J.D., Bushe, B.C. and Hata, T.Y. (1990) *J. Eco. Ento.,* **83** : 1531-1534.

Hata, T.Y. and Hara, A.H. (1992a) *Trop. Pest Manag.,* **38** : 230-233.

Hata, T.Y. and Hara, A.H. (1992b) *Trop. Pest Manag.,* **38** : 152-154.

Hata, T.Y., Hara, A.H., Nagao, M.A. and Hu, B.K.S. (1994) *HortTechnology,* **4** : 159-162.

Henny, R.J. (1999) *HortScience,* **34** : 153-154.

Henny, R.J. and Fooschee, W.C. (1988) *Proc. Fla. Sta. Hort. Soc.,* **101** : 304-305.

Henny, R.J. and Hamilton, R.L. (1992) *HortScience,* **27** : 1328.

Henny, R.J., Norman, D.J. and Chen, J. (2003) *HortScience,* **38** : 133-134.

Herold, F. (1967) *Phytopathology,* **57** : 8.

Hetman, J. and Pudelska, K. (1984) *Prace Instytutu Sadownictwa w Skierniewicach,* B, **9** : 15-24.

Hetman, J., Lashowska, H. and Martyn, W. (1981) *Prace Instytut Sadownictwa w Skierniewicach,* B, **6** : 45-53.

Hew, C.S., Ong, T.K. and Yap, W.P. (1994) *HortScience,* **29** : 1025-1027.

Higaki, T. (1976) *Ph. D. Diss.,* Michigan State Univ., East Lansing.

Higaki, T. (1977) *Proc. Trop. Reg. Amer. Soc. Hort. Sci.,* **21** : 44-45.

Higaki, T. and Imamura, J.S. (1985) *Bull. Res. Series. Hawaii Inst. Trop. Agr. Human Resources,* No. 40, pp. 8.

Higaki, T. and Poole, R.T. (1978) *J. Amer. Soc. Hort. Sci.,* **103** : 98-100.

Higaki, T. and Rasmussen, H.P. (1979) *HortScience,* **14** : 64-65.

Higaki, T., Imamura, J.S. and Paull, R.E. (1992) *HortScience,* **27** : 909-912.

Higaki, T., Rasmussen, H.P. and Carpenter, W.J. (1980b) *J. Amer. Soc. Hort. Sci.,* **105** : 438-440.

Higaki, T., Rasmussen, H.P. and Carpenter, W.J. (1980b) *J. Amer. Soc. Hort. Sci.,* **105** : 441-444.

Higaki, T., Watson, D.P. and Leonhardt, K.W. (1979) *Circular, College Trop. Agric. Human Resources,* No. 420, pp. 19.

Hlatshwayo, M.S. and Wahome, P.K. (2010) *J. Agri. Soc. Sci.,* **6** : 34–38.

Hseu, S.H. and Lin, C.Y. (1998) *Plant Prot. Bull. Taipei,* **40** : 409-417.

Huang, Y.L., Yuan, S.C. and Chen, F.C. (2020) *The Hortic. J.,* **89** : 54-60.

Ignasse, J.E. (1984) *Horticulture Francise,* No. 160, 0. 10.

Imamura, J.S. and Higaki, T. (1984) *Res. Extension Series, Hawaii Inst. Trop. Agr. Human Resources,* No. 47, pp. 17.

Imamura, J.S. and Higaki, T. (1988) *HortScience,* **23** : 353-354.

Islam, M.R., Khan, R., Mamun, A.F.K., Ahmed, G. and Hakim, L. (2009) *Plant Tissue Cult Biostech.,* **19** : 61-69.

Iwata, R.Y., Tang, C.S. and Kamemoto, H. (1979) *J. Amer. Soc. Hort. Sci.,* **104** : 464-466.

Iwata, R.Y., Tang, C.S. and Kamemoto, H. (1985) *J. Amer. Soc. Hort. Sci.,* **110** : 383-385.

Jahan, M.T, Islam, M.R., Khan, R., Mamun, A.N.K., Ahmed, G. and Hakim, L. (2010) *In vitro* Clonal Propagation of Anthurium (*Anthurium andraeanum* L.) Using Callus Culture. *Plant Tissue Culture and Biotechnology,* **19**(1): 61-9. 10.3329/ptcb.v19i1.4961.

Jimenez, F.N., Lennon, A.M. and Umaharan, P. (2012) *Am. Soc. Hortic. Sci.,* **137** : 263–272.

Juárez-Maldonado, A. (2017) *HortScience,* **52** : 1585-1592.

Jungbauer, J. (1973) *Zierpflanzenbau,* **13** : 223-225.

Kagiwata, T. (1990) *Japanese J. Trop. Agric.,* **34** : 289-291.

Kalkman, E.C. (1983) *Vakbladvoor de Bloemisterij,* **38** : 69-71.

Kamemoto, H. (1962) *Hawaii Fm. Sci.,* **11** : 2-4.

Kamemoto, H. and Nakasone, H.Y. (1955) *Hawaii Fm. Sci.,* **3** : 4-5.

Kamemoto, H. and Nakasone, H.Y. (1963) *Tech. Bull. Hawaii Agri. Exp. Stat.,* **58** : 28.

Kamemoto, H. and Sheffer, R.D. (1978) *HortScinece,* **13** : 117-179.

Kamemoto, H., Aragaki, M., Higaki, T. and Kunisaki, J. (1971) *Hawaii Fm. Sci.,* **20** : 12.

Kamemoto, H., Iwata, R.Y. and Marutani, M. (1988) *HITAHR Res. Ext. Ser.,* **8** : 88.

Kanani, M., Ghorbani, N. and Behnamian, M. (2017) In: *International Symposium on Wild Flowers and Native Ornamental Plants.* **1240**, pp. 125-130.

Kaneko, K. and Kamemoto, H. (1979) *J. Heredity,* **70** : 271-272.

Khawlhring, C., Patel, G.D. and Lalnunmawia, F. (2019) *J. App. and Nat. Sci.,* **11** : 240-244.

Khawlhring, N., Thanga, J.L.T. and Lalnunmawia, F. (2012) *J. Hortic. For.*, **4** : 22–26.

Khunle, A.R., Chen, F.C. and Sugii, N. (1995) *HortScience,* **30** : 456-461.

Kim, W.G., Cha, Y.S., Cho, W.D. and Lee, Y.H. (1993) *RDA J. Agric. Sci. Crop. Prot.,* **35** : 327-330.

Klapwijk, D. and Spek, H.J.J. van de (1984) *Vakbladvoor de Bloemisterij,* **39** : 54-55.

Kuanprasert, N., Kuehnle, A.R. and Tang, C.S. (1988) *Phytochem.,***49** : 521-528.

Kuehnle, A.R. and Nan, G.L. (1991) *Physiologia Plantarum,* **82** : A7.

Kuehnle, A.R. and Sugii, N. (1991) *HortScience,* **26** : 919-921.

Kuehnle, A.R., Chen, F.C. and Sugii, N. (1995) *HortScience,* **30** : 456-461.

Kuehnle, A.R., Chen, F.C., Sugii, N., Jaynes, J.M. and Sanford, J.C. (1990) *Proc. 3rd Anthurium Blight Conf., May 23-24, 1990,* Univ. Hawaii, Hilo, pp. 52-54.

Kuehnle, A.R., Fujii, F.C., Chen, A., Alvarez, N., Sugii, R., Fukui, S.L., Aragon and Jaynes.J.M. (2004) *HortScience,* **39** : 1327–1331.

Kuehnle, A., Chen, F.C. and Sugii, N. (1992) *Plant Cell Rep.,* **11** : 438-442.

Kuehnle, A.R., Fujii, R., Chen, F.C., Alvarez, A., Sugii, N., Fukui, R. and Aragon, S.L. (2004) *HortScience,* **39** : 1327–1331.

Kuhle, G. (1960) *Deutsche Gartenb,* **7** : 273.

Kumsingkaew, S. and Akarapisan, A. (2014) *J. Agric. Technol.,* **10** : 755-766.

Kunisaki, J.T. (1980) *HortScience,* **15** : 508-509.

Kunisaki, J.T. and Sagawa, Y. (1971) *Hawaii Fm. Sci.,* **20** : 4-5.

Kuruppu, N.P.U. and Yogaratnam, V. (1989) *Tropical Agriculturist,* **145** : 13-23.

Lamb, A.E. (2002) *U.S. Patent Application No. 09/588,124.*

Lamb, A.E. and Knauss, M.E. (2004) *U.S. Patent Application No. 10/770,217.*

Laws, N. and Galinsky, B. (1996) *Floriculture International,* **6** : 21-23.

Leeuwen, C. van (1980) *Vakbladvoor de Bloemisterij,* **35** : 30-31.

Leeuwen, C. van (1984) *Vakbladvoor de Bloemisterij,* **37** : 48-51.

Leffring, L. (1975) *Acta Hortic.,* **51** : 63-68.

Leffring, L. and Soede, A.C. (1978) *Vakbladvoore de Bloemisterij,* **33** : 25.

Leffring, L. and Soede, A.C. (1979a) *Vakbladvoore de Bloemisterij,* **34** : 40-41.

Leffring, L. and Soede, A.C. (1979b) *Vakbladvoore de Bloemisterij,* **34** : 43.

Leffring, L., Hoogstrate, J. and Braster, M. (1976) *Vakbladvoore de Bloemisterij,* **31** : 14-15.

Li, T., Heuvelink, E., Van Noort, F., Kromdijk, J. and Marcelis, L.F.M. (2014) *Sci. Hortic.*, **179** : 306-313.

Liang, L., Deng, Y., Sun, X., Jia, X. and Su, J. (2018) *J. Am. Soc. Hort. Sci.*, **143** : 3-13.

Lightbourn, G.J. and Prasad, P.V.D. (1990) *Proc. Interamerican Soc. Trop. Hort.*, **34** : 3-5.

Lima, J.D., Nomura, E.S., Fuzitani, E.J. and da Silva, S.H.M.G. (2010) *Scientia Agraria*, **11** : 193-200.

Lima, J.D., Zanetti, S., Nomura, E.S., Fuzitani, E.J., Rozane, D.E. and Iori, P. (2016) *Ciência Rural*, **46** : 440-446.

Lin, H.F., Xiong, J., Zhou, H.M., Chen, C.M., Lin, F.Z., Xu, X.M., and Yeh, K.W. (2019) *BMC Plant Biol.*, **19** : 40.

Liu, L.J. and Serapion, J.L. (1977) *Proc. Amer. Phytopath. Soc.*, **4** : 179.

Maatsch, R. and Bachthaler, E. (1964) *Gartenwelt*, **64** : 273-274.

Maitra, S. and Roychowdhury, N. (2005) *J. Ornam. Hortic.*, **8** : 186-190.

Maitra, S., Ghosh, P.D., Roychowdhury, N., and Satya, P. (2012) *Int. J. Bio Resource and Stress Manag.*, **3**.

Mandhar, S.C., Kumaran, G.S. and Prakash, D. (1998) *Souvenir and Abst. Nat. Sem. On Anthurium Production, 2-3 June, 1998, viv. Ornamental Crops and Central Hort. Expt. Sta.*, IIHR, Bangalore, India, pp. 15.

Marsala, J., Ferraz, M.V., da Silva, S.H.M.G. and Pereira, D.D.M.G. (2014) *Ornam. Hortic.*, **20** : 137-142.

Marutani, M., Sheffer, R.D. and Kamemoto, H. (1993) *Amer. J. Bot.*, **80** : 93-103.

Matsumoto, T.K., Kuehnle, A.R. and Webb, D.T. (1998) *Amer. J. Bot.*, **85** : 1560-1568.

Matysiak, B. and Nowak, J. (1996) *Zeszyty Problemowe Postepow Nauk Rolniczych*, **429** : 241-244.

Matysiak, B., Nowak, J. and Sase, S. (1996) *Acta Hortic.*, **440** : 628-632.

Maurer, M. (1979) *Gartenerborse and Gartenwelt*, **79** : 832-834.

Maurer, M. and Brandes, S. (1979a) *Gartenbauwiss*, **44** : 71-73.

Maurer, M. and Brandes, S. (1979b) *Gartenbauwiss*, **44** : 103-106.

Meerow, A.W. (1995) *HortTechnology*, **5** : 237-239.

Meij, J. (1976) *Vakbladvoor de Bloemisterij*, **31** : 15.

Mirabolfathy, M., Groenewald, J.Z. and Crous, P.W. (2010) *Plant Dis.*, **94** : 278-278.

Miura, N.D.S., Beriam, L.O. and Rivas, E.B. (2013) *Horticultura Brasileira*, **31** : 322-327.

Montes, S., Hernandez, M.M. and Varela, M. (1999) *CultivosTropicales*, **20** : 51-54.

Muraleedharan, A., Sha, K., Sujin, G.S., Karthikeyan, P.K., Kumar, C.P.S., Joshi, J.L. and Nainu, A.J. (2020) *Plant Arch.*, **20** : 4183-4186.

Nakasone, H.Y. and Kamemoto, H. (1957) *Circ. Hawaii Agric. Exp. Stat.*, **53** : 11.

Nakasone, H.Y. and Kamemoto, H. (1962) *Hawaii Agri. Expt. Sta. Tech. Bul.*, 50.

Nhut, D.T., Duy, N., Vy, N.N.H., Khue, C.D., Khiem, D.V., Hang, N.T.T. and Vinh, D.N. (2004) *J. Agric. Sci. Technol.*, **4** : 73–78.

Nikolova, N. and Zafirova, P. (1980) *Nauchni Trudove Vissh Selskostopanski Institut,* **25** : 99-103.

Nishijima, W.T. and Fujiyama, D. K. (1985) *Commodity Fact Sheet,* Hawaii Cooperative Extension Service No. AN-(4) (A), Flower, pp. 3.

Nomura, E.S., Lima, J.D., Rodrigues, D.S., Garcia, V.A., Fuzitani, E.J. and Silva, S.H.M.G.D. (2009) *Ciencia Rural,* **39** : 1394-1400.

Osiecki, M.W. (2002) *U.S. Patent Application No. 09/961,829.*

Owata, R.Y., Tang, C.S. and Kamemoto, H. (1979) *J. Amer. Soc. Hort. Sci.,* **104** : 464-466.

Papenhagen, A. (1986) *Gartenerborse and Gartenwelt,* **86** : 1343-1346.

Paramveer, S., Dhaduk, B.K. and Chawla, S.L. (2011) *Ind. J. Hortic.,* **68** : 86-90.

Paull, R.E. (1982) *HortScience,* **17** : 606-607.

Paull, R.E. (1983) *Res. Extension Series, Hawaii Institute of Trop. Agric. and Human Res.,* No. 037, pp. 225-226.

Paull, R.E. (1987) *HortScience,* **22** : 459-460.

Paull, R.E. and Goo, T. (1982) *J. Amer. Soc. Hort. Sci.,* **107** : 842-844.

Paull, R.E. and Goo, T. T. C. (1985) *J. Amer. Soc. Hort. Sci.,* **110** : 84-88.

Paull, R.E., Higaki, T. and Imamura, J.S. (1992) *Scientia Hort.,* **49** : 125-134.

Penningsfeld, F. (1977) *Obst. and Garsenbauverlag,* pp. 8-13.

Pierik, R.L.M. (1975) *Netherlands J. Agric. Sci.,* **23** : 299-302.

Pierik, R.L.M., Leeuwen, P. van and Rigter, G.C.C.M. (1979) *Netherlands J. Agric. Sci.,* **27** : 221-226.

Pierik, R.L.M., Steegmans, H.H.M., Schaik, W.V. and Eyk-Bos, G.V. (1975) *Vakbladvoor de Bloemisterij,* **30** : 27.

Pignon, C. (1963) *Pyreth. Post. D.,* **7** : 33-36.

Pinheiro, M.V.M., Martins, F.B., Cruz, A.C.F.D., Carvalho, A.C.P.P.D., Oliveira, E.J.D. and Otoni, W.C. (2014) *Acta Scientiarum,* **36** : 87-98.

Pitta, G.P.B., Souza, T.M.W.de, Feichtenberger, E. and De Souza, T.M.W. (1991) *Summa Phytopathologica,* **17** : 159-162.

Pontes, T.A., Van Den Berg, C., Andrade, I.D. and Mayo, S.J. (2014) *Phytotaxa,* **178** : 171-180.

Poole, R.T. and Chase, A.R. (1989) *Foliage Digest,* **12** : 3-5.

Poole, R.T. and McConnell, D.B. (1971) *Proc. Trop. Reg. Amer. Soc. Hort. Sci.,* **15** : 189-195.

Pour, A.S., Chegini, G., Zarafshan, P. and Massah, J. (2018) *Postharvest Biol. Technol.*, **139** : 67-74.

Prakash, D., Aswath, C., Prasad, K.V. and Choudhury, M.L. (1998) *Souvenir and Abst.Nat. Sem. on Anthurium Production, 2-3 June, 1998,* Piv. Ornamental Crops and Central Hort. Expt. Sta., IIHR, Bangalore, India, pp. 16.

Prasad, M. and Fietje, G. (1989) *Acta Hortic.*, **238** : 157-164.

Pritchard, M.K., Hew, C.S. and Wang, H. (1991) *J. Hort. Sci.*, **66** : 209-214.

Promyou, S., Ketsa, S. and van Doorn, W.G. (2012) *Postharvest Biol. Technol.*,**64** : 104–110.

Rachmawati, F. (2007) *Ph.D.* Thesis, pp.146.

Ravinath, D. (2007) Floriculture: a Viable Business. Excel Books India, India.

Reddy, B.S., Yathisha, S.M. and Nataraj, S.K. (2013) *Int. J. Processing and Post Harvest Technol.*, **4** : 114-117.

Rodriguez, S.J., Torres, C.J., Bosaque-Lugo, R. and Semidey-Laracuente, N. (1979) *J. Agric. Univ. Puerto Rico,* **63** : 386-389.

Roni, Z.K., Islam, S., Shamsuzzaman, A.M.M., Uddin, A.F.J., Sahare, H.A. and Alka, S. (2015) *Trends in Biosciences*, **8** : 305-307.

Sakai, W.S., Robertson, S. and Goldstein, M. (1990a) *Proc. 3rd Anthurium Blight Conf., May 23-24, 1990,* Univ. Hawaii, Hilo, pp. 12-13.

Sakai, W.S., Holland, G., Furutani, S., Sewake, K., Imamura, J. and Higaki, T. (1990b) *Proc. 3rd Anthurium Blight Conf., May 23-24, 1990,* Univ. Hawaii, Hilo, pp. 14-17.

Sankari, A., Anand, M. and Arulmozhiyan, R. (2013) *Asian J. Hort.*, **8** : 547-551.

Sato, D. (1983) *Res. Extension Series, Hawaii Inst. Trop. Agric. Human Res.*, No. 037, pp. 245-248.

Schaper, D. and Zimmer, K. (1991) *Gartenbauwiss,* **56** : 163-169.

Schenk, M. and Brundert, W. (1981) *Gartenbau,* **36** : 2064-2065.

Schmidt, K. and Lauterbach, D. (1984) *Deutsche Gartenb,* **38** : 2231-2232.

Schmidt, M., Adam, H. and Patzold, D. (1989) *Gartenbau,* **36** : 108-111.

Schneider, R. and Kiewnick (1974) *PhytopatholgischeZeitschrift,* **79** : 364-367.

Seemanthini, N.S., Chandrashekar, S.Y., Naik, B.H., Shivaprasad, M. and Ganapathi, M. (2018) *Int. J. Curr.Microbiol. App. Sci.*, **7** : 87-91.

Segers, C.M. (1976) *Vakbladvoor de Bloemisterij,* **31** : 45.

Sengupta, D.K., Rabindra-Chettri and Chettri, R. (1989) *Environ and Ecol.*, **7** : 1025-1028.

Sewake, K.T. (1990) *Proc. 3rd Anthurium Blight Conf. May 23-24, 1990,* Univ. Hawaii, Hilo, p. 44.

Sheffer, R.D. and Croat, T.B. (1983) *Amer. J. Bot.*, **70** : 858-871.

Shimasaki, K. (2016) In *Proceedings of the 2016 Annual Meeting of the International Plant Propagators' Society,* **1174** : 137-144.

Shirakawa, T., Bedolph, P.R. and Watson, D.P. (1964) *Proc. Amer. Soc. Hort. Sci.,* **85** : 642-646.

Singh, F. (1998) *Souvenir and Abst.Nat. Sem. on Anthurium Production, 2-3 June, 1998,* Piv. Ornamental Crops and Central Hort. Expt. Sta., IIHR, Bangalore, India, pp. 1-6.

Singh, P., Dhaduk, B.K. and Chawla, S.L. (2011) *Ind. J. Horti.,* **68** : 86-90.

Soleimani Aghdam, M., Jannatizadeh, A., Sheikh-Assadi, M. and Malekzadeh, P. (2016) *Sci. Hortic.,* **202** : 70–76.

Soleimanipour, A. and Chegini, G.R. (2020) *Computers and Electronics in Agriculture,* **174** : 105-109.

Somaya, K.U., Narayanaswamy, P. and Jayaprasad, K.V. (1998) *Karnataka, J. Agril. Sci.,* **11** : 466-470.

Sonneveld, C. and Voogt, W. (1983) *Plant and Soil.,***74** : 41-52.

Sonneveld, C., Voogt, W. and Tattini, M. (1993) *Acta Hortic.,* **342** : 61-67.

Sosa-Flores, V.P., Valdez-Aguilar, L.A., Cartmill, D.L., Cartmill, A.D., Benavides-Mendoza, Spek, H.V.D. and Sonneveld, C. (1985) *Vakbladvoor de Bloemisterij,* **40** : 114-115.

Sreelatha, U., Nair, S.R. and Rajmohan, K. (1998) *J. Ornamental Hort.,* **1** : 48-54.

Srinivasa, V. (2006) *Environ. Ecol.,* **24** : 1034-1036.

Steen, J.A.V.D. (1976) *Vakbladvoor de Bloemisterij,* **31** : 16-17.

Steen, J.A.V.D. (1977) *Vakbladvoor de Boemisterij,* **32** : 24-25.

Steen, J.A.V.D. and Vijvosberg, A.J. (1973) *Vakbladvoor de Bloemisterij,* **28** : 10-11.

Steiner, A.A. and Van Zon, A.A. (1957) *Jversl. Proefstat. Bloem. Aalsmeer,* pp. 30-32.

Straver, N.A. (1980) *Vakbladvoor de Bloemisterij,* **35** : 38-39.

Sumathi, T., Chawla, S.L., Ahlawat, T.R., Patil, S. and Gurjar, R.A. (2018) *Ind. J. Agric. Sci.,* **88** : 1403-1406.

Sumathi, T., Chawla, S.L., Patil, S. and Ahlawat, T.R. (2019) *The Horticultural Society of India,* **76** : 334-337.

Sytsema, W. and Barendse, L. (1975) *Vakbladvoor de Bloemisterij,* **30** : 1921.

Szendel, A.J., Hartman, J. and Laskowska, H. (1982) *Prace. Instytutu Sadownictwa w skierniewicach, B,* **6** : 54-57.

Tai-Chum, P.A. and Ricketts, M.R. (1991) *Proc. Inst. Soc. Trop. Hort.,* **35** : 141-144.

Tesi, R. and Faro, F.L. (1985) *Informatore Agrario,* **41** : 71-76.

Thawiang, N., Buanong, M. and Kanlayanarat, S. (2007) *Acta Hortic.,* 8 : 128-134

Thawiang, N., Buanong, M. and Kanlayanarat, S. (2007) In *International Conference on Quality Management in Supply Chains of Ornamentals,* **755**: 415-418.

Tian, D.Q., Pan, X.Y., Ge, Y.Y., Liu, X.J. and Pan, G.M. (2013) *Jiejiang Agric. Sci. Technol.,* **9** : 1125–1127.

Tian, L.M. and Ma, Y.L. (2008) *J. Anhui Agric. Sci.,***36** : 13977–13978, 13983.

Tombolato, A.F.C., Rivas, E.B., Coutinho, I.N., Bermann, E.C., Imenes, S.D.L., Furlani, P.R., Castro, C.E.F., Matthes, L.A.F. and Saes, L.A. Costa (2002) *Campinas: Instituto Agronômico,* pp.47.

Tombolato, A.F., Saes, L.A., Matthes, L.A.F., Sakai, M., Dias-Tagliacozzo, G., de Castro, C.E.F. and Barbosa, W. (2005) *Acta Hortic.,* **683** : 423.

Toppo, R. and Beura, S. (2018) *Int. J. Curr. Microbiol. App. Sci.,* **7** : 2850-2856.

Tsang, M.M.C., Hara, A.H. and Shintaku, M.H. (2010) *Crop Prot.,* **29** : 525–531.

Turski, R., Flis-Bujak, M. and Martyn, W. (1983) *Prace Instytutut Sadownictwa w Skierniewicach, B,* **8** : 123-132.

Valsalakumari, P.K., Salvi, B.R., Rajeevan, P.K. and Geetha, C.K. (1998) *Souvenir and Abst., Nat. Sem. on Anthurium Production, 2-3 June, 1998, Piv. Ornamental Crops and Central Hort. Expt. Sta,* IIHR, Bangalore, India, pp. 17.

Van der Knaap, L.J.M. (2010) *U.S. Patent Application No. 12/287,359.*

Van Rijn, L.W. (2004) *U.S. Patent Application No. 10/721,989.*

Vargas, T.E., Mejias, A., Oropeza, M. and Garia, de E. Electron. (2004) *J Biotechnol.,* **7** : 282-286.

Vierig, A. (1990) *Deutscher-Gartenbau,* **44** : 2046-2047.

Vonk-Noordegraaf, C. (1969) *Vakbladvoor de Bloemisterij,* **24** : 133.

Voogt, S.J. (1979) *Vakbladvoor de Bloemisterij,* **34** : 26-27, 29.

Wang, G.L., Xu, C.Y. and Wang, G.D. (2010) *J. Plant Res. Environ.,***19** : 26–31.

Wang, K.H., Kuehnle, A.R. and Sipes, B.S. (1998) *Euphytica,* **103** : 23-28.

Wang, K.H., Sipes, B.S. and Kuehnle, A.R. (1997) *Efecto de medios sin sueloen el crecimiento de Anthurium andraeanuminfestado por Radopholussimilis,* **27** : 77-84.

Wang, K.H., Sipes, B.S. and Kuehnle, A.R. (1999) *HortScience,* **34** : 296-297.

Wang, K., Li, Y., Xie, H., Wu, W.J. and Xu, C.L. (2016) *Plant Disease,* **100** : 940-945.

Wang, Y., Liu, S., Tian, X., Fu, Y., Jiang, X., Li, Y. and Wang, G. (2018) *Plant signaling and behavior,* **13** : 148-174.

Wang, Yin Tung and Wang, Y.T. (1999) *HortTechnology,* **9** : 409-412.

Wannakrairoj, S. and Kamemoto, H. (1990) *J. Amer. Soc. Hort. Sci.,* **115** : 169-171.

Winarto, B. (2009) *Ph.D. Thesis,* pp. 235.

Winarto, B. (2011) *J. Agron. Indones.,* **21** : 295–305.

Winarto, B. and Mattjik, N.A. (2009) *J. Agron. Indones.,* **37** : 138–144.

www.ugaoo.com

www.kalliergeia.com

www.en.wikipedia.org

Xu, B., Xin, W., Wang, G., Guo, W., Wen, F. and Jin, J. (2006) *Chinese Bulletin of Botany*, **23** : 698.

Xu, C.Y., Chen, X.X., Xin, W.J. and Wang, G.D. (2010) *Guihaia.* **30** : 117–121

Yu, K.J. and Paek, K.Y. (1995) *J. Korean Soc. Hort. Sci.,* **36** : 684-694.

Zafirova, P. (1982) *Nauchni Trudoe, Vissh Selskostopanski Institut Vassl Kolarov,* **27** : 90-91.

Zens, A. and Zimmer, K. (1986) *Gartenbauwiss,* **51** : 26-31.

Zhang, Y., Li, B., Li, S.F. and Chen, Z.H. (2009) *Acta Laser Biol. Sin.,* **18** : 1–5.

Zhang, Z.S., Li, Y.F., Jiang, L., Li, Y., Wang, Z.Y., Xia, J. and Yi, M.S. (2007) *Acta Hortic.,* **34** : 729–734.

Zheng, F., Zhanchao, W., Duanfen, C. and Guixian, G. (2004) *J. Hebei Agric.Univ.,* **27** : 51-53.

Zimmer, K. (1990) *Gartenerborse and Gartenwelt,* **90** : 870-873.

Gerbera

Ayesha, C. Aswath, Rajiv Kumar, P. Das, P.K. Singh Samanta and V.A. Parthasarathy

1.0 Introduction

Gerbera is an important commercial flower grown throughout the world. It is a perennial herb native to South Africa and Asia, belongs to the family Asteraceae. Gerbera has a wide range of attractive flower colors such as red, pink, orange, peach, maroon *etc.*, which makes it a valuable ornamental species. Gerbera is the fifth most used flower used in cut arrangements and bouquets, after rose, carnation, chrystanthemum and tulip (Dixit, 2016). It is also called Transvaal daisy, Barberton daisy or African daisy. Besides as a cut flower, it is also very ideal for beds, borders, pots and rock gardens. The gerbera plant also used in the preparation of traditional Chinese medicine: tu-er-feng, derived from whole plants of gerbera, is used for curing cold with cough and for rheumatism (Ye *et al.*, 1990).

2.0 Importance and Uses

2.1.0 Aesthetic Uses

It is generally used as a decorative cut flower for bouquets and decoration besides used in pots, flower beds, *etc.* The flowers are popular for floral arrangements due to their wide range of colors with lasting quality. The plant also acts as air purifier which helps in eliminating the harmful toxins. It is also used as a model plant in studying flower formation (https://en.wikipedia.org).

2.2.0 Therapeutic and other Uses

It is primarily the fresh or dried flower heads that are used as a medicine, but the leaves can also be used. Studies have shown that it has anticancer, anti-proliferative, antimetastatic, antispasmodic, gout and ulcers curing properties. In China medicine it is used in the preparation of tu-er-feng for curing colds and treating rheumatism. Plant extract are also used in cosmetic and bath preparations showing skin moisturizing and conditioning effects (www.stuartxchange.org).

The plants act as an herbal remedy for the common cold, bronchitis and other inflammation of the upper respiratory. The herb may be used for loss of appetite as it has a stimulating effect on the digestion system and it has been used as a treatment for many ailments of the digestive tract, such as gastritis, diarrhea, liver and gallbladder complaints and mild constipation. Flowers have been used traditionally for painful menstruation, cystitis and other inflammations of the urinary tract. Additionally, it may have its uses as a natural treatment for atopic dermatitis, gout, and chronic rheumatic conditions. The plant is known for its healing properties and can be used topically on small wounds, sores, and scratches to speed up the healing process. In addition, it has been used externally for dermatitis, rash, eczema, bruises, and boils. A tincture made from the herb can be used for acne, and as a mouthwash or a gargle, it may be used as a remedy for mouth inflammation and sore throat. Also, chewing on the fresh leaves might be helpful as a relief for mouth ulcers. In folk medicine, it was often recommended to eat the fresh leaves to stimulate nutrition uptake, due to the bitter substances found in the herb. The fresh leaves, flowers, buds, and petals have a pleasant taste and can be used in salads

Uses of Gerbera

In Bed — In Border — In Garden

Bouquet — As Cut Flower — In Pot

In Flower Arrangement — In Wedding Decoration — Floral Crown

Floral Rangoli — Floral Veni — In Garland

Source: www.in.pinterest.com, www.flowerexplosion.com, www.indiamart.com, www.bloomsonly.com, www.hilverdaflorist.com, Mokshitha Craft and Creatives, www.mrplantgeek.com, www.suttons.co.uk

or added to soups. Furthermore, the flower heads can be used in vinegar and as a substitute for capers (www.herbal-supplement-resource.com).

3.0 Origin and History

The first official description of the South African species *Gerbera jamesonii* was made by J.D. Hooker in 1889 in Curtis Botanical Magazine (Codd, 1979). The native distribution of this genus, comprising of about 30 species, extends to Africa, Madagascar, and tropical Asia and South America (Bremer, 1994). In India, *Gerbera* spp. are distributed from Kashmir to Nepal at an elevation of 1300 to 3200 m. It was named after German naturalist, Traugott Gerber. *Gerbera jamesonii*, native to Natal and Transvaal, and *G. Viridifolia* from Cape were crossed by Lynch in 1887 after both these species were introduced to Botanic Garden, Cambridge and the hybrid was named as *G. cantabrigensis*. Tourjee *et al.* (1994) revealed that the early development of *G. jamesonii* as a floricultural crop was traced from its collection as a novelty in South Africa to its establishment as a commercial crop in the 1930s. The breeding work was initiated in Europe, and later, the USA for cold hardiness in temperate climates which was an early objective. The relative contribution of *G. jamesonii* and *G. viridifolia* to the modern crop is unknown, but much of the cultivated germplasm can be traced to material that passed through the Botanic Gardens, Cambridge, UK, and La Rosarie, Antibes, France.

4.0 Species and Cultivars

Gerbera is a stemless perennial herb; leaves radical, petiolated, lanceolate, lobed, narrower at the base and wider at tip and are arranged in a rosette at the base; flower heads solitary, showy, yellow, orange, or scarlet, many flowered, the conspicuous ray florets in 1 or 2 rows, those of the inner rows, when present, subtubular and very short and bilipped, disc florets, 2-lipped involucral scales, oblong or lanceolate, closely overlapping. The daisy-like flowers are in various colours. Some bicoloured double cultivars are very attractive on long leafless stalk. Loeser (1986) grouped gerbera into single, semidouble and double cultivars based on flower heads.

There are about 40 species in the genus *Gerbera* which are hardy and perennial, of which only *Gerbera jamesonii* is under cultivation (Brickell, 1997). *Gerbera asplenifolia, G. aurantiaca, G. kunzeana* and *G. viridifolia* are some other important species in the genus.

4.1.0 Species

Gerbera asplenifolia

Leaves leathery, 10-15 cm long, narrow, deeply lobed, roundish; flower heads purple on a hairy scape.

G. aurantiaca

Leaves lanceolate to oblong, acute, 12.5-15 cm long; flower heads orange with yellowish anthers.

G. jamesonii (Barberton Daisy)

Plants hairy, base woody. Leaves runcinate, 12 to 20 cm long; flower heads about 7 to 10 cm across, orange-scarlet.

G. kunzeana

A Himalayan species whose flowers open scarcely.

G. viridifolia

Leaves elliptical or oblong, obtuse, smooth, greenish; flower heads small, dirty white, borne on a short stalk.

G. gossypina

A wild plant of Western Himalayan origin. *G. gossypina* can be used as an ornamental pot plant due to its compact plant size, small light pink flowers and green foliage with woolly fibers on the under surface. The species is free of diseases such as powdery mildew and grey mould. The species is not commonly distributed in Himalayas and is being domesticated for its conservation and sustainable utilization as an ornamental pot plant (Singh and Dhyani 2016).

4.2.0 Cultivars

The criteria considered for the development of new gerbera selections are: ease of growing, uniformity of growth, length of the cultivation period, availability in various colours, uniformity and strength of flower stems, and compactness of growth and suitability to grow under open conditions.

Qualitative and quantitative differences in flower flavonoids have been used for discrimination between varieties (Booy, 1994). Anthocyanins and flavonoid copigments from ray florets have been found to be useful for identification of gerbera cultivars (Booy, 1995).

Steen and Steen (1975) reported *cv.* Sympathie to be the slowest to flower and Zalmora gave the highest yield of cut flowers. Baillot (1976) found that the double pink *cv.* Fredaisy produced more flowers/plant and the flowers were of great diameter with longer stalk. Homogeneity, stability of colour and general quality were excellent. Papenhagen (1978) described the 5 most valuable Friesdorf clones of gerbera– Belero, Adelita, Carita, Rosamunde and Ursula. Lopez *et al.* (1979) described a new *cv.* D-9 which had a satisfactory homogeneity in colour and flower morphology, and a long and narrow ligule. It is early flowering and of high flower productivity. Grabowska *et al.* (1979) found that the most productive cultivars are Fermin, Zalmora, Veronica and Oranje Nassau, and the Bloemfontein strain. The largest proportion of first grade cut flowers was produced by Veronica, Oranje Nassau and Sylvia. Commercially the most profitable cultivars were Veronica, Fermin, Zalmora and Oranje Nassau. Reimherr *et al.* (1980) found that the orange Clementin and the pink Appelblossom were the best out of 10 gerbera cultivars. Schenk and Brundert (1981) tested a red, 2 yellow, 2 orange and 5 pink gerbera cultivars and graded according to their stem length. Sytsema (1981) found that the cultivars Pimpernel, Yellow Moor, Appelblossom, Agnes, Terra Nigra, Symphonie,

Species of Gerbera

Gerbera viridifolia

Gerbera jamesonii

Gerbera aurantiaca

Gerbera kunzeana

Gerbera asplenifolia

Gerbera gossipiana

Gerbera wrightii

Gerbera nivea

Gerbera maxima

Gerbera wrightii

Gerbera ambigua

Gerbera cordata

Source: www.botany.cz, www.zahrady-skalky.sk, www.greathimalayannationalpark.org, www.waterberg-bioquest.co.za, www.inaturalist.org, www.efloraofindia.com, www.ispotnature.org, www.pietrzykowscy.pl, www.redlist.sanbi.org, www.westafricanplants.senckenberg.de

Cultivars of Gerbera

Arka Red | Arka Krishika | Arka Pink

Arka Ashwa | Arka Nesara | Arka White

Rosalin | Dana Ellen | Stanza

Sunway | Savannah | Silvester

Source: www.nestreeo.com, www.iihr.res.in

Cultivars of Gerbera

***Source*: www.nestreeo.com**

Carmen, Uranus, Schlosser, Terra Mix and Petrusa are best having long keeping quality. Eberhardt (1982) described six gerbera cultivars. Enzett Bootes had semi-double pink flowers with a prominent black centre, Enzett Orion was golden, giving 21 blooms/plant, winter yielded of 6 blooms/plant, and a vase life of 10 days; Enzett Perseus was a small flowered double pink cultivar yielding 18 blooms/plant with winter yield of 5.9 blooms/plant and a vase life of 11.3 days; Enzett Auriga Violettosa had long stems and gave a yield of 13 blooms/plant. Enzett Dorado had bright orange red flowers with a vase life of 19.7 days, bearing about 13 flowers/ plant and 3-4 blooms/plant in winter.

Fischer *et al.* (1982) found that Maron Clementine, Yellow Clementine, Red Clemen-tine, Clementine, Xenia and Flamingo were the best cultivars. Maron Clementine produced the highest winter and total yields (434 flowers/m^2). The white *cv.* Delphi produced only 204 flowers/m^2, the dark red Vesta was the least vigorous, the yellow Uranus had the darkest foliage and the pale rose Flamingo the palest. Cream Clementine had the longest vase life (12 days), and Yellow Clementine the shortest (9.4 days). Gergely (1988) evolved four red cultivars *viz.*, Phaedra, Melitta, Anna and Dama; 4 yellow cultivars: Viktoria, Dora, Leila and Terez; one pink cultivar : Ildiko and one lilac cultivar. Buivids *et al.* (1992) found that Jautriite, Ruusinjs, Mikus, Raunis, Cilda and Toms produced flowers in winter and were notable for high quality of the cut flowers. Good ornamental value was the characteristic of Zeltene, Aime, Aira and Lina. Seed-raised selections are available in mixed colours - shades of yellow, apricot, orange, red, pink, scarlet or lavender. The important cultivars are Californian Giant, Pandora series, Parade series, Sunburst series and Tempo series (Brickell, 1996).

Kumar (2013) evaluated the performance of ten genotypes for flower quality characters under naturally ventilated polyhouse, and reported that genotypes Kyllian and Vilassar were found promising for flower quality characters and recommended for growing under naturally ventilated polyhouse. Kumar *et al.* (2012) evaluated 17 exotic cultivars of gerbera under low-cost bamboo made polyhouse and shade net (50 per cent) house under Meghalaya conditions. The cultivars Diego, Lieke, Lion, Pound Sterling, Torbin, Venicie and Wall Street were found promising under low-cost polyhouse and recommended for commercial cultivation for Meghalaya. Kumar and Yadav (2013) found that genotypes Pink Elegance, Piton and Sangria were promising for cut flower production under polyhouse in Meghalaya conditions.

Singh *et al.* (2017) reported that Partrizia was best variety having longest stalk length and a greater number of flowers per plant, while, Szantal and Feliks also exhibited acceptable physical and flowering quality characteristics, so it can also be cultivated under polyhouse for getting better yield and quality flowers.

A new bright yellow gerbera cultivar, Joyful a cross between Chowoo, a semi-double type gerbera cultivar with yellow-orange colored flowers and Focus, a semi-double type cultivar with yellow-colored flowers was developed. Joyful has large, semi-double type flowers with yellow (RHS, 12-A) ray florets and a brown central part with a stable flower shape, strong peduncles, and a vase life of 12.8 days with average flower yield of 49.2 stems per plant per year. Joyful was demonstrated to be based on ploidy tests and RAPD analysis and was registered (Grant No. 4574) to the Korea Seed and Variety Service (KSVS) for commercialization (Mo *et al.*, 2016).

The cultivar Toscana produced maximum stalk thickness and ray florets/ flower, "1133" produced maximum stalk length, Toscana wasmost resistant to fading under high temperature followed by 1133 and Livia. The maximum gross return, net return and B: C ratio was recorded in Fredi followed by Corona, 1133, Toscana and Livia. The cultivars Fredi, Corona and Toscana were recommended for commercial cultivation under naturally ventilated polyhouse (Tameshwer, 2019).

A new bright-pink gerbera cultivar, 'Pink Garden' a cross between 'Pink Light', a semi-double-type gerbera cultivar with pink flowers, and 'Mephysto', a semi-double-type cultivar with purple flowers. 'Pink Garden' has large, double-type flowers with pink (RHS, 55B) ray florets. It also has a steady flower shape, middle-strong peduncles, and a vase life of 13.5 days with average flower yield of '49.5 stems per plant per year. 'Pink Garden' was demonstrated to be based on random amplified polymorphic DNA analysis and was registered (Grant No. 4575) in the Korea Seed and Variety Service for commercialization (Chung and Lee, 2019).

5.0 Genetics and Breeding

5.1.0 Genetics

Cytological investigations indicated that the basic chromosome number of *Gerbera jamesonii* is 2n=50. Changes in chromosome number have influence on the development of phenotypes. Increased plant as well as flower size was reported in tetraploids (Pryor, 1972). The plants of gerbera are grouped into three categories, *viz.* (i) completely self-sterile, (ii) plants with distinct differences in seed set when pollination takes place within an inflorescence, and (iii) when pollination takes place between neighbouring inflorescence on the same plant and plants in which these differences are less pronounced. The third category of plants are less favourable for production of large amount of hybrid seed, since occurrence of selfing within the inflorescence (Maurer and Horn, 1967). Qualitative and quantitative differences in their flavonoid chemical markers distinguished cultivars of very similar colours (Asen, 1984). Inheritance of flower colour is controlled by four factors – *A* causing intense red colour, *a* causing dull red colour, *H* causing *A* and *a* to be lighter in colour, and *B* which became evident in the absence of *A*, causing ivory or yellow colour (Maurer and Horn, 1967).

Productivity is mainly determined by the additive effect of genes. Genes with additive effect have the principal role in flower character inheritance (Mutsenietse *et al.*, 1978a). While studying the inheritance of characters, Mutsenietse *et al.* (1978b) found significant difference in both general and specific combining ability. High heritability for inflorescence diameter and number of ray florets and relatively low for disc diameter and stem length (Schiva, 1973). Self-fertilization resulted in production of less number of viable seeds and of a high proportion of plants with poor characteristics. Heterosis and heritability for ligule length were high and well correlated with the certain other flower characters (Schiva, 1979).

The quality ranking of a sample of gerbera cut flower varies greatly from one person to another hence the selection of a panel to evaluate the flower quality component of a breeding programme is very important (Drennan *et al.*, 1986). Drennan *et al.* (1986) studied heritability, repeatability and components of variance for 68 morphological characters describing the inflorescence, scape and florets. Heritability estimates were generally moderate to high for simple dimension characters and characters which were the mean of several measurements, low to moderate for ratio and composite characters and low for enumeration characters and those describing intraplant variation. Repeatability estimates were generally low for intra plant variation characters, moderate for simple ratio character, and largest for composite and mean characters. A selection index was derived for cut flower yield, consumer rating and vase longevity as function of phenotypic and genetic co-variances of the characters which could be used to select parent in advanced generations. The study indicated that parent-offspring heritability of the index was 0.301 (Harding *et al.*, 1981). For most characters, the variance amongst inflorescence from the same plant was the largest portion of the total phenotypic variance. Aswath *et al.* (1998a) classified the cultivars into three groups based on magnitude of stem bend and recovery. A probable crossing programme between genotypes of class I and II was suggested to achieve structurally strong stem with high water absorption capacity. Percentage inner conduit was found to be governed by dominant and epistatic genes. Water absorption is directly related to percentage inner conduit and is thus considered as an important character for selecting varieties for long transportation periods. Aswath *et al.* (1996) found that only stalk length at bud burst and percentage of culled flowers showed genotype x environmental interaction out of 15 agronomic and floral characters studied in 8 gerbera varieties grown at different N levels. The magnitude of heritability estimates (broad-sense) was very high for many economically important characters except flower number/ plant, indicating that substantial improvement can be made by standard selection criteria. There was a positive correlation between stalk length and days to bud burst, but a negative correlation between flower diameter and days to bud burst (Aswath *et al.*, 1998). Heritable variation for cut flower yield has been reported by many workers (Maurer and Horn, 1967; Borghi and Baldi, 1970; Mutsenietse *et al.*, 1978a,b; Harding *et al.*, 1981). Harding *et al.* (1996) found that the overall estimate of narrow sense heritability is 0.33 based on the least squares and 0.31 based on maximum likelihood. Similar findings were also reported by Mahanta *et al.* (1998) who found that days to visibility of flower bud, days to flower bud opening and days to full flowering exhibited greater genetic variability and high heritability coupled with high genetic advance. It is suggested that these characters can be used as selection criteria for the improvement of gerbera. Broad-sense heritability estimates were very high for all the characters except vase life and size of flowers. The number of days to bending of the stem or wilting of petals of cut flowers increased in 12 parent clones and 80 of their progenies. Genetic variation in days to wilting of petals in progenies lasting for 18 days or individual plants holding over 20 days were noted. It was suggested that choice of parent is the major factor in vase life and 78 per cent variation between progeny means in days to wilt was attributed to the parents chosen (Jong, 1986).

The flower production by seed raised plants were shown to be positively correlated with the number of lateral shoots and with the number of leaves per flower. The most productive plants formed 2.5 lateral shoots during the first year with leaf/flower ratio of 2:4. The number of lateral shoots present at the time the first flower bud became visible was strongly correlated with the number of lateral shoots after one year. There was a positive correlation between the number of lateral shoots and total leaf number, but hardly any between total leaf number and flower production although an increase in lateral shoots resulted in more leaves, the leaf/flower ratio increased with the number of lateral shoots (Leffring, 1973). Meynet (1983) reported that cuttings and *in vitro* plants showed a high genotype x system of propagation interaction. *In vitro* culture mainly modified the ability of the genotype to recover a strong and durable apical dominance. Fredaisy and Fredimo plants differed according to whether they were raised from standard cuttings or by *in vitro* culture and these differences remained stable after several clonal propagation cycles.

Kumar *et al.* (2012) reported that among 19 quantitative traits, high (>20 per cent) phenotypic and genotypic coefficient of variation were observed for number of suckers/plant/year. Leaf breadth and number of suckers/plant/year showed high heritability (95.56 per cent and 96.07 per cent, respectively) coupled with high genetic advance as per cent of mean, indicating presence of additive gene action indicated scope of improving these characters through selection.

Genetic varaibaility studies among 15 quantitative traits revealed that magnitude of phenotypic coefficient of variation (PCV) was higher than genotypic coefficient of variation (GCV) for all the traits, indicating greater genotype and environment interaction. High (>20 per cent) PCV and GCV was observed for number of leaves/plant and leaf width. Heritability estimates ranged from 24.03 per cent (number of suckers/plant/year) to 93.5 per cent (length of ray floret). High heritability (>60 per cent) was observed for all traits except number of suckers/plant/year, flower diameter and flower-stalk diameter. High heritability, coupled with high genetic advance over per cent of mean, was observed for number of leaves/plant, leaf length, leaf width, days to bud-burst, days to first-flower opening, disc diameter, flower-stalk length, number of ray florets per flower head with length and width of ray florets (Kumar, 2013).

High heritability estimates were observed for most of the characters except flower diameter and girth of stalk. High genetic advance coupled with high heritability and comparable genotypic coefficient of variation were observed for number of leaves/plant, disc diameter, weight of harvested stalk and vase life indicating the preponderance of additive gene effects in controlling these characters (Ghimiray and Sarkar, 2015).

Genetic variability, heritability and correlation coefficients for seven flower contributing characters were determined. Phenotypic and genotypic coefficients of variation were highest for the total flower yields per year. Highest heritability recorded for total number of flowers, number of leaves/plant, flower disc diameter and flower stalk length, respectively. The coefficients values of simple correlation exhibited strong and positive association between all the traits, indicative of involvement of linked gene's inheritance pattern. Further, presence of high genetic gain as well as strong positive association between the traits open the ways for the improvement of the traits through selection procedure (Raj *et al.*, 2016).

Kumar *et al.* (2012) reported that flower diameter was significant and positively correlated with disc diameter, stalk length, fresh weight of flower head, durability of single flower, and ray floret length both at genotypic and phenotypic levels.

5.2.0 Breeding

Numerous cultivars of gerbera have been evolved through hybridization. For breeding, flowers of *G. jamesonii* could be grouped into florets narrower than 5 to 5.5 mm and those wider than 5.5 mm. Muller (1959) attempted to evolve varieties for flower form and colour, stem length and stiffness, keeping quality, period of flowering and yield, resistance to biotic and abiotic factors. Hybridization and selection resulted in the development of 12 cultivars of which cultivars Aktur, Migar and Atair showed high resistance of *Phytophthora cryptogea* (Danhardt, 1966). Cantor (1993) reported that the gerbera clones G110 and G44 are useful material for breeding. Leffring (1968) crossed several modern cultivars with wild species to develop a dark centred large flowered hybrid. Further selection and back crossing stabilized the dark centre and seedlings were true to type. Schiva (1980) attempted to evolve varieties for low temperature. Genotypes adapted to cool temperature were crossed with 20 genotypes having good floral characters in pairs to give 10 full sub families. Progeny was raised and assessed for floral attributes. Mean number of flowers produced by first 8 families ranged from 8.09 to 12.90/plant compared to 16.54 and 19.95 for two cool adapted parents. Broad sense heritability was estimated and results were promising. Byrne *et al.* (1987) found that the estimates on genotype x environment (GE) interaction was as much as half of the genetic variance for cut flower yield in gerbera and attributable to GE interactions.

Genetic analysis for flowering time has also been reported by Yu *et al.* (1991, 1993a and 1993b). The yield of 60 clones, including breeding lines ranged between 19 for *cv.* Adrie and 55 for line No. W.30 (Penningsfeld *et al.*, 1976).

Buivids *et al.* (1992) chose parent forms on the basis of combining ability and heritability values for cut flower quality and yield. Inbred lines were produced from 52 cultivars and clones. Inbreeding depression was marked in Aime, Bille, Lelde and Teiksma. No segregation for flower colour was seen in Ede (dark red) and Delphi, Symphonie and clones 35-5, 55-3 and 83-6 (all white). Flower colour may be treated as a composite of three metric characters: hue, chroma and value (McGuire, 1992; Voss, 1992). These characters described the type, namely, red, yellow-green or blue *etc.*, besides intensity and lightness of colour (Tourjee *et al.*, 1995b). The colour distributions in the population of gerbera indicated that they were continuous variables (Tourjee *et al.*, 1993). The variability is largely under genetic control (Tourjee *et al.*, 1995a).

Sparnaaij *et al.* (1975) described a procedure for testing the resistance of gerbera seedlings of *Phytophthora cryptogea*, causing wilt. Results of a diallel cross indicated that in most cases genetic variation was likely to result from additive gene action. In test crosses use of a self-incompatible but highly cross-fertile standard as female partner, was regarded as the simplest method to compare progenies of crosses.

At ICAR-IIHR, Bangalore four lines with double flowers were developed using half-sib selection from locals bearing single flowers. Scoring key for evaluation of these lines, was developed. Among the parameters 'quality' recorded the highest preferential score of 60. Among the individual traits, form of flower and yield recorded the highest score of 15, followed by flower diameter, vase life and pests with 10 each. Among different lines, IIHR 99-2 recorded the highest score X weightage of 96.1, followed by IIHR 99-1, IIHR 99-3 and 99-4 lines with 92.3, 83.1 and 81.1, respectively. All the IIHR lines showed moderate resistance to thrips (Aswath and Rao, 2006).

Cardoso *et al.* (2009) studied the crossing possibility among cultivars and non-commercial accessions through the hybridization between the feminine *cv.* Terra Fame and A8 and masculine genitors, *cvs.* Cariba and Azteca. All accessions presented fifty chromosomes, indicating that the morphological variation in the capitulum (simple, semidouble and double) is not due to the chromosome number mutation or polyploidy. The evaluated accessions displayed high pollen viability, varying from 87.67 per cent to 99.27. The seed formation was 4.46 per cent in the crossings among cultivars, and 50 per cent among cultivar and non-commercial accession of gerbera. The genomic compatibility among the accessions, the high pollen viability and the success in the seed formation among commercial and non-commercial accessions, reveal the possibility of hybrid production with new allelic combination and transference of desirable traits from the non-commercial to the commercial accessions.

Variability was generated and characterized for improvement of floral traits in Gerbera through hybridization program. Considerable variation was observed for the floral traits *viz.*, flower colour, shape, diameter, disc colour, scape length, disc diameter, ray floret width and ray floret number among the parental genotypes and 46 F_1 progeny of an inter-varietal cross IHBT Gr4 x IHBT Gr5. Flower shape and disc colour were observed to be qualitative traits while flower colour, scape length, flower diameter, disc diameter, ray floret width and ray floret number had quantitative inheritance. Analysis of variation revealed high heritability for scape length, flower diameter, disc diameter and ray floret number and moderate heritability for ray floret width. Some of the F_1 genotypes (Gr-05-01-11, Gr-05-01-37, Gr-05-01-42 and Gr-05-01-43) have suitable traits and constitute important germplasm lines to be used in breeding program for development of new varieties (Singh *et al.*, 2009).

Forty lines of potted gerbera (*Gerbera hybrida*) were developed by self-pollination of 80 commercial pot varieties. The lines show wide variation in both qualitative characters (flower and disk color) and quantitative characters (flower diameter, peduncle length and days to flowering). The inbred lines showed varying degrees of inbreeding depression in peduncle length and flower diameter as much as 66-88 per cent and 89 per cent, respectively. As selfed generations were advanced, the quantitative characters of lines became uniform as measured by variance among 30 individual plants per generation. It is postulated that a few number of genes are controlling flower color since most lines of third or fourth selfed generation showed uniform color. Hybrids between developed lines of third selfed generation showed considerable level of heterosis in flower diameter, peduncle length, and number of flowers, and considerably lower degree of heterosis in days to flower (Song *et al.*, 2015).

Gerbera hybrids namely Arka Ashwa and Arka Nesara, were developed through hybridization using half-sib method of breeding where superior lines, IIHR-3 and IIHR-1, were crossed with pollen mixed from different varieties. Both have novel flower colour with double type of flowers and are suitable for cut flower and flower arrangement purposes (Aswath *et al.*, 2017).

5.2.1 Mutation Breeding

Mutation breeding has been successfully used to induce novel variants in case of ornamentals, particularly for flower colour and shape. Walther and Sauer (1986) studied *in vitro* mutagenesis with X-ray doses between 10 and 25 Gy immediately after cutting off from *in vitro* derived microshoots. Shoot productivity increased with time. Similar results were also recorded by Walther and Sauer (1992) using *in vitro* cultures of 5 lines and 1 cultivar with 10-30 Gy X-rays. X-ray irradiation (20 Gy) of explants propagated *in vitro* produced a 10 per cent mutation rate in foliage (number, size and form of leaves) and flowers (stalk length, flower diameter, number and shape of the petals). High intensity of irradiation suppressed the plant growth (Walther and Sauer, 1985).

Ethyl methyl sulphonate (EMS) was more effective than sodium azide in inducing single gene mutations in seeds. Increasing the EMS concentration from 1 per cent to 2 per cent and that of sodium azide from 0.1 mM to 1 mM and the time decreased germination, growth, survival, frequency of flowering plants and seed setting and number of seeds in the M_1 generation and the mutations occurring the M_2 generation. Schiva *et al.* (1984) obtained a large non-lethal M_2 population using low concentration of EMS at shorter duration.

Shoots from *in vitro* plantlets of the pink gerbera *cv.* Rebecca were irradiated with a single dose of 20 Gy gamma rays (9,8 Gy/h) and then micropropagated for two cycles; after rooting, the plantlets were cultivated in the greenhouse and analysed for morphological mutations. Irradiation treatment induced a 25 per cent reduction of propagation rates in each cycle. Out of 247 non-contaminated initial shoots, 129 (52 per cent) produced a clonal progeny showing flower variations and among the 1250 adult plants derived from irradiated shoots, 187 (15 per cent) were variants. The observed variations affected flower morphology (*e.g.* number, length and width of ligulae) and colour. Out of 187 MV2 variant plants, 6 plants (with at least 2 flowers) were presumably solid mutants. The remaining 97 per cent showed chimerism at various levels: 44.4 per cent plants had both normal and uniformly mutant flowers, 4.8 per cent had different uniformly mutant flowers (intra-plant chimerism); 47.6 per cent plants had flower deformations and/or colour sectors (intra-inflorescence chimerism). Out of 424 variant flowers, 147 (35 per cent) were chimeric; of the uniformly variant flowers, 26 per cent showed only single morphological variations, 13 per cent only colour variations and the remaining flowers showed a combination. There were 13 morphological and 6 colour types of variants. In contrast, of 321 control plants derived from 40 non-irradiated shoots, none showed flower variation. Hence, in 'Rebecca' more than two micropropagation cycles are necessary to increase the percentage of solid mutants. In spite of this limitation, the proposed scheme appears to be a quick tool to induce variability without gross genome modification (Laneri *et al.*, 1990).

Jain *et al.* (1998) discussed possible approaches for creating genetic variability for the selection of useful plants, with reference to gerbera using mutagenesis, somaclonal variation and genetic transformation. Somaclonal variation may arise due to genetic or epigenetic changes. The latter being unstable and not transmitted to the progeny have to be identified before plants are multiplied for large-scale production using biochemical (isozymes) and molecular markers (RAPD, RFLP, AFLP). It is suggested that genetic engineering has yet to provide a cost effective and reliable product and, moreover, unstable gene expression, especially for flower colour is a major problem in transgenic plants under field conditions.

Ex vitro derived leaf explants of sixteen cultivars of *Gerbera jamesonii* were irradiated with 5 to 25 Gy doses of gamma rays. For the production of adventitious shoots, leaf explants werecultured on MS medium with 3 mg/l BA, 0.5 mg/l IAA, 100mg/l myo-inositol, 0.4 mg/l thiamine· HC1, 10g/l sucrose and 8 g/l BactoBacto-agar. Shoot formation (frequency and intensity) was dependent on dose ofirradiation and cultivar. Doses of 20 and 25 Gy drastically restricted the regeneration ability of leaf explants, but even 25 Gy dose did not reduce the production of adventitious shoots completely (Jerzy and Lubomski, 1992).

Growth responses of callus were strongly influenced by the radiation dose. Nor *et al.* (2012) studied the effects of gamma irradiation on *in vitro* growth of explants, callus and the formation of shoots and plantlets. A gradual decline was observed in the number of shoots regenerated from irradiated petiole explants. Numbers of shoots regenerated from irradiated petiole explant cultured on Murashige and Skoog medium supplemented with 2.0 mg/l BAP and 0.5 mg/l NAA was reduced to 6.6±0.9 from 7.5±0.4 (control) when explants were exposed to 20 Gray of irradiation dose. Gradual decline was observed in plant height as dose of gamma irradiation increased. A significant decline was observed in the fresh weight of irradiated callus. The fresh weight of callus was reduced when callus tissues were exposed to 20 Gy.

Estimation of mutation frequency on the basis of changes in stomata number per unit area was done. *In vitro* shoots of gerbera 'Regiko' and 'South Pacific' were irradiated with 10, 20 and 30 Gy ^{60}Co gamma rays (100 Gy/min.) and sub-cultured for two cycles, plants raised from 20 and 30 Gy irradiated shoots survived well in the field. Stomata size and number was significantly reduced in plants raised from 30 Gy irradiated shoots. Under 20 Gy dose mutation frequency calculated on the basis of stomata number per mm^2 was close to 0.50 in both cultivars, whereas, it was above 0.60 with 30 Gy dose. This method offers an alternative way to assess effects of gamma irradiation even before the appearance of flowers (Imtiyaz *et al.*, 2014).

In vitro cultures of gerbera established through capitulum explants cultured on 1.0 × Murashige and Skoog (MS) medium supplemented with 5 mg/l benzyladenine (BA) and 0.5 mg/l indole-3-acetic acid (IAA). *In vitro* regenerated shoots were treated separately with different doses of ethyl methyl sulphonate (EMS) or gamma irradiation. The activities of superoxide dismutase (SOD), ascorbate peroxidase (APX), glutathione reductase (GR), catalase (CAT), and polyphenol oxidase (PPO) increased significantly and was maximum at highest dose of either mutagen. Overall, 18.3 per cent and 20.5 per cent reductions in chlorophyll concentration were observed in 2-month-old hardened plants following gamma-irradiation or EMS treatment, respectively. The mutated shoots showed significant increases in total phenolic compound concentrations (36 per cent and 21.7 per cent) and total protein concentrations (17.6 per cent and 17.1 per cent) at the highest dose of gamma rays or EMS, respectively. Increases in the activities of the five enzymes, and in the concentrations of phenolic compounds and proteins, indicated the importance of these biomolecules in protecting gerbera plants grown under abiotic stress conditions (Ghani *et al.*, 2014).

In vitro raised shoot cultures of gerbera cultivar Harley established from petiole explants, were exposed to different doses of gamma rays using a Cobalt-60 source emitting 2.51 kGy/h and ethylmethane sulphonate for 10 min or for 20 min. The LD_{50} values calculated for shoot survival and the induction of mutations were approx. 6.5 Gy for gamma rays and 0.65 per cent (v/v) EMS for 10 min, or = 0.1 per cent (v/v) EMS for 20 min. Investigations revealed a negative correlation between mutagen dose and plant survival, both *in vitro* and after acclimatisation. Morphological variants showing changes in leaf shape, leaf size, scape length, flower diameter, and flower colour were obtained. Significantly, early flowering was induced in all mutated plants compared to non-mutated plants (Ghani *et al.*, 2013).

Gamma irradiation effect in inducing resistance/tolerance to powdery mildew disease was investigated in *cv.* 'Harley'. *In vitro* shoot cultures were established through capitulum explants on Murashige and Skoog medium supplemented with 22.2 μM 6-benzyladenine (BA) and 2.53 μM indole acetic acid (IAA), followed by gamma irradiation of regenerated shoots (3–5 cm). Activity of four antioxidant enzymes *i.e.* superoxide dismutase, ascorbate peroxidase, catalase and glutathione reductase increased and reached to highest level at the most stringent doses of mutagen. Ninety randomly selected irradiated plants (6 months old) and 100 control plants were inoculated with fungal conidial suspension, to screen for tolerance/resistance against powdery mildew. The severity of the disease was recorded on 0–4 scale with '0' indicating highly resistant; '1' indicating resistant; '2' indicating medium resistance; '3' indicating susceptible and '4' indicating highly susceptible. Three plants (3.33 per cent) irradiated with 5 Gy were found to be tolerant to powdery mildew as these plants showed slight and delayed development of fungal colonies on the leaves. The random amplified polymorphic DNA characterization showed that the irradiated plants had DNA patterns that were different from the control and mother plants (Ghani and Sharma, 2019).

6.0 Growth and Flowering

Zayats (1977) described the developmental stages of gerbera from seed germination to maturity. The seedling stage continued for about one month, the juvenile phase for about 4-5 months followed by reproductive phase of 3-4 or more years. In gerbera, like other species of Asteraceae, feminization of marginal florets takes place. During flower development, anthers arrested at intermediate stages of development, result in the formation of non-functional staminodes in the highly zygomorphic ray and intermediate trans florets (Yu *et al.*, 1999). Flowers of Asteraceae lack a leafy calyx, but in gerbera, the base of the corolla is surrounded by pappus bristles or scales, specialized organs that are involved in seed dispersal (Bremer, 1994; Harris, 1995). Kotilaninen *et al.* (1999) made a comprehensive review on the flower development and secondary metabolism in gerbera which includes molecular view of flower organ determination.

6.1.0 Environmental Factors

The performance of cultivars varies between seasons indicating clearly the interaction between genotype and environment. Marked seasonal variation was recorded with regard to flower production (Junges, 1962). The flowering peak reached during the summer but became earlier in successive cropping years and flowers produced during second and third successive years were about the twice that of first year. Pinamonti *et al.* (1996) reported that for gerbera *cv.* Dutch Lion, the yield of gerbera cut flowers from August to October was highest. Venezia *et al.* (1997) found that flower yields were higher on perlite than on rockwool during the winter months. Pascal and Queen Rebecca were the most productive cultivars yielding 184 and 171 marketable flowers/m^2, respectively. Queen Rebecca was the most productive cultivar during the winter months. Steen and Steen (1975) found that productivity was lowest in January except in the case of *cv.* Mandarin, which produced uniformly each month after a slow start. Plants in the centre beds of the glasshouse produced more flowers from December onwards than plants in the side beds.

6.1.1 Light and Temperature

The photoperiod and temperature have been found to influence the growth and flower production. Stinson (1957) reported that plants grown at 12 h daylength produced less number of flowers, while those grown at 8 h day had smaller leaves. Van and Tijskens (1991) found that plants grown under lighting at 10 W/m^2 throughout the duration had a much higher uptake of nitrate and K.

Response to light depends on the genotype. Lin and French (1985) found that Appelblossom was not sensitive to photoperiod; OranjeNasau and Fabiola were promoted by short days (SD) and supplementary lighting with high-pressure sodium (HPS) lamps increased the autumn-winter production of all 3 cultivars. Plants grown under 16 h photoperiod by extending natural daylength with incandescent light produced the least flowers/plant in both summer-autumn and autumn-winter compared with natural daylight. Seasonal variations have also been reported to influence the response of gerbera to light. Rijssel and Visser (1985) reported that the gerbera was less sensitive to low light intensity during the winter months. Different distributions of the same light energy over one day led to a change of about 45 per cent in number of inflorescences (Autio, 2000). A combination of short photoperiod and high light intensity was recommended for obtaining maximum number of inflorescences (Leffring, 1975).

After mid-February when the amount of light was increased, flower production was found to increase in all the side shoots proportionately and a night temperature of 12°C was adequate (Leffring, 1981). When the difference between air and soil temperature was less, flower production was proportionately more, and flower quality was better in cool greenhouse than in warm one (Bulthuis, 1979). Varietal and seasonal variation also influenced the response of gerbera to light. Van Labeke and Dambre (1999a) found that supplementary light increased the number of flowers/m^2.

Soil warming has been reported to improve flower production. Heating has been reported to slightly increase flower stalk length. Flower diameter was not affected by temperature. Response of cultivars to soil heating varies among the cultivars (Vadalie and Laffaire, 1984). Vadalie *et al.* (1985) further reported that with heating flower yield increased by 13 per cent to 27 per cent. Flowing warm water (20°C) through corrugated plastic pipes was best to heat the soil from last week of November to 2nd week of February (San Pedro and Badosa, 1985). However, Lin and French (1985) found that soil warming from 16-20°C to 23°C had no effect on productivity but increased the peduncle length. Vergara-Gongora (1984) recommended preventing air and soil temperatures from dropping below 10 and 17°C, respectively to obtain quality flowers in polyethylene clad greenhouses for winter cut flower production. Warming the root zone by standing the bag on soil-filled benches heated by thermostatically controlled warm water in plastic pipes 30 cm below the soil surface substantially increased the stem length and flower yield in *G. jamesonii* cultivars (Tsujita and Dutton, 1987). Kaewruang *et al.* (1989) found that solarization of soil within plastic bags for 4 weeks increased availability of nutrients such as NH_4^+-N, PO_4^- and K^+ in comparison with bagged soil kept in the shade for the same period.

Flower production of *G. jamesonii* was positively correlated with temperature at the higher light level and the cumulative fresh weight of the flowers was stimulated by temperature fluctuations with a 6-day period at the higher temperature level (Buwalda *et al.*, 2000). Plants grown on warm (increased by 3-5°C) rockwool mats had higher yields, which could be increased even further by CO_2 enrichment (Richter and Matschke, 2000). Sub-surface heating regime resulted in a significant increase in the number of flowers/m^2 (Van Labeke and Dambre, 1999b). Various colour variants of gerberas were found to grow well at temperatures below 15°C which also prevented them from growing too tall (Miske, 1986).

6.1.2 CO_2

Marfa *et al.* (1997) found that CO_2 enrichment increased yields by >10 per cent during December to April. At high N concentration, CO_2 enrichment increased flower diameter in December and January, and at low N concentration, it had increased flower diameter between December and April.

6.2.0 Growth Regulators

Reports pertaining to growth regulators on gerbera are mainly on tissue culture work. Sachs (1968) studied the influence of GA_3 and auxin on intercalary growth of gerbera scape. Inflorescence was removed along with receptacle and involucral bracts, and the cut surfaces were treated with GA_3, IAA, 2,4-D, NAA or kinetin at various concentrations either in a paste or as a solution. Intercalary growth was maintained by applying GA_3 or IAA. IAA normally caused greater elongation than GA decapitated scapes. In scapes from which only the inflorescence was removed, application of GA_3 and/or IAA promoted cell elongation, but there was little or no cell division. In scapes from which only the inflorescence was removed, application of GA_3 and/or IAA promoted cell division as well as cell elongation. Initially both decapitated and deflowered scapes elongated at almost same rate. Kinetin was ineffective either alone or in combination with GA_3 and/or IAA. Lindstorm (1957) reported that 20 µg of GA_3 applied to stem apex of a foliar spray of 10-100 ppm hastened flowering by 10 days to 4 weeks in greenhouse grown gerbera during autumn and winter.

Treatment with low concentration of GA_3 (50 ppm) resulted in early flowering whereas CCC delayed. CCC at 500 ppm promoted flowering in both the seasons while at 750 ppm fewer number of heavy flowers were produced. GA_3 100 ppm accelerated both vegetative growth and flowering characters such as stalk length, number of flowers per plant, flower diameter throughout the crop period in Gerbera (Patra *et al.*, 2015). Gerbera plants were sprayed GA_3 150 ppm produced maximum plant height, number of leaves/plant, plant spread, early flower flowering, flower diameter, stalk length, stalk girth,number of flowers/plant and cut flowers per treatment. However, maximum vase life and number of suckers/plantwere produced by GA_3 100 ppm (Sangama *et al.*, 2017).

Single spray of GA_3 100 ppm at 2 months after planting in pots proved best for growth and flowering, while, GA_3 100 and 150 ppm for yield attributes (Kuotsu *et al.*, 2018). The *cv.* Yeliz achieved the highest yield with GA_3 300 ppm with longest and thickest pedicle length, highest cut flower weight, *cv.* Ula recorded maximum flower diameter and longest vase life (Yazici and Alkac, 2019).

Application of CCC reduced the leaf production, while, it was increased due to application of GA_3. Both the treatments had slight influence on improvement of flower diameter (El-Shafie and Hassan, 1978), Atrinal (dikegulac) spraying at the 7-9 leaf stage induced more number of flowers (Bocion *et al.*, 1977; Laermann, 1980). MH 250 ppm was better in influencing vase life of cut gerbera (Kuotsu *et al.*, 2018).

7.0 Propagation

7.1.0 Seed Propagation

Although gerbera can be propagated by seed, the cultivated gerberas are extremely heterozygous and take more time to come to flowering. The advantage of seed propagation over vegetative propagation is simplicity and absence of pest and disease transmission. It can be produced both in indoor as well as in outdoor plants. Artificial pollination is necessary during warm and sunny days between 10 and 12 hours and each flower should be pollinated 2-3 times (Stupak, 1985). Artificial pollination is necessary for obtaining sufficient seeds and if these are sown immediately after harvesting, 90 per cent germination results (Garthwaite, 1960). Nicolini (1964) reported that seeds retain their full viability for only 15 days after harvesting. According to Franceschetti (1973), the seeds stored at –5.0 to 5.0°C retained high germinability even after two and a half years of storage compared with those stored at room temperature which lost their viability in approximately nine months. Stupak (1985) reported that seeds should be stored at 5°C in a desiccator. The medium for sowing should be light to provide adequate aeration and the optimum temperature for germination was found to be 25°C. Lisiecka (1976a) found that for seed compost 21-29 mg NH_4, 45-53 mg NO_3, 15-20 mg P_2O_5 and 80-90 mg K_2O/l of peat was optimum. The speed of germination was, however, more affected by temperature and was highest between 25 and 28°C (Franceschetti, 1971). The seedlings are pricked at two- to four-leaf stage and are ready for transplanting in about 5-6 weeks.

7.2.0 Vegetative Propagation

7.2.1 Division

In case of asexual propagation rate of multiplication is slow. Among the various methods, multiplication through division of clumps is the most common method. It involves dividing large clumps into smaller units from March to June, when the plants may be set out in the field. Koek (1957) found that division of mother plants at the end of February resulted in 6 plants and they started flowering by May. Trimming of leaves and roots before planting improved establishment and care should be taken that the central part of the sucker does not cover the soil. The transplanted plants should be established by August and late propagation depressed the yield due to low temperature during winter.

7.2.2 Cutting

It can also be multiplied by cuttings. The buds in the leaf axils were detached from the mother plants after withholding water for 3 weeks, pruning of roots, planted in peat substrate for rooting at 25-30°C with 80 per cent RH. About 40-50 plants could be produced in 2-3 months from a single plant through tissue culture (Schiva, 1975). Under intermittent mist young stem cuttings produced roots and shoots much easily.

7.3.0 Micropropagation

Tissue culture multiplication through which a million-fold increase per year of a desired plant can be obtained is an ideal method (Murashige *et al.*, 1974; Pierik *et al.*, 1973; Kaminek *et al.*, 1987). Growing period was much shorter, the product more uniform and better keeping quality in case of tissue cultured plants (Nienhuis, 1989; Loeser and Deiser, 1984).

Different basal media have been reported to be successful for gerbera. Murashige and Skoog's (MS) medium supplemented with various growth substances is commonly used for gerbera tissue culture. Parthasarathy *et al.* (1996) found B_5 to be good out of three commonly used culture media (B_5, MS and White) for *in vitro* multiplication of gerbera. Parthasarathy and Nagaraju (1995) reported success with low levels of BA in contrast to the earlier reports using very high concentrations of kinetin.

Propagation of Gerbera

Seeds

Sowing of Seeds

Germination of Seed

Separation of Suckers

Suckers

Plug Plant

Micropropagation in Gerbera

A: Cultured plants on MS medium containing 1 mg/l BAP and 0.1 mg/l IBA, B: Proliferated plants in cabinet illuminated with cool white fluorescent lamp, C: Proliferated plants in cabinet illuminated with 100 per cent blue LED lamps, D: Rooted plants in cabinet illuminated with 100 per cent blue LED lamps, E: Acclimatization stages of the plants (1 and 21 days after planting, respectively), F: Acclimatized plants (42 days after planting in pots)

Source: San *et al.* (2016).

Bed Preparation | Gerbera in Greenhouse | Cultivation in Pot

Source: www. agritech.tnau.ac.in, www.agricultureguruji.com, www.gerardomercurio.com, www.ellepot.com, www.youtubeGardening Upbeat, www.gardenworldimages.com, www.gardengatemagazine.com

All the steps in the micropropagation of gerbera *i.e.* seed germination, multiple shoot differentiation, rooting and hardening is highly genotype dependent, sterilized seeds of nine gerbera genotypes were germinated on MS medium supplemented with BAP 3.0 mg/l, *in vitro* grown seedlings after 30 days of seed culture were transferred onto MS medium containing BAP 2.0 mg/l + NAA 1.0 mg/l to allow multiple shoot formation, elongated shoots were then rooted on MS medium supplemented with IBA 1.0 mg/l for rooting, the genotype 'Kristen' showed highest germination frequency and number of leaves/plant, Aladin exhibited higher shoot regeneration, Silvester and Wink showed greater response in *in vitro* shoot elongation. The genotype Pepe showed maximum root formation and root elongation. Kristen, Aladin and Alabandaare the suitable gerbera plant types for *in vitro* propagation through seed explant (Islam *et al.*, 2017).

Gerbera hybrid seeds obtained by half-sib mating raised in half strength MS medium. The earliest seed germination, first leaf development and rooting were recorded line IIHR1-1, IIHR5-4 and IIHR1-1, IIHR1-2, IIHR1-5, IIHR3-2 and IIHR5-4. The response of *in vitro* leaf explants to callusing and regeneration was recorded when cultured on full strength Murashige and Skoog's media supplemented with different concentrations of 2,4-D or BAP or IAA and Kinetin. Leaf explants cultured on MS medium fortified with equal concentration of BA (1 mg/l) and 2, 4-D (1 mg/l) produced green and greenish white granular callus within 25 days, however, leaf explants cultured on MS medium supplemented with 2,4-D produced yellowish callus. Plant regeneration was also found earliest in the same media within 14 days (Mahindrakar *et al.*, 2018).

Gerbera seeds were used as primary source of explant.IAA served as the best medium for shoot proliferation from *in vitro* shoots as it produced maximum number of shoots with leaves and shoot length at three weeks of inoculation. For root regeneration, MS media supplemented with 2 mg/l IBA recorded best rooting media as regenerated roots early with maximum number of roots and root length (Vijayalakshmi *et al.*, 2019).

7.3.1 Bud and Shoot Proliferation

Shoot tips (Murashige *et al.*, 1974; Huang and Chu, 1985; Parthasarathy and Nagaraju, 1995; Parthasarathy *et al.*, 1996; Nagaraju *et al.*, 1998), inflorescence buds (Pawlowska, 1977), capitulum (Pierik *et al.*, 1973; Arello *et al.*, 1991), leaf explants (Reynoird *et al.*, 1993; Nagaraju *et al.*, 1998), midribs (Pierik and Segers, 1973), petioles (Elomaa *et al.*, 1993) and leaf blades (Hedtrich, 1979) have been employed as explants for regeneration and multiplication.

Satisfactory rate of increase of gerbera divisions was achieved by employing MS medium containing 0.5 mg/l IAA and 10 mg/l kinetin (Murashige *et al.*, 1974). Jerzy and Lubomski (1991) reported that direct regeneration from petiole explants was better than indirect shoot formation through callus phase. Raalte (1978) reported meristem culture for the propagation of gerbera. Hedtrich (1979) successfully regenerated adventitious shoots by using leaf blade explants on modified MS media containing 1 mg/l BAP and 0.1 mg/l GA. Reynoird *et al.* (1993) standardized the protocol for regeneration of plants from leaf explants of several *Gerbera* species.

Pinto *et al.* (1993) reported the best propagation rate when the $NH_4^+ : NO_3^-$ ratio was optimum. Kataeva *et al.* (1991) found that high humidity in the culture vessels and excess BA in the medium were the major factors influencing vitrification. Zeatin was also found to stimulate shoot multiplication *in vitro* (Blakesley *et al.*, 1986).

Akter *et al.* (2012) opined that flower bud and flower stalk were suitable and superior explants for callus induction and subsequent regeneration of *in vitro* shoots when cultured on MS supplemented with 5.0 mg/l BAP and 1.0 mg/l NAA. However, highest number of multiple shoots were obtained when the flower bud derived callus was subcultured on MS supplemented with 2.0 mg/l BAP. Variety with red petal showed the best response in producing multiple shoots. Best root induction was observed on half the strength of MS supplemented with 0.2 mg/l IBA.

7.3.2 Haploids

Haploidy has an important role in improvement of gerbera. Bappadocia *et al.*, (1988) used *in vitro* culture of unpollinated ovules and ovaries for the production of haploid plants. Seventy-six per cent of the regenerated plants were haploid and 24 per cent were diploid. Gynogenetic plants of pot gerbera using unpollinated ovules *in vitro* (Miyoshi and Asakura, 1996; Cagnet-Sitbon, 1981). Meynet and Sibi (1984) produced haploid plants using unfertilized ovules of gerbera *cv.* Fresultane. Ahmim and Vieth (1986) successfully regenerated haploid plants using IAA (0.1 mg/l) and BA (0.2 mg/l).

Ovule-derived haploid culture is an effective and important method for genetic study and plant breeding. *Gerbera hybrida* is a highly heterozygous species, and the lack of homozygous lines presents a challenge for molecular genetic research. Li *et al.* (2020) studied haploid induction through unpollinated ovule culture and evaluated the effects of several important factors on this culturing procedure in *G. hybrida*, including genotype, low temperature, and the development seasons of the ovules. Among 45 *G. hybrida* cultivars analyzed, 29 cultivars exhibited adventitious bud induction via *in vitro* unpollinated ovule culture with significant different responses, indicating that the genotype of donor plants was a vital factor for inducibility. Ovules extracted in the summer consistently had the highest induction rates, and even the non-induced cultivar included in the analysis could be induced at low levels when ovules from summer were used. Low temperature treatment could also promote adventitious bud induction, and in particular, a strong and significant effect was detected after 7 days of cold treatment. Ploidy level measurements by flow cytometry revealed that 288 ovule-derived regenerants were haploid (55.17 per cent) and 218 lines were diploid (41.76 per cent). Genetic stability analysis of the regenerants indicated 100 per cent similarity to the marker profile of the mother plant. This report of ovule-derived haploids in *G. hybrid* facilitates the development of homozygous lines for molecular research and plant breeding.

7.3.3 Capitulum Regeneration

Pierik and Segers (1973) and Pierik *et al.* (1975, 1982) regenerated adventitious shoots from capitulum or midrib explants and also from the axils of involucral bracts. Arello *et al.* (1991) also obtained plantlets from capitulum explants of the varieties Appelblossom, Marleen, Clementine and Pimpernel when cultured on modified MS medium supplemented with BA, IAA or kinetin.

The one-fourth pieces of young capitula of tight flower stage and half-strength MS medium containing 0.25 mg/l BAP was the suitable for initiation and produced higher number of shoots per explant up to 3.8 shootsanditwas improved by culturing the one-fourth piece of 01.092 capitulums on MS medium fortified by 0.2 mg/l BAP and 0.02 mg/l NAA producing the highest shoot formation up to 8.5 shoots per explant with 28.7 leaves/explant and 2.1 cm leaf length. Highmultiple shoots were determined in third to fourth subculture periods and reduced thereafter with high multiplication rate noted on 01.092 clone. Shoots were easily rooted on half-strength MS medium supplemented with 2 g/l activated charcoal. Plantlets were transferred to *ex vitro* condition with 96.4 per cent survivability of 03.045 clone using Cycas rumphii bulk and cocopeat (1:1, v/v) under spraying 1 g/l Growmore (32N:10P:10K) solution once week periodically. The route has high potential applied in qualified plantlet production for other Gerbera's due to high shoots produced up to 35 shoots per whole young capitulum used (Winarto *et al.*, 2020).

Capitulum segments were used as explants in micro propagation of Orange and Malibu cultivars of gerbera plants. The capitulum segments of the gerbera cultivar Orange and Malibu cultivars cultured *in vitro* as explants led to the direct proliferation of adventitious shoots, significant superiority of the Orange cultivar in its response rate to direct shoot regeneration, which reached 86.72 per cent after 12 weeks of culturing. While the Malibu cultivar was significantly superior in the number and length of the shoots that were formed from direct organogenesis, which reached 10.48 shoots explant and 1.78 cm, respectively. The concentration of 1.0 mg IBA was significantly superior to the other concentrations in the response of explants to direct shoot proliferation and the number and length of shoots recorded 99.22 per cent and 15.90 shoots explant and 1.78 cm, respectively. While the explants that cultured on the MS medium that supplied with 5.0 mg L BA recorded the lowest response to direct shoot proliferation and the number and length of shoots, which reached 27.50 per cent, and 2.00 shoots explant and 0.35 cm, respectively(Ibrahim and Yassin, 2020).

7.3.4 Rooting and Transplantation to Soil

Mariska *et al.* (1989) obtained best rooting with 0.1 mg/l IAA. Use of NAA resulted in short and swollen roots. Improved root growth was reported by Khunachaka *et al.* (1987) by incorporating 1 mg/l ancymidol in the culture media. Zakharova (1987) added 1 mg/l IAA to agar media for good rooting.

Acclimatization of micropropagated gerbera genotypes was studied, after 90 days of acclimatization, high survival rate of plants was found and the genotypes 'Igor', GA and GR showed higher growth than 'Pacific', 'Igloo', 'Mephisto' and GL at young stage (Couto and Araujo, 2018).

8.0 Soil and Climate

8.1.0 Soil

For successful production of quality flowers, a well-drained, rich, light, neutral or slightly alkaline soil is ideal. They prefer an organic soil with high nitrate content and tolerate lime in the soil (Garthwaite, 1959). Loeters (1968) studied the root system of gerbera in various mineral and peaty soils. In soils with good structure and drainage the root system reaches very deep and drainage appeared to be the most important factor. Flower diameter is not influenced by the pH of the soil (Stinson, 1953). Soils with a pH of 5.0 to 7.2 produced higher number of flowers with long flower stem than when grown on acidic soil with pH below 5.0. Gotz (1983) suggested that the substrate pH must not exceed 6.5.

Sonneveld *et al.* (1997) found that chlorosis was suppressed as pH was reduced. Increasing the Mn supply reduced chlorosis only at the highest pH. The number of flowers produced at the lowest pH was higher than at the highest pH. The effect of pH on the absorption of macroelements was insignificant, but on micronutrient absorption was substantial.

Moulinier and Montarone (1978) observed better growth in friable soil than in compact soil. Vergara-Gongora (1984) recommended lowering the soil pH (from 8.2 to 5.5) and preventing air and soil temperatures from dropping below 10 and 17°C, respectively for winter cut flower production.

8.1.1 Soilless Culture

Soilless culture like hydroponics creates a number of superiorities against custom methods. Farajollahzadeh *et al.* (2014) devised a hybrid technique that could possess more benefits and fewer problems compared to both traditional soil culture and Hydroponics. In this hybrid method, the typical nutrient solution cycling system is similar to conventional closed hydroponic systems while the soil remains to be the source of nutrient. Nitrogen-fixing bacteria in an inline bioreactor were responsible for supplying plant nitrogen demand. After six months the results showed that even though the chemical system was superior in leaf count, chlorophyll content, shoot fresh weight and fresh biomass but could not surpass the organic system in root parameters (volume, fresh and dry weight) and dry biomass accumulation. The plants could manage to extract the necessary elements from the soil indirectly. The outcome suggests considering further optimizations towards an organic hydroponic system that could be considered sustainable.

8.2.0 Climate

In tropical and subtropical climate, gerberas are grown in the open, but in temperate climate, they are protected from frost and cultivated in greenhouse. They like sunny situation in cool weather but during the summer months they should be lightly shaded if left out in beds. Poor light during winter adversely affects the flower production. A night temperature of 12°C was found to be optimum (Leffring, 1981).

9.0 Cultivation

9.1.0 Substrate and Planting

The influence of season and climate on growth and flowering has been discussed earlier. Extending the growing season by 6-14 weeks substantially increased the yield (Matous, 1985). The plants should be set with the crowns well above the soil surface. Steen (1975) grew gerberas on raised beds in rows 30 cm apart. Laurie *et al.* (1968) recommended to set the plants 30-37.5 cm apart. The spacing in the rows ranged from 20 to 40 cm giving from 9.4 to 4.7 plants/m^2. Spacing does not affect quality. Closer spacing is suitable only for annual crop, while for perennial crops wider spacing is advisable. Growing in raised beds improves drainage and aeration. David (1977) suggested that closer spacing increased winter production by 15-20 per cent.

Substrates used for planting depend on the type of planting materials. Rockwool or peat/perlite have been reported to be good substrates (Scholten and Prakash, 1991; Maloupa *et al.*, 1999). Growing of gerbera in open hydroponic system resulted in high total marketable yield and also during winter months (Venezia *et al.*, 1997). Strojny and Nowosielski (1980) obtained satisfactory growth in 30 cm high containers than 15 cm containing high peat and composted bark substrate with 2 plants/container.

Rice husk, corn cob, peat and sand in the ratio of 3:2:1:1 appeared the best soilless formula in terms of plant growth, flowering time, cut flower yield and quality of cut flowers. Media with 2 portion of peat and 1 portion of perlite ranked the second followed by rice husk, corn straw, peat and sand in the ratio 3:2:1:1 and rice husk, corn straw, corn cob, peat and sand in the ratio 2:1:1:1:1 (Meng *et al.*, 2012).

Cocopeat: perlite mix is a common substrate in gerbera commercial production. Plantlets grown in mushroom compost: perlite and then sandy substrate had the highest leaf number, shoot and root dry weight. The early flowering and the highest number of flowers were obtained in these two-culture media. Plants treated with sawdust: perlite and perlite did not show desirable vegetative and reproductive growth and due to high substrate EC in leaf mold: perlite, the percentage of leaf chlorosis and necrosis was significant (Kharrazi, 2019).

CPV (cocopeat: perlite: vermiculite) amended with PHG (Pusa hydrogel 0.25 per cent) and VC (vermicompost 20 per cent) coupled with vermiwash spray provide congenial conditions for gerbera plant growth and flower production while also simultaneously reducing irrigation water use considerably (Verma *et al.*, 2019).

9.2.0 Irrigation

Gerbera needs thorough irrigation but water logging should be avoided. Production of flowers was high in summer when the water regime was high and low in winter with a constant water supply. Autumn-planted gerberas respond favourably to a hydrothermic quotient of 2 in summer and 1 in winter. It was also observed that the difference in flower number due to seasonal differences in light intensity could partly be reduced by irrigation and stem length increased with the water supply at flower initiation (Junges, 1964).

Mineral contents of irrigation water also affect the yield and quality. Gerberas are not specifically sensitive to chloride in the water but are susceptible to high total salt content in irrigation water and response of cultivars varies (Voogt, 1976b). Salinity above 8 mM NaCl reduced the number of flowers, flower weight and flower area (Baas *et al.*, 1995). Hanger (1984) obtained optimum growth of seedlings when solutions had electrical conductance values between 2 and 3.9 mS/cm in rockwool under hydroponic conditions.

Filella and Penuelas (1994) found that chlorophyll content was highly correlated with the wavelength of the red edge peak and leaf area index (LAI) was correlated with the amplitude of the peak, but the area of the red edge peak was the best estimator of LAI. Schmidt and Exarchou (2000) developed a device for measuring transpiration flows on several leaves in a plant stand to test the possibility for controlling the water supply of greenhouse plants.

Blomme and Dambre (1982) evaluated the response of pot-grown gerbera *cv.* Peter to intermittent sub-irrigation, irrigation by means of capillary matting + capillary tubes, drip irrigation or sprinkler irrigation and found that the differential treatment effects on flower production were slight, but capillary matting + capillary tubes gave the highest percentage of healthy plants. Olivella *et al.* (2000) found that flooding reduced leaf gas exchange parameters (stomatal conductance, net photosynthetic assimilation rate, and leaf transpiration rate) in five, 1-year-old gerbera cultivars under greenhouse conditions. The marked intolerance of gerbera to flooding could be a serious barrier to its culture under an aerobic conditions, and hence careful irrigation management is required.

Utilization of treated waste water for irrigation in the region is profitable and better results are always attained once supplementary irrigation is provided (Darwish *et al.*, 1999). Subsurface drip irrigation have significantly higher flower production, for all substrates and water qualities used, than surface drip irrigation (Maloupa *et al.*, 1999). Lhoste (1998) compared the disinfection of recycled irrigation water contaminated with *Phytophthora cryptogea* with Javel water (4 and 20-30 ppm active chlorine) in a greenhouse with gerbera *cv.* Testarossa. Disinfection with 4 ppm chlorine resulted in healthy plants after 90 days compared to 100 per cent plant mortality after the use of non-disinfected water.

Regarding flower and dry matter production, there are no differences between irrigation systems, but there were significant differences between substrates and an interaction between irrigation system and substrates. The vermiculite/perlite mixture gave the best results on average. With sub-irrigation, yields were highest in fine perlite/coarse perlite mixture and with drip irrigation they were highest in the peat/perlite mixture (Lenzi *et al.*, 2000).

9.3.0 Manuring and Fertilization

Organic and microbial fertilizers have been found to be useful for gerbera. Pinamonti *et al.* (1997) evaluated compost prepared from bark and sewage sludge as an amendment for 4 growing media. Compost amended growing media offered clear advantages in both the quantitative and qualitative characteristics of production. The physical and physiochemical properties of compost combined with those of white peat to create growing media offered higher production. Strojny *et al.* (1976) found that composted pine bark with N and Mg 250 and 150 mg/l, respectively was effective for quality flowers. Flower yields were in all seasons highest with 25 per cent Forestina humus (Fischer and Forchthammer, 1984). Rodriguez Navarro *et al.* (2000) found that the plant length, chlorophyll content, and number, length and diameter of inflorescences increased with vermicompost 20 per cent and the incidence of *Rhizoctonia solani* (in young plants), *Phytophthora drechsleri* and *Fusarium oxysporum* (both in mature plants) was also low. This was a confirmation of the earlier findings by Orlikowski (1999).

Giri and Beura *et al.* (2020) developed an INM model for cut flower production of gerbera in open field condition. Application of 75 per cent RDF (15:10:30 g NPK/m^2) + Vermicompost (25 g/m^2) + Azospirillum/Azotobacter (20 g/m^2) + PSB+ macro and micro element spray increased vegetative parameters and number of flower/plant. It is concluded that reducing level of chemical fertiliser to 75 per cent RDF along with application of different organic source of nutrient can improve flower yield of gerbera in open field condition.

Investigations have been carried out on the nutrient requirement of gerbera and some recommendations are presented in Table 1.

Table 1: Fertilizer Recommendation for *Gerbera*

Fertilizer Dose	*Reference*
60:48:96 kg/1000 m^2	Volpi and Farina (1987)
140-280 mg N, 30-60 mg P, 175-350 mg K/l of peat	Kacperska (1985)
30 mg N:50 mg P_2O_5 : 50 mg K_2O/100 g of 1:2 loam:peat	Skalska (1985)
6.8 g N:1.24 g P_2O_5:15 g K_2O:4.54 g CaO:1.25 MgO per plant	Riviere and Alegre (1984)
1.5:0.8:2.5 of N : P_2O_5:K_2O at 2g/l substrate	Skalska (1978)
2 g N:4 g P_2O_5:1 g K_2O/plant	Kamel (1975, 1977)
10g N (+ 5 top dressing of 5 g N); 5 top dressing of 6 g K_2O/m^2	Blomme and Dambre (1980)
281-293 mg N : 127-133 P_2O_5 and 360-440 mg K_2O/l of peat substrate	Lisiecka (1976b)
4-10 g N : 2.4-8 g P_2O_5/plant	Gotz (1983)
47 g N for 10 months in every 3rd irrigation	Skalska (1983)
100 kg N/ha in 3 split application	Aswath *et al.* (1995)
20:30:40 g NPK/m^2	Ayemi *et al.* (2017)
12.5 g P m^2 and 15 g K_2O m^2	Amin *et al.* (2015)
15:10:30 g NPK/10 plants	Giri and Beura *et al.* (2020)
15:10:30 g of NPK/m^2	Palanisamy *et al.* (2015)

Application of nutrient solution at the highest concentration resulted in salt accumu-lation and the salt content was highest at the top of the pot, and lowest in the centre (Vass, 1986). Flower numbers/m^2 were greater with the nutrient solution method compared with other methods (Blomme and Dambre, 1984). Baillot (1976) studied the influence of plantora (14:10:14) or with coic nutrient solution using the double pink gerbera Fredaisy. Plantora tended to improve the quantity and coic solution the quality of flowers. Response of cultivars varied in their reaction to fertilizer application (Fischer and Forchthammer, 1983). Oszkinis and Lisiecka (1971) obtained best flower production with 'Flora' (a complete fertilizer) applied monthly from January to August at 75 g/m^2. Vass and Hargitai (1986) found that during vegetative growth the soluble N content of the soil gradually decreased even when nutrient solutions were applied.

Aswath *et al.* (1996) found variation between varieties for their response to nitrogen. Filella and Penuelas (1994) found chlorophyll content to be correlated with N availability. Mantrova *et al.* (1982) found that N at 0.025 g/kg soil applied 15-17 days after planting, N : K_2O at 0.1 : 0.15 g/kg at the stage of rapid vegetative growth, N : P_2O_5 at 0.1 : 0.1 g/kg at bud formation, and N:P:K at 0.1:0.1:0.23 g/kg at the beginning of flowering were optimum for plant development and flower production. Skalska(1980) obtained the best results (>40 flowers/year) when the N level in spring, summer, autumn and winter was 40, 20, 10 and 5 mg/100 g substrate, respectively, the P level in winter as 20-25 mg/100 g and the K level in autumn and spring was 30-40 mg/100 g. Flower production was lowest in winter. Blomme and Dambre (1979) observed that when the N dose for top-dressing was higher than 15 g/m^2 (in the range 5-25 g) the quality of the flowers was adversely affected.

Graf and Kress (1996) found about 200 mg higher N uptake at 4 weeks after planting from plants grown in 30 per cent flax shavings, which stabilized after a further 5 weeks. However, some plants showed mild chlorosis due to N-fixation and Fe deficiency as a result of an increase of 0.5 in pH value. Guba (1994) reported that the greater the proportion of NH_4 the lowest was the pH of the rooting medium, with a mean, pH of 4.9 at the highest NH_4 : NO_3 ratio. Flower quality tended to deteriorate with increasing NH_4 : NO_3 ratio. Kreij (1989) reported that gerbera *cv.* Joyce supplied with a nutrient solution with a low K : Ca ratio (0.8-0.9 : 3.0-4.0 mmol/l) showed symptoms of K deficiency, whereas use of a high ratio solution (7.0-8.0 mmol/l) showed no symptoms of K deficiency. At both ratios, plants supplied with 0.6 mmol Mg/l were of better quality than those supplied with 1.7 mmol Mg/l.

Gerbera plants received 12.5g P/m^2 and 15 K_2O g/m^2 recorded more plant height, leaves/plant, flowers blooming period, plant canopy, flowers/plant, flower stalk length, flower diameter and vase life (Amin *et al.*, 2015).

Studies carried out by Palanisamy *et al.* (2015) on the effect of fertigation regimes, foliar spray of micronutrient mixture and humic acid on gerbera plants revealed that plant height, number of leaves, leaf area, plant spread, suckers number and plant density had progressive increase at 100 per cent RDF+0.004 per cent MN Mixture + 0.2 per cent humic acid, while, 75 per cent RDF + 0.004 per cent MN Mixture + 0.2 per cent humic acid induced first flowering. The 100 per cent RDF+0.004 per cent MN Mixture + 0.2 per cent humic acid recorded highest flower yield, longest flower stalk and stalk girth and largest flower diameter. Hence these treatment combinations could be recommended for commercial cultivation of cut gerbera.

Micronutrients also play an important role in flower production. Muthumanickam *et al.* (1999) sprayed *cv.* Ibiza monthly with 2 per cent $MnSO_4$, $FeSO_4$, $ZnSO_4$ alone and in combination. Treatment with all 3 micronutrients resulted in the highest leaf Fe, Mn and Zn contents, and gave the highest flower yield, stalk length, flower diameter, stalk girth, plant height and vase life. Lisiecka (1976a) reported that 60 g $CuSO_4$ + 2 kg $CaCO_3$/$m_3$peat improved the commercial quality of flowers and the quantity/m^2 increased by 46.5 per cent. Fe citrate at 20 g/m_3 or Na_2MoO_4 at 2.5-5.0 g/m_3 slightly improved seedling fresh weight. Spraying of 0.1-0.2 per cent $MnSO_4$ every 2-4 weeks corrected interveinal chlorosis (Voogt, 1976a). Gutmane and Grivina (1975) reported that raising the salt concentrations from 2 to 4 g/l improved the growth and increased N, P and K accumulation in the leaves and roots. Fischer and Forchthammer (1987) suggested supplementation of the normal fertilizer treatment from the 4th to 7th month of growth with a weak solution of $CuSO_4$ (5-6 mg/plant), at intervals of 6-8 weeks for plants grown in small pots during winter.

Macro nutrients NPK (19:19:19) applied at the rate of 0, 30, 45 g per m^2 per month at 15 days interval resulted maximum duration of flowering, number of flowers per plant and stalk thickness in *var*. Dune. Maximum number of petals/ flower, petal length and flower diameter were recorded in *var*. Rich when treated with 45g/m^2/month of NPK. The *var*. Zingard recorded quickest to bud opening with highest vase life at 45g/m^2/month of NPK (Srivastava *et al.*, 2017).

Foliar spray of ferrous sulphate (0.6 per cent) recorded the maximum plant height, highest flowers/plant with flower diameter, whereas, best result in terms of leaves and suckers/plant, plant spread and flower stalk length was recorded in magnesium sulphate in *var*. Rosaline. Vase life of flower was found maximum in plant treated with 0.6 per cent magnesium sulphate. Application of ferrous sulphate, magnesium sulphate and zinc sulphate were found to significantly increase the growth and flowering of plant. However, boric acid was found to be less effective (Sarkar *et al.*, 2020).

Klossowski and Strojny (1983) found nutrient analysis of leaf blade was more reliable than petiole and obtained good flower yields when the leaves contained 2.7-3.13 per cent N, 0.19-0.35 per cent P, 3.06-3.64 per cent K, 1.66-2.18 per cent Ca and 0.3-0.48 per cent Mg. Hwang and Yoon (1994) reported that the optimum concentrations of N, P_2O_5, K_2O, CaO, MgO and Na_2O were 2.70, 0.40, 3.55, 2.75, 1.15 and 0.45 per cent, respectively at 90 days after transplanting in shoots for maximum biomass yields.

9.4.0 Weed Control

Mixtures with dual (metolachlor) + 1 or 2 kg adol or + 6-8 kg tenoran/ha markedly improved control of *Digitarias anguinalis*. Nitrofen at 8 lb *a.i.*/acre gave good control and a combination of 4 lb nitrofen and 2 lb oryzalin/acre was also effective in controlling *Oxalis corniculata* (Pyeatt and Elmore, 1975).

The integrated weed management practices in gerbera using different pre-emergence herbicides, mulching materials and cultural practicesrevelaed that Pendimethalin at 1.0 kg *a.i.*/ha and black polyethylene sheet and alachlor at 1.5 kg *a.i*/ha showed better results (Shalini and Patil, 2006). Black polyethylene mulch produced tallest plant with maximum leaves and suckers/plant, flower size, flowers/clump, flower stalk and earlier flower bud visibility in *cv*. Red Gem (Sarmah *et al.*, 2014).

9.5.0 Pot Gerbera

Gerbera is highly suitable for pot culture. Bruckner and Elgner (1989) compared the growth of the seedling cultivar Feuerfalter and the clonal cultivar Cyclop in 12 cm pots containing 5 different substrates and supplementary fertilizer application was necessary in all substrates because of high N demand. Hermann (1978) obtained flower yields in pots similar to those grown in beds. A pot compost which is porous, rich in organic matter and with high moisture retentive capacity suits the requirements of gerbera. Plants grew better in peat to which no lime or lime at only 1 kg/m_3 has been added than in peat with higher addition of lime. Jesiotr *et al.* (1975) suggested a mixture of 90 per cent (by volume) pine bark and 10 per cent birch bark. It is suggested to compost the bark for at least 4-6 months before it is used. Excess of tannins in bark can be reduced by treatment with hot water. Addition of urea at 3 kg/m_3 of bark and 10-15 per cent (by volume) of active sludge has also been found beneficial in order to compensate the nitrogen depletion that occurs during compos-ting. A mixture of composted pine bark and sphagnum peat moss (1:1 by v/v) is also being used by some growers. When the major portion of the media consists of soil or soil composts, addition of wetting agents like hydrogell or hydramull (about 20 per cent by volume) improves the moisture retention in such media. Richter (1996) reported that for pot plants of *cv.* Jaguar Dunkelrosa EC values of 1.8-2.4 mS/cm in the nutrient solution and irrigation at –100 and –200 hPa provided controlled growth and good pot plant quality. Hell and Ter (1996) found that 30 per cent shading resulted in the highest number of flowers and longest stems, while 70 per cent shading gave the poorest results in *cv.* Feuerfalter. Plants grown under 30 per cent shading showed the best keeping quality under simulated living-room conditions. Unshaded plants showed significant growth depression in the form of very small leaves and low fresh weight/plant compared with those grown with 30 or 50 per cent shade.

Steamed or non-steamed bark, composted for 35 days with urea added at 2.8-5.6 kg/m^3 was used to grow the plants in 7.5 litre plastic pots. Compared with plants grown in a 90:10 per cent peat : perlite mixture, the bark-grown plants gave poor results and this was attributed mainly to an excessively large air volume (Pivot, 1985).

In a trial with two types of rockwool (grodan and cultilene) plants grown on grodan were generally better than in cultilene. With heating, cultilene rockwool gave increased yield of 13 per cent in *cv.* Clementine and 22 per cent in Valentine. With grodan, flower quality was better without heating (Vidalie *et al.*, 1985). Charpentier *et al.* (1986), however, reported that two types of rockwool gave similar results. It is important to point out that the quality of flowers grown in pots declines considerably after the first year. So it would be beneficial to repot the plants dividing clumps immediately after flowering is over.

Effect of substrates with different physical and chemical characteristics on potted gerbera cultivars (Cherry and Red) was investigated.Plants grown on substrate with pH above 7.0 had a reduction in iron absorption, resulting in lower intensity of green coloration in the leaves. Plants grown on substrate with pH below

5.0 showed toxic levels of manganese and lower dry mass. The characteristics of the substrate, especially the pH, influence the nutrients absorption and the production of potted gerbera, altering the final plant quality (Ludwig *et al.*, 2014).

Using different growing media instead of single media in potted gerbera is very useful and more profitable. The highest benefit cost ratio (B:C ratio) (3.17 and 3.37 in two years) was obtained from flowers grown with cocopeat 40 per cent +red soil 30 per cent + F.Y.M. 20 per cent + sand 10 per cent combination. The treatment cocopeat 40 per cent +red soil 30 per cent + F.Y.M. 20 per cent + sand 10 per cent realized maximum net returns of Rs.7,01,236 and Rs.7,11,124 in 560 m^2 area for two years (Kulkarni *et al.*, 2017).

10.0 Diseases and Pests

10.1.0 Diseases

Diseases can be major factor limiting gerbera production. Important diseases and their control measures are given below.

Fungal Diseases

10.1.1 Root Rot

It is a soil-borne disease caused by *Pythium irregularae* and *Rhizoctonia solani*. The infection results in stunted growth and ultimately drying of entire plants. *Rhizoctonia solani* caused more losses than *P. irregularae* and can attack older plants (Krober and Plate, 1975; Plate and Krober, 1975). Soil sterilization gives good control of the disease.

10.1.2 Foot Rot

This disease is also caused by soil-borne pathogens and *Phytophthora cryptogea* is very common attacking gerbera. Foot rot of *Gerbera jamesonii* (grown under glass in the Netherlands) was found by isolation and inoculation experiments, to be caused by *Fusarium oxysporum*. The short stems blacken and rot and the leaves and flowers die (Vonarx, 1952). Soil sterilization with vapam at 100 ml/m^2 was found very effective to control the disease (Koek, 1958; Pag, 1960). Warming of soil (26°C) also reduced the incidence of disease (Scholten, 1972). Bigre (1980) suggested cultural methods to control *Phytophthora cryptogea* such as avoiding infected substrate, excess water, and temperatures which favour the pathogen. Fongacid and prothiocarb may be used for chemical control, beside growing resistant cultivars like Sacha and Palas. Schickedanz (1976) compared the effects of certain fungicides and reported that prothiocarb at 0.15 per cent gave better control of *P. cryptogea*on young plants than fenaminosulf at 0.04 per cent. Watering immediately after planting was more effective than dipping the roots before planting. The treated plants remained disease free for about 3 months. Pielka and Bartynska (1974) controlled *P. cryptogea* by using topsin at 0.05 per cent and Garibaldi (1974) by drenching of dexon or difolatan at 1 g/m^2. In 3 glasshouse trials, the disease was controlled by monthly soil drenching of 0.4-1.6 g metalaxyl + 0.8-3.2 g folpet/m^2. Good control was also recorded by using 2 g furalaxyl or 5 g of urace. Results were variable with 6.2 cm_3 captafol and 4 g cymoxanil. Results were poor with 6.6 cm^3 propamocarb, 8 g fosetyl aluminium and 4.2 g fenaminosulf. Treatments with 2-3 g benalaxyl, 0.8 cm^3 hexachlorophene and 2.2 g etridiazole + 2.7 g captan were effective (Pasini *et al.*, 1984).

10.1.3 Sclerotium Rot

Sclerotium rolfsii was identified as the cause of this serious fungal disease affecting the entire above ground parts of the plants. The incidence may be as high as 90 per cent during August and September when the atmospheric relative humidity is high and the temperature ranged between 30° and 34°C (Singh, 1973).

10.1.4 Blight

This disease, also known as gray mould, is caused by *Botrytis cinerea* which kills young growing tissues. Kerssies (1993) found that the incidence of *Botrytis cinerea* was low in spring and early summer whereas at other times of the year many lesions appeared on the cut flowers. Garibaldi (1968) reported that flower heads of gerbera growing in humid conditions under glass developed very small black spots on the ray florets. An evidence of a distinct blocking reaction of the host cells, restricting the development of the fungal mycelium to very small areas of the petals was also noted. Deep planting, bad drainage, poor ventilation and predispose plants to infection should be avoided. Watering should be discontinued from mid-October until the end of March (Garthwaite, 1963). Infected parts should be removed and destroyed. Spraying of 0.7 per cent captan, 1 per cent TMTD or 0.7 per cent euparen at 8- to 10-days interval was also suggested to control the disease (Kulibaba, 1972). Orlikowski *et al.* (1974) inoculated seeds with *Botrytis cinerea* causing damping off in 5 to 90 per cent plants which was controlled by treatment with 0.1 per cent benlate or thiram.

10.1.5 Powdery Mildew

Two fungi, *Erysiphe cichoracearum* (Pirone *et al.*, 1960) and *Oidium erysiphoides* f. sp. gerbera (Folk *et al.*, 1984), producing a white powdery coating on the foliage have been reported on gerbera. Pirone *et al.* (1960) suggested spray with wettable sulphur or karathane to control *E. cichoracearum*. To prevent *O. Erysiphoides* f. sp. gerbera, the relative humidity should be reduced especially in spring and autumn, old leaves should be removed and the house ventilated. Good control was obtained before flowering with, elemental sulphur or dinocap (as karathane FN 570.05 per cent + karathane IC 0.05 per cent) or the systemic fungicides like benomyl (chinoinfundazol 50 WP 0.1 per cent), bupirimate (nimrod 25 CE 0.1 per cent), fenarimol (rubigan 12 CC 0.03 per cent), pyrazophos (afugan 0.1 per cent) or thiophanatemethyl (topsin M WP 0.08 per cent). On flowering plants, only benomyl, bupirimate, fenarimol or pyrazophos should be used (Folk *et al.*, 1984). Treating plants with 0.5-0.7 per cent colloidal sulphur, 0.7 per cent thiovit (sulphur) or 0.15 per cent karathane (dinocap) is also suggested (Kulibaba, 1972).

Baiswar *et al.* (2010) reported first record of powdery mildew caused by *Podosphaera* sp. on *G. jamesonii* in India. Kumar *et al.* (2013) studied fungicidal and varietal trial in gerbera. Least disease severity (5.23 per cent) was observed in spray of wettable sulphur at 2.5g/l of water followed by carbendazim at 2g/l of water foliar spray (7.23 per cent). The highest flowers/m^2/year (195.00) and suckers/ plant/year (6.60) were also recorded in wettable sulphur at 2.5g/l of water foliar spray. The var. Teresa showed moderately resistant reaction (24.57 per cent) against powdery mildew infection. Varieties Pia and Tecala were highly susceptible to powdery mildew which showed per cent disease index of 65.30 per cent and 54.27 per cent, respectively.

10.1.6 Leaf Spot

It is also a common disease caused by *Phyllosticta gerberae, Alternaria* spp. and others. In greenhouse and field tests, Bordeaux mixture gave little control of *Alternaria* leaf spot, but captan was found promising. Kulibaba (1972) recommended spraying of 1 per cent Bordeaux mixture, 0.5-7 per cent maneb, 0.5 per cent zineb, 0.4 per cent polyram or 0.5 per cent ziram to control the disease.

10.1.7 Downy Mildew

The disease for the first time had been reported from Poland and the causal organism was identified to be *Bremia lactucae* (Leski, 1984). The disease can be effectively controlled with the use of ridomil 25 WP (metalaxyl) at 0.1-0.15 per cent by spraying 3-6 times at fortnightly interval or by soil drenching at 0.7 to 1.0 g/plant twice at 3-4 weeks' interval. Dithane M-45 (mancozeb), sanspor 50 EC (captafol) or captan at 0.3 per cent used 8 times at 9-10 days interval also proved useful in controlling the disease.

10.1.8 Other Diseases

Some other fungi have also been reported to attack gerbera plants. Symptoms of the disease caused by *Sclerotinia sclerotiorum* and conditions for the growth of the pathogen were described by Hoffmann and Menzinger (1963). Strict hygiene and careful choice of preceding crops were suggested to prevent the spread of the disease. *Verticillium dahliae* caused heavy losses in commercial gerbera cultivation in Israel by severe stunting and wilting of plants, rather than immediate death. Fumigation with methyl bromide, chloropicrin or mixtures (70 : 30) of the two gave good control of the disease. Schlang and Sikora (1978) found that when *G. jamesonii* was inoculated simultaneously with *Meloidogyne arenaria* and *Verticillium dahliae*, or with *V. dahliae* 40 days after *M. arenaria*, there was a reduction of 40 to 50 per cent in the dry weight of the roots but there was no significant effect on the verticillium-wilt index.

Bacterial Disease

10.1.9 Bacterial Blight

A severe leaf blight disease which limited the gerbera production in the USA was characterized by small to large, circular to irregular, brownish black leaf spots with or without concentric rings. Large brown areas, extending from the margins and narrowing as they reached the mid-veins of the leaf, often occurred (Miller and Knauss, 1973).

Diseases and Pests of Gerbera

Powdery Mildew

Alternaria Leaf Spot

Wilt

Root Rot

Root Rot

Root Knot Nematode

White Fly

Leaf Miner Injury

Mite Injury

Diseases and Pest of Gerbera

Spider Mite

Leaf Miner

Aphid

Ring Spot Virus

Virus Mottling of Foliage

Source: www. extension.umd.edu, www.flowerweb.com, www.entnemdept.ufl.edu, www. gerbera.org, www.ag.umass.edu, www.omafra.gov.on.ca, www.biobestgroup.com, www. pestnet.org, www.amazevegegarden.com, www. dfr.icar.gov.in, www.semanticscholar. org, www. pnwhandbooks.org, www.link.springer.com/article/10.1007

Viral Diseases

10.1.10 Tobacco Rattle Virus

This disease was first reported from Florida. Yellow or black annulated ring spots were observed on the foliage of field-grown gerbera. Necrotic lesions were produced on 6 species of indicator plant (Stouffer, 1965). The most reliable control method was found to be soil steaming (Schmelzer, 1966). The occurrence of this virus was also reported from the Netherlands. It was suggested that the normal soil steaming before each crop should destroy the nematode vectors such as *Trichodorus* spp. and prevent the disease (Hakkaart, 1968).

10.1.11 Mosaic

The mottling caused by gerbera mosaic virus was described by Verma and Singh (1980). They reported that the disease can be transmitted by grafting.

10.2.0 Pests

Among different insect pests, white flies, aphids, leaf miners, spider mites, caterpillars and root-knot nematodes are common attacking gerbera plants. Pollution caused by the use of pesticides has become a major concern in today's horticultural production. Insect-proof nets over vents can lower the use of insecticides in greenhouses, yet they hinder air exchange, causing high temperature problems. A com-promise between acceptable natural ventilation and pest protection can be found if the area of the greenhouse openings is increased. Montero *et al.* (1999) found that the insect-proof mesh reduced the population of whiteflies, heliothis and leaf miners in greenhouses with a significant reduction in the number of pesticide applications. However, the insect net was not an effective barrier against the thrips *Frankliniella occidentalis*.

An integrated pest management (IPM) programme for gerbera integrating bioagents along with reduced risk pesticides, soil-solarization and cultural practices is beneficial. Insect monitoring devices like yellow sticky cards, mass trapping products such as sticky trapscan also used for detection and management of insects (Sabir *et al.*, 2012).

10.2.1 White Fly

In greenhouse, gerbera is subject to heavy infestation of whitefly (*Trialeurodes vaporariorum*), a sucking insect. Control measures include biological control with *Encarsia formosa* and malathion sprays (Anon., 1968). Sutterlin *et al.* (1991) investigated the prospects of biological control of the aleyrodid with the parasitoid *Encarsia formosa*, the dispersal and distribution of *T. vaporariorum* on gerbera in greenhouse experiments. In intact plants, many more aleyrodids were found on younger than older leaves. Sutterlin *et al.* (1999) studied the foraging behaviuor of *Encarsia formosa* and found that it is independent of the cultivar. Reist (1997) investigated the effectiveness of chemical and integrated methods for control of aleyrodids, thysanoptera, leaf miners and aphidoidea on gerbera. The IPM treatment was calculated at about 35 per cent more than the chemical treatment. Albert (1997) found biological control was either expensive or not always successful. Lupa (1987) reported the possibility of replacing insecticidal control of *T. vaporariorum* on gerbera by biological control using *Encarsia formosa*. The effectiveness of this control was found to depend on the method of rearing the parasitoid, the date of its introduction and the average daily temperature in the greenhouse. In greenhouse experiments, Cresswell *et al.* (1994) reported that mortality of *T. vaporariorum* was related to the concentration of acephate (orthene) in nutrient solution applied to gerbera grown in perlite. A single irrigation with 80 mg/l acephate killed all immature *T. vaporariorum* in 10 days with no apparent damage to the plants. Dirlbek *et al.* (1989) found that for control of *T. vaporariorum*, the use of boverol (containing conidia of *Beauveria bassiana*) applied alone or in combination with chemicals gave favourable results when compared with chemicals alone and is promising especially in view of the reduced chemical consumption and contamination. Boverol at 0.3 per cent kept numbers of adults at low levels for about 28 days, with slightly better results in combination with deltamethrin at 0.016 per cent. Baker (1987) found that application of morestan [chinomethionat] gave the greatest reduction of nymphs and also appeared to be toxic to the eggs of *T. vaporariorum*. Pasini *et al.* (1997) found spraying 6 times at 7-day intervals with MYX 400/l gave control of white flies *T. Vaporariorum* and *Bemisia tabaci* with the same efficiency of buprophezine (buprofezin) (applaud) and etophenprox (etofenprox) (trebon). An integrated control with the use of yellow-orange plexiglass sticky traps was also suggested by Nucifora and Vacante (1981) and Quaglia (1983). Baranowski and Blaszak (1996) determined the influence of colour and smell on the detection of *T. vaporariorum* infesting gerbera and found that yellow plates were more attractive than blue, and the addition of benzaldehyde or benzyl acetate increased the effectiveness of traps.

Four foliage thrips predators (*Transeius montdorensis* (Schicha), *Orius armatus* (Gross), *Mallada signata* (Schneider) and *Neoseiulus cucumeris* (Oudemans) and three soil predators *Geolaelaps aculeifer* (Canestrini), *Steinernema feltiae* (Filipjev) and *Dalotia coriaria* (Kraatz) were studied to determine their ability to reduce the numbers of WFT on gerbera. Of the foliage predators, *T. montdorensis*, *O. armatus* and *N. cucumeris* performed best, significantly reducing the numbers of adult and immature thrips on flowers and foliage by 30–99 per cent (Manners *et al.*, 2013).

10.2.2 Leaf Miner

The leaf mining fly, *Liriomyza trifolii*, is a serious pest of gerbera. Another species, *L. soncho*, was also recorded by Vrie (1976). Adults of these small winged insects lay eggs on the leaf. The larvae bore into the leaf and make irregularly shaped tunnels or blotches which are generally light yellowish tan to brown in colour. As these larvae mature, they fold the leaf together with threads and feed on the inner surface. Park and Park (1996) found that the developmental threshold (DT) and effective accumulative temperature (ET) of *L. trifolii* on gerbera leaves were 11.2°C, 33.9 DD (degree-days) in eggs, 10.3°C, 38.6 DD in larvae, 10.7°C, 152.3 DD in pupae, and 10.8°C, 222.8 DD from egg to pupa. Provisional advice, based on trial results, for the control of *L. soncho* in greenhouse crops is to apply the insecticide ambush (permethrin) at 50 ml per 1000 m^2 by swingfog or pulsfog atomizers, the application being repeated if necessary. Dengler (1984) stated that gerberas do not tolerate all insecticides. Twentyone different sprays were applied 3 times at weekly interval to the cultivars Clementine, Marleen and Farina on 3 sites at 8 and 20°C. Ekamet (etrimfos), hostathion (triazophos) and vydate (oxamyl) controlled *L. trifolii* but frequent sprays damaged plants, particularly when mixed with mildew controlling fungicides. Damage was severe at 20°C, particularly to Farina. Nimrod (bupirimate) and fungaflor (imizalil) alone or in combination were compatible with the 3 cultivars. Rubigon (fenarimol) and F 238 (dodemorph) damaged plants when applied separately and were incompatible in mixture with insecticides. Saprol (triforine) alone did not damage plants but was incompatible when mixed with other products. Vande *et al.* (1988) found that applications of buprofezin at 0.75 g *a.i.*/10 litres applied twice with an interval of 8 days gave effective control of *L. trifolii* on gerbera. Natskova and Karadzhova (1990) found that 0.03 per cent permethrin (ambush 25 EC), 0.02 per cent cypermethrin (ripcord 40 EC), and oxamyl (vydate 10 G) and aldicarb (temik 10 G), both at 2 g/plant against adults of *L. trifolii* and *L. bryoniae*. Against larvae, the most effective compounds were cypermethrin, 0.1 per cent methamidophos (tamaron 59 EC), 0.03 per cent cyromazine (trigard 75 WP), 0.1 per cent diflubenzuron (dimilin 25 WP), 0.1 per cent triflumuron (alsystin 25 WP) and aldicarb. Calabretta and Nucifora (1985) reported the possibilities of biological and integrated control against leaf miner.

Control tests in Sicily showed that combination of yellow trap against the adults and aqueous sprays of pyrazophos against the larvae within the leaf miners were very effective (giving up to 100 per cent larval mortality). Yathom *et al.* (1988) conducted a greenhouse study to determine the optimal height, position (angle) and direction of yellow sticky traps for the attraction of the agromyzid *L. trifolii*. Catches on vertical traps were greater than those on horizontal and 45° angle traps. Free-hanging vertical traps caught more flies than fixed ones.

Jones *et al.* (1986) reported that *L. trifolii*, easily develops resistance to insectcides. Biological control is more practicable. *Diglyphus intermedius*, released at the rate of about 1000 parasites/week, resulted intrap catches of *Liriomyza* adults lower than those in a nearby greenhouse in which chemical and microbial treatment had been applied. Natural immigration of the eulophid *Diglyphusisaea* into the area afforded about 90 per cent parasitism of *L. trifolii* and effectively protected the crop without artificial control measures. When no parasitoids were released, the population level of the aleyrodids was 100 times higher at the end of an 11-week observation period and the rates of damage to leaves of gerbera by *L. trifolii* was 17.9-63.3 (Hong *et al.*, 1996).

10.2.3 Mites

Greenhouse grown gerberas often attacked by *Hemitarsonemus latus* and *Steneotarsonemus pallidus*. The development of leaves and flower buds was adversely affected, the flowers were malformed and unsaleable. Symptoms of damage by *Polyphagotarsonemus latus* were described by Szentgyorgyl *et al.* (1984). Elemental sulphur gave 100 per cent mite mortality in laboratory trials and when applied to crop plants at 20 kg/ha, gave 97-98 per cent control on forced plants. Flowers were, however, soiled and dusting with sulphur was recommended only for non-flowering plants. Flowering plants should be treated with actellic (pirimiphos methyl), hostathion (triazophos), mitac (amitraz), rospin (chloropropylate), torque (fenbutatin oxide) or vydate (oxamyl).

10.2.4 Thrips

Lacasa *et al.* (1993) studied the incidence of *Frankliniella occidentalis* and estimated the damage and occurrence of tomato spotted wilt tospovirus. *F. occidentalis* was the most frequent and abundant thrips, although *Thrips abaci, T. simplex* and *T. angusticeps* were also present. Red-flowered varieties (Beauty and Monika) had small thrips population densities and were not infected by virus. The white-flowered variety Ansophie had an average density of 30.5 thrips/flower with 90.5 per cent viral infection and losses of 2.23 flowers/plant. Charlotte (orange), Delphi (white) and Princessa (yellow) had similar densities (15.2, 12.5 and 12.1 thrips/flower, respectively) but different disease levels (19.04, 19.94 and 64.28 per cent) and flower losses (0.24, 1.08 and 2.42 flowers/plant). Horizon (yellow) had a low population density, losses of 0.25 flowers/plant and 11.9 per cent viral infection. Labanowski (1990) found that application of seedox 80W [bendiocarb] at 0.1 per cent 4 times at intervals of 4-5 days resulted in a 96 per cent reduction in pest population of thrips (*F. occidentalis*) in greenhouse grown gerbera varieties.

Vestergaard *et al.* (1996) assessed the efficacy of *Metarhizium anisopliae* against *F. occidentalis* in a greenhouse, by mixing fungal spores into peat containing gerbera plants. Two of 3 monitoring methods showed a reduction in the thrips population 4-6 weeks after treatment with *M. anisopliae*, after which the thrips population increased rapidly.

Sorensson and Nedstam (1993) introduced the predators *Amblyeius cucumeris* (*Neoseiulus cucumeris*) and *Orius insidiosus* alone and together on gerbera infested with *F. occidentalis* in greenhouses in Sweden. Both predators established well and reduced thrips numbers on all 4 crops, the reduction being greatest when both predators were present.

10.2.5 Aphids

This insect infests young leaves and buds and causes injury by sucking the sap which results in distortion of tissue. Life history and damage caused by *Myzus persicae* and *Aphis fabae* have been described by Tusnadi (1984a). For the control of both the species spraying or fumigation are recommended. As sprays, 0.1 per cent dichlorvos, 0.1 per cent methomyl, 0.1 per cent oxamyl, 0.05 per cent pirimicarb, 0.04 per cent permethrin + tetramethrin, 0.04 per cent decamethrin or 0.04 per cent cypermethrin was suggested. Fumigation with sulfotep smoke cartridges at 200 m^3 or pirimicarb smoke candles at 700 m^3 may also be used. Among the other insect pests that occasionally infest gerbera are *Heliothis armigera* (Jefferson, 1952), *Mamsestra brassicae* (Tusnadi, 1984b) and *Spodoptera liltoralis*.

10.2.6 Integrated Pest Management

Sabir *et al.* (2012) investigated two modules (IPM and Non- IPM) in pest management of gerbera. IPM components included soil drenching with 2 per cent carbosulfan, soil application of bioagents mixture of *Pseudomonas fluorescens* and *Trichoderma harzianum* and spraying with combined mixtures of biorationals like azadirachtin and Agrospray (0.5 per cent). The non-IPM treatment included soil drenching with 2 per cent formaldehyde and spraying of imidacloprid (0.02 per cent) for controlling whitefly and aphid, spinosad (0.02 per cent) for controlling thrips, bitertanol (0.05 per cent) for controlling powdery mildew and white rust. In IPM treatment, the effects of biotic stress was significantly reduced in comparison to non-IPM treatment. Total pest-affected (disease, insect and nematode) plants recorded were 13.4 per cent in IPM module as compared to 37.56 per cent in non-IPM module. Combination of azadirachtin and agrospray (0.5 per cent) was the most effective component of IPM for controlling the sucking pest of gerbera although it required repeated application. IPM treatment including the organic amendments enriched with combination of bioagents gave significant increase in yield (33 per cent) and improved quality of flowers (65 per cent grade I flowers).

10.2.7 Nematodes

Root-knot nematode is occasionally serious in gerbera. Stokes *et al.* (1974) recommended the use of dasanit (fensulfothion) at 800 or 1000 ppm or mocap (athoprophos) at 400 or 800 ppm, the latter being more effective. Hoogstrate (1982) compared M 10 granules (experimental product), vydate (oxamyl) as granules or liquid and temik (aldicarb) granules for controlling soil infestations of *Meloidogyne* spp. under glass. All 4 nematicides considerably reduced nematode population, resulted in improved root system and increased cut flower number compared with plants in untreated infested soil. Vydate treatment (10 per cent granule) at 0.5 kg/100 m^2 incorporated before planting resulted in a better root system than M 10 (10 per cent granules at 2 kg/100 m^2).

11.0 Harvesting and Postharvest Management

11.1.0 Harvesting

The flowers of gerbera are generally harvested when the outer two rows of disc florets are perpendicular to the stalk. The stage of the harvest is very critical as the flowers should not be cut before the outer whorl of flowers show pollen (Auman, 1980). Generally, recutting of stem before placing in water or preservative solution or at all stages of marketing is necessary to prolong the vase life.

11.2.0 Yield

Skalsha (1980) obtained more than 40 flower/clumps/year at different levels in the substrate. The highest yield (244.86 flowers/m^2) was produced by a substrate concentration of 30 mg/100 g (Skalsha, 1985). Muthumanickam *et al.* (1999) reported maximum number of cut flowers (46.7/plant/year) in *cv.* Ibiza.

11.3.0 Postharvest Management

If the flower stem is not fully developed before harvest, a point about 10 cm below the capitulum becomes very susceptible to the disorder 'stem break'. It was shown that the zone of curvature differentiated slowly and in parallel to

Harvesting and Postharvest Management

Harvesting

AR Glass for Determination of Harvesting Stage

Robotic Harvesting

Grading and Bunching

Grading

Individual Packing

Bunching

Packing in CFB

Box Packaging

Packaging of Flowers

Wet Packaging

Transportation

Source: www.gerjahelden.nl, www.gerjahelden.nl, www.hortibiz.com, www. web. fleurametz.com, www.indiamart.com, www.gerbera.co.nz, www. in.pinterest.com, www. floraldaily.com

the opening of the disc florets. When only 2 rows of these florets had opened (the stage when flowers were usually cut) the weak section contained few conductive xylem elements and hardly any supportive elements. In contrast, 16 per cent of the underlying section (in the middle of the stem) consisted of supportive tissues and this prevented bending. Cultivars which have sturdier stems, the zone of curvature differentiates relatively rapidly hence it is better to avoid cutting flowers too early and also better to use a vase life preservative when flowers are stood in water before marketing (Dubuc Lebreux and Vieth, 1985).

11.3.1 Packaging

The flower heads packed in insulated boxes were protected from cold or freezing temperature. In the Netherlands, plastic coated metal grids measuring 50 x 70 cm and 2 x 2 cm mesh size were used to pack gerbera flower heads (Gardner, 1975). The flower heads are supported by the grid which is suspended above a plastic tray measuring 48 x70 x30 cm, at a height which can be adjusted as per the length of flower stalk so that the stalk ends are immersed in water up to 15 cm depth in the tray. Nijlunsing (1976) used non-returnable cardboard cups or minisleeves in place of metal grids to protect the cut blooms.

Polypropylene (24 µ) and HDPE (24 µ) films used for MAP at 5°C cold storage temperature found promising in maintaining flower quality of gerbera during and after storage. The quality of cold stored flowers (without MAP) deteriorated with reduced flower size and increased scape bending. MA (PP and HDPE) stored gerbera flowers at 5°C temperature showed significantly lower physiological loss in weight and scape bending after storage and maintained higher water uptake with improved flower size, petal length and width during vase life after seven days of storage as compared to cold stored (without MA) gerbera. Membrane stability index (MSI) of the petal tissue was also significantly higher of MA (PP and HDPE) cold (5°C) stored gerbera on 6th day of vase life which delayed the petal senescence and enhanced the vase life as compared to cold stored (without MA) gerbera (Patel and Singh, 2009).

The cut gerbera flowers pulsed with $AgNO_3$ 20 ppm + sucrose 5 per cent + RT + 24 h and packed with plastic sleeves with different ventilations for flower head only and whole flower scape in corrugated fibre board boxes (CFB) with newspapers (NP) and tissue papers (TP) as bedding materials, showed significant results pertaining to postharvest quality. After 4 days of storage at room temperature (RT) as prevailing in local or domestic market, the physiological loss in weight and wilted flowers were highest in the 60 per cent ventilated sleeves for flower head only and more spoiled flowers were recorded in zero per cent ventilated plastic sleeves for whole scape. Among the best treatments evaluated for vase life, the treatment plastic sleeves for whole scape + 20 per cent ventilation + CFB + TP and NP recorded highest vase life (2.71 and 2.68 days respectively) due to more water balance (2.101 and 1.972 g/f respectively) compared to all other treatments. Further, these treatments also maintained moderate levels of anthocyanins compared to other treatments (Prashanth and Chandrasekar, 2010).

11.3.2 Storage

Gerberas are not suitable for long term storage as the flowers lose 40 per cent of their vase life and 17°C is optimum for short term storage (Staby *et al.*, 1978), while they can be stored up to 8 days at 4°C temperature (Auman, 1980).

Sang *et al.* (1998) observed increased water absorption with increasing period of dry storage. Fresh weight decreased with increasing period of dry storage. Stem curvature angle after dry storage increased with increasing storage period. Aswath *et al.* (1998) stored flowers of 23 varieties of *G. hybrida* in dry storage for 24 and 48 h. These varieties were classified into three groups based on magnitude of stem bend and recovery. Water absorption is directly related to percentage inner conduit and is thus considered an important character for selecting varieties for long transportation periods. Anuradha and Gowda (1999) studied the effects of dry storage by subjecting flowers to 24 h dry storage immediately after harvesting, measuring the curvature of the stem before and after storage, and after turgor was regained when the flowers were placed in water.

Muniz *et al.* (2016) opined that, the inflorescences of gerbera *cv.* 'Intenza' maintained at 4°C, 8°C and 12°C showed a greater postharvest life, highlighting those kept at 4°C which presented less loss of mass and higher soluble sugar content; thus, maintaining the quality of gerbera for a longer period, with vase life of approximately 15 days.

11.3.3 Senescence of Flower and Vase Life

The vase life also depends on the genotype and the seasons (Eberhardt, 1982; Fischer *et al.*, 1982; Brundert and Schmidt, 1983; Shin *et al.*, 1994; Yoon *et al.*, 1996a). Cells from ligulae of inflorescences held in preservative solution maintained their ultra structure adequately with that of fresh inflorescences compared to low cytoplasm content, small mitochondria and elongated chromoplasts in water after 8 days (Burzo *et al.*, 1992; Amariutei *et al.*, 1995a, 1995b). Garibaldi and Deambrogio (1989) found that petal water content was generally higher in flowers in preservative, and it rose with time. Ion release was higher in the water control.

Put (1990) isolated a wide variety of microorganisms from freshly harvested cut flower stems and from vase water. After 10 days of vase life, fungal growth increased markedly in vase water. Van (1980) found that the increase of ion leakage (IL) from gerbera petal cells during aging of cut flowers was accelerated in isolated petals by decreasing the water potential and thereby the pressure potential of the tissue. The increase of IL was delayed by placing flower stems in KNO_3 solution to increase pressure potential. It is suggested that differences in the IL time curves between cut and intact flowers are initiated by differences in pressure potential. Van Meeteren and Van Gelder (1980) concluded that cytokinin activities were not responsible for differences in the semipermeability of petal cells of cut and intact flowers.

Proper harvesting, postharvest handling and use of floral preservatives improve the keeping quality of cut flowers of gerbera. Several preservatives have been found to be effective in prolonging the vase life. The results of experiments on vase life of gerbera are summarized in Table 2. Van *et al.* (1991) developed a computer model for calculating the potential vase life of cut flowers using various factors *viz.*, time and temperature between harvest and placement in water, dry storage or dry transport, fungal infection, bacterial infection, low temperature, ethylene, and flower preservatives in the vase water.

Table 2: Effect of Sugar and Chemicals on Vase Life of Gerbera

Varieties	Treatment	Vase Lfe (Days)	Reference
–	300 ppm 8 hydroxy-quinoline SO_4 (HQS) + 300 ppm sodium benzoate + 10^{-4} aminooxyacetic acid + 10^{-4} 3,4,5-T all in 2 per cent sucrose solution	4-5 days	Accati and Jona (1989)
Rebecca	200 mg HQS + pulsing with 80 g/l sucrose	10.5 days	Deambrogio *et al.* (1991)
–	50 per cent coconut water	14.1 days	Nair *et al.* (1998)
Campitano Dino Testarossa	Injection of 1.0 per cent $CaCl_2$ followed by dipping in 1.0 per cent $CaCl_2$	4-5 days increase over control	Gerasspoulos and Chebli (1999)
–	Injection of ACC in scapes	9-10 days	Gerasspoulos and Chebli (1998)
Beauty	2 to 3 application of 1,4 - naphtho-quinone	Low scape deformation	Yoon *et al.* (1996b)
–	40 mg/l $AgNO_3$ + 50 mg/l sodium benzoate + 30 mg/l sugars + 7.5 mg/l $Al_2(SO_4)_3$ + 2 mg/l kinetin	Extends vase life	Yildirim *et al.* (1995)
–	200 mg/l $AgNO_3$ (or) 200 mg/l 8-hydroxyquinoline citrate (HQC) with 100 g/l sucrose	2 weeks	Nowak (1989)
–	Deionized water with 200 mg/l 8-HQC + 20g/l sucrose	3 days increase	Tjia *et al.* (1987)
Richard Symphonie	20 per cent sucrose + 0.02 per cent HQS + 0.0025 per cent $AgNO_3$	8 days	Amariutei *et al.* (1986)
Peter	25mg/l AgNO3 + 500 mg/l Na_2 EDTA (hold for 6 hours) followed by distilled water	2 weeks	Nowak and Plich (1981)
–	200mg/l HQS + 50 mg/l $AgNO_3$ + 5 per cent sucrose (hold for 24 hr.)	Increase by 4-5 days	Lupi (1977)
–	0.05 per cent chlorine water changed every week	Kills bacteria	Barendse (1978)
–	Flower preservative "Proflovit – 72"	7-21 days	Rudnicki and Nowak (1976)
Alp	Chrysal solution	30 days	Brundert and
Terraflorida		9 days	Schmidt (1983)

Steinitz (1983) reported that dry weight was more than double within 11 days and fibre and lignin increased by 40-50 per cent when the flowers were kept in 6 per cent solution of sucrose immediately after harvest. Placing flowers immediately after harvest in fresh water or chrysal-VB solution improved vase life of cut gerbera flowers (Van, 1977; Schenk and Brundert, 1981). Flowers of quilled cultivar lasted longer in water containing sugar and silver nitrate (Byrne, 1973).

Javad *et al.* (2011) found that gerbera cultivars 'Tropic Blend', 'Cacharelle' and 'Aventura' had maximum vase life, while, stem bending varied from 0 to 100 per cent among different cultivars.

All the cut gerbera of *cv.* Savannah were precooled at 5°C for 6 hours and followed by pulsing with sucrose 20 per cent + sodium hypochlorite 50 ppm for 12 hours and then kept in preservative floral solutions *i.e.* neem extract, rosemary oil, calcium chloride and their combinations along with sucrose 4 per cent. Solution of neem extract 1 per cent + calcium chloride 0.1 per cent + sucrose 4 per cent significantly maintained water relations and reduced scape bending curvature. Anthocyanin content in ligules and total soluble solids in flower stalk of cut gerberas kept in neem extract 1 per cent + calcium chloride 0.1 per cent + sucrose 4 per cent solution were also highest *i.e.* 6.66 mg Congo Red/g f wt and 10.13 °Brix respectively. The cut gerbera kept in solution of neem extract 1 per cent + calcium chloride 0.1 per cent + sucrose 4 per cent solution recorded lowest optical density (0.046) which was attributed due to lowest microbial count (3.42 x 10^5 cfu/ml) and resulted in highest vase life of 13.37 days (Murthy and Subbaiah, 2020).

Stems in $AgNO_3$ solution softened, lost rigidity and bent within 1 day, while, stems in $AgNO_3$ + sucrose increased rigidity and mechanical stability and flowers remained erect up to 10 days, even after the petals wilted. Histochemical observations of stem sections showed a strong postharvest cell wall thickening and lignification of phloem cells in sucrose-treated flowers (Steinitz, 1982). Aswath *et al.* (1998) found a correlation between total sugars of the fresh stem and vase life, indicating the beneficial response to external application of sugar. Nowak (1979) correlated flower senescence in vase with a decrease of carbohydrate content in the flower and flower stalk.

Gerbera scapes sprayed preharvest with 1.0-1.5 per cent $CaCl_2$ had the longest flower longevity and the lowest bending incidence. Scape calcium concentration also increa-sed resulting in a 3 to 4 day increase in vase life and a 3 to 5 day delay in bending incidence (Gerasspoulos and Chebli, 1999).

Injecting the scapes of gerbera with ACC at harvest, then holding at 20°C for 15 days in a preservative solution reduced bending incidence, inhibited flower scape elongation, enhanced firmness of the flower scapes, advanced the maturation of stamen in disc florets and increased vase life (Gerasspoulos and Chebli, 1998). Similarly, ethephon treatment induced early maturity, short scapes, advancement of full maturity of healthy flowers (Papadopoulou *et al.*, 1996). Flowers sensitive to ethylene require a solution inhibiting ethylene emission whereas those less sensitive require a solution to enhance vase life (Jacquemont, 1985).

11.3.4 Stem Break

Since various cultivars exhibit different osmotic pressures which is related to longevity, osmotic pressure of the preservative solution would influence the stem osmotic pressure (Accati and Jona, 1989). Drooping is caused by water shortage as a result of vessel blocking due to attack by microorganisms and reducing their growth in the vase water could prevent drooping using of a preservative in the vase and by placing flowers, which had a stem medullary cavity visible at the cut surface (Van, 1977). Stem break can be prevented and longevity of cut inflorescences increased by using a preservative solution containing an antimicrobial agent, an acidifying agent, and sucrose applied either continuously or as a pulsing treatment (Abdel Kader and Rogers, 1986). Jona *et al.* (1989) reported that the preservative solutions prevented pectic decomposition, decreased ion leakage, and increased vase life from 6.7 days in the control to 12.95-17.2 days. Van (1978) investigated cut flowers of gerbera in relation to stem break, a sudden bending of the stem, which is a common disorder. The fresh weight decreased sharply 3 days before stem break occurred, and this was accompanied by a decline in absorption of water by the flowers. The petal water potential (psi) decreased in these flowers whereas it remained constant in flowers without stem break. Stem break could be prevented by pretreatment of the stems with sodium hypochlorite or silver nitrate, by adding silver nitrate or dichlorophen to the base water and by handling the stems in such a way that water could enter into the cavity of the stem. Barendse (1981) found a minimum of 4 h storage in water before marketing was necessary for gerberas and a maximum of 24 h was possible without stem damage occurring. Presence of hollow stems did not affect keeping quality (Sytsema, 1981).

Effects of salicylic acid (SA) on phenylalanine ammonia-lyase enzyme activity (PAL), as an important enzyme in phenylpropanoids pathway, consequently lignin formation and stem bending disorder of cut gerbera, with different pulse treatment of SA (0,0.25,0.5,0.75 and 1 mM) on two gerbera cultivars with different stem bending percentages was studied.'Double dutch' a resistant cultivar to stem break showed 2 times higher PAL activity than 'Ecco' with high stem bending. SA effect depending on cultivar and concentration, had significant effect on PAL, lignification and stem bending; so that, the highest enzyme activity in 'Double dutch' and 'Ecco' cultivars was observed at 0.5 and 0.25mM concentrations, respectively. Hence, SA in low concentrations with increasing stem lignification leads to decreasing stem bending disorder in both cultivars (Deljou *et al.*, 2015).

During the vase period in gerbera flowers, gibberellic acid maintains significantly more favourable fresh weight, water uptake, water loss, as compared to benzyl adenine solution. The highest values of measured parameters were found more significant with gibberellic acid at 50 ppm solution. The result revealed that the quality attributes and vase life of cut gerbera flowers were improved by the use of gibberrelic acid treatment (Sangma and Chaurasiya, 2019).

Chemical preservative 8-HQS treatment prevents the stem bending, increases fresh weight and prolongs the vase-life of gerbera flowers. Using solution of 8-HQS (20 mg/l) alone or in combination with 4 per cent sucrose, significantly increases water uptake, vase life and decreases stem bending during all shelf life periods (3rd, 6th, 9th and 12th days).The chemical solutions $CaCl_2$ (10mg/l) supplemented with 4 per cent increased fresh weight and considerably reduces the bending deviation, respectively on 9th and 12th day (Sharma *et al.*, 2018).

It was observed that the gerbera vase life was significantly prolonged in the presence of titanium ions. Titanium ions are known to reduce the entire bacterial counts as well as altered the absolute abundance of different bacterial species in the vase solution. Therefore, titanium ions could extend effectively the longevity of gerbera and possess the broad-spectrum antibacterial properties (Li *et al.*, 2019).

12.0 Biotechnology

Biotechnological research in gerbera is very advanced. Parthasarathy *et al.* (2001) reviewed the entire aspect of gerbera biotechnology. The following is the brief summary of the research on this aspect.

12.1.0 Somatic Embryogenesis

Callus initiates readily in gerbera but organ regeneration occurs only inconsistently and in small numbers. Ruffoni and Sulis (1988) reported callus formation in 16 clones of gerbera of which plantlet regeneration was reported only in 2 clones. Jerzy and Lubomski (1991) also recorded very poor shoot regeneration from callus. Ruffoni and Massabo (1991) cultured apex, petiole and leaf explants on MS medium with 27 combinations of plant growth factors. Apex explants produced most plants, followed by petiole explants. Axillary bud cultures on MS with 5-10 µM BA induced limited amount of callus in *G. aurantiaca* (Meyer and Van-Staden, 1988). From the proximal end of the main leaf vein bud initiation was achieved after a phase of callogenesis in modified MS media containing BA and NAA (Reynoird *et al.*, 1993).

Pale green and friable callus was obtained from emerging leaves and brown callus when fully expanded leaf used as an explant. The younger leaves from the middle of rosette have more potential to regenerate and it gradually decreased with the development of the leaf. Parthasarathy *et al.* (1997) obtained embryogenic callus from leaf explants cultured on MS medium supplemented with BAP 0.75 mg/l, NAA 1 mg/l and IBA 0.75 mg/l, while absence of BAP in the medium resulted in only rooting.

Mohanty *et al.* (2005) standardized protocol for shoot proliferation and somatic embryogenesis in gerbera *cv.* Alsmeera. Explants like shoot tips, leaf discs and petioles were cultured on MS medium with varied combinations of phytohormones. Further, the calli were subjected to basal and half-strength MS media and exposed to X-ray radiation for induction of somatic embryos. MS medium supplemented with BAP (1 ppm) was found to the best for shoot proliferation, whereas, MS medium fortified with 2,4-D (5 ppm) recorded maximum callusing from leaf explants, kept in diffused light conditions.MS medium supplemented with kinetin (0.5 ppm), GA (0.5 ppm) and NAA (0.5 ppm) and exposure to X-ray irradiation (1–2 KVP) for 2 sec recorded highest percentage of somatic embroyos.

Cell suspension cultures were established from embryogenic callus induced from leaf explants of *Gerbera jamesonii.* Embryogenic callus was induced when leaf explants were cultured on MS medium containing 1.0 to 2.0 mg/l 2,4-D. Cream friable callus was formed within two weeks. Proliferated callus was transferred to MS liquid medium containing 2,4-D with a small concentration of NAA and subcultured at 2 weeks interval. Induction of somatic embryos (globular, heart and torpedo) was observed after 2 weeks of culture. Somatic embryos developed in MS suspension medium containing 1.0 to 2.0 mg/l 2,4-D with 0.1 or 1.0 mg/l NAA and globular embryos were further differentiated into the cotyledonary phase embryos. The addition of amino acids (L-glutamine or L-proline, 5.0 mg/l, respectively) to the culture media, in the range of concentrations tested, yielded higher enhancement of the embryo growth and development. Transfer of individual embryos onto a freshbasal medium with no plant growth regulators was able to achieve complete maturation. Relatively, only a few number of embryos developed shoots and roots when transferred to MS medium supplemented with 2.0 mg/l BAP and 0.5 mg/l NAA in addition to 3 per cent (w/v) sucrose and 0.8 per cent (w/v) agar containing medium. About 11 per cent of somatic embryos were converted to true-to-type fertile plants (Hasbullah *et al.*, 2011).

Mature embryos were used as explant sources and cultured on MS medium amended with diverse auxins and cytokinins in varying concentrations as sole as well as in different combinations ofculture media (MS + 3.0 mg/l 2,4D + 0.5 mg/l BA + 30.0 g/1 sucrose + 7.5 g/l agar powder) proved well for callus initiation. Culture media (MS + 2.0/3.0mg/l NAA + 0.5 mg/l 2-ip + 30.0 g/l sucrose + 7.5 g/l agar) exhibited higher response *i.e.* number of shoot proliferating explants and number of shoot(s) per explant. However, shoot of higher length was recorded on nutrient medium (MS + 2.0/3.0 mg/l BA + 30.0 g/l sucrose + 7.5 g/l agar). Higher rooting response (root proliferating efficiency, number of roots and mean root length) was exhibited by rooting medium (MS + 0.1 mg/l IBA + 15g/l sucrose + 7.5 g/1 agar). Phenotypic normal plants were recovered when transferred to pots (Bhatt *et al.*, 2016).

Nodular, greenish brown callus obtained from leaf sections when cultured on MS basal medium supplemented with 2.0 mg/l 2,4-D. Somatic embryo maturation and shoot regeneration was obtained in MS medium supplemented with 2.0 mg/l BAP and 3.0 mg/l IAA. Multiple shoot proliferation from the regenerated shoots was obtained under the influence of BAP (2.0 mg/l) and Kn (3.0 mg/l) followed by root induction using Indole-3-acetic acid (IAA 3.0 mg/l) (Monica and Vandita, 2017).

12.2.0 Molecular Studies

The ability to transform gerbera genetically is one of the major advantages of using this species as a new model system. Thus, gerbera is a powerful system for functional analysis of the general regulation of various aspects of flower development, organ differentiation and metabolism (Kotilainen *et al.*, 1999). Molecular studies of floral organ differentiation in compositae were investigated by isolating characterizing corolla-abundant genes (Kotilainen *et al.*, 1994). Organ identity genes and modified patterns of flower development using newly isolated *MADS* box genes were reported by Yu *et al.* (1999). Elomaa *et al.* (1998) studied the control of floral organ differentiation and the regulation of flavonoid biosynthesis at molecular level. The biosynthetic pathway to anthocyanins, flavones and flavonols is well known. The gerbera chalcone synthase gene family consists of 7-10 members and Elomaa *et al.* (1993) isolated two non-allelic cDNA in a clone encoding gerbera chalcone synthase (*gchs 1*, *gchs 3* and one cDNA encoding gerbera dihydroflavonol-4-reductase), and found that in *cv.* Terra Regina expression of *gchs 3* is detectable in sepals and petals. They also deve-loped a regeneration method and an *Agrobacterium*-mediated genetic transformation using petiole explants of cultivar Terra Regina. Helariutta *et al.* (1993) studied the regulation of flavonoid pathway genes during inflorescence development in gerbera and cloned a *dfr* cDNA from a ray floret corolla cDNA library of *Gerbera hybrida cv.* Regina by a PCR technique. Chemogenetic investigations of gerbera flowers by Martens *et al.* (2000) indicated two blocks at different steps of the pathway leading to acyanic flowers. One block most probably concerns flavanone 3-hydroxylase (FHT) and the other dihydroflavonol 4-reductase (DFR). The formation of flavones is assumed to be catalyzed by flavone synthase II (FNS II). Chemogenetic investigations indicated that this step is controlled by a gene named *Fns*. Up to now, no mutant was found for regulation of the enzyme activity of flavonol synthase (FLS). The formation of cyanidin instead of pelargonidin depends on the activity of the enzyme flavonoid 3´-hydroxylase (F3´H). This hydroxylation reaction is assumed to be controlled by the gene named *F3´h*. Surprisingly, in some lines the formation of cyanidin was found to be suppressed by an as yet not identified dominant factor. Helariuta *et al.* (1995a) studied the expression of genes encoding (naringenin) chalcone synthase (*chs*) and dihydroflavonol-4-reductase (*dfr*) (dihydrokaempferol-4-reductase) in five gerbera varieties with spatially restricted anthocyanin pigmentation patterns. Elomaa and Holton (1994) reviewed the modification of flower colour using genetic engineering.

During the commercial production of gerbera seedlings, flower color is one of the main morphological aspects that have an agronomic interest and becoming an important feature in genetic breeding programs. The use of molecular markers may serve to direct crossings, new hybrids and mutants, besides confirm and identify new genotypes for commercial purposes. In that context, this work aimed to analyze the genetic divergence among six cultivars of *Gerbera jamesonii* ('Jaguar Yellow', 'Jaguar Cream', 'Jaguar Lemon', 'Jaguar Salmon Pastel', 'Jaguar Red', 'Jaguar Deep Rose'). The genetic divergence among cultivars of gerbera was carried out with 21 primers, which amplified 37 DNA polymorphic fragments, used to estimate the Jaccard index and presented an average of 0,38, ranging from 0,28 to 0,56. The genetic structure among cultivars was estimated by UPGMA and revealed two distinct groups, at 38 per cent genetic similarity. The largest genetic similarity found (56 per cent) was between cultivars 'Jaguar Yellow' and 'Jaguar Lemon'. The results showed that the RAPD is a fast, relatively inexpensive and useful technique for genetic divergence characterization between different cultivars of *Gerbera jamesonii* Rezende *et al.* (2009).

Rashmi *et al.* (2018) assessed the genetic fidelity of *in vitro* raised gerbera plants multiplied through micropropagation upto fifteen *in vitro* subcultures by using inter-simple sequence repeat (ISSR) markers. 40 ISSR primers screened, 20 ISSR primers produced a total of 54 ISSRs clear, distinct and reproducible amplicons, which were monomorphic across all micropropagated plants studied. All banding profiles from micropropagated plants were monomorphic and similar to those of the mother plant. These results indicate that the micropropagation protocol developed for rapid *in vitro* multiplication is appropriate and suitable for clonal propagation gerbera.

The transcriptome sequences provide a valuable resource in revealing the molecular mechanism underlying stem bending of cut flower and offer novel genes that can be used to guide future studies for ornamental plant breeding. Ge *et al.* (2019) studied high-throughput transcriptome sequencing of gerbera during stem bending using the Illumina sequencing technology. Three cDNA libraries constructed from mRNAs of gerbera stem at stem bending stage 0, 2 and 4 were sequenced. More than 300 million high quality reads were generated and assembled into 96,492 unigenes. Among them, 34,166 unigenes were functionally annotated based on similarity search with known protein. Sequences derived from plants at different stem bending stages were mapped to the assembled transcriptome, and 9,406 differentially expressed genes (DEGs) were identified. Through Kyoto Encyclopedia of Genes and Genomes (KEGG) enrichment analysis, specific pathways were identified during the stem bending process, such as phenylpropanoid biosynthesis pathway, phenylalanine metabolism pathway, starch and sucrose metabolism pathway, and plant hormone signal transduction pathway. A total of 211 transcription factors (TFs), including TF families involved in plant senescence, such as NAC, MYB, WRKY, and AP2/ERF members, as well as TFs related to water stress tolerance, were shown to be regulated during stem bending. Gene Onotology (GO) functional enrichment analysis indicated that key genes involved in responses to osmotic and oxidative stresses were also varied in expression during this process. Furthermore, analysis of DEGs involved in the hormone signaling pathways and determination of endogenous abscisic acid (ABA) content showed that stem bending may be an ethylene-independent process, but regulated by ABA. In short, the findings suggests that the stem bending of cut gerbera may be caused by the involvement of water stress and regulation of ABA during the postharvest life.

12.3.0 Genetic Transformation

Like any ornamental plant, gerberas are produced exclusively for their aesthetic values. Thus the improvement of quality attributes such as flower colour, longevity and form, plant shape and architecture and the creation of novel variation are important economic goals. Conventional breeding of gerbera by crossing and selection has generated several clonally propagated elite genotypes that have desirable traits, such as colour, shape, vase life and resistance against pests and diseases. However, one of the disadvantages of traditional breeding is the limited genepool of any single species. An important application of genetic engineering to the ornamental plant is the possibility of changing flower colour. Flavanoids, which are common flower pigments, are synthesized via the phenylpropanoid pathway. Cloning and transferring into plants the genes affecting flavonoid synthesis has led to alter flower colour or patterns.

Zuker *et al.* (1998) and Elomaa and Teeri (2001) reviewed genetic transformation in gerbera. Orlikowska and Nowak (1997) found that young leaf petiole was not efficient enough to enable successful transformation. Nagaraju *et al.* (1998) reported *Agrobacterium*-mediated genetic transformation protocol for *G. hybrida* using petiole, leaf and shoot tip explants. Genetic transformation depends upon regeneration efficiency. Studies by Reynoird *et al.* (2000) have revealed differences in bud regeneration capacity from *G. hybrida* leaf explants excised from various genotypes for transformation using *A. tumefaciens* strain 82139.

Plants of four gerbera cultivars transformed with nucleocapsid N-gene of Tomato Spotted Wilt Virus were evaluated in terms of resistance to the virus and several phenotypical traits. Sixteen out of 33 transformed genotypes (with transgenic plant status confirmed by PCR with specific primers for N and npt II genes) survived when transferred to the greenhouse. After mechanical inoculation with TSWV, typical symptoms of viral infection appeared in the control plants after two to four weeks. No disease symptoms were observed at that time in any of the infected transgenic plants. Assessment of other phenotypical traits of gerbera confirmed lack of significant differences between transformed and control plants. Except for one genotype of 'Prince' and one genotype of 'Zuzanna', all of the transformed plants can be potentially good breeding material (Korbin, 2006).

Stem bending is a major disorder in cut gerbera flowers caused mainly by the lack of mechanical support in the flower stem. Lignin is the component of plant cell walls that provides strength. Three major enzymes comprise the pathway of lignin anabolism including peroxidase (POD), cinnamyl alcohol dehydrogenase (CAD), and cinnamoyl-CoA reductase (CCR). In this study, three overexpressed mutants for each of these major genes were generated and analyzed for their role in stem bending. Comparing the pod transgenic plants with ccr and cad mutants indicated that pod plants had the highest lignin content. Compared to wild-type (WT) plants, the highest vase life, membrane stability and solution uptake were obtained in gerbera pod mutants. Therefore, it appears that higher lignin contents can decrease stem bending and improve cell wall rigidity of cut gerbera flowers. These results also suggest that POD is a key gene in the biosynthesis of lignin in gerbera (Hamedan *et al.*, 2019).

13.0 References

Abdel-Kader, H. and Rogers, M.N. (1986) *Acta Hortic.*, **181**: 169-176.

Accati, E.G. and Jona, R. (1989) *Acta Hortic.*, 261 : 63-68.

Ahmim, M and Vieth, J. (1986) *Can. J. Bot.*, **64** : 2355-2357.

Akter, N., Hoque, M. and Sarker, R. (2012) *Plant Tissue Cult. and Biotech.*, **22** : 143-152.

Albert, R. (1997) *Nachrichtenblatt des Deutschen Pflanzenschutzdienstes*, **49** : 137-142.

Amariutei, A., Craciun, C., Burzo, I. and Fjeld, T. (1995a) *Acta Hortic.*, **405** : 108-116.

Amariutei, A., Alexe, C., Burzo, I. and Fjeld, T. (1995b) *Acta Hortic.*, **405** : 372-380.

Amariutei, A., Burzo, I. and Alexe, C. (1986) *Acta Hortic.*, **181**: 331-337.

Amin, N., Sajid, M., Qayyum, M.M., Shah, S.T., Wahid, F. and Hashmi, R.S. (2015) *Int. J. Biosci.*, **7** : 1-11.

Anonymous (1968) *Adv. Leafl. Minist. Agric.*, London, No. 86 : 5.

Anuradha, S. and Gowda, J.V.N. (1999) *Mysore J. Agric. Sci.*, **33** : 312-316.

Arello,E.F., Pasqual, M., Pinto, J.E.B.P. and Barbosa, M.H.P. (1991) *Pesquisa Agropecuaria Brasileira*, **26** : 269-273.

Asen, S. (1984) *Phytochem.*, **23** : 2523-2526.

Aswath, C., Parthasarathy, V.A. and Bhowmik, G. (1995) *J. Orn. Hort.*, **3** : 59-64.

Aswath, C., Parthasarathy, V.A. and Bhowmik, G. (1996) *Crop Res.*, **12** : 349-352.

Aswath, C., Parthasarathy, V.A. and Bhowmik, G. (1998) *J.Maharashtra Agrl. Univ.*, **23** : 140-142.

Aswath, C., Parthasarathy, V.A. and Bhowmik, G. (1998a) *J. Orn. Hort.*, **1**: 55-60.

Auman, C.W. (1980) In : *Introduction to Floriculture* (ed. R.A. Larson), Academic Press, New York.

Autio, J. (2000) *Acta Hortic.*, **515** : 91-98.

Ayemi,T.J., Singh, D. and Fatmi, U. (2017) *Int. J. Curr. Microbiol. App. Sci.*, **6** : 1049-1056.

Baas, R., Nijssen, H.M.C., Van den, B.T.J.M. and Warmenhoven, M.G. (1995) *Scientia Hort.*, **61** : 273-284.

Baillot, F. (1976) *Horticulture Francaise*, **72** : 9-6.

Baiswar, P., Chandra, S., Kumar, R. and Ngachan, S.V. (2010) *Plant Pathol.*, **59** : 802. doi: https: //doi.org/10.1111/j.1365-3059.2009.02240.x.

Baker, J.R. (1987) *North Carolina Flower Growers' Bull.*, **31**: 3,47.

Bappadocia, M., Chretien, L and Laublin, G. (1988) *Can. J. Bot.*, **66** : 1107-1110.

Baranowski, T. and Blaszak, D. (1996) *Roczniki Nauk Rolnic Zych.*, **25** : 59-61.

Barendse, L.V.J. (1978) *Vakbladvoor de Bloemisterij.*, **33** : 48,25.

Barendse, L. (1981) *Vakbladvoor de Bloemisterij.*, **36** : 29.

Bhatt, D., Tripathi, M.K., Sankar, M.V., Patel, R.P. and Jatav, R. (2016) *Int. J. Agric. Sci.*, **8** : 2542-2547.

Bigre, J.P. (1980) *Revue Horticole*, **205** : 13-15.

Blakesley, D., Lenton, J.R and Horgan, R. (1986) *Int. Congr. Plant Tiss. Cell Cult.*, 6 Meet., pp. 375.

Blomme, R. and Dambre, P. (1979) *Verbondsnieuws voor de Belgische Sierteelt.*, **23** : 175-178.

Blomme, R. and Dambre, P. (1980) *Verbondsnieuwsvoor de Belgische Sierteelt.*, **24** : 551-554.

Blomme, R. and Dambre, P. (1982) *Verbondsnieuwsvoor de Belgische Sierteelt.*, **26**: 195-199.

Blomme, R. and Dambre, P. (1984) *Verbondsnieuwsvoor de Belgische Sierteelt.*, **28** : 684-687.

Bocion, P.F., Silva W.H. de., Walter, H. and Graf, H.R. (1977) *Gb-+-Gw.*, **77** : 634-636.

Booy, G. (1994) *Plant Var. Seeds,* **7** : 183-191.

Booy, G. (1995) *J. Hort. Sci.*, **70** : 135-146.

Borghi, B. and Baldi, V. (1970) *Sementielette,* **6**.

Bremer K. (1994) In : *Asteraceae: cladistics and classification.* Timber Press : Portland, Oregon.

Brickell (1996) *A-Z Encyclopedia of Garden Plants*, Dorling Kinder Sley Limited, London, p. 467.

Broer, S. (1968) *Vakbladvoor de Bloemisterii,* **23** : 853.

Bruckner, U. and Elgner, N. (1989) *Deutscher Gartenbau*, **43** : 280-284.

Brundert, W. and Schmidt, K. (1983) *Gb-+-Gw.*, **83** : 1270-1271.

Buivids, K., Ozolinya, L., Muceniece, G. and Marga, G. (1992) *Eksperimentine Biologija*, **3-4** : 61-62

Bulthuis, J. (1979) *Vakbladvoor de Bloemisterij*, **34** : 31-33.

Burzo, I., Amariutei, A. and Craciun, C. (1992) *Lucrari Stiintific eInstitutul Agronomic Nicolae Balcescu BucurestiSeria B Hortic.*, **35** : 103-114.

Buwalda, F., Eveleens, B. and Wertwijn, R. (2000) *Acta Hortic.*, **515** : 141-149.

Byrne, T., Drennan, D. and Harding, J. (1987) *J. Amer. Soc. Hort. Sci.,* **112** : 724-727.

Byrne, T.G. (1973) Flower Nursery Rep., November, 10-11.

Cagnet-Sitbon, M. (1981) *Agronomie,* **1** : 807-812.

Calabretta, C. and Nucifora, A. (1985) *Accademia Nazionale Italiano di Entomologia*, pp. 807-813.

Cantor, M. (1993) *Buletinul Universitatii de Stiinte Clujapoca. Seria Agricultura siHorticultura*, **47** : 81-89.

Cardoso, R.D.L., Grando, M.F., Basso, S.M.S., Segeren, M.I., Augustin, L., Suzin, M. (2009) *Hortic. Bras.*, **27** : 40-44.

Charpentier, S., Laffaire, M., Riviere, L.M. and Vidalie, H. (1986) *Revue Horticole*, **271**: 47-54.

Chung, Y.M. and Lee, J.S. (2019) *Korean J. Hortic. Sci.,* **37** : 404-414.

Codd, L.E. (1979) *Veld and Flora* (December), 114-115.

Couto, T.R.D and Araujo, J.S.D.P (2018) *Rev. Agr. Acad,* **1** : 52-63.

Cresswell, G.C., Macdonald, J.A. and Allender, W.J. (1994) *Pesticide. Sci.*, **42** : 13-16.

Danhardt, W. (1966) *Arch. Gartenb.*, **14** : 17-24

Darwish, M.R., El Awar, F.A., Sharara, M. and Hamdar, B. (1999) *Applied-Engineering in Agrl.,***15**: 41-48.

David, H. (1977) *Gartenbau*, **24** : 370-372.

Deambrogio, F., Dolci, M. and Accati, E. (1991) *Advances Hort. Sci.,* **5** : 135-138.

Deljou, M.J.N., Khalighi, A., Arab, M., Karamian, R. and Hamedan, H.J. (2015) *Int. J. Hort. Sci. Tech.*, **46** : 279-290.

Dengler, R. (1984) *Dtsche Gartenb.*, **38** : 1005-1006.

Dirlbek, J., Dirlbekova, O., Veldova, I. and Dobrovodsky, I. (1989) *Sbornik-UVTIZ,-Ochrana-Rostlin*, **25** : 289-298.

Dixit, S. (2016)https: / /blog.nurserylive.com / 2016 / 09 / 15 / do-you-know-gerberas-are-fifth-most-popular-cut-flowers-in-the-world-and-gardening-in-india.

Drennan, D., Harding, J. and Byrne, T.G. (1986) *Euphytica*, **35** : 319-330.

Dubuc Lebreux, M. A. and Vieth, J. (1985) *Acta Botanica Neerlandica*, **34** : 171-182.

Eberhardt, M. (1982) *Gartenbau*, **29** : 314-316.

Elomaa, P. and Holton, T. (1994) *Biotechnol. and Genetic Engineering Rev.*,**12** : 63-87.

Elomaa, P. and Teeri,T.H. (2001) In : *Biotechnology in Agriculture and Forestry* (ed. Y.P.S. Bajaj), Transgenic crops, **48** : 139-154.

Elomaa, P., Mehto, M., Kotilainen, M., Helariutta, Y., Nevalainen, L. and Teeri, T.H. (1998) *Plant J.*, **16** : 93-99.

Elomaa, P., Honkanen, J., Puska, R., Seppanen, P., Helariutta, Y and Mehto, M. (1993) *BioTech.*, **11** : 508-511.

El-Shafie, S.A. and Hassan, H.A. (1978) *Arch. Gartenb.*, **26** : 333-342.

Farajollahzadeh, Z., Hadavi, E. and Mirkohi, A.K. (2014) *Acta Hortic.*,**1037** : 1075-1082.

Filella, I. and Penuelas, J. (1994) *Intnatl. J. Remote Sensing.*, **15**: 1459-1470.

Fischer, P., Forchthammer, L. and Kalthoff, F. (1982) *Deutscher-Gartenbau*, **36** : 138-141.

Fischer, P. and Forchthammer, L (1983) *Deutscher Gartenbau*, **37** : 1792-1794.

Fischer, P. and Forchthammer, L (1984) *Deutscher Gartenbau*, **38** : 2176-2179.

Fischer, P. and Forchthammer, L. (1987) *Deutscher Gartenbau*, **41** : 674-678.

Folk, G., Tusnadi, C.K. and Bruckner, A. (1984) *Kerteszet es Szoleszet*, **33** : 7.

Franceschetti, U. (1971) *Coltivatore e Giornale Vinicolo*, Italiano, **1-2** : 8.

Franceschetti, U. (1973) *Sementi-Elette.*, **19** : 7-10.

Gardner, U. (1975) *Dtsche Gartenb.*, **29** : 1643.

Garibaldi, A. (1968) *Phytopath. Medit.*, **7** : 91-92.

Garibaldi, A. (1968) *Rivistadella Ortoflorforutticoltura Italiana*, **58** : 108-120.

Garibaldi, E.A. and Deambrogio, F. (1989) *Informatore Agrario.*, **45** : 103-106.

Garthwaite, J.M. (1959) *Exp. Hort.*, **2** : 12-15.

Garthwaite, J.M. (1960a) *Exp. Hort.*, **3** : 65-68.

Garthwaite, J.M. (1960b) *Exp. Hort.*, **8** : 114.

Ge, Y., Lai, Q., Luo, P., Liu, X. and Chen, W. (2019) *BMC Genomics*,**20** : 600.

Gerasopoulos, D. and Chebli, B. (1998) *J. Amer. Soc. Hort. Sci.,* **123** : 921-924.

Gerasopoulos, D. and Chebli, B. (1999) *J. Hortic. Sci. Biotechnol.,* **74** : 78-81.

Gergely, I. (1988) *Kerteszet es Szoleszet,* **37**: 4.

Ghani, M. and Sharma, S.K. (2019) *Physiol. Mol. Biol. Plants.,* **25** : 159-166.

Ghani, M., Kumar, S and Thakur, M (2013) *J. Hortic. Sci. Biotechnol.,* **88** : 585-590.

Ghani, M., Kumar, S. and Thakur, M. (2014) *J. Hortic. Sci. Biotechnol.,* **89** : 301-306.

Ghimiray, T.S. and Sarkar, I. (2015) *Int. J. Biores Sci.,* **2** : 81-83.

Giri, B. and Beura, S. (2020) *Int. J. Curr. Microbiol. App. Sci.,***9** : 1680-1688.

Gotz, W (1983) *DeutscherGartenbau,* **37** : 1898-1900.

Grabowska, B., Jarosz, B. and Rudnicka, L. (1979) *PraceInstytutu Sadownictwai Kwiaciarstwa w Skierniewicach,* **4** : 7-18.

Graf, W. and Kress, O. (1996) TASPO-Gartenbau, 5 : 8-10.

Guba, W. (1994) *J. Fruit Orn. Plant Res.,* **2**: 143-155.

Gutmane, L. and Grivina, D. (1975) *Izvestiya AN LatvSSR,* **11** : 22-26.

Hakkaart, F.A. (1968) *Neth. J. Plant Path.,* **74** : 28-29.

Hamedan, H.J., Sohani, M.M., Aalami, A. and Nazarideljou, M.J. (2019) *Sci. Hortic.,* **245** : 274-279.

Hanger, B.C (1984) *Combined Proc. Intnl. Plant Propagators Soc.,* **34** : 124-134.

Harding, J., Byrne, T. and Nelson, R.L. (1981) *Euphytica,* **30** : 653-657.

Harding, J., Huang, H.Z. and Byrne, T. (1996) *Euphytica,* **88** : 55-60.

Harris, E.M. (1995) *Botanical Rev.,* **61** : 93-278.

Hasbullah, N.A., Saleh, A. and Taha, R.M. (2011) *Afr. J. Biotechnol.,* **10** : 13762-13768.

Hedtrich, C.M. (1979) *Gartenbauwissenschaft,* **44** : 1-3.

Helariutta, Y., Elomaa, P., Kotilainen, M., Griesbach, R.J., Schroder, J. and Teeri, T.M. (1995a) *Plant Molecular Biol.,* **28** : 47-60.

Helariutta, Y., Elomaa, P., Kotilainen, M., Seppanen, P andTeeri, T.M. (1993) *Plant Molecular Biol.,* **22** : 183-193.

Hell, T.B. and Ter, H.B. (1996) *Nachrichtenblatt des Deutschen Pflanzenschutzdienstes,* **27** : 150-152.

Hell, T.B. and Ter, H.B. (1996) *TASPO Gartenbau,* 5 : 13-14.

Hermann, P. (1978) *Erwerbsgartner,* **27** : 2260-2261.

Hoffmann, G.M. and Menzinger, W. (1963) *Gartenwelt,* **63**: 236-237.

Hong, K. J., Han, M.J., Kim, I.S., Ahn, S.B. and Lee, M.H. (1996) *RDA J. Agric. Sci. Crop Protn.,* **38** : 539-544.

Hoogstrate, J.G. (1982) *Vakbladvoor de Bloemisterij,* **37** : 23

Huang, M. and Chu, C.Y. (1985) *J. Japanese Soc. Hort. Sci.,* **54** : 94-100.

Hwang, K.S. and Yoon, J.H. (1994) *RDA J. Agric. Sci. Soil Fert.*, **36** : 268-272.

Ibrahim, M.A. and Yassin, M.M. (2020) *Plant Cell Biotechnol. Mol. Biol.*, **21** : 1-9.

Imtiyaz, Nazki, T., Siddique, Rather, Z.A. and Nelofar. (2014) *Indian J. Hortic.*, **71** : 77-81.

Islam, M.M., Mamun, A.A., Dash,P.K., Kundu,R.R. and Akter. (2017) *Int. J. Curr. Res.*, **9** : 55326-55329.

Jacquemont, R. (1985) *C.R.Academied'Agriculture de France.*, **71**: 1219-1223.

Jain, S.M., Vitti, D., Tucci, M., Grassotti, A., Rugini, E. and Saccardo, F. (1998) *Adv. Hort. Sci.*, **12** : 47-53.

Javad, N.D.M., Ahmad, K., Mostafa, A. and Roya, K. (2011) *Afr. J. Biotechnol.*, **10** : 560-566.

Jefferson, R.N. (1952) *J. Econ. Ent.*, **45** : 137-138.

Jerzy, M. and Lubomski, M. (1992) *Acta Hortic.*, **349** : 269-274.

Jerzy, M. and Lubomski, M. (1991) *Scientia Hort.*, **47** : 115-124.

Jesiotr, E., Strojny, Z. and Saniewski, M. (1975) *Acta Hortic.*, **51** : 103-106.

Jona, R., Fronda, A., de, A.F., Accati, G.E. and De, A.F. (1989) *Acta Hortic.*, **246** : 313-318.

Jones, V.P., Parrella, M.P., Hodel, D.R. (1986) *California Agric.*, **40** : 10-12.

Jong, J De (1986) *Acta Hortic.*, **181** : 353-735.

Junges, W. (1962) *Arch. Gartenb.*, **10** : 434-444

Junges, W. (1964) *Arch. Gartenb.*, **12** : 469-482.

Kacperska, I. (1985) *Prace Instytutu Sadownictwai Kwiaciarstwa w Skierniewicach B Rosliny Ozdobne*, **10** : 105-114.

Kaewruang, W., Sivasithamparam, K. and Hardy, G.E. (1989) *Biol. Fert. Soils.*, **8** : 38-47.

Kamel, H.A., Bispara, A.P., Ibrapim, A.A. and Nada, M.A.K (1975) *Agrl. Res. Rev.*, **53** : 97-103.

Kamel, H.A., Ibrapim, A.A., Bispara, A.P. and Nada, M.A.K (1977) *Technical Bull. Qubba Bot. Garden.*, **1** : 78.

Kaminek, M., Vanek, T., Kalendova, K.A. and Pilar, J. (1987) *Scientia Hort.*, **33** : 281-289.

Kataeva, N. V., Alexandrova, I.G., Butenko, R.G and Dragavtceva, E.V. (1991) *Plant Cell Tiss. Organ Cult.*, **27** : 149-154.

Kerssies, A. (1993) *Plant Pathol.*, **42** : 754-762.

Kharrazi, M., Sharifi, A., Zadeh, S.N., Khadem, A., Moradian, M. (2019) *J. Agric. Sci. Sust. Produc.*, **9** : 277-291.

Klossowski, W and Strojny, Z. (1983) *Prace Instytutu Sadownictwai Kwiaciarstwa w Skierniewicach B Rosliny Ozdobne*, **8** : 111-121.

Koek, P.C. (1957) *Jversl. Proefst. Bloem. Aalsmeer*, pp. 49-52.

Koek, P.C. (1958) *Tijdschr. Plzeikt.*, **64** : 121.

Korbin, M. (2006) *J. Fruit Ornam. Plant Res.*, **14** : 85-93.

Korner, H.J. (1983) *Nachrichtenblatt fur den Pflanzenschutz in der DDR*, **37** : 8-11.

Kotilainen, M., Helariutta, Y., Elomaa, P., Paulin, L. and Kerssies, A. (1994) *Plant Pathol.*, **42** : 754-762.

Kotilainen, M., Albert, V.A., Elomaa, P., Koskela, S., Mehto, M., Pollanen, E., Yu, D. and Teeri, T.H. (1999) *Flowering Newsl.*, **28** : 20-31.

Kotilainen, M., Helariutta, Y., Elomaa, P., Paulin, L. and Teeri, T.H. (1994) *Plant Mol. Biol.*, **26** : 971-978.

Kreij, C. (1989) *Vakbladvoor-de Bloemisterij*, **44** : 46-47.

Krober, H. and Plate, H.P. (1975) *Nachrichtenblatt des Dentschen Pflanzenschutzdienstes*, **27** : 150-152.

Kulibaba, Y.F. (1972) *Byulleten Glavnogo Botanicheskogo*, **8** : 115-119.

Kulkarni, K.V., Thawal, D.W., Kharbade, S.B., Shaikh, A.A. and Jagtap, K.B. (2017) *Int. J. Agric. Econ. Manag.*, **7** : 1-6.

Kumar R. and Yadav, D.S. (2005a) *J. OrnamHortic.*, **8**(3) : 212-215.

Kumar, R. (2013) *Asian J. Hortic.*, **8**(2) : 680-682.

Kumar, R. (2013) *J. Hortic. Sci.*, **8** : 111-113.

Kumar, R. and Yadav, D.S. (2005) *J. Ornam Hortic.*, **8**(4) : 275-277.

Kumar, R. and Yadav, D.S. (2013) *Hort Flora Res. Spect.*, **2** : 244-246.

Kumar, R., Deka B.C., War, M. and Venugopalan, R. (2012) *Indian J. Agric. Sci.*, **82**(6) : 543–547.

Kumar, R., Deka, B.C. and Venugopalan, R. (2012) *Indian J. Agric. Sci.*, **82**(7): 615–619.

Kumar, S., Krishna, S.T. and Shakywar, R.C. (2013) *Ann. Hortic.*, **6** : 158.

Kuotsu, N.R., Keditsu, R. and Hemanta, L. (2018) *Int J. Curr. Microbiol. Appl. Sci.*, **7** : 1537-1541.

Labanowski, G. (1990) *Materialy Sesji Instytutu Ochrony Roslin*, **30** : 121-128.

Lacasa, A., Gonzalez, A., Martinez, M.C., Torres, J. and Fernandez, J. (1993) *Boletin-de-Sanidad-Vegetal,-Plagas*, **19** : 193-209.

Laermann, H.T. (1980) *Gesunde Pflanzen*, **32** : 225-230.

Laneri, U., Franconi, R and Altavista, P. (1990) *Acta Hortic.*,**280** : 395–402.

Laurie, A., Kiplinger, D.C. and Nelson, K.S. (1968) In : *Commercial Flower Forcing*, McGraw Hill Book Co., New York.

Leffring, L. (1968) *Vakbladvoor de Bloemisterij*, **23** : 1375.

Leffring, L. (1973) *Scientia Hort.*, **1** : 221-229.

Leffring, L. (1975) *Acta Hortic.*, **51** : 263-265.

Leffring, L. (1981) *Vakbladvoor de Bloemesterij*, **36** : 108-109.

Lenzi, A., Oggiano, N., Rinaldi, M. and Maletta, M. (2000) *Special issue: Fiori. Colture-Protette*,**29** : 85-91.

Leski, B. (1984) *Annals Warsaw Agril. Univ. SGGW-AR, Hort.*, **12**: 47-51.

Lhoste, A. (1998) *PHM Revue Horticole.*, **396** : 28-29.

Li, C.X., Fan, Y.F., Luan, W., Dai, Y., Wang, M.X., Wei, C.M., Wang, Y., Tao, X., Mao, P. and Ma, X.R. (2019) *Microb Ecol.*, **77** : 967-979.

Li, F., Cheng, Y., Zhao, X., Yu, R., Li, H., Wang, L., Li, S. and Shan, Q. (2020) *Sci. Rep.*, **10** : 1702.

Li, L., Zhang, W., Na, Z.L.L., Yaqin, J.P., Wang., Zhong, C., Yang, Y., Sun, S., Liang, S. and Wang, X. (2015) *Front. Plant Sci.*, **6** : 168.

Lin, W.C. and French, C.J. (1985) *HortScience*, **20**: 271-273.

Lindstorm, R.S. (1957) *Quart. Bull. Mich. Agric. Stat.*, **39** : 673-681.

Lisiecka, A. (1976a) *Zesz. Probl. PostepowNauk Rol.*, **179** : 323-326.

Lisiecka, A. (1976b) *Roczniki Akademii Rolniczej w Poznaniu, Ogrodnictwo*, **85** : 87-93.

Loeser, H. (1986) *Dtsche Gartenb.*, **40** : 1461-1464.

Loeser, H. and Deiser, E. (1984) *Deutscher Gartenbau*, 38 : 2284-2288.

Loeters, J.W. (1968) *Meded. Dir. Tuinb.*, **31** : 487-96.

Ludwig, F., Fernandes, D.M., Guerrero, A.C. and Bôas, R.LV. (2015) *Ornam. Hortic*, **21** : 177-184.

Ludwig, Fernanda, Fernandes, D., M., Guerrero, A., C. and Boas, R.L.V. (2013) *Hortic. Bras.*, **31** : 622-627.

Lupi, R., Accati, E.gG and Schiva, T. (1977) *Annali dell' Istituto Sperimentale per la Floricoltura*, **8** : 1-13.

Mahanta, P., Choudhury, S., Paswan, L. and Talukdar, M.C. (1998) *South Indian Hort.*, **46** : 43-46.

Mahindrakar, A., Kumar, R., Aswath, C. and Munikrishnappa, P.M. (2018) *Indian J. Agric. Sci*, **88**(6): 860-864.

Mahindrakar, A., Kumar, R. and Aswath, C. (2017) *Environ. Ecol*, **35** : 1117-1122.

Maloupa, E., Gerasopoulos, D and Tuzel, Y. (1999) *Acta Hortic.*, **491** : 433-438.

Maloupa, E., Traka Mavrona, K., Papadopoulos, A., Papadopoulos, F., Pateras, D. and Papadopoulos, A.P. (1999) *Acta Hortic.*, **481** : 603-607.

Manners, A.G., Dembowski, B.R. and Healey, M.A. (2013) *Aust. J. Entomol.*, **52** : 246–258.

Marfa, O., Biel, C., Penuelas, J., Montero, J. I., Guri, S. and Save, R. (1997) *Horticultura Revista de Hortalizas Flores y Plantas Ornamentales*, **118** : 68-69.

Mariska, I., Gati, E. and Sukmadjaja, D. (1989) *Bull. Penelitian Hortikultura*, **17** : 34-43.

Martens, S., Forkmann, G. and Cadic, A. (2000) *Acta Hortic.*, **508** : 39-44.

Matous, J.E. (1985) *Zahradnictvi*, **12** : 67-72.

Maurer, J. and Horn, W. (1967) *Gartenwelt*, **67** : 63-64.

McGuire, R.G. (1992) *HortScience*,**27** : 1254-1255.

Meng, H., Xie, Z., Cheng, Z. and Su, L. (2012) *Acta Hortic.*, **937** : 499-510.

Meyer, H.J and van-Staden, J. (1988) *Plant Cell Tissue Organ Cult.*,**24** : 25-30.

Meynet, J and Sibi, M. (1984) *Agronomie*, **3** : 839-846.

Miller, J.W and Knauss, J.F. (1973) *Plant Disease Rep.*, **57** : 504-505.

Miske, T. (1986) *Zierpflanzenbau*, **26** : 1008-1010.

Miyoshi, K and Asakura, N. (1996) *Plant Cell Rep.*, 16 : 1-5.

Mo, C.Y., Soo, L.J., Jeong, L.B. and Chang, K.O (2016) *Korean J. Hortic. Sci. Technol.*, **34** : 966-976.

Mohanty, B. K., Kumar, S., Srivastava, R. and Chand, S. (2005) *J. Ornam. Hortic.*, **8** : 196-200.

Monica, J and Vandita, B. (2017) *Int. J. Sci. Res. Rev.*, **6** : 244-251.

Moulinier, H. and Montarone, M (1978) *Pepinieristes Horticulteurs Maraichers*, **188** : 13-18.

Muller, B. (1959) *Dtsche Gartenb.*, **6** : 72-76.

Muniz, Corrêa, A.C., Galati, Cury, V., Marques, Magalhães, K., Mattiuz, Machado, C.F., Mattiuz and Ben-Hur (2016) *Cienc. Rural.*, **46** : 1945-1951.

Murashige, T., Serpa, M. and Jones, J.B. (1974) *HortScience*, **9**: 175-180.

Murthy, K.C.B. and Subbaiah, K.V. (2020) *J. Pharmacogn. Phytochem.*, **9** : 900-903.

Muthumanickam, D., Rajamani, K. and Jawaharlal, M. (1999) *J. Orn. Hort., New-Series*, **2** : 131-132.

Mutsenietse, G.Y., Rashal, I.D. and Dishler, V.Y. (1978a) *Genetica*, **14** : 238-241.

Mutsenietse, G.Y., Rashal, I.D. and Dishler, V.Y. (1978b) *Genetica*, **14** : 779-783.

Nagaraju, V., Srinivas, G.S.L and Lakshmi Sita, G. (1998) *Curr. Sci.*, **10** : 630-634.

Nair, S.A., Attri, B.L. and Sharma, T.V.R.S. (1998) *Indian Coconut J.*, **29** : 5-6.

Natskova, V. and Karadzhova, O. (1990) *Rasteniev"dni-Nauki.*, **27** : 96-102.

Nicolini, G. (1964) *Ital. Agric.*, **101** : 833-841.

Nijlunsing, W. (1976) *Vakbladvoor de Blemisterij*, **31** : 44-45.

Nor, H.A., Rosna, T., Saleh, M., Azani. and Noraini, M. (2012) *Hortic. Bras.*, **30** : 252-257.

Nowak, J. (1979) *ActaHortic.*, **91** : 105-110.

Nowak, J. (1989) *Prace Instytutu Sadownictwai Kwiaciarstwa w Skierniewicach B Rosliny Ozdobne*, **14** : 169-174.

Nowak, J. and Plich, H. (1981) *Prace Instytutu Sadownictwai Kwiaciarstwa w Skierniewicach B Rosliny Ozdobne*, **6** : 89-97.

Nucifora, A. and Vacante, V. (1981) *Colture Protette*, **10** : 33-36.

Olivella, C., Biel, C., Vendrell, M. and Save, R. (2000) *HortScience*, **35** : 222-225.

Orlikowska, T and Nowak, E. (1997) *Acta Hortic.*, **447** : 619-621.

Orlikowski, L., Hetman, J. and Tjla, B. (1974) *HortScience*, **9** : 239-240.

Orlikowski, L.B. (1999) *Proc. 51st Internatl. Symp. crop protection*, Gent, Belgium, 4 May 1999.

Oszkinis, W. and Lisiecka, A. (1971) *Roczniki Wyzszej SzkolyRolniczej w Poznaniu*, **50** : 137-147.

Pag, H. (1960) *NachrDtsch., Pflanzenschutzdientes Braunschweig*, **12** : 74-77.

Palanisamy, Kannan, D., Sharma, R., Bhatt, S.S. and Singh, A. (2015) *Int. J. Trop. Agric.*,**33** : 29-34.

Papadopoulou, E., Gerasopoulos, D. and Maloupa, E. (1996) *Advances Hort. Sci.*, **10** : 29-32.

Papenhagen, A. (1978) *Gb-+-Gw.*, **78** : 340-341.

Park, J. D. and Park, J.D. (1996) *Korean J. Appl. Ento.*, **35** : 302-308.

Parthasarathy, V.A and Nagaraju, V. (1995) *Ann. Plant Physiol.*, **10** : 10-12.

Parthasarathy, V.A., Nagaraju, V. and Parthasarathy, U. (2001) In : *Biotechnology of Horticultural Crops*, Naya Prakash (Eds. V.A. Parthasarathy, T.K. Bose and P. Das), Calcutta, pp. 169-185.

Parthasarathy, V.A., Parthasarathy, U., Nagarju, V. and Mishra, M. (1997) *Folia Hort.*, **9** : 83-86.

Parthasarathy, V.A., Parthasarathy, U. and Nagaraju, V. (1996) *Ann. Plant Physiol.*, **11** : 34-39.

Pasini, C., D'Aquila, F. and Bianchini, C. (1984) *Colture Protette.*, **13** : 63-66.

Pasini, C., D'-Aquila, F. and Gandolfo, M. (1997) *Colture Protette.*, **26** : 73-76.

Patra, S.K, Beura, S. and Shasani, T. (2015) *Agric. Sci. Digest.*, **35** : 173-177.

Pawlowska, H. (1977) *Hod. Rosl. Aklimat, Narieman*, **21** : 177-181.

Penningsfeld, F., Forchthammer, L., Fischer, P. and Kalthoff, F. (1976) *Deutscher-Gartenbau*, **30** : 1123.

Pielka, J. and Bartynska, M. (1974) *Prace Institutu Sadownictwa w Skierniewieaca E.*, No. 4, pp. 22-27.

Pierik, R.L.M., Steegmans, H.H.M., Verhaegh, J.A.M. and Wouters, A.N. (1982) *Netherlands J. Agril. Sci.*, **30** : 341-346.

Pierik, R.L.M., Steegmans, H.M.M and Marelis, J. J. (1973) *Scientia Hort.*, **1** : 117-119.

Pierik, R.L.M., Jansen, J.L.M., Maasdam, A. and Binnendijk, C.M. (1975) *Scientia Hort.*, **3** : 293-297.

Pierik, R.L.M.andSegers. T.A. (1973) *Z. Pflanzenphysiol.*, **69** : 204-12.

Pinamonti, F., Zanella, T. and Zorzi, G. (1996) *Informatore Agrario.*, **52** : 47-52.

Pinamonti, F., Stringari, G. and Zorzi, G. (1997) *Compost Sci.Utilizn.*, **5** : 38-46.

Pinto, J.E., Sato, A.Y and Pinto, C.A. (1993) *Bulletin-OILB-SROP*, **19** : 240-246.

Pirone, P.P., Dodge, B.O. and Ricket, H.W. (1960) *Diseases and Pests of Ornamental Plants*, Constable and Company Limited, London.

Pivot, D. (1985) *Acta Hortic.*, **172** : 211-222.

Plate, H.P. and Krober, H. (1975) *Gartenwelt*, **75** : 340-342.

Prashanth, P. and Chandrasekar, R. (2010) *Indian J. Agric. Res.*, **44** : 66-69.

Pryor, R.L. (1972) *HortScience*, **7** : 197-198.

Put, H.M.C. (1990) *Scientia Hort.*, **43** : 129-144.

Pyeatt, L.E. and Elmore, C.L. (1975) *Flower and Nursery Report.* 1975, July-August, 9-11; 1 pl.

Quaglia, F. (1983) *Atti XII Congresso Nazionale Italiano di Entomologia*, Roma 5-9 novembre 1980. Volume II. 325-326.

Raalte, D. Van. (1978) *DeutscherGartenbau*, **32** : 1754-1755.

Raj., Kumar, P.A., Singh, D.B., Mer, M.S., Kishor, A. and Devnath, S. (2016) *Biotech Today*, **6**(2) : 29-33.

Rashmi, R., Aswath, C., Dhananjaya, M.V. and Reddy, D.C.L. (2018) *Int J. Curr. Microbiol. Appl. Sci.*, **7** : 508-514.

Reimherr, P., Kneissl, B., Loeser, H., Miske, T., Papenha gen, A. and Schenk, W. (1980) *Gb-+-Gw.*, **80** : 1002.

Reist, A. (1997) *Revue Suisse de Viticulture d'Arboriculture et d'Horticulture*, **29** : 333-335.

Reynoird, J.P., Chriqui, D., Noin, M., Brown, S. and Marie, D. (1993) *Plant Cell Tissue Organ Cult.*, **33** : 203-210.

Reynoird, J.P., Dewitte, W., Prinsen, E., Van Onckelen, H., Noin, M., Chriqui, D. and Cadic, A. (2000) *Acta Hortic.*, **508** : 261-263.

Rezende, Silva, R.K., Vilela, L.P., Renato, P., Antonio, C.J., Paula, T., Pereira, Masetto and Elisa, T. (2009) *Cienc. Rural.*, **39** : 2435-2440.

Richter, M. (1996) *TASPO Gartenbaumagazin*, 5 : 39-41.

Richter, M. and Matschke, J. (2000) *TASPO Gartenbaumagazin*, **10** : 11-12.

Rijssel, E-van and Visser, A.J-de (1985) *Groenten Fruit.*, **41** : 27.

Riviere, L. M. and Alegre, J (1984) *P. H. M. Revue Horti Cole.*, **246** : 11-13.

Rodriguez-Navarro, J.A., Zavaleta-Mejia, E., Sanchez-Garcia, P. and Gonzalez-Rosas, H. (2000) *Fitopatologia*, **35** : 66-79.

Rudnicki, R.M. and Nowak, J. (1976) *Acta Agrobot.*, **29** : 289.

Ruffoni, B. and Massabo, F. (1991) *Acta Hortic.*, **289** : 147-148.

Ruffoni, B. and Sulis, S. (1988) *Annalidell'IstitutoSperimentale per la Floricoltura*, **19** : 73-81.

Sabir, N., Sumitha, R., Deka, S and Sindhu, S.S (2012) Integrated Pest Management of Gerbera under Protected Cultivation.

Sachs, R.M. (1968) *Amer. J. Bot.*, **55** : 62-68.

San Pedro, M. De and Badosa, R. (1985) *Communicaciones de la III reunion de Ornamentals Jornadasteinicas 1985*, pp. 202-205.

Sang, C.K., Choi, B.J. and Koh, J.C. (1998) *J. Korean Soc. Hort. Sci.*, **39** : 833-837.

San, B., Gok, K.M. and Bayhan, A.K. (2016) *Süleyman Demirel Üniversitesi Fen Bilimleri Enstitüsü Dergisi*, **20**(3):10. DOI:10.19113/sdufbed.13860 .

Sangma, L.B. and Chaurasiya, A.K. (2019) *Int. J. Curr. Microbiol. Appl. Sci.*, **8** : 1380-1392.

Sangma, Z.C.N., Singh, D. and Fatmi, U. (2017) *Int. J. Curr. Microbiol. Appl. Sci.*, **6** : 468-476.

Santos, F.T., Ludwig, F., Costa, L.A.D.M. and Costa, M.S.S.D.M. (2015) *Ornam. Hortic*, **21** : 251-258.

Sarkar, A., Kumar, T. and Sadhukhan, C.R. (2020) *Int. J. Curr. Microbiol. Appl. Sci.*, **9**: 2042-2051.

Sarmah, D., Mahanta, P., Talukdar, M.C. and Das, R. (2014) *Res. Crops*, **15** : 211-214.

Schenk, M. and Brundert, W. (1981) *Gb-+-Gw.*, **81** : 275-276.

Schickedanz, F. (1974) *Gartenwelt.*, **74** : 32.

Schiva, T. (1973) *Annalidella Facolta di Scienze Agrariedella Universitadeglistudi di Torino*, **8** : 355-366.

Schiva, T. (1975) *Annali dell' Instituto Sperimentale per la Floricultura*, **6** : 133-513.

Schiva, T. (1979) *EUCARPIA Meet. on Genet. Carnation and Gerbera*, April, 1978, Italy, pp. 211-221.

Schiva, T. (1980) *Annali dell' Istituto Sperimentale per la Floricucoltura*, **11** : 65-75.

Schiva, T., Ruffoni, B. and Vaccarino, R. (1984) *Annali dell' Istituto Sperimentale per la Floricucoltura*, **15** : 29-71.

Schlang, J. and Sikora, R.A. (1978) *Zeitschrift fur Pflanzenkrankheiten und Pflanzenschutz.*, **85** : 203-209.

Schmelzer, K. (1966) *Arch. Gartenb.*, **14** : 89-92.

Schmidt, U. and Exarchou, E.S.O. (2000) *Acta Hortic.*, **537** : 487-494.

Scholten, H. and Prakash, J. (1991) *Current Plant Sci. and Bioech.*, **12** : 367-370.

Scholten, G. (1972) *Vakblad-voor-de-Bloemisterij*, **27**: 22.

Shalini, M. and Patil, V.S. (2006) *J. Farm Sci.*, **19** : 144-147.

Sharma, V., Karma, G., Thakur, R. and Kaur, R. (2018) *Int. J. Innov. Pharm. Sci. Res.*, **6** : 1-9.

Shin, H.K., Kim, J.Y., Shin, R.W., and Ko, J.Y. (1994) *RDA J. Agric. Sci. Hort.*, **36** : 444-452.

Singh, P., Bhardwaj, P., Kumar, R. and Singh, D. (2017) *Int. J. Curr. Microbiol. App. Sci.*, **6** : 112-116.

Singh, S. and Dhyani, D. (2016) *Int. J. Plant Sci. Eco.*, **2** : 10-14.

Singh, S., Dhyani, D. and Karthigeyan, S. (2009) *J. Ornam. Hortic.*, **12** : 225-231.

Singh, S.J. (1973) *Sci. Cult.*, **39** : 133-134.

Skalska, E. (1978) Sbornik UVTIZ Zahradnictvi, **5/6** : 123-130.

Skalska, E. (1980) *Sbornik UVTIZ Zahradnictvi*, **7** : 145-150.

Skalska, E. (1983) *Sbornik UVTIZ Zahradnicti*, **10** : 237-245.

Skalska, E. (1985) *Acta Pruhoniciana*, **50** : 43-54.

Song, C.Y., Moon, J.Y., Song, E.K. and Oh, D.G. (2015) *Korean J. Breed. Sci.*, **47** : 11-19.

Sonneveld, C., Voogt, W. and Roeber, R.U. (1997) *Acta Hortic.*, **450** : 139-147.

Sorensson, A. and Nedstam, B. (1993) *Bull. OILB-SROP*, **16** : 129-132.

Sparnaaij, L.D., Garrestsen, F. and Bekker, W. (1975) *Euphytica*, **24** : 551-6.

Srivastava, R., Kumari, J., Negi, N. and Chand, S. (2017) *Int. J. Chem.*, **5** : 1888-1891.

Staby, G.L., Robertson, J.L., Kiplinger, D.C. and Conover, L.A. (1978) *Chain of life*, Ohio State Univ., Columbus, Ohio.

Steen, J.A van-der. (1975a) *Vakbladvoor de Bloemisterij*, **30** : 12-13.

Steen, J.A.van-der, (1975b) *Vakbladvoor de Bloemisterij*, **30** : 16-17.

Steinitz, B. (1982) *Gartenbauwissenschaft*, **47** : 2, 77-81.

Steinitz, B. (1983) *Gartenbauwissenschaft*, **48**: 2, 67-71.

Stinson, R.F. (1953) *Proc. Amer. Soc. Hort. Sci.*, **62** : 487-491

Stinson, R.F. (1957) *Quart. Bull. Mich. Agric. Stat.*, **40** : 393-396.

Stokes, D.E., Mc Farlin J.R. anmd Taylor, A.L. (1974) *Proc. Fla. St. Hort. Soc.*, **86** : 378-80.

Stouffer, R.F. (1965) *Phytopathology*, **55** : 501.

Strojny, Z., Saniewski, M. and Jesiotr, E. (1976) *Prace Instytutu Sadownictwai Kwiaciarstwa w Skierniewicach B*, **2** : 87-95.

Strojny, Z. and Nowosielski, O. (1980) *Prace Instytutu Sadownictwai Kwiaciarstwa w Skierniewicach B*, **5** : 103-113.

Stupak, N.N. (1985) *Rrferativnyi Zhurnal*, **55** : 10.55.802.

Sutterlin, S., van, L.J.C and. van-Lenteren, J.C. (1999) *J. Insect Behavior.*, **12** : 105-122.

Sutterlin, S., van-der, M.G.J. and van-Lenteren, J.C. (1991) *Proc. Section Exptl. Appl. Ent. Netherlands Ent. Soc.*, **2** : 115-120.

Sytsema, W. (1981) *Vakbladvoor de Bloemisterij*, **36** : 50-51.

Szentgyorgyl, L., Nemestothy, K., Budal, C. and Csolle, I. (1984) *Kerteszet es Szoleszet*, **33** : 5.

Tameshwer, J., Sharma, G., Jangde, B. and Banjara, N.C. (2019) *J. Pharmacogn. Phytochem*, **8** : 2112-2114.

Tjia, B., Marousky, F.J. and Stamps, R.H. (1987) *HortScience*, **22** : 896-897.

Tourjee, K.R., Harding, J. and Byrne, T.G. (1993) *HortScience*, **28** : 735-737.

Tourjee, K.R., Harding, J. and Byrne, T.G. (1995a) *Heredity*, **74** : 303-310.

Tourjee, K.R., Harding, J. and Byrne, T.G. (1995b) *Heredity*, **75** : 290-296.

Tourjee, K.R., Harding, J. and Byrne, T.G. (1994) *HortTechnology*, **4** : 34-40.

Tsujita, M.J. and Dutton, R.G. (1987) *Canadian J. Pl. Sci.*, **67** : 585-587.

Tusnadi, C.K. (1984a) *Kerteszet es Szoleszet*, **33** : 5.

Tusnadi, C.K. (1984b) *Kerteszet es Szoleszet*, **33** : 6.

Vadalie, H. and Laffaire, M. (1984) *Revue Horticole*, **246** : 15-19.

Vadalie, H., Laffaire, M., Riviere, L.M. and Charpentier, S. (1985) *Revue Horticole*, **262**: 13-18.

Van Labeke, M.C. and Dambre, P. (1999b) *Verbondsnieuws*, **43** : 34-36.

Van Labeke, M.C. and Dambre, P.S.O. (1999a) *Verbondsnieuws.*, **43** : 27-29.

Van Meeteren, U. and van Gelder, H. (1980) *Scientia Hort.*, **12** : 273-281.

Van, D.W.G., Tijskens, L.M.M. and Van, D.W.G. (1991) *Agric. Systems.*, **35** : 111-127.

Van, M.U. (1977) *Vakbladvoor de Bloemisterij*, **32** : 46-47.

Van, M.U. (1978) *Scientia Hort.*, **8** : 65-74.

Van-de, V.M., Vacante, V., Van-de, V. M. and De-Veire, M. V. (1988) *Boletin-de Sanidad Vegetal Fuera-de-Serie.*, **17** : 425-435.

Vass, E. (1986) *Kerteszeti Egyetem Kozlemenyei*, **50** : 289-295.

Vass, E. and Hargitai, L. (1986) *Kerteszeti Egyetem Kozlemenyei.*, **50** : 281-287.

Venezia, A., Martignon, G., Schiavi, M. and Casarotti, D. (1997) *Colture Protette.*, **26** : 129-135.

Vergara-Gongora, J. (1984) *Boletin Informativo Centro de Investigacion y Desarrollo Horticola*, **8** : 69-77.

Verma, A.K., Sindhu, S.S., Singh, A., Kumar, A., Singh, A. and Chauhan, V.B.S. (2019) *Ecol. Eng.*, **132** : 23-30.

Verma, V.S. and Singh, S. (1980) *Gartenbauwiss.*, **45** : 61-62.

Vestergaard, S., Gillespie, A.T. and Eilenberg, I. (1996) *Proc.I Meet. Insect Pathogens and Insect Parasitic Nematodes*, **19**: 240-246.

Vidalie, H., Laffaire, M., Riviere, L.M. and Charpentier, S. (1985) *Revue Horticole*, **262**: 13-18.

Vijayalakshmi, C.L., Babu, P., Bagali, A.N., Soregaon, C.D. and Wadageri, A.H. (2019) *Int. J. Curr. Microbiol. Appl. Sci.*, **8** : 2610-2619.

Volpi, L. and Farina, E. (1987) *Colture Protette.*, **16** : 100-102.

Vonarx, J.A. (1952) *Tijdschr. Plzeikt.*, **58** : 5-9.

Voogt, S.J. (1976a) *Vakbladvoor de Bloemisterij.*, **31** : 20-21.

Voogt, S.J. (1976b) *Vakbladvoor de Bloemisterij.*, **31** : 28-29.

Voss, D.H. (1992) *HortScience*, **27** : 1256-1260.

Vrie, M. van De (1976) *Vakbladvoor de Bloemisterij*, **31** : 24-25.

Walther, F. and Sauer, A. (1985) *Deutscher Gartenb.*, **39** : 2097-2098.

Walther, F. and Sauer, A. (1986) *Proc. Internatl. Conf. Nuclear Techniques for Plant Improvement*, IAEA, Vienna, 1986, pp.113-120.

Walther, F. and Sauer, A. (1992) *Gartenbauwissenschaft.*, **57** : 68-71.

Winarto, B., Yuniarto, K. and Soehendi, R. (2020) *Not. Sci. Biol*, **12** : 264-276.

www.stuartxchange.org

www.stuartxchange.org

www.herbal-supplement-resource.com

Yathom, S., Marcus, R., Chen, M. and Tal, S. (1988) *Phytoparasitica*, **16** : 217-224.

Yazici, K and Alkaç, O.S. (2019) *JAFAG*, **36** : 45-54.

Ye, J.N., Wang, T.Z., Cai, S.Q., Komatsu, K., Mikage, M. and T. Namba (1990) *J. Pharmac. Soc. Jpn.*, **110** : 374-382.

Yildirim, T.B., Hatipoglu, A., Zager, B. and Gulgun, B. (1995) *Ege Universitesi Ziraat Fakultesi Dergisi*, **32** : 175-182.

Yoon, H.S., Sang, C.K. and Choi, B.J. (1996a) *J. Korean Soc. Hort. Sci.*, **37** : 831-835.

Yoon, H.S., Sang, C.K. and Choi, B.J. (1996b) *J. Korean Soc. Hort. Sci.*, **37** : 593-597.

Yu, D., Kotilaninen, M., Pollanen, E., Mehto, M., Elomaa, P., Helariutta, Y., Albert, V.A., and Teeri, T.H. (1999) *The Plant J.*,**17** : 51-62.

Yu, Y., Harding, J., Byrne, T. and Famula, T. (1993a) *Theor. Appl. Genet.*, **86** : 234-236.

Yu, Y., Byrne, T. and Harding, J. (1991) *Euphytica*, **53** : 19-23.

Yu, Y., Harding, J. and Byrne, T. (1993b) *Euphytica*, **70** : 97- 103.

Zakharova, J.K. (1987) *Nauchnye Trudy Moskovskii Lesotekhnicheskii Institut.*, **188** : 72-75.

Zayats, T.V. (1977) *Byulletin Glavnogobotanicheskogo Sada*, **103** : 54-59.

Zuker, A., Tzfira,T. and Vainstein, A. (1998) *Biotechnol. Adv.*, **16** : 33-79.

3

Antirrhinum

Anamika Gurung, Rajiv Kumar, K. Dutta,
J.S. Arora and P. Pal

1.0 Introduction

Antirrhinum, commonly known as snapdragon, is a perennial plant but treated as an annual or biennial. The name is derived from Greek *anti*-like, *rhinos*-snout, referring to the curiously shaped flowers.

2.0 Importance and Uses

2.1.0 Aesthetic Uses

Antirrhinum is one of the principal cut flower crops grown in many parts of the world, in suitable environment during the growing period, in tropical, subtropical and temperate countries. Magnificent flowers are borne on terminal long spikes of many colours and numerous shades. It has good keeping quality and flowers remain fresh for several days. These qualities make it as an excellent cut flower. It is also highly suitable for various landscape plans as bedding, in rockeries, pots or herbaceous borders. The variations in plant height in different types and groups provide wide scope of using antirrhinum for different purposes in gardens. In renaissance period, antirrhinum flowers were worn on sleeves by subjects to welcome the king. Theophrastus stated that wearing antirrhinum endowed a man with grace, humour and glory in the eyes of both poor and rich (Coats, 1968).

2.2.0 Therapeutic and other Uses

In language of flowers, antirrhinum stands for presumption. In old days, there was a common belief that plant of antirrhinum saves from being bewitched. Russians used to grow on large-scale to extract oil from its seeds which was equal to olive oil (Genders, 1969). Earlier, it was recommended to wash the face with the water in which antirrhinum seeds were soaked if a woman wanted to restore her beauty and youthfulness. In addition to the value of their showy and fragrant blooms, snapdragon leaves and flowers are also used in the treatment of tumours, ulcers, inflammation, and hemorrhoids.

Antirrhinum majus is commercially available as edible flower and it contains bioactive compounds like flavonolgly-cosides (*e.g.* quercetin and kaempferol glycosides) and anthocyanins, such as cyanidin and pelargonidin glycosides. It contains 29 μg/mg total carotenoids, and lutein is the dominant compound (González-Barrio *et al.*, 2018).

It emerged as a model plant during early studies of inheritance and mutation because of its diploid inheritance, ease of cultivation, and variation in morphology and flower color.

3.0 Origin and History

Snapdragon (*Antirrhinum majus*) is native of the west Mediterranean area, and its genetic centre is southern Spain (Schwarz-Sommer *et al.*, 2003), with 22 of 25 species in the Iberian Peninsula (Vargas *et al.*, 2009). It was believed to have reached Britain with the Roman at an early date where it has been naturalized on

Uses of Antirrhinum

In Bed

In a Garden

As Border in a Garden

In Hanging Baskets

As Specimen in Garden

In Pathways

In Pot

In Flower Arrangement

In Garden Decoration

mountainous regions. From Britain it spread to the different parts of the world. *Antirrhinum glandulosum* is native of California.

Mateu-Andres (1999) evaluated the 3 wild species *A. lopesianum*, *A. mollissimum* and *A. microphyllum*, endemic to the Iberian peninsula, for 8 enzymes using electrophoresis. Alleles unique to individual species were found and indicated ancient diversification and supported a geographical model of specification.

Cultivars were first grown in the United States for flowering during long day periods, but in about 1926 Frank Volz of Chevoit, Ohio, introduced the cultivar, Chevoit Maid which produced flowers in winter (Lindstrom, 1966). The first F_1 hybrid Christmas Cheer was introduced by Fred and Helen Windmiller in Columbus, Ohio, in 1938 (Ball, 1957) and was followed by others, like Maryland Pink, Mary Ellen and Dorcas Jane, George J. Ball, Inc. and Yoder Brothers, Inc. have been the main introducers of new F_1 hybrids since about 1950.

The *Antirrhinum* species group comprises approximately 20 morphologically diverse members that are able to form fertile hybrids (Hudson *et al.*, 2008). Liberal González *et al.* (2014) reconstructed the evolutionary history of four taxa Antirrhinum (*A. majus* subsp. majus, *A. majus* subsp. striatum, *A. molle*, and *A. sempervirens*) in the Pyrenees inferred from phylogeographic analyses. It revealed enormous influence of the glacial history of the mountain ranges on the current configuration of intra- and inter-specific genetic diversity in Antirrhinum, as well as the importance of periglacial areas for the survival of species during glacial periods of the Quaternary.

Genetic diversity has generally been found to be lower in smaller populations, and the distribution of genetic variants within and between populations has been correlated with their level of self-incompatibility.

4.0 Morphology, Species and Cultivars

4.1.0 Morphology

Plants are erect or climbing herbs or even half-shrubs having usually opposite and entire leaves; flowers terminal racemes or solitary and axillary, corolla are saccate or gibbous at base, but not spurred, personate or closed at the throat, lobes of upper lip erect, lower lip spreading; stamens are four and not winged; fruits 2-celled capsule, the upper cell opening by 1, the lower by 2 many-toothed pores. Closely allied to *Linaria*, from which it differs in the spurless flowers.

The flower is zygomorphic and has one plane of symmetry, coinciding with the dorsoventral axis (Luo *et al.*, 1995). During the vegetative phase, the plants display decussate phyllotaxis, where the two leaves formed per node are positioned at opposite sides of the stem. After the transition to the reproductive phase, Antirrhinum plants switch to a spiral phyllotaxis, producing small leaf-like organs (bracts) at each node along the main (inflorescence) stem. Flowers arise in the axils of bracts and consist of four types of organs arranged in a whorled phyllotaxis. Five sepals in the calyx constitute the outer (first) whorl, followed by five petals (second whorl), four stamens and a stamenoid (third whorl), and two fused carpels in the inner (fourth) whorl (Masiero *et al.*, 2004).

Greenhouse types have been classified into 4 groups depending on growth and flowering responses to temperature and day length. Cultivars in Group I are highly reproductive and flower early at 10°C night temperature and short days of mid-winter. Group II cultivars flower with good quality blooms at 10°C night temperature but require more time for growth and flowering than the cultivars in Group I. Group III cultivars are very slow to flower at 10°C but grow and flower well at 15°C night temperature under longer day and higher light intensities. Cultivars in Group IV are reproductive only at night temperature of 15°C or more and the plants are blind at 10°C (Duffett, 1960).

The species group has traditionally been divided into three morphological subsections. (1) Subsection Antirrhinum comprises species with large leaves and flowers and includes the wild ancestor of cultivated *A. majus*. (2) Subsection Kickxiella comprises species with small leaves and flowers, including *Antirrhinum charidemi*, which is endemic to a dry coastal desert in southeastern Spain and has the smallest leaves and flowers within the species group, about one tenth of the area of *A. majus*. (3) Subsection Streptosepalum consists of two species with organs that are approximately intermediate in size between those of subsections Antirrhinum and Kickxiella (Webb, 1971 and Sutton, 1988).

4.2.0 Species

There are about 32 species of *Antirrhinum* (Chittendan and Synge, 1956). Several species have arisen after hybridization between subsections, and that some species continue to hybridize. However, all potential hybrids appear to have retained a phenotype similar to one of their ancestors, suggesting that ancestral combinations of characters are maintained by selection at many different loci (Wilson and Hudson, 2011). Some of the important species are described below.

Antirrhinum majus

It is native to the Mediterranean region where they grew as tender perennials. Plants erect, 20 to 120 cm high, branches, produced from base; leaves racemes, 3-5 cm long, oblong to lanceolate and glabrous; flowers 2.5-5.0 cm long, racemes upright, various colour.

A. molle

Branches decumbent, slender, woolly; leaves opposite, small, glandular, racemes, few flowers, produced at end of branches. It is native of Pyrenees.

A. glandulosum

Plants annual, about 60 cm, branches reddish, glandular; raceme leafy; flowers rose-pink, lower portion whitish yellow.

A. sempervirens

Plants woody, evergreen, prostrate; leaves opposite, narrow, oval; flowers whitish. It is native of Pyrenees.

Species of Antirrhinum

Antirrhinum majus

Antirrhinum molle

Antirrhinum hispanicun

Antirrhinum sempervirens

Antirrhinum glutinosum

Antirrhinum siculum

Cultivars of Antirrhinum

Madame Butterflies

Rocket Hybrid

Colossal Tip Top

F_1 Coronette

Tetra Ruffled

Sonnet Mix

Magic Carpet

Tahiti Series

A. glutinosum

Plants perennial, woody, branches fragile; leaves alternate, oblong to ovate; flowers large, erect, solitary in leaf axil. It is native to Spain.

Some other species are *A. asarina, A. hispanicun, A. siculum, A. turtuosum* and *A. maurandioides*.

4.3.0 Cultivars

Antirrhinum is broadly divided into three groups (i) tall, (ii) semi-tall, and (iii) dwarf varying in plant height. Since names of cultivars of annual flowering plants vary with seedsmen in different countries, flower colour has been mentioned in various strains and cultivars.

4.3.1 Tall (Maximum)

The height of the plant varies from 70-120 cm in this group, the important types are.

Antirrhinum majus maximum

Flowers have one or different shades of colour including white, scarlet, rose, lilac, crimson-purple, orange-scarlet, coppery, yellow and many other colours; used for bedding and cut flower.

Coronette Series

Suitable for bedding and cut flower, resistant to weather and rust.

F_1 Butterfly Type

Tall plants (70-75 cm) with open-faced flowers. This type has two distinct strains, Bright Butterflies (single flowered), Madame Butterflies (double azalea flowered). These strains are also known as Pentstemon flowered.

Rocket Hybrids (F_1)

Long spikes are produced on sturdy stems, recommended for bedding and cut flower. The colours in different cultivars include white, pink, rose-red, golden yellow, cherry-red, bronze.

Rocket Series

Tall (1.2 m), vigorous, excellent for cut flowers.

Tip Top

Branches arise from base, good for cut flower. Many cultivars vary in flower colour – orange-rose, scarlet, rose-pink, dark red, canary-yellow and pure white.

4.3.2 Semi-tall (Intermediate)

The height of plants varies from 50-75 cm and has several types.

A. majusnanum

Popular type for bedding and cut flower having many colours – pure white, rose, dark red, bright scarlet, fiery red, dark yellow, orange-yellow, orange-scarlet.

F_1 Carioca

Branches arise from base, uniform and vigorous growth, ideal for bedding with several colours.

F_1 Coronette

Sturdy growth, ideal for bedding.

Sonnet series

Bushy in habit, wet-weather tolerance, early and free flowering and produce bronze, pink, caramine red, crimson, burgundy, white and yellow.

Tetra Ruffled

Tetraploid antirrhinum, flowers are large and ruffled. Spikes are long and compact on strong stems and flowers in many attractive colours.

4.3.3 Dwarf or Tom Thumb (*Pumilum*)

Dwarf snapdragons are bushy plants, 15-30 cm in height, producing large number of spikes. They are very useful for edging, bedding, rockery and in boxes and pots.

Bells

Dwarf, early-flowering with long lasting, open-faced or hyacinth-like flowers with various colours.

Chimes

Exceptionally dwarf and compact, produce flowers in a wide colour range.

Floral Showers

Early-flowering, tolerates wet-weather.

F_1 Floral Carpet

Dwarf, ball-shaped plants bearing masses of flowers.

F_1 Kolibre

Early flowering, in many colour.

Little Darling

Bushy habit, flowers face upwards.

Magic Carpet

Very dwarf, useful for edging and in rockeries.

Tahiti Series

Rust resistant. Antirrhinum hybrids (F_1) suitable for growing in early and late winters have been developed. Rust-resistant cultivars and F_2 hybrids with attractive colours in tall and semi-tall groups are also available.

5.0 Genetics and Breeding

5.1.0 Genetics

The diploid chromosome number of *Antirrhinum majus* is 2n=16 and is easily self- and cross-pollinated (Hudson *et al.*, 2008). Recently, as suggested by sequence analyses, Antirrhinum was moved from the Scrophulariaceae to the Plantaginaceae family (Veronicaceae) (Olmstead *et al.*, 2001). Tetraploid forms are also available. Baur (1910, 1924) studied large number of its types including species and inter-specific hybrids and reported about 200 genes. The large majority of these genes were found to be recessive mutations leading to number of changes in the phenotype. Some major changes were reported pertaining to habit of plant, flower colour and its pattern, leaf shapes, chlorophyll deficiency and pollen incompatibility.

Two major classes of genes are regulating meristem and organ identity (Simon *et al.*, 1994). The fimbriata (*fim*) gene of *A. majus* acts between these two classes in a sequence of gene activation. The *fim* gene, originally described in 1930, was cloned by transposon tagging and encodes a product with no detectable homology to other proteins. Mutations in *fim* result in partial homeotic transformations of floral organs and in reduced determinacy of the meristem. Expression and function of *fim* depends on the activity of meristem identity genes, and *fim* in turn controls the spatial and temporal expression of organ identity genes. The pattern of *fim* expression defines a new domain of the floral meristem that changes with time in a complementary manner to those of the meristem identity gene *floricaula* and the organ identity gene *plena*. Weigel and Meyerowitz (1994) reviewed the ABCs of floral homeotic genes of *A. majus* under the following main heading : present models for flower development (a summary of the ABC model, where A, B and C represent 3 classes of organ identity function genes); homeotic genes at the molecular level; gain-of-function situations : further refinement of the ABC model; regulation of organ identity genes; additional functions of organ identity genes; and comparison of *fly* and plant homeotic genes.

Matkowski and Adler (1999) analysed main shoot phyllotaxtic patterns during vegetative and generative phases of growth of *A. majus* mutant lines *Phantastica ambigua* (M250) and *Graminifolia constans* (M146), and wild-type plants of Sippe 50 and W108 by using scanning electron microscopy. The results suggest a possible role of genetic factors in determining the regularity of spatial arrangement of organs.

In antirrhinum, asymmetry of the flower and its component organs requires *cyc* and *dich* gene activity (Luo *et al.*, 1999). It showed that, like *cyc*, the *dich* gene encodes a product belonging to the TCP family of DNA-binding proteins that is first expressed in the dorsal domain of early floral meristems. However, whereas *cyc* continues to be expressed throughout the dorsal regions, expression of *dich* eventually becomes restricted to the most dorsal half of each dorsal petal. The effects of *dich* mutations and ectopic *cyc* expression on petal shape, provided an indication that plant organ asymmetry can reflect subdomains of gene activity. The results indicate that plant organ asymmetry can arise through a series of steps during which early asymmetry in the developing meristem is progressively built upon.

Semi-dominant *macho* alleles of the *plena* gene display carpelloid sepals and staminoid petals of *A. majus*, but the two inner flower whorls of stamens and carpels are normal and produce fertile gametes. In the recessive *plena* mutant, the two outer whorls are normal whereas the stamens are largely or entirely petaloid and the carpels sepaloid, thus producing weakly male-fertile or fully sterile lines. Two new *plena* and two new *macho* alleles were induced in transposon tagging experiments. Genetic and molecular analyses revealed that the two contrasting mutant phenotypes were caused by mutations in one and the same gene (Lonnig and Saedler, 1994).

In Antirrhinum, the C-function genes PLENA (PLE) and FARINELLI (FAR) make an unequal contribution to the specification of male and female organs (Davies *et al.*, 1999).

During flower development, *PLE* (PLENA) expression is first detected in the central region of the floral meristem while later it is found in stamen and carpel primordia where was maintained throughout their development (Bradley *et al.*, 1996). In the recessive *plena* mutant, perianth organs replace sex organs, therefore *ple* flowers have sepals in the first whorl, petals in the second and third whorls, and carpels are substituted by petaloid or sepaloid organs. Moreover, *ple* mutant results in a proliferation of whorls within the flower, therefore extra perianth organs develop inside the fourth whorl. In contrast, ectopic expression of the Antirrhinum FAR gene in Antirrhinum converts petals to stamens but does not alter sepal identity (Causier *et al.*, 2005). The expression of PLE in the outer whorls of Antirrhinum promotes both male and female organs, whereas FAR can only specify male organs.

Two mutants of *A. majus*, *deficiens* and *globosa*, have a homeotic conversion of the floral organs in whorls 2 and 3, namely petals to sepals and stamens to carpels (Okamoto *et al.*, 1994). The class B MADS box transcription factors DEFICIENS (DEF) and GLOBOSA (GLO) of *Antirrhinum majus* together control the organogenesis of petals and stamens (Saedler and Zachgoa, 2004). These proteins could also be involved in the direct repression of the expression of another MADS-box gene, namely DEFH125, and could thereby contribute to the formation of mature male gametophytes (Lauri *et al.*, 2006). Higher plant shoot meristems are multicellular structures that are the site of postembryonic organogenesis. Analysis of chimaeric plants has indicated that cells in different regions of the meristem can interact with each other so that their activities are coordinated during developmental processes. Correlations have not been demonstrated between events at a molecular level and the interactions observed at a phenotypic level in chimaeras.

Szymkowiak (1995) reported that expression of the *floricaula* gene in one region of chimaeric snapdragon (*A. majus*) meristems is sufficient to promote the transition from inflorescence to floral development, and to induce the expression of the downstream organ identity genes *deficiens* and *plena* throughout the meristem.

The meristem-identity gene *floricaula* (*flo*) determined floral meristem identity, control of flower development and gene expression across cell layers by *flo* of *Antirrhinum* (Becraft, 1995). The homologous genes *floricaula* (*flo*) regulate the formation of determinate floral meristems. Transcripts of single-copy gene is confined to floral meristems and some floral organs that subtend antirrhinum flowers (Kelly *et al.*, 1995).

Masiero *et al.* (2004) proposed that *INCO* represses prophyll development in Antirrhinum, and show that absence of this control results in impaired floral architecture. Also, *INCO* controls Antirrhinum floral meristem identity revealing the genetic interactions between *inco* and mutants of *FLORICAULA*, a gene that controls the inflorescence to floral transition, together with SQUAMOSA.

In the developed paracorolla (a petaloid organ fused with stamens at the base in *Antirrhinum majus*), Yamaguchi *et al.* (2010) observed that the expression of the A-class gene SQUAMOSA (SQUA) and the B-class genes DEFICIENS (DEF) and GLOBOSA (GLO) were high, while those of the C-class genes PLENA (PLE) and FARINELLI (FAR) were very low. In contrast, in the paracorolla primordia, expressions of SQUA and PLE were not detected. Thus, the expression patterns of homeotic genes in the paracorolla were basically the same as those of the petal, and this pattern was established during paracorolla development. In the stamen, which fused with the paracorolla at the base, DEF, GLO, PLE, and FAR were expressed in the anther. However, in the filament, SQUA was also expressed at a higher level in addition to these genes, indicating that the expression pattern shifted from the stamen type in the anther, to the petal type in the paracorolla, through the intermediate type in the filament.

5.1.1 Floral Scents

Genes encoding the enzymes involved in the production of floral scents monoterpenes (myrcene, *E*-β-ocimene, and linalool) and phenylpropanoids (methylbenzoate, acetophenone, dimethoxytoluene, *cis*-methylcinnamate, and *trans*-methylcinnamate) have been identified from *A. majus* and used to study the regulation of scent production and the effects of variation in scent composition on pollinator attraction. Results showed that honeybees are capable of using all of the floral volatiles to discriminate subtle differences in scent (Wright *et al.*, 2005).

5.1.2 Flower Colour

Investigations by Baur (1919, 1924) showed that certain colours exhibited simple Mendelian inheritance. Thus, the cross between red and ivory-flowered plants gave F_1 with pink flowers and the F_2 population segregated as 1 red : 2 pink : 1 ivory. The F_3 progeny of pink F_2s segregated in a similar manner. Red flowers differed from ivory flowers by single gene (*F*) and from yellow flowers by two independent genes (*F* and *C*). Baur also demonstrated the independent assortment of some of the flower genes. The cross of red peloric x ivory normal gave a modified dihybrid ratio in the F_2, and the cross of red peloric x yellow normal yielded a F_2 with a more complex segregation, as expected. Baur identified the genes for colours as *C* for ivory or yellow, *F* for red, *G* for modifying red and *D* for controlling colour of tube. These genes were redesignated by Baur (1930) as *N* for albino, *P* for pink, *M* for modifier and *Y* for yellow. Later work added other genes to the list and changed the gene nomenclature as nivea (*Niv*), sulfurea (*Sulf*=*C*), incolorata (*Inc*=*F*), eosinea (*Eos*), pallida (*Pal*), diluta (*Dil*) and eluta (*El*). Factorial analysis revealed that these genes determine different colours due to synthesis of different pigments. The main seven colours are white, yellow, ivory, bronze, pink, crimson and magenta (Stubbe, 1966). The genotypes corresponding to phenotypes are presented in Table 1.

Table 1: Genotypes for Flower Colour in *Antirrhinum majus**

Colour	*Genotype for Four Genes*			
White	*niv/niv*			
Yellow	*niv*$^+$	*sulf/sulf*	*inc/inc*	
Ivory	*niv*$^+$	*sulf*$^+$	*inc/inc*	
Bronze	*niv*$^+$	*sulf/sulf*	*inc*$^+$	*eos/eos*
Pink	*niv*$^+$	*sulf*$^+$	*inc*$^+$	*eos/eos*
Crimson	*niv*$^+$	*sulf/sulf*	*inc*$^+$	*eos*$^+$
Magenta	*niv*$^+$	*sulf*$^+$	*inc*$^+$	*eos*$^+$

The plus superscript as in *niv*$^+$ indicates dominant wild type allele.

* Stubbe (1966).

It is evident from Table 1 that homozygous recessive *niv/niv* blocks pigment formation and results in white flowers, whereas dominant *niv*$^+$ permits pigment formation. The colour is yellow in the presence of *niv*$^+$ and the homozygous recessive conditions for the genes *sulf* and *inc*. In gene *inc* which exercises a switching action between yellowish anthoxanthin and reddish anthocyanin, a dominant *inc*$^+$ allele is necessary for anthocyanin formation. The colour becomes pink when the first three genes are represented by dominant alleles and magenta with all four dominant alleles.

Three additional colour genes are *Pal, Dil* and *El*. *Pal* is complementary to *Inc* and both dominant allele *Inc*$^+$ and *Pal*$^+$ must be present for anthocyanin formation. *Dil* is a modifier which intensifies the colouration. Different allele combinations of *Inc, Pal* and *Dil* determine fine shades of bronze or pink. *El* causes leaching of the reddish colour, and depending on the dosage of the dominant *El* allele different shades of magenta are obtained. The incompletely dominant allele *El* in single dose makes the flower colour one shade lighter (Stubbe, 1966). Flower colour becomes progressively lighter with increasing dose of *El* (Table 2).

Table 2: Dosage Effect of the Dominant Allele *El* on Flower Colour in *Antirrhinum majus**

Genotype	*Flower Colour*	*Colorimetric Value for Anthocyanin Concentration*
+ + + +	Intense magenta	103.8
El + + +	Dark magenta	60.2
El El+ +	Light magenta	43.9
El ElEl +	Very light magenta	31.3
El ElElEl	Nearly white	21.8

* Stubbe (1966).

Single gene heterosis is exhibited by several mutant genes of *Antirrhinum majus*. The heterozygotes of the mutant types –spectabilis, lurida, vitellina, virescens are superior to the homozygotes in size of various plant parts (Stubbe, 1953, 1966).

De Vries (1901) reported that the strains of *A. majus* produce flowers showing magenta flakes on an ivory ground and later on it was reported by Baur (1924). The gene responsible for this variegation was identified as *Pallida-recurrence*. Instability of this gene established from the occasional appearance of fully coloured plants among the progeny was derived by self-fertilization. Kuchuch and Schuck (1930) described other alleles of *pal*, responsible for various types of variegation in flower colour, such as :

RUBRA : pal^{rub}/pal^{rub}	plants have reddish pink flowers.
CARNEA : pal^{car}/pal^{car}	plants have flesh-coloured flowers.
RHODE : pal^{rhode}/pal^{rhode}	plants have light flesh-coloured flowers.
MALACEA : pal^{mal}/pal^{mal}	plants have pink tinged (pale magenta) flowers.
TUBOCOLORATA : pal^{tub}/pal^{tub}	plants develop the red colour only in a very small zone at the base of flower tube.
TINCTA : pal^{tin}/pal^{tin}	plants have ivory-coloured flowers.
MACULOSA : pal^{mac}/pal^{mac}	plants have flowers with irregularly described indistinct spots in which the red colour corres sponds to some extent to the tone of malacea flower.
VARICOLOR : pal^{var}/pal^{var}	plants have a concentration of pigment at the base of corolla tube (Fincham and Harrison, 1967).
$pal^{rec\ low}$	It has much higher mutation frequency (Sastry, 1976).

The anthocyanin biosynthetic pathway is responsible for the production of anthocyanin pigments in plant tissues and shares a number of enzymes with other biochemical pathways. The six core structural genes of this pathway have been previously cloned and characterized in snapdragon. Rausher *et al.* (1999) reported that upstream genes in the anthocyanin pathway have evolved substantially more slowly than downstream genes. In addition, regulatory genes associated with the anthocyanin pathway tend to evolve more rapidly than the structural genes they regulate, suggesting that adaptive evolution of flower colour may be mediated more by regulatory than by structural genes.

Harrison and Fincham (1964) observed the effects of different temperatures on the frequency of somatic and germinal mutation of pal^{rec}. It was noted that mutations, both somatic and germinal, were much more at 15°C than at 25°C and that the timing of mutation could be controlled by manipulation of the temperature at which plants were cultured. A number of phenotypically distinct mutants was observed with more or less impaired anthocyanin pigmentation carrying mutable allele *pal*^{rec}. The stable patterns were very little affected by temperature changes between 15°C and 25°C but the mutational rates were suppressed at 25°C. All the *pal* mutants produced the same anthocyanin, though in amount, they varied widely from allele to allele (Fincham and Harrison, 1967; Harrison and Fincham, 1968). The instable recurrence alleles of nivea and pallida have been compared by Harrison and Carpenter (1973) who found that *niv*^{rec} was less sensitive to temperature but somatic instability was still inversely correlated with temperature. There is no difference in genetical instability at two temperatures for *niv*^{rec}. Another new type of genetic instability was found to be associated with special allele of *pal*, an anthocyanin producing gene. This allele has been denoted as *pal*$^{rec\ low}$ to distinguish it from *pal*^{rec} which has a much higher mutation frequency (Sastry, 1976). Cyclic changes at the *pal* locus of an unusual allele *pal*$^{rec\ low}$ have been reported by Aslam and Sastry (1979). This allele is colourful and in heterozygous individual which undergoes cyclic shifts from a relatively colourless stable condition to an unstable state. The instability of another gene perluteavariabilis (*perl*^{var}) has been studied by Stubbe *et al.* (1981). There was varying degree of reversion in somatic tissue and generative reversion to *perl*$^{+}$. The generative to *perl*$^{+}$ was not influenced by the degree of somatic reversion but it depends upon specific gene.

Several major loci involved in natural flower color variation. These include the linked ROSEA (ROS) and ELUTA (EL) loci, affecting the intensity and pattern of magenta anthocyanin pigment, and SULFUREA (SULF), affecting the distribution of yellow aurone pigment (Schwinn *et al.*, 2006; Whibley *et al.*, 2006). A number of mutations that affect the activity of regulatory genes that control the expression of the structural genes of floral pigment production have also been described, including delila (del), Eluta (El), rosea (ros), Venosa (Ve), and mutabilis (mut). Mutations in the regulatory genes do not abolish pigmentation but change the pattern of pigmentation within the flowers; Delila affects pigmentation in the corolla tube; Mut affects pigmentation in the corolla lobes; Rosea affects the pattern and intensity of pigmentation in both lobes and tubes; and Venosa affects pigmentation of the epidermis overlying the veins in both lobes and tubes (Schwinn *et al.*, 2006).

Structural genes encoding most enzymes involved in the biosynthesis of magenta anthocyanins and the structurally related yellow aurones have been identified and isolated through a combination of genetics and biochemistry. Several regulatory genes encoding MYB or basic helix-loop-helix (bHLH) transcription factors are known to affect the intensity or pattern of pigmentation (Schwinn *et al.*, 2006).

The *nivea* mutation also affects petal colour, as it consists of a deletion of the single Antirrhinum gene encoding chalcone synthase. This enzyme is necessary for the first committed step of the flavonoid biosynthetic pathway (Wienand *et al.*, 1982). Therefore, this mutant line is lacking not only the purple anthocyanins visible to the human eye but also all of the flavonoid pigments capable of absorbing light of any wavelength, including ultraviolet (UV) A radiation (Glover and Martin, 1998). These flowers not only appear white to human eyes, but will also be 'beewhite', as a result of the reflection of all incident light from cell walls (Kevan *et al.*, 1996).

5.1.3 Flower Form

Crosses between penstemon (non-snap) flower and bunny face produce a bunny face in F_1 with 3:1 segregation bunny : penstemon in F_2. Bergbusch (1999) investigated the manipulation of flower shape and symmetry in *A. majus* in relation to the use of symmetry as an indicator of organization and possible prepatterns in plant development.

The flower form is controlled by a single dominant gene (Watts, 1980). Endress (1999) observed that the *cycloidea* gene in antirrhinum produces monosymmetric flowers with the adaxial side of the androecium reduced, taxa also occur in which the androecium is reduced on both sides, adaxial and abaxial. Actinomorphic flowers have arisen from zygomorphic ones during evolution. Theissen (2000) reported that this mutant carried a defect in *LCYC*, a homology of the *CYC* gene, which controls zygomorphy in *Antirrhinum majus*. The mutant phenotype is not due to changes in the *LCYC* nucleotide sequence, but rather to an extensive, heritable methylation of the gene. A second gene controlling zygomorphy in snapdragon, *DICH*, has also been shown to be a *CYC* homolog and both genes share significant sequence similarity with *TBI*.

Zygomorphy, considered to have coevolved with insect pollination, is apparent in the different morphologies of the dorsal, lateral, and ventral petals of Antirrhinum and in stamen development (Hudson *et al.*, 2008). The dorsal and lateral petals are specified by the paralogous TCP transcription factor genes CYCLOIDEA (CYC) and DICHOTOMA (DICH), which act partly by activating dorsal expression of RADIALIS (RAD). RAD is a protein with a single MYB transcription factor repeat that is thought to compete with the two-repeat MYB protein encoded by the DIVARICATA (DIV) gene to antagonize its ventralizing effect (Almeida and Galego, 2005). However, DIV is not involved in determining the overall dorsoventral asymmetry of the flower and its effects on regional asymmetries depend on interactions with CYC/DICH (Almeida and Galego, 2004). At early stages of corolla development, DIV influences specifically the growth of ventral and lateral petals. However, DIV is transcribed in all primordia irrespective of their dorsoventral positions. It is unlikely that the lack of effect of DIV in dorsal petals is due to redundancy, because inactivation of a DIV-like gene (DVL) either on its own or in combination with a div null mutant has no effect on corolla morphology (Galego and Almeida, 2002).

Asymmetric growth can be caused by gain-of-function mutations in two paralogous homeobox genes, HIRZINA (HIRZ) and INVAGINATA (INA). Ectopic expression of either HIRZ or INA in developing petals results in the duplication of the petal tube as an outgrowth from the existing tube, which produces a structure that resembles the spurs (Golz *et al.*, 2002). Rolland-Lagan *et al.* (2003) observed that asymmetric petal shape might result from a rotation of the petal lobe relative to the principal direction of growth, if growth direction was maintained parallel to the proximodistal axis of the flower, presumably by a long-range signal.

Cells of the petal epidermis of Antirrhinum, like those of many flowering plants, have a conical shape that is dependent on activity of the MYB transcription factor encoded by MIXTA. These conical cells intensify flower color by reducing reflection, as revealed by MIXTA mutants with flat epidermal cells, and are important in attracting pollinating bees (Noda *et al.*, 1994). In addition to having a role in the reproductive ecology of Antirrhinum, the misexpression of MIXTA or related genes can give rise to the formation of epidermal hairs (trichomes), indicating a common regulation of these two cell types and a role for other MIXTA-like genes in regulating multicellular trichome development in Antirrhinum.

Whitney *et al.* (2011) suggested that petal cell shape does significantly influence the physical properties of petals. The commonly found conical petal epidermal cells make a great deal of difference to wettability. Thus, besides providing visual and tactile cues to pollinating animals, conical petal cells may also act to prevent waterlogging and to optimise shedding of dirt and pathogens.

5.1.4 Leaf Shape

Leaves, like petals, are dorsoventrally asymmetric in their morphology and anatomy. Analysis of the classical Antirrhinum mutant *phantastica* (*phan*) indicated that interaction between cells in the dorsal and ventral parts of each developing organ might be needed to establish lateral growth of the leaf blade or petal lobe (Waites and Hudson, 1995).

Once growth of the leaf blade has been initiated, it must be coordinated between cell layers and in different regions to result in a flat leaf with a characteristic shape. This coordination is lost in cincinnata (cin) mutants of Antirrhinum, which show increased leaf growth, particularly towards the margins, which are therefore curled. This is consistent with CIN acting to promote sensitivity to a growth-arrest signal (Heidmann *et al.*, 1998).

The growth and asymmetry of leaves in *Antirrhinum majus* involves the related YABBY transcription factors GRAMINIFOLIA (GRAM) and PROLONGATA (PROL). GRAM is expressed in abaxial margins of organ primordia where it promotes lateral growth and abaxial cell fate. GRAM, however, is not needed for abaxial fate in the absence of adaxial cell specification, suggesting that it promotes abaxial fate by excluding adaxial identity. Although GRAM expression is abaxially restricted, it functions redundantly with its abaxially expressed paralogue, PROL, and with the ubiquitously expressed PHANTASTICA gene to promote adaxial identity via intercellular signalling. This non cell-autonomous behaviour is consistent with the ability of GRAM in only the abaxial most cell layer to direct normal development of more adaxial cells. The contrasting roles of GRAM in promoting and inhibiting adaxial identity might serve to reinforce and maintain the distinction between adaxial and abaxial domains in the growing leaf primordium (Golz *et al.*, 2004).

5.1.5 Resistance against Lodging

Wisniewski (1998a,b) concluded that lodging resistance snapdragon plants are largely under genetic control. The length of the root collar and root diameter at a half of its length may be used in selection of breeding material to improve lodging resistance. He (1998c) also reported that morphological characters like plant height and the main shoot and inflorescence lengths should be included in selection to improve its lodging resistance.

Antirrhinum majus cultivar 'Apollo Cinnamon' was found to have resistance against lodging in the field in Punjab, Pakistan region (Shafique *et al.*, 2011).

5.1.6 Resistance against Rust Disease

The resistance against rust disease was found to be dominant and controlled by a single gene (Emsweller and Jones, 1934). Most of the mutations are recessive and do not express themselves in the heterozygous condition. This necessitates raising of the self progenies of individual plants so that homozygous recessive genotypes can be identified. On account of this very reason the mutation frequencies ranging from 1.23 to 7.0 per cent have been observed in F_2 and F_3 progenies.

The isolates of races of rust, *Puccinaantirrhini* urediospores were inoculated on the leaves of *Antirrhinum majus.* Two races of *P. antirrhini* were identified (α and β). None of the cultivars challenged exhibited resistance to isolates of Race β. Some cultivars were resistant to Race α. Resistance was inherited in a manner consistent with the existence of a dominant, nuclear allele at a single locus (Aitken *et al.*, 1989).

5.2.0 Breeding

Breeding priorities change time to time. Earlier, emphasis was given on breeding against rust disease which is very common in greenhouses. Emsweller and Jones (1934) reported the development of cultivars resistant to rust *Puccinia antirrhini.*

The back cross method was used to combine the resistance with the flower colour and plant habit of standard commercial cultivars. Therefore, resistant parent lines were used widely for F_1 hybrid production with large number of other susceptible parents (Watts, 1980).

Nebel and Ruttle (1938) made crosses between tetraploid Velvet Beauty and tetra-ploid Red Shades and obtained large ruffled flowers of deep colour and heavier stems. Triploid Velvet Beauty produced by tetraploid Velvet Beauty was crossed with diploid Velvet Beauty. This triploid was partially fertile. Diploid Red Shades x Velvet Beauty resulted into small flowers, less ruffled and less deep in colour.

Antirrhinum majus cultivars and some wild species are self-fertile, but they require self-pollination for high seed yields. Although self-fertile, *A. majus* showed unilateral incompatibility and can only be crossed to other self-incompatible species as the female parent (Hudson *et al.*, 2008).

In Antirrhinum, pollinator constancy and preference and poor hybrid seed viability are the most important reproductive isolating mechanisms. Average pre-zygotic isolation was greater than post-zygotic isolation, in accordance with the trend observed in flowering plants in which reproductive isolation is principally caused by pre-zygotic mechanisms. However, average post-zygotic isolation was also high, in contrast to what was expected among Antirrhinum spp. (Carrió and Güemes, 2014).

5.2.1 Interspecific Hybrids

The interspecific hybrids between *A. majus* and *A. molle* have been reported. There was intergrading variation in characters such as plant height, habit of branching and corolla form. From the progeny of this specific hybrid, new class of antirrhinum has been developed. These hybrids have creeping habit and are highly suitable for covering the ground as carpet and hence, the group is called as miniature or magic carpet. These hybrids are also suited to rock gardens. The fertile hybrids between *A. majus* x *A. linkianum* have also been produced.

The reproductive barriers between the rare *Antirrhinum pulverulentum* and its common congener *A. litigiosum* have been analysed under the reproductive isolation index (RI). *Antirrhinum pulverulentum* had a high value of total RI, indicating that there are barriers to gene flow from *A. litigiosum* towards this species, and also had a high value for the intrinsic RI, reflecting a low inherent capacity for production of hybrid plants. In addition, the possibility of successful backcrosses between this species with hybrids produced from *A. litigiosum* ovules were low, given the high intrinsic RI of *A. litigiosum*. These data indicate the current existence of strong and permanent barriers to hybridization between the two species, suggesting that hybridization does not seem to be a serious problem for the conservation of *A. pulverulentum* in the studied population, nor for the near future (Carrió and Güemes, 2019).

5.2.2 Male Sterility

An apetalous pollen sterile form with a thicker fasciated pistil appears quite commonly. Although it sets seed but the yield is low and has not proved to be useful (Goldsmith, 1968).

5.2.3 Self-incompatibility

Cultivated *A. majus* and some wild species *e.g.*, *Antirrhinum valentinum*, *Antirrhinum subbaeticum* and *Antirrhinum siculum* are self-fertile. The majority of Antirrhinum species show gametophytic self-incompatibility that is determined by a single, complex S locus. Individuals of self-incompatible species reject pollen carrying an S allele that corresponds to one of their own alleles and are therefore obligate outbreeders (Hudson *et al.*, 2008). Mapping and sequence analysis of the active self-incompatibility (S) locus of *Antirrhinum hispanicum*, followed by expression and functional studies, identified the likely pollen component to be an F-box protein involved in targeting the RNase for degradation (Lai *et al.*, 2002).

5.2.4 F_1 Hybrid Production

Present era is of F_1 hybrids in antirrhinum. First F_1 hybrid was produced in 1953 for greenhouse forcing. Since then, many hybrids from different groups and types have been introduced. Bright Butterflies which have been produced by recombination of peloric Juhwa and strong normal flowered cultivars (Goldsmith, 1968). Limited use has been made of single gene pollen steriles either in segregating pollen fertile, pollen sterile population or vegetatively propagated male sterile lines. Pollination is done by hand.

5.2.5 Mutation Breeding

The rate of mutation is considerably accelerated by the use of X-rays, chemicals, heat and other mutagenic agents. The total mutation rate for all genes in the complement will naturally be much higher (Baur, 1924). Large differences between strains in spontaneous mutation rates in *A. majus* have also been found by Baur (1924). Strain A_2 had an overall mutation rate of about 5 per cent. On the contrary, the highly inbred strain 50 had a mutation rate of 0.74 per cent. Growth and vigour of the standard strain 50 was studied with those of six mutant types in normal and modified greenhouse environments. It was observed that the wild type was superior to all mutants in the normal greenhouse environment. In abnormal greenhouse environments, however, the wild type was observed to be inferior. The adaptive value of a mutant allele is a function of the environment as well as the intrinsic properties of new allele (Briicher, 1943).

Year to year variation in mutation rates ranges from 0.50 to 0.86 per cent and give overall average rate of 0.6 per cent. A number of macromutation has been isolated in *A. majus* by Stubbe (1952, 1959, 1966). These mutations are : Transcendens – number of stamens reduced from normal four to two; Hemiradialis – five stamens approach to radial symmetry of corolla; Neohemiradialis – six anthers; Radialis – radially symmetrical flowers; Hirzina – floral spurs, approaching *Linaria*; Fistulata – reduction of corolla limb and tubular corolla; Subconnata – one cotyledon. Three mutants of Sippe 50 were recessive to wild type (Delool and Tilney-Bassett, 1986).

Although there have been molecular studies of morphological mutants but it is unclear how these are related to mutants in natural populations, where the constraints of natural selection and breeding structure are quite different. Cubas *et al.* (1999) reported that the mutant carries a defect in *Lcyc*, a homologue of the *cycloidea* gene which controls dorsoventral asymmetry in antirrhinum. The *Lcyc* gene is extensively methylated and transcriptionally silent in the mutant. This modification is heritable and co-segregates with the mutant phenotype. Occasionally the mutant reverts phenotypically during somatic development, correlating with demethylation of *Lcyc* and restoration of gene expression. This indicates that epigenetic mutations may play a significant role in evolution than has hitherto been suspected.

Mutants on M_1 population were detected when *Antirrhinum majus* seeds were treated with ethyl methanesulfonate (EMS) at 0.5 per cent, 0.75 per cent, and 1.0 per cent concentrations for 10 h exposure period. Morphological changes in the M_1 population showed dwarfism, chlorophyll deficiency, and leaf morphology abnormality (Heffron *et al.*, 2006).

5.2.6 Use of Polyploid

Nebel and Ruttle (1938) and Pryor (1951) reported the value of tetraploid antirrhinums which were developed by the use of colchicine (0.2 per cent). This brought revolution due to much increase in flower size. Tetraploid cultivars are widely grown and known for large flower. Triploids have been produced by crossing tetraploid and diploid. Normally, the triploids should have been tidier in appearance because of the absence of seed pods and seeds but all the triploids produce normal number of capsules. This is probably due to the fact that ovules and pollen grains are produced in such abundance, failures in fertilization passed unnoticed (Watts, 1980).

The tetraploids of snapdragon could be successfully induced by 0.1 per cent ~ 0.3 per cent colchicines for 12-36 h and the best treatment 0.2-colchicines for 24 h. Comparing with the diploid the obtained tetraploid had the features of big thick flower-long rough stem. The net photothesis rate of tetraploid snapdragon was higher than that of the diploid (Xiu, *et al.*, 2004).

Kondo *et al.* (2020) obtained tetraploid plants of *Antirrhinum majus* from the seeds of the F_1 variety 'Maryland True Pink' by amiprophos-methyl (APM) treatment and subsequent *in vitro* culture. The highest adventitious shoot formation frequency reached in non-primed seeds treated with 10 mg/l APM for 7 d. Tetraploid frequency increased with higher concentration and/or longer duration of APM treatment and ranged from 0 to 65.9 per cent in the non-primed seeds and 10.3 per cent to 100 per cent in the osmoprimed seeds. Osmopriming of seeds accelerated the induction of tetraploids by APM treatment at the same concentration and duration by 10.3–39.0 per cent.

5.2.7 Genetic or Inherent Factors

Haughn *et al.* (1995) reported that genes in *Antirrhinum majus* play key roles in : (i) the transition from the vegetative to reproductive phase, (ii) the activation of floral development in specific shoots, and (iii) the unique arrangement of floral organs have been identified and in many cases cloned. Many of the genes appear to encode transcription factors that act to select specific developmental programmes of division and differentiation for groups of primordial cells. Other genes may be involved in detecting environmental conditions.

6.0 Growth and Flowering

6.1.0 Environmental Factors

6.1.1 Light

Both duration and intensity of light influence the growth and flowering in snapdragon. Investigations made by Roberts and Struckmeyer (1939), Post (1942), Hanney (1953) and Rogers (1958) clearly indicated that plants responded to day length and flowered earlier under long days. Day length treatments were found to exert quantitative effects and snapdragons were classified as quantitative long day plants (Hedley, 1974). Long days hastened flowering in most cultivars and short days retarded it but did not completely prevent flowering in most cases with a few exception. Maginnes (1960) and Maginnes and Langhans (1960) showed that photoperiod does not influence flowering over the entire period of growth from germination to flower bud initiation but rather during a definite period just prior to flower bud initiation. Langhans and Maginnes (1962) described this as the sensitive stage, a period of growth 40-65 days after germination when the plant has 5 to 10 pairs of leaves. Maginnes and Langhans (1967) also reported that a 4-h continuous or 4-h flash or cyclic light break at night was more effective in shortening the time of flowering and hastening the change from leaf to flower bud production.

The effect of long day cycles on early cultivar Pink Ice and late cultivar Orchid Rocket was studied by Hedley and Harvey (1975). The early cultivar was found to be more responsive to long days than late cultivars. One hour light break given mid-way during 16-h dark period resulted in a decrease in the number of leaves in *cv.* Pink Ice. Supplementary lighting at the same intensity level during the day time had little or no beneficial effect on the growth of the plants. Sanderson and Link (1967) reported that plants receiving supplementary light in the middle of the night flowered earlier than those grown on an un-supplemented natural day or 9-h day.

The effects of duration and intensity of light on growth and flowering were also studied by Flint (1960). Warm-white fluorescent lamps were used and light intensities of 200, 400, 800 and 1600 ft-candle were maintained for 6, 12 and 24 hours. The results of the experiment showed that a given amount of light applied over a long daily duration produced more dry matter than when applied over shorter period. Longer duration at all total light levels also resulted in greater leaf area and stem elongation. Plant grown under 6-hour light did not produce flower buds after 5 months (Table 3). Increased mid-winter light intensities developed by reflecting additional sunlight onto greenhouse benches increased stem length, fresh and dry matter. The plants receiving the reflective illumination flowered 5-7 days earlier but no difference in spike length was found (Carpenter, 1964). Miller (1960) concluded that adjusting night temperature upwards during the growing period after bright winter day resulted in earlier flowering but was associated with slightly smaller size of plant compared with those grown at normal temperature.

Table 3: Flowering Dates, Number of Leaves Formed Prior to Flower Bud Formation and General Appearance of Snapdragon Plants under different Light Duration Treatments with Equal Total Light Originally (1600 ft-candle hours/day)*

Daily Light Duration (Hours)	*Flowering Dates*	*Days from Sowing to Flowering*	*No. of Leaves Formed Prior to Flower Buds*	*Growth Habit*
6	–	More than 150	More than 90	Vegetative on Nov. 21. No axillary growth
12	Sept. 24	92±3	60	Most vigorous growth. Some axillary growth.
24	Sept. 1	69±3	32±1	Rapid thin growth. Profuse axillary growth.

* Flint (1960).

Snapdragons markedly respond photosynthetically to supplementary lighting at intensities and durations higher than needed to elicit photoperiodic responses. Marked increase in dry weight of young plants following all-night supplementary lighting with fluorescent lamps at 86 to 108 W/m^2 with the tubes 20 to 25 cm above the plants was reported by Peterson (1955).

Reductions in flowering time and increase in fresh weight and stem length with supplemental high pressure sodium (HPS) irradiation at a quantum flux density (QFD) of 105±15 $\mu E\ m/s^2$ (Table 4) was also noted by Stefanis and Langhans (1982).

Table 4: Days to Flower, Fresh Weight and Stem Lengths for 4 Snapdragon Cultivars Grown under Natural Irradiation or Natural Plus Supplemental HPS Irradiation at Two Plant Spacings during the Winter*

Cultivar (Response group)	*Natural Plus HPS Irradiation*		*Natural Irradiation*	
	(8 x 10 cm)	*(10 x 13 cm)*	*(8 x 10 cm)*	*(10 x 13 cm)*
Snowman (II)				
Days to flower	92 a[z]	94 a	101 b	108 c
Fresh weight (g)	45 a	45 a	23 b	19 b
Stem length (cm)	99 a	92 ab	83 bc	78 c
Double Azalea Bronze (II)				
Days to flower	99 a	97 a	123 b	122 b
Fresh weight (g)	79 a	76 a	47 b	40 b
Stem length (cm)	130 a	122 ab	113 b	106 b
Phoenix (II)				
Days to flower	95 a	98 a	108 b	106 b
Fresh weight (g)	45 a	49 a	19 b	21 b
Stem length (cm)	100 a	101 a	88 b	88 b
Rocky Mountain (II)				
Days to flower	93 a	95 a	107 b	111 b
Fresh weight (g)	70 a	72 a	37 b	46 b
Stem length (cm)	110 a	108 ab	98 b	106 ab

z Mean separation by row by HSD, 1 per cent level.

*Stefanis and Langhans (1982).

A range of antirrhinum cultivars shown to be quantitative long day plants. Adams *et al.* (2003) observed that snapdragon seedlings were effectively insensitive to photoperiod when very young (juvenile). However, after the end of the juvenile phase, SD delayed flowering and increased the number of leaves below the inflorescence. Plants transferred from LD to SD showed a sudden hastening of flowering and a decrease in leaf number once sufficient LD had been received for flower commitment. Photoperiod had little effect on the rate of flower development.

Analysis of color filter's effect on presumed photoreceptors to exist indicated that antirrhinum plant height was regulated by the action of a blue acting photoreceptor (BAP) and not the phytochrome. There was no evidence for an effect of phytochrome or BAP on time to flowering; however, increasing temperature levels effectively decreased the time to flowering (Khattak and Pearson, 2005).

Munir *et al.* (2010) assessed the stages of photoperiod sensitivity in antirrhinum, flowering time and leaf numbers data of antirrhinum cultivars. Plants at juvenile phase (initial phase of development) were insensitive to photoperiod in both inductive (LD) and non-inductive (SD) environment. After juvenile phase when plants were transferred from LD to SD, they recognised the stimulus under inductive environment and induced flowering. However, plants transferred from non-inductive environment to inductive showed a continuous phase of photosensitivity. Rate of flower development was less sensitive to photoperiod. The duration of photoperiod sensitive phases varied with the cultivars. These cultivars require 4-8 days of photoperiod at critical phase to flower.

Use of shade cloth increased total length of snapdragon flower stems (at the first harvest) compared with no-shade control. Red shade cloth resulted in longer flower stems of snapdragon (at the second harvest), however, shade cloth delayed blooming of snapdragon for 1 week compared with no-shade control along with reduced transpiration rate (Li *et al.*, 2017).

Snapdragon treated with 10 h white light, 3 h red light, 3 h blue light (WRB) promoted stem elongation, inflorescences length, size and number of florets (Xiang *et al.*, 2020).

6.1.2 Temperature

This is another important factor influencing the growth and flowering of antirrhinum. Miller (1960) reported that earlier maturity occurred when plants were subjected to various temperature treatments. Reducing night temperature after cloudy days had no significant effect on size or quality of flower spikes while increasing the night temperature from normal (10°C) to 15°C after bright days resulted in slight earlier flowering, but produced slightly smaller flower spikes. The use of difference in temperature regimes (DIF) drop, with a drop in temperature at sunrise, was investigated by Sach *et al.* (1995) in *A. majus cv.* Coronet Chery as a means of controlling plant height. A 6°C drop for 2 h was used and temperature regimes in the greenhouse were adjusted to give an average diurnal temperature of 16°C. The plants showed variable responses to treatments, depending on sowing date and the plug size used, with reductions in plant height compared with controls ranging from 0 to 43 per cent.

The effects of differences (DIF) between day temperature (DT) and night temperature (NT) and constant temperature regimes (15, 20 or 25°C) on the elongation and development of *A. majus* seedlings was studied by Ito *et al.* (1997a). The 25/15°C (DT/NT) treatment (positive DIF) promoted plant height and internode and petiole lengths compared with a constant 20°C (zero DIF) regime, whereas the 15/25°C treatment (negative DIF) suppressed growth compared with zero DIF. However, leaf length and leaf unfolding rate responded significantly differently to different constant temperature, while not being significantly affected by DIF treatments. It was concluded that taking advantage of DIF effects might facilitate the production of compact seedlings but that short-term control of elongation may be more easily achieved by altering average temperature than by changing DIF.

Under positive DIF, seedling height, internode length and 1st-leaf petiole length were greater than in plants grown under negative DIF. When seedlings were switched from the negative DIF regime to the positive DIF regime (or vice versa) on day 13, significant differences were observed in the final (*i.e.*, 10 days later) plant height, 1st internode length and 1st leaf petiole length between the corresponding treatments. It was concluded that DIF during the early seedling stage had a small effect on subsequent growth (Ito *et al.*, 1997b).

Miller (1962) observed that as the size of the plant increased there was a rapid drop in the optimum growing temperature for dry weight accumulation. Tayama (1963) found that the maximum net assimilation rate, the dry weight accumulation of both tops and roots and the leaf dry weight/total dry weight ratio decreased as the plant age increased. He concluded that relatively high temperatures favour growth of young plants whereas low temperatures favour growth of mature plants.

Seeley (1965) grew greenhouse cultivars Jackpot and Margaret in both heated and unheated soil and did not find any significant difference in the number of flowering stems per plot, weight of 75 cm stems and number of days from planting to harvesting by heating the soil between 68° and 74°F (20.8° and 23.3°C). Very little advantage with soil heating was also recorded by Berninger (1978).

Plants of an early flowering cultivar 'Chimes White' of antirrhinum were subjected to three temperature regimes (0, 10 and 20°C) at 6-leaf pair stage to observe their effects on the flowering time and plant quality. Plants at higher temperature (20°C) flowered earlier than the lower ones (10°C and 0°C). However, maximum flower numbers were counted in plants at 10°C followed by 20°C. Plants at 0°C produced lowest flower buds. The quality of plants was improved at 10°C temperature; whereas, it was poor at 0°C (Munir *et al.*, 2004). Flowering time and rate of progress to flower of antirrhinum *cv.* Chimes White was accelerated by subjecting them to high temperatures, and 28°C is appeared to be a ceiling temperature as above it (30°C) no supra-optimal effect of temperature was observed. Antirrhinum plants can be grown at lower temperatures (18 to 22°C) to enhance their quality and then moved to ceiling temperature to hasten flowering (Munir *et al.*, 2015).

6.1.3 Carbon Dioxide (CO_2)

Carbon dioxide enrichment has been found to be less effective in snapdragon. When plants were grown at high CO_2 (1000 µl/l) leaf ribulose-1, 5-bisphosphate carboxylase/oxygenase content declined but was not affected in the mannitol-producing species of snapdragon (*Antirrhinum majus*). The total leaf carbohydrates increased 60-100 per cent, but mannitol metabolism did not function as an overflow mechanism for the increased accumulation of carbohydrate (Moore *et al.*, 1997).

Shaw and Rogers (1964) recorded little difference in stem length, stem weight or flower spike length due to CO_2 enrichment at either 10 or 15°C night temperature and higher day temperatures. Flowering was, however, accelerated by about 10 days compared to control plants. Slight benefit from CO_2 enrichment at 400 to 500 ppm was also reported by Koths (1964). Lindstrom (1966) grew cultivars belonging to different response groups at CO_2 level of 1500 and 4000 ppm at about 15°C night temperature and observed that cultivars of later response groups, *viz.* (i) late spring and fall group and (ii) summer group flowered earlier and produced better quality flowers when grown under conditions of enriched CO_2. There was, however, not enough additional benefit from 4000 ppm CO_2 over 1500 ppm. Duffett (1960) recommended the use of 15°C night temperature and CO_2 enrichment to 800 ppm for northern Ohio. In North Carolina, CO_2 enrichment at either 750 or 1500 ppm was found to be little beneficial for the production of this crop (Nelson and Larson, 1969). Thus it is concluded that CO_2 enrichment to about 800 ppm and warmer than normal temperature are beneficial for snapdragon production only when cultivars of later response groups are grown.

6.1.4 Light and Temperature

In an experiment on the effect of temperature and photoperiod, Sanderson and Link (1967) observed that the opening of first floret was earliest when the plants were grown at 70°F (21.2°C) night temperature (than at 62°F (16.6°C) or 52°F (11.1°C). The largest number of flowers were, however, produced at 52°F (11.1°C) (Table 5).

Table 5: Influence of Temperature (°F) and Photoperiod on the Mean Society of American Florists Grade[x] of a Winter (cv. Winter Green) and a Summer (cv. Florida) Variety of Snapdragon*

Photoperiod[y]	*Cultivar*				*Cultivar*			
	Minimum Night Temperature (°F)							
	Winter Green				*Florida*			
	70°	*62°*	*52°*	*cv. Mean*	*70°*	*62°*	*52°*	*cv. Mean*
Experiment I								
Natural day (ND)	0.6	1.0	1.7	1.1	2.5	3.8	3.3	3.2
ND plus 2 hr	0.1	0.4	1.8	0.8	1.4	3.4	3.0	2.6
ND plus 4 hr	0.1	0.3	1.2	0.5	0.5	3.3	3.0	2.3
ND plus 6 hr	0.1	0.1	1.1	0.4	0.5	2.6	2.5	1.9
Experiment III								
Natural day	1.1	2.0	2.9	2.0	3.0	3.7	2.9	3.2
9 hr day	2.0	2.4	2.2	2.2	2.8	3.5	3.2	3.2
9 hr plus 2 hr	0.3	1.0	2.0	1.1	1.6	3.2	2.9	2.6
9 hr plus 4 hr	0.3	1.2	1.2	0.9	1.4	3.6	2.4	2.5
9 hr plus 6 hr	0.1	0.9	1.0	0.7	1.2	2.6	3.5	2.4

x Mean SAF grade determined by assigning these numerical values to the SAF grading system for snapdragons : Special=4, Fancy=3, Extra=2, First=1. Flowers failing to meet a grade were scored zero.

y Natural day length occurring at College Park, Maryland was approximately 9.5 to 14.5 hr (December-April) in Experiment I and 9.5 to 15.5 hr (December-May) in Experiment III. 9-hour day achieved by covering plants with black cloth from 4.30 pm to 8.0 am. Natural day or 9-hour day plus 2, 4, or 6-hour of supplementary light applied in the middle of the night.

* Sanderson and Link (1967).

Rutland and Pallas (1972) in an experiment on effect of temperature on transpiration observed increased transpiration rate as the day time temperature was increased in the range of 20 to 35°C while vapour pressure deficit and night temperature were held constant in a controlled environmental chamber. Transpiration rates declined when day temperature was 20 or 25°C after night of 15°C but remained steady during days of 30 or 35°C following 15°C night. The slower rate of warming with 20 and 25°C night influenced the rate of transpiration by causing morning wilting and by increasing stomatal resistance. Marked resistance to passage of water through roots occurred at temperature lower than 25°C (Kramer,

1940). Edwards and Goldenberg (1976) stated that temperature was an important environmental factor when day length was kept constant at 16 hour. At 25°C, inbred lines An 105 and 106 were earlier to flower than An 101 while at 12°C the flowering was much later.

In pinched antirrhinum plants, flowering of the stem was accelerated by high night temperature, but was not accelerated by long day treatment. Yields of cut flowers of variety Light Pink Butterfly II were the highest when grown under a combination of night temperature of 11°C and long day treatment. Cut flower quality was not affected by long day treatment. In non-pinched plants, the whole growth period was accelerated by a high night temperature, but only the flower development was accelerated by long day treatment (Inaba and Horiuchi, 2003).

Long-day conditions anticipated flowering of 40 days compared to the short-day treatment and reduced both stem length and leaf number at the anthesis. Thermal treatments did not affect the flowering time and the main cut stems characteristics (Paradiso *et al.*, 2007).

The long-day treatment combined with 16°C night temperature was effective for decreasing the number of days to flowering and increasing the yield of cut flowers of 'Light Pink Butterfly III' in winter (Inaba *et al.*, 2008).

Munir *et al.* (2017) quantified the flowering and leaf number response of *Antirrhinum majus* L. *cv.* Chimes White to different photoperiods, night temperatures and light integrals using photo-thermal model. Plants flowered earlier at long photoperiod (17 h/d), higher mean diurnal temperature (19.2°C in February and 23.4°C in June) and high ambient light integrals (8.26 $MJ/m^2/d$) and *vice versa*.

6.2.0 Growth Regulators

Growth substances have been used to modify the plant growth and regulate flowering in antirrhinum. Srivastava and Singh (1972) observed that single application of IAA at 20 ppm at 17 days after transplanting improved the vegetative growth and delayed the flowering whereas treatment with 60 ppm depressed the growth but enhanced the flowering. Nijs and Boesman (1981) noted remarkable increase in plant growth by the application of GA_3 at 5-100 ppm. Similarly, Sarhan and El-Sayed (1983) recorded considerable increase in plant height and flower stalk length with the use of GA_3 at 50-200 ppm.

Treatment with ethepon at 1000 ppm reduced the height of antirrhinum plants. Higher doses were toxic and delayed flowering (Munshi *et al.*, 1980). Sarhan and El-Sayed (1983) recorded the largest number of flowers per spike when plants were sprayed with 50-1200 ppm CCC. Incorporation of granular ancymidol to the soil at 311-1243 mg/m^3 also influenced the plant growth (Murray *et al.*, 1986). Application of SADH (250-2000 ppm) 4 weeks after planting and again a month later reduced the plant height and delayed flowering. However, there was an increase in the number of inflorescence and florets per inflorescence (Shedeed *et al.*, 1986).

Brown and Konuk (1994) reported that nebularine, a purine riboside antibiotic compound inhibited the growth of seedlings of antirrhinum and caused mitotic aberrations in root tips because of the potent effect of this purine derivative on cell division, due to cytokinin antagonist effect. Blahut-Beatty *et al.* (1998) applied N6-benzylaminopurine (BAP) to young inflorescences of wild-type *Antirrhinum majus* which resulted in the formation of chimaeric filamentous structures in the dorsal region of the third whorl. In addition, BAP induced similar filamentous structures in between the first and second whorls, and this response was concentration dependent. These observations suggested that cytokinins may be site-specific factors involved in the regulation of floral organ identity genes or genes that control floral symmetry, *i.e.*, the *CYCLOIDEA* gene.

Mannitol 10 to 500 mM markedly promoted flower bud development and spike elongation accompanied by an increase in the number of nodes in snapdragon. Although glucose or sucrose at 250 mM promoted the flower opening more than mannitol, these carbohydrates promoted the bud development and spike elongation only slightly. The treatment with mannitol increased the concentration of mannitol in terminal buds more than those of other carbohydrates. On the contrary, the treatments with glucose and sucrose markedly increased the concentrations of glucose, fructose and sucrose, but only slightly that of mannitol in these organs. These results show that mannitol in snapdragon, has a specific physiological action, which is observed neither with glucose, sucrose nor sorbitol (Ichimura *et al.*, 2003).

The maximum diameter of the floret, leaf area, chlorophylls a, b and total and shoot dry weight was obtained of with gibberellic acid 400 mg/l of (Chehrazi *et al.*, 2017). However, Malik *et al.* (2019) observed maximum growth of vegetative and reproductive characteristics when gibberellic acid 200 ppm was applied. The application of 5 nutrient sprays proved best for vegetative and flowering parameter except for minimum days to first inflorescence initiation, first inflorescence opening and 50 per cent flowering which were obtained by 3 nutrient sprays.

7.0 Propagation

7.1.0 Seed Propagation

Antirrhinum is normally propagated by seeds which are sown in seed boxes, pans, pots or on raised beds. It takes about 3-4 days to start germination. Optimum germination of snapdragon seeds occurs at temperature of 18-21°C and is enhanced by light (Cathey, 1969).

Seeds are very small with approximately 6,000 to 9,000 seeds per gram and can be grown in 3-4 inch pots using rich, well-drained soil with a pH between 5.5 and 5.8. A higher pH can cause iron deficiency in snapdragons. Snapdragons are sensitive to high 'salt' levels in the growing medium, and especially to ammonium. The electro-conductivity should be less than 0.75 mS/cm, and ammonium levels less than 5 ppm (Creel and Kessler, 2007). Seeds when sown in July and September, plant heights exceeded sufficiently 100 cm to satisfy the highest quality of cut flowers (Inaba *et al.*, 2010).

Seeds in pots can be exposed to air without soil cover, but soil should be kept moist during seed germination. To prevent the soil from drying out, pots can be placed in a tray with a plastic cover. Seed germination can be improved either by seed stratification, *e.g.* chilling the seeds suspended in 0.01 per cent agar or water

Propagation of Antirrhinum

Seed Pods

Seeds

Seedlings in Protray

Cultivation of Antirrhinum

Cultivation in Open

Cultivation in Greenhouse

Staking in Antirrhinum

Harvesting Stage and Flowering in Greenhouse

in a refrigerator for 3–5 days before sowing or chilling pots with seeds in a cool room 4–8°C for 3–5 days. Seed germination will be more uniform after a chilling treatment. Germination should occur in 7–10 days at 17–20°C (Hudson *et al.*, 2008). The optimum concentration of growth regulator for the promotion of seed germination was found to 250 uM of GA_3 (Kang and Choi, 2006).

Seed dressing with thiram (0.3 per cent) and captan (0.5 per cent) proved beneficial in increasing the percentage of germination (Zamorski and Milczarek, 1977). Seedlings are highly susceptible to *Pythium* and *Rhizoctonia* and, therefore, seeds should be sown in pathogen-free media. A loose, open, well aerated medium rich in organic matter is necessary for good root growth. Spacing and depth of sowing are also very important. Seeds should not be sown too deep and seedlings not be very crowded. Delworth (1946) recommended about 3000 seedlings/m^2. When seedlings attain the height of 2-3 cm or the first set of two leaves are developed they are ready for transplanting. They can be stored for up to 6 weeks at 2-4°C by providing fluorescent light at 2.7 Klx for 14 hours daily (Kumpf *et al.*, 1966).

Seed priming is a treatment to improve seed quality and plant growth upon germination and is applied widely to crop species to improve uniformity and yield. The maximum germinability was observed when seeds were treated with KNO_3 for 2 days where germination was increased by 14 per cent and shortened the day to germination by 3.5 days. The optimum temperature for germination was 15~25°C. The germination frequency was decreased to lower than 20 per cent at 30°C, which showed that seeds of snapdragon germinates better at low temperature than high temperature. The germination frequency was different at light condition; it was low at dark condition, but was increased by 15~20 per cent at red light condition (Kang and Choi, 2006).

Different priming treatments like hormonal priming with GA_3, chemical priming with KNO_3 and KH_2PO_4 increased the germination, growth, flowering and seed attributes of antirrhinum selections *i.e.* UHFSA-9 and UHFSA-13. Improved vegetative and flowering parameters of plants were observed when seeds were bio-primed with *Trichoderma harzianum* at 1×10^5 cfu/ml as compare to non-primed plants (Bhargava *et al.*, 2015).

The field performance of 24 exotic cultivars of snapdragon at three sowing dates (November 10, March 10 and October 10) was assessed. November 10 sowing was most suitable compared to the two other sowing dates in terms of flowering parameters in Pakistan (Nawaz *et al.*, 2017).

7.2.0 Vegetative Propagation

Shrubby species of *Antirrhinum* are propagated by softwood cuttings, taken in September and rooted in a cold frame. When cuttings are taken in spring, rooted in gentle heat (Chittendan and Synge, 1956; Brickell, 1996). All Antirrhinum species can be propagated clonally from cuttings. It also readily forms grafts within and between species. Cuttings can be prepared by cutting 20–50 mm immediately below the leaf node. Dip the cut end in a solution containing 1 per cent Indole-3 butyric acid or 0.5 per cent 1-naphthaleneacetic acid. Insert cuttings into rooting substrate or soil and maintain them at high humidity (100 per cent) or mist every 30 minutes in an enclosed container or chamber at 17–20°C. After approximately 2–3 weeks, rooted cuttings can be transplanted (Hudson *et al.*, 2008).

7.3.0 Micropropagation

Pfister and Widholm (1984) were successful in the regeneration of plants from cultures initiated from root and shoot tissues. Complete plants with root placed in soil developed into fertile plants. Excised hypocotyls from 8 commercial cultivars, 3 inbred lines and an F_1 hybrid were cultured on MS medium containing 2.22 µM BA to assess genotypic effects on adventitious shoot formation. Ten-day-old seedlings grown under a 16 h photoperiod was optimum for culture, on medium containing 2.22 µM BA for 3 weeks for adventitious shoot formation. Adventitious shoot formation was genotype-dependent (Schroeder and Stimart, 1999).

A rapid and efficient regeneration system has been developed for inter-specific self-incompatible hybrids between *Antirrhinum majus* × *A. hispanicum*. Multiple shoots were induced from hypocotyls explants on Murashige and Skoog (MS) medium or Loblolly pine medium supplemented with a low concentration of 3-indoleacetic acid (0.05 mg/l) and a high concentration of 6-benzyladenine (1 mg/l). Rooting took place in 1/2 MS medium without growth regulators, by using this procedure, it takes about 5–6 weeks from seed to plantlet (Wang *et al.*, 2004).

Shoot tips of *Antirrhinum majus* were excised from field grown mature plants and transferred to hormone-free MS medium. The addition of 1.5 mg/l kinetin and 0.5 mg/l benzyl adenine (BAP) produced the highest multiplication rate per explant. The explants were successfully stored without serious losses by using MS medium supplemented with an appropriate concentration of 3 per cent sucrose, 3-6 per cent sorbitol or 6 per cent glucose at 24±concentration for up to 32 weeks (Sheyaib *et al.*, 2010).

For the direct micropropagation in snapdragon, MS medium supplemented with BAP at 0.50 and 1.00 mg/l produced the highest shoots number from the expalnts (Hypocotyl, cotyledon and cotyledonary node). Ninety per cent of the microshoots were rooted on MS medium supplemented with 0.50 and 1.0 mg/l IBA but rooting which achieved on medium fortified with NAA at 0.50 mg/l significantly recorded the highest roots number. For indirect micropropagation, explants successfully formed callus by using MS medium supplemented with Thidiazuron (TDZ) at 1.00, 2.00 or 4.00 mg/l. The highest shoots number derived from callus was recorded for callus cultured on MS medium fortified with 2.00 mg/l TDZ + 0.50 mg/l NAA (Hamza *et al.*, 2013).

8.0 Soil and Climate

8.1.0 Soil

Antirrhinum needs a porous and well-drained medium, rich in organic matter. Soil having pH 6.0 to 6.5 and good water holding capacity is ideal for proper growth and flowering. As snapdragons need good aeration, proper preparation of growing media prior to planting is important (Miller, 1957). Soil compaction causes stunted growth, weak root system and delayed flowering (Carpenter, 1963). Excellent growth and flowering were recorded in a media ranging from 1:1 mixture of peat and perlite to 1:1:1 mixture of peat: soil: perlite or in 3:3 soil and sand mixture provided aeration, moisture and nutrition were maintained properly (Hanan and Langhans, 1962). Secerov-Fiser and Vujosevic (1998) obtained best root system of antirrhinum when grown in 1:1:1 peat: sand: leaf mould mixture.

Snapdragons were grown on saline soil (0.00 to 0.9 per cent) in Yemen. The highest salinity caused all plants to die. In salinity up to 0.6 per cent NaCl, the plants did not reach the flowering stage. Plant height, number of branches/plant, flower spike length, fresh and dry weights of both herbage and roots decreased significantly as the level of NaCl increased. The uptake of N, P, K, Ca, Na and Cl tended to decrease as soil NaCl increased (Kandeel *et al.*, 1999).

In studies to find the non-fumigant alternatives to methyl bromide for management of nematodes, soil-borne disease, and weeds in production of snapdragon by McSorley *et al.* (2009) found that solarization alone provided results similar to methyl bromide for managing weeds, nematodes, and plant disease mortality.

8.1.1 Growing Media

Mehmood *et al.* (2013) suggested that leaf mould can be recommended as a standard potting media for *Antirrhinum majus*. For commercial cut antirrhinum production, substrates with locally available components of coco-coir (CC), rice hulls (RH) and sugarcane pressmud (PM) can be recommended. The highest quality plants in terms of shortest production time, tallest plant height, optimal leaf chlorophyll content, highest leaf area, fresh and dry weight, longest vase life were recorded when plants were grown in 50 per cent CC + 40 per cent RH +10 per cent PM (Ahmad *et al.*, 2020).

8.2.0 Climate

Snapdragons are cool season herbaceous perennials but grown as annuals, especially in sub-tropical regions (Huo and Chen, 2018). Sowing of seeds of *cvs.* Navajo and Delworth from February to August at 15-day interval reduced the flowering time until May, while inhibition of lateral shoot formation became more pronounced. With sowing from May until 15 July, the time required for development and lateral shoot formation remained fairly constant, sowing in August took a long time to flower (Maatsch and Runger, 1960).

9.0 Cultivation

9.1.0 Transplanting

As mentioned earlier, seedlings should be transplanted when one or two sets of true leaves are developed. Soil or growing medium is to be thoroughly prepared before this operation. A distance of 7.5 x 15 cm or 10.0 x 12.5 cm per plant is recommended for winter crops and 7.5 x 10.0 cm or 7.5 x 12.5 cm for summer crops (Fries, 1962; Sanderson, 1975a). Holley (1966) recorded increased flower production with greater plant density. Plants should be watered immediately after transplanting.

Decrease in weight of cut flowers and the number of axillary buds was recorded at the higher planting densities. Days to anthesis and the length of cut flowers of 'Maryland Pink' decreased when planted in deep type containers. The growth of 'Light Pink Butterfly II' was not affected by the container size (Inaba *et al.*, 2010).

Under sub-tropical region, antirrhinum had acceptable raceme length, quality, yield and vase life when seeds were sown from 15 September to 15 October and transplanted at 22.5 × 22.5 cm spacing. Stem quality was greater with increased plant spacing (30 × 30 cm), but the benefits were reduced by longer crop time and reduced number of harvestable stems (Abdullah *et al.*, 2020).

9.2.0 Irrigation

Antirrhinums need adequate moisture in soil for all stages of growth. Irrigation schedule mainly depends on the soil and weather conditions. Five crops of snapdragons were subjected to 3 irrigation regimes – watering when soil suction was 30 cm, 300 cm and 600 cm or higher, between 8 and 9 a.m. Reducing frequency of irrigation generally reduced mean grade of cut flower, stem length, number of florets per spike and fresh and dry weight and increased the per cent dry weight of flower. Usually the higher the solar radiation during initial growth, the greater was the reduction in plant growth when watered at 300 and 600 cm suction. The amount of water necessary to produce 1 g dry weight of plant material decreased at higher suction levels and with increasing solar radiation (Hanan, 1965).

The effect of soil moisture content in 4 greenhouse soils, on growth and flowering of snapdragon was studied by Hanan and Langhans (1964). Maximum growth and flower quality were observed when water equalled about 39 per cent of the total soil volume. At moisture content above 34 per cent or below 24 per cent, mean grade stem length and fresh weight decreased. Dry weight of cut flowers increased as the soil became dry. Per cent dry weight was least between moisture levels of 34 and 24 per cent. They also found that major oxygen flux, measured 7 cm below the soil surface, was found to occur when 55 per cent of the total pore space was filled by air (45 per cent filled with water).

Snapdragons can be produced from seed when irrigated with saline water upto 14 dS/m because germination was recorded at 92 per cent or greater. Overall, quality of stems produced with saline waters ranging from 2.5 to 11 dS/m was very high ('special'). Irrigation with more saline water (14 dS/m) resulted in a slight reduction in quality and stems were rated as 'fancy' depending on the cultivar. Both cut flower cultivars can be produced for commercial use under saline conditions up to at least 14 dS/m (Carter and Grieve, 2008).

Saavoss *et al.* (2016) used wireless sensor networks to automatically control irrigation in commercial hydroponic *Antirrhinum majus* cut-flower production in a greenhouse with tight control over environmental conditions. The time to first harvest and time to last harvest were reduced for all cultivar groups, indicating that the plants grew faster using sensor networks. Production time per crop was decreased, allowing 2.5 additional crops per year. Electricity usage was also reduced. Increases in irrigation efficiency achieved by the use of sensor networks have important environmental benefits as well, such as reducing pressure on water supplies, reducing greenhouse gas emissions by reducing energy and fertilizer use and reducing nutrient leaching into waterways.

9.3.0 Manuring and Fertilization

Nutritional deficiency symptoms in snapdragons have been extensively studied by Oertli (1970). Nitrogen deficiency causes stunted growth, and yellowing of leaves. The plants become thin, woody and few short axillary branches arise. Phosphorus deficiency also suppressed growth, young leaves become dark green in colour and the tip may curve downward and inward toward the central stem.

Deficiency of potassium results in interveinal chlorosis of young leaves similar to iron deficiency followed by necrosis of tips and margins of older leaves.

In calcium deficient nutrient solution the root system in snapdragon is first affected. The roots of seedlings become thin, less branched and poorly developed and die very quickly. Older plants also begin to wilt in 1 or 2 weeks and die. The symptoms of magnesium deficiency are interveinal chlorosis and necrosis of older leaves on young plants. The tips of older leaves curl downwards and younger leaves upwards.

Sulphur deficiency causes pale, yellow-green upper leaves in which veins are lighter than the rest of the leaf blade. Interveinal chlorosis appears on the older leaves.

Iron chlorosis appears first as interveinal yellowing of the youngest leaves and gradually there is complete loss of green colour. Axillary shoots on such plants are small and almost white in colour.

Boron deficiency results in death and abortion of terminal growing points and terminal flower spikes, followed by the growth of axillary shoots in the lower part of the stem. Boron deficiency symptoms can be accelerated by over application of calcium containing fertilizer or lime due to the antagonistic reaction between calcium and boron (Carmichael, 1968).

The nutritional requirement of antirrhinum is relatively low compared to rose and chrysanthemum. Rogers (1951) did not find significant difference in growth in soils having a range of 1 to 10 ppm P or 5 or 30 ppm K; higher nitrate levels, however, stimulated growth. Jeong *et al.* (1992) reported that *A. majus cv.* Rocket Mixed grew best with 25:75 NH_4 : NO_3. In sand culture, Flint and Asen (1953) found nutrient solution containing 160 ppm N, 32 ppm P_2O_5, 240 ppm K_2O, 176 ppm Ca and 64 ppm Mg optimum for growth and development; higher concentrations of nutrients caused leaf chlorosis, low dry matter production, shorter stem and fewer flowers. Boodley (1962) suggested application of N at the early stage to ensure vegetative growth. Excessive N, on the other hand, should be avoided. Increased P concentration (15 versus 3 μg/ml) significantly increased plant fresh weight and the total number of flowers per spike (Besmer and Koide, 1999). In case of constant liquid fertilization, Sanderson (1975b) recommended about half the normal concentration of 200 ppm each of N and K.

In an experiment, Bunt (1968) concluded that magnesium ammonium phosphate was promising as a base fertilizer and hoof and horn was less satisfactory than 45 *a.i.*

urea-formaldehyde. In John Innes No. 1 potting compost, increasing rates of hoof and horn (0.6 kg, 1.2 kg and 2.4 kg/m^3) increased the flower number in all the early and late cultivars used (Hedley *et al.*, 1977). Bevacqua and Mellano (1994) reported that seedlings established more vigorously when sewage sludge compost was mixed with soil but yield (plant FW) was not affected. Soil analysis of compost treated plots revealed lowered pH and increased levels of organic matter, primary nutrients, soluble salts and heavy metals. Rainbow *et al.* (1998) developed a process to convert different composts into highly effective growing media by application of 1.7 to 3.0 g ammonium nitrate and 50 ml phosphoric acid (5.4 per cent m/m P_2O_5) per litre of compost, followed by dilution with coir in a ratio of between 1 and 5 parts by volume to 1 part by volume of compost. Treated compost promoted better growth of *A. majus.*

Besmer and Koide (1999) reported that mycorrhizal colonization in a soilless medium significantly increased flower vase life of *A. majus* and decreased flower ethylene production. The reduction in ethylene production caused by mycorrhizal colonization was as large as the variation in ethylene production among snapdragon cultivars. It was, therefore, concluded that mycorrhizal colonization may be a viable alternative to toxic ethylene inhibitors such as silver thiosulfate. Mycorrhizal colonization did not significantly influence response to exogenous ethylene.

Memon and Khetran (2014) observed that under higher humic acid level (350 g/sub-plot) in addition to calcium chloride (180 g/sub-plot), maximum flowers/ spike were achieved with minimum blooming period and early emergence of first flower. Calcium chloride when applied in addition to humic acid at higher level, the growth and flower production of Snapdragon improved substantially. Hence, it is suggestible that for growing successful Snapdragon, the plants may be fertilized with 350 g humic acid + 180 g calcium chloride.

Mycorrhizal inoculation (*Rhizophagus intraradices*) at seedling stage of snapdragon improved plant height, length and basal diameter of flowering stems. It also increased starch, phenolics, chlorophylls and carotenoids but decreased soluble sugars and proline in leaves (Tognon *et al.*, 2016).

The mycorrhizal treatment alleviated the deleterious effects of salt stress in snapdragon plants. Under saline conditions, snapdragon plants require mycorrhiza, not only for acclimation purposes, but also for continuing nutrient uptake during the growth stages (El-Nashar, 2017). Adding of *gyttja* (organic material) with nitrogen fertilizer to the calcareous and infertile soil will enhance the morphological attributes of plants (Mjeed and Ali, 2017).

Nitrogen and phosphorous at 60 kg/ha resulted in maximum duration of flowering, length of spike, number of spike/plant and per hectare and vase life, however, maximum number of florets per spike was obtained by nitrogen at 90 kg/ha (Verma *et al.*, 2018).

The spraying tryptophan at 200 mg/l and zinc sulfate at 4 g/l improved vegetative and flowering growth in snapdragon. The interaction treatment between the spraying with tryptophan at 200 mg/l and zinc sulfate at 4 g/l was found best for vegetative and flowering traits (Hadi and Saeed, 2019).

9.4.0 Mulching

In the wavelength zone of 300 nm to 1100 nm, 85 per cent of the total light source was reflected by reflective film (RF) mulch, while over 95 per cent was absorbed by black polyethylene film (BL). RF significantly increased stem hardiness and transpiration rate, and significantly reduced the lodging percentage, compared to BL mulching. RF mulching also significantly reduced the number of days to full bloom (by approximately 12 days) compared to BL, and produced good cut flower quality (Kim *et al.*, 1998).

A transparent biodegradable film and a transparent biodegradable spray coating were compared with commercial transparent low-density polyethylene film mulch in snapdragon. The capacity to reduce the long wave infrared thermal losses was higher for the biodegradable mulches. The increase of the growing medium temperature allowed the harvest of the snapdragon grown in the mulched trays 7 days before that of the flowers grown in the un-mulched trays. At the end of the crop cycle, the disposal of the biodegradable mulching materials was carried out by fragmenting and mixing them with plant residue and the growing medium. The residue biodegraded in almost 1 month for the spray coating and in about 12 months for the biodegradable film (Schettini *et al.*, 2007).

9.5.0 Weed Control

Control of weeds, particularly at early stages of plant growth, is essential for the normal growth of plants. Generally 4-5 weedings are required for successful raising of crop. Crag 1 at 4 lb/acre and simazine at 1 lb/acre gave good weed control when applied in July and caused no injury to snapdragons (Warren and others, 1960). Alachlor at 4 lb, chloramben 4 lb and oxidiazon 4 lb/acre were very effective in controlling weeds when applied after transplanting of seedlings (Bing, 1981). Haramaki and Kuhns (1985) recorded complete control of weeds with the application of oryzalin. The chemical did not show any harmful effect on the plants.

Solarization or mulching alone reduced weed numbers but integration of solarization and mulching provided the most effective control of weeds in *Antirrhinum majus* 'Potomac Pink' (Gill and McSorley, 2010).

9.6.0 Pinching

Pinching is a method that can be used to produce well proportioned plants, suitable for selling as high quality pot-plants and also as a method of controlling the time of flowering, when plants are required for selling them flowered at a certain time and occasion. In antirrhinum, pinching is done to enhance the number of branches per plant and obtain greater number of flowers. The operation should be done at the early stage of plant growth so that branches become quite strong before flowering. Wainwright and Irwin (1987) observed that pinching reduced the plant height but produced greater number of flower spikes. Flowering was, however, delayed with this practice.

The plant quality of antirrhinum cultivar 'Chimes White' was much improved in plants pinched at 5-7 leaf-pair stage at the expense of flowering time. Though control plants flowered earlier than the pinched ones, they produced less number of flower buds. Flower time and rate of progress to flowering in pinched plants increased linearly and significantly. The quality of pinched plants regarding branch numbers, leaf area, plant height, plant fresh weight *etc.* was significantly improved in all treatments (Munir and Naz, 2006).

The suitable positions for pinching the main stems of *Antirrhinum majus* were just above the second or third node for 'Maryland Pink', and the second node for 'Light Pink Butterfly IF (Inaba *et al.*, 2010). Pinching the plants at 3-leaves stage and treated with paclobutrazol (PBZ) caused a remarkable reduction in plant height. The plants treated by 150 mg/l PBZ and pinched at 7-leaves stage, the shortest inflorescence length was obtained (Shanan and Soliman, 2011).

9.7.0 Greenhouse

In an experiment, Sullivan and Pasian (2000) produced cut flower of 3 antirrhinum cultivars (Potomac Pink, Winter White and Potomac Light Pink), by growing in trays and ground beds in a greenhouse. The quality of flowering shoots was lower when plants were grown in raised trays rather than in on ground trays or in ground beds. Flowering shoot grade (a subjective quality indicator) correlated well with the ratio of shoot dry-weight to shoot length (an objective quality indicator). Flower quality of plants grown on ground trays was equal to that of those grown in ground beds (Table 6).

Table 6: Changes in Length, Grade, Dry Weight and the Ratio of Dry Weight to Length of Flowering Shoots of Three Cultivars of Snapdragon as Affected by Growing Method and Culture Time*

		Culture Period, 1996-1998				
Positions	*Tray*	*22 Aug.- 1 Jan.*	*24 Sept.- 2 Jan.*	*8 Feb.- 7 May*	*5 May- 13 July*	*25 Aug.- 19 Jan.*
		Cultivar				
		Potomac Pink	*Winter White*	*Pot. Light Pink*	*Potomac Pink*	*Potomac Pink*
		Flowering shoot length (cm)				
Ground	None	154.9 az	109.4 a	146.0 a	97.4 a	131.2 a
	Small	155.4 a	93.4 b	150.3 a	99.4 a	127.3 a
	Large	153.2 a	98.0 b	140.3 a	99.2 a	133.9 a
Raised	Small	124.7 b	70.0 d	119.2 b	76.3 b	106.8 b
	Large	131.5 b	79.9 c	125.1 b	79.4 b	104.0 b
		Flowering shoot gradey				
Ground	None	1.14 b	1.21 b	1.31 a	1.10 b	1.00 a
	Small	1.27 ab	1.53 ab	1.00 a	1.11 b	1.08 a
	Large	1.24 ab	1.28 ab	1.28 a	1.10 b	1.00 a
Raised	Small	1.40 ab	1.86 a	1.35 a	1.75 a	1.32 a
	Large	1.44 a	1.61 ab	1.37 a	1.83 a	1.22 a

Culture Period, 1996-1998						
Positions	*Tray*	*22 Aug.- 1 Jan.*	*24 Sept.- 2 Jan.*	*8 Feb.- 7 May*	*5 May- 13 July*	*25 Aug.- 19 Jan.*
		Cultivar				
		Potomac Pink	*Winter White*	*Pot. Light Pink*	*Potomac Pink*	*Potomac Pink*
Flowering shoot dry weight (g)						
Ground	None	15.9 a	6.6 a	13.0 b	7.6 a	15.4 a
	Small	15.4 a	5.2 ab	16.7 a	7.8 a	13.7 a
	Large	13.6 a	5.6 a	10.6 bc	8.3 a	13.6 a
Raised	Small	9.7 b	2.8 c	10.2 bc	4.2 b	8.6 b
	Large	10.4 b	3.8 bc	9.3 c	4.3 b	8.0 b
Ratio of flowering shoot dry weight to flowering shoot length (g/m)						
Ground	None	10.2 a	6.0 a	8.9 b	7.8 a	12.0 a
	Small	10.0 ab	5.6 ab	11.0 a	7.8 a	10.8 a
	Large	8.8 ab	5.7 ab	7.5 b	8.3 a	10.6 ab
Raised	Small	7.7 b	4.0 c	8.5 b	5.5 b	8.0 bc
	Large	7.9 ab	5.0 bc	7.4 b	5.4 b	7.7 c

z Mean separation with columns and observations by Tukey's pairwise comparison at $P \leq 0.05$.

y Numbers close to 1 represent highest quality and numbers close to 3 represent poorest quality.

* Sullivan and Pasian (2000).

The results of the morphological parameters revealed that the protected cultivation promotes higher growth and development rates, when compared to the open field (Bhargava *et al.*, 2016).

10.0 Diseases and Pests

10.1.0 Diseases

Fungal Diseases

10.1.1 Damping Off

The disease is characterized by the damping off of the seedlings in the nursery, leading to their complete death. The major pathogens associated with this disease have been reported to be *Rhizoctonia solani* (Sharma and Mahmud, 1950) and *Pellicularia praticola* (Saksena, 1960). Disease symptoms of root rot in snapdragon caused by *Pythium* spp. include crown and root rot, seed and seedling damping-off, rot of the stem base, and stunting (Garzon *et al.*, 2011). The disease is more severe under high soil moisture and temperature conditions. This can be checked to some extent by nursery sanitation, thin sowing and regulating the water supply. The raising of seedlings in sterilized soil and application of fungicides like brassicol (0.3 per cent) also proves beneficial. Dickerson (1999) suppressed damping off (*Pythium ultimum*) of snapdragon seedlings by using 25 per cent compost + 25 per cent peat moss in potting mixtures in greenhouse. A compost was produced from biosolids, yard trimmings, horse track manure and source-separated landscape residuals and residential solid waste. The 50 per cent peat moss mix resulted in the lowest suppression.

Table 7: Effect of Growing Conditions on Growth, Flowering and Seed Production of *Antirrhinum* Selection UHFSA-13*

Treatment	*Plant Height (cm)*	*Plant Spread (cm)*	*Number of Spikes*	*Spike Length (cm)*	*Days Taken to Visible Bud Formation*	*Days to Flowering*
Polyhouse	81.42 ± 4.36a	42.84 ± 2.06a	12.29 ± 1.00a	63.44 ± 8.66a	90.79 ± 8.43a	116.00 ± 7.42a
Open	65.67 ± 1.77b	40.15 ± 1.09b	9.79 ± 0.30b	55.83 ± 3.89b	103.92 ± 6.72b	123.29 ± 8.66b
	Number of Florets per Spike	*Duration of Flowering (Days)*	*Total Number of Pods per Spike*	*Number of Seeds per Pod*	*Test Weight of Seeds (mg)*	*Seed Yield per Plant (g)*
Polyhouse	23.96 ± 3.83a	46.29 ± 8.08a	20.08 ± 2.00a	321.25 ± 6.60a	165.58 ± 9.78a	3.51 ± 0.18a
Open	17.89 ± 2.99b	30.00 ± 4.48b	17.22 ± 0.63b	310.63 ± 15.03b	157.75 ± 8.60b	3.17 ± 0.11b
	Cost of Cultivation (Rs./m²)	*Gross Income (Rs./m²)*	*Net Returns (Rs./m²)*	*Benefit Cost Ratio*		
Polyhouse	25.87	141.03	115.16	4.45		
Open	18.62	74.30	55.63	2.99		

Similar letter with in a column do not differ significantly by Fischer *t* test (*P* B 0.05).

* Bhargava *et al.* (2016).

Aluminum as $Al_2(SO_4)_3$ was applied as a drench at 0.75, 1.9, or 3.75 g/150 ml water to the surface of infested medium in 650-cm^2 plug trays (1300-cm^3 tray volume). All concentrations of aluminum were effective in controlling pre-emergence damping-off of snapdragon. At 4 days after seeding and drenching with aluminium sulfate, exchangeable aluminum was 0, 0.5, and 2.03 meq Al^{+3}/100 g medium, respectively, for the three concentrations used. Control of damping-off of snapdragon and vinca with 10 meq Al^{+3}/100 cm^3 medium with no detectable exchangeable aluminum 4 days after application suggests that *P. parasitica* was suppressed by aluminum early in the host–pathogen interaction, whereas petunia was susceptible to damping-off for a longer period before seedling emergence. Aluminum was not phytotoxic to vinca, snapdragon, or petunia grown in a limed medium.

Strategies commonly used to control Pythium root rot include using soilless media, disinfesting irrigation water in recirculating systems (Daughtrey and Benson, 2005), scouting crops, rogueing symptomatic plants, and applying fungicides and/or biological control agents (Moorman and Kim, 2004). When 12 snapdragon cultivars were inoculated with *Pythium aphanidermatum* or *Pythium irregulare*, no cultivar was completely resistant to either pathogen, but Twinny White and Candy Showers Yellow were least susceptible to both. Integrating effective fungicides with less susceptible cultivars can limit disease caused by *Pythium* spp. Two fungicides, fenamidone and mefenoxam were the best treatments against *P. aphanidermatum* and *P. ultimum* in snapdragon. Integrated uses of *Trichoderma harzianum*, fenamidone and potassium phosphite were the best treatment to control *P. irregulare* infecting snapdragon (Múnera and Hausbeck, 2015).

10.1.2 Anthracnose

This is the most common disease of greenhouses and caused by *Colletotrichum antirrhini* (Pirone *et al.*, 1960), and *C. fuscum* (Sohi, 1984). The symptoms appear on stem and leaves as pale yellowish-green to grey spots with a narrow border. Spots on old stem are sunken, oblong and pale yellowish green to grey with narrow brown borders. Anthracnose of snapdragon caused by *Colletotrichum destructivum* was reported by Tomioka *et al.* (2010). Water-soaked or chlorotic lesions initially appeared on leaves. The lesions gradually enlarged, coalesced, turned brown and dried, resulting in early blight of the affected leaves. Diseased plants with numerous lesions withered rapidly. Spraying with dithane M-45 or dithane Z-78 or captan at 0.2 per cent or bavistin (0.1 per cent) is useful in controlling the disease. The infected plant debris should be destroyed.

10.1.3 Blight

This disease is caused by *Phyllosticta antirrhini*. It is characterized as creamish to brownish spots on the foliage. Ashy-grey spots are also produced on stems. Pale brown dead tissues covered with gray fungi appear on shoots, leaves, and flowers when plants are infected with *Botrytis* blight (Huo and Chen, 2018). The disease, which is more serious in summer (Sohi, 1984) can be controlled by taking measures as mentioned in case of anthracnose. Spray the shoots with suspended sulfur or other proprietary fungicide (Huo and Chen, 2018).

10.1.4 Leaf Spot

In Europe, *Septoria antirrhini* has been reported to cause small circular spots with a pale centre and bordered by a purplish brown margin appearing on the leaves (Pirone *et al.*, 1960). Leaf spot caused by *Cercospora antirrhini* is a serious problem in North Carolina (Porter and Aycock, 1967). Removal of dried infected leaves, which are inoculum sources, is recommended as a primary control measure.

10.1.5 Rust

It is observed both in field and greenhouses. The affected plants show brown pustules on leaves, stem and sepals (Pirone *et al.*, 1960). The pathogen causing this disease is *Puccinia antirrhini*. The disease is more severe under dry conditions and severely attacked plant gets crippled. Several cultivars are resistant to rust. Saniewska (1995) reported that cultivars Anna, Ajak, Eldorado, Eris, Irene, Marlena and White Spire were the most resistant to rust (*Puccinia antirrhini*). The disease susceptible cultivars can be controlled by spraying fungicides namely thiram or a mixture of sulphur and zineb (0.2 per cent) at regular intervals. *Bacillus polymyxa* (*Paenibacillus polymyxa*) and their filtrate inhibited sporulation of rust (*Puccinia antirrhini*) on leaves of antirrhinum (Saniewska *et al.*, 1998).

10.1.6 Powdery Mildew

The typical symptoms of the disease include the formation of a white powder coating on both leaf surfaces and young stems. It is caused by *Oidium* sp. and the attack is more severe under conditions of high humidity and dry soil. The disease can be checked by spraying wettable sulphur or karathane at 0.2 per cent. Application of benomyl as a spray (100 to 300 ppm) or as a soil drench (100-150 ppm) gave good control of powdery mildew caused by *Podosphaera leucotricha* (Raabe *et al.*, 1970).

10.1.7 Downy Mildew

It is caused by fungus *Peronospora antirrhini* which appears on the lower side of the leaves in mealy white patches. General stunting and crippling of plants are other symptoms. The seedlings are worst affected by the disease (Sohi, 1984). It can be controlled by spraying dithane M-45 (0.2 per cent) at fortnightly intervals. Paulus *et al.* (1983) controlled downy mildew successfully by spraying subdue (metalaxyl) at the rate of 0.75 pints/100 gal water and manzate 80 W at 1.5 lb/100 gal water.

Hourly concentrations of airborne sporangia of *Peronospora antirrhini* at a commercial snapdragon farm were investigated by Byrne *et al.* (2005). Minimum daily temperatures <10.0°C appeared to have a moderate limiting effect on atmospheric sporangial concentrations, whereas temperatures <6.0°C had more severe limiting effects. Maximum daily temperatures ≥30.0°C limited concentrations of atmospheric sporangia. Long dew periods (≥6 h) were associated with relatively large sporangia releases. Consecutive days with short leaf wetness periods were associated with low atmospheric sporangial concentrations.

Cultivar 'Potomac Rose', 'Potomac Apple,' and 'Potomac Dark Orange' are significantly less susceptible to downy mildew (Byrne *et al.*, 2000).

10.1.8 Wilt

The disease results in dropping of leaves which become chlorotic and may ultimately lead to the death of plants. It is attributed to different strains of *Fusarium solani* and *F. solani* var. *merittii* (Srivastava, 1969). High temperature (29-35°C) and high soil moisture are predisposing factors for the disease development. Seed treatment with captan (0.2 per cent) or bavistin (0.1 per cent) is useful in checking the disease.

Alternaria and *Helminthosporium* species have also been reported to attack snapdragon flowers in Florida (Engelhard, 1971). Stem rot caused by *Sclerotinia sclerotiorum* is another disease attacking snapdragon. Hamdollah-Zadeh (1993) observed that the defoliant strains of *Verticillium dahliae* caused more severe wilt symptoms on antirrhinum.

10.1.9 Basal Stem and Root Rot Disease

It is caused by Alternaria, Aspergillus, Cladosporium, Fusarium, Macrophomina, Penicillium and Stemphyllium. *Fusarium oxysporum* was the most frequent fungi approached 36.0 x 10^2 cfu/gm soil. *F. oxysporum* reduced chlorophyll content of infected plants, and the percentage of its inhibition increased up to 81.88 in severe infections. The infection caused significance stunting of plants when plant height inhibited by 58.09 per cent in dead plants. Antagonism ability of Trichoderma species as *T. harzianum, T. viride,* and *T. reesei* against *F. oxysporum* showed control efficiency of.1.666, 2.333, and 2.666, respectively (Ramadan, 2018).

Viral Diseases

Seedlings of antirrhinum in nursery are attacked by leaf mosaic which has been found to be a strain of cucumber mosaic virus due to which the plants remain stunted and leaves become small. In nursery, the disease is transmitted by aphids (Lastra and McCain, 1974).

Senthilraja *et al.* (2018) first reported Tomato Spotted Wilt Virus (TSWV) infecting *A. majus* in the Nilgiris district of Tamil Nadu. Snapdragon plants exhibited typical symptoms of tospovirus infection, including severe chlorotic and necrotic rings in leaves accompanied with stunting.

10.2.0 Pests

Antirrhinum is relatively free of insect pests. However, several types of insect pests may attack the plants. Aphids, spider mites and looper larvae are the most common.

10.2.1 Aphids

Aphids suck the sap on the underside of the leaves and on the tender tip ends. *Aphis gossypii* is often troublesome in greenhouses and gardens. The peach aphid *Myzus persicae* also attacks and can cause leaf discolouration and distortion similar to virus-like symptoms. Foliar spray of endosulphan (thiodan) or oxydemetonmethyl (metasystox-R) has also been suggested (Rogers, 1980). Harbaugh and Mattson (1973) observed effective control of aphids on greenhouse snapdragons by predatory activities of lacewing (*Chrysopa carnea*).

10.2.2 Spider Mite

Red spider mite (*Tetranychus telarius*) may be very abundant and injurious during the dry seasons. Spray with miticide such as aramite, dimite, ovex or kelthane will control the mites.

10.2.3 Looper Larvae

This is also one of the common pests of snapdragon. Foliar spray with trichlorfon has been suggested by Rogers (1980). In India, gram caterpillar or gram pod borer, *Heliothis* (*Helicoverpa*) *armigera*, has been found to cause severe damage to the plants. It eats away foliage, flower and pods and causes serious damage in terms of quality and yield of seed. It has been observed by Singh and Arora (1980)

Diseases and Pests of Antirrhinum

Damping Off   Anthracnose   Blight

Rust   Downy Mildew   Spider Mite

Harvesting and Marketing of Antirrhinum

Stage of Harvesting   Harvesting of Flowers   Harvested Flowers

Cut Flowers in Market   Pots Plants in Market

that the pest caused highest (38.24 per cent) damage to the pods. The control of this pest should be done by spraying thiodan at 3 ml/l, or ripcord at 4.5 ml/l to avoid loss in quality of blooms and seed yield.

Cyclamen mite (*Steneotarsonemus pallidus*) and red-and-black stink bug are other pests noticed on snapdragon.

10.2.4 Nematode

Root-knot nematode (*Meloidogyne incognita*) occasionally attacks the roots of young seedlings and the plants become stunted. McSorley and Frederick (1994) reported that *A. majus cv.* First Ladies, was highly susceptible (mean gall rating ≥ 4.2 and ≥14,500 eggs/plant) to *M. incognita*.

Steaming was more effective than solarization, and was comparable to methyl bromide (MeBr) fumigation for root-knot nematode (*Meloidogyne arenaria*) control and yield enhancement in cut flower production (Qiao *et al.*, 2011).

11.0 Harvesting and Postharvest Management

11.1.0 Harvesting Stage

Snapdragons are generally cut when nearly one-third of the florets are open. Some growers, having bud opening facilities, harvest their crop when one or two florets on the spike are open (Staby *et al.*, 1978). The best harvesting stage of cut *Antirrhinum majus cv.* Fujinoyuki flowers were when 7-9 florets/spike opened (Lee *et al.*, 1995).

Stems of snapdragon should be harvested when the lowermost floret starts expanding, for longest vase life and for better visual quality, harvesting should be done when two to three snapdragon florets have opened (Ahmad and Dole, 2014).

11.2.0 Postharvest Management

11.2.1 Grading

The flowers are graded according to spike length, stem weight and number of open florets per stem. The standard grades established by the Society of American Florists are most commonly used by growers in the USA. Under these grades, the material must have straight stems, clean and uniform foliage, free of diseases and pests. Foliage on the lower third of the stem should be removed.

11.2.2 Pulsing

Pulsing with 0.2 mM STS for 16 h improved flower quality and prolonged vase life and prevented floret abscission due to ethylene (Lee *et al.*, 1995). Flowers pre-treated with STS and held in a preservative solution (2 per cent sucrose + 150 ppm 8-HQC + 25 ppm $AgNO_3$) at 1°C for 1, 2 or 4 weeks had good flower quality and long vase life. The addition of 150 ppm daminozide in the solution prevented excessive elongation of spike.

Snapdragon stems should be pulsed with 5 per cent sucrose along with isothiazolinone for 24 h, and maintained continuously in vase solution with Chrysal

Clear Universal Flower Food or Floralife Clear Professional Flower Food, but without floral foam, which reduced vase life (Ahmad and Dole, 2014).

Mannitol treatment delayed senescence of spike tips, resulting in marked stem growth and promotion of bud development in cut snapdragon flowers. Degradation of nuclei, a parameter of programmed cell death (PCD), as well as carbohydrate levels of spike tips were suppressed by mannitol treatment. Water potential and osmotic potential in the spike tips was modestly decreased by the mannitol treatment, resulting in the suppression of senescence (Ichimura *et al.*, 2016).

11.2.3 Storage

The temperature suitable for short duration storage is –0.6 to 1.7°C and chilling injury is evident below –0.94°C. Storing the flowers in dark by using preservative solution was found to be beneficial (Staby *et al.*, 1978). Yee *et al.* (1980), however, observed deterioration in flower quality by storing at 5°C. With the use of dry pack storage at –1°C cut spikes could be held for longer period (Mastarlerz, 1953). Hypoberic storage is another development. Burg (1973) observed that cut flowers of snapdragons could be held at 40 mm Hg pressure for up to 6 weeks in good condition, compared with the 3-week maximum for the best methods used previously.

Treatment with 0.5 µL/1-MCP, sealed with a gas permeable film, and stored at 5°C resulted in slower changes in package headspace O_2, CO_2, and C_2H_4 partial pressures, maintained higher overall quality of both flower species and reduced electrolyte leakage and abscission in edible flower of snapdragon. All samples prepared with MAP had significantly reduced dehydration and higher overall quality compared to flowers packaged commercially in plastic clamshell containers. Treatments with controlled release of 1-MCP and MAP significantly extended storage life of edible snapdragon flowers (Kou *et al.*, 2012).

The respiration of cut snapdragon flowers increased exponentially as the temperature increased from 0 to 20°C. The vase life of flowers of *cv.* 'Potomac Pink' stored dry at 0°C was 10.8 days, similar to that of freshly harvested controls (10.6 days), and 4.4 days longer than that of flowers stored at 7.5°C. When spikes were placed horizontally at 20°C, growth became negatively gravitropic within 20 min. Bending was significantly higher than controls (stored vertically) in all flowers stored horizontally at temperatures above 5°C. Vase life of flowers stored for 5 days at a range of temperatures then placed in an interior environment was directly correlated with respiration rate at the storage temperature. Wet storage of cut snapdragon flowers reduced the loss of quality at storage temperatures above 5°C but the vase life of flowers stored in water at 12.5°C was less than half that of flowers stored dry at 0°C. Ethylene treatment caused 100 per cent floret abscission which was prevented by pre-treatment either with 1-methycyclopropene (1-MCP) or with silver thiosulfate (STS), but neither of these inhibitors prevented gravitropic bending (Çelikel *et al.*, 2010).

11.2.4 Bending of Inflorescence

There are two difficulties in handling of snapdragons, abscission of florets (shattering) and negative geotropic curvature of spikes (Rogers, 1980). Flower shattering is caused by endogenous ethylene (Wang *et al.*, 1977) or presence of ethylene in ambient air. This can be prevented by pre-treating the spikes with rhizobitoxin (an inhibitor of ethylene synthesis) or spraying of flowers with 5×10^{-4}M benzothiadiazole (TH 6241). Treatment with STS (silver thiosulphate) also delayed the flower abscission in snapdragons (Farnham *et al.*, 1967; Wang *et al.*, 1977 and Staby *et al.*, 1978). Attempts are being made to develop shattering resistant lines.

The salicylic acid (SA) applied at 0.5/1 mM inhibits the upwards gravitropic bending in snapdragon by interfering specifically with downstream processes of the gravitropism signaling pathway (Friedman *et al.*, 2003).

Friedman *et al.* (2003) suggested that both the epidermal and cortical cells play an important role in gravitropic shoot bending of snapdragon. At the onset of bending, a negative relative growth rate of the upper flank was detected, followed by increased relative growth rate in both lower and upper flanks. Consequently, a differential stem growth pattern was obtained during gravistimulation, which was significantly and specifically abolished by calcium antagonists reported previously to inhibit stem curvature of snapdragon. The differential growth patterns resulted from dynamic modifications of the cell dimensions in the epidermal and cortical stem layers. Bending started with both shrinking and widening of the epidermal cells and a parallel decrease in length and height of cortical cells at the upper stem flank. These changes were accompanied with a concomitant increase in length and height of the cortical cells on the lower stem flank, followed by a growth increase of epidermal cells.

Screening of various cultivars to select those with non-bending spikes following gravistimulation could not be used to predict the bending responses of their corresponding mature spikes, as the bending response of various snapdragon cultivars was strongly dependent on their growth conditions. The use of 20 mM $LaCl_3$ (a calcium channel blocker), or buffers with a high pH could successfully inhibit the gravitropic response of snapdragon spikes, without reducing flower quality (Friedman *et al.*, 2005).

11.3.0 Vase Life

Cut spikes of most cultivars have a vase life of about 1 week in tap water or distilled water. Use of the most suitable preservative can double or triple the vase life of snapdragons. A solution containing 300 ppm 8-HQC (8-hydroxy quinoline citrate) and 1.5 per cent sucrose was found to be the most useful preservative to enhance the vase life, number of open florets and spike length after cutting (Larsen and Scholes, 1966; Raulston and Marousky, 1970). Larsen and Scholes (1966) also found that addition of 25 ppm SADH was beneficial. The best result recorded by Johnson (1972) was from a solution containing 300 ppm 8-HQC, 0.5 per cent sucrose and by Halaba*et al.* (1986) those containing 300 ppm 8-HQC + 50 ppm CCC and 5 per cent sucrose. Sucrose at 50 g/l + 200 mg 8-HQS (8-hydroxyquinoline sulphate) markedly promoted flower opening and extended the vase life from 8.2 days (control) to 20.7 days of *cv.* Yellow Butterfly (Ichimura and Hisamatsu, 1999). The onset of climacteric ethylene production was delayed by sucrose treatment suggesting that the effects of sucrose on promoting bud opening and inhibiting flower senescence are attributed to an increase in sugar concentrations and inhibition of ethylene synthesis. Continuous treatment with sucrose (20 g/l) plus 200 mg 8-hydroxyquinoline/l was also effective in improving the vase life of cut flowers (Ichimura, 1998). The longevity of cut flowers was also increased by dipping in 1000 ppm $AgNO_3$ for up to 40 minutes (Awad *et al.*, 1986). Mugge (1983) reported that treatment with AKN (0.08 per cent $Al_2(SO_4)_3$ + 0.03 per cent KCl + 0.02 per cent NaCl) at 20 ml/l extended the vase life of cultivars Alaska and Die Rose by 4.8 and 3.6 days, respectively, over control.

Among the different preservatives used on snapdragon *cvs.* Super Tetra Crimson Giant, Proflovit (0.39 g-hydroxyquinoline + 0.05 g CCC + 50 g sucrose/l) showed the best effect with regard to vase life and flower quality (Nowak and Rudnick, 1975).

Floret drop of snapdragons (cv. Houston) induced with 150 ppm ethephon in the holding solution was prevented and vase life was extended by 12-h pre-treatment in silver thiosulphate solution prior to ethephon treatment (Anderson *et al.*, 1993). The longevity and quality of freshly cut spikes kept under simulated transport conditions were almost doubled by 20-h treatment with a solution containing STS plus sucrose compared with control (Nowak, 1981).

Light and floral preservatives markedly influenced the development of floret colour that opened after harvest. Spikes kept in dark produced little anthocyanin and were poorly coloured whereas those in light and held in 8-HQC + sucrose solution produced much more intensely coloured florets (Rogers, 1980). Effect of artificial light of different wavelengths (supplied by blue, red, warm white, cool white, fluorescent and incandescent lamps) and treatment with 0.1, 1.0 and 10.0 ppm ethephon, uniconazole, BA, IAA, ABA and GA on the coloration of the florets of cut snapdragon flowers were investigated by Sang *et al.* (1992). Anthocyanin content was highest under fluorescent light and blue light and lower under red and incandescent lamps. GA and IAA decreased anthocyanin content, especially at higher concentrations. Anthocyanin content increased slightly at the lowest concentrations of BA and ABA, ethephon and uniconazole. With the addition of sucrose (5 per cent) to the preservative solution, anthocyanins increased to some extent under all light regimes. It was concluded that the most effective artificial light for good coloration was the fluorescent lamp with wavelengths of 470 and/or 600 nm.

Pre-treatment for 6h with low concentrations of 1-MCP (1-methylcyclopropene, formerly designated SIS-X), a cyclic ethylene analogue, inhibited the normal wilting response of cut antirrhinum. In the presence of ethylene at 1 µl/l, the vase life of 1-MCP-treated flowers was up to 4 times that of the control (Serek *et al.*, 1995).

Cut flowers kept in water showed inflorescence bent, while those pre-treated with 1 per cent $CaCl_2$ for 16 h did not show such symptoms. STS pre-treatment prevented floret wilting but calcium did not. Calcium pre-treatment enhanced inflorescence elongation and open florets. When vase water contained 5 per cent glucose and germicide the calcium and STS pre-treated flowers produced inflorescence with larger florets, resulting in extended vase life (Fukai and Uehara, 2006.).

Cut stems snapdragon did not appear to be sensitive to handling procedures or external ethylene; however, pre-treatment with 0.2 mMSTS extended longevity of snapdragon stems. Snapdragon stems can be stored dry for up to 2 weeks (Ahmad and Dole, 2014).

12.0 Biotechnology

A number of basic researches have been done on the biotechnology of antirrhinum particularly pertaining to molecular biology of flower colour and tissue culture.

12.1.0 Tissue Culture

12.1.1 Protoplast Culture

Research on tissue culture of snap dragon is well documented. In one of the earliest reports, Poirier Hamon *et al.* (1974) isolated protoplasts enzymatically from the leaves of Kimosy Blanc antirrhinums and cultured on a synthetic medium *in vitro*. Cultured protoplasts increased size, regenerated walls, divided and formed small colonies and embryoids. Cell wall formation and cell division required both auxin and cytokinin in the medium. Isolated stem segments of the same species, when cultured on a synthetic medium with an added auxin, yielded rapidly growing callus tissues in which differentiation of embryoids again occurred. Okubo *et al.* (1991) observed highest adventitious shoot formation in the basal hypocotyl segments by the addition of 0.1 mg BA/l (9.36 shoots/section). Endogenous cytokinins were higher in the basal hypocotyl segments than in the 2 upper hypocotyl segments, whereas auxin content was higher in the 2 upper than in the basal hypocotyl segments. Ratios of cytokinins to auxin were also the highest in the basal hypocotyl segment (200 : 1). Gonzalez Benito *et al.* (1996) successfully micropropagated two wild taxa [*A. Majus* subsp. *barrelieri* (*A. barrelieri*) and *A. microphyllum*] and the commercial cultivar Floral Carpet. Seedling nodes were multiplied on MS medium supplemented with 5 µM BA (or 4.4 µM for the cultivar). Rooting took place in MS medium without growth regulators. In general, the addition of NAA to the medium did not improve the rooting of shoots. High survival percentages (>80 per cent) were obtained when plantlets were transferred to soil.

The highest callus fresh weight as well as callogenesis frequency were observed in Murashige and Skoog (MS) medium containing 4.52 µM 2,4D from hypocotyls explant. MS medium supplemented with 4.44 µM 6 benzyl amino purine (BAP) plus 0.49 µM 3indolebutyric acid (IBA) along with 30 µM sodium nitroprusside (SNP) had the highest shoot organogenesis frequency and shoot number from callus. MS medium containing 60 µM SNP was the best treatment for root induction (Zafarghandi and Rahmati-Joneidabad, 2020).

12.2.0 Molecular Markers

Ottenschlager *et al.* (1999) isolated immature pollen of snapdragon at different developmental stages, bombarded with plasmids containing green fluorescent protein (*GFP*) genes *GFP4, GFP5ER* and *GFP4S65C* and cultured *in vitro* for several days until maturity. GFP expression was monitored every day during *in vitro* maturation, germination and pollination, as well as after *in situ* pollination. GFP-expressing pollen from snapdragon developed *in vitro* for several days until maturity and germinated *in vitro* as well as on the surface of stigmata, strongly suggesting that all 3 GFPs are not toxic for the development of functional pollen. GFP can be used as a non-destructive marker for plant reproduction biology and development. The *UFO* gene showed extensive homology with fimbriata (*Fim*), a gene mediating between meristem and organ identity genes in antirrhinum. All three *Ufo* mutant alleles that were sequenced are predicted to produce truncated proteins. Phenotypic, genetic and expression pattern comparison between *Ufo* and *Fim* suggested that they are cognate homologues which play a similar role in mediating between meristem and organ identity genes. However, some differences in the functions and genetic interactions of *Ufo* and *Fim* were apparent, indicating that changes in partially redundant pathways have occurred during the evolutionary divergence of antirrhinum (Ingram *et al.*, 1995). Sequence analysis of *ndhD* and the corresponding cDNA showed that editing of an *ACG* triplet in snapdragon leads to the creation of an *AUG* codon. Conservation of translation starts codons of plastome-encoded genes can, therefore, be achieved by editing of transcripts, and the definition of plastome-encoded *ORFs* would take the potential editing events into consideration (Neckermann *et al.*, 1994).

Cytochrome P450 (unspecific monooxygenase) cDNAs, AFNS2, and TFNS5, were isolated from snapdrogon (*Antirrhinum majus*) petal cDNA libraries, based on sequence homology with licorice CYP93B1 cDNA encoding (2S)-flavanone 2-hydroxylase. They were expressed in yeast and identified to encode flavone synthase II catalysing direct conversion of flavonoids to flavones, probably via 2-hydroxy-flavanones (Akashi *et al.*, 1999).

A molecular linkage map has been constructed mainly on the basis of transcribed genes (Schwarz-Sommer *et al.*, 2003). This has an average interval between markers of 2.5 cM at present. However, most of the protein-coding loci are aggregated into clusters, which reduce the effective distance between genes, and further markers continue to be added.

A molecular recombination map, currently comprising more than 250 loci with an average distance between loci of ~2 cM, has been produced for Antirrhinum using hybrids between *A. majus* and wild species (Hudson *et al.*, 2008). It has been aligned with a map of classical mutants and with the eight Antirrhinum chromosomes using fluorescence *in situ* hybridization (FISH) (Zhang *et al.*, 2005). Maps constructed using hybrids between different Antirrhinum species are largely collinear, suggesting an absence of extensive chromosomal rearrangements, although distorted transmission of some genomic regions in interspecies hybrids can hinder genetic mapping (Schwarz-Sommer *et al.*, 2003). Recombinant inbred lines (RILs) and near-isogenic lines (NILs) have been produced from hybrids between different Antirrhinum species, allowing those genes that underlie differences between species to be identified.

Molecular linkage map of *A. majus* based on segregation of markers in the F_2 population of two inbred lab strains of *A. majus*. The resulting map consisted of over 300 markers in eight linkage groups, which could be aligned with a classical recombination map and the *A. majus* karyotype. The differences varied in magnitude and direction between chromosomes, suggesting that they had multiple causes. The map, which covered an estimated of 95 per cent of the genome with an average interval of 2 cM. The molecular recombination map further indicates that the IDLE MITE transposons are distributed throughout the genome and are relatively stable (Schwarz-Sommer *et al.*, 2010).

12.3.0 Genetic Transformation

Snapdragon has been well studied for expression of various genes and has been used as a model system too. *Agrobacterium* mediated gene transfer has been well documented in snapdragon.

Hoshino and Mii (1997) induced hairy roots of snapdragon by a wild-type strain of *A. rhizogenes* on media containing various concentrations of a phosphinothrcin-based herbicide, bialaphos (bilanafos) or plant growth regulators. Adventitious shoot regeneration from hairy roots was observed at a low frequency (10 per cent) on half-strength MS medium. Addition of α-NAA in combination with BA, thidiazuron or zeatin to the medium had no effect on shoot regeneration from hairy roots. Although bialaphos at 0.9 mg/l or more was toxic to hairy roots, at 0.5 mg/l it significantly increased the shoot regeneration frequency by up to 56 per cent. In contrast, non-transformed roots and leaves regenerated no shoots on media with or without bialaphos. Regenerated shoots detached from host roots readily developed roots on gellangum-solidified medium. Regenerated plants were successfully transferred to the greenhouse, but did not produce seed. Later they (Hoshino *et al.*, 1998) obtained transgenic herbicide resistant *A. majus* plants by co-cultivation of leaf explants with *A. rhizogenes* harbouring *Ri* plasmid and pARK5. The transgenic nature of the plants was confirmed by opine, PCR and southern analyses. Regenerated plantlets were successfully established in the growth chamber. Acclimatized plants exhibited altered phenotypes known as *Ri* syndrome. These included reduced apical dominance with highly branched stems and short internodes. These plants also showed resistance to an applied standard level of commercial herbicides.

Hairy roots were produced following the co-cultivation of *Agrobacterium rhizogenes* cells with hypocotyls of five varieties of *Antirrhimum majus*. The use of a strain containing a binary plasmid with T-DNA bearing the β-*glucuronidase* reporter gene resulted in the co-transformation of some root clones. Regeneration of shoots from hairy roots occurred only with variety Golden Monarch. Regenerated plants, some of which were *GUS*-positive, exhibited abnormal morphology common among hairy root regenerants; they were dwarf with altered leaf shape, a poor root system and flowering was delayed (Senior *et al.*, 1995).

In one of the earliest work on snapdragon Fritze *et al.* (1991) studied the developmental and UV light regulation of the snapdragon (naringenin-) chalcone synthase (CHS) promoter linked to the β-glucuronidase reporter gene in transgenic tobacco. The pattern of expression of the chimaeric gene was compared with the expression of the endogenous *CHS* genes in tobacco and snapdragon. It was demonstrated that expression of the CHS promoter is controlled in both an organ-specific and tissue-specific manner. The highest level of expression was observed in immature seeds. Deletions were used to define regions of the promoter required for expression in roots, stems, leaves, seeds and petals of transgenic plants. Spena *et al.* (1992) subsequently characterized the promoter activity of a 2.2 kb DNA fragment for the 5′ flanking sequence of the *tap1* (tapetum-specific) gene of snapdragon in tobacco with the β-glucuronidase reporter gene. The *tap1* promoter conferred to the reporter gene an expression that was limited to the early stages of anther development. Expression of the root induction *rolB* gene of *A. rhizogenes* under the control of the *tap1* promoter impaired the development of tobacco flowers. Tobacco plants transgenic for the *tap1-rolB* chimaeric gene showed altered anthers and a reduction in whole flower growth. The perturbation of anther development would in turn affect gibberellin production and content, which is reflected in a reduced elongation of flowers. Dennis and Bowman (1993) also characterized the genes that control flower development in snapdragon. Two types of homeotic gene have been identified, the first associated with the identity of the meristems (including the *floricaula* and *squamosa* genes of snapdragon) and the second controlling the identity of the floral organs (the *agamous* and *apetala* 2 genes of Arabidopsis). Results of genetic transformation and mutational studies showed how 3 classes of homeotic gene function could specify organ identity in each of the 4 whorls (sepals, petals, stamens and carpels).

Stable transformants in *Antirrhinum majus* were obtained by Cui *et al.* (2003) by inoculating explants with *A. tumefaciens* strain GV2260 harboring the binary vector pBIGFP121, which contains the neomycin phosphotransferase gene (NPT II) as a selectable marker and the gene for the Green Fluorescent Protein (GFP) as a visual marker. Putative transformants were identified by selection for kanamycin resistance and by examining the shoots using fluorescence microscopy. PCR and Southern analyses confirmed integration of the GFP gene into the genomes of the transformants. The transformants had a morphologically normal phenotype. This improved method requires only a small number of seeds for explant preparation, and three changes of medium; the overall transformation efficiency achieved, based on the recovery of transformed plants after 4-5 months of culture, reached 8-9 per cent. This success rate makes the protocol very useful for producing transgenic *A. majus* plants.

Particle bombardment and Agrobacterium infiltration of developing petals have been used successfully to obtain transient expression of pigment genes and suppression by RNA interference (RNAi) (Schwinn *et al.*, 2006; Shang *et al.*, 2007).

12.4.0 *In vitro* Conservation

Gonzalez Benito *et al.* (1998) were successful in the cryopreservation protocol, by encapsulation-dehydration of nodal explants of the Spanish endemic plant *Antirrhinum microphyllum.* The explant was pre-cultured on MS medium for 1 day and preserved the protocol by encapsulation, culture of beads on 0.75 M sucrose for 19 h, desiccation to 20 per cent moisture content, fast freezing and thawing.

13.0 References

Abdullah, B., Ahmad, I. and Dole, J.M. (2020) *Sci. Hortic.*, **261** : 108909.

Adams, S.R., Munir, M., Valdes, V.M., Langton, F.A. and Jackson, S.D. (2003) *Ann. Bot.*, **92**(5) : 689-696.

Ahmad, A., Ahmad, I. and Ferdosi, M.F.H. (2020) *Sarhad J. Agric.*, **36**(2) : 586-592.

Ahmad, I. and Dole, J.M. (2014) *Sci. Hortic.*, **172** : 308-316.

Aitken, E.A.B., Newbury, H.J. and Callow, J.A. (1989) *Plant Pathol.*, **38**(2) : 169-175.

Akashi, T., Fukuchi-Mizutani, M., Aoki, T., Ueyama, Y., Yonekura-Sakakibara, K., Tanaka, Y., Kusumi, T. and Ayabe, S.I. (1999) *Plant Cell Physiol.*, **40** : 1182-1186.

Almeida, J. and Galego, L. (2004) *Int. J. Dev. Biol.*, **49**(5-6) : 527-537.

Almeida, J. and Galego, L. (2005) *Int. J. Dev. Biol.*, **49** : 527-537.

Anderson, R.D., Sanderson, K.C., Smith, D. and Williams, J.C. (1993) *Plant Growth Regulator Soci. Amer. Quar.*, **21** : 144-150.

Aslam, K.M. and Sastry (1979) *Heredity*, **42** : 227-278.

Awad, A.R.E., Meawad, A., Dawh, A.K. and El-Saka, M. (1986) *Acta Hortic.*, **181** : 177-182.

Ball, V. (1957) *The Ball Red Book*, 9th ed., Geo. J. Ball. Inc. West Chicago, Illinois.

Baur, E. (1910) *Z. Indukt. Abstamn*, VererBungal, **4** : 81-102.

Baur, E. (1919) *Einfuhrung in die experimontelleVerer bungs lehre*, Bountragar, Berlin.

Baur, E. (1924) *Bibil. Genet.*, **4** : 1-170.

Baur, E. (1930) *Z. Bot.*, **23** : 676-702.

Becraft, P.W. (1995) *Trends Genetics*, **11** : 253-255.

Bergbusch, V.L. (1999) *Ann. Bot.*, **83** : 483-488.

Berninger, E. (1978) *PepinieristsHorticulteursMaraichers*, **191** : 13-19.

Besmer, Y.L. and Koide, R.T. (1999) *Mycorrhiza*, **9** : 161-166.

Bevacqua, R.F. and Mellano, V.J. (1994) *Communications Soil Sci. Plant Analysis*, **25** : 395-406.

Bhargava, B., Gupta, Y.C. and Sharma, P. (2016) *Proc. Natl. Acad. Sci. India Sect. B Biol. Sci.*, **86**(1) : 65-69.

Bhargava, B., Gupta, Y.C., Dhiman, S.R. and Sharma, P. (2015) *Natl. Acad. Sci. Lett.*, **38**(1) : 81-85.

Bing, A. (1981) *Proc. Northeastern Weed Sci. Soc.*, **35** : 262-266.

Blahut-Beatty, L.M., Bonham-Smith, P.C. and Sawhney, V.K. (1998) *Canadian J. Bot.*, **76** : 1828-1834.

Boodley, J.W. (1962) In : *Snapdragons* (Ed. R.W. Langhans), *N.Y. State Flower Grs. Assoc., Inc.*, Ithaca, New York.

Bradley, D., Carpenter, R., Copsey, L., Vincent, C., Rothstein, S. and Coen, E. (1996) *Nature*, **379**(6568) : 791-797.

Brickell, C. (1996) *A-Z Encyclopedia of Garden Plants*, Dorling Kindersley, Limited, London, pp. 124-125.

Briicher, H. (1943) *Z. Bot.*, **39** : 1-47.

Brown, E.G. and Konuk, M. (1994) *Phytochemistry*, **37** : 1589-1592.

Bunt, A.C. (1968) *Acta Hortic.*, **8** : 56-61.

Burg, S.P. (1973) *HortScience*, **8** : 202-205.

Byrne, J.M., Hausbeck, M.K. and Sconyers, L.E. (2005) *Plant Dis.*, **89**(10) : 1060-1066.

Carmichael, O.E. (1968) *Master's Thesis*, University of Missouri, USA.

Carpenter, W.J. (1963) *Flor. Rev.*, **132** : 26.

Carpenter, W.J. (1964) *Proc. Amer. Soc. Hort. Sci.*, **84** : 624-629.

Carrió, E. and Güemes, J. (2014) *Bot. J. Linn. Soc.*, **176**(2) : 159-172.

Carrió, E. and Güemes, J. (2019) *J. Nat. Conserv.*, **48** : 47-53.

Carter, C.T. and Grieve, C.M. (2008) *HortScience*, **43**(3) : 710-718.

Cathey, H.M. (1969) *Flor. Rev.*, **144** : 26-29, 75-77.

Causier, B., Castillo, R., Zhou, J., Ingram, R., Xue, Y., Schwarz-Sommer, Z. and Davies, B. (2005) *Curr. Biol.*, **15**(16) : 1508-1512.

Çelikel, F.G., Cevallos, J.C. and Reid, M.S. (2010) *Sci. Hortic.*, **125**(3) : 429-433.

Chehrazi, M., Hosseini, H.R., Dehkordi, E.H. and Vafa, K.A. (2017) *Iranian J. Hortic. Sci.*, **48**(2) : 265-273.

Chittendan, F.J. and Synge, P.M. (1956) *Dictionary of Gardening*, The Clarendon Press, Oxford.

Coats, A.M. (1968) *Flowers and Their Histories*, Adam and Charles Black, London.

Creel, R. and J.R. Kessler. (2007) *Alabama Cooperative Extension System. ANR*, **1312** : 1-5.

Cubas, P., Vincent, C. and Coen, E. (1999) *Nature*, London, **401** : 157-161.

Cui, M.L., Handa, T. and Ezura, H. (2003) *Mol. Genet. Genom.*, **270**(4) : 296-302.

Daughtrey, M.L. and Benson, D.M. (2005) *Annu. Rev. Phytopathol.*, **43** : 141-169.

Davies, B., Motte, P., Keck, E., Saedler, H., Sommer, H. and Schwarz Sommer, Z. (1999) *The Embo J.*, **18**(14) : 4023-4034.

De Vries, H. (1901) *Die Mutations Theorie*, Leipzig.

Delool, R.A.H. and Tilney-Bassett, R.A.E. (1986) *J. Heredity*, **77** : 236-240.

Delworth, C.I. (1946) *Flor. Rev.*, **99** : 35-36.

Dennis, E. and Bowman, J.L. (1993) *Current Biol.*, **3** : 90-93.

Dickerson, G.W. (1999) *BioCycle*, **40** : 62-63.

Duffett, W.E. (1960) *Ohio Flor. Ass. Bull.*, **371** : 4-5.

Edwards, K.J.R. and Goldenberg, J.B. (1976) *Ann. Bot.*, **40** : 1277-1283.

El-Nashar, Y.I. (2017) *Photosynthetica*, **55**(2) : 201-209.

Emsweller, S.L. and Jones, H.A. (1934) *Hilgardia*, **8** : 197-211.

Endress, P.K. (1999) *Intern. J. Plant Sci.*, **160** : 5.

Engelhard, A.W. (1971) *Proc. Fla. St. Hort. Soc.*, **83** : 455-457.

Fincham, J.R.S. and Harrison, B.J. (1967) *Heredity*, **22** : 211-224.

Flint, H.L. (1960) *Proc. Amer. Soc. Hort. Sci.*, **75** : 769-773.

Flint, H.L. and Asen, S. (1953) *Proc. Amer. Soc. Hort. Sci.*, **62** : 481-486.

Friedman, H., Vos, J.W., Hepler, P.K., Meir, S., Halevy, A.H. and Philosoph-Hadas, S. (2003) *Planta*, **216**(6) : 1034-1042.

Friedman, H., Zhang, Z., Meir, S., Halevy, A.H. and Philosoph-Hadas, S. (2005) *J. Hortic. Sci. Biotechnol.*, **80**(4) : 433-438.

Fries, H. (1962) In : *Snapdragons* (Ed. R.W. Langhans), *N.Y. State Flower Grs. Assoc. Inc.*, Ithaca, New York.

Fritze, K., Staiger, D., Czaja, I., Walden, R., Schell, J. and Wing, D. (1991) *Plant Cell.*, **3**: 893-903.

Fukai, S. and Uehara, K. (2006) *Hortic. Res.*, **5**(4) : 465-471.

Galego, L. and Almeida, J. (2002) *Genes Dev.*, **16**(7) : 880-891.

Garcý´a-Barriuso M., Ferna´ndez-Castellano C., Rocha J., Bernardos S. and Amich F. (2012) *Plant Biosyst.*, **146** : 291-301.

Garzon, C. D., Molineros, J. E., Y ´ anez, J. M., Flores, F. J., Del Mar Jim´enez-´Gasco, M., and Moorman, G. W. (2011) *Plant Dis.*, **95** : 1233-1238.

Genders, R. (1969) *The Cottage Garden and the Old Fashioned Flowers*, Petham Books, London.

Gill, H.K. and McSorley, R. (2010) *In Proc. Fla. State Hort. Soc.*, **123** : 308-311.

Glover, B.J., Perez-Rodriguez, M. and Martin, C. (1998) *Development*, **125** : 3497-3508.

Goldsmith, A.G. (1968) *HortScience*, **3** : 269-271.

Golz J.F., Roccaro M., Kuzoff R., Hudson A. (2004) *Development*, **131** : 3661-3670.

Golz, J.F., Keck, E.J. and Hudson, A. (2002) *Curr. Biol.*, **12**(7) : 515-522.

Gonzalez Benito, M.E., Tapia, J., Rodriguez, N. and Iriondo, J.M. (1996) *J. Hort. Sci.*, **71** : 11-15.

González-Barrio, R., Periago, M.J., Luna-Recio, C., Garcia-Alonso, F.J. and Navarro-González, I. (2018) *Food Chem.*, **252** : 373-380.

Gonzalez-Benito, M.E., Nunez-Moreno, Y. and Martin, C. (1998) *Cryo Letters*, **19** : 225-230.

Hadi, H.A.I. and Saeed, A.K.A.M. (2019) *Euphrates J. Agric. Sci.*, **11**(4) : 57-69.

Halaba, J., Kwiatkowski, M. and Rudnick, R.M. (1986) *Prace Inst. Sadownictwa w Skierniewicach*, B, **8** : 185-190.

Hamdollah-Zadeh, A. (1993) *Iranian J. Plant Pathol.*, **29** : 53-54.

Hamza, A.M., El-Kafie, A., Helaly, A.A. and EL-Mongy, M.S. (2013) *J. Plant Prod.*, **4**(11) : 1621-1637.

Hanan, J.J. (1965) *Proc. Amer. Soc. Hort. Sci.*, **86** : 681-692.

Hanan, J.J. and Langhans, R.W. (1962) *Bull. N.Y. St. Flower Grs.*, **198** : 1-2, 6.

Hanan, J.J. and Langhans, R.W. (1964) *Proc. Amer. Soc. Hort. Sci.*, **84** : 613-623.

Hanney, W.J. (1953) *National Snapdragon Soc. Bull.*, **2** : 1-3, 12.

Haramaki, C. and Kuhns, L.J. (1985) *Proc. 39th Annual Meeting of North Eastern Weed Sci. Soc.*, pp. 231-234.

Harbaugh, B.K. and Mattson, R.H. (1973) *J. Amer. Soc. Hort. Sci.*, **98** : 306-309.

Harrison, B.J. and Carpenter, R. (1973) *Heredity*, **31** : 309-323.

Harrison, B.J. and Fincham, J.R.S. (1964) *Heredity*, **19** : 237-258.

Harrison, B.J. and Fincham, J.R.S. (1968) *Heredity*, **23** : 67-72.

Haughn, G.W., Schultz, E.A. and Martinez-Zapater, J.M. (1995) *Canadian J. Bot.*, **73** : 959-981.

Hedley, C.L. (1974) *HortScience*, **9** : 105-112.

Hedley, C.L. and Harvey, D.M. (1975) *Ann. Bot.*, **39** : 257-263.

Hedley, C.L., Aurther, A.E. and Rabinowitch, H.D. (1977) *Euphytica*, **26** : 755-760.

Heffron, L., Blowers, A. and Korban, S. (2006) *HortScience*, **41**(4) : 1021A-1021.

Heidmann, I., Efremova, N., Saedler, H. and Schwarz Sommer, Z. (1998) *Plant J.*, **13** : 723-728.

Holley, W.D. (1966) *Bull. Colo. Flower Grs. Assoc.*, **200** : 1-4.

Hoshino, Y. and Mii, M. (1997) *Plant Cell Rep.*, **17** : 256-261.

Hoshino, Y., Turkan, I. and Mii, M. (1998) *Scientia Hort.*, **76** : 37-57.

Hudson, A., Critchley, J. and Erasmus, Y. (2008) *CSH Protoc.*, 1-7.

Huo, H. and Chen, J. (2018) *EDIS*, 1-5.

Ichimura, K. (1998) *Japan Agril. Res. Quart.*, **32** : 275-280.

Ichimura, K. and Hisamatsu, T. (1999) *J. Japanese Soc. Hort. Sci.*, **68** : 61-66.

Ichimura, K., Kohata, K. and Yamada, K. (2003) *In VIII International Symposium on Postharvest Physiology of Ornamental Plants*, **669** : 119-124.

Ichimura, K., Yoshioka, S. and Yamada, T. (2016) *Postharvest Biol. Technol.*, **113** : 20-28.

Inaba Z., Kato., C., Murakami S. and Ishii C. (2008) *Hort. Res.* (Japan), **7**(3) : 393-398.

Inaba, Z. and Horiuchi, M. (2003) *Hort. Res.* (Japan), **2**(3) : 199-203.

Inaba, Z., Kato, C., Horiuchi, M. and Ohtsuka, H. (2010) *Hort. Res.* (Japan), **9**(2) : 165-170.

Inaba, Z., Kato, C., Murakami, S., Ishii, C., Baba, F., Horiuchi, M. and Ohtsuka, H. (2010) *Hort. Res.* (Japan), **9**(3) : 351-356.

Ingram, G.C., Goodrich, J., Wilkinson, M.D., Simon, R., Haughn, G.W. and Coen, E.S. (1995) *Plant Cell,* **7** : 1501-1510.

Ito, A., Hisamatsu, T., Soichi, N., Nonaka, M., Amano, M. and Koshioka, M. (1997a) *J. Japanese Soc. Hort. Sci.,* **65** : 809-816.

Ito, A., Hisamatsu, T., Soichi, N., Nonaka, M., Amano, M. and Koshioka, M. (1997b) *J. Japanese Soc. Hort. Sci.,* **65** : 817-823.

Jeong, B.R., Lee, C.W., Hayashi, M., Kano, A. and Goto, E. (1992) *Acta Hortic.,* **319** : 505-510.

Johnson, C.R. (1972) *Flor. Rev.,* **149** : 95-97.

Kandeel, A.M., El-Ramah, S.O. and Al-Qubati, A.A. (1999) *Arab Univ. J. Agril. Sci.,***7** : 261-271.

Kang, J.S. and Choi, I.S. (2006) *J. Life Sci.,* **16**(3) : 493-499.

Kelly, A.J. Bonnlander, M.B. and Meeks-Wagner, D.R. (1995) *Plant Cell,* **7** : 225-234.

Kevan, P., Giurfa, M. And Chittka, L. (1996) *Trends Plant Sci.,* **1** : 280-28.

Khattak, A.M. and Pearson, S. (2005) *J. Zhejiang Univ. Sci.,* **6**(2) : 119.

Kim, W.S., Huh, K.Y., Cho, I.H. and Woo, Y.H. (1998) *Korean J. Hort. Sci. Tech.,* **16** : 350-351.

Kondo, H., Deguchi, A. and Miyoshi, K. (2020) *Plant Cell, Tissue Organ Cult.,* **142**(1) : 157-166.

Koths, J.S. (1964) *Michigan Flor.,* **399** : 15.

Kou, L., Turner, E.R. and Luo, Y. (2012) *J. Food Sci.,* **77**(5) : 188-193.

Kramer, P.J. (1940) *Plant Physiol.,* **15** : 63-67.

Kuchuch, H. and Schuck, R. (1930) *Die Erbfaktorenbei Antirrhinum majus and ihreBezeichnung,* Z.A.I.V., **56** : 51-83.

Kumpf, J., Horton, F. and Langhans, R.W. (1966) *Bull. N.Y. St. Flower Grs.,* **244** : 103.

Lai Z., Ma W., Han B., Liang L., Zhang Y., Hong G. and Xue Y. (2002) *Plant Mol. Biol.,* **50** : 29-42.

Langhans, R.W. and Maginnes, E.A. (1962) In : *Snapdragons,* (Ed. R.W. Langhans), *N.Y. State Flower Growers Assoc., Inc.,* Ithaca, New York.

Larsen, F.E. and Scholes, J.F. (1966) *Proc. Amer. Soc. Hort. Sci.,***89** : 694-701.

Lastra, R. and McCain, A.H. (1974) *Flower and Nursery Report.,* pp. 3-4.

Lauri, A., Xing, S., Heidmann, I., Saedler, H. and Zachgo, S. (2006) *Planta,* **224**(1) : 61-71.

Lee, J.S., Kim, Y.A. and Sin, Y.M. (1995) *J. Korean Soc. Hort. Sci.,* **36** : 926-942.

Li, T., Bi, G., LeCompte, J., Barickman, T.C. and Evans, B.B. (2017) *HortTechnology.*, **27**(6) : 860-867.

Liberal González, I., Burrus, M., Suchet, C., Thébaud, C. and Vargas, P. (2014) *BMC Evol. Biol.*, **14** : 146.

Lindstrom, R.S. (1966) *Flor. Rev.*, **139** : 18-19, 51-54.

Lonnig, W.E. and Saediler, H. (1994) *Mol. General Genetics*, **245** : 636-643.

Luo, D., Carpenter, R., Copsey, L., Vincent, C., Clark, J. and Coen, E. (1999) *Cell Cambridge*, **99** :

367-376.

Luo, D., Carpenter, R., Vincent, C.A., Copsey, L. and Coen, E. (1995) *Nature*, **383** : 794-799.

Maatsch, R. and Runger, W. (1960) *Gartenbawiss*, **25** : 231-235.

Maginnes, E.A. (1960) *Dissert. Abst.*, **23** : 1471.

Maginnes, E.A. and Langhans, R.W. (1960) *Bull. N.Y. St. Flower Grs.*, No. 171.

Maginnes, E.A. and Langhans, R.W. (1967) *Bull. N.Y. St. Flower Grs.*, No. 261.

Malik, S.A., Neelofar, Z.A., Nazki, I.T., Mir, S.A., Khan, F.A. and Pukhta, M.S. (2019) *Int. J. Soil Plant. Sci.*, **28**(1) : 1-6.

Masiero, S., Li, M.A., Will, I., Hartmann, U., Saedler, H., Huijser, P., Schwarz-Sommer, Z. and Sommer, H. (2004) *Development*, **131**(23) : 5981-5990.

Mastarlerz, J.W. (1953) *National Snapdragon Soc. Bull.*, No. 3 : 1, 3.

Mateu-Andres, I. (1999) *Bot. J. Linnean Soc.*, **131** : 187-199.

Matkowski, A. and Adler, K. (1999) *Bot. J. Linnean Soc.*, **131** : 235-248.

McSorley, R. and Frederick, J.J. (1994) *J. Nematol.*, **26** : 773-777.

McSorley, R., Wang, K.H., Rosskopf, E.N., Kokalis-Burelle, N., Hans Petersen, H.N., Gill, H.K. and Krueger, R. (2009) *Int. J. Pest Manag.*, **55**(4) : 265-273.

Mehmood, T., Ahmad, W., Ahmad, K.S., Shafi, J., Shehzad, M.A. and Sarwar, M.A. (2013) *Univers. J. Plant Sci.*, **1**(3) : 104-111.

Memon, S.A. and Khetran, K. (2014) *J. Agric.*, **10**(6) : 1557-1569.

Miller, R.O. (1957) *Bull. N.Y. St. Flower Grs.*, **140** : 1-3.

Miller, R.O. (1960) *Proc. Amer. Soc. Hort. Sci.*, **75** : 761-768.

Miller, R.O. (1962) *Proc. Amer. Soc. Hort. Sci.*, **81** : 535-543.

Mjeed, A.J. and Ali, M.A. (2017) *Kurdistan J. Appl. Res.*, **2**(1) : 1-7.

Moore, B.D., Palmquist, D.E. and Seemann, J.R. (1997) *Plant Physiol.*, **115** : 241-248.

Moorman, G. and S. Kim. (2004) *Plant Dis.*, **88** : 630-632.

Mugge, A. (1983) *Gartenbau*, **30** : 218-219.

Múnera, J.D.C. and Hausbeck, M.K. (2015) *HortScience*, **50**(9) : 1319-1326.

Munir, M. and Naz, F. (2006) *J. Appl. Hortic.*, **8**(1) : 25-28.

Munir, M., Hadley, P., Carew, J., Adams, S. and Pearson, S. (2017) *Pak. J. Bot.*, **49**(1) : 181-186.

Munir, M., Hadley, P., Carew, J., Adams, S.T, Pearson, S.I. and Sudhakar, B. (2015) *Pak. J. Bot.*, **47**(5) : 1717-1720.

Munir, M., Hadley, P., Carew, J., Zubair, M.U., Adams, S.T, Hussain, S.B, Baloch, J.U., Hussain, N.A. and Amanullah, M. (2010) *Pak. J. Bot.*, **42** : 1931-40.

Munir, M.U., Jamil, M.U., Baloch, J. and Khattak, K.R. (2004) *Int. J. Agric. Biol.*, **6** : 173-178.

Munshi, P.S., Nayak, K.C. and Mukherjee, D. (1980) *Indian Agric.*, **24** : 153-159.

Murray, G.E., Sanderson, K.L. and Williams, J.C. (1986) *HortScience*, **21** : 120-122.

Nawaz, M.A., Shafique, A., da Silva, J.A.T., Ahmed, W. and Saleem, B.A. (2017) *Int. J. Agric. Biol.*, **19**(1) : 23-28.

Nebel, B.R. and Ruttle, M.L. (1938) *New York Agric. Exp. Stn., Geneva, Circ.*, No. 183.

Neckermann, K., Zeltz, P., Igloi, G.L., Kossel, H. and Maier, R.M. (1994) *Gene*, **146** : 177-182.

Nelson, P.V. and Larson, R.A. (1969) *N.C. Agric. Stn. Tech. Bull.*, **194** : 1-15.

Nijs, L.P.D. and Boesman, G. (1981) *Medelimgin Vans de Facultiel Land BeuwetanSchappan Ri Rijkanverdite Genet.*, **46** : 236-240.

Noda K., Glover B.J., Linstead P. and Martin C. (1994) *Nature*, **369** : 661-664.

Nowak, J. (1981) *Scientia Hort.*, **15** : 255-262.

Nowak, J. and Rudnick, R.M. (1975) *Prace Inst. Sadownictwa w Skierniewicach B*, **1** : 173-179.

Oertli, J.J. (1970) *Flor. Rev.*, **146** : 20-21.

Okamoto, H., Yano, A., Shiraishi, H., Okada, K. and Shimura, Y. (1994) *Plant Mol. Biol.*, **26** : 465-472.

Okubo, H., Wada, K. and Uemoto, S. (1991) *Plant Cell Rep.*, **10** : 501-504.

Olmstead, R.G., DePamphilis, C.W., Wolfe, A.D., Young, N.D., Elisons, W.J. and Reeves, P.A. (2001) *Am. J. Bot.*, **88** : 348-361.

Ottenschlager, I., Barinova, I., Voronin, V., Dahl, M., Heberle-Bors, E. and Touraev, A. (1999) *Transgenic Res.*, **8** : 279-294.

Paradiso, R., Aronne, G. and De Pascale, S. (2007) *In International Symposium on High Technology for Greenhouse System Management: Greensys*, **801** : 1399-1406.

Paulus, A.O., Besemer, S. and Nelson, J. (1983) *Flower and Nursery Report, Summer and Spring*, **102**, University of California.

Peterson, H. (1955) *Bull. N.Y. St. Flower Grs.*, **122** : 2-3.

Pfister, J.M. and Widholm, J.M. (1984) *HortScience*, **19** : 852-854.

Pirone, P.P., Dodge, B.O. and Rickett, H.W. (1960) *Diseases and Pests of Ornamental Plants*, Constable and Co. Ltd., London.

Poirier Hamon, S., Rao, P.S. and Harada, H. (1974) *J. Expt. Bot.*, **25** : 752-760.

Porter, D.M. and Aycock, R. (1967) *N.C. Agric. Exp. Stn. Tech. Bull.*, **179** : 1-31.

Post, K. (1942) *New York (Cornell) Agric. Exp. Stn. Bull.*, **787** : 56.

Pryor, R.L. (1951) *Proc. Amer. Soc. Hort. Sci.*, **57** : 406-407.

Qiao, K., Shi, X., Wang, H., Ji, X. and Wang, K. (2011) *J. Agric. Food Chem.*, **59**(6): 2362-2367.

Raabe, R.D., Hurlimann, J.H. and Sciaroni, R.H. (1970) *California Agric.*, **24** : 8.

Rainbow, A., Wilson, N. and Szmidt, R.A.K. (1998) *Acta Hortic.*, **469** : 79-88.

Ramadan, N. (2018) *Mesopotamia J. Agric.*, **46**(1) : 271-278.

Raulston, J.C. and Marousky, F.J. (1970) *HortScience*, **5** : 355.

Rausher, M.D., Miller, R.E. and Tiffin, P. (1999) *Molecular Biol. and Evolution*, **16** : 266-274.

Roberts, R.H. and Struckmeyer, B.E. (1939) *J. Agric. Res.*, **59** : 699-709.

Rogers, M.N. (1951) *Master's Thesis*, Univ. Missouri, USA.

Rogers, M.N. (1958) *Mo. State Florists News.*, **18** : 3-7.

Rogers, M.N. (1980) In : *Introduction of Floriculture* (Ed. R.A. Larson), Academic Press, New York.

Rolland-Lagan, A.G., Bangham, J.A. and Coen, E. (2003) *Nature*, **422**(6928) : 161-163.

Rutland, R.B. and Pallas, J.E. Jr. (1972) *J. Amer. Soc. Hort. Sci.*, **97** : 34-37.

Saavoss, M., Majsztrik, J., Belayneh, B., Lea-Cox, J. and Lichtenberg, E. (2016) *Irrig. Sci.*, **34**(5) : 409-420.

Sach, L., Hendriks, L. and Ueber, E. (1995) *Acta Horticuturae*, **378** : 81-86.

Saedler, H. and Zachgoa, S. (2004) *Plant Cell*, **16**(12) : 3197-3215.

Saksena, H.K. (1960) *Indian Phytopath.*, **13** : 165-167.

Sanderson, K.C. (1975a) *Flor. Rev.*, **156** : 58.

Sanderson, K.C. (1975b) *Flor. Rev.*, **156** : 17, 59-61.

Sanderson, K.C. and Link, C.B. (1967) *Proc. Amer. Soc. Hort. Sci.*, **91** : 598-611.

Sang, C.K., Kim, H.Y. and Yun, H.S. (1992) *J. Korean Soc. Hort. Sci.*, **33** : 79-86.

Saniewska, A. (1995)*Ochrona Roslin*, **39** : 4, 14.

Saniewska, A., Orlikowski, L.B. and Wojdyla, A.T. (1998) *Prog. Plant Protection*, **38** : 198-203.

Sarhan, A.Z. and El-Sayed, A.A. (1983) *Annals Agric. Sci. Moshtohor*, **19** : 419-426.

Sastry, G.R.K. (1976) *Heredity*, **36** : 316-329.

Schettini, E., Vox, G. and De Lucia, B. (2007) *Sci. Hortic.*, **112**(4) : 456-461.

Schroeder, K.R. and Stimart, D.P. (1999) *HortScience*, **34** : 736-739.

Schwarz-Sommer, Z., Davies, B. and Hudson, A. (2003) *Genetics*, **4** : 657-666.

Schwarz-Sommer, Z., de Andrade Silva, E., Berndtgen, R., Lönnig, W.E., Müller, A., Nindl, I., Stüber, K., Wunder, J., Saedler, H., Gübitz, T. and Borking, A. (2003) *Genetics*, **163**(2) : 699-710.

Schwarz-Sommer, Z., Gübitz, T., Weiss, J., Gómez-di-Marco, P., Delgado-Benarroch, L., Hudson, A. and Egea-Cortines, M. (2010) *BMC Plant Biol.*, **10**(1) : 275.

Schwinn, K., Venail, J., Shang, Y., Mackay, S., Alm, V., Butelli, E., Oyama, R., Bailey, P., Davies, K. and Martin, C. (2006) *Plant Cell*, **18** : 831-851.

Secerov-Fiser, V. and Vujosevic, A. (1998) *Review Res. Work, Faculty of Agri.*, Belgrade, **43** : 87-92.

Seeley, J.G. (1965) *Proc. Amer. Soc. Hort. Sci.*, **86** : 693-694.

Senior, I., Holford, P., Cooley, R.N. and Newbury, H.J. (1995) *J. Exp. Bot.*, **46** : 1233-1239.

Senthilraja, C., Renukadevi, P., Malathi, V.G., Nakkeeran, S. and Pappu, H.R. (2018) *Plant Dis.*, **102**(8) : 1676-1676.

Serek, M., Sisler, E.C. and Reid, M.S. (1995) *Plant Growth Regulation*, **16** : 93-97.

Shafique, A., Maqbool, M.U., Nawaz, M.A. and Ahmed, W.A. (2011) *Pak. J. Bot.*, **43**(2) : 1003-1010.

Shanan, N.T. and Soliman, A.S. (2011) *Am. -Eurasian J. Sustain. Agric.*, 150-158.

Shang, Y., Schwinn, K.E., Bennett, M.J., Hunter, D.A., Waugh, T.L., Pathirana, N.N., Brummell, D.A., Jameson, P.E. and Davies, K.M. (2007) *Plant Methods*, **3**(1) : 1-12.

Sharma, P.P. and Mahmud, A.K. (1950) *Mag. Agric. Coll. Nagpur*, **25** : 47-48.

Shedeed, M.R., El-Gammasy, K.M., Hashim, M.E. and Almula, A.M.N. (1986) *Annals Agric. Sci.*, Ain Shams University, **31** : 691-705.

Sheyaib, S., Shatnawi, M.A., Shibli, R.A., Obeidat, M., Shadiadeh, A.N., Alhussaen, K.M. and Abu-Zahara, T. (2010) *Jordan J. Agr. Sci*, **6**(2) : 171-182.

Simon, R., Carpenter, R., Doyle, S. and Coen, E. (1994) *Cell Cambridge*, **78** : 99-107.

Singh, J. and Arora, J.S. (1980) *J. Res. Punjab Agric. Univ.*, **17** : 258-261.

Sohi, H.S. (1984) *Diseases of Ornamental Plants and Their Control*, ICAR Publication, New Delhi.

Spena, A., Estruch, J.J., Prinsen, E., Nacken, W., Van Onckelen, H. and Sommer, H.S.O. (1992) *Theor. Appl. Genet.*, **84** : 520-527.

Srivastava, O.P. (1969) *J. Indian Bot. Soc.*, **48** : 176-189.

Srivastava, O.P. and Singh, S.N. (1972) *Punjab Hort. J.*, **12** : 195-199.

Stefanis, J.P. and Langhans, R.W. (1982) *HortScience*, **17** : 601-603.

Stubbe, H. (1952) *Abhandl. Sanchsisch. Akad. Wiss. Math. Natur. Kl.*, Leipzig, **47** : 2-23.

Stubbe, H. (1953) *Z. Indukt. AbstamumVererbungs.*, **85** : 450-478.

Stubbe, H. (1959) *Cold Spring HarborSymp. Quant. Biol.*, **24** : 31-40.

Stubbe, H. (1966) *Genetik Und Zytologic Von Antirrhinum L. Sect. Antirrhinum*, Jena : Gustav Fisher.

Stubbe, H., Berg, W., Meyer, M. and Hegemann (1981) *BiologicalesZentralblatt*, **100** : 383-385.

Sullivan, K.J. and Pasian, C.C. (2000) *HortScience*, **35** : 25-27.

Sutton, D.A. (1988) A revision of the tribe Antirrhineae. (Oxford, UK: Oxford University Press).

Szymkowiak, E.J. (1995) *BioEssays*, **17** : 387-390.

Tayama, H.K. (1963) *Dissert. Abstr.*, **24** : 4340.

Theissen, G. (2000) *BioEssays*, **22** : 209-213.

Tognon, G.B., Sanmartín, C., Alcolea, V., Cuquel, F.L. and Goicoechea, N. (2016) *Plant Growth Regul.*, **78**(3) : 389-400.

Tomioka, K., Nishikawa, J., Moriwaki, J. and Sato, T. (2011) *J. Gen. Plant Pathol.*, **77**(1) : 60-63.

Vargas, P., Carrió, E., Guzmán, B., Amat, E. and Güemes, J. (2009) *J. Biogeogr.*, **36**(7) : 1297-1312.

Verma, S., Kumar, A. and Kumar, A. (2018) *Plant Arch.*, **18**(1) : 62-64.

Wainwright, H. and Irwin, H.L. (1987) *J. Hort. Sci.*, **62** : 401-404.

Waites, R. and Hudson, A. (1995) *Development*, **121**(7) : 2143-2154.

Wang, C.Y., Baker, J.E., Hardenburg, R.E. and Lieberman, M. (1977) *J. Amer. Soc. Hort. Sci.*, **102** : 517-520.

Wang, H., Ouyang, J. and Xue, Y. (2004) *Plant Cell, Tissue Organ Cult.*, **76**(1) : 61-65.

Warren, G.C. and others (1960) *Res. Rep. East. Sect. Nat. Weed Committee*, Canada, pp. 83-85.

Watts, L. (1980) *Flower and Vegetable Plant Breeding*, Grower Books, London.

Webb, D.A. (1971) *Bot. J. Linn. Soc.*, **64** : 271–275.

Weigel, D. and Meyerowitz, E.M. (1994) *Cell Cambridge*, **78** : 203-209.

Whibley, A.C., Langlade, N.B., Andalo, C., Hanna, A.I., Bangham, A., Thebaud, C. and Coen, E. (2006) *Science*, **313** : 963-966.

Whitney, H.M., Poetes, R., Steiner, U., Chittka, L. and Glover, B.J. (2011) *PLoS One*, **6**(3) : 7576.

Wienand, U., Sommer, H., Schwarz, Z., Shepherd, N., Saedler, H., Kreuzaler, F., Ragg, H., Fautz, E., Hahlbrock, K., Harrison, B. and Peterson, P.A. (1982) *Mol. Gen. Genet.*, **187**(2) : 195-201.

Wilson, Y. and Hudson, A. (2011) *Plant J.*, **66**(6) : 1032-1043.

Wisniewski, J. (1998a) *Roczniki Akademii Rolniczej w Poznaniu Ogrodnictwo*, **26** : 145-151.

Wisniewski, J. (1998b) *Roczniki Akademii Rolniczej w Poznaniu Ogrodnictwo*, **26** : 129-144.

Wisniewski, J. (1998c) *Roczniki Akademii Rolniczej w Poznaniu Ogrodnictwo*, **26** : 113-128.

Wright, G.A., Lutmerding, A., Dudareva, N. and Smith, B.H. (2005) *J. Comp. Physiol.*, **191**(2) : 105-114.

Xiu, H., Sixiang, Z. and Xun, G. (2004) *J. Yunnan Agric. Univ.*, **19**(5) : 524-527.

Yamaguchi, H., Niki, T., Niki, T. and Nishijima, T. (2010) *J. Jpn. Soc. Hortic. Sci.*, **79**(2) : 192-199.

Yee, S.W., Suh, J.K. and Roh, S.M. (1980) *J. Korean Soc. Hort. Sci.*, **21** : 193-197.

Zafarghandi, M.S.R. and Rahmati-Joneidabad, M. (2020) *Adv. Hortic. Sci.*, **34**(1) : 105-111.

Zamorski, C. and Milczarek, K. (1977) *Acta Hortic.*, **30** : 335-340.

Zhang D., Yang Q., Bao W., Zhang Y., Han B., Xue Y., Cheng Z. (2005) *Genetics*, **169** : 325-335.

Petunia

Debmala Mukherjee and Subhrajyoti Chatterjee

1.0 Introduction

Petunias are one of the best summer flowering annuals for mass display. *Petunia* is genus of 30 species of flowering plants of South American origin. The popular flower of the same name derived its epithet from the French, which took the word *'petun'*, meaning 'tobacco', from a Tupi- Guarani language. An annual, most of the varieties seen in gardens are hybrids (*Petunia* × *atkinsiana*, also known as *Petunia* × *hybrida*). Petunia is a genus in the family Solanaceae, subfamily Petunioideae. Most petunias are diploid with 14 chromosomes and are interfertile with other petunia species. The basic petunia flower is funnel shape, but hybridizers have created many variations including singles and doubles with petals that have wavy or fringed margins. Many patterns are available in stripes, speckles and borders in an extensive colour palette that includes purple, mauve, lavender, pink, red, white and yellow.

2.0 Importance and Uses

2.1 Aesthetic Uses

Petunias are versatile annuals. They can be used for colour masses, borders, containers, hanging baskets or as a seasonal groundcover and also used in private garden, and building decoration. Some petunias are more suited to container growing, while others are better for massing in the garden. Petunias are a representative bedding plant known for their tolerance for hot and dry climates (Hoang and Kim, 2018) also. The showy trumpet-shaped flower of *Petunia* × *atkinsiana* is an excellent choice for summer flower beds and window boxes.

2.2 Therapeutic and other Uses

A mild flavour, they can be eaten raw in salads or be used as a garnish on other foods. The Maya and Inca believed that the scent of petunias had the power to ward off underworld monsters and spirits. Their flower-buds were bunched together for magical drinks. According to New Age folklore, Petunias will only thrive where there is 'positive energy' and will not grow in places where there is 'negativity'. When given as a gift, petunias have (in the language of flowers) two mutually exclusive meanings, symbolising on the one hand being comfortable with someone, and on the other anger and resentment.

3.0 Origin and History

The petunia, carries a close relation to tobacco, tomato, and the deadly nightshade. Spanish explorers in the 16th century discovered a low growing, white-flowered form the Tupi-Guarani Indians called 'Petun', which in their language translated to the 'worthless tobacco plant'. Considered 'ugly' the Spanish explorers did not send samples of the 'petun' back to Spain. The Mayan and Incas believed that the fragrance of the petunia could chase away the underworld monsters and spirits.

In 1823, King George III sent Joseph Bonaparte, (Napoleon's Brother) back to explore Argentina. During his journey, samples were collected and sent back to Spain where botanists confirmed the petunias relationship to tobacco. In 1831, a collector

Uses of Petunia

In Pot

As Border

In Hanging Basket

Flower Bed in Garden

In Decoration of Wall

Entrance of Home

On Wall Decoration

On Roof Top

In Flower Vase with Other Flowers

Petunia in Colour Combination

Window Basket

Uses of Petunia

Petunia in Decoration

On the Edge of Water Pool **Terrace Garden** **In Roundabout**

In Flower Bed **In Road Median** **In Trellis**

Garden Design

Source: **www.in.pinterest.com, www.1zoom.me, www.gardensonline.com.au, www.in.pinterest.com, www.swallowtailgardenseeds.com, NDMC, New Delhi, www.123rf.com, www.tropical.theferns.info, www.lushome.com**

for the Edinburgh Botanical Gardens, Scottish Explorer John (James) Tweedie came across another petunia, purple in colour, while exploring the Americas.

The first single petunia was found by Commerson in Argentina, on the banks of La Fata river, and sent by him to Jussieu, who named it Petunia nycta-ginaflora. Introducing it into France in 1823. This plant had an upright habit with thick sticky leaves and long-tubed fragrant white petunias flowers. The second species was sent by Tweedie from Buenos Ayres to the Glasgow Botanical Gardens in 1831, this plant had a decumbent habit, small violet purple flowers and short tube and was named Petunia vlolacea. From these two species all varieties of petunias have been bred. They have been freely crossed with each other: hence the garden varieties now go under the general name of Petunia hybrida.

In 1803, Jussieu established the genus *Petunia* (Solanaceae), later referred to as *Petunia sensu* Jussieu. In the 1980–1990s *P. sensu* Jussieu was divided into two genera: *Petunia* and *Calibrachoa*. Ando *et al.* (2005) performed a detailed phylogenetic analysis of *P. sensu* Jussieu and demonstrated that the separation of *Petunia* and *Calibrachoa* into diverent genera is supported by chloroplast DNA RFLP data. The study of the garden petunia, termed *Petunia* × *hybrid* (Hook.) Vilm. and now commonly known as *Petunia hybrida*, was decisive for the future of the taxonomy of the genus. The garden petunia was first obtained by hybridization in 1834 by Atkins of Northampton, a British nurseryman, and it soon spread to European gardens (Sink, 1984). Today, it is cultivated all over the world and is one of the most important Solanaceae utilized for ornamental purpose. Several clades in the Petunia phylogenetic trees were found to correspond with geographic distribution, suggesting that recent speciation occurred independently in diverent regions. To date, around 30 *Petunia* (sub) species have been described. The geographic origin of petunia is the southern or central part of South America, and various species have been documented from collections made in Argentina, Brazil, Paraguay, and Uruguay (Ando *et al.*, 2005).

It is native of South America and is popularly known as the garden petunia, a highly cultivated hybrid for ornamental purposes derivated from a cross between *Petunia axillaris* (Lamarck) Britton, Sterns and Poggemburg and *P. integrifolia* (Hooker) Schinz and Thellung (Wijsman, 1982). Fries (1911) mentioned the two genetically distinct groups designated by Wijsman and Jong (1985) as two genera: *Petunia* (2n=18) and *Stimoryne* Rafin. (2n=14). According to this work, the garden petunia (2n=14) was included in the genus *Stimoryne*. However, as the name *Petunia* was previously established among horticulturists, Wijsnands *et al.* (1986) suggested the conservation of *P. nyctaginiflora* Juss. (2n=14) as the nomenclatural type of the genus *Petunia* (Brummitt, 1989) and revalidated the genus *Calibrachoa* (2n=18). Twelve species remained in the genus *Petunia* (Ando and Hashimoto 1996) and 24 were recombined to *Calibrachoa* (Wijsman, 1990; Stehmann and Semir, 1997).

4.0 Distribution

The genus Petunia is endemic to South America, with subtropical distribution ranging from 22° to 39°S. The major species richness is found in Brazil (13), followed by Argentina (5), Uruguay, Paraguay, and Bolivia (2). All species occur in Brazil, except for *P. occidentalis*, which has a disjunct distribution restricted to the Sub-Andean mountains in northwestern Argentina and southern Bolivia. There are two major disjunct areas of occurrence of petunia: the Serra da Mantiqueira in Minas Gerais, Brazil, and the Sub-Andean region in Argentina and Bolivia. In the first, only *P. mantiqueirensis* was found, an endemic species phylogenetically related to the Brazilian highland group. The geographic barrier corresponds to the Atlantic rainforest and savanna that cover almost all the Sao Paulo state. The Sub-Andean region is separated from the core ˜ Petunia distribution by the Chaco, a large and drier region, and inhabited by two taxa: *P. axillaris subsp. subandina*, which grows from San Luis, Argentina, toward Tarija, Bolivia (Ando, 1996; Kokubun *et al.*, 2006), and *P. occidentalis*, with distribution restricted to northern Argentina and adjacent areas in Bolivia (Tsukamoto *et al.*, 1998).

5.0 Morphology

Petunia comprises annual or perennial herbs, up to 1 m tall, with erect, ascendant, decumbent, or procumbent stems, rarely rooting at the nodes. The leaves are sessile or petiolate, with blades elliptic, ovate or obovate, more rarely rounded or linear, membranaceous, somewhat juicy, flat, and usually without marked venation. The leaves are sessile (*e.g.*, lacking a petiole, or leaf stem) and are usually oval-shaped with smooth margins; some feature are having fine sticky hairs. Although technically a perennial, the common garden petunia is most often grown as an annual; its flowers bloom profusely from early summer until frost. The innumerable horticultural varieties fall into two general types: the compact erect type, reaching 15–25 cm and adapted for summer garden beds, and the sprawling long-stemmed balcony petunia, which grows to about 46 cm and is often potted in hanging baskets and window boxes.

Members of the Solanaceae family show a wide range of floral morphologies. Some species have zygomorphic or monosymmetric flowers, while others have actinomorphic or radially symmetric flowers (Knapp, 2002b). Petunia plants, as other Solanaceae species, exhibit determinate inflorescences composed of scorpioid or cincinnus cymes (Weberling, 1989). The flowers are funnel-shaped, consisting of five fused or partially fused petals and five green sepals. The flowers range from pure white to deep crimson or purple and are often speckled or veined in contrasting colours. There are single- and double-bloom varieties. At the base of each flower, two bracts are formed, each with a dormant (vegetative) meristem in its axil. A wild type flower contains five sepals, five petals, five stamens, and two carpels arranged in four concentric whorls. The five petals are fused. Stamen filaments are partly fused to the tube of the flower. The petunia flower is zygomorphic in all floral whorls, which is partly due to the whorled or irregular arrangement of sepals and petals in the floral bud before it opens (Knapp, 2002b). The calyx is green, deeply lobed, with a very short tube and long linear or spatulate lobes with usually inconspicuous ribs. The corolla is funnelform, campanulate, or salverform, with imbricate aestivation, tube 1.5–7 cm long, and purple, white, red (bright or orange), or pink limb. The five stamens are variously adnated to the corolla tube, usually included and arranged in. Petunia species are mostly annual herbs.

Ovules have a single integument (Angenent *et al.*, 1995). The fruit is a capsule, containing variable amounts of seeds for different petunia species (Gunn and Gaffney, 1974; Sink and Power, 1978). Seed capsules are conic, tapering to the apex and widest at the base. The mature fruit is surrounded by an enlarged, glandular hairy calyx composed of five lobes (the sepals) which are equal to or longer than the capsule, depending on the species (Gunn and Gaffney, 1974).

5.1 Pollination in Petunia

The typical example of petunia pollination syndrome is by Hawkmoth in *Petunia axillaris* and bee in *P. integrifolia.* These two representatives of two groups of petunia species have a complex set of morphological and physiological traits that are adapted to their respective pollinators. *P. exserta* is pollinated by hummingbirds. *P. axillaris* has white flowers, with long petal tubes that exactly fit the length of the tongue of the hawkmoths that pollinate them (*Manduca contracta* and *M. diyssa* ssp. *Petuniae*). Moreover, for the nocturnally active hawkmoths a coloured flower is not as important as a strong scent, and in accordance with that *P. axillaris* has white, nocturnally scented flowers (Ando *et al.*, 2001; Stuurman *et al.*, 2004). *P. integrifolia* has unscented purple coloured flowers, with a wide and short petal tube.

P. integrifolia exhibit diurnal opening and closing movements synchronous with the activity period of the bee, *Hexantheda* sp. Amounts of floral nectar in *P. axillaris* and *P. integrifolia* are within the range of hawkmoth and bee pollinated flowers, respectively (Ando *et al.*, 2001; Stuurman *et al.*, 2004). A thorough study in which the Petunia pollination syndromes have been dissected into their most important phenotypic and genetic components has been conducted by Stuurman *et al.* (2004). Natural hybrids of *P. axillaris* and *P. integrifolia* have never been reported, even though artificial crosses can produce fertile hybrid offspring. Therefore, Ando *et al.* (2001) studied the reproductive isolation between the two species. Differential insect visitation of *P. integrifolia* and *P. axillaris* in sympatric populations was observed, suggesting an important biological meaning of the floral differences in colour, scent, and amount of nectar. Insect visitation is not the only reproductive barrier among Petunia species, as genetic incompatibilities between ecotypes can also be considerable (Ando *et al.*, 2001). Still, the extensive divergence in the Petunia floral pollination syndromes indicates that insect visitation has certainly had a huge impact on the evolutionary history of the Petunia species (Stuurman *et al.*, 2004).

6.0 Genetics and Breeding

Petunias are popular seed and vegetatively propagated bedding plants with numerous phenotypes, series and cultivars available on the world market. The history of petunia development as a crop is highlighted with implications for a wide variety of other floricultural crops. Hybrid seed production transformed the crop; continued domestication by flower breeders produced a wide range of flower colour and flower patterns (Griesbach, 2006). In 1944, a severe summer drought killed record number of bedding plants throughout the United States. Petunias became national news because of their drought tolerance (Haughton, 1978). This tolerance established their widespread popularity, which lead to a major effort to develop new cultivars with novel flower colours and improved garden performances.

Breeding method for petunia radically changed during the 1930's. Prior to 1930, most cultivars were developed through mass selection and superior clones were vegetatively propagated (Ewart, 1984). After 1930, breeding focused on individual plant selection, inbreeding and seed production of superior strains.

Warner and Walworth (2010) reported that the leaf unfolding rate (*i.e.*, development rate) and the number of nodes forming prior to floral initiation were two factors determining production times of petunia. Wild relative species of the cultivated petunia (*Petunia* × *hybrida*) that exhibited faster development rates than modern cultivars and may therefore be useful genetic sources to develop cultivars with decreased production time were identified. Three interspecific F_2 families, *Petunia exserta* × *P.axillaris*, *P.* × *hybrida* 'Mitchell' × *P. axillaris*, and *P. axillaris* × *P. integrifolia*. All exhibited transgressive segregation for development rate and node number below the first flower. Development rate and time to flower segregated independently in all families. Leaf number below the first flower was positively correlated with leaf unfolding rate in all families except *P. axillaris* × *P. integrifolia*. Time to flower was positively correlated with flower bud number in the *P.* × *hybrida* 'Mitchell' × *P. axillaris* and *P. axillaris* × *P. integrifolia* families only. Based on these results, wild petunia germplasm should be useful for developing petunia cultivars with reduced crop production times, but some negative effects on crop quality parameters may need to be overcome.

Cartolano *et al.* (2007) showed that the 'blind' (bl) mutant of *Petunia hybrida* and the 'fistulata' (fis) mutant of *Antirrhinum majus*, which had similar homeotic phenotypes, were recessive alleles of two homologous miRNA-encoding genes. The BL and FIS genes controlled the spatial restriction of homeotic class C genesto the inner floral whorls, but their ubiquitous early floral expression patterns were in contradiction with a potential role in patterning C gene expression. They provide genetic evidence for the unexpected function of the MIRFIS and MIRBL genes in the centre of the flower and proposed a dynamic mechanism underlying their regulatory role.

Que and Jorgensen (1998) used flower colour transgenes to address the hierarchical operation of both mechanisms *i.e.* transcript homology and co-suppressionin plants carrying two transgene copies in petunia. It was shown that co-suppression of homologous, endogenous flower colour genes by single-copy transgenes required that the transgene be driven by a strong promoter and that the degree of co-suppression was highly sensitive to increasing transgene dosage. Together, these observations suggested that co-suppression should be a sensitive reporter of epigenetic changes in transgene transcription, such as might be caused by paramutation-like interactions between transgene loci. Intercrosses bringing together two homologous transgene loci, one a known epimutable reporter and the other a transgene inverted repeat, resulted in complete loss of co-suppression in some outcross progeny and a qualitative change in morphology-based patterns of co-suppression in other outcross progeny. This paramutation-like behaviour suggested that the transgenes may be altered at the transcriptional level, eliminating co-suppression altogether or changing the spatial pattern of transgene transcription to produce a new pattern of co-suppression.

Van der Krol *et al.* (1988) reported that Chalcone synthase (Chs) gene had been used for production of pink, white and variegated flowers (sense and antisense genes were used) in petunias. Meyer *et al.* (1987) reported the first application of genetic engineering to modify flower colour which led to the production of an orange pelargonidin-producing Petunia variety, which produced flowers with pale brick colour. This was achieved by the expression of dihydroflavonal-4-reductase (dfr) gene from maize in a petunia line. Kashikar and Khalatkar (1981) treated genetically pure and physiologically similar seeds of uniform size with either ethyl methyl sulphonate (EMS) or gamma radiations. Treated seeds were sown in the field to raise the M_1 generation. The screening of M_1population indicated several flower colour changes in the EMS treatments, while very few colour alterations were obtained with gamma radiation. Muszynski (1976) started mutation breeding programme with an old variety known as 'Purpurea' which was having tall and single stem and purple flowers. Seed samples of 1000 seeds each were treated with thermal neutron flux of 2.9-3.4 × $10^9 n_{th}/cm^2$.sec and irradiation time of 30-60-90-120 minutes. In the M_1 generation some interesting chlorophyll as well as some morphological changes was observed.

6.1.0 Self Incompatibility in Petunia

In *Petunia*, 19 different S alleles have been described (Ai *et al.*, 1991). Phylogenetic analysis suggested that the different alleles arose through intra-genic combination. The 'S' alleles within the pistil encode S-RNAses which have two highly variable regions (HV_a and HV_b) and five conserved regions (C_1 through C_5). The exact nature of pollen S alleles have not yet been elucidated; however they are physically linked to the pistil 'S' alleles (Mccubbin*et al.*, 2000). A potential pollen 'S' allele encoded protein has been identified (Sims and Ordanic, 2001). This protein ($PhSBP_1$) is only expressed in pollen and binds to S-RNAses with a high degree of specificity. $PhSBP_1$ contains a RING-HS domain which functions as and E_3 ubiquitin ligase. This suggests that $PhSBP_1$ might function by degrading S-RNAses.

6.2.0 Male Sterility in Petunia

Maureen Hanson's group at Cornell University has cloned a mitochondrial gene that encodes Cytoplasmic Male Sterility (CMS) in petunia. The gene encodes an abnormal protein that disrupts mitochondrial activities. A nuclear gene (the Restorer fertility or Rf gene) is known to interact with this mutant mitochondrial gene, reducing its expression and thereby restoring normal fertility to male sterile genotypes. The group has identified the petunia Rf gene by cloning candidate genes from map position and demonstrating that such a gene is able to confer fertility to CMS Petunia lines. Subsequently genes highly similar to the petunia Rf have been identified as fertility restorers in *Brassica* and rice (Gerats and Vandenbussche, 2005).

6.3.0 Interspecific Hybridization

Within the two subgenera (*Pseudonicotiana* and *Eupetunia*), geographic isolation is the primary reproductive isolating mechanism. Within a subgenus, the species are allopatric producing fertile progeny between any interspecific combination. The species in the different subgenera are syntopic. The range of the *Pseudonicotiana* and *Eupetunia* species overlaps, but hybrids are not found commercially. All *Pseudonicotiana* × *Eupetunia* hybrids are fertile but vary in pollen viability. Even within a subgenus, pollen viability of interspecific hybrids can vary from 100 per cent to less than 50 per cent that of a sibmated species control. *Petunia axillaris* × *Petunia integrifolia* is the most studied *Pseudonicotiana* × *Eupetunia* hybrid. The cultivated petunia (*Petunia hybrida*) was derived from this hybrid. Significantly higher success rate is obtained when *Petunia axillaris* is used as the female parent. Morphologically, the hybrid is intermediate in between the parents except flower colour (Griesbach, 2006).

6.4.0 Somatic Hybridization

Petunia is being considered as one of the model plants for tissue culture. Unlike in most plants, petunia protoplasts are easily regenerated into whole plants. Izhar and Tabib (1980) obtained two types of cytoplasmic hybrids by protoplast fusion. These contained one or the other original parental nucleus and heteroplasrnon, a mix of plasmons inducing cytoplasmic male sterility and fertility. In subsequent generations, following selfing, stable male sterile and male fertile lines segregated from single fertile cytoplasmic hybrid plants. These demonstrated the existence of a heteroplasmic state in the somatic hybrids and the occurrence of cytoplasmic segregation of the heteroplasmon into homoplasmons following the first and the second meiotic cycles.

7.0 Species

Petunia axillaris and *P. integrifolia*, parents to the familiar garden hybrid, have the largest distribution. Several species are narrow endemics, for example, *P. bajeensis*, *P. bonjardinensis*, *P.exserta*, *P.mantiqueirensis*, *P. reitzii*, *P. saxicola* and *P. secreta*. Species of petunia can be found in sunny, partially shaded, or completely shaded sites. *Petunia altiplana*, *P. axillaris*, *P. inflata*, and *P. integrifolia* are easily found in disturbed places such as roadside slopes, especially in rocky ground; *P. scheideana* and *P. mantiqueirensis* occur in partial shade at edges of Araucaria forests; and *P. exserta* grows in the shaded relief from shallow caves sculpted by the wind in sand stone (Stehmann *et al.*, 2009).

Petunia altiplana

This species is easily recognized by its rooting stems usually broadly, spatulate leaves, purple corolla and stigma located below the anthers of the longest pair of stamens. Its repent habit of usually forming a round mat is unique to the genus. The radial growth pattern plus the massive blooming permit the use of species for ornamental purposes. This species is distributed in the highlands of the Santa Catarina and Rio Grande do Sul, Brazil, in altitude of 800-1200 m and grows in outcrops or exposed roadside slopes (Ando and Hasimoto, 1993).

P. exserta

This is a rare member of the genus *Petunia*, endemic to the Serras de Sudeste in southern Brazil. It was first described in the year 1987, only fourteen plants were found in the wild during an expedition in 2007. In the wild, the plant is

Species of Petunia

Petunia exserta

Petunia axillaris

Petunia integrifolia

Petunia × atkinsiana

Petunia altiplana

Petunia littoralis

Types of *Petunia × hybrida*

Grandiflora petunia

Picobella miliflora petunia

Multiflora petunia 'Primetime'

Spreading Petunia, Wave Type

Multiflora petunia, Carpet Series

Grandiflira petunia (Ultra Series) 'Rose Star'

found growing only in shaded cracks on sandstone towers. It is the only *Petunia* species that is naturally pollinated by hummingbirds. *Petunia exserta* has a bright red corolla with distinctive exserted stamens and stigma, the latter which are typically associated with hummingbird pollination. It shares the erect habit, and similar corolla, pollen, and stalk attributes of *Petunia axillaris* and *Petunia secreta* (Stehmann *et al.,* 2009). The flowers have no fragrance, which is not necessary for attracting hummingbirds, which have little sense of smell. The red colour is due to several anthocyanin pigments, which are different from the cause of red colour in modern hybrid petunias. In addition to its very limited range, *P. exserta* is also threatened because it so easily hybridizes with other *Petunia* species in its native range, especially with *Petunia axillaris,* thus producing offspring that are hybrids rather than the species. The plant is increasingly being sold in horticulture, so the threat to its survival is only in its native habitat.

P. axillaris

This large white petunia, wild white petunia or white moon petunia, is an annual herbaceous plant. It is native to temperate South America. The plant's flowers, the only white ones found in the *Petunia* genus, are 3 to 7 cm long (Turchetto, 2014). The commonly-grown garden petunia is a hybrid of *P. axillaris* and *P. integrifolia*. *P. axillaris* has three subspecies: *P. axillaris axillaris, P. axillaris parodii,* and *P. axillaris subandina*. *P. axillaris axillaries* and *P. a. parodii* are found in flat parts of the Pampas. Meanwhile, *P. subandina* is found in highlands near the Andes (Turchetto, 2014).

Petunia integrifolia and *Petunia violacea* have both been known by the vernacular name of violet petunia, both being species of wild petunia with violet-coloured blooms. *Petunia integrifolia* is native to Argentina. *P. integrifolia* bears flowers approximately 4 cm in diameter and the plant is typically smaller and harder to cultivate than the well-known hybrid bedding Petunia now known correctly as *Petunia × atkinsiana*. *Petunia violacea* Lindl. has been reported to be used as a hallucinogen in Ecuador, where the plant has the vernacular name shanín. The drug is said to cause sensations of levitation and flight - a type of hallucination often associated with the use of the more toxic hallucinogenic plants of the deliriant type *e.g.* the tropane containing Atropa and Hyoscyamus -active constituents of the witches' flying ointments of Medieval and Early Modern Europe (Schultes, 1976).

P. bajeensis

This species is characterized by its viscid vestiture, foliose stem forming a cushion like structure, ovate, elliptic or oblong leaves with prominent primary and secondary veins, purple funnel form corolla, adnated more than 7 mm to the base of the corolla tube, stigma located below the anthers of the longest pairs of stamens and deflexed pedicel in fruit stage. Vegetatively, the individuals of the species roughly resemble more robust plants of *P. bonjardinensis*, but the morphology of the flower does not differ from that of *P. integrifolia* except for the larger size of the floral parts.

P. bonjardinensis

The decumbent habit, with very fragile stems, villose vestiture, campanulate and purple corolla, stigma positioned above the anthers of the longest pair of stamens, and deflexed peduncles in fruit stage are characters permitting clear identification of the species. *Petunia bonjardinensisis* endemic to a small area near to the border of the southern Brazilian plateau.

P. integrifolia

This species is recognized by the following morphological characters: decumbent stems, elliptic or obovate leaves (linear in some coastal populations), funnel form purple corolla, anthers completely opened (flat) after dehiscence, showing the bluish pollen, stigma placed below the anthers of the longest pair of stamens, and deflexed pedicels in the fruiting state. *Petunia integrifolia* inhabits the Pampas province and occurs in Argentina, Uruguay, and southern Brazil (from Rio Grande do Sul to thecoast of Santa Catarina) (Stehmann *et al.*, 2009).

8.0 Horticultural Types of Petunia

Grandiflora

The grandiflora petunia type was developed in early 1950 with the first F_1, hybrid grandiflora, 'Ballerina', introduced in 1952. Petunia grandiflora hybrids are a versatile annual available in a wide range of colours. They are prized in the home garden for their large, bright blooms and low maintenance. This type is most popular and is having the largest flowers of 10 cm or more. They may be singles or ruffled doubles. Some have a pendulous, cascading habit that makes them more suitable for hanging baskets and window boxes, but most are upright plants that develop over the summer into large mounds of flowers which are 30-35 cm tall. Although the flowers are spectacular, their tendency is to get tired and spindly in mid- summer. Grandiflora petunias perform best in moderate summers without excess humidity or moisture. Older cultivars are large plants, but the recent trend is to develop more compact cultivars. Double-flower forms have multi-petaled flowers, resembling carnation flowers. Traditionally, grandifloras are the best selling petunias, but the flowers do not hold up well under adverse garden conditions in the south. Grandiflora petunias series include: Ultra, Dream, Storm, Daddy, Supermagic, Supercascade. 'Sugar Daddy' (*Petunia* Daddy Series), which sports purple flowers with dark veins. 'Rose Star' (*Petunia* Ultra Series), whose flowers look striped because of its rose-pink flowers with a white centre.

Multiflora

The multiflora petunia type was developed in late 1940 with the first F_1, hybrid multiflora/Comanche', introduced in 1953. Multiflora petunias have greater numbers of somewhat smaller flowers and are often more compact. The stems are strong, which makes multiflora petunia varieties suitable for windy climates. The blooms tend to hold up a bit longer than Grandiflora petunia varieties, especially during rainy weather. Multiflora petunias are available in both single and double varieties. Popular Multiflora petunias include: Primetime, Celebrity, Carpet, Horizon, Mirage. Typically, they're massed together to create big splashes of colour in the garden.

Cultivars of Petunia

'Sweet Sunshine Dark Violet'

Grandiflora *Petunia* 'Blue Daddy'

'Hula Hoop Red'

Grandiflora *Petunia* 'Sugar'

'Ultra Blue Star'

'Blue Frost'

'Red Pirouette'

Grandiflora *Petunia* 'Bravo Salmon Veined'

'Limelight'

'Cascadias Rim Magenta'

Fortunia Early Blue Vein

'Mini Rose Blast Pink'

'Milliflora'

This type of petunias developed from a chance genetic mutation from *Petunia × hybrida* resulting in a true miniature plant. Miliflora petunias are miniature with even smaller flowers and plant size. Milliflora petunias are compact, and produce abundant quantities of perfect, small flowers. Mature size of the plants is generally about 20 cm. tall and wide. Milliflora petunias bloom early and are often grown in containers or hanging baskets. They make good edging plants and are also attractive when mixed with other flowering annuals in containers, where they may be viewed close-up. They are low-maintenance plants that require no dead heading. Milliflora petunias include 'Picobella' and 'Fantasy'.

'Spreading Types'

Spreading or Wave petunia is a recent addition with blooms typically measuring about 5 cm across. The plants, which typically spread up to 100 cm by the end of the season, look great in containers and work well as ground covers. They tolerate heat and drought fairly well and generally require no dead heading. Wave petunias include: Easy Wave, Shock Wave and Avalanche. This is the fast growers that can fill in large spaces. Plants spread so rapidly that they cover a huge area over one growing season, provided they're watered and fertilized frequently. This makes them ideal for scrambling down a hillside garden, or planting atop a retaining wall. They're also amazing in hanging baskets and window boxes, where they trail 20 cm or more over the summer. When grown in full sunlight, they are so covered by flowers that you hardly see any foliage.

'Purple Wave' was the first cultivar in the class of spreading petunias. It produces large blooms of deep rose-purple. It is tolerant of summer heat, drought and rain damage. 'Purple Wave' remains under 4 inches tall. 'Wave' series petunias are available in a multitude of colours. Most are not quite as ground-hugging as the original. They are weather-tolerant, disease resistant and heavy-blooming.

'Floribunda'

This type resulted from recent hybridization between grandifloras and multifloras. General characteristics fall in between the two in terms of plant growth habit and flower size, but may offer increased disease resistance.

8.1.0 Cultivars of Petunia

'Limelight': 'Limelight' petunias show the trendy magenta-and-lime-green colour combination. This is a nursery hybrid; parent species are native to South America. The tidy, compact growth habit of this petunia variety is suited at the front of a border. This flourishes well at full sun to partial shade.

'Cascadias Rim Magenta': This variety is having vigorous mounding plant with a trailing growth habit. Height is approximately 30-50 cm and well suited to full sun to partial shade. Cascadias Rim Magenta has deep purple flowers with striking white edging. This is a trailing plant that does well hanging in baskets.

'Fortunia Early Blue Vein': The Fortunia series of petunias was developed as a day-neutral breed, which means that gardeners can expect an earlier blooming time. It has a more compact growth habit than many other petunias. 'Fortunia Early Blue Vein' has an almost silvery appearance, which pairs well with other silver plants, such as dusty miller. It is also a nursery hybrid and parent species are native to South America.

'Rose Star' (*Petunia* Ultra Series) is one of the spreading petunias that presents a striped appearance thanks to rose-colored flowers featuring a white centre.

'Carpet Series' is a compact, mound-forming multiflora variety that makes an excellent edging plant or potted plant. This variety is compact, early blooming with 4-5 cm blooms that come in a wide variety of colours, and ideal for ground cover also.

'Primetime' series is a variety of multiflora type, compact and uniform, covered with 2¼-inch flowers.

'Heavenly Lavender' is an early, compact, double, deep lavender blue multiflora petunia with 3-inch blooms on 12-to 14-inch plants.

'Sugar Daddy' (*Petunia* Daddy Series) is a fairly durable Grandiflora producing big, showy dark purple flowers ornamented with dark veins. Unlike many of its brethren, the blossoms of this plant do stand up well to rain.

'Surprise Lime' bears large, white flowers with yellow centre. It is a wonderful choice for a moon garden as the blossoms look very much like Moon Flowers and are very attractive to hummingbird moths.

'Petunia Plus Red' is an excellent choice if you want to attract hummingbirds and other beneficial pollinators. This Grandiflora features big, bold, red flowers that stay bright and vibrant even in hot, dry weather.

'Suncatcher Pink Lemonade' is a pretty, trailing early bloomer in shades of mingled pink and yellow.

'Limelight' presents a striking magenta coloration with lime-tinted edges. This compact plant is an excellent choice for a border.

'Prism Sunshine's' flowers are deep yellow in the centre with creamy yellow edges.

'Sweetunia White Merlot' produces masses of tidy lavender/rose blossoms with deep magenta centre. The plants are compact and present a neat, rounded shape that looks great in a hanging basket or container.

'Mini Rose Blast Pink' has rosy blossoms with splashes of magenta. This is a compact plant that tends to stand erect. It does not trail. It's a nice choice as a bedding or border plant.

'Purple Pirouette' is a double flowering variety in deep purple with frilly white edging. It's a strikingly beautiful plant.

'Wave Blue' features masses of electric blue flowers and very little foliage. It's a beautiful choice for a focal point such as a window box.

'Fantasy': This variety comes under multiflora type and it is a series of nine hybrids, colours are crimson pink, red with veins and a white throat, violet-blue, salmon, white, crimson red, pale pink with a creamy throat, pale salmon with dark veins, *etc.*

'Celebrity' series petunias are compact and rain-tolerant. The flowers reach 6-8 cm across.

'Madness' series petunias have big, 8 cm flowers in many veined and solid colours. They are compact and bloom until frost. They bounce back well after rain.

'Double Madness' petunias are compact and floriferous with big, 8 cm flowers all through the summer. Like their single counterparts, 'Double Madness' petunias bounce back within hours of a rainstorm.

9.0 Growth and Flowering

In the vegetative phase, leaf primordia are produced alternately along the stem and axillary buds develop which do not elongate until flowering starts. Flowering is controlled by several endogenous factors and environmental stimuli, including day length and vernalization. The identity of floral organs is determined by homeotic genes. Homeotic mutants develop organs at positions normally occupied by other organs. To-date no petunia A type gene has been isolated, but the phenotype of the petunia blind mutant resembles the phenotype of weak alleles of Arabidopsis A–type mutants (Colombo, 1997). Flower development can be divided into several distinct phases: (1) transition to flowering, (2) inflorescence/flower meristem formation, and (3) floral organ patterning. Flowering in Petunia is photoperiodically controlled, and long day conditions or a night interruption with artificial light promote early flowering (Adams *et al.,* 1999). Long photoperiods (>13hours) result in earlier flowering and taller, relatively unbranched plants. Short photoperiods (8-10 hours) delay flowering, retard elongation of the main stem, and promote lateral branching. This response to photoperiod, however, is influenced by temperature. At average daily temperatures less than 20°C, plants are compact and well branched regardless of photoperiod, but flower faster under long days. At average daily temperatures greater than 20°C with short days, plants have more branches than under long days, but flowering is delayed. The most rapid flowering occurs at average daily temperatures greater than 20°C with long photoperiods. However, the central stem is elongated, lateral branching is restricted, and leaf size is smaller. This relationship between temperature and photoperiod is apparent during spring production. In early spring, petunias are compact and slow to flower. As the season progresses, temperature and photoperiod increases resulting in plants that are elongated and flower faster.

10.0 Soil and Climate

10.1.0 Soil and Growing Media

Petunia can grow in a range of soil texture but it prefers sandy or loamy fertile soils. It reacts well on decomposed manure or humus. The site should be open and sunny and well- drained. Fertile soil with pH 6-7 is desirable for petunia. Grandifloras need only enough water to keep the soil moist. In high pH soil, Fe deficiency is a problem. They grow best in rich soil with good drainage. They bloom best with regular fertilization and will continue to flower all seasons. In the flowering stage, the highest photosynthesis rates (8.612 µmol $CO_2/m^2/s$) as well as leaf area (1.766 dm^2) of petunias were obtained on growing media with 60 per cent biolan peat, 30 per cent acid peat and 10 per cent perlite (BP60-AP30-P10) (Popescu and Popescu, 2015).

10.2.0 Climate

This fast growing half-hardy annual grow well in all zones. It prefers bright sunlight for optimum growth but some cultivars prefer light shady condition. Although these plants can tolerate several hours of shade each day, but it needs at least 5 or 6 hours of good sunlight; they'll perform even better when located in full sun all day. The plant grows well in temperate climates also but does not tolerate shade. In shady condition, the numbers of flowers become less. The plants are frost-sensitive, but with the proper precaution, they can survive the winter undamaged. Petunias arc highlight plants requiring as much light as possible early in the spring to flower quickly and to prevent stretching. The lower the available light, the lower the temperature should be to produce high quality plants.

11.0 Propagation

11.1.0 Seed Propagation

Grandiflora petunia hybrids are commonly propagated from seed, which is best sown indoors for at least four weeks before moving outside. Seeds should be sown uncovered in a moist, well-drained soil with a pH of 6 to 7. When planting outdoors, care should be taken that soil temperatures should reach at least 15°C, and there is no chance of frost. The soil for planting should be loose, light and nourishing and the growing media should consists about two parts of leaf or sod soil, two parts of rotted humus, one part of sand, and two parts of peat. The soil needs to be well watered a day before sowing. Before sowing, the seeds are mixed with dry sand and sprinkled on the top of soil, leveled, sprayed, covered with glass or foil and kept at the temperature of 20-22°C to germinate. In a week the seedlings will germinate and will require spraying twice a day. In greenhouse condition, high air humidity as well as air circulation must be ensured. Care should be taken for controlling root rot disease at this stage. Therefore, daily as soon as condensate appears on the glass or foil, the glass should be turned up and the foil should be wiped out. In addition, from time to time, the seedlings need to be sprayed light solution of potassium permanganate along with irrigation. When the first leaf of seedling appears, the glass can be removed, the soil should be lightly sprinkled with sand, and watering should be reduced.

11.2.0 Vegetative Propagation

Petunia is commercially propagated by soft wood cutting or terminal cutting. About 3 inch length cuttings with at least three leaf joints on stem are planted in soil after removing all leaves from lower half of stem. Removal of open flowers, floral buds and lower leaves are necessary for the encouragement of roots rather than shoots. The planted cuttings should be placed in a warm area that receives full sun

Propagation and Growing

Cuttings Taken

Preparation of Cuttings

Planting of Cuttings

Petunia Seeds (Pellated)

Planting in Portray

Plug Plant

Pollination of Petunia

Harvesting of Pods

Harvesting of Seeds

Micropropagation in Petunia

(1) Seeds of *P. hybrida*; (2) Plantlets derived from seeds cultured on MS medium without PGRs; (3) Auxiliary buds used as explants; (4) Explants on MS medium containing PGRs; (5) Auxiliary buds proliferation on medium enriched with 0.50 and 2.00 mg/l BA without NAA; (6) A plantlet grown on medium enriched with 0.50 mg/l BA ready for rooting; (7) Plantlets with roots produced on medium supplemented with 0.01 mg/l NAA without BA; (8) Roots produced at the base of shoot on MS medium containing 0.01 mg/l NAA; (9) Hardened plants, acclimated in a greenhouse using pots filled with peat and perlite (1:1). (*Source*: Mehri *et al.,* 2018).

Open Cultivation **Greenhouse Production**

Pot Petunia Production **Production in Open**

Cultivation in Protected Structures

***Source*: www.greenhousecanada.com, www.asergeev.com, www.shutterstock.com, www.cannygardening.com, www.harrisseeds.com**

or at least five to six hours of sunlight per day. Keeping plants under fluorescent light is also helpful for root growth. Root will form within 6 weeks.

11.3.0 Micropropagation

The highest shoot length (5.75 cm) was recorded from the MS medium supplemented with 0.2 mg/l GA_3 +0.2 mg/l BAP. Rooting of regenerated shoots were achieved on MS medium supplemented with 0.1-1 mg/1 IBA and NAA. The regenerated shoots with well developed roots were successfully acclimatized and established in pots containing sterilized peat moss and grown under laboratory conditions with 70 per cent survival rates (Habas *et al.*, 2019). The appropriate solid culture medium of crowd shoots induction is MS+6-BA 3.2mg/l +NAA 0.1mg/l and MS+6-BA 1.6mg/l+ NAA 0.2mg/l, while the proper culture medium for rooting is 1/2 MS without any hormones was standardized by Guang-rong, 2004. From the study of Mehri, 2018, the highest number of shoots (8.44) and the maximum average shoot length (13.16 cm) were recorded on MS medium supplemented with 0.50 mg/l BA without NAA. Root length (5.20 cm) and root number (8.77) were the maximum in the medium containing 0.10 mg/l NAA. The plantlets regenerated in vitro with well-developed shoots and roots were successfully established in pots containing peat and perlite and grown in a greenhouse within 4 weeks with a 100 per cent survival rate.

12.0 Cultivation

12.1.0 Planting and Care

When shoots get 3-4 true leaves, they are pricked out. The soil in the container should be pre-moistened. Then a seedling is pricked out into a peat or plastic pot with the same soil mixture. Transplanted plants are watered and covered by a paper or some other protection fabric for two or three days. At this stage of growth petunia requires an attentive care, and the most important point is to keep the soil in a state of constant moderate humidity.

Fertilizers are applied weekly in a week after pricking out the seedlings alternating between leaf and root feedings by nitrophosphate and other combined water-soluble fertilizers at the rate of approximately 25-34 g/l of water. The varieties of multiflora petunia bloom in two and a half months after being planted, the varieties of grandiflora petunia bloom in three months. Before planting in open ground it is desirable to hold the already grown seedlings in a cool place or expose them to fresh air for some time during a day for acclimatization. In temperate climate, planting out of petunia seedlings is carried out when all the spring frosts have gone (approximately in the second half of May), and it is better to do this in a cloudy day or in the evening.

12.2.0 Cultivation in Greenhouse

Greenhouse or protected cultivation of petunia comprises with plug tray sowing of seeds. Seeds are sown in plug flats containing a well-drained media without covering the seed. Light is not essential for germination in this structure. The most effective way to sow petunia seeds is with an automatic seeder. Because petunia seed are very small, they are available in pelleted form to accommodate handling by an automatic seeder. pH of growing media should be 5.5 to 6.0 and with an electrical conductivity level of less than 0.75 mmhos/cm based on the 2:1 extraction method. Temperatures for seed germination should be 23-25°C for the first three to five days, 20-23°C when cotyledons unfold and 18-21°C when true leaves appear. Use of bottom heat is sometimes recommended if needed to maintain a minimum 21°C media temperature. High temperatures, greater than 32°C, will inhibit germination. Petunia seed require near 100 per cent relative humidity for rapid, uniform germination. This can be accomplished using timed mist, tenting the flats with clear polyethylene or using a germination room. Keep the germination media moist, but not saturated. Germination should begin two to three days after sowing and be completed in 10 days. Seedlings must receive sufficient light after germination to prevent unwanted stretching. Maintanance of a minimum of 2500 footcandles after germination is must. This level can be raised to 5000 footcandles as seedlings mature provided temperature can be controlled. Weekly applications of 50 to 75 ppm nitrogen from a base-forming fertilizer [15-0-15 or calcium nitrate (15.5-0-0) and potassium nitrate (13-0-44)] can be applied beginning when radicles emerge. Increase the rate to 100-150 ppm nitrogen and applied on a constant liquid fertilization basis when cotyledons expand can be practiced. Generally, minimize use of phosphorous and ammonium forms of nitrogen fertilizer to prevent stretching and soft growth. However, a high ammonium and phosphorous fertilizer such as 20-10-20 can be applied every other fertilization if seedling growth slows. Watering of seedlings is done in plug flats to maintain a turgid condition, but avoid excessive moisture that can lead to root diseases and seedling stretching. Night temperatures can be lowered to 17-18°C during the last few weeks in plug flats to tone the seed lings in preparation for shipping and transplanting. Generally, petunia seedlings are ready to transplant when four to five true leaves are present and the root ball pulls from the plug intact.

12.3.0 Manuring and Fertilization

Before planting, well decomposed organic matter is to be spread atleast at a depth of 5 cm. At planting, application of 900 gm of a 5-10-5 fertilizer, or 450 gm of 10-10-10 fertilizer per 100 square feet of soil is given. In July, spreading around a half-rate application of fertilizer is recommended, if needed. Too much fertilizer can reduce the number of blooms on grandiflora petunias, and excessive vegetative growth. Fertilize petunias monthly with a balanced fertilizer to support their rapid growth and heavy blooming. Double-flowered cultivars like a biweekly dose of fertilizer. Too high doses of nutrients can damage the roots and cause excessive growth of the shoots. Liquid fertilizer should only be administered with the pouring water. If the trailing petunia is wintered, the last fertilizer is delivered at the beginning of August. Liquid fertilizer is specially feed on container planting and window box planting.

A biofertilizer study on petunia by Hoda and Mona, (2014) revealed that, using *Azospirillum lipoferum* + *Bacillus polymxa* plus 5 g per plant of the chemical fertilizer produced the highest significant values of all growth (*i.e.* plant height, branches number, leaf area, dry weights of shoots and roots), flowering parameters (*i.e.* flowering date, number of flower/branch and flowering period) and chemical analysis (chlorophyll, total carbohydrates, N and P percentages) compared with the control. It was clear that biofertilizers beside its ability to improve the nutrient supply in the soil, they also increased the efficiency of added chemical fertilizer.

12.4.0 Irrigation

Petunias are tolerant to heat so it does not need to be watered them regularly. A thorough watering once a week is sufficient unless there are prolong periods of drought in that area. Avoid watering shallowly as this encourages shallow roots. The spreading types of petunias and those in containers require more frequent watering. Water the soil to a depth of 15 cm when the soil becomes dry. Irrigation in mid- day is not recommended rather it should be applied in morning or late afternoon. But over- watering may lead to stagnation of water and thus rotting of roots may occur. This flowery beauty cannot cope with a dry substrate. Even short-term drought can cause considerable damage on the trailing petunias. The danger increases with pot plants in the summer. The soil in the pots can absorb only a small amount of moisture, which quickly evaporates in the heat. For this reason a regular checking of the soil is recommended.

12.5.0 Pinching

Pinching is a special practice followed in petunia. Pinching the plants can increase bloom stems and discourage vegetative growth. Pinching a flower means to snap off the last inch of the main growing stem during the stage prior to the first flower buds forming, when growth is vigorous. This removes the growth bud at the end of the main stem and encourages more buds to bloom. Removing dead blooms when they begin to fade also encourages re-blooming. Pinching is practiced in case of grandifloras or multifloras, when these grow about 15 cm tall. But pinching of milliflora or 'spreading' petunia is not followed.

12.6.0 Pruning

In midsummer, most petunias get leggy and thus producing blossoms at the tips of long, leafless stems. To keep petunias tidy and flowering, pruning the shoots back to about half their length is often practiced. This will encourage more branching and flowers. After pruning, fertilization and irrigation must be provided to force out new growth and flowers. The plants may look raggedy at first but they'll rebound with more colour and blooms. Older garden petunia plants can be pruned prune hard (within a few cm of the base) to re-encourage vigour, especially in cooler climates, but keeping the remaining leaves. Faded, old or dead blossoms should be removed time to time for attractiveness, to improve bloom quality especially for the larger-flowered petunias.

13.0 Disease and Pests

13.1.0 Diseases

13.1.1 Botrytis Blight

Botrytis or gray mold of petunia is caused by *Botrytis cinerea*. It can occur on spent blossoms and cause flower blight. Flowers have small translucent, or dead spots. Infected flowers can spread the disease by dropping onto leaves of healthy plants. Controlling weeds and remove plant debris, maintaining spacing between plants to allow good air circulation, reducing humidity within the canopy, and minimizing leaf wetness may be helpful to prevent the spread of the disease. Fungicides appropriate for *Botrytis* include Carbendazim 50 per cent WP at 2.5 gm/10 l of water, Diethofencarb can be helpful.

13.1.2 *Phytophthora* Crown Rot

Phytophthora crown rot of petunia is capable of wiping out a bed of petunia within a few weeks. Plants are infected at the stem and rapidly wilt and die. The interesting thing about this disease is that the root system on a freshly killed plant will look perfectly healthy. The fungus doesn't affect the root system. The fungus *Phytophthora nicotianae* is long lived in the soil and will attack and kill susceptible plants next season. Young plants are quickly killed. Pasteurized potting mix or soilless mixes are known to be free of pathogens. Discards flats containing infected plants. Crop rotation to marigold, zinnia or another resistant bedding plant is the best option to combat this disease. Application of metalaxyl 35 per cent WS at 1.5 gm/l of water, Captan 50 per cent WP at 3 gm/l of water is very effective in controlling *Phytophthora* diseases. It is used as a drench and works better as a regular preventative application.

13.1.3 Powdery Mildew

Powdery mildew fungi cause the development of white, powdery growth on the foliage of petunia. The powdery mildew organism prefers moderate temperatures between about 15-26°C and shady conditions. Powdery mildew can be controlled by planting the petunias under full sun, providing the plant with adequate air circulation and avoiding excessive fertilizer applications. Protectant fungicides offer control of powdery mildew, if applied before or soon after the disease appears.

13.1.4 Bacterial Fasciation

It is another problem which causes significant plant deformities in petunias. This disease spreads through plant injuries and water. Infected petunias have flattened stems, and sometimes the stems appear to have fused together. The leaves are smaller than normal but are produced in large numbers, and the shoots may be swollen or undersized. Abnormal growths called galls occasionally form on the crown. An organism called a phytoplasma causes aster yellows. This disease, which spreads through feeding insects called leafhoppers, causes yellowed leaves and greenish, deformed flowers.

13.1.5 Verticillium Wilt

Verticillium wilt causes an overall decline in plant vigour, often causing older leaves to wilt before younger ones, or only part of a plant to die at first. There is no cure for verticillium wilt, so pull plants and try again in pots. In some areas, the soil can be heated enough through soil solarization to kill the fungal pathogen.

Diseases and Pests of Petunia

Botrytis Blight or Grey Mould

Phytophthora Crown Rot

Powdery Mildew on Petunia Leaves

TMV Affected Petunia

Verticillium Wilt of Petunia

Thrips Infestation

Aphids Infestation

Spider Mites Infestation

13.1.6 Virus

Several viruses have been a problem on petunia including *Cucumber mosaic virus* (CMV), *Tobacco mosaic virus* (TMV), *Beet curly top virus*, and *Impatiens necrotic spot virus*. Other viruses reported on petunia include *Potato virus Y, Tomato mosaic virus, Alfalfa mosaic virus, Tomato ringspot virus* (ToRSV), *Tobacco streak virus, Petunia vein clearing virus*, and *Broad bean wilt virus 1*. A newer virus causing leaf mottling in 2012 was provisionally named *Petunia chlorotic mottle virus* (PCMoV). CMV is spread by aphids, TMV by infected plant to healthy plant contact, workers with contaminated hands who handle the plants and with contaminated tools during propagation. TMV is very stable and may survive on greenhouse benches, tools and other surfaces for months. Curly top is widespread, fairly common, and vectored by the beet leafhopper. Plants infected with TMV have mottled green and yellowish green deformed leaves. Necrotic tip dieback, strapped shaped or rat-tailing of leaves and yellowing along the leaf veins is also possible. Blossoms usually are deformed and mottled and could also have a colour break. Plants are stunted. Many infected plants may show no symptoms at all. Plants with beet curly top usually are stunted and have mottled, deformed leaves. Small patches on the leaf are surrounded by a dark halo. Infected plants are destroyed to check further spread of the virus. Good sanitation practices and ensuring proper soil moisture can prevent or minimize disease by encouraging plant health.

13.2.0 Pests

13.2.1 Thrips

Western flower thrips (*Frankliniella occidentalis*) are pests of petunia. These feed on leaves, petals and terminals, where they prefer rapidly growing tissue and cause the appearance of tiny light scars, or stippling. Thrips can carry viruses to petunias, and may cause leaves to turn papery or flowers to develop white spots, known as 'colour break'. This thrips also serves as a vector for diseases that impact petunias like impatiens necrotic spot virus and tomato spotted wilt virus. Knocking thrips off of plants with a strong blast of water, removing spent flowers, providing with adequate water and avoiding the use of excessive nitrogen fertilizer can limit thrips presence and impact. Dimethoate 30 per cent EC at 1 ml/l, Imidacloprid 17.8 per cent SL at 3 ml/10 l, Fipronil 5 per cent EC at 1.5 ml/l, Spinosad 45 per cent SC at 3.2 ml/10 l or neem oil application result better to control thrips.

13.2.2 Spider Mites

Spider mites are common landscape pests that are most prevalent when conditions are hot and dusty. Mites are nearly microscopic pests that suck the juices directly out of petunia cells causing the development of tiny light dots on leaves. Heavy feeding can cause leaf yellowing and drop and the spider mites' characteristic webbing may appear. Drought-stressed petunias are particularly vulnerable to this pest. Spider mites also leave thin webs behind where they feed. Spray your petunias with neem oil once a week until all signs of mites are gone. Spraying the petunia with water knocks the spider mites off and reduces the amount of dust on the plant.

13.2.3 Aphids

Aphids are tiny green insects that congregate on the stem joints and undersides of leaves, damage by sucking out the plant juices. Spraying of organophosphate, Cypermethrin or carbamate group insecticides can help control aphid. Application of neem oil or other horticultural oil along with soap or sticker is also beneficial. Spray the plants every seven to 10 days as needed.

13.2.4 Caterpillars

Caterpillars chew through foliage and buds, sometimes causing extensive damage in no time. They're easy to see if you pick through thick foliage. The best solution is to remove them by hand and drown them in a bucket of water daily, but if you can't bring yourself to do it, sprays of *Bacillus thuringiensis* applied weekly should knock them out quickly. Emamectin benzoate 5 per cent SG at 4 g/10 l, Novaluron 10 per cent EC at 7.5 ml/10 l of water may effectively check lepidopteran infestation.

13.3.0 Nutrient Deficiencies

Multiple nutrient deficiencies could lead to light-coloured spots or similar foliage symptoms. An iron deficiency first appears as yellowing between leaf veins, while a severe problem results in young leaves turning completely white or light yellow. Where zinc is inadequate, leaves develop chlorosis, causing them to twist. This chlorosis turns into tan or white dead patches. When plants don't get enough potassium, mature petunia leaves develop yellow-green patches between veins that become tan dead spots before the entire leaf withers. Proper application of a balanced or supplemental fertilizer and, if necessary, taking steps to alter the pH, usually addresses nutrient deficiencies. The petunia plant may fail to flower or may produce fewer leaves or blossoms than normal; spotted foliage is also a common symptom of a nutrient disorder. Petunias that receive too much or too little water may suffer from wilted or discolored foliage.

13.3.1 Stunt

This is a physiological disorder due to lack of boron in plant. Young plants are stunted, thick and are manifested as hard, distorted, and mottled upper foliage, proliferation of lateral shoots, and terminal bud abortion. This problem can be caused by high media pH or low boron concentration. Maintain media pH at 5.5 to 6.0 and apply supplemental boron once or twice during production. As a supplement, apply Borax as a drench at 10 kg/ha. Water test should be done before application in soil. Excessive amounts of sodium or calcium in the soil can also cause young plants to appear thick and stunted.

14.0 Harvesting and Postharvest Management

For harvesting, the flower which is not fully opened but just started to open is to be selected. Harvesting should be started after the dew dries but before afternoon. The stem is cut 6 mm above a set of leaves and the harvested flowers are immediately put into container with water and preservative, otherwise, the flowers may wilt.

14.1.0 Harvesting of Seeds

Once the bloom fades and dies, the seed pod will begin to develop. The seedpod is enclosed in the calyx, a bulbous structure at the flower's base. The calyx turns from green to brown and it swells accordingly. The seedpods are stored in dry area out of direct sun for about a week to dry completely with proper air circulation. Then the seedpod is cut off when it begins to crack and fully matured seeds are harvested. Any soft or damp pods should be rejected. The faint rattling sound of pod indicates the ripened seeds inside. Otherwise, in order to extend flowering period, pinching of the flowers can be done or else, a few of them are allowed to dieback naturally for seed harvesting. When the blooms dry out, after pinching at the base, and stored them in an airtight container. The pods will open and release seeds. This way also harvesting of petunia seeds can be practiced. Finally, the harvested seeds are stored in plastic food storage bag.

14.2.0 Postharvest Management

The postharvest life of large number of flowers is determined by the onset of petal senescence. Petal senescence is an active physiological process through several biochemical changes. Harvesting time of flowers creates an impact of petal senescence and ethylene also is a primary plant hormone involved in this aspect. The petals exhibit a characteristic 'in-rolling' behaviour during senescence, and also in response to exogenously supplied ethylene. This 'in-rolling' can be delayed, or even prevented, by treatments (such as the application of silver ions or cytokinins) that block the biosynthesis or action of ethylene (Netam, 2018). Inhibition of the synthesis or action of ethylene delays the onset of senescence symptoms and increases the vase life of the flowers. As per the study of Shvarts *et al.,* 1997, Petunia (*Petunia hybrida, cv.* 'VR') flowers produced at 32/27°C (day/night) developed from Stage 1, when the corolla is barely visible, to full anthesis within 4 days, as compared with 9 days for flowers produced at 17/12°C (day/night). However, the flowers produced at high temperatures were approximately half the size of the latter at anthesis, owing to differences in both fresh and dry weights. Corolla pigmentation was enhanced at the lower growth temperatures, owing to higher anthocyanin content. Northern analyses revealed that both growth and pigmentation are regulated by temperature at the molecular level. In contrast to its beneficial effect on size and pigmentation, growth at lower temperatures resulted in a shorter vase life for the cut flowers, which was associated with enhanced ethylene production. Cut petunia flowers were treated with aminooxyacetic acid, silver thiosulfate or 1-aminocyclopropane carboxylic acid. The results supported the contention that ethylene is indeed involved in the regulation of the growth temperature effects on petunia flower vase life. In this study, petunia did not behave according to the common conception that lower growth temperatures lead to better quality cut flowers.

14.3.0 Improving Vase Life through Molecular Approach

Generation of transgenic plants with suppressed production or action of ethylene is an excellent way to lengthen the longevity of ornamental plants. Mutated ethylene receptor genes have frequently been used to convey resistance to ethylene in transgenic floral species. Bleecker *et al.* (1988) showed that etr1-1 is a dominant mutation that confers ethylene insensitivity in other plants. Ethylene-insensitive petunia was created by Wilkinson *et al.* (1997) when they introduced etr1-1 under the control of the constitutive CaMV35S promoter. Transfection of petunia with etr1-1 placed under the control of floral-specific promoters FBP1 (floral binding protein) and AP3 (involved in floral organ development) resulted in a vase life of up to five times that of non-transformed flowers (Cobb *et al.*, 2002). The transgenic petunia plants were insensitive to ethylene, and produced flowers that were larger and had a longer vase life than those from non-transformed plants. They found, however, that disease resistance was compromised. Earlier, Clark *et al.* (1999) observed reduced adventitious root formation in ethylene- insensitive transgenic petunia and concluded that ethylene plays an important role in the response of roots to environmental stimuli. More recently, Clevenger *et al.* (2004) reported that transgenic ethylene-insensitive petunias carrying the etr1-1 transgene had a decrease in pollen viability, root mass, seed weight, and seed germination. While seed germination and weight are maternally regulated, the transgene was found to be completely dominant in its effect on flower senescence. Shibuya *et al.* (2004) have recently isolated PhEIN2, a petunia homolog of the Arabidopsis EIN2 gene, and constructed transgenic petunia plants with reduced PhEIN2 expression, and significantly delayed flower senescence. Some undesirable traits similar to those seen in transgenic etr1-1 plants were observed including inhibition of adventitious and seedling root hair formation, increased hypocotyl length, and premature death. A more dramatic reduction in ethylene sensitivity was achieved with the expression of both the etr1-1 and PhEIN2 transgenes (Shibuya *et al.*, 2004).

15.0 Biotechnology

15.1.0 Genetic Transformation

Flower colour modification of Petunia hybrida commercial varieties by metabolic engineering study has been conducted by Tsuda *et al.*, 2004. Petunia is a good model for the study of flavonoid/anthocyanin biosynthesis because of its genetics and molecular biology (Holton and Cornish 1995). Many of the genes involved in the pathway have been cloned from petunia for the first time, such as, for example, flavonoid 3, 5-hydroxylase (F35H, Holton *et al.*, 1993a), flavonol synthase (FLS, Holton *et al.*, 1993b), anthocyanidin 3-glucoside rhamnosyltransferase (3RT, Brugliera *et al.*, 1994), and flavonoid 3-hydroxylase (F3H, Brugliera *et al.*, 1999).

Flower colour modification by metabolic engineering is useful to breed novel floricultural crops, as reviewed (Tanaka *et al.*, 2004). The flower colour of commercial varieties of *Petunia hybrida* was successfully modified by the suppression of endogenous flavonoid biosynthetic genes, the expression of a hetelorogous flavonoid biosynthetic gene, and the combination of both. Flower colour changed from purple to almost white or from purple to red by the suppression of the endogenous gene expression, from red to orange by the down-regulation of the flavonoid 3- hydroxylase gene and the expression of the rose dihydroflavonol 4-reductase gene, and from violet to pale violet by the expression of the flavonol synthase or flavone synthase gene. From the study, it was clearly indicated that the usefulness of metabolic engineering of the flavonoid biosynthetic pathway to modify flower colour but only a few of the transgenic petunia exhibited phenotypic stability. For commercialisation, it is necessary to generate many independent transgenic lines, select elite lines with stable phenotypes and maintain them in tissue culture.

A mutant receptor, etr1-1, identified from *Arabidopsis* was shown to be unable to bind ethylene. Plants harbouring this mutation have a greatly decreased sensitivity to ethylene since the ethylene response continues to be suppressed even in the presence of the hormone (Schaller and Bleecker, 1995). When the *Arabidopsis* etr1-1 is constitutively expressed in other species, such as petunia, a decrease in ethylene sensitivity is also observed. Since ethylene signaling is involved in many developmental processes, plants constitutively expressing etr1-1 show a variety of defects (Wilkinson *et al.*, 1997). Transgenic petunias expressing an antisense fragment of petunia EIN2, a gene that encodes a positive regulator of ethylene action, show similar ethylene insensitivity and delayed flower senescence (Shibuya *et al.*, 2004).

Research on extending the life of ethylene-insensitive flowers was focused by Reid and Jhiang, 2012. Floricultural crops with ethylene-insensitive flowers seem a very appropriate target for deploying transgenic strategies to extend postharvest life. Although there is still no clear demonstration of the initiation signal for senescence in these flowers, their senescence can be delayed substantially by treating the flowers with cycloheximide, an inhibitor of protein synthesis (Jones *et al.*, 1994; Pak and van Doorn, 2005). The use of this metabolic poison is not commercially feasible, but these results suggest a molecular strategy for extending floral life by using amolecular approach to inhibiting protein synthesis. It was hypothesized that targeted expression of an antisense sequence to a protein from the ribosome should have the same effect as cycloheximide. Accordingly, we transformed petunia plants with the GVG system driving an antisense construct for RPL2, one of the ribosomal subunits. The plants grew normally, but when flowers were exposed to dexamethasone their life was considerably extended.

Petunia plants were transformed with the glucocorticoid inducer system driving an antisense construct of PBB2. Transformed plants grew normally, and their flowers showed normal longevity when placed in water. However, when the flowers were treated with low concentrations of the dexamethasone inducer, their longevity was greatly increased (Stier *et al.*, 2010).

Another study on cytokinin by Clark, (2004) was to produce transgenic petunia (*Petunia* × *hybrida*) plants that over-produced endogenous cytokinins in a senescence specific manner. Cytokinins have been shown to delay the onset of leaf senescence. This was achieved by transforming plants with the IPT (isopentenyl transferase) gene driven by the senescence-associated transcriptional promoter, P SAG12. Two independent transgenic events produced T_1 and T_2 generation seedling lines that demonstrated the desired nonsenescent phenotype in progeny trials. These lines were used to evaluate the horticultural performance of PSAG12-IPT petunia plants in terms of delayed senescence, rooting of vegetative cuttings, lateral branch growth, flower number, floral timing, and fruit set. Although both lines displayed a delayed senescence phenotype the two PSAG12-IPT transgenic lines differed from each other in regard to other horticultural traits. In addition to delayed leaf senescence, line I-1-7 also demonstrated a decrease in adventitious rooting and an increased number of branches during plant production. Line I-3-18 also demonstrated a delayed leaf senescence phenotype; however, plants of this line were not greatly altered in any other horticultural performance traits in comparison to wild-type 'V26'. IPT transcript was detected in young fully expanded leaves of both lines, although mRNA levels were higher in I-1-7 plants. A greater than 50-fold increase in IPT transcript abundance was detected in leaves of transgenic plants following drought stress. These results demonstrate that it is possible to use PSAG12-IPT to produce transgenic plants with delayed leaf senescence but differences in plant morphology between transgenic lines exist, which may alter horticultural performance characteristics.

16.0 References

Adams, S.R., Pearson, S., Hadley, P. and Patefield, W.M. (1999) *Ann. Bot.*, **83** : 263-269.

Ai, Y., Kron, E. and Kao, T.H. (1991) *Molec. Gen. Genet.*, **230** : 353-358.

Ando, T. and Hashimoto, G. (1993) *J. Linnean Soc. Bot.*, **111** : 265-280.

Ando, T., Nomura, M., Tsukahara, J., Watanabe, H., Kokubun, H., Tsukamoto, T., Hashimoto, G., Marchesi, E. and Kitching, I.J. (2001) *Ann. Bot.*, **88** : 403-413.

Ando, T., Kokubun, H., Wantabe, H., Tanaka, N., Yukawa, T., Hashimoto, G., Marchesi, E., Suarez, E. and Basualdo, I. L. (2005) *Ann. Bot.*, **96** : 289-297.

Ando, T. (1996) *Acta Phytotax. Geobot.*, **47** : 19-30.

Angenent, G.C., Franken, J., Busscher, M., van Dijken, A., van Went, J.L., Dons, H. and van Tunen, A.J. (1995) *Plant Cell.*, **7** : 1569-1582.

Bleecker, A.B., Estelle, M.A., Somerville, C., Kende, H. (1988) *Science.*, **241** : 1086-1089.

Brugliera, F., Barri-Rewell, G., Holton, T.A. and Mason, J.G. (1999) *Plant J.*, **19** : 506-602.

Brugliera, F., Holton, T.A., Stevenson, T.W., Farcy, E., Lu, C.Y. and Cornish, E. (1994) *Plant J.*, **5** : 81-92.

Brummitt, R.K. (1989) *Taxon.*, **38** : 301.

Cartolano, M., Castillo, R., Efremova, N., Kuckenberg, M., Zethof, J., Gerats, T., Schwarz-Sommer, Z. and Vandenbussche, M. (2007) *Nature Genet.*, **39** : 901-905.

Clark, D.G., Gubrium, E.K., Barrett, J.E., Nell, T.A., Klee, H.J. (1999) *Plant Physiol.*, **121** : 53-59.

Clevenger, D.J., Barrett, J.E., Klee, H.J., Clark, D.G. (2004) *J Amer. Soc. Hort. Sci.*, **129**(3) : 401-406.

Cobb, D., Schneiter, N., Guo, S., Humiston, G.A., Harriman, B., Bolar, J. (2002) *XXVIth International Horticultural Congress, Toronto, Canada. Book of abstracts;* 83.

Colombo, L., van Tunen, A.J., Dons, H.J.M. and Angenent, G.C. (1997) *Adv Bot Res.*, **26** : 229-250.

Dowd, P.E., Mccubbins, A.G., Wang, X., Verica, J.A., Tsukamoto, T., Ando, T. and Kao, T.H. (2000) *Ann. Bot.*, **85** : 87-93.

Ewart, L. (1984) *Plant breeding. Petunia* (Ed. Sink, K.C.). Springer-Verlag, Berlin. pp.180-202.

Facciola, S. (1998) Cornucopia II, Kampong Publications, California, ISBN: 0-9628087-2-5.

Fries, R.E. (1911) *Kungliga Svenska Vetenskapsaka-demiens Handlingar.*, **46** : 1-72.

Gerats, T. and Vandenbussche, M. (2005) *Trends Plant Sci.*, **10** : 251-256.

Griesbach, R.J. (2006) Petunia. *Flower Breeding and Genetics* (Ed. Anderson, N.O.). Springer, The Netherlands. pp. 301-336.

Gunn, C.R. and Gaffney, F.B. (1974) *United States Department of Agriculture Technical Bulletin,* **1471** : 1-32.

Habas, R.R., Turker, M. and Ozdemir, F.A. (2019) *TURJAF.*, **7**(10) : 1554-1560.

Haughton, C.S. (1978) Green Immigrants. Harcourt Brace Jovanovich, New York, pp. 269-272.

Hoang, L. H. N. and Kim, W. S. (2018) *Hortic. Sci. Technol.*, **36**(1) : 10-19.

Hoda, E. El-M. and Mona, S. (2014) *Am. J. Plant Physiol.*, **9** : 68-77.

Holton, T.A, Brugliera, F., Lester, D.R., Tanaka, Y., Hyland, C.D., Menting, J.G.T., Lu, C.Y., Farcy, E., Stevenson, T.W. and Cornish, E.C. (1993a) *Nature.*, **366** : 276-279.

Holton, T.A., Brugliera, F. and Tanaka, Y. (1993b) *Plant J.*, **4** : 1003–1010.

Holton, T.A., Cornish, E.C. (1995) *Plant Cell.*, **7** : 1071-1083.

Holton, T.A., Kusumi, T., Tanaka, Y. (2004) *Plant Biotechnol J.*, **21**(5) : 377-386.

Izhar, S. and Tabib, Y. (1980) *Theor. Appl Genet.*, **57** : 241-245.

Jones, R.B., Serek, M., Kuo, C.-L., and Reid, M.S. (1994) *J. Am. Soc. Hort. Sci.*, **119** : 1243-1247.

Kashikar, S.G. and Khalatkar, A.S. (1981) *ActaHortic.*,**111** : 35-40.

Kessler, J.R. (1999) *Southeastern Floriculture,* March/April, pp. 15-19.

Knapp, S. (2002b) *Floral diversity and evolution in the Solanaceae*. In "Developmental Genetics and Plant Evolution" (Cronk, Q.C.B., Bateman, R.M. and Hawkins, J.A. eds.), pp. 267-297. Taylor and Francis, London.

Kokubun H, Nakano M, Tsukamoto T, Watanabe H, Hashimoto G, Marchesi E, Bullrich L, Basualdo IL, Kao T, Ando T. (2006) *J. Plant Res.*, **119** (5) : 419-430.

McCubbin, A.G., Wang, X. and Kao, T.H. (2000) *Genome,* **43** : 619-627.

Mehri, S., Ilkaee, M. N. And Saeedzadeh, F. (2018) *J. Ornm and Hortic Pl.*, **8**(4) : 255-263.

Meyer, P., Heidmann, I., Forkmann, G. and Saedler, H. (1987) *Nature,* **330** : 677-678.

Muszynski, S. (1976) *Acta Hortic.*, **63** : 203-204.

Netam, N. (2018) *J. Pharmacogn. Phytochem.*, **7**(2) : 1687-1691.

Olmstead, R.G. and Bohs, L. (2007) *A summary of molecular systematic research in Solanaceae: 1982–2006.* In: D.M. Spooner, L. Bohs, J. Gionannoni, R.G. Olmstead and D. Shibata (Eds.), Solanaceae VI: Genomics Meets Biodiversity. *Inter. Soc. for Hort. Sci.*, Brugge, pp. 255-268.

Pak, C. and van Doorn, W.G. (2005) *New Phytol.*, **165** : 473-480.

Popescu, G.C., and Popescu, M. (2015) *Chil. J. Agric. Res.*, **75**(1) : 21-26.

Que, Q. and Jorgensen, R.A. (1998) *Dev. Genet.*, **22** : 100-109.

Reid, M.S. and Jiang, C-Z. (2012) *Hortic Rev.*, **40** : 3-56.

Robbins, T.P., Harbord, R.M., Sonneveld, T. and Clarke, K. (2000) *Ann. Bot.*, **85** : 105-112.

Schaller, G.E. and Bleecker, A.B. (1995) *Science.*, **270** : 1809-1811.

Shibuya, K., Barry, K.G., Ciardi, J.A., Loucas, H.M., Underwood, B.A., Nourizadeh, S., Ecker, J.R., Klee, H.J., Clark, D.G. (2004) *Plant Physiol.*, **136** : 2900-12.

Shvarts, M., Weiss, D. and Borochov, A. (1997) *Sci. Hortic.*, **69** (3/4) : 217-227.

Sims, T.L. and Ordanic, M. (2001) *Plant Molec. Biol.*, **47** : 771-783.

Sink, K.C. and Power, J.B. (1978) *Euphytica.*, **27** : 725-730.

Sink, K.C. (1984) *Petunia*. In: K.C. Sink (Ed.), *Petunia: Monographs on Theoretical and Applied Genetics* 9. Springer-Verlag, Berlin, pp. 3-9.

Stehmann, J.R. and Semir, J. (1997) *Novon.*, **7** : 417-419.

Stehmann, J.R., Lorenz-Lemke, A.P., Freitas, L.B. and Semir, J. (2009) *The Genus Petunia,* In: Gerats, T. and Strommer, J. (eds.), Petunia, Evolutionary, Developmental and Physiological Genetics, Springer Science and Business Media, pp. 1-28.

Stier, G.N., Kumar, P., Jiang, C.-Z., and Reid, M.S. (2010) *HortScience.*, **45** : S147.

Stuurman, J., Hoballah, M.E., Broger, L., Moore, J., Basten, C. and Kuhlemeier, C. (2004) *Genetics.*, **168** : 1585-1599.

Tanaka, Y., Katsumoto, Y., Brugliera, F. and Mason J (2005) *Plant Cell Tiss Org Cult.*, **80** : 1-24.

Tsukamoto, T., Ando, T., Kokubun, H., Watanabe, H., Tanaka, R., Hashimoto, G., Marchesi, E. and Kao, T. (1998) *Acta Phytotax. Geobot.*, 49 : 115-133.

Turchetto, C., Fagundes, N.J., Segatto, A.L., Kuhlemeier, C., Solís Neffa G., Speranza, P.R., Bonatto, S.L. and Freitas, L.B. (2014) *Mol Ecol.*, **23** (2) : 374-389.

Van der Krol, A.R., Lenting, P.E., Veenstra, J., Van der Meer, I.M., Koes, R.E., Gerats, A.G.M., Mol, J.N.M. and Stuitje, A.R. (1988) *Nature*, **333** : 866-869.

Warner, R.M. and Walworth, A.E. (2010) *J. Hered.*, **101**: 308-316.

Weberling, F. (1989) *Cambridge University Press, Cambridge.*

Wijsman, H.J.W. and Jong, J.H. (1985) *Acta Bot. Neerl.*, **34**(3) : 337-349.

Wijsman, H.J.W. (1982) *Acta Bot. Neerl.*, **31** : 477-490.

Wijnands, D.O., Bos, J.J., Wijsman, W.J., W, Schneider, F., Brickell, C.D. and Zimmer, K. (1986) *Taxon.*, **35** : 748-749.

Wijsman, H. J. W. (1990) *Acta Bot. Neerl.*, **39**(1) : 101-102.

Wilkinson, J.Q., Lanahan, M.B., Clark, D.G., Bleecker, A.B., Chang, C., Meyerowitz, E.M., Klee, H.J. (1997' A'z) *Nat Biotech.*, **15** : 444-447.

Ranunculus

Madhubala and Reeta Bhatia

1.0 Introduction

Ranunculus is a herbaceous perennial plant known for their thin, papery and rose like petals and long lasting cut flowers. The word *'Ranunculus'* is derived from the Latin words *'rana'* and *'unculus'*, meaning 'frog' and 'little', respectively because it is found growing abundantly along the streams during spring season similar to little frogs which appear during the same season. *Ranunculus* is commonly known as buttercup, Persian buttercup, or Coyote eyes, widely used as ornamental plants and for medicinal purpose. *Ranunculus asiaticus* L., commonly known as 'Persian buttercup' or 'Rose of Spring' is the only cultivated species having ornamental value. Double-flowered forms of this species are a popular ornamental plant in gardens, and widely used in floristry. It offers a multitude of colors ranging from yellow to white, orange, red, pink, and purple blooms and have versatile ornamental values. In the year 2014, ranunculus was the fifth top-selling summer flowers with about 60 million stems being sold, and accounted for 0.4 per cent of the total turnover of cut flowers and foliage at the Dutch Flower Auction (Beruto *et al.*, 2018). It is commercially cultivated in different Mediterranean countries including, Italy, France and Israel, the Netherlands, South Africa, California, and Japan.

2.0 Importance and Uses

2.1.0 Aesthetic Uses

Ranunculus are popular cut flowers highly valued for their rose like blooms, brilliantly colored and ruffled petals, tall stems, and long vase life. They are the choicest flowers among the florist, widely used for making flower arrangements and wedding bouquets owing to their luxuriant floral display, long stems, attractive and long lasting flowers. The exquisite flowers of ranunculus are in great demands in high-end flower shops and wedding bouquets. In the language of flowers, the ranunculus is the flower symbolises seduction, charm, and beauty. The flower is viewed as a compliment, loudly proclaiming, "you are charming." Some genetically dwarf varieties are suitable for pot culture, can be planted in containers. Now days it is becoming popular for bedding decoration, borders, gardening and landscaping purposes. Flowers can be single, double or frilled type, among these double type are commonly used for cut flower purpose. This flower offers a multicolour flowers of varied hues ranging from purple, yellow to white, orange, red, pink that makes it more suitable as gifting flower for ornamental purposes in the home and celebrations like birthdays, anniversaries, and graduations, among others.

2.2.0 Therapeutic and other Uses

Most of the species in the genus *Ranunculus* have medicinal value against rheumatism, intermittent fever and redness of the skin. Bulbous buttercup (*Ranunculus bulbosus*) is used in homeopathic medicines for treatment of skin diseases, arthritis, nerve pain, flu (influenza), and other conditions. In more concentrated doses (*i.e.*, non-homeopathic amounts), bulbous buttercup can be unsafe.

Uses of Ranunculus

Cut Flowers

Garden Display

Pot Plant

Floral Arrangements

Vase Display

Centrepiece for Table

Bridal Bouquet

Floral Crown

Button Holes

Exquisite Display of Ranunculus in Landscaping

Flower Beds

Pot Display

Bedding Plant

Herbaceous Border

Window Boxes/Patio Planting

Containers

3.0 Origin and History

The genus *Ranunculus* consists of about 600 species (Tamura 1995) distributed through the two hemispheres. They are found distributed on every continent, from tropical to the Arctic and Sub-Antarctic regions. Ranunculus were widely grown in the Mediterranean regions and in the cool oceanic areas of Western Europe (Flanders, England) in the 18th and 19th centuries (Meynet, 1993). The cultivation of *Ranunculus asiaticus* goes back more than 350 years. According to the literature by Timothy Clark in a the Royal Horticultural Society journal 'The Garden', the ranunculus plants were brought to Europe from the Levant (the old name for regions bordering the eastern Mediterranean) in the 17th century by bulb traders who were also importing the first tulips and anemone. The plants were so popular with "gentlemen florists" that before World War I, it was estimated that for every tulip cultivar, there were ten Ranunculus cultivars. Ranunculus plants survive in various environments, from low-lying wetlands to the cold alpine mountains. This genus varies in morphology and physiology, embodying a strong adaptability. Most of the species originated from Central Asia and eventually reached Europe and other parts of the world. *Ranunculus asiaticus* L is the only cultivated species, native of Northern Africa, South-western Asia, Southern Europe and Eastern Mediterranean regions.

4.0 Taxonomy

The buttercup family, Ranunculaceae, has a worldwide distribution and comprises of more than 2500 species belonging to 59 genera. *Ranunculus* is one of the largest and most cosmopolitan genus among the family Ranunculaceae consisting of about 600 species of flowering plants (Almerekova, 2020). The Initial classifications of *Ranunculus* were based on descriptions of achenes, flowers, roots and fruit anatomy. Later, Tamura revised the classification of the genus based on a reassessment of the achene structure (Tamura, 1993). Tamura's classification sub-divided the genus into seven subgenera: *Pallasiantha, Coptidium, Ficaria, Batrachium, Crymodes, Gampsoceras*, and *Ranunculus*. In this classification, the subgenus *Ranunculus* was further sub-divided into 20 sections. Recent taxonomic reports by Almerekova (2020) suggest that the genus has a monophyletic origin, divided into two subgenera, and consists of 17 sections. The taxonomical studies of the *Ranunculus* genus based on molecular markers have suggested that the molecular taxonomy of the genus is not congruent with the classifications based on botanical studies.

The plants of this genus are mostly herbaceous perennial, occasionally annual or biennial, aquatic or terrestrial in habitat. The leaves are produced in a rosette at the base of the stem. In many perennial species, runners are produced that will develop into new plants with roots and rosettes at the distanced nodes. The leaves are palmately veined, entire, more or less deeply incised, or compound, lack stipules, have petioles. The flowers are hermaphrodite, borne either solitary or in a cyme, have usually five (but occasionally as few as three or as many as seven) mostly green sepals and, five yellow, greenish or white petals that are sometimes flushed with red, purple or pink (but the petals may be absent or have a different, sometimes much higher number). The fruits (achenes), are smooth or hairy, winged, nobby or have hooked spines.

5.0 Species and Cultivars

5.1.0 Species

The genus *Ranunculus,* one of the largest genus among the family Ranunculaceae, comprises of about 600 species of flowering plants (Almerekova, 2020). Some of the commercially important species have been described as under:

Ranunculus asiaticus

Also known as 'Persian Buttercup', this species can grow up to 60 cm in both height width, flowering in May and June. It produces flowers in hues like yellow, purple, white orange, red, pink that open in the late spring. The leaves are green like parsley and can be planted in containers, borders, and flowerbeds. The foliage is fuzzy and bright green in colour. The flowers are like daisy or multi-petaled borne on 12- to 18-inch hollow stems in spring. The flower petals are tissue-paper thin and translucent. They can be used for long-lasting fresh flower arrangements.

R. aconitifolius

Commonly known as 'bachelor's buttons', is native to central Europe. The plants are herbaceous perennial, growing up to 60 cm in height and 40 cm in spread. The leaves are palmate, slightly hairy up to 20 cm long, flowers are produced in loose panicles of white, saucer-shaped during spring. This species become dormant when summer's heat arrives. Flowers are cheerful white in colour during May. This species is naturalised in damp woodland and in ditches and by streams.

R. abortivus

It is known by several names including, 'Early wood buttercup' or 'Kidney-leaf buttercup' or 'Little-leaf buttercup' or 'Small-flower crowfoot'. The stems are erect, 10 to 60 cm tall. The leaves are variable in shape, the basal leaves are kidney-shaped to circular in shape. Both stems and leaves are hairless with each stem can bear up to 50 small flowers with drooping, yellow petals. The flowers are less showy than other buttercups. Each flowers has five petals (1.5 to 3.5 mm long), with a ring of stamens around a round cluster of green carpels.

R. acris

Commonly known as 'Meadow buttercup' 'Tall buttercup', 'Common buttercup', or 'Giant buttercup', is among the common buttercups found growing across Europe and temperate Eurasia. The plant grows up to a height of 60-70 cm. Five petaled flowers yellow in colour bloom from May to August. The flowers are glossy yellow, about 25 mm in diameter.

R. bulbosus

Also known as 'Bulbous buttercup', due to its distinctive perennating organ, a bulb-like swollen underground stem or corm. The stems are erect, branching, and slightly hairy can grow 20–40 cm tall and have a swollen corm-like base. Flowers are of five petaled bright-yellow in colour.

Species of Ranunculus

Ranunculus asiaticus | Ranunculus ficaria | Ranunculus repens

Ranunculus bulbosus | Ranunculus alpestris | Ranunculus acris

Ranunculus asiaticus | Ranunculus aconitifolius | Ranunculus lyallii

Ranunculus cortusifolius | Ranunculus auricomus | Ranunculus nivalis

Source: www.homestratosphere.com; www.thespruce.com; www.petalrepublic.com; www.flowerglossary.com

Cultivars of Ranunculus

Bloomingdale | Amandine Rose | Clooney Hanoi

Accolade | Cloni Dark Orange | Hanoi

Venere | Purple Picotee | Rhone Pink

Tango | Tecolote White | Tecolote Pink

Source: www.thespruce.com; www.gardenia.net

Cultivars of Ranunculus

Delano Yellow | Delano Red | Tecolote Purple

Elegance Red | Elegance White | Elegance Orange

Tomer White | Tomer Yellow | Tomer Picotee

Tecolote Orange | Tecolote White | Tecolote Yellow

Source: www.thespruce.com; www.gardenia.net

R. ficaria

Commonly found along streams, in grassy meadows, and in woodland areas. This species is a common wildflower in North America.

R. fascicularis

Known as early buttercup or tufted buttercup, flower in April and May and can grow up to 60 cm. It looks beautiful in perennial flower beds and wildflower gardens.

R. repen

Also called as 'Creeping buttercup, it is native to Europe, Asia and north-western Africa. It is a herbaceous, stoloniferous perennial plant growing up to 50 cm height. It has both prostrate running stems, which produce roots and new plants at the nodes, and more or less erect flowering stems. In moist and shady areas it blooms early in the month of April and May. The leaves are shaped like snowflakes, gold or lime green on colour. The petals are bright-yellow in colour.

5.2.0 Cultivars

A number of cultivars/hybrids were developed by amateur and professional breeders, which are very popular among growers. A list of commercial cultivars available is presented below:

Cultivars	*Description*
Accolade	Flowers are bicolour with pink yellow colour, flowering will continue for three weeks in a cool and sheltered spot.
Amandine Rose	High in demand having many ruffled layers of rich pink petals. Bloomingdale Plants grow up to only 20 to 25cm in height. This cultivar is best suitable for late winter to early spring. Flowers are large about 9-10 cm in diameter, densely petaled that resemble small peonies.Available in eight colours: Tangerine, Gold, Pink, Yellow Red, Rose, White and Purple.
Cloni Dark Orange	The 'Cloni' series resemble English roses with their many layers of petals and big sized flowers. 'Cloni Dark Orange' is used to brighten up a rainy spring day. Delano Series Plant height is about 30 cm, rose like flowers 12 cm in diameter having multiple layers of delicate silky petals with dark corners. Available in different colours like, red, white, yellow and pink.
Elegance Series	Plant height is 30-35 cm, rose like flowers of 7-15 cm in diameter with multiple layers of delicate silky petals with dark centers. Flowers colour may be white, pink, orange, red *etc.*
Hanoi	Flowers are pink in colour used for bouquet making.
Purple Picotee	'Purple Picotee' creates a lavish display in combination with peonies, larkspur, or lisianthus flowers. Flowers are purple in colour and used in bouquets.

Cultivars	*Description*
Rhone Pink	This variety is of Japanese types, imported and sold in flower markets.
Tango	Tango looks like rose with clustered flowers having multiple layers of bright red-orange petals. Tecolote Series Plants grow up to 30-35 cm tall. Flowers are large in size like roses, about 7-15 cm in diameter with multiple layers of silky petals and dark centers. Available in Pink, red, white, yellow, pink and orange colours.
Tomer Picotee	Plant grows up to 25-30 cm in height, flower are large, like rose, white in colour with multiple layers of delicate, silky petals with purple tips. Venere Flowers are not quite pink and not quite coral, used for bouquet making.

6.0 Genetics and Breeding

6.1.0 Genetics

The basic chromosome number of *R. asiaticus* has been established as n= 8, resulting in 16 R-type chromosomes in a diploid genome. According to Smith and Bennett 1975, the genome size of the genus *Ranunculus* is highly variable, the nuclear DNA amount (2C) differs from 3.9 pg in *R. lateriflorus* to 51.6 pg in *R. lingua*). Recently, Zonneveld *et al.* (2005) measured the genome size (2C) for *R. asiaticus* as 16.5 pg. Ranunculus is a cross pollinated crop due to the presence of Selfincompatible and hence show severe inbreeding depression. Most of the commercial cultivars in this crop are hybrids as it exhibits heterosis for various traits (Meynet 1993).

6.2.0 Breeding

The breeding work in Persian buttercup has been carried out either by small research groups or private companies in France, Italy, Netherlands, Japan, Israel, and the United States. Breeding efforts in ranunculus has led to the development of wide range of cultivars that are propagated primarily through seeds. Genetically homogeneous planting material in this crop can be produced through sib mating technique. Alternatively, the true-to-type F_1 hybrids can be produced by using double haploids as parental lines. Meynet and Duclos (1990) have reported the development of doubled haploids in ranunculus using *in vitro* anther culture (androgenesis) techniques However, double haploid plants reported to have low vigour and seed production.

For cut flower production, the major focus of breeding *R. asiaticus* is on flower morphology, diameter and colour, peduncle length, precocity, and flower longevity. While, the major breeding objectives for pot production in ranunculus includes, compactness, prolonged flowering duration, uniformity in flower colour and flowering time. Besides, *R. asiaticus* is highly susceptible to several diseases including, *Fusarium oxysporum* f. sp. *ranunculi*, powdery mildew, and bacterial disease (*Pseudomonas* spp). Hence, breeding new cultivars resistant to biotic stresses is an important breeding goal in ranunculus.

Recently, novel lines known as "PONPON" having bicoloured and jagged petals have been developed by Biancheri Creations through breeding and micro propagation techniques. These lines represent novel new colour and altered flower shape that offer the consumers innovative Ranunculus lines having "green center" (Beruto *et al.*, 2018).

6.2.1 Hybridization

Meynet (1974) successfully established a breeding program for the development of homogeneous lines within this species, which is highly sensitive to inbreeding. He tried to create hybrid lines using parents that were partially inbred, and revealed the loss of vigor after three to four generations of selfing. With the aim to develop homozygous inbred lines, Meynet and Duclos (1990) conducted preliminary studies on development of double haploids using *in vitro* anther culture approach. In 1983, Sakata Seed America, Morgan Hill, Calif., introduced the seed propagated F_1 hybrid series Bloomingdale. More recently, Italian breeders developed innovative *R. asiaticus* lines, which are well appreciated in the international markets. F1 hybrids have been developed by crossing two homozygous parental lines obtained through *in vitro* anther culture techniques in Persian buttercup (Ferrero, 2006). Beruto *et al.* (2018) carried out the studies to identify the most receptive stage of female flower for pollination. They classified the flower development of *R. asiaticus* into six flower stages. The best ratio of the pollen tube to the total style length, as confirmed by aniline blue staining, was during flower stage 4 of anthesis. They further observed that, the stamens were ripe at flower stage 5 of anthesis. The maturity of the stamens lagged behind the optimal receptivity of the carpels, indicating the protogynous nature of the flower. A difference 4 to 14 days was noticed in female and male maturity in *R. asiaticus*, which was significantly influenced by the growing period. The carpels remain receptive for 3–5 days, and the anthers dehisce their pollen for about 2–3 days. The results of aniline blue staining revealed that the fertilization probably take place between 3 and 4 days after pollination.

Interspecific or intertribal crosses have been attempted to improve the agronomic traits and to create the genetic diversity in this crop. The interspecific crosses between *R. asiaticus* x *R. cortusifolius* has to led to the development of hybrids with larger leaves along with fragrant petals. Similarly, intertribal crosses between *R. asiaticus* and *Anemone coronaria* were performed to enrich the ranunculus gene pool (Dhooghe, 2009). Inhibition of pollen tube growth in the stylar region was observed as major prezygotic barriers, while, normal pollen capture, adhesion, hydration, and germination was observed in these crosses. Among the post-zygotic barriers, seed abortion was recognized as major hurdle and was successfully circumvented using *in vitro* embryo rescue techniques.

While studying the fertilization barriers in *R. asiaticus*, Beruto *et al.* (2018) selfed and crossed three cultivars namely, 'Alfa', 'Bianco Strie' and 'Krisma' with each other. The results revealed that both 'Bianco Strie' and 'Krisma' successfully produced fruitlets without exhibiting any fertilization barrier on selfing or crossing. However, selfing in *cv.* 'Alfa' lead to either no fruitlet, or occasionally one to five fruitlets were formed per pollinated flower. Self-pollinated flowers of 'Alfa' showed

Hybrid Seed Production in Ranunculus

Hand Pollination: (a) Gently brushing against the stamen to pick up the pollen; (b) Emasculated flower ready to receive the pollen; (c) Recovering the female with a bag.

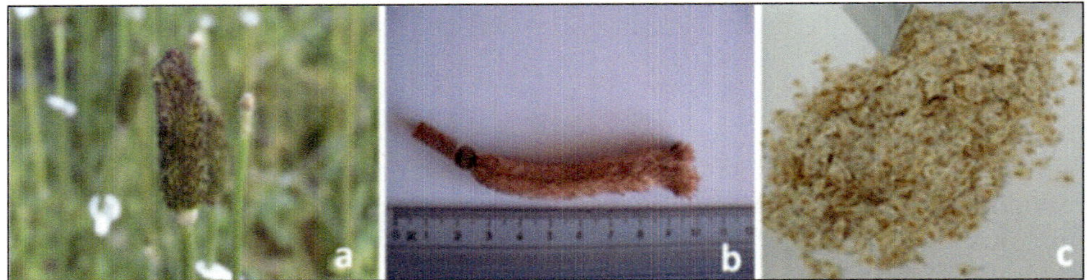

Successful Hand Pollination: (a) Fruit growing; (b) Mature ear from which the different achenes are released; (c) Seeds ready to be commercialized.

Novel Ranunculus Lines: PONPON (a–c) and lines with green canters (d–f); developed by Biancheri Creations, Camporosso, IM, Italy.

Source: Beruto *et al.* (2018). *Ornamental Crops, Handbook of Plant Breeding*, 649-67.

twisted and inhibited pollen tube growth in the styles when stained with Aniline blue and hence, confirmed self-incompatible. Cultivar, 'Alfa' was identified as a good pollinator or acceptor when used in cross-pollinations.

6.2.2 Polyploidy

Polyploidization is an important breeding tool for creation of variability in ranunculus. The diploid chromosome of *R. asiaticus* is 2n=2x=16. Efforts have been made to alter the polyploidy and production of homozygous diploid genotypes through androgenesis approach in *R. asiaticus* (Ferrero *et al.,* 2006). Dhooghe *et al.* (2009) studied the effects of different antimitotic agents including, colchicine, oryzalin and trifluralin for inducing polyploidization in *Ranunculus asiaticus* cv. 'Alfa' using *in vitro* shoots. The highest percentage of polyploidization was achieved by using trifluralin (2 μM, solid medium, exposure time of 8 weeks), although all antimitotic agents were successful in inducing polyploidization.

7.0 Growth and Flowering

Flowering in ranunculus is a complex process, strongly influenced by both the temperature and photoperiodism (Carillo *et al.,* 2019). *R. asiaticus* L. has been reported as a quantitative long-day type geophyte, and photoperiod strongly influence its growth and flowering. The duration of the vegetative period and the number of flowers in ranunculus vary with the genotypes and is greatly influenced by the size of the tuberous root. Larger tuberous roots often results in early floral induction and more number of flowers per plants as compared to the smaller tuberous roots. This may be due to the greater quantity of reserve starch available for flower stem development in the larger tuberous roots (Meynet, 1993).

7.1.0 Temperature

R. asiaticus is highly photo and thermo sensitive crop and both temperature and photoperiod influence growth and flowering of this crop (Karlsson 2003; Meynet 1993). *R. asiaticus* exhibits a low temperature requirement (night/day regime: 5–10/12–25°C, optimum day temperature: 16°C) and has a medium to high light intensity requirement. Vernalization is essential in this crop to break the summer vegetative dormancy in tuberous roots and have been commercially exploited during vegetative propagation (Kamenetsky *et al.,* 2005). Vernalization of rehydrated tuberous roots have been reported to induce early sprouting, leaf rosette formation and flowering in this crop. The time for flowering and the sensitivity to cold treatment, in terms of flower induction, varies with the genotypes.

Cold pre-treatment of the hydrated tuberous roots at 8–10 °C for 5 weeks have proved effective in inducing early flowering in commercial cultivars of *R. asiaticus* (Beruto *et al.,* 2009). This treatment significantly shortened the number of weeks required to flower and improved the average number of flowers produced per plant compared to the untreated tuberous roots.

7.2.0 Light

It is considered as a quantitative long day (LD) plant (Horovitz, 1985). According to Ohkawa (1986), long-day treatment hasten the flower initiation in ranunculus, but often results in less number of flowers as compared to the plants grown under natural and short-day conditions. Similarly, it was observed that the ranunculus seedlings grown for pot production promoted leaf growth under long-day conditions. Vernalization and a short day (SD) stimulate meristematic activity in tuberous roots, thus increasing the number of buds sprouting, as well as the number of leaves and flowers (Meynet, 1993). However, tuberization and flowering show a direct antagonistic response to SD treatment. The long day (LD) increases the growth of tuberous roots and induce early flowering compared to SD, however it reduces the number of flowers as it promotes the growth and development of already formed buds (Meynet, 1993). The temperature and photoperiod interactions also play a crucial role in flowering, a LD promotes flowering at low temperatures, while LD is ineffective at high temperatures (Farina *et al.*, 1985). Long days (greater than 13.5 hours) with high irradiance (15 – 20 mols) are helpful in inducing and enhancing flowering. However, after flower initiation, while buds are immature, short days should me given to improve flower size and maintain floral uniformity.

The photoperiodic light quality has been reported to affect flowering of long day plants, by influencing the phytochrome photoequilibria (PPE) at plant level. Modarelli *et al.* (2020) studied the influence of light spectrum of three light sources, with different induced PPE (natural day length (NL): white fluorescent light (PPE 0.84), light emitting diodes (LEDs) with red: far red (R:FR) light at 3:1 ratio (PPE, 0.84) and LEDs with R:FR light at 1:3 ratio (PPE 0.63) on photosynthesis, metabolic profiling, plant growth and flowering in two hybrids of *R. asiaticus* L *viz.* MBO (early flowering) and MDR (medium earliness). It was revealed that compared to NL, photoperiodic treatments did not affect net photosynthesis, while they promoted the quantum yield of PSII and reduced the non photochemical quenching. Under NL, plant growth was greater in MBO, while flowering initiated earlier in MDR and flowers characteristics were similar in both the hybrids. Despite the greater sensitivity of MDR plants in terms of metabolism, photoperiodic lighting improved plant growth and reduced the flowering time only in MBO, with a stronger effect under R:FR 3:1 light. MDR plants were characterized by higher soluble sugars, polyphenols, photosynthetic pigments and proteins, while MBO plants by higher starch and amino acid content. Hence, the morphological effects of photoperiodic light quality and the hybrid-specific response should be taken into account to optimize lighting protocols in commercial farms.

7.3.0 Growth Regulators

Plant hormones like, GA_3 and BA plays an eminent role in plant growth development and flowering in ornamental plants. Ranunculus can be forced to produce flowers for export markets through application of various plant growth regulators such as, GA_3 and though alteration of growing conditions (Armitage, 1992). The effects of Gibberellic acid (GA_3) and benzyladenine (BA) either individual or in combinations as soil drench on growth and flowering in eight cultivars of *R. asiaticus* have been assessed by Kwak *et al.* (2018). Parameters like plant height and first flower size was increased and days to flowering was hastened by 5 to 9 days with GA_3 application. However, contrasting results were obtained with

BA treatment. The combined application of GA_3 + BA decreased the flowering of ranunculus by 5 to 6 days (Kwak *et al.*, 2018). The flower quality parameters like plant height and cut stem length were improved with application of GA_3 in shade. GA_3 at 100 mg/l significantly increased number of flowers per plant and both GA_3 and BA have also been reported to affect days to flowering in *Ranunculus* sp. (Mayoli *et al.*, 2009).

8.0 Propagation

The commercial method of propagation is through seeds or vegetative propagation.

8.1.0 Vegetative Propagation

Ranunculus grows from unusual spider-like tuberous roots that remain viable for as long as a year when stored under cool and dry location. The most common method of propagation of ranunculus is through division of the tuberous roots in autumn season. Underground storage organs like tuberous roots flowers very faste and more profusely. The optimum size of the tuberous roots for commercial flower production is about 10 cm (4 inches). The tuberous roots must be soaked before planting for better sprouting. Vegetative propagation through tuberous roots is preferred because time to flowering is shorter and the number of flowers per stem is higher than in seed propagated plants (Meynet, 1993). The age of the tuberous roots and storage conditions has strong influence on sprouting of tuberous roots in ranunculus. Generally, elevated moisture and temperature are detrimental to long-term viability of this crop (Cerveny and Miller, 2011). According to Beruto *et al.* (2009), the sprouting was significantly affected when the tuberous roots were stored at room temperature for 1 year. While, the tuberous roots stored at low temperature (2°C) for 1 year under open or modified atmosphere (2 per cent O_2 and 4 per cent CO_2) retained their sprouting ability. Dormancy in tuberous roots of Ranunculus has also been reported (Dole and Wilkins, 1999; Dole, 2003). Cold treatment of ranunculus tuberous roots/bulbs at 5°C for 4 weeks was found effective in early initiation of flower primordia Ohkawa (1986).

8.2.0 Seed Propagation

The commercial hybrids of ranunculus are propagated through seeds. A flower normally produces 200–700 seeds. The seeds are dormant at harvesting and require 3-4 months of rest period before they are ready for sowing (80-85 per cent germination). The seeds retain their viability over a period of one year time, seeds are dormant; after 3–4 months they become reactive (80–85 per cent germination). Its seed can be stored at 15–16 °C and 45 per cent RH, and no special seed treatments are required to improve the germination and seedling growth. Seeds can be sown in plug trays using commercially available media. In area with cool, moist summer climate, seeds can be sown in May for flowering in early November (Meynet 1993). While, under hot and dry summer conditions, sowing should either be done in July to August, or in early autumn, when embryos have naturally overcome the dormancy period and attained physiological maturity. The seeds sown during September to October will produce flowering in the month of March to April (Beruto *et al.*, 2018).

Propagation of Ranunculus

Tuberous Roots | Tuberous Roots after Soaking | Planting

Sprouting after 10-14 Days | At Bud Formation Stage | Seeds

Source: **www.longfield-gardens.com, https://www.brecks.com/blog/growing-ranunculus**

Micropropagation of *Ranunculus* through axillary bud stimulation. (a) Tuberous roots with sprouted buds ready for *in vitro* inoculation; (b) adult plant collected to take the axillary buds; (c) *in vitro* multiplication phase; (d) TC shoots in vitro rooted ready to be in vivo acclimatized.

Source: Beruto ***et al.*** (2018). ***Ornamental Crops, Handbook of Plant Breeding***, 649-67.

Seeds may be covered with a thin layer of medium-grade vermiculite to maintain moisture levels. The radicle emergence occur in 14 – 21 days. Light is not essential during germination phase. However, from day 15 – 29, a light intensity of 1,500 – 2,000 foot candles (15,000 – 20,000 lux) should be maintained. For obtaining good quality seedlings, the temperature should be maintained at 10 – 13°C. It has been observed that increasing the temperature from 13.5 – 18.5°C leads to drastic reduction in seedling quality. For quality seedling production, application of liquid fertilizers at 14-4-14 or 17-5-17 with 60 – 75 ppm of nitrogen is recommended. The seedlings are ready transplanted 8 to 10 weeks after sowing.

8.2.1 Effect of Growth Regulators on Germination of Seeds

Ranunculus asiaticus seeds germinate within 4-7 weeks at 20°C (Kim *et al.*, 1991). Morphological and physiological dormancy have been reported earlier (Atwater, 1980; Geneve, 1998). Application GA_3 at 0.289 m mol/l and KNO_3 at 0.1 mol/l have synergistic effect on seed germination of acid scarified seeds (Young *et al.*, 1992). Although no dormancy has been reported for cultivated ranunculus (Dole and Wilkins, 1999; Hartmann *et al.*, 2002), but there are some variabilities in some species or cultivars and wild varieties regarding seed dormancy conditions. The germination percentage (98.75 per cent) and average daily germination (10.76 per cent) were recorded when scarified seeds were treated with GA_3 at 200 mg/l. These treatments improved germination of seeds, but the hypocotyls grew only up to 2 mm in length and then growth was ceased. Hypocotyls length was increased significantly when moist stratification was done at 5°C for one month (Karami *et al.*, 2007)

8.3.0 Micropropagation

Micropropagation is an important technique for the production of virus free propagation material and large scale multiplication of ranunculus (Beruto and Debergh 2004; Beruto 2010). This techniques offers the possibility to fix new varieties obtained through sexual reproduction particularly when unstable ornamental characteristics are envisaged through seed propagation (Beruto *et al.*, 2018). Compared to seedlings, the tissue cultured plantlets exhibited earlier flowering, increased number of shoots, and better tuberous root yield per plant. Further, it was revealed that the tuberous roots obtained through cloned material have been proved to be more vigorous and produce more flower buds with larger flowers of better quality and greater uniformity compared to the conventionally obtained tuberous roots (Beruto *et al.*, 2018). Axillary buds taken from healthy mother stock plants or from the sprouted tuberous roots are considered ideal explants for initial culture establishment (Beruto, 2010). It is recommended to use buds from plants at least one month before opening of flowers. Care should be taken while taking buds from sprouted tuberous roots. Positive results were obtained when hydrated tuberous roots pre conditioned at 12°C, 50 μmol/m^2/s along with 8 h of light. MS (Murashige and Skoog 1962) medium supplemented with different concentrations of growth regulators BA (benzyladenine), kinetin, and NAA (naphthalene acetic acid) was used for multiplication stage. Each shoot can produce on an average

1.5–4.5 new shoots under *in vitro* conditions, after every 4 weeks depending on the genotype (Beruto and Debergh, 2004).

Various factors including type of sugars, growth regulators, water status, nutrients availability and agar impurities significantly influence the mass multiplication (Beruto 1997; Beruto and Curir 2006; Beruto *et al.*, 2001). Among the different sugars, fructose has proved beneficial during *in vitro* multiplication of *Ranunculus* (Beruto and Portogallo, 2000). The tissue cultured shoots are used to induce the tuberous roots under *in vivo* conditions. The low temperature treatment can slow down the growth of *in vitro* raised shoots, and thus shoots can be stored at 2°C for about 3 months at multiplication and rooting stage (Beruto *et al.*, 2011).

9.0 Soil and Climate

9.1.0 Soil

Well-drained, light soil with medium moisture holding capacity are preferred for ranunculus. Over moist soil should be avoided as corms and roots may rot. Soil can be mixed with addition of organic matter if required. Loosening of soil before planting with improve drainage. After planting, watering along with thin layer of mulching with organic matter such as bark, straw or coco hulls will help to retain moisture and coolness in the soil. Ranunculus are prone to root rot under wet conditions, so unless the soil dries out, do not water again until you see sprout. For quality pot plant production, good quality soilless porous media such as peat moss, compost, ground bark or decomposed manure can be used. The pH of media should ranges 5.8 – 6.0 along with EC level of 1.25 – 1.75 dS/m.

9.2.0 Climate

Ranunculus can grow well in full sun shine. It can tolerate shade in the morning, in the afternoons full sun shine required to bloom. Ranunculus is a cool season flower that can grow well in springs with temperatures of about 12°C. Temperatures of 12 to 16°C is ideal for its optimum growth and high number of leaves per plant. Increasing temperature from 8 to 20°C, improved the plant height, hasten flowering, but reduced flower number and quality. Temperature above 16°C is detrimental for its growth. Yellowing of lower foliage can be seen if the temperature is above 27°C. Short-day conditions and cool nights are optimal for high-quality flower production and high yields. Long days accelerate floral development and increase plant height but reduce flower quality (Dole and Wilkins, 2005). For commercial production ranunculus are usually grown in greenhouse so that high and low temperatures can be maintained to produce flawless flowering for wholesale markets. Ranunculus is grown outdoors during the fall and winter for spring blooming in Mediterranean climates.

10.0 Cultivation

10.1.0 Planting and Spacing

In warm climate zones, the tuberous roots can be planted in the fall, which will come in to bloom in late winter or early spring. Climate zones where winters are not hardy, planting can be done in spring for flowering in late summers. For cut flower production, a planting density of 45 to 65 plants/tubers/m^2 is maintained.

10.2.0 Manuring and Fertilization

The stored food in the tuberous roots is considered sufficient with a starter solution at planting; however, after establishment, a nitrogen level of approximately 100 ppm can be used. Ranunculus crop should be fertilized with a water-soluble fertilizer diluted to half strength every 10 days during the foliage production period. Fertilizer should be withheld once blooming begins. Too much fertilizer will support foliage at the cost of bloom production. Fertilize the bulbs in the spring with a 10-10-10- fertilizer by diluting 6 tablespoons of fertilizer into 1 gallon of water. Thoroughly drench the soil around the plants, taking care not to splash the leaves. The treatment with biofertilizers like, Trichoderma + Mycorrhiza + Basils improved growth and flowering parameters of ranunculus, the use of humic acid to improve the growth also helped to improve the interaction between fertilizers and humic acid for increased growth of ranunculus, flowers production and vase life (Jabbar and Hussein, 2020).

Bernstein *et al.* (2011) studied the effects of three levels of K fertigation (60, 120, or 180 mg K/l) under 50 mg N/l application, and the intermediate level of 120 mg K/l, with three levels of N applications (50, 100, or 150 mg N/l). They revealed better flower production and lower incidences of stem-toppling with the two lowest K and N treatment. They suggested that *R. asiaticus* L. has a low nutritional requirement for K and N. It was revealed that stem toppling occur under high application of N and K due to reduced availability of Ca to the expanding tissue of the flower stem.

10.3.0 Irrigation

Ranunculus requires well-drained media. Overwatering or anaerobic medium conditions increase disease incidence. There has been growing concern over the sustainability of ranunculus cultivation when plants irrigated with water of marginal quality because the viability of the tuberous roots may be compromised. Valdez-Aguilar *et al.* (2009) studied the effect of saline irrigation waters, with and without pH control, on the growth of plants and tuberous roots of ranunculus. Four irrigation water solutions with increasing concentration of Ca^{2+},Mg^{2+}, Na^{+}, SO_4^{2-}, and Cl^- to meet an electrical conductivity (EC) of 2, 3, 4, and 6 dS/m and pH 6.4. Salt sensitivity of ranunculus, as determined by growth of the flowering stems and viability of the tuberous roots, was increased by irrigation with alkaline waters, which was associated with additional increases in Na^+ and Cl^-tissue concentration and decreased iron accumulation. Hypersensitivity to salinity makes ranunculus crop a poor candidate for water reuse systems.

11.0 Diseases and Pests

11.1.0 Diseases

11.1.1 Powdery Mildew

This fungal disease is very common in ranunculus that causes small, circular, gray, black, white or pinkish powdery spots on older leaves, which eventually spreads to cover the entire leaves and stems. Symptoms appear will be wilting of plants and browning of leaves. This problem can be controlled through proper ventilation of growing area and make sure that plants get adequate sun and air circulation. Remove the affected parts of the plant using sterile knife. Water the plants through drip irrigation to avoid more humidity around the plants.

11.1.2 Root Rot

The symptoms appear as wilting and drooping of the plant even after adequate watering. It is a fungal disease and spread very fast through spores in the soil. Discard the infected plants immediately after observing the incidence. Use well-drained soil and planting should be done in the area with adequate sunlight and air movement.

11.2.0 Insect-Pests

11.2.1 Aphids

The most common pests of ranunculus, very small in size, green in colour that suck the sap from leaves, flowers and stems. Symptoms appear to be wilting and discoloration of leaves. Aphids can be removed from the plants by spraying with water, mixture of mild dish soap or with essential oils.

11.2.2 Birds

Young shoots in ranunculus are soft and tender and become tasty food for birds. The growing area should be covered with net until the shoots grow up to 15 cm to keep the birds away. After growing these tall plants are no longer attractive to the birds.

12.0 Harvesting and Postharvest Management

12.1.0 Harvesting

Cut flowers are harvested at bud stage or half-open stage. The flowers keep on opening for several days when the stem are kept in clean and cool water. Always use sharp cutting tool to cut flowers as low as possible to the ground without damaging the foliage. The plant needs the foliage to create energy for next year's growth. Harvesting of flowers on time from the plant will produce more flowers.

12.2.0 Postharvest Management

Flowers close at night and harvesting, grading, and packing are best done early in the morning when flowers are still closed. Flowers are stored and shipped at 1 to 2°C. If stored for long periods, the temperature should be 0 to 1°C and good-quality water should be used. Cut stem should be kept in clean vase with fresh and lukewarm water. Water level should be enough to submerge 3 to 4 inches of the cut stems. The length of flower vase should be at least T! the length of the stems. The cut stems should be re-cut, change the water, and add flower food every two days t o increase vase life of cut stems.

Pre-treatment of flowers harvested at half-open stage with cycloheximide (0.01 and 0.05 mM) before transfer to normal water improved the vase life and prolonged membrane integrity (Scariot *et al.*, 2009). Treatment with AOA (amino-oxyacetic acid) and STS (silver thiosulfate) can enhance the cut flower longevity. The effect of ethylene in cut stems to regulate senescence in Ranunculus is not an important aspect as reported by Kenza *et al.* (2000). Flower senescence in Ranunculus is visible in the form of changes in petal and anther colour and by increased pistils dimension which appear like well diverged from the stamens. Flower maturation and senescence have also been reported to be caused by reduction in total soluble carbohydrates and phenol levels in petal tissues that result in increase in ion leakage and protein degradation (Shahri and Tahir, 2011).

12.3.0 Maintenance of Plants after Harvesting

Once flowering is over, leave the plants as such in foliage stage. The foliage will absorb sunlight that will provide reserve food material to next year. Water as needed during active growth periods. Ranunculus prefers not to be watered when dormant. As summer comes to an end, ranunculus leaves will yellow and die back as the plant becomes dormant. Foliage can be removed at this point. The tuberous roots are harvested after shoot senescence, stored during the summer for dormancy release, and marketed for garden plantations or flower beds for the next season.

13.0 Biotechnology

13.1.0 Somatic Embryogenesis

In vitro callus were induced from thalamus of *Ranunculus asiaticus* L. on Murashige and Skoog's medium supplemented with different concentrations of 2,4-D (0.1, 0.2, 0.4, 0.8, 1.6 mg/l) for 75 days. Calli induced on 2,4-D at 1.6 mg/l concentration exhibited embryoid differentiation, 75 days after subculture onto hormone-free medium. The calli maintained at lower concentrations of 2,4-D at 0.05 mg/l also showed the embryoid differentiation when transferred onto hormones free medium one month after subculture. All the somatic embryos developed into plants, and 96 per cent survival was observed after transplanting to *in vivo* growth conditions (Beruto and Debergh, 1992).

13.2.0 Molecular Markers

The hybrids of *Ranunculus asiaticus* are susceptible to several viruses and the common are Tomato spotted wilt virus (TSWV), Ranunculus white mottle virus (RWMV), Ranunculus leaf distortion virus (RLDV) and others. Leaf tissues were used to evaluate these viruses by double step RT-PCR assays and specific primers. The analysed samples showed the virus species, either in single or in mixed infection and it was confirmed that the molecular markers can be easily used to detect viruses for production of virus-free Ranunculus plants on mass scale propagation (Sacco *et al.*, 2015).

SSR and SNP markers were identified on the basis of 151,136 petal transcriptome-derived unigenes with MISA, GATK3 and annotation tools. The results revealed that SSRs had high polymorphism; 586 SSRs located in coding regions and mostly were tri-nucleotide motif. Among the 116 unigenes of 8 families of enzyme genes of anthocyanin biosynthesis, 19 SSRs and 127 SNPs were identified in 17 unigenes and 49 unigenes, respectively. All the SSRs were in non-coding region and 53.13 per cent SNPs were in coding region including 41 synonymous types. The problem of information on absence of genome or EST in *Ranunculus asiaticus* L. to develop SSRs and SNPs was solved in this study which can be useful for future researches (Liu *et al.*, 2018).

13.3.0 Protoplast Culture

Ranunculus sceleratus L. plants were grown from seeds under semi controlled greenhouse conditions to produce leaf material. Plants were maintained at average temperature of 17°C, 50 per cent R.H. and low light intensity of approx. 1500 lux throughout the growing period. Viable protoplasts were isolated from leaves by a one-step mixed enzyme treatment. Cultured protoplasts regenerated cell walls and cell divisions were observed critically and calli formed from the cell colonies differentiated into plantlets which were then grown as complete plants (Dorion *et al.*, 1975).

14.0 References

Almerekova, S., Shchegoleva, N., Abugalieva, S. *et al.* (2020) *PLoS ONE*, **15**(10): e0240121

Armitage, A.M. (1992) *In*: R.A. Larson (ed.). *Introduction to Floriculture*. Academic Press, San Diego, California. pp.160-191.

Atwater, B. (1980) *Seed Sci. Technol.*, **8**: 523-573.

Bernstein, N., Ioffe, M., Luria, G., Bruner, M., Nishri, Y., Philosoph-Hadas, S., Salim, S., Dori, I. and Matan, E. (2011) *Pedosphere*, **21**(3): 288-301.

Beruto, M. (1997) *PhD Thesis*, Gent, Faculteit Landbouwkundige en Toegepaste Biologische Wetenschappen, University Gent, Belgium, pp.21–44, 125–184.

Beruto, M. (2010) In: Jain SM, Ochatt SJ (eds), vol 589. Humana Press, New York, pp.29–37

Beruto, M. and Curir, P. (2006) Global Science Books, Isleworth, **2**: 277–284

Beruto, M. and Debergh, P. (1992) *Plant Cell, Tiss. Organ Cult.*, **29**: 161-165.

Beruto, M. and Debergh, P. (2004) *Plant Cell Tiss. Org.cult.*, **77**: 221–230

Beruto, M. and Portogallo, C. (2000) *Acta Hort*, **520**: 163–170.

Beruto, M., Fibiani, M., Rinino, S. *et al.* (2009) *Israel J. Plant Sci.*, **57**: 377–388.

Beruto, M., La, Rosa. C., Portogallo, C. (2001) *Acta Hort*, **560**: 399–402.

Beruto, M., Rinino, S., Bisignano, A. and Fibiani, M. (2011) *Acta Hort.*, **908**: 391–403.

Beruto, M., Rabaglio, M., Bisignano, Viglione, S., *et al.* (2018) In: Ornamental Crops Book, pp. 649-671.

Carillo, P., Arena, C., Modarelli, G.C., *et al.* (2019) *Front. Plant Sci.*, **10**: 241.

Cerveny, C.B., Miller, W.B., BjOrkman, T. *et al.* (2012) *Hort. Sci.*, **47**(2): 212-216.

Dhooghe, E. (2009). *Ph.D. Thesis*, Ghent University, pp.244.

Dhooghe, E., Denis, S., Eeckhaut, T., *et al.* (2009) *Euphytica*, **168**: 33–40.

Dole, J.M. and Wilkins, H.F. (1999) *Floriculture: Principles and Species*. Prentice Hall, Upper Saddle River, N.J., USA.

Dole, J.M. (2003) *Hort Sci.*, **38**: 341-346.

Dorion, N., Chupeau, Y and Bourgin, J.P. (1975) *Plant Sci. Letters*, **5**(5): 325-331.

Farina, E., Paterniani, T., and Suffia, G. (1985) *Annali dell'Istituto Sperimentale per la Fioritura San Remo*. **16**: 93–99.

Ferrero, F., Duclos, A., Ottenwaelder, L., *et al.* (2006) *Postharvest Biol Technol.*, **19**(3): 287–290.

Geneve, R. (1998) *Seed Technol.*, **20**: 236-250.

Geneve, R. (2003) *Hortic. Sci.*, **38**: 336-341.

Hartmann, H.T., D.E. Kester., F.T. Davies., *et al.* (2002) *Plant Propagation: Principles and Practices*, 7th Edn., Prentice Hall, USA., pp.199-236.

Jabbar, A.A. and Hussein, J.K. (2020) *Plant Archives*, **20**: 2201-2207.

Kamenetsky, R., Peterson, R. L., Melville, L. H. *et al.* (2005) *New Phytologist*, **166**(1): 193-204.

Karami and Khosh-Khui, M. (2007) *Int. J Agricl Res.*, **2**: 97-101.

Karlsson, M. (2003) *Greenhouse Product News*, 44–48.

Kenza, M., Umiel, N. and Borochov, A. (2000) *Acta Hort.*, **714**: 119–128.

Kim, Y.S., E.H. Park, S.O. Yoo, *et al.* (1991) *J. Korean Soc. Hortic. Sci.*, **30**: 401-410.

Kwak, H.G., Young, R.L., Youn, J.C., *et al.*(2018) *Flower Res J.*, **26**(4) : 179-186.

Liu, Y., Wang, J., Zhang, J., *et al.* (2018) *Nanoscience and Nanotechnology Letters*, **10**(2): 267-273(7)

Mayoli, R.N., Isutsa, D. K. and Tunya,G. O. (2009) *Afr J Hort Sci.*, **2**: 13-28.

Meynet, J. (1974) *Acta Horti.*, **43**: 191–195.

Meynet, J. (1993) In: De Hertogh A, Le Nard M (eds) The physiology of flower bulbs. Elsevier Science Publishers, Amsterdam, pp.603–610.

Meynet, J. and Duclos, A. (1990) *Agronomie.*, **10**: 213–218.

Modarelli, G. C., Arena, C., Pesce, G., *et al.* (2020) *Physiol Plant.*, **170**(2): 187-201.

Murashige, T. and Skoog, F. (1962) *Physiol Plant.*, **15**: 473–497.

Ohkawa, K. (1986) *Acta Hort.*, **177**: 165–177.

Sacco, E., Borghi, C., Laura, M. *et al.* (2015) *Acta Hort.* **1072**, 165-171.

Scariot, V., Larcher, F., Caser, M., *et al.* (2009) *Europ J Hort Sci.*, **74**(3): 137–142.

Shahri, W. and Tahir, I. (2011) *J Fruit Ornamental Plant Res.*, **19**(2): 123–131.

Smith, J.B., Bennett, M.D. (1975) Heredity, **35**: 231–239.

Tamura, M. (1995) *Ranunculaceae*, Duncker and Humblot, Berlin, **2**: 220–519.

Tamura, M. (1993) *In* En Kubitzki, K. Rowher, J.G. and Bittrich, V. (Eds). *The families and genera of vascular plants*. Springer-Verlag, pp.563–583.

Valdez-Aguilar, L. A., Grieve, C. M., Poss, J., *et al.* (2009) *Hort. Sci.*, **44**(1), 138-144.

Young, J.A., Martens, E. and West, N.E. (1992) *J. Range Manage.*, **45**: 358-362.

Zonneveld, B.J.M., Leitch, I.J. and Bennett, M.D. (2005) *Ann. Bot.*, **96**(2): 229-244.

Marigold

S. Sangeetha Priya, Shyama Kumari, J.S. Arora and P. Pal

1.0 Introduction

Marigold gained popularity amongst gardeners and flower growers on account of its easy culture and wide adaptability. Its habit of free flowering, short duration to produce marketable flowers, wide spectrum of attractive colour, shape, size and good keeping quality attracted the attention of flower growers. In India, it is one of the most commonly grown flowers and extensively used on religious and social functions, in one form or other.

2.0 Importance and Uses

2.1.0 Aesthetic Uses

In India, flowers are sold in the market as loose or after making into garlands. Other than cut flower, marigold especially is used for beautification and also used in landscape plans due to its variable height and colour of flowers. It is highly suitable as a bedding plant, in a herbaceous border and is also ideal for newly planted shrubberies to provide colour and fill the spaces. French marigold is most ideal for rockery, edging, hanging baskets and window boxes.

2.2.0 Therapeutic and other Uses

Both leaves and flowers are equally important from medicinal point of view. Leaf paste is used externally against boils and carbuncles. Leaf extract is good remedy for ear ache and have fungicidal effect (Grewal and Grewal, 1988). Park *et al.* (1998a,b) reported that high levels of dietary lutein from marigold extract can inhibit mammary tumour development and increase immunity against cancer in BALB/c mice. It also has potential benefit in the prevention of macular degeneration, development of visual and nervous system of fetus and antioxidant properties. Despite these advantages, incorporation of lutein into food is limited because of its low water solubility, chemical instability and poor oral bioavailability. To overcome this, Steiner *et al.* (2018) explored the lutein-loaded encapsulation systems like emulsion-based, nanoparticle and microparticle based and molecular inclusion encapsulation systems which are also cost efficient and stable.

Rojas-Martinez *et al.* (1999) observed less injury from *Alternaria solani* in tomato when grown in association with marigold (*T. erecta*) than monocropped tomato. Single whorl orange coloured *Tagetes erecta* was grown along the borders of a tobacco field and a long every 30th row of tobacco, trapped a significantly higher number of *Helicoverpa armigera* eggs and larvae (Shivayogesh-wara *et al.*, 2000). Also, marigold flower extract effectively controlled canker (62.82 per cent), early blight (61.53 per cent), wilt (18.42 per cent), fruit spot (27.41 per cent), blossom end rot (50.43 per cent) and sun scald (26.44 per cent) in tomato (Nahak and Sahu, 2017). Goel *et al.* (2015) reported that marigold flower extract significantly enhanced the activities of peroxidase (POX), polyphenol oxidase (PPO) and catalase (CAT) in tomato. Further, it induced additional isoforms of POX, PPO and lipoxygenase (LOX) which plays vital role in host resistance against several biotic and abiotic stresses.

Uses of Marigold

In Pot | In Flower String | In Garland

Loose Flower | Flower Rangoli | In Decoration

In Landscaping | In Flower Bed | Along Path

Flower Border | In Garden | Flower Show using Marigold

Source**: www.in.pinterest.com, www.123rf.com, www.reddeer.ca, www.flickr.com, www.bhubaneswartourism.in, www.thehindu.com**

The essential oil present in different species of *Tagetes* can find a use in the perfume industry. Dhingra and Dhingra (1956) reported that all parts of *Tagetes patula* contain essential oil which can readily be extracted by steam distillation. The oil has a pronounced odour and acts as a repellent to flies. Kapelev (1971) while evaluating the different species like *T. patula, T. tenuifolia, T. lucida* and *T. signata* for essential oil contents, found that *T. signata* contained higher essential oil (4.25 per cent of dry matter) of very good quality which can be used in perfume industry. Tamut *et al.* (2017) reported that essential oil 0.02-0.09 per cent in the floret tissues of *Tagetes patula*. Maximum content was recorded with Orange Winner followed by Bolero Red, Boy-O-Boy and Pusa Arpita. The essential oil comprised of β-caryophyllene, terpinolene, caryophyllene oxide, (Z)-β-ocimene, piperitenone, piperitone, (E)-ocimenone, (Z)-ocimenone, limonene and germacrene-D. In addition, *Tagetes* has also been reported to be a good source of helenine, which varies with flower colour, *i.e.*, deep yellow flower had higher content than dark red colour (Tarpo and Cucu, 1961). Illuminated marigold (*T. erecta*) flower meal containing oleoresin with a higher concentration of all-trans-zeaxanthin and xanthophyll pigments of air dried petal are used for poultry ration to increase the colour of egg yolk (Medina *et al.*, 1993; Delgado-Vargas *et al.*, 1998; Armstrong *et al.*, 2000; Willoughby *et al.*, 2000) and for colouring dairy product and dog food (Willoughby *et al.*, 2000). Marigold flower extract as feed additive at 15 ppm to quails improved production performance, egg quality with reduced egg cholesterol and increased yolk colour (Nuraini *et al.*, 2017). Inclusion of marigold extract in the sorghum-based diets of laying hens reduced the percentage and thickness of the eggshell thus improving the egg quality (Maria *et al.*, 2017).

Shahane and Sidhaye (2018) reported that environment-friendly pathway for the synthesis of reduced graphene oxide nanostructures by using *Tagetes erecta*. Also, marigold had developed defense mechanism against the heavy metals, lead and nickel by increasing the anti-oxidant activity, root length, proline content and dry weight thus proving its efficacy in phytoremediation for soil conservation (Bhurat *et al.*, 2017). Padalia *et al.* (2015) synthesized silver nanoparticles (AgNP) by reducing aqueous silver ions to flower broth which could be used in the production of biomedical products. The four flavonols extracted from *T. minuta*, 6-hydroxy quercetin 7-O-β-(6″-galloyl glucopyranoside), 6-hydroxy quercetin 7-O-β-(6″-caffeoyl glucopyranoside), 6-hydroxy kaempferol 7-O-β-(6″-galloyl glucopyranoside) and 6-hydroxy kaempferol 7-O-β-(6″-caffeoyl glucopyranoside) could be potentially utilized for antibacterial drug discovery (Shahzadi and Shah, 2015).

The yield of astaxanthin, a pigment obtained from red marigold was found maximum with the use of Hexane: Isopropanol (50.21 ± 0.32 µg/g) followed by DMSO: Acetone (43.21 ± 0.25 µg/g) and acetone (29.95 ± 0.18 µg/g). This pigment also inhibited *E. coli* from spoiled milk in the range of 12.10 ± 0.9 mm to 12.41 ± 0.44 mm (Sugan, 2017).

2.3.0 Nematodes Management

Nematodes are very harmful eel worms to many vegetables, fruits and ornamental plants and cause substantial reduction in yield and quality. These nematodes can be controlled by the use of chemicals like dichloropropene dichloropropane (DD) or nemagon though costly, it is used and required in many crops. It has been noticed that by growing certain plants, the population of nematodes can be checked. Amongst these plants, marigold has been found to be highly effective in keeping the population of nematode under control. The earliest report about the resistance of *Tagetes* to nematode (*Meloidogyne*) infestation was made by Tyler (1938) and Steiner (1941). It was many years later that, Meijneke and Oostenbrink (1958) reported a marked reduction of *Pratylenchus penetrans* and other types of nematodes by growing *T. erecta* and *T. patula* for 3-4 months. Visser and Vythilingam (1959) observed a considerable decrease in population of meadow nematodes *Pratylenchus coffeae* and root-knot nematode (*Meloidogyne javanica*) by growing *T. erecta* and *T. patula*. Cayrol (1960) recommended nematode control in bulb crops by green manuring or intercropping with African marigold. Growth of *Prunus avium*, *Rosa* spp., and apple seedlings was improved when grown after a crop of marigold. Roots of nursery plants of apple and rose were almost free from *Pratylenchus* sp. when grown in a two-year rotation with *Tagetes* sp. (Oostenbrink, 1960). Preplanted marigold also reduced galling and numbers of second-stage juveniles in subsequent tomato (Ploeg, 1999). Akhtar and Alam (1990) observed that amendment of agro-waste of *T. erecta* at 100 g/kg of pot mixture significantly reduced the *M. incognita* and population of other plant parasitic nematodes. Addition of dried chopped portions of *T. erecta* at 10, 20 and 30 g/pot was effective to reduce the number of egg masses and gall of *M. incognita* on roots of sunflower (Abadir *et al.*, 1994, 1996). Incorporation of chopped marigold leaves at 40 and 80 kg/ha significantly reduced the number of galls and egg masses of *M. incognita* in tomato and brinjal plants and final juvenile population in soil (Walia and Gupta, 1997). Five pairs of papaya intercropped with 5 rows of marigold showed greater reduction in nematode population. As it was not economical, 5 pairs of papaya with 3 rows of marigold were recommended for nematode control (Selvam *et al.*, 2015). Kurt and Askin (2018) revealed that highest gall index was observed with only marigold flowers whereas lowest gall index was recorded with application of green parts of plant. Thus, it was insisted that crop rotation with marigold could reduce the root-knot nematode damage considerably.

3.0 Origin and History

Kaplan (1960) established that marigold is native of Central and South America, especially Mexico. From Mexico it spread to different parts of the world during early part of the 16th century. The name *Tagetes* was given after Tages, a demigod, known for his beauty.

African marigold was first introduced into Spain early in the 16th century and became popular in southern Europe under the name, Rose of the Indies. This plant was reintroduced into Europe in 1535 under the name, *Flos Africonus* by Emperor Charles V who launched an expedition to free Tunis from the Moors where it was naturalized on Algerian coast. It continued to be so-called until well up to the 18th century, although it was correctly figured as *Plantas Tagetes indica* in *Fuch's Herbal* of 1542 (Genders, 1971).

Another well-known type of marigold, *i.e.*, French marigold (*T. patula*) followed a similar route from Mexico to that of African type. Parkinson mentioned that it was called as *Tanacetum peruvinum* as if it grew in Peru. James Justice believed that it is a Chinese plant and it was first sent to the Royal Garden at Paris from where it was distributed under the name of French marigold (Coats, 1968).

Bailey (1963) mentioned that *T. erecta* was put into cultivation in 1596 and *T. patula* in 1573 in Europe. Tubular form of floret of single hollow, French marigold was reported by Parkinson (1629) in the *Paradiscus*. The chrysanthemum types of flower heads of African marigold was illustrated in the *Plant of Mexico* by Francesco Hornandez in 1645 (Iltis, 1945).

4.0 Species, Types and Cultivars

4.1.0 Species

There are about 33 species of the genus *Tagetes* (Rydberg, 1915). The characters of important species (Chittendan and Synge, 1956) are given below.

Tagetes erecta (African Marigold)

The plant hardy, annual, about 90 cm tall, erect and branched. Leaves pinnatisect, leaflets are lanceolate, serrated; flowerheads single to fully double, lemon-yellow to yellow, golden yellow or orange, large-sized; florets 2-lipped or quilled.

T. patula (French Marigold)

A hardy annual, 30 cm tall, bushy. Foliage dark green; stem reddish; leaves pinnatisect, leaflets linear lanceolate, serrated; flower heads small, single or double, yellow to mahogany-red, peduncles long.

T. tenuifolia (syn. *T. signata*)

Annual with a branching habit. Leaves pinnatisect, segments into 12, oblong, linear, sharply serrated; flowers 5 rays, yellow, roundish, obovate. *T. signata cv. pumila* dwarf, bushy, less than 30 cm; flowers bright yellow, small, numerous.

T. lucida (Sweet Scented Marigold)

The plants are annuals. Leaves sessile, small, lanceolate; flowers 2-3 rayed, in dense, terminal corymbs, with more agreeable odour than other species.

T. lacera

It was discovered in California. The plant up to 120-150 cm in height, flowers profuse, yellow.

T. lemmonii

It is a shrubby plant, grows up to 60-70 cm. Leaves slender, opposite; leaflets about 2-3 cm long; flowers showy, 2-3 cm diameter.

The other species grown in gardens are *T. minuta, T. pusilla* and *T. corymbosa*. In India, however, the cultivation of *T. erecta* and *T. patula* dominates.

Species of Marigold

Tagetes erecta *Tagetes patula* *Tagetes tenuifolia*

Tagetes lucida *Tagetes lemmonii* *Tagetes micrantha*

Tagetes minuta *Tagetes coronopifolia* *Tagetes pusilla*

Source: www.en.wikipedia.org, www.commons.wikimedia.org, www.magicgardenseeds.com, www.discoverlife.org, www.conabio.gob.mx, www.swbiodiversity.org

4.2.0 Types and Cultivars

In the two main groups of marigold– African and French, there are many types and strains varying in plant height, growth habit, flower shape and size. African marigolds have yellow and orange colours in various shades – light yellow, canary yellow, golden yellow, bright yellow, cadmium-orange, golden orange, deep orange, bright orange. A strain with white and almost white flowers has been developed. In French marigold, besides the colours mentioned above, the crimson and mahogany red colours are very attractive. Some cultivars have bicolour flowers like light yellow with maroon blotches, deep crimson-edged yellow, gold and red bicolour.

Like many other flowers of commercial importance, marigold has also been classified differently by floriculturists and seedsman in different countries. Therefore, no standard and accepted classification of marigold is available. However, attempts have been made to include the major groups but the cultivars are not always known or widely accepted in different countries.

4.2.1 African Marigold

4.2.1.1 Carnation Flowered

The plants grow to a height of 75 cm, flowers up to 10 cm across. This type has orange, golden yellow, deep orange or lemon-yellow flowers.

Carnation flowered dwarf – grows up to 40 cm, large flowering.

4.2.1.2 Chrysanthemum Flowered

Tall Double chrysanthemum flowered – Plants 60-80 cm tall, large, chrysanthemum-like flowers in various shades of yellow and orange.

Dwarf Double chrysanthemum flowered – The height is 20-30 cm. It has some strains, like.

(i) Luxor Series – improved Cupid type, and

(ii) Rexor Series – improved Spun type.

4.2.1.3 Tall F_1 Hybrids

Plants about 3 m high, large, fully double, flowers up to 12 cm across.

(i) F_1 Gold Coin Series, and

(ii) F_1 Climax Series.

4.2.1.4 Semi-tall F_1 Hybrids

This is also known as hedge type because of uniform and compact growth, 50 cm high. Double flowers are 10 cm across with lemon and golden yellow light orange.

4.2.1.5 Dwarf F_1 Hybrids

Plants are 15 to 40 cm high, and of compact growth; many flowers appear at one time.

Cultivars of Marigold

Pusa Narangi Gainda | Pusa Basanti Gainda | Pusa Arpita

Space Hopper | Arka Bangara 2 | Arka Agni

Arka Bangara | Arka Alankar | Bidhan Basanti

Bidhan Lakshmi | Bidhan Marigold 3 | Kesar Gold

Source: www.nurtureplant.in, www.iihr.res.in, www.iandbseeds.com, www.suttons.co.uk

Cultivars of Marigold

Arka Honey, Pusa Deep, Pusa Arpita

Bonanza Deep Orange, Bonanza Bee, Bonanza Harmony

Bonanza Flame, Bonanza Bolero, Bonanza Yellow

Red Brocade, Narayanpuri 2, Safari Red

Source: www.iihr.res.in, www.davesgarden.com, www.pixels.com, www.iandbseeds.com, www.plants.thegrowingplace.com, www.sites.google.com, www.panamseed.com

(i) Inca Series – Flowers are large, fully double and compact,

(ii) Space Age Series – Early flowering, dwarf and uniform, and

(iii) Galore Series – Long flowering duration, uniform.

4.2.1.6 F_1-triploid

Early and very free flowering, large and golden yellow flowers.

4.2.2 French Marigold

4.2.2.1 Dwarf Double

Plants 20-30 cm high, colours – yellow, orange, reddish brown, mahogany-red, golden yellow, bicolour.

4.2.2.2 Dwarf Double-scabious Flowered

Flowers with crested centre, wide range of colour – golden yellow, golden orange with red marking, golden yellow with red, brownish red with orange centre, golden yellow with spotted red, *etc.*

4.2.2.3 Dwarf Double Petite

Very dwarf plants, 15-20 cm high, bear numerous attractive flowers with golden yellow, orange, yellow and maroon colour.

4.2.2.4 French Dwarf Single

Plants 20-35 cm high, compact flowers, single, golden yellow, yellow, golden yellow with maroon eye, mahogany-red with yellow centre, yellow with brown.

4.2.2.5 Dwarf Triploid F_1 Hybrid

Plants 25-40 cm high, extremely early, profuse flowering; colour – yellow, golden yellow, orange and brownish-red.

4.2.2.6 Dwarf Double

(i) Dwarf All Saints – About 20 cm high, good for bedding, wide range of colour. (ii) Dwarf Chrysanthemum – Plants are 20 cm high, compact and bushy, chrysanthemum-flowered, rose, crimson.

4.2.2.7 Tetraploid Hybrid

French-type plants, early flowering, brilliant orange, fully double, carnation-type blooms, 6-7 cm across.

4.2.3 Single Signet (*T. tenuifolia*)

It is another dwarf type of marigold. Plants grow 30-35 cm tall, bushy with fine lacy foliage and covered with small single flowers of lemon-yellow, and orange colour. These are grown for pot culture, edging or rock garden.

4.2.4 Interspecific Hybrids (Afro-French marigold)

The interspecific hybrids between African marigold (*T. erecta*) and French marigold (*T. patula*) have been produced in the USA which are intermediate in characters. These hybrids are early flowering, medium in height, growing about 60 cm tall. Plants are bushy and produce double flowers with delightful colour combination of red and gold and are called 'Red and Gold' hybrids. Nugget type with yellow colour, and Show Boat with golden yellow flowers have been released. Another good interspecific hybrid, Red Seven Star has also been released. The fully double flowers, 12 cm in diameter are produced on 35 cm tall plant.

Several trials of marigold have been conducted to find out suitable cultivars for different purposes. Burney (1961) suggested suitable cultivars for growing in Switzerland. In North India, Sharga *et al.* (1972-73) and Sharga (1975) recommended those cultivars of African marigold which produce either 50 flowers or 800 g of flowers by weight for commercial growing. For this purpose, Alaska, Fire Glow, Golden Jubilee and Yellow Fluffy were selected. In case of French marigold, preference was given to the cultivars which produce either 100 flowers or 200 g flowers per plant and Brownie Scout, Burpees Nugget, Cupid Yellow, Petite Gold and Yellow Pygmy were good yielders.

Trials conducted at the Punjab Agricultural University revealed that Giant Double African Orange, Climax and Bangalore Selection from African group were suitable for growing commercially. In French group, cultivars like Rusty Red, Butter Scotch and Red Brocade were suitable for commercial purpose. Cultivars like Valencia, Sussana, Tetraploid were excellent for bedding (Arora and Singh, 1980a). Among 8 cultivars of African marigold, the promising cultivars were Giant Double African Orange and Giant Double African Lemon, at Konkan region of Maharashtra (Mehta *et al.*, 1995).

Shivakumar *et al.* (2015) reported that the genotype Nilakkotai Local Orange performed well for plant height, number of primary and secondary branches per plant, stem girth, flower diameter, flower weight, number of petals per flower and number of flowers per plant. However, genotype Coimbatore Local Yellow resulted in maximum leaf area and leaf area index. Maximum total chlorophyll content was recorded with Darmapuri Local.

ICAR-IIHR, Bengaluru has developed and released three marigold hyrids *viz.*, Arka Bangara, Arka Bangara 2, Arka Agni which are vegetatively propagated. The salient features of these hybrids have been given here.

Arka Bangara: Flowers are of yellow gold colour, compact and medium in size. The variety comes to flowering by 40-45 days. Its flowering duration is 65 -70 days. Flower size 5-6.5 cm in diameter. Yield potential 8 tons/acre in winter and 5 tons/acre in summer.

Arka Bangara-2: Flowers are gold yellow in colour. Flowering starts 40-45 days after planting and continues to flower for 60 days. Flowers are compact and large in size (7-7.5cm). Yield potential is 8-10 tons/acre.

Arka Agni: Flowers are orange in colour. Flowering starts 40-45 days after planting and continues to flower for 60 days. Flowers are compact and large in size (7.5-8 cm). Yield potential is 8-8.5 tons/acre.

Recently ICAR-IIHR, Bangalore released four F_1 hybrids on male sterile background resulting in male sterile hybrids that are high yielding and with compact flowers having good shelf life. The salient features of these F_1 hybrids are given here.

Arka Vibha: F_1 hybrid of African marigold identified for its attractive flower shape, highly compact, attractive orange colour (RHS colour orange group, N25C), with a shelf life of 8-9 days, and yield potential of 10-12 tons/acre.

Arka Abhi: F_1 hybrid of African marigold identified for its attractive radiant lemon yellow colour (RHS colour yellow group, 5A), large flowers (7-8 cm), with good shelf life (6-8 days) and high yield of 10-11 tons/acre.

Arka Bhanu: F_1 hybrid of marigold identified for its attractive flower shape and golden yellow colour (RHS colour yellow group 12 B), compact, with a shelf life of 7-8 days. Yield potential is 10-11 tons/acre.

Arka Shubha: F_1 hybrid of African marigold was identified for its carotenoid content. Plants are spreading type with a yield potential of 12-14 tons/acre. Flowers are orange in colour (RHS colour orange group N25C), with shelf life of 5-6 days. Arka Shubha is rich in carotenoid content (3.25 per cent).

Two varieties in French marigold namely Arka Madhu and Arka Pari are also released.

Arka Madhu: It is photo insensitive, dwarf with spreading habit and floriferous, flowering throughout the year, flower initiation 25 days after planting, flowering duration 6 weeks, flowers diameter 4.5-5 cm and flower yield 5-6 tons/acre.

Arka Pari: It is photo insensitive, dwarf plant habit and floriferous nature, flowering throughout the year, flower initiation 30 days after planting, flowering duration 9 weeks, flowers diameter 4-4.5 cm, flower yield 4-5 tons/acre.

ICAR-IARI, New Delhi also developed and released high yielding varieties of African marigold namely Pusa Narangi Gainda, Pusa Basanti Gainda, Pusa Bahar and French marigold varieties Pusa Arpita, Pusa Deep.

5.0 Genetics and Breeding

5.1.0 Genetics

A very few studies have been conducted on the cytological status of marigold. Chromosome number of several species of *Tagetes* have been compiled by Towner (1961) as follows :

Diploid Species-2n	***Tetraploid Species-4n***
Tagetes erecta-24 (Eyster, 1939)	*T. patula*-48 (Eyster, 1939)
T. tenuifolia-24 (Eyster, 1939)	*T. minuta*-48
T. elongata-24	*T. remotiflora*-48
T. lemmonii-24	*T. mendocina*-48
T. jaliscensis-24 (Towner, 1962)	*T. biflora*-48
T. lucida-22	

Towner (1956, 1958) and Bolz (1961) proposed that *T. patula* is an allotetraploid (2n=48) species. Towner (1961) suggested that *T. patula* is an allotetraploid of *T. erecta* and *T. tenuifolia*. When they grew the seeds from the interspecific cross, most of the hybrid seedlings died before attaining the reproductive stage, and seedlings which survived were weak. The bivalent frequency per cell was 0 to 11 and averaged 4.4 at meiosis in the diploid hybrid. The pollens were essentially non-viable and seed set was less than 1 per cent. A true breeding, fertile allotetraploid *T. erectatenuifolia* was synthesized by colchicine doubling of the chromosome number of the F_1 hybrid. In the synthetic allotetraploid, the average chromosomal association was 23.97 biva-lents plus 0.05 univalents and 97.4 per cent of MI cells had bivalents. There were no multivalent, bridges or fragments. Synthetic hybrid showed striking morphological similarity to *T. patula*. Natural as well as synthetic allotetraploids were intermediate between *T. erecta* and *T. tenuifolia* for nearly every character. Further studies carried out by Towner (1962) revealed that although *T. jaliscensis* and *T. erecta* are morphologically distinct, F_1 hybrids between them had normal meiosis and were fertile. Segregation in F_2 was undisturbed for marker genes like tubular ray, chlorophyll defect and disc anthocyanin, on 4 pairs of chromosomes. From the plant morphology it was postulated that the sub genome *A* of amphiploid *T. patula* was derived from *T. erecta* and not from *T. jaliscensis*. Sub genome *B* was derived from *T. tenuifolia*.

Galica-Fuentes (1995) observed wide diversity of inflorescences in marigold germplasm, especially in *T. erecta*. Kennedy (1997) noted genetic variability for seed characters – vigour index, germination percentage and seedling dry weight and high heritability coupled with low genetic advance for 1000-seed weight and seedling length. The coefficient of variation was lowest for duration of flowering and highest for flower yield (kg/plot) in French marigold, while, in African marigold it was lowest for plant height 60 days after transplanting and highest for flower dry weight. Heritability estimates for all the characters were generally very high. Singh and Sen (2000) observed high heritability along with high genetic advance for flower yield per plot and per hectare in both *T. erecta* and *T. patula*.

A line × tester crossing programme in *T. erecta* was carried out using 3 male sterile lines, ms7, ms8 and ms12, and a set of 11 diverse pollinators, selections 7, 8, 14, 19, 21, 22, 27, 28, 29, 31 and 56, as testers. The hybrids (33 F_1s) along with parents were evaluated by Gupta *et al.* (2001) during winter and summer. The morphological characters during winter and summer showed that F_1 plants performed better than their parents during both seasons. Analysis of variance revealed that genotypic differences were highly significant for all the characters during both cropping seasons, except for seed number per head during winter. Significant heterosis was observed in all traits except for stalk length and number of seeds during winter. The highest heterosis was observed in flower yield, which was 310.2 per cent for the winter crop and 535.7 per cent for the summer crop.

5.1.1 Male Sterility

Male sterility in marigold is of two kinds, (i) apetalous and (ii) double-flowered (Goldsmith, 1968). Apetalous sterility is preferred to full double flowers because latter type is prone to break down and it gives rise to a few disc florets at later stages which may be due to either age of plants or environmental factors. However, apetalous flowers are less attractive to pollinating insects. Male sterility is governed by recessive gene and is incorporated into seed parent. It is maintained by crossing the heterozygous plants with the sterile ones.

Male sterile apetalous lines, MS7 and MS8, were crossed with a male fertile petalous line (Selection-21). F_1 and F_2 (selfed F_1) plants (*T. erecta*) were grown to investigate segregation of petalous and apetalous flowers. In F_2 generation, segregation of male fertile (petalous) and male sterile (apetalous) plants in a 3:1 ratio indicated that the apetalous sterile trait was governed by a single recessive gene (Gupta *et al.*, 1999b). Santhosh *et al.* (2018) assessed the contribution of two types of male sterile lines (3 petaloid, 4 apetaloid) for exploitation of heterosis of biochemical components using 3 fertile pure lines. Results showed that total carotenoids, zeaxanthin and lutein were found to be governed by non-additive gene action. Moreover, parent IIHRMO 9-7 was observed as good general combiner for lutein and total carotenoid. Maximum positive heterotic values were recorded with hybrid combination IIHRMO 9-8 x IIHRMO 12-12 relative to the better parent.

The inheritance studies reported for some of the characters are discussed below :

5.1.1.1 Mode of Inheritance

Double Flower vs Single Flower

Punnet (1924 a,b) reported that double flowering habit was monogenic dominant to single. Towner (1961, 1962) while studying the F_2 crosses of *T. erecta* × *T. tenuifolia* observed that the double flowering habit was monogenically dominant over single flowering. In diallel crosses including 7 parents, Singh (1970) observed that double flowering was dominant over single flowering and F_2 segregated into 3:1. The monogenic nature of dominance was confirmed in the test crosses.

Types of Flower

Singh and Swarup (1973b) while working on the inheritance of different types reported that in a cross between carnation and cupid types, the carnation type was dominant over cupid. In F_2, two types of flowers segregated in a ratio of 9 carnation : 7 cupid, thus indicating the presence of two complementary genes governing this character. Whereas the carnation type was found to be dominant over chrysanthemum in a cross between these two types. The F_2 population segregated in a ratio of 15 carnation : 1 chrysanthemum which could be due to duplicate epistasis. Two recessive genes were reported to be responsible for chrysanthemum type. The crossing between cupid and chrysanthemum type resulted in intermediate flowers. The F_2 population segregated in a ratio of 9 intermediate : 3 cupid : 4 chrysanthemum. This showed that cupid and chrysanthemum types were governed by two genes. Based upon these studies, the following genetic makeup of different types was proposed (Singh, 1970).

Carnation (double-flowered)	: *RR TuTu CC*
Carnation (single-flowered)	: *rrTuTu CC*
Cupid	: *rrTuTuCuCu*
Chrysanthemum	: *rr tutu cc*
	: *RR tutu cc*
Intermediate	: *RR tutu cucu*

Flower Colour

Punnet (1937), in an investigation on African marigold, reported that the flower colour was governed by two or more factors. The flower colour of interspecific hybrid between *T. erecta* and *T. jaliscensis* revealed that disc corolla anthocyanin from *T. jaliscensis* segregated as simple dominant over the absence in *T. erecta*. The dominant allele for the presence of anthocyanin in the lobes of disc corolla is given here by the gene symbol R (red) (Towner, 1962).

The inheritance of yellow, tangerine and orpiment colour was studied by Singh and Swarup (1973b). Orpiment and tangerine are orange colours but former is deeper in shade. In a cross between yellow and tangerine coloured parents, F_1 was found to be tangerine and F_2 population segregated in a ratio of 12 tangerine : 3 yellow : 1 orpiment. This suggested that flower colour was governed by two dominant genes. In another cross between yellow and orpiment coloured flowers, orpiment colour was observed to be dominant over yellow and F_2 population segregated in a ratio of 13 orpiment : 3 yellow. It also showed that two genes were responsible for inheritance of colour.

In the cross between two orange colours, *i.e.*, orpimentitangerine, orpiment was dominant over tangerine and F_2 population segregated into 13 orpiment : 3 tangerine ratio, which also suggested the involvement of two genes. On the basis of above results, following genetic constitutions have been proposed for different parental colours (Singh, 1970).

Yellow	: *tt ii YY*
Tangerine	: *TT ii YY, TT ii yy*
Orpiment	: *TT II YY, tt II yy, tt ii yy*

Flat vs Tubular Flowers

Flatness of florets was found to be monogenic dominant over tubular florets (Punnet, 1924a). In a cross between *T. erecta* × *T. jaliscensis*, Towner (1962) also observed that flat floret was dominant over tubular ray floret. The dominant and recessive alleles of *T. jaliscensis* and *T. erecta* were assigned *Tu* and *tu*, respectively.

Leaf Colour

In healthy plants, foliage is green whereas virescent plants have chlorotic leaves. It was observed by Kantor (1938) and Towner (1962) that the duplicate recessive genes were responsible for virescent character. However, Little *et al.* (1940) in *T. erecta* crosses found that virescent character was monogenic recessive and partially lethal whereas a few plants died due to chlorophyll deficiency. Early flowering plants were observed to be somewhat weaker than late flowering ones, indicating that the character may be acting as a partial lethal.

Days Required for First Flowering

From a diallel cross between seven parents, Singh and Swarup (1973a) reported that two crosses showed significant additive gene effects whereas in three crosses dominance was significant. The epistatic gene effects in all crosses were more pronounced than additive and dominance gene effects. However, *I* effects contributed more to the inheritance of earliness. Duplicate epistasis was present in cross between Sun Giant × Cupid Orange Mum whereas complementary epistasis was present in other six crosses. Transgressive of segregation of earliness was observed.

Flower Weight

From the studies conducted by Singh and Swarup (1973a), it was revealed that dominance played a more important role than the additiveness. Six crosses showed significant dominance effect, their magnitude was greater than additive gene effects and in one cross it exceeded the mean effect also. Only one cross had significant additive effects. In six crosses, the mean effects were greater than *d* and *h*. The absolute magnitude of the total epistatic gene effects was observed to be larger than mean effects in six crosses. Epistasis was significant in four crosses. The contributions of *j* and *I* were greater than those of *1*. Complementary epistasis was observed in two crosses. Transgressive segregation in F_2 generation resulted in an increase between 0.03 and 36.17 per cent in flower weight over better parent.

Number of Flowers

It was reported by Singh and Swarup (1973a) that the dominance effects were significant. The additive gene effects were significant in three crosses and significant interaction effects were observed in six crosses. The absolute magnitude of the total epistatic gene effects of these six crosses was observed to be larger than the mean effects, thus indicating major contribution of the epistatic effect to variation in these crosses. The dominance × dominance had more contribution to the inheritance of total number of flowers. Duplicate epistasis was also observed. Transgressive segregation in the F_2 generation was observed for this trait also. The increase in the number of flowers in the transgressive segregates varied from 19.51 to 17.75 per cent over better parent. Giri *et al.* (2018) reported that flower yield per ha had positive correlation with plant height, plant spread, dry weight of flower, number of flowers per plant, flower diameter and flower yield per plant. Thus, to improve flower yield of marigold, selection should be done based on these traits.

5.2.0 Breeding

Hybridization between distantly related types is the most effective and commonly employed tool to induce variation so as to improve the existing cultivars and evolve new, high yielding and better quality of genotypes. Different principles of breeding have successfully been used in marigold which resulted in new cultivars and present-day F_1 hybrids.

5.2.1 Inter-varietal Hybridization

In a well-planned breeding studies, Singh and Swarup (1972) observed appreciable heterosis. The highest percentage of heterosis was observed in a cross between Alaska × Cupid Orange Mum for early flowering, Local from Katrain × Cupid Orange Mum for total number of flowers and between Alaska × Hawaii for flower size and weight. Attempts were made to develop F_1 hybrids in marigold because F_1 hybrids are considerably uniform and capable of producing large blooms with high yield potential. In addition, they are characterized by their semi-tall nature with excellent large full double flowers. Goldsmith (1968) discussed the advances in production of F_1 hybrid seeds of important flower. For about last 20 years F_1 hybrid seeds of *T. erecta* are available. Climax was first F_1 hybrid to be released. A number of other F_1 hybrids are also available in the market.

Gupta *et al.* (2016) evaluated the performance of 14 French marigold genotypes which were maintained as inbred lines. Among those genotypes, Safari Queen recorded early flowering; Harmony Boy had maximum plant height and spread; Spray Max produced maximum number of flowers per plant with greater flower yield and Cupid Lemon produced bigger flowers. Thus, it was suggested that var. Sunkist, Spray Boy, Safari Queen, Spray Max and French Bonita-6 could be used as parents in hybridization programme for higher yield.

5.2.2 Inter-specific Hybridization

The interspecific hybridization between *T. erecta* × *T. tenuifolia* and *T. erecta* × *T. jaliscensis* has been done by Towner (1961, 1962). Recently, interspecific hybridization between *T. erecta* × *T. patula* has been carried out which finally led to evolution of Red and Gold hybrids. These hybrids are dwarf like French marigold but produce bigger flowers like that of African marigold in large number.

5.2.3 F_1 Hybrid Seed Production

The technique used in F_1 hybrid seed production involves the use of male sterility. The F_1 hybrid seed is produced in field by using alternate rows of male sterile line and a pollen parent. The male sterile line gives rise to normal and sterile plants in a ratio of 50 : 50 and normal plants are rogued out. Only sterile plants are kept for pollination and seed production.

5.2.4 Mutation Breeding

Sarkar *et al.* (2016) identified 20 and 15 Gy dose of ^{60}Co gamma radiation as LD_{50} for *in vivo* and *in vitro* conditions with the survival rate of 53.24 and 52.55 per cent, respectively. Under *in vivo* study, treatment with 20 Gy induced early flower bud compared to control. Under *in vitro* condition, yellow flowers (10 YR) were obtained at 15 Gy instead of orange control (5 YR). In M_1 generation, two mutants for flower form, one each in colour and earliness were obtained under *in vitro* condition, while, 5 mutants in flower colour and 2 in flower form under *in vivo* condition. Among them, putative mutant-3 (pm3) produced early flowering and pm-4 gave light orange coloured flowers (10 YR).

Majumder *et al.* (2018) developed distinct putative mutants of Pusa Narangi Gainda through gamma irradiation and studied variability as well as correlation in M_2 generation. Greater heritability was observed with days to flower harvest. Number of flowers per plant and days to first flower harvest showed maximum positive phenotypic correlation, whereas, seed yield and number of flowers per plant showed maximum positive genotypic correlation. On further assessment of putative mutants using ISSR, putative mutant – 3 (pm3) showed highest dissimilarity from parent (0.58). Moreover, the mutants pm5 and pm6 were found similar to each other (0.65) but distinct from parent. The mutants selected above were multiplied for 6 generations to attain homozygosity for large scale field evaluation.

5.2.5 Polyploidy

Sajjad *et al.* (2013) studied the effect of colchicine on *in vitro* polyploidy induction in African marigold. Nodal segments were cultured on MS medium supplemented with different doses of colchicine. After treatment, nodal segments were shifted to colchicines free MS medium for normal growth. Highest mortality rate (72.74 per cent) and restricted shoot growth (3.80 cm) was observed in 0.05 per cent colchicine treatment. Stomatal size was increased in 0.01 per cent (811.34 µm) and 864.70 µm in 0.05 per cent and 289.30 µm in control. Stomatal number was reduced to 9 in 0.05 per cent colchicine treatment to 16 in control. Thus, 0.01 and 0.05 per cent were found effective in inducing polyploidy in African marigold (Table 1).

Table 1: Effect of Colchicines on *In vitro* Polyploidy Induction in African Marigold*

Parameters	*0.00 per cent*	*0.001 per cent*	*0.01 per cent*	*0.05 per cent*
Mortality (per cent)	2 d	12.17 c	40.14 b	72.74 a
Regeneration (per cent)	98 a	87.83 b	59.86 c	27.26 d
Shoot proliferation (per cent)	44.0 c	71.67 b	94.0 a	11.67 d
Shoot length (cm)	9.93 a	8.26 b	7.63 b	3.80 c
Stomatal size (µm)	289.30 b	322.77 b	811.34 a	864.70 a
Stomatal number	16 a	15 a	10 b	9 b

* Sajjad *et al.* (2013).

Patil *et al.* (2017) revealed that white African marigold plants survived in all concentrations of colchicines (0, 0.5, 1.0, 1.5, 2.0 and 2.5 per cent) when treated to the seeds, whereas, plants exposed only upto 1.5 per cent colchicine survived on root treatment. Frequency of stomata and guard cell decreased with increase in colchicine dose while, stomata size, guard cell size and pollen grain size increased with increasing concentration. However, seeds soaked in colchicine at 0.5 per cent concentration for 12 h were found most effective in inducing more number of variant types.

6.0 Growth and Flowering

Growth and flowering in marigold are generally governed by day length and temperature (Roberts and Struckmeyer, 1939; Merritt and Ting, 1995). Certain growth substances also influence this process (Murray *et al.*, 1986; Shedeed *et al.*, 1986). Factors like soil moisture (Kurup *et al.*, 1994) and nutrition influence both plant growth and flower production.

6.1.0 Environmental Factors

6.1.1 Light and Temperature

The effect of temperature and day length was first investigated by Roberts and Struckmeyer (1939). They reported that African marigold flowered under long day at temperature of 21.1-23.9°C, 17.2-18.3°C and 12.8°C, while the plants failed to grow under short days at higher temperature. Intermediate classes, however, grew and produced flowers under short days at 12.8°C. Allard and Garner (1940) opined that all of the cultivars except those of African type (*T. erecta*) are intermediate in their day length requirement. The African cultivars showed a slight delay in flowering under long days. Marigold, earlier said to be long-day plant was, however, found to be a short day plant (Duda, 1967). In a well-designed experiment, Tsukamoto *et al.* (1968) observed that the weakest requirement for short photoperiod was exhibited by *T. erecta* whereas *T. patula* was intermediate and *T. tenuifolia* had the strong requirement. Eight *cvs.* of *T. erecta* and 5 *cvs.* of *T. tenuifolia* were grown under short and long day conditions from seed germination whereas *T. patula cvs.* from the age of 4 weeks. Long day plants *T. erecta* and *T. tenuifolia* produced more number of nodes and delayed (3 weeks) appearance of terminal flower bud. In *T. patula*, only Orange Winner and Tanga cultivars responded to long day during summer but did not respond to long day during winter (Zimmer, 1989). The flowering of *T. erecta cv.* Golden Eagle, a short-day type was unaffected by daily application of abscisic acid (ABA). With this treatment, the stem growth was inhibited without promoting flowering (Cathey, 1968). The investigation carried out by Tsukamoto *et al.* (1971) revealed that French marigold *cv.* Butter Ball flowered under 10 h light at 30°C but not under 18 or 24 h light. At 20°C, however, the plants flowered at all day length but less profusely and slowly than at 30°C under 10 h day length. Repeated application of 25 ppm GA_3 slightly improved flowering but TIBA did not affect flowering of African marigold. Under greenhouse, 27°C day temperature for 9 h and 18°C night temperature produced early flowers by 3 weeks in cultivar Perfection Yellow (*T. erecta*) and 2 weeks in Aurora Fire (*T. patula*) and significantly higher crop productivity (27 g dry weight/m^2 compared to low night temperature in both cultivars (Kim and Merritt, 1991). *T. patula* and *T. erecta* × *T. patula* also produced 2 to 17 days early flowering in 26°C day/17°C night with a 21°C mean daily temperature (Merritt and Ting, 1995). Stem elongation of *T. erecta cvs.* Excel Gold, Golden Jubilee and Sovereign was recorded with far red light. The duration of far red light and time of application (day/night) had pronounced effect on stem elongation. Both red and blue light reversed the effect of far red light. Red light was effective in day time (Jatzokoski and Zimmer, 1994). Pot of Gold cultivar was not markedly affected by either day length or temperature. Carlson *et al.* (1981) reported that when marigold plants were grown at 21.1°C (70°F) (night temperature) the plants had higher fresh weight but the postharvest keeping quality of flowers was best with plants grown at 10-15.5°C.

6.1.2 Growth Regulators

Marigolds are excellent pot plant, especially the cultivars of *T. patula* and *T. tenuifolia* which are grown in pots for window garden, *verandah* or other situations. The cultivars of *T. erecta* usually grow taller and growth retarding chemicals have been recommended for reducing the height and improving the appearance of the pot plant. Wilson and Holcomb (1979) observed reduction in plant height, and dark green colour of the foliage in *cvs*. Moon Shoot, Yellow Galore, First Lady when the plants were sprayed with SADH at 2500 ppm.

Novoselova *et al.* (1985) observed that spraying or watering *T. patula*, grown as a pot plant in winter and spring, with CCC at the start of bud formation had a marked effect on growth and flowering, best results were obtained when 2-5 per cent solutions were used as soil drench on 2 dates. The treated plants were more uniform in growth, dwarf and bushy with more branches and produced larger number of flowers in comparison with the control plants. Spraying African marigolds (*T. erecta*) with cycocel (chlormequat) (750 ppm) resulted in the highest flower weight (10.8 g/flower), and number (10.84/plant) and reduced the number of days to flowering (43.68). Maleic hydrazide (MH) reduced plant height and the number of days to flowering (Yadav, 1997). Various rates of ancymidol as granular incorporated, granular broadcasted, soil drench or foliar spray treatments were also tested by Murray *et al.* (1986). The plant height of *Tagetes erecta* was controlled most effectively by 311 to 622 mg ancymidol/m^3 as a drench or granular incorporated. During two successive seasons, *T. erecta* was treated with GA_3 at 100-400 ppm or SADH at 250-2000 ppm at 4 weeks from planting out and again a month later. GA_3 generally increased plant height whereas SADH decreased it. Flowering time was not affected by GA_3 whereas SADH delayed it (Shedeed *et al.*, 1986). Syamal *et al.* (1990) sprayed GA_3 (100 or 200 ppm) and MH (200 or 400 ppm) on *T. erecta* 15 days after transplanting and twice more at 10 days intervals. GA_3 at 200 ppm recorded highest flower and seed yield, whereas MH suppressed vegetative growth and yields of flowers and seeds (Table 2).

Table 2: Effect of GA_3 and MH on Growth, Flowering and Seed Yield of Marigold*

Treatments	*Height of Plants (cm)*	*Number of Branches/ Plant*	*Number of Leaves/ Plant*	*Number of Flowers/ Plant*	*Seed Yield/ Plant (g)*
MH 200 ppm	40.3	13.6	277.6	68.3	18.9
MH 400 ppm	37.3	12.6	262.4	56.3	13.8
GA_3 100 ppm	84.7	16.0	379.2	136.6	27.0
GA_3 200 ppm	94.7	16.1	491.9	152.0	29.4
Control	44.0	13.6	265.6	76.6	19.5
C.D. at 5 per cent	3.72	0.97	17.74	15.93	4.38

* Syamal *et al.* (1990).

African marigold *cv.* Sunset Gaint was sprayed with GA_3 and ethrel (100-400 ppm) 22 days after transplanting. Higher concentration of GA_3 increased growth, number of flowers/plant, weight of flowers and duration of flowering. Flower yield was more than double with 400 ppm GA_3 application. Lower concentrations of ethrel enhanced the flowering time and also increased duration of flowering (Singh *et al.,* 1991). Davis (1991) applied uniconazole at 5-25 mg/l and noted controlled growth of seedling without adverse effect on subsequent growth and flowering of African marigold. Seeds of *T. patula* 'Bonanza Gold' were soaked in solutions of 0, 500 or 1000 mg paclobutrazol (PB)/l for 16 hours at 25°C or they were primed (–0.5 MPa for 7 days at 20°C) in Grade 5 exfoliated vermiculite moistened with 0, 500 or 1000 ppm PB solutions. Soaked and primed seeds were dried for 1 day at 19°C and 25 per cent relative humidity. Soaking marigold seeds in 1000 ppm PB failed to decrease shoot height below that of plants from non-treated seeds at 32 days after planting (DAP). However, exposure to 1000 ppm PB during priming of marigold seeds resulted in shoot height suppression (13 per cent) similar to that of the growth medium drench, and similar shoot dry weight reduction (21 per cent) to the shoot spray. Suppression of shoot growth by this seed treatment was short-term since by five weeks after transplanting into 15 cm pots (Pill and Gunter, 2001). Marigold seeds *var.* Pusa Narangi Gainda and Pusa Arpita primed with water for 12 h resulted in higher germination percentage, establishment percentage, plant height, number of branches per plant, flower yield per plant, number of days for seed maturity, seed yield per plant, 1000-seed weight followed by osmotic potential of -0.5 MPa polyethylene glycol for 24 h. Minimum improvement was observed in 2 per cent KNO_3 primed seeds for 24 h (Damabi *et al.,* 2014). Pot marigold seeds primed with 0.1 µM foliar spray + 0.1 µM priming with 2, 4- epibrassinolide resulted in maximum shoot weight (79 per cent), root (33 per cent), fresh weight of plant (28 per cent) and dry weight (26 per cent). Moreover, higher increase in chlorophyll a, b, total and carotenoid (2.4, 3.1, 2.6 and 2.7 per cent, respectively), protein (35 per cent) and reduced sugar (18.3 per cent) were recorded with the same treatment (Kara *et al.,* 2016). Seed treatment with magnetic field for 5 min increased marigold leaf area while, seeds treated with laser irradiation for 10 min produced more number of leaves (Mirshekari *et al.,* 2015).

7.0 Propagation

There are two common methods of propagation of marigold- (i) by seeds, and (ii) by cuttings. Plants raised from seeds are tall, vigorous and heavy yielder and hence, seed propagation is preferred to cuttings.

7.1.0 Seed Propagation

Seeds of wide range of cultivars of *T. erecta*, *T. patula* and *T. tenuifolia* are easily available and germinate quickly. It is, therefore, an easy method to propagate marigold through seeds which germinate well at temperature from 18 to 30°C. For raising seedlings, seeds can be sown in pots, seed boxes, on flats or raised nursery beds. Nursery beds are thoroughly prepared by digging and incorporating well rotten farmyard manure. Seeds should be sown thinly and be covered with light soil or sand or sieved leaf mould and watering should be done through fine rose. For the entire period, the nursery bed should be kept moist. Boodley (1961) reported that seedlings of marigold attained transplanting stage earlier if seedlings are raised

Propagation and Growing of Marigold

Seeds | Seedlings in Plug Tray | Ready for Transplanting

Planting in Pots | Flat Bed Preparation | Raised Beds with Mulching

Standing Crop | Commercial Cultivation | Cultivation in Open

Intercropping with Bean | Intercropping with Sugarcane | Intercropping with Papaya

Source: www.gardenerspath.com, www.indiamart.com

at a temperature of 18.3 to 21.1°C (65-70°F) under 18 h light with a minimum light intensity of 500-800 ft candles. In *T. erecta cv.* Spungold, pre-sowing seed treatment with 0.75 per cent potassium nitrate for 16 h registered 89.5 per cent germination and good seedling vigour (Selvaraju and Selvaraj, 1994). Seeds inoculated with *Beauveria bassiana* resulted in earlier germination with higher germination percentage and compact plant architecture (Heinz *et al.*, 2018).

7.1.1 Sowing Time

Marigold can be raised three times a year - rainy, winter and summer seasons. The seasons of sowing and transplanting of seedlings for obtaining flowers at different seasons of the year are given below:

Flowering Season	*Sowing Time*	*Transplanting Time*
Late rains	Mid-June	Mid-July
Winter	Mid-September	Mid-October
Summer	January-February	February-March

7.2.0 Vegetative Propagation

7.2.1 Cutting

This method is followed for perpetuation of all characters of a particular plant or cultivar. The emergence of adventitious roots from the stem helps in establishment of cuttings. About 6-10 cm long cuttings are made from the apical portion of the shoot and treated with root promoting chemicals. Roy *et al.* (1973) reported that rooting of marigold cuttings was enhanced by IAA and IBA treatment. The synergism between IAA and p-hydroxy benzoic acid and gallic acid was also observed. Cuttings planted in sand or vermiculite root easily and the rooted cuttings are transferred in bed or pots for flowering.

7.2.2 Grafting

Cleft grafting of 4 weeks old *Tagetes erecta* scion onto 6 weeks old rootstock showed good graft survival percentage compared to splice and hole insertion grafting. However, grafting on four week old rootstock showed no survival, irrespective of the grafting methods (Anupama *et al.*, 2018).

7.3.0 Micropropagation

Plantlet regeneration of *T. erecta* from hypocotyl-derived calluses occurred on MS medium containing 0.2-1.0 mg NAA and 0.5-5.0 mg BA/l, whereas leaf-derived calluses produced mainly roots and green spots in some plant growth regulator combinations. The leaf explants also produced large yellow calluses on MS medium containing 5.0 mg NAA and 0.2-5.0 mg BA/l (Belarmino *et al.*, 1992).

Kothari and Chandra (1984) developed protocol using leaf segments. Regeneration of a large number of adventitious shoot buds was observed on MS medium with BA (3-5 mg/l) + IAA (1 mg/l). Shoots with sufficiently developed internodes were rooted on filter paper bridges in culture tube on MS liquid medium with IBA (0.5 mg/l) + GA_3 (0.5 mg/l) to obtained complete plantlets with well-developed root and shoot systems. Misra and Dutta (2001) developed a protocol from leaf segments of white marigold (*T. erecta*). GA played a significant role in shoot bud induction as well as in suppressing callus formation. Shoot buds were induced directly in MS medium supplemented with 14.43 µM GA and 4.44 µM 6-BA in the absence of any auxin. Shoot buds (2-5) differentiated from the margins as well as leaf lamina of the lower petiolar segment within 4 weeks of incubation. Differentiated shoots grew well and proliferated in the MS medium having 1.1 µM BA and 29.41 µM $AgNO_3$, as it had a beneficial effect on the growth and proliferation of shoots. Shoots were excised, rooted in 0.27 µM NAA.

Kumar *et al.* (2017) standardized an *in vitro* protocol for mass multiplication of petaloid male sterile *cvs.* Siracole Orange and Siracole Yellow. Nodal segments were choosen as explants and were pretreated with carbendazim 0.2 per cent + metalaxyl (0.2 per cent) + 8-HQC 200 mg/l for 60 min followed by surface sterilization with 0.1 per cent $HgCl_2$ for 4 min. MS medium + 0.5 mg/l BAP + 0.05 mg/l NAA resulted in higher culture establishment (82.2 per cent) with minimum of 3.88 days for bud emergence. Maximum proliferation with healthy shoots (6, 28, 122 and 404 shoots/ explants at 30, 60, 90 and 120 days after culture, respectively) were obtained with MS + 0.5 mg/l BAP + 0.1 mg/l NAA + 2.5 mg/l $AgNO_3$. However, maximum elongation (2.1 cm) was noted with hormone free MS medium. Further, maximum rooting percentage (96.5 per cent), more number of roots (23.37), quicker root induction (5.25 days) and greater *ex vitro* survival (91.25 per cent) were recorded with ½ MS medium + 0.5 mg/l IBA. Highest plant survival (98.10 per cent) and better plant growth resulted with low-cost polypropylene glasses rather than traditional glass bottle system.

Nodal segments of marigold var. Pusa Narangi Gainda and Pusa Arpita were pretreated with carbendazim 0.2 per cent + Metalaxyl 0.2 per cent + 8-HQC 200 mg/l for 60 mins followed by surface sterilization with 0.1 per cent $HgCl_2$ for 4 minutes. Explants cultured on MS medium + BAP 2 mg/l + NAA 0.05 mg/l resulted in higher culture establishment of 69.44 per cent and earliest bud emergence in 4.45 days. Complete proliferation occurred in MS medium without hormones, MS + 0.5 mg/l Kinetin+ 0.1 mg/l NAA and MS+ 0.5 mg/l BAP + 0.1 mg/l NAA + 2.5 mg/l $AgNO_3$. MS + 0.5 mg/l BAP + 0.1 mg/l NAA + 0.25 mg/l $AgNO_3$ produced maximum number of quality shoots per explants (4.3, 18.8, 64.2 and 208.2 shoots at 30, 60, 90 and 120 days after culture, respectively) (Kumar *et al.*, 2018).

Kumar *et al.* (2018b) reported that anthers of marigold cultured on MS medium fortified with 1 mg/l BAP + 1 mg/l 2,4-D + 45 g/l sucrose obtained highest callus (38.74 per cent) induction in the period of 16 days. However, the embryogenic callus induction was significantly more with Pusa Arpita (27.7 per cent) than Pusa Basanti Gainda (23.37 per cent). Further, maximum adventitious shoot buds were induced on MS medium + 2 mg/l BAP + 0.5 mg/l NAA + 30 g/l sucrose with 1.09 buds per callus in a minimum duration of 19.47 days. Shoot bud induction was also found higher with Pusa Arpita (3.81 per cent) over Pusa Basanti Gainda (2.9 per cent). The whole regeneration took only 30 days in case of the French genotype whereas, 40 days in African genotype.

Mehraj *et al.* (2018) cultured the *in vitro* regenerated doubled haploids line of marigold derived through ovule culture to the rooting medium consisting of ½ MS basal medium supplemented with IBA (0.5, 1 mg/l), NAA (0.5, 1.0 mg/l) and 0.25 mg/l of IBA and NAA each along with 60 g/l of sucrose. The medium comprising IBA 0.5 mg/l produced a maximum of 33.54 roots per shoots with the rooting percentage of 93.67 per cent in a minimum of 3.44 days for root initiation. Though longer roots (9.64 cm) were observed with MS medium devoid of hormones, ½ MS + 0.5 mg/l IBA produced healthy, white, thick, vigorous roots but of shorter root length (3.43 cm). Further, these rooted plants on hardening in media comprising cocopeat: vermiculite: perlite in 1:1:1 ratio showed good survival percentage at 15 days (96.8 per cent), 30 days (95.0 per cent) and 45 days (95.0 per cent) with 18.6 leaves per plant and shoot length of 12.50, 15.46 and 23.40 cm, respectively. Thereafter for *ex vitro* hardening, maximum plant survival (94 per cent), plant height (39 cm), plant spread (30.5 cm), stem diameter (0.97 cm), shoot fresh weight (0.75 g), root fresh weight (0.46 g), leaf length (10.27 cm), root length (28.50 cm), leaf width (6.82 cm) and greater colonization (60.78 per cent) were observed with media comprising soil+ vermicompost (3:1) + 20 g VAM consortia. Clonal fidelity of doubled haploid marigold lines was assessed using SSR markers which obtained same banding pattern between *in vitro* leaf derived regenerants and donor mother plant insisting that the tissue culture derived plants were monomorphic and true to type (Mehraj *et al.*, 2019a,b).

Thaneshwari *et al.* (2018) obtained better callus induction and shoot differentiation when the flower buds were pretreated at 45°C for 2 h. MS medium fortified with 4 per cent sucrose + 4.44 µM BAP + 4.52 µM 2,4-D incubated in dark for 4 weeks at 25°C recorded maximum callus induction (89.66 per cent). Further sub-culturing on MS + 4 per cent sucrose + 4.44 µM BAP + 1.07 µM NAA exposed to 16/8 h photoperiod resulted in higher shoot differentiation rate (23.35 per cent). Thereafter, the shoots were rooted on hormone free MS media.

Aguilar *et al.* (2000) obtained tissue cultured plants in ventilated containers. These plants exhibited greater control of leaf water loss and increased survival in the field than plants cultured in sealed containers. Increased field survival of plants cultured in ventilated containers was attributed to higher levels of endogenous abscisic acid (ABA).

8.0 Soil and Climate

8.1.0 Soil

Marigold is adaptable to different types of soil conditions and thus can be grown successfully in a wide variety of soils. However, a deep, fertile, friable soil having good water holding capacity, well-drained and near to neutral in reaction (pH 7.0-7.5) is most desirable. An ideal soil for marigold cultivation is fertile sandy loam.

The growth and flowering of marigold (*T. erecta*), were evaluated by Yahya *et al.* (1997) in different growing media comprising a range of percentage of coconut dust (coir) and tropical peat (100:0, 75:25, 50:50, 25:75 and 100 per cent v/v). Plants grown in 100 per cent peat were the shortest and accumulated the least amount of dry matter. Marigolds grown in media containing coconut dust produced more flowers than those grown in peat alone. Incorporation of sand into the coconut dust-based growing media to improve media wettability up to 33 per cent helped maintain good moisture content and air-filled porosity. Secerov-Fiser and Vujosevic (1998) reported that the peat:sand:leaf mould mixture was best to obtain better root system in pot grown *T. erecta*. Marigold plants var. Marvel Mixture and Taishan Orange (*T. erecta*) and Durango Red and Bonanza Flame (*T. patula*) grown on peat and peat-coconut showed maximum plant height, internodal length, leaf length and stem diameter compared to compost, flower soil and low moor peat (Malgorzata and Renata, 2017).

In a study to assess the effect of sodicity on chlorophyll content in leaves, six marigold *var*. Local Selection, Pusa Narangi Gainda, Pusa Basanti Gainda, African Tall Orange, Sunset Giant and Inca-Hybrid were grown in sodic soils having 9.9 (control), 20, 30 and 40 ESP levels. Chlorophyll 'a' content reduced by 2, 16 and 30 per cent; chlorophyll 'b' by 13, 27 and 34 per cent and total chlorophyll by 8, 23 and 30 per cent at 20, 30 and 40 ESP levels, respectively. Among the 6 varieties, Inca-Hybrid and Sunset Giant recorded maximum and minimum chlorophyll content, respectively in the sodic soils (Trivedi *et al.*, 2017). Increasing salinity stress level from 0 to 8 dS/m reduced plant height, fresh yield of flowers, yield of flower dry weight, number of flowers per plant, grain yield, petals and seed weight by 43, 60, 56, 17, 64, 15 and 30 per cent, respectively. It was also reported that incorporation of bacteria could minimize the effect of stress on marigold (Tina *et al.*, 2017).

8.2.0 Climate

Marigold requires mild climate for luxuriant growth and flowering. In very high temperature plants cease to grow, flower production is affected adversely and flower size is also reduced to a great extent. During severe winter, plants and flowers are damaged by frost. Therefore, depending on environment, planting is done in three seasons, *i.e.*, rainy, winter and summer and seeds are sown accordingly. Hence, flowers of marigold can be obtained during larger part of the year.

The environmental conditions after the seedlings are transplanted markedly influence the growth and flowering in both African and French marigold (Table 3). Mild climate during growing period (14.5-28.6°C) greatly improved flowering while higher temperature (26.2-36.4°C) adversely affected flower production (Yadav, 1983). Sowmeya *et al.* (2017) recorded maximum plant spread increased single flower weight, increased flower yield (346.08 g per plant in L_3 hybrid) and higher number of harvests (in all three varieties) during September-October. Moreover, it had been insisted that marigold cultivation under polyhouse was more beneficial than open cultivation for off-season production.

Table 3: Influence of Planting Time on Flower Production in African and French Marigold*

Planting time	African Marigold		French Marigold	
	No. of Flowers per Plant	Weight of Flowers (g) per Plant	No. of Flowers per Plant	Weight of Flowers (g) per Plant
P_1 (15 February)	22.3	162.9	26.3	32.3
P_2 (15 April)	15.6	106.3	23.8	28.9
P_3 (15 June)	18.5	146.6	30.2	37.1
P_4 (15 August)	29.5	225.8	36.8	47.2
P_5 (15 October)	35.0	319.1	37.9	87.3
P_6 (15 December)	26.2	208.6	40.0	56.7
C.D. at 5 per cent	1.00	9.04	2.33	3.29

* Yadav (1983).

9.0 Cultivation

9.1.0 Transplanting of Seedlings

Marigold seedlings are easily transplanted and established in the field without much mortality. At the time of transplanting, they should be stocky and bear 3-4 true leaves. Thin and long seedlings do not make a good plant. Very old seedlings are also not desirable. Yun *et al.* (1996) suggested that 4 weeks old seedlings are optimum for transplanting. Mohanty *et al.* (1993, 1997) recommended that September planting was most profitable. Transplanting should be done in well prepared land and soil is pressed around root zone to avoid air pocket. After transplanting, a light irrigation or watering with rose should be done.

9.2.0 Storage of Seedling

The effects of seedling age (2- to 6-weeks-old) and storage for 1-15 days at room temperature or 7°C on the growth of marigold *cv.* Inca Yellow was investigated by Yun *et al.* (1996). The optimum period for seedling growth was 4 weeks. Storage of seedlings at room temperature had a tendency to decrease the ratio of DW:FW due to overgrowth; seedling quality was maintained following storage at 7°C. Marigold seedlings could be stored for 7 days at room temperature, and for 12 days at 7°C.

9.3.0 Spacing

Proper spacing between plants is required for better development of plant and higher flower yield. Two species of marigold which are commonly cultivated for commercial production of flowers require different spacings. *Tagetes erecta* requires wider spacing than *T. patula*. Experiments conducted by Singh and Arora (1980) and Yadav (1983) revealed that in *cv.* Giant Double African Orange of *T. erecta* flower yield was highest by planting at the distance of 40 × 30 cm.

Kobza (1993) planted *T. erecta*cvs. Clinton, Zitronenprinz and Mesic at 40 × 40 cm, 40 × 30 cm or 40 × 20 cm and *T. patula* cvs. Burpee Melody, Sonnenglanz,

Oranzovy Paprsek and Mars at 30 × 30 cm, 30 × 20 cm, 30 × 15 cm or 30 × 10 cm. Increased plant density by 30 to 100 per cent increased seed yield by 30-40 per cent in *T. erecta*, whereas in *T. patula*, 56 and 200 per cent rise in plant population improved the seed yield by 20 and 35 per cent respectively. In *T. erecta cv.* African Yellow, higher number of flowers per plant was recorded with 30 × 20 cm spacing (Mohanty *et al.*, 1993). Maximum plant height and secondary branches per plant were observed at closer spacing of 50 cm x 40 cm. However, spread, diameter of plant, primary branches per plant, duration of flowering and yield per plant (0.41 kg) were recorded maximum with wider spacing of 50 cm x 50 cm. But, yield per ha was obtained maximum (202.8 q/ha) in closer spacing (Chauhan, 2016).

9.4.0 Irrigation

Marigold is herbaceous plant and makes vegetative growth rapidly in early stages. It takes about 55-60 days to complete vegetative growth and enter into reproductive stage. At this stage terminal flower buds appear which break the apical dominance and side branches start growing. At all stages of vegetative growth and during flower production, sufficient amount of moisture in soil is essential. Moisture stress at any stage of development adversely affects normal growth and flowering. Stressed plants exhibited higher peroxidase activities and also contained higher concentrations of free phenolic compounds, due to enhanced biosynthesis and formation of lignin, than control plants (Kurup *et al.*, 1994).

T. erecta cv. Apollo was grown in a growth chamber for 3 weeks on pine bark filled 3.8 l container and received 12 fertigations. It was applied continuously or cyclically (3 applications each of 1/3 of the total volume with 1 h between applications) methods. Cyclic method increased root growth and dry weight and N content in both plant and roots and reduced 43 per cent leaching loss of total nitrogen compared to continuous fertilization method (Karam *et al.*, 1994). Razin and Omer (1994), reported that watering *T. patula* every 10 days with 0.5 litres of water/pot, produced highest petal dry weight, volatile oil and carotene yields/plant.

Ram *et al.* (1999) studied the effect of irrigation regimes and N fertilization on growth, herb and oil production, consumptive use of water (CUW) and water use efficiency (WUE) of *T. minuta* during winter season. Plant height and number of primary branches/plant were highest when the crop was irrigated at 0.4 IW: CPE (Irrigation water : Cumulative Pan evaporation) and fertilized with 225 kg N/ha. Irrigation at 0.4 IW: CPE and 225 kg N/ha produced the highest herb yield, while the highest oil yield was obtained with irrigation scheduled at 0.2 IW: CPE and fertilized with 150 kg N/ha. The differential response of irrigation and N levels on the herb and oil production was due to decreases in herb oil content with higher levels of either of the factors. At 0.2 IW: CPE and 150 kg N/ha, CUW was 338.5 and 423.6 mm, respectively. The WUE values at the same levels of irrigation and nitrogen fertilization were 0.31 and 0.26 kg oil/mm/ha, respectively. For optimum exploitation of *T. minuta* in the foot hills of Uttar Pradesh, irrigation at 0.2 IW-CPE ratio and N at 162 kg/ha is recommended. Riaz *et al.* (2013) revealed that overall plant quality of varieties, Super Giant and Inca F_1 decreased with the progression of drought stress where 70 per cent field capacity (FC) could be considered appropriate for acceptable plant quality and Inca F_1 performed well compared to Super Giant for all the attributes observed (Table 4).

Table 4: Effect of Drought on Growth of Marigold*

Parameters	*80 per cent FC*		*70 per cent FC*		*60 per cent FC*	
	Super Giant	*Inca F_1*	*Super Giant*	*Inca F_1*	*Super Giant*	*Inca F_1*
Plant height (cm)	4.17	1.90	6.70	4.43	5.70	3.20
Number of leaves per plant	4.33	8.34	3.00	5.00	4.33	3.00
Leaf area (cm^2)	7.33	6.00	12.33	12.33	9.67	4.33
Plant quality	0.40	0.77	0.60	0.23	0.67	0.43
Shoot fresh weight (g)	10.00	11.73	5.53	6.46	3.76	3.90
Shoot dry weight (g)	5.20	6.03	3.76	4.70	2.60	3.40
Root fresh weight (g)	2.56	5.26	2.26	2.63	2.26	1.46
Root dry weight (g)	2.33	3.96	2.03	2.43	2.00	1.33

* Riaz *et al.* (2013).

9.5.0 Manuring and Fertilization

Although marigold responds to fertilization, information on its requirement of macro and micronutrient is scanty. In an investigation on the nutrient uptake by *Tagetes patula cv.* Tina, Hoffman and Komora (1977) reported that uptake by 10 plants was 2.31 g N, 0.87 g P, 7.16 g K, 2.94 g Ca, 0.59 g Mg and 0.35 g Na. The rate of nutrient utilization from soil was 57.5 per cent of available N, 20.1 per cent P, 94.4 per cent K, 8.6 per cent Ca, 23.0 per cent Mg and 6.4 per cent Na. It was observed that fresh and dry weight of plants increased for 70-75 days until flowering and need for N and Mg lasted up to this stage, K was required up to 94 days (until seed formation), Ca and Na up to 108 days (until seed ripening) and P up to the end of seed production. Joiner and Grius (1961) observed that weekly application of the lowest amount (1 lb/100 sqft) of 8-0-8 fertilizer for 4 weeks was sufficient for satisfactory growth and flowering. The effects of N, P and K were also studied by Tsurushima and Date (1977), on nutrient content and dry weight of *Tagetes patula* and it was noticed that omitting N resulted in a low N content without altering the dry weight. By omitting P, both dry weight and P content were reduced and omitting K showed no effect either on dry weight or on K content.

Anuradha *et al.* (1988) found that with the increase in N and P application, plant N, P and K content and uptake increased. In a pot experiment, Yadav and Singh (1997) reported that lower rates of FYM (1 per cent) and 180 ppm N improved growth and flowering of African marigold. The number of days to 50 per cent flowering of potted African marigold was significantly reduced with increasing N and P rates. The number of flowers/plant, plant height, and number of branches/plant were increased up to 240 ppm N, whereas plant spread decreased beyond 180 ppm N. Beyond 180 ppm N, no beneficial effect was observed on flower yield. Increasing P rates significantly improved the growth parameters and yield of marigold (Dahiya *et al.*, 1998). Yadav *et al.* (1999) studied the effect of N and farmyard manure on leaf chlorophyll and nutrient contents of winter season African marigolds at the flower bud initiation stage. Total leaf chlorophyll content increased with N application up to 180 ppm N and remained almost constant beyond this level. Leaf N content increased consistently with N rate but other nutrient concentrations (P, K, Ca and Mg) were highest at the 180 ppm N application rates. Leaf chlorophyll decreased and nutrient content increased with increasing application of FYM. Singh *et al.* (2016) noted significant improvement in flowering attributes *i.e.* early flower bud initiation, 50 per cent flowering, longer duration of flowering, grater yield with the application of 50 per cent RDF + 50 per cent poultry manure.

Table 5: Response of Nitrogen and Phosphorus to Flower Production in African and French Marigold*

Planting time	*African Marigold*		*French Marigold*	
	No. of Flowers per Plant	*Weight of Flowers (g) per Plant*	*No. of Flowers per Plant*	*Weight of Flowers (g) per Plant*
Nitrogen levels				
N_0 (0 kg/ha)	19.6	153.0	30.2	44.1
N_1 (100 kg/ha)	24.9	194.0	34.6	52.4
N_2 (200 kg/ha)	30.4	262.0	42.7	68.1
N_3 (300 kg/ha)	37.2	342.1	48.4	84.1
N_4 (400 kg/ha)	37.7	349.0	50.7	91.6
C.D. at 5 per cent	1.70	16.14	2.02	4.31
Phosphorus levels				
P_0 (0 kg/ha)	28.1	235.6	38.3	58.8
P_1 (100 kg/ha)	30.9	270.4	40.4	69.8
P_2 (200 kg/ha)	30.8	271.2	43.0	74.9
P_3 (400 kg/ha)	29.9	262.9	41.7	68.7
C.D. at 5 per cent	1.52	14.44	3.55	3.85

* Yadav (1983).

Experiments were conducted at Punjab Agricultural University, Ludhiana for two years to find the optimum rates of N and P for marigold (*T. erecta*) *cv.* Giant Double African Orange and to study the effects on growth and flower production (Arora and Singh, 1980b). With the increase of N from 0 to 40 g/m^2, the number of flowers per plant and flower yield (weight) continued to increase whereas plant height responded only up to 30 g/m^2. The response to P was inconsistent on growth and flower production. In both *T. erecta* and *T. patula* the maximum plant growth and flower yield with the application of 400 kg N and 200 kg P per hectare were recorded by Yadav (1983) (Table 5). Watanabe *et al.* (1984) reported that growth rate in French marigold increased steadily with increasing P concentration from 1.1 to 11.3 ppm. The number of flower buds and florets increased in proportion to P levels and P concentration increased with plant growth rate. Pradhan *et al.* (2015) stated that photosynthetic pigments (1.594, 0.788, 2.382 and 0.789 mg/g of Chlrophyll-a, b, total chlorophyll and carotenoid, respectively) increased when 10 per cent fly ash was amended with soil on 40th day of growth.

Arulmozhiyan and Pappaiah (1989) reported that application of 120 kg/ha nitrogen along with foliar application of 1000 ppm ascorbic acid 30 days after transplanting on *T. erecta* was beneficial. Anuradha *et al.* (1990) obtained increased number of flowers/plant and weight of individual flowers of *T. erecta* with the application of 60 kg N in combination with 90 kg P_2O_5 and 60 kg K_2O/ha (Table 6). Hameed and Sekar (1999) recorded early flowering and maximum flower diameter, single flower weight and yield per plant when *T. erecta cv.* Salem and Dindi-gul were treated with 150 kg N/ha + 120 kg P_2O_5/ha in 2 split doses (basal and top-dressing). Durga *et al.* (2017) indicated that 0.46 kg N, 0.07 kg P and 0.49 kg K were required to produce 1 quintal of marigold flowers in inceptisols of New Delhi. Custic and Poljak (1994) stated that two formulations of liquid fertilizers fertina-1 (N:P:K:Mg at 4:6:8:2 + B, Cu, Fe, Zn) and fertina-2 (N:P:K:Mg at 10:5:4:2 + B, Cu, Fe, Zn, Mn) were very useful for *T. patula*. Fertina-1 at 1.5 and 6 per cent and fertina-2 at 0.5, 1.0 and 1.5 per cent produced significantly more flowers/plant in comparison with control. Plants (*T. erecta*) treated with azotobacter + phosphorus solubilizing bacteria (PSB) + 75 per cent N gave the highest flower yield per ha (58.46 q/ha) which was significantly higher than the control (19.23 q/ha) (Chandrikapure*et al.*, 1999; Gupta *et al.*, 1999a). Liquid feeding of nitrogen, phosphorous and potassium at the rate of 2 per cent each at 15, 30 and 45 days after transplanting resulted in floriferous and productive crop yield in African marigold *cv.* Double African Yellow (Narayan, 2018).

Table 6: Effect of N and P_2O_5 on the Yield and Attributing Characters in Marigold (*T. erecta*)*

Treatments	*No. of Days for 50 per cent Flowering*	*No. of Flowers per Plant*	*Weight of Single Flower (g)*	*Size of Flower (cm)*	*Yield of Flowers (q/ha)*
Nitrogen levels					
N_0 (0 kg/ha)	56.3	15.8	6.2	5.7	81.9
N_1 (30 kg/ha)	46.2	26.0	8.0	5.8	87.2
N_2 (60 kg/ha)	40.3	39.1	8.9	6.7	96.9
N_3 (90 kg/ha)	35.2	50.9	11.9	7.0	130.3
C.D. at 5 per cent	1.9	1.9	NS	0.3	1.3
Phosphorus levels					
P_0 (0 kg/ha)	51.7	28.2	7.0	5.7	99.5
P_1 (30 kg/ha)	47.4	31.4	7.8	6.1	97.6
P_2 (60 kg/ha)	41.4	34.1	9.1	6.3	102.6
P_3 (90 kg/ha)	37.6	38.0	11.3	6.9	107.6
C.D. at 5 per cent	1.8	1.7	NS	0.3	1.3
C.D. at 5 per cent N × P	9.2	NS	NS	NS	0.7

* Anuradha *et al.* (1990).

Ezawa and Yoshida (1994) studied the effect of phosphorus on vesicular-arbuscular mycorrhizal fungus on *T. patula cv.* Bonanza and found that mycorrhizal plants without P-fertilization increased shoot fresh weight and higher ISPase (infection-specific phosphatase) activity. Suzuki *et al.* (2000) studied the effect of arbuscular mycorrhizal (AM) colonization on the uptake of trace elements in marigold (*T. patula*) using a multitracer consisting of radionuclides of 7Be, 22Na, 46Sc, 51 Cr, 54 Mn, 59 Fe, 56Co, 65Zn, 74As, 75Se, 83Rb, 85Sr, 88Y, 88Zr and 95mTc. The uptake of 7Be, 22Na, 41Cr, 59Fe, 65Zn and 95mTc was higher in the mycorrhizal marigolds than in the non-mycorrhizal ones, while that of 46Sc, 56Co, 83Rb and 85Sr was lower in the mycorrhizal marigolds than in the non-mycorrhizal ones. Dhami *et al.* (2015) stated that *Azotobacter* + Phosphorous solubilizing bacteria + vermicompost recorded maximum available nitrogen, phosphorous and potassium (406.00, 38.75 and 260.33 kg/ha during flowering and 380.66, 34.31 and 198.33 kg/ha during harvesting, respectively). Moreover, maximum organic carbon content in soil during flowering (1.83 per cent) and after harvesting (1.33 per cent) was also observed with the same treatment. Ramalakshmi and Bharathiraja (2015) reported that inoculation of AM fungi + phosphobacteria + 100 per cent N and K along with 75 per cent P significantly enhanced growth and yield of marigold than with 50 per cent P.

Onsley *et al.* (1994) suggested that *Trichoderma* strains (1 per cent w/v) inoculated peat/girt compost may be used for *T. patula cv.* Orange Winner cultivation. Each strain increased either number or weight of flowers and also dry matter content of shoots.

Application of sea weed (*Ecklonia maxima*) concentrate kalpak at 1 ml/plant on *T. patula cv.* Jaine as foliar spray or root drench at transplanting, improved both vegetative and reproductive growth and also increased seed production up to 50 per cent (Staden *et al.*, 1994). Application of poultry manure (PM) + chemical fertilizers (CF) - 100: 60: 60 kg/ha resulted in superior growth, chlorophyll-a (45 per cent increase), chlorophyll-b (35 per cent increase) and oil yield in marigold. Thus, the nitrogen fertilization and organic amendments linearly correlated with biomass and oil yield. Maximum nitrogen uptake was found in CF (28 per cent) followed by PM + CF (22 per cent). Antioxidant activity (DPPH) was higher (14 per cent) in FYM + CF and PM + CF than control (Pandey *et al.*, 2015). Application of vermicompost to marigold grown on podzolic soil produced stronger sprouts, tall plants with greater stem diameter, flower diameter and yield than control (Kovshov and Iconnicov, 2017), while use of 100 per cent sewage sludge showed superior yield in *T. erecta* (Solanki *et al.*, 2017). Khaleghi *et al.* (2018) revealed that root length and proline content increased with higher level of drought stress. The maximum proline content was observed with 15 day irrigation interval without the application of humic acid, whereas, the minimum proline content was recorded with 5-day irrigation interval along with 250 g/l humic acid. Though drought stress decreased the number of leaves, flowers, chlorophyll, carotenoid and calcium content in plants, application of higher levels of humic acid (250 mg/l) overcame the effect of drought stress. Thus, it was concluded that humic acid at 250 mg/l not only reduced the negative effect of drought stress, but also improved the physiological and morphological characteristics.

T. erecta cvs. Discovery and Inca, *T. patula cv.* Sophia and *T. erecta* × *T. patula* cv. Nugget Supreme grown in a soilless medium amended with phosphorus charged alumina had adequate nutrition and found similar or superior plant height, branch numbers, flower production, fresh and dry weight of shoot in comparison with plants grown in commercial fertilizer (Lin *et al.*, 1996). Borch *et al.* (1998) also reported that in *T. patula cv.* Janie Tangerine, with reduced P using Al-P, P leaching was greatly reduced and plant quality was improved. Al-P treated plants had more flowers and fewer wilted flowers. Roots were confined to a small volume beneath the drip tube in control plants, while roots of Al-P plants were well distributed through the medium. Pal and Ghosh (2010) reported that sources of potassium (MOP and Potassium sulphate) failed to influence the plant characters, whereas, different levels of potassium showed positive response. Yield of flowers increased with increasing quantity of potassium from 0 to 200. Furthermore, plant height, number of branches, plant spread, flower size, nutrient and chlorophyll content of leaf and anthocyanin content in petals were directly related to potassium level (Table 7).

Table 7: Effect of different Sources and Levels of Potassium on Vegetative and Flowering Traits of African Marigold *cv.* Siracole*

Parameters	*Sources*				*Levels (kg/ha)*						
	MOP	*K_2SO_4*	*S.Em*	*C.D (5 per cent)*	*0*	*50*	*100*	*150*	*200*	*S.E.m*	*C.D (5 per cent)*
Plant height (cm)	57.13	57.10	0.91	NS	51.58	54.74	57.90	59.39	61.97	1.60	4.97
Primary branches/ plant	5.07	5.13	0.04	NS	4.23	4.75	5.17	5.40	5.65	0.14	0.43
Secondary branches/ plant	18.67	19.15	0.10	NS	15.20	16.83	19.25	20.43	22.25	0.22	0.64
Plant spread (cm)	50.29	51.77	0.62	NS	41.95	47.84	51.98	55.52	57.85	1.58	4.73
Days to flower bud emergence	46.79	46.99	0.67	NS	43.25	45.96	47.16	48.27	49.80	0.45	1.35
Flower diameter (cm)	5.46	5.48	0.02	NS	4.60	5.49	5.62	5.74	5.84	0.09	0.27
Weight of individual flower (g)	5.09	5.13	0.01	NS	4.43	4.91	5.22	5.43	5.56	0.07	0.23
Flower yield (kg/5.76 m^2)	19.84	19.96	0.22	NS	13.64	17.96	21.25	22.72	23.94	0.53	1.59

* Pal and Ghosh (2010).

Containerized *T. patula cv.* Orange Boy plants were fed with nutrient solution containing 0.25, 0.5, 1, 2, 3, 4, 5, or 6 mM B, Cu, Fe, Mn, Mo or Zn and toxicity development was monitored (Lee *et al.*, 1996). The threshold micronutrient concentrations that induced visible foliar toxicity symptoms were 0.5 mM B, 4 mM Cu, 4 mM Fe, 2 mM Mn, 1 mM Mo and 5 mM Zn.

Iron deficiency stress influences physiology of iron acquisition in *T. erecta cv.* First Lady (Albano and Miller, 1996). Albano *et al.* (1996) also reported specific physiological disorder, bronze speckle in *T. erecta* cultivars First Lady and Voyager by nutrient solutions containing Fe-DTPA concentrations greater than 0.018 mM (1 ppm) applied to a soilless medium. The disorder was characterized visually by speckled patterns of chlorosis and necrosis, and downward curling and cupping of leaves. The percentage of total leaf DW affected with symptoms generally increased with increasing Fe-DTPA concentration. Symptomatic leaf tissue had a greater Fe concentration than corresponding asymptomatic leaf tissue. Albano and Miller (1998) observed that symptoms of Fe toxicity (bronze speckle disorder in marigold characterized by chlorotic and necrotic speckling and downward leaf cupping and curling) developed only in the high-Fe treatment.

Table 8: Percentage of Plants Affected with Bronze Speckle, and Symptom Severity Rating for the High-Fe (0.36 mmol/l Fe-DTPA) Treatment, Plant Height and Leaf Dry Weight as an Average of both the High-Fe and Low-Fe (0.018 mmol/l Fe-DTPA) Treatments, and Total Leaf Fe and Mn Concentration for Seven Marigold Cultivars as Affected by Fe-DTPA Concentration in the Nutrient Solution*

Cultivar	*Affected Plants (per cent)^z*	*Symptom Severity^x*	*Height (cm)*	*Leaf DW (g)*	*Fe (µg/g)*		*Mn (µg/g)*	
					Low	*High*	*Low*	*High*
Orange Jubilee	99 a^y	3.5 a	31.8 a	1.12 a	120d	666 b***	95 c	98b^{NS}
Gold Lady	54 bc	2.1 cd	14.5 b	0.94 b	286a	828 a***	138 b	139a^{NS}
Yellow Galore	61 b	2.2 c	14.4 b	0.89 b	157cd	364 c***	89 d	88 b^{NS}
Orange Lady	71 b	2.5 bc	14.3 bc	0.92 b	229b	753ab***	151a	144a^{NS}
Marvel Gold	40 c	1.9 d	13.3 bc	0.88 b	180c	548 c***	103 c	127a**
First Lady	71 b	2.6 b	13.1 c	0.75 c	255ab	679 b***	137 b	131a^{NS}
Discovery Orange	24 d	1.4 d	7.0 d	0.57 d	241b	406 d**	157 a	132a**

z_n = 72 plants per cultivar. yMean separation within columns by LSD, P < 0.05.

x_1: Normal, 2: Chlorotic speckling, 3: Chlorotic and necrotic speckling, 4: Chlorotic and necrotic speckling, and downward leaf cupping and curling.

NS,*,**,***Non-significant or significantly different from low Fe-DTPA concentration at P = 0.05, 0.01 and 0.001, respectively, by pairwise *t* test.

* Albano and Miller (1998).

The concentration of Fe in leaves in the high-Fe treatment was 5.6 and 1.7 times as great as in the low-Fe treatment for Orange Jubilee and Discovery Orange, respectively. Based upon the percentage of plants affected and leaf symptom severity, relative cultivar susceptibility to Fe toxicity was Orange Jubilee > First Lady > Orange Lady > Yellow Galore > Gold Lady > Marvel Gold > Discovery Orange (Table 8). Snehitha *et al.* (2019) revealed that application of 100 per cent RDF in the form of water soluble fertilizers through drip irrigation increased the number of flowers, flower surface area, fresh weight of flowers, number of pickings and flower yield compared to control (no fertilizers), conventional fertilizers and 75 and 50 per cent RDF through water soluble fertilizers.

Sridhar and Rengasamy (2010) reported that 1.0 per cent seaweed liquid fertilizer of *Ulva lactuca* along with 50 per cent of recommended fertilizers showed maximum growth characteristic, number and fresh weight of flowers (Table 9).

Table 9: Effect of different Proportions of Seaweed Liquid Fertilizer and Chemical Fertilizers on Growth and Flowering Characteristics of African Marigold*

Treatment	*Plant Height (cm)*	*Shoot Height (cm)*	*Root Height (cm)*	*Total Fresh Weight (g)*	*Total Dry Weight (g)*	*No. of Branches*	*No. of Flowers/ 10 Plants a Row*	*Fresh weight of Flowers/ 10 Plants/ Row (kg)*
100 per cent RDF	34.84 ± 1.82	29.40 ± 1.49	5.44 ± 0.51	72.96 ± 3.84	30.88 ± 3.05	3.40 ± 0.54	118	1.95
75 per cent RDF + 1.0 per cent SLF	37.28 ± 1.59	31.64 ± 1.31	5.64 ± 0.48	78.10 ± 3.35	33.05 ± 2.67	4.20 ± 0.83	125	1.76
50 per cent RDF + 1.0 per cent SLF	41.80 ± 0.60	34.86 ± 0.55	6.94 ± 0.58	87.60 ± 1.26	37.07 ± 0.97	4.60 ± 0.54	142	2.60
25 per cent RDF + 1.0 per cent SLF	35.38 ± 1.33	20.58 ± 1.20	5.80 ± 0.40	74.14 ± 2.85	31.38 ± 2.27	3.80 ± 0.33	137	1.93
1.0 per cent SLF only	34.22 ± 1.14	28.72 ± 1.30	5.50 ± 0.25	71.66 ± 2.43	30.33 ± 1.95	2.80 ± 0.83	120	1.69
Water only	29.80 ± 1.45	24.90 ± 1.27	4.90 ± 0.27	62.40 ± 3.10	26.41 ± 2.45	2.40 ± 0.54	90	1.25

* Sridhar and Rengasamy (2010).

9.6.0 Weed Control

In marigold production, control of weeds is an important operation. If the weeds are not removed in time, a great loss would occur in terms of growth and productivity of marigold. In India, 3-4 manual weeding are required for the entire period. Chemical weed control in marigold has also been recommended by various workers. Haramaki and his co-workers (1962, 1965, 1966, 1968) tried various herbicides and recorded satisfactory weed control by using EPTC at 8-10 lb, triflurabin and diphena-mid at 4 lb per acre. Schubert (1968) found the use of amiben at 4 lb, propachlor and diphenamid at 6 lb/acre as safe and satisfactory in controlling weeds of marigold. Ashley (1971) obtained full-season weed control in marigold with alachlor at 3 lb/acre and chloroxuron at 4-8 lb/acre. In *T. patula* field, Hurt and Vencill (1994a) reported that foliar application of pursuit (imazethapyr) at 560 g/ha successfully controlled purple nutsedge (*Cyperus rotundus*) and yellow nutsedge (*C. esculentus*). Application of manage (halosulfuron) alone and with a surfactant (X-77) at 9 and 18 g/ha controlled 87-91 per cent purple nutsedge and 79-84 per cent yellow nutsedge (Hurt and Vencill, 1994b). Four weeks after treatment crops exhibited >10 per cent injury with manage.

9.7.0 Pinching

In tall cultivars of *Tagetes erecta*, development of axillary branches and flower production are influenced by the presence of apical dominance. It has been observed that such cultivars first grow upward to their final height and later on produce terminal flower buds. After the formation of terminal flower bud, axillary branches develop which also bear flowers. However, if the apical portion of shoot is removed early, large number of axillary shoots arises resulting in well-shaped bushy plants bearing more number of uniform flowers. Experiment on pinching revealed that removal of shoot apices 40 days after transplanting enhanced the flower yield, late pinching at 50 or 60 days proved less effective in this respect (Singh and Arora, 1980).

10.0 Diseases and Pests

10.1.0 Diseases

Fungal Diseases

10.1.1 Damping Off

It is caused by *Rhizoctonia solani* and appears as brown necrotic spots girdling the radicle, later on extends to plumule and causes pre-emergence mortality. Post-emergence symptoms appear on lower part of hypocotyl as water soaked brown, necrotic ring leading to collapse of seedling. When infected seedlings are pulled, the root system appears partially or fully decayed (Subramanyam *et al.*, 1975).

10.1.2 Collar Rot

This disease is caused by a number of pathogens and the common ones are *Rhizoctonia solani*, *Phytophthora* sp., *Pythium* sp. and *Sclerotium rolfsii*. These pathogens cause collar rot in nursery stage or in grown up plants depending upon soil type, moisture conditions and other factors (Sohi, 1984).

10.1.3 Leaf Spot and Blight

Various species of *Alternaria*, *Cercospora* and *Septoria* are known to cause leaf spot or blight of marigold. *Alternaria tagetica* and *A. tenuissima* damage the foliage. The symptoms appear as minute, brown, circular spots which enlarge at later stage of infection (Choudhury, 1969; Some and Mustaffee, 1966; Sohi, 1984). Indofil M-45 (mancozeb) (0.25 per cent) significantly controlled the disease (Mazumdar, 2000). Bulb extract of garlic was the most effective treatment against *Alternaria tenuissima* on *Tagetes* sp., followed by mint and datura extract (Barnwal *et al.*, 1998). Luo *et al.* (2018) reported *A. jacinthicola* causing leaf spot disease on *T. erecta* in China for the first time.

Alternaria tagetica was isolated for the first time from seeds of diseased marigold (*T. erecta*) in Taiwan. Seed-borne *A. tagetica*, treating with *Bacillus azotoformanis* (1 × 109 CFU/ml) was found effective and mancozeb highly effective (Wu *et al.*, 2001). Mamta *et al.* (2016) revealed that tebuconazole and propicanozole (1 mg/l) resulted in 98.89 and 97.78 per cent mycelia growth inhibition followed by mancozeb, copper oxychloride and chlorothalonil. Also, fungicide application along with seed treatment of tebuconazole (0.1 per cent) and propiconazole (0.1 per cent) reduced disease infection up to 12.84 and 15.71 per cent on leaves, 8.65 and 10.27 per cent on flowers under field condition.

Three species of *Cercospora, i.e. C. tagetis, C. tagetiserecta* and *C. tageticola* infect various species of *Tagetes.* The lesions form on lamina and scatter to coalesce irregularly. They are blackish to brownish grey on lower surface, pale tan on upper surface. Stroma are many; olivaceous brown, small globular to irregular (Mitra, 1935; Thirmulachar and Govindu, 1956; Kar and Mandal, 1973).

Septoria tageticola also infects the foliage of marigold and is characterized by smoky grey to black spots. At later stage, black, minute pycindia develop on necrotic lesions. It was observed that cultivars of *T. erecta* were more susceptible while those of *T. patula* were resistant or immune (Sohi, 1984). Septoria leaf spot on marigold was first reported from Himachal Pradesh by Shukla and Thakur (2018). The initial symptoms were small, water-soaked circular spots on older leaves which gradually turned to gray, tan and dark brown margins.

The blight of marigold (*T. erecta*) caused by *Colletotrichum capsici* has been reported from India (Saksena and Singh, 1959). The infection appears on petiole, peduncle and branches.

10.1.4 Inflorescence Blight

This disease is caused by *Alternaria zinnae*. Elongated lesions are formed on inflorescence. Symptoms may appear as large irregular blotches on the leaves, which are light tan to dark brown with zonations (Edward, 1957; Sohi, 1984). This can effectively be controlled by dithane sprays (Edward, 1957).

10.1.5 Flower Bud Rot

It is caused by *Alternaria dianthi*. The disease appears mainly on young flower buds and results in dry rotting of the buds with brown scorched, necrotic discolouration of sepals and peduncles. The ray and disc florets also turn brown. At later stages, buds are shrivelled, turn deep brown and dry up. Symptoms are less prominent on matured buds but these buds also fail to open. On older leaves few deep brown necrotic spots develop (Mondal and Chaudhury, 1976).

10.1.6 Powdery Mildew

Oidium sp. and *Leveillulat aurica* (*Oidiopsis taurica*) have been reported to cause powdery mildew of marigold. First, whitish, tiny, superficial spots appear on leaves, later on the whole aerial parts of plant is covered with whitish powder, *i.e.*, conidia and conidiophores of the pathogens (Grover, 1952; Sreevamulu, 1953).

10.1.7 Rust

In California, African marigold has been reported to be infected by rust caused by *Coleosporium madiae*. Another rust fungus, *Puccinia tageticola* also occurs on *Tagetes* (Pirone *et al.*, 1960). Removal and destruction of infected plants have been suggested.

10.1.8 Gray Mould

Gray mould of *Tagetes erecta* caused by *Botrytis cinerea* has been recorded in China, Australia, and the U.S.A and recently in Korea. According to the field observations carried out by Aktaruzzaman *et al.* (2018), gray mould was active at low temperatures (16-24°C) and lesions expanded rapidly in humid conditions. Water-soaked, brown or gray spots followed by abundant mycelium with conidia occurred on the affected flower heads which reduced the aesthetic value of flowers. Leaves and stems were not affected.

10.1.9 White Mould

Rahman *et al.* (2015) first reported white mould on marigold caused by *Scletotinia sclerotiorum* with the symptoms initiating on petals as fluffy white mycelia, expanding to whole flower and lower part of the plant. Dark brown lesions with necrotic tissues appeared on infected leaves and stems. Seriously infected plants showed flower dropping and branch wilting.

Other fungal diseases reported on marigold include charcoal rot caused by *Macrophoma phaseoli* and wilt caused by *Verticillium alboatrum*.

Viral Diseases

Marigold is infected by cucumber mosaic virus (*Marmor cucumeris*). Aster yellow virus is also common. Rouging of diseased plants and spraying with malathion at frequent intervals to control aphids and leaf hoppers which spread the disease, are suggested (Pirone *et al.*, 1960; Singh *et al.*, 1999).

Alp *et al.* (2016) reported naturally occurring 'Candidatus phytoplasma Solani' as alternate host for the first-time affecting *T. erecta* with the symptoms of shoot proliferation, dwarfing and reddening in ornamental gardens of Van Province (Turkey).

10.2.0 Pests

10.2.1 Red Spider Mite (*Tetranychus* sp.)

These mites sometime appear near flowering time. Plants give dusty appearance. It can be controlled by spraying kelthane.

10.2.2 Hairy Caterpillar (*Diacrisia obliqua*)

It is a polyphagus insect and the caterpillar eats away foliage. This pest can be controlled by nuvon or thiodan at 1 ml/l of water.

Diseases and Pests of Marigold

Damping-off
Stem Rot
Leaf Spot
Powdery Mildew
Flower Blight
Gray Molds
Impatiens Necrotic Spot Virus (INSV)
Aster Yellow
Spider Mite
Japanese Beetles
Thrips
Leaf Miner

Diseases and Pests of Marigold

Leaf Hopper Thrips Aphids

Mealy Bugs Slug Caterpillar

Source: www.pnwhandbooks.org; www.ijcmas.com; www.forestryimages.org; www.ag.umass.edu; www.forestryimages.org; www.britannica.com; www.invasive.org; www.hgtv.com; www.nzacfactsheets.landcareresearch.co.nz; www.ag.umass.edu;www.no-dig-vegetablegarden.com; www.extension.unh.edu; www.agritech.tnau.ac.in

Harvesting and Postharvest Management

Harvesting Stage Field Harvesting Transport in Crates

Packing in Sac Packing in Gunnybag In Market

Source: www.freepik.com, www.communityfarm.in

10.2.3 Japanese Beetle (*Popillia japonica*)

It is a common pest on marigold and feeds on flowers and upper leaves.

10.2.4 Leaf Hopper (*Empoasca fabae* and *Macrosteles fascifrons*)

The symptoms of infestation are cupping or rolling of leaves, wilting of shoot tips and leaflets. French marigold is more susceptible than the African. Spraying with malathion controls this pest.

10.2.5 Tarnished Plant Bug (*Lygas lineolaris*)

Marigold is also attacked by this insect causing distortion of flowers and leaves. Malathion has proved effective to control the pest.

10.2.6 Thrips

Neohydatothrips samayunkur is recorded for the first time from Panama, Brazil on *Tagetes patula* (Monteiro *et al.*, 1999).

10.2.7 Aphids

Stoltz *et al.* (1997) observed aphid (*Myzus persicae, Macrosiphum euphorbiae* and *Aphis gossypii*) infestation in *T. erecta* (1.6 per cent) in Alaska. They concluded that increased seed production in this area would require aphid management strategies.

10.2.8 Mealy Bugs

Pattanaik and Misra (2019) observed complete control (100 per cent) of mealy bug in marigold even after 15 days of spraying thiamethozam 0.5 g/l whereas, propargite 50 per cent + Bifenthrin 5 per cent, tolfenpyrad and bifenthrin were next effective treatments in reducing the number of mealy bugs (6.33-8.67).

10.2.9 Slugs

Limax maximum and *Milaxgagates* feed on the leaves of marigold. Infected areas should be dusted with 15 per cent metaldehyde.

Plusia orichalcea (*Thysanoplusia orichalcea*) is reported for the first time from marigolds (*Tagetes* spp.) in India. Larvae were found feeding on leaves and shoots during August and September (Ali *et al.*, 1998). Weaver *et al.* (1997) reported that floral extract of Mexican marigold (*T. minuta*) was toxic to adult male and female *Sitophilus zeamais*, under illuminated and dark conditions. The original extract and certain fractions probably contained photoactive thiophenes, which are toxic to certain larval *Lepidoptera*.

10.2.10 Leaf Miners

In California, under greenhouse condition, Heinz and Parella (1990) controlled *Liriomyza trifolle* population with the help of parasitic wasp *Digliphus leginii* in African marigold. They also controlled the population of *Lepidoptera* species by spot application of *Bacillus thuriengiensin*.

11.0 Harvesting and Postharvest Management

11.1.0 Harvesting

Marigold flowers are plucked by hand when they have attained the full size. Plucking of flowers should be done in cool hours of the day either in the morning or evening. Field should be irrigated before plucking so that flowers keep well for longer period after harvest. Flower stalk is hollow which breaks off easily when twisted between thumb and fingers. Productivity of plants increases considerably by regular plucking of flowers. Willoughby *et al.* (2000) developed a mechanical harvester for marigold flowers. Two different header systems were tried. One uses extendable fingers mounted in a drum and the other uses solid-mounted fingers to pluck the flower from the plant. Marigold flowers are mainly used for making garlands and, hence, plucked flowers are collected in polythene bags or bamboo baskets for carrying to market.

11.2.0 Yield

The yield of flowers in African and French marigold not only varies with cultivar but cultural practices including planting, spacing and fertilization showed a marked variation in flower number from 6 to 8 millions and from 1.5 to 2.5 millions/ha in French and African marigold, respectively, the corresponding ranges of variation in the weight of flowers were 8 to 12 tonnes/ha in French marigold and 11 to 18 tonnes in African marigold (Yadav, 1983). Singh and Singh (1997-98) evaluated the yield of 4 varieties each of African and French marigolds. Yield of flowers/plant as well as yield/ha were highest for Giant Double African (lemon). Average yields for African and French marigolds were 164.889 and 130.278 q/ha, respectively.

11.2.1 Herbage and Oil Yield

The application of nitrogen at 200 kg/ha, irrigation at 0.5 IW : CPE ratio (irrigation water : cumulative pan evaporation) and autumn planting produced significantly higher herbage and oil yields of South American marigold (*T. minuta*). Cis-beta-ocimene and cis-tagetenone concentrations were higher in the main crop and concentrations of limonene, dihydrotagetenone and cis-tagetenone were higher in the ratoon crops (Singh, 2001).

11.2.2 Yield of Carotenoid

Kopsell *et al.* (2016) studied the impact of genotype and light on carotenoid content in *T. tenuifolia*. Cultivars 'Lemon Gem', 'Tangerine Gem' and 'Red Gem' were cultivated in nutrient solution under greenhouse. Results revealed that 'Red Gem' produced maximum concentration of α-carotene, β-carotene, lutein, zeaxanthin, antheraxanthin, violaxanthin and neoxanthin. The petal carotenoid content was found increasing when 'Tangerine Gem' and 'Lemon Gem' were grown under LED light and 20 per cent blue light.

Fresh marigold treated with enzymes showed a higher susceptibility to pigment extraction than untreated samples, and the highest carotenoid yields were obtained using the enzyme Econase-cep. This enzyme at 0.1 per cent w/w increased extraction from 1.7 to 7.4 g/kg of marigold flower in dry weight; such treatments enhanced carotenoid extraction at the industrial level as well. Total carotenoids ranged from 11.4 to 17.4 g/kg and 18 to 24.7 g/kg of control and treated marigold meal, respectively. The concentration of carotenoids was highest when a 5 per cent level of treated meal was used (Delgado-Vargas and Paredes-Lopez, 1997 a,b). Vandekar *et al.* (2015) reported ethanol as best solvent on sonicator, 1:20 ratio of natural source: solvent at 80 and 100°C as optimum on heating solvent, 50°C on ultrasound probe, 320 watt on microwave reactor and 1.5 h as optimum extraction time on soxhlet apparatus. Solar drying was found economical which resulted in maximum moisture loss and retention of anti-oxidant activity, β-carotene, fibre, ash and protein content (Ahluwali *et al.*, 2018). Fu *et al.* (2018) obtained 71.14 per cent higher yield of lutein with microwave and enzyme co-assisted aqueous two-phase extraction (MEAATPE) technique over enzyme assisted ATPE, microwave assisted ATPE and Soxhlet extraction methods. Indrawati *et al.* (2019) insisted that the extraction of lutein using vegetable oil was 1.9 times more than that of with acetone. Moreover, the encapsulated lutein incorporated into milk-tea was highly liked by the sensory panelists.

11.3.0 Postharvest Management

11.3.1 Flower Storage

Jadhav (2018) reported that flowers kept inside cold storage after harvesting showed minimum weight loss, shrivelling percentage, rotting percentage, better appearance and brighter colour than at room conditions. Moreover, flowers stored in black polythene bags (BPB) in cold storage had greater retention and lesser shrivelling/rotting with better quality compared to air-stored flowers. BPBs inside cold storage had a maximum shelf life of 8 days which was only 3.5 days in case of ordinary room storage. Marigold petals dried through microwave assisted convective dryer (MACD), packed with polyethylene lined aluminium foil and stored at room temperature maintained good retention of lutein and total antioxidant activity upto 75 days of storage (Duhan *et al.*, 2016).

11.3.2 Seed Storage

T. erecta seeds (7 ± 0.5 per cent moisture content) treated with thiram + cytozyme resulted in the highest vigour and viability after 8 months storage is sealed polythene bag. Seeds could be stored for up to 6 months in cloth bags but for a longer period in polyethylene bags.

12.0 Biotechnology

Phytoplasmas associated with Mexican marigold (*T. erecta*) phyllody proliferated on micropropagated plants grown in MS medium supplemented with different concentrations of growth regulators kinetin (1.0 and 1.5 mm) and indole-3-acetic acid (1.0 and 1.5 mm) (Rojas-Martinez and Villegas, 1999).

12.1.0 Somatic Embryogenesis

Embryogenic callus and somatic embryos were induced from cotyledonary explants of *Tagetes erecta* (Bespalhok and Hattori, 1998). Cotyledons were first cultured on MS medium supplemented with 2.0 mg 2,4-D and 0.2 mg kinetin/l. Friable embryogenic callus developed when the compact embryogenic callus was transferred to medium containing 2,4-D and subcultured every 2 weeks. Somatic embryo development was obtained when friable embryogenic callus was transferred to MS medium supplemented with 3 mg ABA and 60 g sucrose/l. The addition of 10-30 mM glutamine improved somatic embryo development.

12.2.0 Molecular Markers

A marigold flower cDNA library was constructed and two cDNA (*hmg1* and *hmg2*) encoding 3-hydroxy-3-methylglutaryl coenzyme A reductase (hydroxymethyl-glutaryl-CoA reductase) (*HMGR*) were isolated and sequenced (EMBL/GenBank accession numbers AF034761 and AF034760, respectively). The *hmg1* cDNA encodes *HMGR1*, an enzyme consisting of 574 amino acid residues with a putative molecular weight of 61.9 Ku, while *hmg2* encodes a truncated form of the enzyme lacking the 90 amino acid residues present at the C-terminus of *HMGR1*. The two cDNAs have identical 3′- and 5′-untranslated regions. Southern blot analysis revealed the presence of 3 or more genomic sequences of marigold showing homology to the *hmg1* cDNA, while northern blot analysis detected at least 2 size patterns of *HMGR mRNAs* in flowers, leaves and roots (Villar-Martinez *et al.*, 1999).

Characterization of 22 marigold genotypes was carried out by Panwar *et al.* (2017) with 26 RAPD markers which amplified 241 amplicons, out of which 238 amplicons were polymorphic (98.8 per cent) at a rate of 9.27 amplicon per primer. Jaccard's similarity coefficient varied from 0.18 to 0.80 which suggested wide genetic base among the genotypes. The polymorphic information content (PIC), resolving power (RP) and marker index (MI) of RAPD primers were 0.38, 8.36 and 3.44, respectively.

Panwar *et al.* (2018) assessed the genetic diversity among 22 genotypes of marigold using 19 Inter Simple Sequence Repeats (ISSR) and 12 Universal Rice Primers (URP) markers. Out of 184 amplicons amplified with ISSR primers, 171 (92.73 per cent) were found polymorphic with PIC ranging between 0.23 and 0.47 and an average of 0.34. Moreover, the RP was observed between 7.0 and 16.0 with 11.20 being average, whereas, the average MI was 2.93 ranging between 0.77 and 4.83. In case of URP markers, polymorphism of 93.08 per cent (*i.e.* 122/131), average PIC content of 0.32 (0.17 to 0.39), average RP of 13.75 (8.82 to 22.45) and average MI of 3.17 (0.55 to 5.87) was observed. Further, the Jaccard's similarity coefficient was found in the range of 0.33-0.81 in ISSR markers and 0.33-0.86 for URP markers thus insisting the presence of wide genetic divergence among the studied marigold genotypes. UPGMA cluster method grouped the 22 genotypes into 2 clusters at 0.44 similarity coefficient in both the markers in which the two male sterile lines formed a distinct group.

12.3.0 Production of Secondary Metabolites

Tagetes spp. was excellent source of thiophenes which could control nematodes, insects and pathogens. Marigold hairy root cultures developed through *Agrobacterium rhizogenes* mediated transformation produced 4 types of thiophenes *viz.*, butenenyl bithiophene (BBT), hydroxyl butenenyl bithiophene (BBTOH), acetoxy butenenyl bithiophene (BBTOAc) and α-terthienyl (α-T) with the total thiophene content (0.31 – 0.96 mg/gfw) being 4-folds greater than wild type hairy root cultures (Gupta *et al.*, 2016).

12.4.0 Analysis of differentially Expressed Male Sterile Genes

Hou *et al.* (2016) first studied the differential expression of male fertility related gene pools by construction of cDNA libraries using Suppression Subtractive Hybridization (SSH) strategy. Three SSH cDNA libraries were developed between male fertile and sterile marigold flower buds. Out of 1920 clones, 506 positive clones were analysed by dot-blot hybridization. Further, 286 non-redundant ESTs were produced in which 192 ESTs corresponded to known functional proteins. Fifteen candidate genes were detected through GO function annotation and were found having role in hormone pathways, cell cycles, transcription factors and protein kinases responsible for male sterility. Further, five genes: CDBK2-1 in cell division, AMS in anther wall tapetum development, LAP3 in pollen exine formation, ACOS5 and CYP703A2 in sporopollenin biosynthetic process were identified through quantitative real time PCR. This could be used to understand the mechanism of male reproductive development in marigold.

13.0 References

Abadir, S.K., Ismail, A.E. and Kheir, A.M. (1994) *Annals Agril. Sci. Moshtohor*, **32** : 1027-1033.

Abadir, S.K., Ismail, A.E. and Kheir, A.M. (1996) *Pakistan J. Nematol.*, **14** : 95-100.

Ahluwalia, P., Kaur, A., Kumar, A. and Sachdev, P.A. (2018) *Agril. Res. J.*, **55**(2) : 324-330.

Akhtar, M. and Alam, M.M. (1990) *Pakisthan J. Nematol.*, **8** : 25-28.

Aktaruzzaman, M., Afroz, T., Kim, B.S. and Shin, H.D. (2018) *Plant Dis.*, **102** : 1656.

Albano, J.P. and Miller, W.B. (1996) *J. Amer. Soc. Hort. Sci.*, **121** : 438-441.

Albano, J.P. and Miller, W.B. (1998) *HortScience*, **33** : 1180-1182.

Albano, J.P., Miller, W.B. and Halbrooks, M.C. (1996) *J. Amer. Soc. Hort. Sci.*, **121** : 430-437.

Ali, S.S., Rane, A.E., Patil, B.N. and Neha-Chopde (1998) *J. Soils Crops*, **8** : 111.

Allard, H.A. and Garner, W.W. (1940) *US Dept. Agr. Tech. Bull.*, Washington, D.C., No. 727.

Alp, S., Usta, M., Sipahioglu, H.M. and Guller, A. (2016) *Turk. J. Agric. For. Sci.*, **40**(3) : 311-318.

Anupama, T.V., Sreelatha, U., Baburaj, A. and Narayanankutty, C. (2018) *J. Tropical. Agric.*, **56**(2) : 176-178.

Anuradha, K., Pampapathi, K. and Narayana, N. (1990) *Indian J. Hort.*, **47** : 353-357.

Anuradha, K., Pampapathy, K. and Narayana, N. (1988) *South Indian Hort.*, **36** : 209-211.

Armstrong, P.R., Brusewitz, G.H., Stone, M.L. and Maness, N.O. (2000) *Transactions ASAE*, **43** : 379-384.

Arora, J.S. and Singh, J. (1980a) *Proc. Nat. Sem. Prodn. Techn. Comm. Flower Crops*, Tamil Nadu Agricultural University, Coimbatore, pp. 81-82.

Arora, J.S. and Singh, J. (1980b) *Proc. Nat. Sem. Prodn. Techn. Comm. Flower Crops*, Tamil Nadu Agricultural University, Coimbatore, pp. 83-84.

Arulmozhiyan, R. and Pappaiah, C.M. (1989) *South Indian Hort.*, **37** : 169-172.

Ashley, R.A. (1971) *Proc. North Eastern Weed Sci. Soc.*, **25** : 233-239.

Bailey, L.H. (1963) *The Standard Cyclopedia of Horticulture*, Vol. III, MacMillan Co. Ltd., New York.

Barnwal, M.K., Jha, D.K. and Dubey, S.C. (1998) *J. Res.*, Birsa Agril. University, **10** : 221-223.

Belarmino, M.M., Abe, T. and Sasahara, T. (1992) *Japanese J. Breeding*, **42** : 835-841.

Bespalhok, F.J.C. and Hattori, K. (1998) *Plant Cell Rep.*, **17** : 870-875.

Bhurat, K.B., Sharma, S., Mishra, Y. and Patankar, C. (2017) *Rendiconti Lincei*, **28**(4): 673-678.

Bolz, G. (1961) *Zeit Pflanzenzucht*, **46** : 169-211.

Boodley, J.W. (1961) *Bull. N.Y. St. Flower Grs.*, No. 181, pp. 3-4.

Borch, K., Brown, K.M. and Lynch, J.P. (1998) *HortTechnology*, **8** : 575-579.

Burney, M. (1961) *Rev. Hort. Swiss.*, **34** : 534.

Carlson, W., Nelson, T. and Armitage, A.M. (1981) *Amer. Veg. Gr.*, **29** : 28, 30, 32.

Cathey, H.M. (1968) *Proc. Amer. Soc. Hort. Sci.*, **9** : 693-698.

Cayrol, J.C. (1960) *Rev. Hort. Paris*, **132** : 2276.

Chandrikapure, K.R., Sadawrte, K.T., Panchbhai, D.M. and Shelke, B.D. (1999) *Orissa J. Hort.*, **27** : 31-34.

Chauhan, S.K. (2016) *Ann. Agril. Res.*, **37**(2).

Chittendan, F.J. and Synge, P.M. (1956) *Dictonary of Gardening*, The Clarendon Press, Oxford.

Choudhry, S.R. (1969) *Sydowia*, **23** : 46-53.

Coats, A.M. (1968) *Flowers and Their Histories*, McGraw Hill Book Co., New York, pp. 240-242.

Custic, M. and Poljak, M. (1994) *Polgoprivredna Znanstvena Smotra*, **59** : 105-112.

Dahiya, S.S., Singh, N. and Singh, S. (1998) *Environ. Ecol.*, **16** : 855-857.

Damabi, S.M., Singh, K.P., Pandey, C.D. and Swaroop, K. and Raju, D.V.S. (2014) *J. Ornam. Hortic.*, **17**(3-4).

Davis, T.D. (1991) *Hort. Technology*, **1** : 49-52.

Delgado-Vargas, F. and Paredes-Lopez, O. (1997a) *Food Chemistry*, **58** : 255-258.

Delgado-Vargas, F. and Paredes-Lopez, O. (1997b) *Plant Foods Human Nutrition*, **50** : 163-169.

Delgado-Vargas, F., Paredes-Lopez, O., Avila-Gonzalez, E. and Avila-Gonzalez, E. (1998) *J. Agri. Food Chemistry*, **46** : 698-706.

Dhami, V., Rao, V.K. and Sachan, S. (2015) *J. Ornam. Hortic.*, **18**(1-2): 40-46.

Dhingra, S.N. and Dhingra, D.R. (1956) *Perfum. Essent. Oil Rec.*, **47** : 391-394.

Duda, M. (1967) *Biologie Brastislema*, **22** : 227-234.

Duhan, S., Kar, A. and Bhan, C. (2016) *Adv. Life Sci.*, **5**(11): 4560-4565.

Durga, M.L., Raju, D.V.S., Pandey, R.N., Singh, K.P., Kumar, P., Pandey, R., Gopalakrishnan, S. and Chandra, S. (2017) *Indian J. Agric. Sci.*, **87**(4): 500-504.

Edward, J.C. (1957) *Sci. and Cult.*, **22** : 883-884.

Eyster, W.H. (1939) *Proc. 7th Int. Genet. Congr.*, 117.

Ezawa, T. and Yoshida, T. (1994) *Soil Sci. Plant Nutrition*, **40** : 255-264.

Fu, X., Ma, N., Sun, W. and Dang, Y. (2018) *Ind. Crops Prod.*, **123** : 296-302.

Galicia-Fuentes, S.S. (1995) *Revista Fitotecnia Mexicana*, **18** : 43-53.

Genders, R. (1971) *Pelham's New Gardening Annual*, Pelham Books Ltd., London, pp. 53-55.

Giri, T.K., Kumar, M., Kumar, S., Sharma, V.R., Naresh, R.K. and Malik, S. (2018) *Biotech Today*, **8**(2) : 41-45.

Goel, N., Anukrati, K., Singh, A. and Paul, P.K. (2015) *J. Chem. Pharm. Res.*, **7**(7) : 466-475.

Goldsmith, A.G. (1968) *HortScience*, **3** : 269-271.

Grewal, P.S. and Grewal, S.K. (1988) *Phytopathologia Mediterranea*, **27** : 112-114.

Grover, R.K. (1952) *Bull. Bot. Soc.*, Univ. of Saugar, **4** : 9-13.

Gupta, N.S., Sadavarte, K.T., Mahorkar, V.K., Jadhao, B.J. and Dorak, S.V. (1999a) *J. Soils Crops*, **9** : 80-83.

Gupta, V., Shanker, K. and Rahman, L.U. (2016) *Afr. J. Biotech.*, **15**(17) : 706-713.

Gupta, Y.C., Dhiman, S.R., Sharma, P., Sharma, B., Sharma, P. and Bhatia, S. (2016). *J. Ornam. Hortic.*, **19**(1-2): 13-18.

Gupta, Y.C., Raghava, S.P.S. and Misra, R.L. (1999b) *J. Orna. Hort. New Series*, **2** : 65-66.

Gupta, Y.C., Raghava, S.P.S. and Misra, R.L. (2001) *Indian J. Genet. Pl. Breed.*, **61** : 65-68.

Hameed, A.S. and Sekar, K. (1999) *South Indian Hort.*, **47** : 339-341.

Haramaki, C. (1962) *Proc. North East Weed Control Conf.*, pp. 212-217.

Haramaki, C. and Atmore, S. (1965) *Pa. Flower Grs. Bull.*, **173** : 1-2, 9-10.

Haramaki, C. and Atmore, S. (1966) *Proc. 20th North East Weed Control Conf.*, pp. 195-199.

Haramaki, C. and Koch, G. (1968) *Proc. 22nd North East Weed Control Conf.*, pp. 259-263.

Heinz, K.M. and Parella, M.P. (1990) *Environ. Ento.*, **19** : 825-835.

Heinz, K.M., Harding, P.A., Ek-Ramos, M.J., Hernandez, H., Krauter, P.C. and Sword, G.A. (2018) *HortScience*, **53**(12) : 1791-1798.

Hoffman, M. and Komora, A. (1977) *Pro. Komis, Nauk, Rol. iKomis, Nauk. Lesn. PTPN.*, **43** : 69-81.

Hou, Z., Tang, D. and Tang, N. (2016) *Pak. J. Bot.*, **48**(6) : 2423-2431.

Hurt, R.T. and Vencill, W.K. (1994a) *J. Environ. Hort.*, **12** : 135-137.

Hurt, R.T. and Vencill, W.K. (1994b) *J. Environ. Hort.*, **12** : 131-134.

Iltis, H. (1945) *J. Hered.*, **36** : 291-295.

Indrawati, R., Kurniawan, J.M., Wibowo, A.A., Juliana, Gunawan, I.A., Heriyanto, Brotosudarmo, T.H.P. (2019) *CYTA: J. Food*, **17**(1): 121-127.

Jadhav, P.B. (2018) *Int. J. Curr. Microbiol. Appl. Sci.*, **7**(12) : 832-843.

Jatzokoski, M. and Zimmer, K. (1994) *Gartenbauwissenschaft*, **59** : 270-274.

Joiner, J.N. and Grius, J.T. (1961) *Proc. Fla. St. Hort. Soc.*, **74** : 445-447.

Kantor, J.H. (1938) *Master's Thesis*, Bucknell Univ., Lewisburg, pa., USA.

Kapelev, I.G. (1971) *Rastitelnye Resursy*, **77** : 571-574.

Kaplan, L. (1960) *Econ. Bot.*, **14** : 200-202.

Kar, A.K. and Mandal, M. (1973) *Indian Phytopath.*, **25** : 685-690.

Kara, Z.S., Saffari, V.R. and Tavassolian, I. (2016) *J. Crop Improv.*, **18**(3) : 557-567.

Karam, N.S., Niemiera, A.X. and Leda, C.E. (1994) *J. Environ. Hort.*, **12** : 208-211.

Kennedy, R.R. (1997) *Madras Agri. J.*, **84** : 328-330.

Khaleghi, A., Solgi, M. and Dalvand, M. (2018) *J. Sci. and Tech. Greenhouse Cul.*, **9**(2): 67-80.

Kim, S.K. and Merritt, R.H. (1991) *J. Korean Soc. Hort. Sci.*, **32** : 223-230.

Kobza, F. (1993) *Zahradnictvi*, **20** : 103-110.

Kopsell, D., Belisle, C., Lowery, H., Whitlock, C. and Sams, C.E. (2016) *Acta Hortic.*, **1134.**

Kovshov, S.V. and Iconnicov, D.A. (2017) *Indian J. Agril. Res.*, **51**(4): 327-332.

Kumar, K.R., Singh, K.P., Jain, P.K., Raju, D.V.S., Kumar, P., Bhatia, R. and Pawar, S. (2018) *Indian J. Agril. Sci.*, **88**(6): 970-977.

Kumar, K.R., Singh, K.P., Raju, D.V.S., Panwar, S., Bhatia, R., Jain, P.K. and Kumar, V. (2017) *Indian J. Agril. Sci.*, **87**(10): 31-38.

Kumar, R., Singh, K.P., Raju, D.V.S., Panwar, S., Bhatia, R., Kumar, S. and Verma, P.K. (2018) *Int. J. Curr. Microbiol. Appl. Sci.,* **7**(1) : 2768-2781.

Kurt, N. and Askin, M.A. (2018) *Int. J. Agri., Fores. and Life Sci.,* **2**(2).

Kurup, S.S., Nalwadi, U.G., Basarkar, P.W. and Feucht, W. (1994) *Acta Hortic.,* **381** : 488-493.

Lee, C.W., Choi, J.M. and Pak, C.H. (1996) *J. Plant Nutrition,* **19** : 901-916.

Lin, Y.L.P., Holcomb, E.J. and Lynch, J.P. (1996) *HortScience,* **31** : 94-98.

Little, T.M., Kantor, J.H. and Robinson, Jr. B.A. (1940) *J. Hered.,* **31** : 73-78.

Luo, H., Jia, G.G., Ma, D.F., Sun, Z.X. and Deng, J.X. (2018) *Plant Dis.,* **102**(11) : 2375.

Majumder, J., Singh, S.K., Kumari, M. and Verma, M. (2018) *Plant Tissue Cul. and Biotech.,* **28**(2) : 223-236.

Malgorzata, M. and Renata, M. (2017) *Folia Horticulturae,* **29**(2) : 189-198.

Mamta, Singh, R. and Mehta, N. (2016) *J. Mycol. and Plant Pathol.,* **46**(4) : 362-367.

Maria, O., Silva, C.D., Da, W.D., Castro, O.H., Barros, M.E., Larissa, F., Sousa, g.Y., Junior, S. and Proto de, M.A. (2017) *Braz. J. Animal Health and Prod.,* **18**(2).

Mazumdar, N. (2000) *Plant Dis. Res.,* **15** : 28-33.

Medina, A.L., BeMiller, J.N. and Janick, J. (1993) *Proc. Second Nat. Symp. New Crops Exploration, Res. Commercialization,* Indianapolis, Indiana, October 6-9, pp. 389-393.

Mehraj, U., Panwar, S., Singh, K.P., Namita and Sagwan, S. (2019a) *Int. J. Curr. Microbiol. Appl. Sci.,* **8**(1) : 1393-1403.

Mehraj, U., Panwar, S., Singh, K.P., Namita, N., Pandey, R., Solanke, A.U., Mallick, N. and Kumar, S. (2019b) *Indian J. Agril. Res.,* **89**(7): 1162-1166.

Mehraj, U., Panwar, S., Singh, K.P., Namita, Pandey, R., Solanke, A.U. and Mallick, N. (2018) *Indian J. Hort.,* **75**(4) : 651-656.

Mehta, S.H., Nadkarni, H.R. and Rangawala, A.D. (1995) *Indian J. Agri. Sci.,* **65** : 810-812.

Meijneke, C.A.R. and Oostenbrink, M. (1958) *Meded. Dir. Tuinb.,* **21** : 283-290.

Merritt, R.H. and Ting, K.C. (1995) *Scientia Hort.,* **60** : 313-324.

Mirshekari, B., Tabrizi, M.R.G. and Kouchebagh, S.B. (2015) *Biol. Forum – An Int. J.,* **7**(1) : 1070-1074.

Misra, P. and Datta, S.K. (2001) *In vitro Cellular Dev. Biol.,* **37**: 466-470.

Mitra, M. (1935) *Report Imperial Mycologist, Agric. Res. Inst.,* Pusa, 1933-34, pp. 134-147.

Mohanty, C.R., Behera, T.K. and Samantaray, D. (1993) *J. Orna. Hort.,* **1** : 55-60.

Mohanty, C.R., Behera, T.K. and Samantaray, D. (1997) *South Indian Hort.,* **45** : 41-44.

Mondal, N. and Choudhury, S. (1976) *Curr. Sci.,* **45** : 75.

Monteiro, R.C., Zawadneak, M.A.C. and Mound, L.A. (1999) *Anais da SociedadeEntomologica do Brasil*, **28** : 323-326.

Murray, G.E., Sanderson, K.C. and Williams, J.C. (1986) *HortScience*, **21** : 120-122.

Nahak, G. and Sahu, R.K. (2017) *Pak. J. Biol. Sci.*, **20**(1) : 12-19.

Narayan, S. (2018) *HortFlora Res. Spectrum*, **7**(3) : 217-220.

Novoselova, A.N., Serova, O.K. and Pyatitskaya, L.F. (1985) *Byulleten Glavnogo Botanicheskogo Sada*, No. 138, pp. 41-44.

Nuraini, Mirzah and Djulardi, A. (2017) *Int. J. Poultry Sci.*, **16**(1) : 11-15.

Onsley, M.A., Lynch, J.M. and Whipps, J.M. (1994) *Scientia Hort.*, **59** : 147-155.

Oostenbrink, M. (1960) *Meded. LandbHogesch Gent.*, **25** : 1065-1075.

Padalia, H., Moteriya, P. and Chanda, S. (2015) *Arab. J. Chem.*, **8**(5) : 732-741.

Pal, P. and Ghosh, P. (2010) *Indian J. Nat. Prod. and Resources*, **1**(3) : 371-375.

Pandey, V., Patel, A. and Patra, D.D. (2015) *Ind Crops Prod.*, **76** : 378-385.

Panwar, S., Singh, K.P., Janakiram, T., Bharadwaj, C., Namita, N., Sonah, H., Deshmukh, R.K. and Sharma, T.R. (2017) *Indian J. Agril. Sci.*, **87**(5).

Panwar, S., Singh, K.P., Namita, Janakiram, T., Sonah, H. and Sharma, T.R. (2018) *Indian J. Hort.*, **75**(1) : 105-110.

Park, J.S., Chew, B.P. and Wong, T.S. (1998b) *J. Nutrition*, **128** : 1802-1806.

Park, J.S., Chew, B.P., Wong, T.S. and Park, J.S. (1998a) *J. Nutrition*, **128** : 1650-1656.

Patil, S.R., Kuchanwar, O.D., Deotale, R.D., Jiotode, D.J., Kalamkar, V.B. and Rathod, A. (2017) *Int. J. Res. Biosci. Agric. Tech.*

Pattanaik, M. and Misra, H.P. (2019) *Int. J. Chem. Studies*, **7**(4) : 36-37.

Pill, W.G. and Gunter, J.A. Jr. ((2001) *J. Environ. Hort.*, **19** : 11-14.

Pirone, P.P., Dodge, B.O. and Rickett, H.W. (1960) *Diseases and Pests of Ornamental Plants*, Constable and Co. Ltd., London.

Ploeg, A.T. (1999) *J. Nematol.*, **31** : 62-69.

Pradhan, A., Dash, A.K., Mohanty, S.S. and Das, S. (2015) *Ecol., Environ. and Conserv.*, **21** : 369-376.

Punnet, R.C. (1924a) *Studia Mendeliana*, Burno, pp. 185.

Punnet, R.C. (1924b) *Studia Mendeliana*, Burno, pp. 187-199.

Punnet, R.C. (1937) *Yearbook of Agriculture* 1937, U.S. Dept. Agri.

Rahman, Hossain, Dey, Sarker, Nonaka, M. and Harada, N. (2015) *J. Plant Pathol.*, **97**(2) : 398.

Ram, P., Kumar, B., Singh, A., Mohd-Yaseen, Kothari, S.K. and Kumar, B. (1999) *J. Medicinal Aromatic Plant Sci.*, **21** : 361-366.

Ramalakshmi, R. and Bharathiraja, S. (2015) *Int. J. Curr. Res.*, **7**(1) : 12264-12266.

Razin, A.M. and Omer, E.A. (1994) *Egyptian J. Hort.*, **21** : 195-202.

Riaz, A., Younis, A., Taj, A.R., Karim, A., Tariq, U., Munir, S. and Riaz, S. (2013) *Pak. J. Bot.*, **45**(S1) : 123-131.

Risugan (2017) *Int. J. Adv. Life Sci.*, **10**(2) : 279-289.

Roberts, R.H. and Struckmeyer, B.E. (1939) *J. Agr. Res.*, **59** : 699-723.

Rojas-Martinez, R.I. and Villegas, M.A. (1999) *Fitopatologia*, **34** : 69-76.

Rojas-Martinez, R.I., Zavaleta-Mejia, E. and Gomez, R.O. (1999) *Fitopatologia*, **34** : 83-89.

Roy, B.N., Basu, R.N. and Bose, T.K. (1973) *Plant Sci.*, **5** : 117-122.

Rydeberg, P.A. (1915) *North Amer. Flora*, **34** : 148-159.

Sajjad, Y., Jaskani, M.J., Mehmood, A., Ahmad, I. and Abbas, H. (2013) *Pak. J. Bot.*, **45**(3) : 1255-1258.

Saksena, H.K. and Singh, B.B. (1959) *Plant Dis. Reptr.*, **43** : 670-673.

Santhosh, N., Tejaswini, Shivashankar, K.S. and Gadre, A.K. (2018) *Int. J. Chem. Studies*, **6**(5) : 955-962.

Sarkar, J., Singh, S.K., Singh, K.P. and Guha, S.K. (2016) *Indian J. Agril. Sci*, **86**(7).

Schubert, O.E. (1968) *Proc. 22nd North East Weed Control Conf.*, pp. 264-271.

Secerov-Fiser, V. and Vujosevic, A. (1998) *Rev. Res. Work Faculty Agril.*, Belgrade, **43** : 87-92.

Selvam, S.I.K., Devaraj, R.A. and Rani, C.S. (2015) *Indo-American J. Pharm. Res.*, **5**(10).

Selvaraju, P. and Selvaraj, J.A. (1994) *Madras Agril. J.*, **81** : 496-497.

Shahane, S. and Sidhaye, D. (2018) *Int. J. Modern Phy. B.*, **32**(19) : 1840068.

Shahzadi, I. and Shah, M.M. (2015) *Frontiers in Pharmacol.*, **6**.

Sharga, A.N. (1975) *Farmer and Parliament*, **10** : 9-10 and 28.

Sharga, A.N., Motial, V.S. and Basario, K.K. (1972-73) *Prog. Hort.*, **4** : 71-76.

Shedeed, M.R., El-Gamassy, K.M., Hashim, M.E. and Almulla, A.M.N. (1986) *Ann. Agric. Sci., Ain Shams Univ.*, **31** : 677-689.

Shivakumar, Srinivasa, Praneeth, Y.S., Jayasheela, D.S. and Gokavi, N. (2015) *Ecol., Environ. Conserv.*, **21**(4) : 1817-1821.

Shivayogeshwara, B., Gowda, B.L., Shankar, S. and Patil, N.M. (2000) *Insect-Environ.*, **6** : 4.

Shukla, A. and Thakur, R. (2018) *Int. J. Curr. Microbiol. Appl. Sci.*, **7**(1) : 1744-1748.

Singh, B. (1970) *Ph.D. Thesis* submitted to Agra University.

Singh, B. and Swarup, V. (1972) *Indian J. Genet.*, **31** : 407-415.

Singh, B. and Swarup, V. (1973a) *Indian J. Genet.*, **33** : 172-175.

Singh, B. and Swarup, V. (1973b) *Indian J. Genet.*, **33** : 373-377.

Singh, C.V., Shankar, M.V., Kasana, B.S. and Bobede, A. (2016) *Int. J. Agril. Sci.*, **8**(4) : 1000-1002.

Singh, D. and Sen, N.L. (2000) *J. Orna. Hort. New-Series*, **3** : 75-78.

Singh, D., Naqvi, Q.A., Garg, I.D. and Singh, D. (1999) *Indian Phytopath.*, **52** : 114-117.

Singh, J. and Arora, J.S. (1980) *Proc. Nat. Sem. Prodn. Techn. Comm. Flower Crops*, Tamil Nadu Agricultural University, Coimbatore, pp. 85-87.

Singh, M. (2001) *J. Hort. Sci. Biotech.*, **76** : 52-54.

Singh, M.P., Singh, R.P. and Singh, G.N. (1991) *Haryana J. Hort. Sci.*, **20** : 81-84.

Singh, R., and Singh, A.R. (1997-98) *Recent Hort.*, **4** : 89-91.

Snehitha, R., Sneha, C.H., Kalyani, D.L., Sravani, P. and Sumanth, T. (2019) *Pharma Innovation*, **8**(8) : 169-173.

Sohi, (1984) *Disease of Orna. Plants*, ICAR, New Delhi.

Solanki, P., Kalavagadda, B., Akula, B., Sharma, S.H.K. and Reddy, D.J. (2017) *Int. J. Curr. Microbiol. Appl. Sci.*, **6**(6) : 825-831.

Some, S.K. and Mustafee, T.P. (1966) *Curr. Sci.*, **35** : 370-371.

Sowmeya, Kannan and Ranjana (2017) *Int. J. Curr. Res.*, **9**(12).

Sreevamulu, T. (1953) *Sci. and Cult.*, **18** : 540-541.

Sridhar, S. and Rengasamy, R. (2010) *J. Phytol.*, **2**(6) : 61-68.

Staden, J. Van., Uphold, S.J. and Drewes, F.E. (1994) *J. Appl. Phycology*, **6** : 427-428.

Steiner, B.M., Mcclements, D.J. and Davidov-Pardo, G. (2018) *Trends in F. Sci. Tech.*, **82** : 71-81.

Steiner, G. (1941) *Proc. Biol. Soc. Wash.*, **54** : 31-34.

Stoltz, R.L., Gavlak, R.G. and Halbert, S. (1997) *J. Veg. Crop Production*, **3** : 27-36.

Subramanyam, N., Reddy, N. and Rao, A.S. (1975) *Indian Phytopath.*, **28** : 516-519.

Suzuki, H., Kumagai, H., Oohashi, K., Sakamoto, K., Inubushi, K., Enomoto, S. and Ambe, F. (2000) *Soil Sci. Plant Nutrition*, **46** : 283-289.

Syamal, M.M., Rajput, C.B.S., Upadhyay, R.K. and Singh, J.N. (1990) *Indian J. Hort.*, **47** : 439-441.

Tamut, O., Singh, K.P., Raina, A.P., Namita, Panwar, S., Kumar, P. and Verma, P.K. (2017) *Indian J. Hort.*, **74**(1) : 97-102.

Tarpo, E. and Cucu, V. (1961) *Pharmazie*, **16** : 486-488.

Thaneshwari, Aswath, C. and Chaithra, N. (2018) *Int. J. Curr. Microbiol. Appl. Sci.*, **7**(3) : 3216-3226.

Thirmulachar, M.J. and Govindu, H.C. (1956) *Sydowia*, **16** : 284-288.

Tina, A., Pezhman, M. and Deris, Z. (2017) *Int. J. Environ., Agric. Biotech.*, **2**(3) : 1109-1112.

Towner, J.W. (1956) *Ph.D. Thesis*, Univ. Calif., Los Angles.

Towner, J.W. (1958) *Proc. 10th Inter. Cong., Genetics*, **2** : 296.

Towner, J.W. (1961) *Amer. J. Bot.*, **48** : 743-751.

Towner, J.W. (1962) *Amer. J. Bot.*, **49** : 1064-1067.

Trivedi, H., Shekhar, C. and Singh, N. (2017) *HortFlora Res. Spectrum*, **6**(3): 206-208.

Tsukamoto, Y., Imanishi, H. and Yahara, H. (1968) *J. Japanese Soc. Hort. Sci.*, **37** : 213-219.

Tsukamoto, Y., Imanishi, H. and Yahara, H. (1971) *J. Japanese Soc. Hort. Sci.*, **40** : 401-406.

Tsurushima, H. and Date, N. (1977) *Bull. Tokyo Agric. Exp. Sta.*, **10** : 43-52.

Tyler, J. (1938) *Plant Dis. Reptr., Suppl.*, **109** : 133-151.

Vandekar, V.D. (2015) *J. Nat. Prod. Plant Resources*, **5**(2) : 43-52.

Villar-Martinez, A.A. del., Ma, R. and Paredes-Lopez, O. (1999) *J. Plant Physiol.*, **155** : 205-211.

Visser, T. and Vythilingam, M.K. (1959) *Tea Quart.*, **30** : 30-38.

Walia, K.K. and Gupta, D.C. (1997) *Indian J. Nematol.*, **27** : 18-23.

Watanabe, Y., Kimura, N. and Shimada, N. (1984) *Tech. Bull. Fac. Hort.*, Chiba Univ., **34** : 15-19.

Weaver. D.K., Zettler, J.L., Wells, C.D., Baker, J.E., Bertsch, W. and Throne, J.E. (1997) *J. Eco. Ento.*, **90** : 1678-1683.

Willoughby, R.A., Solie, J.B., Whitney, R.W., Maness, N.O. and Buser, M.D. (2000) *ASAE Annual Intern. Meeting*, Milwaukee, Wisconsin, USA, July 9-12, pp. 1-14.

Wilson, M. and Holcomb, S.J. (1979) *Bull. Pennyslvania Flower Grs.*, No. 311, p. 12.

Wu, W.S., Chou, H.H., Lin, S.M. and Wu, H.C. (2001) *J. Phytopath.*, **149** : 91-96.

Yadav, L.P. (1983) *Ph.D. thesis*,Bidhan Chandra Krishi Viswavidyalaya, Kalyani.

Yadav, P.K. (1997) *Curr. Agri.*, **21** : 113-114.

Yadav, P.K. and Singh, S. (1997) *Environ. Ecology*, **15** : 849-851.

Yadav, P.K., Singh, S., Dahiya, S.S. and Singh, S. (1999) *Environ. Ecol.*, **17** : 188-190.

Yahya, A., Mohd, R. I. and Roeber, R.U. (1997) *Acta Hortic.*, **450** : 31-38.

Yun, T., Kim, T.J., Hong, S.M., Cho, J.T., Jeong, I.M., Lim, K.B. and Paek, K.Y. (1996) *RDA J. Agri. Sci. Hort.*, **38** : 604-608.

Zimmer, K. (1989) *Gesischer Gartenbau*, **43** : 961-963.

7

China Aster

Rajiv Kumar, R.S. Dhuna, P. Pal and S.L. Chawla

1.0 Introduction

China aster (*Callistephus chinensis*) is a half hardy flower annual. Plants are erect having hispid-hairy branches bearing alternate, broadly ovate or triangular-ovate, deeply and irregularly toothed leaves. Flowers are solitary, single, semi-double or double and indigo-blue, lavender, rose and white are the prominent colours.

2.0 Importance and Uses

2.1.0 Aesthetic Uses

It is one of the most popular of all garden annuals grown throughout the world. They can easily be grown in the open fields and cloth/net houses for the production of cut flowers. Cut flowers last long and are used in vases and floral decoration. Asters also make very showy bedding plants when grown in large masses and are valuable for filling up the gap in mixed herbaceous border, the dwarf types are put in front and the taller behind. Some strains are used as pot plants. Dwarf cultivars are also suitable for edging and window boxes. Several countries produce seeds in large quantities due to its wide popularity. In tropical and subtropical countries aster is extensively grown for bedding and cut blooms during the winter months and spring.

The crop is mainly grown in India both by marginal and small farmers in Karnataka, Tamil Nadu, Andhra Pradesh, Maharashtra and West Bengal. Under Bengaluru conditions, this crop is being grown throughout the year. The crop is found to be suitable for intercropping in coconut, arecanut, jackfruit and mango orchards to utilise the interspacing and fetch more returns.

2.2.0 Therapeutic and other Uses

The aster root has been used for centuries in Chinese medicine. The flowers and leaves can be eaten fresh or dried when eating aster plants. The Native American people harvested wild aster for a multitude of uses. The roots of the plant were used in soups and young leaves were cooked lightly and used as greens. The Iroquois people combined aster with bloodroot and other medicinal plants to make a laxative. The Ojibwa used an infusion of aster root topically to aid with headaches. Portions of the flower were also used to treat venereal diseases.

3.0 Origin and History

The genus *Callistephus* derived its name from two Greek words *kalistos*, most beautiful, and *stephos*, a crown, referring to the flower. Cassini described the China aster as *C. hortensis*. It was first named by Linnaeus as *Aster chinensis*, and Nees subsequently changed this name to *Callistephus chinensis*.

The evolution of China aster is a history of remarkable variation. The present day asters have been developed from single form of wild species, *C. chinensis*, a plant native to China. The plant was introduced in Europe in 1731 by a Jesuit missionary, R. P. d'Incarville (Bailey, 1963). The original plants had single flowers

Uses of China Aster

In Hanging Basket In Pots As Cut Flowers

Loose Flower As Garland Floral Crown

Floating Floral Arrangement Floral Rangoli In Flower Beds

Along Path As Borders In Rock Gardens

Source: www.floretflowers.com, www.gardenchronicle.com, www.graden.lovetoknow.com, www.bigstockphoto.com, www.i.pinimg.com, www.peakpx.com, www.cf.ltkcdn.net/garden, www.getflowersdaily.com, www.media.istockphoto.com, www.amazon.co.uk www.etsy.com

with two to four rows of blue, violet or white ray florets. The first change in the flower type had been the prolongation or development of the central florets and the production of quilled flowers. French florists contributed much in the development of double forms. German growers also raised many double cultivars. These asters were then known as German asters because the first great advances in the evolution of the plant were made in Germany and the seeds then used came largely from that country. The comet types, introduced in 1886 or 1887 subsequently replaced the quilled types. With the introduction of new branching types by James Vick in 1893, the USA became the main centre of development of this plant as well as for the production of seeds. Since its introduction into Europe, it is grown in almost all the continents of the globe as a garden annual.

4.0 Species, Types and Cultivars

China aster is the member of the family Asteraceae and genus *Callistephus* having a single species *C. chinensis*. Nechansky and Jirasek (1967) suggested that the most suitable character for the classification of aster was the shape of the ray florets.

4.1.0 Types and Cultivars

Annual asters have many types and cultivars varying in growth habits, flower shape, size and appearance of florets; colours of the flowers are also very attractive in different cultivars in a particular type. Double flowered asters are classified broadly into 3 groups – tall, medium-tall and dwarf.

4.1.1 Tall

The height of the plants under this group varies from 70 to 90 cm. The types and colour in different cultivars are given hereunder.

4.1.1.1 American Branching

Long-stemmed flowers have wide range of colour in different cultivars – dark blue, light blue, rose, salmon, scarlet, violet, white and pink.

4.1.1.2 Bouquet Powderpuffs

Medium-sized flowers on rigid stems have also the colour range in different cultivars as mentioned above.

4.1.1.3 Chikuma Strain

Fully double, ball-shaped pompon-type flowers, ideal as cut flower. Many cultivars have attractive range and shades of colour – red, rose, purple, white, red with golden yellow centre, yellow and salmon-rose.

4.1.1.4 Compliment Series

Long-stemmed double flowers are in salmon-pink, light blue and white, good for cut flowers, var. Compliment Light Blue – ray-florets are pale violet-blue, disc florets yellow.

Cultivars of China aster

Arka Archana

Arka Shashank

Arka Aadya

Phule Ganesh White

Phule Ganesh Pink

Phule Ganesh Purple

Arka Violet Cushion

Arka Poornima

Arka Kamini

Arka Advika

Arka Nirali

Arka Shubhi

Cultivars of China aster

Matsumoto Pink — Matsumoto Yellow — Bonita Rose

Serenade Blue — Fellowship Asters — Barr's Pink

Purple Cloud — September Ruby — Kickin New England

Ada Ballard — Chatterbox New York — Royal Ruby

***Source*: www.sakataornamentals.com, www.leafyplace.com**

4.1.1.5 Duchesse Series

Flowers incurved have wide range of colours from yellow to red and purple, good for cut flowers, var. Matsumoto – long-stemmed semi-double flowers with yellow disc-florets. Plants are weather-resistant and partially wilt resistant. Good for cut flower.

4.1.1.6 Giant Massagno

Large flowers with quilled petals about 10 cm across are borne on bushy plant. Rose, scarlet, light blue, silvery blue, golden yellow are the flower colours in different cultivars.

4.1.1.7 Giant of California

Large and late flowering type having good colour range.

4.1.1.8 Mutsumoto

Plants erect, semi-double flowers with yellow disc florets, good for cut flower. Both weather and wilt resistant.

4.1.1.9 Paeony-flowered

Upright growth, and chrysanthemum-like flowers have various attractive colours. Good as cut flower.

4.1.1.10 Princess Series

Incurved, semi-double, quill-petalled flower heads in a wide range of colour. Suitable for cut flowers, partial wilt-resistant. var. Giant princes – semi-double flowers on long woody stems, red-purple, disc-florets tipped with yellow. This type is much branched and ideal for cut flowers.

4.1.1.11 Totem pole

Large bouquet-type flowers, 10-12 cm across, rose-pink, blue, scarlet-cerise and white in colour.

4.1.2 Medium-tall

This group of aster has many types, height of which varies from 49-60 cm.

4.1.2.1 Early Burpeeana

Early mum-like flowers about 7 cm across with semi-incurved petals, in blue, rose, white and scarlet.

4.1.2.2 Giant Comet

Much branched, hardy plants, flowers with curled petals.

4.1.2.3 Giant Grego

Bushy and vigorous growth, flowers have curled petals. Colour of flower includes salmon, scarlet, white, shell-pink, deep rose, light and deep blue.

4.1.2.4 Kyoto Pompon

Small button like double flowers, wide colour orange. Weather-resistant.

4.1.2.5 Ostrich Plume Series

Long stem, spreading habit, flowers feathery, reflexed, double, pink and crimson, ideal cut flowers, partial wilt-resistant.

4.1.2.6 Pompon

Globular-shaped flowers with quilled petals having wide range of colour – white, yellow, dark blue, lilac, scarlet, rose, light blue and also with white centre.

4.1.2.7 Unicum

Large-flowered with quilled petals.

4.1.3 Dwarf Double

The plant height in this group ranges between 20 and 40 cm. It has also several types varying in growth habit and flower characteristics.

4.1.3.1 Colour Carpet

Dwarf, uniform globular plants covered with mum-like flowers.

4.1.3.2 Comet Series

Dwarf, spreading growth, early flowering, double flowers with quill-petalled and wide range of colour – white, pink, blue, purple, scarlet and yellow. Suitable for containers.

4.1.3.3 Dwarf Chrysanthemum

Compact, bushy growth, good range of flower colour.

4.1.3.4 Milady Series

Sturdy-branched, round-shaped double-flowers have rose-pink, rose-red, scarlet, blue, white and mixed colours, ideal for bedding and partial wilt-resistant.

4.1.3.5 Pinocchio Series

Large number of incurved, round, double flowers appear on dwarf and compact plants, excellent for bedding, edging and window boxes.

4.1.4 Asterix

Single-flowered, dwarf cultivars are useful for bedding and cut flowers. Whereas, single-flowered tall cultivars are good for cut flowers. Several types and cultivars of aster are resistant to wilt.

In India, six improved varieties of China aster *viz.*, Arka Aadya, Arka Archana, Arka Kamini, Arka Poornima, Arka Shashank and Arka Violet Cushion were developed by ICAR-IIHR, Bengaluru. Four varieties *viz.*, Phule Ganesh Pink, Phule Ganesh Purple, Phule Ganesh Purple and Phule Ganesh White from Regional Fruit

Research Station (M.P.K.V., Rahuri) Ganeshkhind, Pune were developed. Varietal descriptions of these varieties are given here.

Arka Aadya: It was developed through Individual Plant Selection from OP population of Line No.173. It is spreading type, floriferous, early flowering (66.44 days), high loose flower yield (178.56 g/plant), extended blooming period (16.11 days) and tolerance to lodging. The flower are pink coloured (RHS 62.A, Red Purple Group) and semidouble type. It is recommended for loose flowers (flower with short stalk) and for bedding purpose.

Arka Archana: It was developed through Individual Plant Selection from selfed population of Line No. 15. It is spreading type, floriferous, early flowering (69.40 days), high loose flower yield (222.62 g/plant), extended blooming period (18.44 days) and tolerance to lodging. The Flowers are white coloured (RHS NN155.C, white group) and semi-double type. It is recommended for loose flowers (flower with short stalk) and bedding purpose.

Arka Kamini: It was derived by crossing two pure lines (AST 6 x AST 36) which was developed through pedigree method of breeding. This cultivar produces deep pink colour flowers that are more attractive than local pink cultivars. It takes 120-140 days to flower production and attains a height of 60 cm. Its flowers are 5.5-6.5 cm in diameter and weigh 2 g each. Each plant produces about 40-50 flowers. It has vase life of 8 days.

Arka Poornima: It is developed through advanced pedigree selection of the cross between AST 29 and AST 3. Flowers are pure white, powdery puff type and superior to the local cultivars. It takes 120-140 days to flower and grows to about 50-60 cm tall. Flower diameter is 5.5-6.5 cm. It has a stalk length of 25 cm and a vase life of 7 days. The plant produces 20-25 flowers with an individual flower weight 3.5 g.

Arka Shashank: It was developed by crossing a local variety and a pure line (Local Pink x AST 2). Flowers are creamy white, powder puff type and more attractive than local cultivars. Plants are 60-80 cm tall. Flower diameter is 4.5-5.5 cm and weighs 2.5 g each. It bears about 45 flowers per plant with a stalk length of 25 cm and a vase life of 9 days.

Arka Violet Cushion: This variety was developed through pedigree selection from a cross between Local Pink x AST 2. Flowers are violet in colour and superior to local violet cultivars. It produces pompon type flowers having 4-5 rows of ray florets with tubular disc florets. The flower diameter varies from 5.0-5.5 cm and an individual flower weighs about 2.5g. It has a stalk length of 20 cm and a vase life of 8 days. It produces 35 to 50 flowers per plant.

Phule Ganesh Pink: It is an early, semi-erect and superior quality variety. The flowers are attractive, double chrysanthemum type, big size and rosy pink colour. It produces 43.03 lakh flowers per hectare.

Phule Ganesh Purple: It is a semi-erect variety. It bears bright purple colour flowers with a yield potential of 46.83 lakh flowers per hectare.

Phule Ganesh Violet: It is early, semi-erect and superior in growth and flower yield characters. It produces medium size violet flowers. The flowers are double chrysanthemum type with a yield potential of 60.47 lakh flowers per hectare.

Phule Ganesh White: It is a medium, tall, erect growing late variety. It produces snow white flowers. The flowers are double chrysanthemum type with stalk length 48.6 cm. Yield potential is up to 47.07 lakh flowers per hectare.

Recently, three novel China aster varieties have been identified for release by ICAR-IIHR, Bengaluru during 2021. They are Arka Advika, Arka Nirali and Arka Shubhi. The salient features of these new China aster varieties are given here.

Arka Advika: It was developed through natural mutation from the variety Arka Poomima. This mutant variety has unique semi-double flower with short pseudo ray florets. Flowers are white in colour (White group, NN155D, Fan 4). It produces average 36 flowers per plant, stalk length (36 cm) and vase life (11 days). It is suitable for cut flower, flower arrangement and bouquet making.

Arka Nirali: It was developed through natural mutation from the variety Arka Violet Cushion. This mutant variety has unique semi-double flower with short pseudo ray florets. Flowers are violet in colour (N81A, Purple group, Fan 2). It produces average 43 flowers per plant, stalk length (42 cm) and vase life (11 days). It is suitable for cut flower, flower arrangement and bouquet making.

Arka Shubhi: It was derived from the cross 'Arka Kamini' × 'Local White' following pedigree selection. Flowers are unique light pink colour (Red Purple group, 65D, Fan 2). It has average flower diameter (5.04 cm), number of flowers per plant (56.00), stalk length (47 cm) and vase life (10 days). It is suitable for cut flower and flower arrangement.

5.0 Genetics and Breeding

5.1.0 Genetics

Most of the present-day China asters are diploids. The behaviour of this plant as a diploid (2n=18) was reported by Huziwara (1954). Schwantiz and Schwantiz (1950) studied the sexuality of polyploid forms and found that in tetraploid forms the proportion of green or rudimentary flowers was much higher than in diploid plants. The poor functioning of the sexual organs in polyploid forms was due to deterioration in transport of reserve materials caused by the increase in the cell size.

Based on the segregation in the progenies of rogues, Fleming (1937) concluded that purple colour is dominant to red and deep pink; red to white and white are recessive to all. Witt (1937) found that colour is controlled by a series of multiple alleles. He reported that *R* is responsible for delphinidin, *r′* for cyanidin and *r* for pelargonidin. Forkmann and Seyffert (1977) while analyzing six genotypes of China aster, also concluded that multiple alleles govern flower colour. Beale (1941) however, contradicted the result of Witt (1937) and ruled out the possibility of multiple alleles. Negi and Raghava (1990) investigated the flower colour in 17 cross combinations involving 10 parents. Violet was dominant over all other colours; blue was dominant over creamy white; deep pink was dominant over pink and white, but incompletely dominant over light pink; and pink was dominant over light pink and white. Four genes, designated *C, B, B* and *P*, governed flower colour. It is proposed that *P* has a dominant allele *PD* (for deep pink) and a recessive allele *P* (light pink). Guzenwski (1976) determined anthocyanins in eight cultivars in their F_1 hybrids and in backcross and F_2 progenies. He found 3 anthocyanins, namely, delphinidin, cyanidin and pelargonidin. The flower colours of the cultivars determined by the combinations of anthocyanins were violet, pink-violet, violet-red, bluish red-pink and pink. Seyffert (1962) gave an account of interaction between genes affecting the anthocyanin pigment molecule structure in *Callistephus chinensis*. Petrenko (1975) studied the pigments in the flowers of 13 different colours. Pelergonidin, peonidin, cyanidin and delphinidin were found to be present in the colouration of flowers and cyanidin was observed to be the most occurring pigment.

The cDNA clones were used for characterization of wild-type line 06 and the genetically stable anthocyaninless mutant lines 07, 08 and 10 h. Southern blot analysis revealed that the genes encoding chalcone isomerase (CHI), naringenin 3-dioxygenase (FHT), dihydrokaempferol 4-reductase (DFR) and anthocyanidin synthase (ANS) are single copy genes. Northern blot analyses confirmed that yellow coloured mutant lines 07 and 10 h are CHI-deficient, and line 10 h is additionally DFR-deficient. Lines 07 and 08 are ANS-deficient (Henkel *et al.*, 2000).

Selfed progenies of plants arising by cross pollination among the varieties Irma, Turandot and Srdce Francie (Heart of France) were examined by Novotna (1990). Segregation resulted in a wide range of flower and foliage colours, flower shapes and plant habits. In the F_2 of Srdce Francie, 7 flower colours were distinguished : 46.55 per cent were purple, 22.41 per cent white, 10.34 per cent dark bluish-purple, 8.62 per cent reddish purple, 8.62 per cent red, 1.72 per cent dark purple and 1.72 per cent carmine red. Single flowers accounted for 75.87 per cent of the flower shapes studied and 15.5 per cent of flowers were full. In the F_3, purple and dark bluish-purple flower colours predominated (32.95 and 30.72 per cent, respectively). Single and semi-full flowers accounted for 79.32 and 18.98 per cent of all flower structures encountered, respectively.

The inheritance of flower doubleness and flower types in China aster was studied in 7 crosses involving 9 parents. Flower doubleness (DD) was observed to be monogenically dominant over singleness (dd). Pompon flower type with one row of ray florets and tubular disc florets was dominant over both chrysanthemum and fluffy types, while chrysanthemum type was dominant over fluffy type. Three independent genes Tu, D and C were found to govern these three different flower types. Another genotype having pompon flower type with four rows of ray florets and tubular disc florets was dominant over pompon with one row of ray florets and tubular disc florets. When this pompon type was crossed with chrysanthemum and fluffy types, it was found that the tubular (showy) disc florets were incompletely dominant over normal (non-showy) disc florets. Thus, in this pompon type, there is a dominant allele C′ of the gene C which does not allow the conversion of ray florets into tubular disc florets. Tu alone or C or C′ alone fail to produce tubular disc florets (Raghava and Negi, 2001).

Khangjarakpam *et al.* (2014) recorded high (>20 per cent) GCV and PCV for plant height, number of branches and leaves per plant, flower diameter, number of ray and disc florets/flower head, stalk length, and, number and weight of flowers/plant however, it was recorded high for plant height and plant spread (Rai *et al.*, 2017).

Kumari *et al.* (2017) recorded higher genotypic and phenotypic coefficient of variation was recorded for number of leaves/plant and number of ray florets/ flower head. High heritability (>60 per cent) was recorded for all traits. The genetic advance ranged from 0.97 (flower head diameter) to 143.01 (number of leaves/ plant). High genetic gain was recorded for number of leaves/plant and number of ray floret/flower head.

Harishkumar *et al.* (2017) assessed and estimated the magnitude of variation among the F_2 population of two crosses, *i.e.* AAC-1 × Arka Poornima and Arka Kamini × P G Purple. High heritability along with high genetic advance existed the crosses AAC-1 × Arka Poornima and Arka Kamini × PG Purple for plant height, number of branches, number of leaves, flower stalk length, flower diameter, disc diameter, number of flowers per plant, individual flower weight and flower yield per plant. Thus, these characters could be improved through simple selection procedure due to the presence of an additive type of gene action.

Naikwad *et al.* (2018) recorded high phenotypic and genotypic coefficients of variation for test weight, yield per plant, flower bud initiation, days to flowering, plant height, leaf area, leaf area index and secondary branches. High heritability estimates coupled with high genetic advance over mean was observed for all the traits except duration of flowering and dry matter.

Ramya *et al.* (2019) evaluated genetic variability in F_2 population of cross Arka Archana × AAC-1 in China aster and found high (>20 per cent) phenotypic coefficient of variation (PCV) and genotypic coefficient of variation (GCV) was recorded for number of branches per plant, plant spread East-West, number of flowers per plant, disc diameter, individual flower weight and flower yield per plant.

Rai *et al.* (2017) found high heritability (>80 per cent) was recorded for most of the traits except number of harvests of loose flower, however, Bhargav *et al.* (2019) observed high (>20 per cent) PCV and GCV for plant height, number of leaves per plant, plant spread, flower stalk length, 100 flower weight, number of flowers per plant and weight of flowers per plant. Heritability estimates ranged from 92.32 per cent (number of leaves per plant) to 99.84 per cent (100 flower weight). High heritability coupled with high genetic gain as per cent of mean was recorded for all traits.

Rai *et al.* (2017) reported high heritability (>80 per cent) for all traits, except number of harvests of loose flower. High values of genetic gain were recorded for plant height (69.81 per cent), followed by weight of flowers/plant (50.97 per cent).

Ramya *et al.* (2019) found that high heritability (>60 per cent) coupled with high genetic advance as per cent over mean (>20 per cent) were recorded for plant height, number of branches per plant, plant spread, flower diameter, disc diameter, flower stalk length and flower yield per plant. This indicated that the high heritability is due to additive gene effects which can be utilized for further crop improvement programme. Hosalli *et al.* (2019) revealed that high heritability along with high genetic advance existed in cross *viz.*, AAC-1 × Arka Poornima and Arka Kamini × P. G. Purple for number of flowers per plant, individual flower weight and flower yield per plant. Thus, these characters could be improved through simple selection procedure due to the presence of additive type of gene action.

Naikwad *et al.* (2018) found that number of leaves and secondary branches/ plant and plant spread showed significant positive correlation with yield suggesting good scope for improvement of yield. Path analysis indicated that 10 flower weight, number of secondary branches/plant and number of flowers/plant directly influenced the flower yield and had maximum direct effect. These traits can be directly used to improve the yield of China aster. Rai *et al.* (2017) found that weight of flowers/plant was positively correlated with plant height, plant spread, flower head diameter, number of flowers/plant and individual flower weight.

Miteva and Vasileva (2015) studied the correlations between the vegetative and decorative indicators in *Callistephus chinensis* L. by type of cultivation. Khangjarakpam *et al.* (2015) found that number and weight of flowers/plant were significant and positively correlated with plant height, number of branches/plant, number of leaves/plant and flower stalk length. Kumari *et al.* (2017) reported that weight of flowers/plant was positively correlated for earliness, duration of flowering, and number of flowers per plant, 100 flower weight and shelf life. Rai *et al.* (2017) reported that weight of flowers/plant is significantly and positively correlated both at genotypic and phenotypic level to plant height, plant spread, flower head diameter, number of flowers/plant and individual flower weight. Ramya *et al.* (2019) carried out phenotypic correlation coefficients and path coefficient analysis for F_2 segregating population of cross Arka Kamini × PG Violet in China aster. The results revealed that flower yield per plant had positive correlation with the individual flower weight, flower diameter, disc diameter, number of flowers per plant and number of branches per plant. Bhargav *et al.* (2019) observed that the weight of flowers per plant was significant and positively correlated with all the economic traits, except for shelf life.

Rai *et al.* (2017) reported path coefficient analysis using correlation coefficients revealed that number of flowers/plant contributed highest positive direct effect on weight of flowers/plant, followed by individual flower weight and plant height. Path coefficient analysis using correlation coefficients revealed that number of flowers/plant contributed highest positive direct effect on weight of flowers/ plant, followed by individual flower weight and plant height. Therefore, cultivars with superior traits can be utilised in hybridization to transfer desirable traits into a single cultivar (Rai *et al.*, 2017). Kumari *et al.* (2017) found that path coefficient analysis using correlation coefficient revealed that 100 flowers weight contributed highest positive direct effect on weight of flowers/plant followed by number of leaves/plant and number of flowers/plant.

Ramya *et al.* (2019) carried out path analysis revealed that number of branches per plant, plant spread North-South, plant spread East-West, days to first flowering, duration of flowering, flower diameter and disc diameter, number of flowers per plant, individual flower weight had maximum positive direct effect on flower yield per plant.

Bhargav *et al.* (2019) found that path coefficient analysis using correlation coefficients revealed that number of flowers per plant contributed highest positive direct effect on weight of flowers per plant, followed by weight of 100 flowers. Therefore, the selection on the basis of traits, *viz.* plant height, number of leaves per plant, plant spread, flower stalk length, 100 flower weight, number and weight of flowers per plant will be more effective for improvement of traits in breeding of China aster.

Stalk length and flower weight had positive significant correlation between themselves and with flower size, plant height and days taken to flower. Flower size was observed to have positive significant correlation with plant height and days taken to flower. Number of flowers per plant and days taken to flower showed significant association with number of branches per plant.

Patil and Rane (1994a,b) also studied heterosis and combining ability on 12 yield components in 9 *C. chinensis* lines and their 72 F_1. Heterobeltiosis for weight of flowers was highest (99.25 per cent) in Alandi selection x Kashmir local selection, and for flowers/plant it was highest (74.8 per cent) in Rase selection x Otur selection.

Bhargav *et al.* (2018) estimated heterosis over mid parent in thirty crosses of China aster involving six lines and five testers. Results revealed that the cross Matsumoto Scarlet x Phule Ganesh Violet exhibited highest positive relative heterosis for plant height, number of branches per plant, and duration of flowering. The cross Matsumoto Scarlet x IIHRJ3-2 exhibited maximum negative relative heterosis for days to first flowering, however, the cross Matsumoto Scarlet x IIHRG13 recorded the maximum positive relative heterosis for flower head diameter and 100 flower weight. The cross Matsumoto White x Local white recorded the maximum relative heterosis for flower yield per plant and flower yield per hectare.

Kumari *et al.* (2018) carried out heterosis analysis with line × tester set of 8 parents involving five females and three pollen parents. They revealed that the cross Arka Poornima x Arka Shashank exhibited highest positive significant heterosis for plant height over mid parent, better parent as well as commercial check *cv.* Arka Kamini. For days to first flowering, cross Arka Aadya x Arka Shashank exhibited maximum negative heterosis over commercial check. The cross (Arka Poornima x Arka Shashank) exhibited highest positive significant heterosis for duration of flowering over mid parent, better parent as well as commercial check. For flower stalk length, the highest positive significant heterosis over mid parent, better parent as well as commercial check was observed in cross Arka Archana. The cross Arka Aadya x Local Violet recorded highest heterosis for flower yield over better, mid and commercial check.

Bhargav *et al.* (2019) estimated heterosis over better parents in thirty crosses of China aster. They revealed that the cross Matsumoto White x Local exhibited highest positive significant heterobeltiosis for plant height, flower stalk length, number of flowers per plant, weight of flowers per plant and flower yield per hectare. The cross Matsumoto Pink x IIHRJ3-2 exhibited maximum negative heterobeltiosis for days to first flowering. However, Matsumoto Scarlet x IIHRG13 recorded the maximum positive heterobeltiosis for flower head diameter and 100 flower weight.

Nataraj *et al.* (2020) evaluated ten parents (without reciprocals) involved in diallel cross and forty-five F_1 hybrids of China aster for commercial heterosis for economic traits. The overall maximum positive heterosis for yield in terms of number of flowers per plant was observed in Local Pink × Local White, Local Violet × AAC-1 for cut flowers, weight of flowers per plant PG Violet × Arka Poornima for loose flower over commercial heterosis. Negative heterotic crosses Local Pink × Local Violet, for days to 50 per cent flowering are important to exploit heterosis for earliness in China aster. The F_1 hybrids Local Violet x AAC-1, Local Pink × Local White and PG Violet × Arka Poornima recorded high yield potential.

A line x tester analysis was carried out involving 11 divergent genotypes (6 lines and 5 testers) and 30 F_1 hybrids of China aster for assessing the combining ability for 13 economic traits. The genotypes Matsumoto White and Phule Ganesh Violet showed good general combining ability for most of the economical traits. The cross combinations Matsumoto Scarlet x Phule Ganesh Violet, Matsumoto White x Phule Ganesh Violet and Matsumoto Red x IIHRJ3-2 showed best specific combining ability for various traits. The variance of *sca* was the highest for all the traits except for plant spread and days to first flowering which indicated the dominance of additive gene effects. Hence, it was inferred that specific combining ability can be exploited for the creation of novel flower colours and phenotypes (Bhargav *et al.*, 2019).

5.2.0 Breeding

China aster is a self-pollinated crop (Strube, 1965; North, 1979 and Watts, 1980). Fleming (1937) estimated approximately 10 per cent natural crossing. Strube (1965) described floral biology of China aster. According to him, its flower head consists of both pistillate ray florets and perfect disc florets. The proportion of ray florets and disc florets is a measure of doubleness of the flowers. As a rule, the stamens and pistils do not mature simultaneously in the individual flowers. The stigma of the individual flower unfolds after the pollen is discharged from the flower. The China aster was, therefore, mentioned as geitonogamous. Novotna (1989) reported that the number of seeds produced by the tubular flowers was several times higher than the number produced by the ligulate flowers in the same anthodium. This difference was probably due to poor pollination of the ligulate flowers.

The breeding objectives of China aster includes : creation of compact habit, tolerance of different climatic conditions, abundant flowering, horizontal placement of flowers on the stalks, long and stiff stalks, size, shape and colour of flowers, keeping quality of flowers, transportability of cut flowers and resistance to fusarium (Varbanova, 1995, Janakiram and Rao, 1996).

While crossing, it should be remembered that the florets in the capitulum become sexually mature from outside towards inside. The outer marginal flowers are only female and are therefore, used for artificial pollination. For this at colour break stage, the ray florets are trimmed up to the level of stigma with scissors or sharp blade and disc florets are removed completely. Care should be taken while removing disc florets without damaging receptacle. The emasculation should be done between 10 am to 2 pm. Third day after the emasculation, pollination is done. The flowers which will provide the pollen are put in the sun for 1 to 2 hours and then beaten out on the trimmed flower. The flowers are covered with butter paper bags soon after emasculation and pollination. The seed will mature after one month of crossing.

Nataraj *et al.* (2020) screened 45 China aster hybrids and their parents developed by crossing in diallel mating (without reciprocals) for growth, flowering, yield and quality traits. All the hybrids out performed with their parents indicating high amount of heterosis present in them. The crosses *viz.*, Local Pink × Local White and Local Violet × AAC-1 and PG Violet × Arka Poornima were early with medium height with erect branching and stalk length, higher number of flowers per plant, number of primary branches and higher flower yield per plant and these hybrids are suitable for cut flower purpose. The cross combinations *viz.* Arka Archana × AAC-1 and Arka Aadya × Arka Poornima are suitable for landscape gardening by witnessing more spreading habit of growth and maximum flowering duration.

Bhargav *et al.* (2018) characterized 42 genotypes for 21 traits as per DUS test guidelines. Among these 21 traits, two were monomorphic, four dimorphic and 15 were polymorphic indicating their potential for varietal characterization. This characterization clearly established distinctness among the different genotypes for various traits. It will be useful for breeders/farmers to identify and differentiate China aster genotypes and to pursue protection under PPV and FR Authority, New Delhi.

Khangjarakpam *et al.* (2014) carried out seed setting studies in China aster genotypes and their cross combinations. Among genotypes, Kamini recorded maximum seed yield per plant (6.23 g) followed by Poornima (3.46 g) and Shashank (2.98 g); maximum 1000 seed weight was recorded in genotype Shashank (2.61), however, maximum number of seeds per flower head was recorded in Kamini (154.20), followed by Shashank (139.60) and Poornima (96.20); highest seed setting per cent was recorded in Kamini (61.19), followed by Shashank (42.00) and Poornima (28.38). Among cross, maximum seed setting per cent was recorded in Kamini x Poornima (43.15) followed by Poornima x Kamini (43.15) and Violet Cuhsion x Shashank (4.87).

The plant apex of *cv.* Gerda and Princess Armida treated with 1.5 and 1.0 per cent colchicine for 5 days increased the number of polyploid plants (Hanzelka and Kobza (2001).

The brief description of China aster genetic stocks registered with ICAR-NBPGR, New Delhi for their unique traits

- ☆ IIHR-42 (IC0610423; INGR14051): A China aster (*Callistephus Chinensis* (L.) Nees.) pure line with creamy white flower colour and powderpuff flower type, resistant to rootknot nematode (*Meloidogyne Incognita* Race 1) (Rao *et al.*, 2015).
- ☆ IIHR-35 (IC0610422; INGR14050): A China aster (*Callistephus chinensis* (L.) Nees.) pure line with violet (83.a) flower colour and pompon flower type (Negi *et al.*, 2015).

- ✰ IIHRCA H13A (IC0610420; INGR14048): A China aster (*Callistephus chinensis* (L.) Nees.) pure line with early flowering and higher number and weight of flowers/plant (Kumar *et al.*, 2015).
- ✰ IIHRCA J17 (IC0610421; INGR14049), a China aster (*Callistephus chinensis* (L.) Nees.) pure line with early flowering, higher number and weight of flowers/plant (Kumar *et al.*, 2015).
- ✰ IIHRV1 (IC0624189; INGR 19099), a China aster [*Callistephus chinensis* (L.) Nees.] germplasm with white flower colour (NN155D), semi-double flower form and pseudo ray florets (Anon., 2019).
- ✰ IIHRV2 (IC0624190; INGR 19100), a China aster [*Callistephus chinensis* (L.) Nees.] germplasm with violet flower colour (N81A), semi-double flower form and pseudo ray florets (Anon., 2019).

6.0 Growth and Flowering

6.1.0 Environmental Factors

Apart from the genetical factors, the major environmental factors influencing growth and flowering of China aster are light and temperature. Certain growth substances have also been used to regulate these processes.

6.1.1 Light

The duration of light has a remarkable effect on growth and development of aster. A day length of less than 14 hours was found to cause rosetting of leaves (Post, 1935). Asters can be flowered year round provided the seedlings are lighted during short days from the time of emergence to flowering (Laurie *et al.*, 1958).

Supplementary light causes early flowering. Lin and Donald (1950) recorded flowering 15 days earlier under high temperature long days than high temperature short days and 30 days earlier at low temperature long days than low temperature short days. Floral initiation occurred more or less at the same time under low temperature long days and high temperature long days. Kofranek (1951) also observed that plants flowered 3 months earlier than normal by giving 4 hours of supplementary light. The investigations carried out by Payne and Kays (1954) revealed that long photoperiods caused axis elongation followed by flower bud initiation but reversal to the short photoperiods after this stage resulted in quicker flower development than when long photoperiods were continued up to flowering. The number of leaves produced by plants before flower initiation was much greater in short day conditions than in the long days. In long days, floral induction was completed about 5 weeks after sowing. Inflorescence development was faster in short days than in long days and in continuous short days or long days plants flowered almost simultaneously. Plants transferred from long days to short days 6 weeks after sowing flowered about a month earlier than those kept in continuous long days (Doorenbos, 1959). Similarly, Gruis and Joiner (1960) noted that plants receiving 7 weeks of long days flowered 12 and 19 days earlier than those receiving 10 and 13 weeks, respectively. Earliness in flowering was, however, associated with shorter stems and smaller and less number of flowers (Table 1). Shedeed *et al.* (1990a,b) treated about 10 cm high seedlings under different lighting treatments : natural daylength (controls), an additional 2 or 4 h light/day for 3 or 6 weeks or they were covered for 2 or 4 h/day with black fabric for 3 or 6 weeks. August produced plants which took the shortest time to flower. The additional light accelerated flowering and increased the total carbohydrates, RNA, DNA and also that of gibberellin and auxin-like substances; total phenols were decreased.

Table 1: The Effect of Levels of Fertilizer and 3 Periods of Long Days on the Average Number of Flowers per Plant in *Callistephus chinensis* 'All Saints'*

Fertilizer treatment	*Light Treatments*			
	L1 (7 weeks)	*L2 (10 weeks)*	*L3 (13 weeks)*	*Fertilizer Means*
F1 (2 lbs)	6.8	7.8	8.8	7.8
F2 (4 lbs)	6.1	9.6	10.7	8.8
F3 (8 lbs)	5.4	7.9	9.5	7.6
Lighting means	6.1	8.4	8.8	
L.S.D.			0.05	0.01
Between lighting means			0.8	1.2
Between fertilizer means			0.5	0.7
Between means within table			0.9	1.2

* Gruis and Joiner (1960).

Hughes and Cockshull (1964) reported that red light interruption for 1 hour in the middle of the 16 h dark period increased the area of leaf plus cotyledons, leaf area ratio, specific leaf area and the dry weight of leaf plus cotyledon and also the total dry weight compared with short days with no interruption. An hour's exposure to red fluorescent light during each 16 h night caused flower induction as well as concomitant stem extension, but transfer back to 8 h days after an inductive period accelerated further flower development and restricted the stem elongation of both the main axis and laterals (Hughes and Cockshull, 1965). It was also indicated that the partition of available dry weight between leaves, stems and roots followed a definite pattern, dependent only upon total plant dry weight. The diversion of dry weight into flowers was strongly accelerated by transfer into 8 h days after induction, but the remaining dry matter still appeared to be distributed between the vegetative parts along the general pattern determined by total vegetative weight. Hughes and Cockshull (1966) stated that China aster was a long day plant and a night break of 2 h improved vegetative growth and flowering and also induced earlier flowering even when the plants received a natural day length of 14 h.

China aster *cv.* Kometa Pink were grown in 8 greenhouse compartments in autumn (natural day length maximum 13 h) under six lamp types : incandescent, fluorescent cool white and blue, compact discharge (SL), low pressure sodium (SOX), light emitting diodes (LED, red) and 2 provisional light sources, far-red and a combination of far-red and fluorescent red. The light levels were applied in a gradient from 0.1 to 2.0 µmol/m^2/s (400-800 nm) from the edge to the centre of the chamber. Light treatments started at sunset and continued until an 18 h photoperiod had been achieved. All lamp types were effective for photoperiodic control except for far-red. Blue was far less effective than the other types. Positive relationship between light intensity and the earliness of flowering was noted (Graaf-van-der-Zande *et al.*, 1992).

6.1.2 Temperature

Temperature is another factor influencing the growth and flowering. Asters at a night temperature of 10°C produced fully double flowers with strong stems. At higher temperatures they will grow reasonably well but will have weaker stem and less number of petals (Biebel, 1937; Laurie *et al.*, 1958).

According to Runger and Wehr (1967), flower initiation took place at a temperature of 15°C and above in both long and short days, although, it was retarded by latter. Floral induction increased sharply with photoperiods between 13 and 15 h, but it was irregular in short days. However, flower development was enhanced by short days and retardation by long days greatly increased with photoperiods between 13 and 15 h; stem elongation and the number of bracts increased with longer days.

6.2.0 Growth Regulators

Plant growth regulators play an important role in morphology and physiology of the plants and their effect varies with plant, species, variety, concentration used, method of application, frequency of applications *etc.*

Lindstrom *et al.* (1957) reported that gibberellic acid (20 mg) applied to stem apices or as foliar spray (10-100 ppm) hastened flowering in China asters. Induction of early flowering and increased stem length, branching and flower size with the application of this substance at 0.01 per cent was also noted by Tamberg (1963). Under long days, GA_3 did not affect the flower initiation but reduced the number of leaves appeared before the emergence of inflorescence in short days (Doorenbos, 1959). Foliar application of GA_3 at 200 ppm on *C. chinensis*, 15 days after transplanting and twice more at 10-day intervals increased the plant height, number of branches and leaves, number of flowers/plant and seed yield (Syamal *et al.*, 1990).

Foliar sprays of 50 ppm GA_3 applied to potted China aster *cv.* Heart of France twice, at 30 and 45 days after planting increased flower diameter, stem length and flowering period and reduced the number of small and deformed flowers (Jayanthi *et al.*, 1990). Potted *C. chinensis* seedlings were sprayed (3 times at 30-day intervals) with GA_3 at 150 ppm hastened flower opening by approximately 5 days, increased plant height (by 5-6 cm) and dry weight (DW) and flower diameter compared with unsprayed controls. Kinetin treatments generally reduced plant height and flower stalk length. The shortest plants were obtained with 10 ppm kinetin + 10 ppm IAA. Both GA_3 and kinetin treatments generally reduced inflorescence DW and the number of flowers per plant (Shedeed *et al.*, 1991a). GA_3 also increased plant N, P, K, DNA and RNA contents, but reduced total phenols content (Shedeed *et al.*, 1991a,b).

Foliar application of SADH resulted in compact growth. The best time to apply the retardant was when the stem began to elongate (Cathey *et al.*, 1963). In a trial with GA_3, MH and NAA each at varying concentrations up to 1000 ppm, Reddy

(1978) observed that the number of flowers/plant and the duration of flowering was enhanced by GA_3 at 200 or 300 ppm, MH delayed flowering while NAA had little effect on growth and flowering. Syamal *et al.* (1990) reported that MH at 400 ppm suppressed vegetative growth, flowering and seed yield. Flower bud appearance was delayed by 12.40, 7.54 and 3.54 days by spraying MH, cycocel and alar at 1500 ppm, respectively, as compared to controls. Alar and cycocel each at 1500 ppm increased the number of flowers but reduced flower diameter. Plant N content was highest with 1500 ppm alar, while P and K contents were highest with 1500 ppm cycocel (Aswath *et al.*, 1995a).

Preharvest spray with alar at 1500 ppm, MH at 500 ppm and cycocel at 500 ppm on flowers of China aster increased the vase life by 2.6, 2.2 and 2 days respectively, when placed in a 2 per cent sucrose solution (Aswath *et al.*, 1995b).

Application of GA_3 (200 ppm) followed by Boron (0.1 per cent), MH (500 ppm) and TIBA (100 ppm) increased the seed germinability and seedling vigour. Among the interactions, maximum seed germination and seedling vigour were recorded in 240: 160: 80 kg N: P_2O_5: K_2O/ha either with GA_3 (200 ppm) (94.1 per cent) or Boron (0.1 per cent) (92.3 per cent) or MH (500 ppm) (90.3 per cent) or mepiquat chloride (500 ppm) (89.1 per cent) spray once at 30 days after transplanting and another at ray floret initiation stage (Doddagoudar *et al.*, 2004). Foliar spray of salicylic acid (200 mg/l) is recommended for increased growth, flower and seed yield in China aster *cv.* Kamini as it is cheaper than GA_3 (Kumar *et al.*, 2015).

Foliar spray of GA_3 at 250 ppm significantly increased plant height, leaf area per plant and number of leaves per plant, induced early emergence of flower bud to full blooming of flowers, longer duration of flowering, improved flower longevity, flower fresh and dry weight, flower diameter and enhanced flower yield in terms of number of flowers per plant and flower weight per plant (Sharma and Joshi, 2015), however, shelf life, vase life, flower weight and flower diameter was improved by Azospirillum (8 per cent) and Biovita (1 per cent) (Vinutha *et al.*, 2017).

Vijayakumar *et al.* (2017) observed that foliar spray of GA_3 at 150 ppm at 45 days and 60 days after transplanting increased plant height, plant spread, number of branches and nodes plant per plant, longer internodal length and leaf total chlorophyll, chlorophyll 'a' and 'b' contents. Therefore, foliar application of GA_3 at 150 ppm would be effective to improve the growth and yield of China aster; GA_3 (150 ppm) at 30 days after transplanting resulted in early flowering and increased number of flowers per plant, fresh weight and dry weight of flowers (Maurya *et al.*, 2017), while, GA_3 (150 ppm) sprayed twice on three varieties of China aster resulted in maximum leaf area, stem girth and girth of flower stalk (Mishra *et al.*, 2018). GA_3 at 50 ppm applied thrice resulted in maximum flower diameter, maximum fresh weight and dry weight of flower (Mishra *et al.*, 2018).

Spraying of GA_3 at 200 ppm produced maximum plant height, number of secondary branches per plant and plant spread, earliest first flower bud initiation and flower bud opening and duration of flowering, stalk length, flower diameter, flower weight, number of flowers per plant and per plot, test weight, seed yield per plant and plot, gross return, net return and benefit cost ratio (Kumar *et al.*, 2003;

Munikrishnappa and Chandrashekar, 2014; Kumar *et al.*, 2015; Gupta *et al.*, 2015; Vinutha *et al.*, 2017; Rawat *et al.*, 2018; Sindhuja and Prasad, 2018) (Table 2).

Table 2: Effect of different Plant Growth Regulators on Yield and B : C Ratio of China aster *cv.* Shashank*

Treatments	*No. of Flowers/ Plant*	*Flower Yield/ Plant (g)*	*Total Flower Yield (t/ha)*	*Benefit Cost Ratio*
Control	33.89	99.275	3.17	1.4:1
GA_3 @ 50 ppm	63.5	300.63	9.62	4.19:1
GA_3 @ 100 ppm	65.50	320.65	10.26	4.41:1
GA_3 @ 150 ppm	67.01	339.40	10.86	4.62:1
GA_3 @ 200 ppm	71.54	393.42	12.58	5.30:1
NAA @ 50 ppm	62.09	289.08	9.25	4.06:1
NAA @ 100 ppm	65.49	307.14	9.82	4.02:1
NAA @ 150 ppm	61.63	281.35	9.00	3.92:1
NAA @ 200 ppm	59.16	258.46	8.27	3.59:1
CCC @ 500 ppm	50.58	183.59	5.87	2.58:1
CCC @ 1000 ppm	55.4	215.94	6.91	3.03:1
CCC @ 1500 ppm	57.29	232.94	7.45	3.26:1
CCC @ 2000 ppm	53.62	201.29	6.44	2.81:1
SEd(±)	0.86	30.49	0.98	
C.D. at 5 per cent	1.78	62.94	2.01	

* Sindhuja and Prasad (2018).

Application of Asahi SL (sodium ortho and para-nitro phenolate, sodium 5 nitro guaiacolate) applied during flower bud development on tertiary stems or during flowering of flowers on tertiary capitula, increased seed yield, germination and vigour and did not modify vegetative growth of plants (Górnik and Grzesik, 2005).

Foliar spray of CCC at 1500 ppm recorded highest flower yield and flower quality. Soil drench of paclobutrazol at 200 ppm delayed flowering, while, early flowering was observed with 25 ppm. Application of MH at 1000 ppm resulted in minimum plant height, inter nodal length, number of leaves/plant and leaf area in China aster (Munikrishnappa and Chandrashekar, 2014). Different concentrations of MH were effective in reducing plant height and improving various quality traits in China aster *cv.* Poornima (Rawat *et al.*, 2018).

Pinching at 25 DAT with fertilizer application (270:180:100) and foliar spray of MH (500 ppm) followed by CCC (200 ppm) on same day of pinching resulted in marked increase in number of branches, leaves and flowers per plant with higher seed yield and quality (Gnyandev *et al.*, 2014).

7.0 Propagation

Aster is propagated by seeds (12000 seeds/28 g) and will germinate in a week at about 21°C (Auman, 1980). Seeds of asters have no dormancy and freshly

Propagation and Growing of China aster

Seeds

Seed Germination

Seedlings in Protrays

Polyhouse Production

Intercrop with Brinjal

Open Field Cultivation

Source: **www.shutterstock.com, www.fi.pinterest.com, www.floretflowers.com, www.s1.1zoom.me, www.iihr.res.in, www.i.ytimg.com, www.gardeningknowhow.com, Jyothi and Thomas (2016)**

Micropropagation

Production under Retractable Tunnels

Source: www.shutterstock.com, www.fi.pinterest.com, www.floretflowers.com, www.s1.1zoom.me, www.iihr.res.in, www.i.ytimg.com, www.gardeningknowhow.com, Abraham and Thomas (2016)

harvested seeds have a narrower temperature requirement for germination than older ones (Ichihara, 1954). Domsch (1955) observed that healthy seeds produced vigorous plants at temperatures of 12°, 16° or 20°C after 4 days of sowing whereas seeds infected with *Botrytis cinerea* germinated slowly at lower temperature and produced poor quality plants than at higher temperatures. In Japan, Yasuda (1955) recorded 70-80 per cent germination in local aster seeds. According to Massante (1963), the seeds germinate at a temperature range of 10-35°C, the optimum being 21°± 4°C (Thompson and Cox, 1979). They also observed that at temperature between 11°C and 26°C the reduction in germination was found to be 7 per cent, while at 31°C germination was less by nearly 20 per cent. Carneiro (1996) obtained highest percentage (78.28 per cent) and early germination at 20°C. No germination was observed at 35°C. Seed germination was limited by temperature conditions and the requirement of light was of little practical importance but water stress immediately following sowing reduced seedling production.

Simulated acid rain (pH 1, 2, 3 or 4) inhibited the germination of *C. chinensis* seeds. Germination percentage and rate decreased with increasing acidity. Seeds did not germinate at pH 1 while pH 2 lowered the flower quality of China aster. This effect was observed 1-2 days after treatment (Lee *et al.*, 1996).

The main differences in seed production technology are time of seeds sowing, risk of killing plants by winter, healthy status of plants, number of off-type, frequency of seed harvesting, seed yields and germination capacity of seeds. The profitability of China aster was 865 z³ (1 euro=3.2 z³)/0.1 ha (Houbowicz and Borysiak, 2008).

Among 7 China aster genotypes evaluated for seed production, highest seed yield/plant was recorded in Arka Poornima (3.45 g), whereas, lowest seed yield/plant was recorded in Arka Aadya (1.87 g) (Chakraborty *et al.*, 2019).

Padmini *et al.* (2014) reported that effect of boric acid on improving seed yield in powder puff type could be attributed to increased seed setting, improved pollen viability, increased average seed weight/head and number of filled seeds/head.

Mathad *et al.* (2009) found that the seeds harvested at 42 days after flowering (DAF) recorded maximum capitulum diameter, capitulum weight, 1000 seed weight, seed weight per capitula, seed germination, rate of germination, root length, shoot length, seedling dry weight and seedling vigour index. These parameters increased from 7 to 42 DAF then onwards gradually decreased up to 56 DAF. Moisture content in seeds gradually decreased from 7 to 56 DAF.

Seeds primed with GA_3 (200 ppm) found best for shoot length, root length, leaf length, leaf width, fresh and dry weight of seedlings, while, unprimed seeds failed to exert any significant effect on seedling parameters. Thus, priming with GA_3 (200 ppm) could be used as an effective method to enhance seedling growth in China aster *cv.* Kamini (Sidana *et al.*, 2019).

Seed priming (PEG 8000 at -1.25 MPa) and fungicide treatment combined with PEG on the germination of two seed samples of China aster at 30°C, 20°C and 10°C revealed that priming seeds in PEG alone and combined with fungicide significantly reduced mean germination time and increased germination index at all three temperatures. Combining fungicide treatments with PEG primary caused a significant decline in the percentage of seeds infested with fungi compared to PEG priming alone (Zhao *et al.*, 2004).

Effect of different priming methods on seed quality, biochemical changes and storability of China aster by using two different ages of the seed *viz.*, one year old seed and half year-old seed and four treatments *viz.*, hydro priming, osmo-priming, halo priming, unprimed as control as another factor was carried out. Seed samples were primed and kept in poly ethylene bag and stored for six months at ambient condition. Six months old seed recorded maximum germination percentage, field emergence, speed of germination, seedling length, seedling dry weight and seedling vigor index and least was observed in one year old seed. Priming of seed with KNO_3 at 0.5 per cent resulted in best performance of the seed regarding all the physiological and biochemical parameters followed by hydro priming. However, storage of the six months old KNO_3 treated seed up to six months was found to be good compared to one year old seed regarding all the quality parameters (Vimala and Pratap, 2014).

Rosinska *et al.* (2009) divided seeds of 3 commercial China aster lots into 3 grades with different colours: dark brown, brown and light brown, and 2 grades with a different size: length below 3.9 mm (small) and above 3.9 mm (large). Results showed that the size of seeds had no effect on their germination. The dark brown seeds germinated better than the light brown ones. Removing light brown seeds from the China aster seed lot improved their germination.

Han (2010) studied the effects of microwave on China aster seed germination, vigour and infestation with fungi, at temperature 20°C. The seeds were treated in microwave oven (power 850 W heating power) in dry conditions and in water for 10, 20, 30, 40, 50, 60, 90, 120 and 180 s, respectively. Treating seeds in dry condition for 120 and 180 s and in water for 180 s, resulted in loss of seed viability. Dry microwave seed treatment for time longer than 20 s negatively influenced total number of germinating seeds. Seeds treated in water maintain high values of maximum germination up to 90 s of exposition. In general, microwave treatment had no positive effect on seed vigour. Treating seeds with Kaptan Zawiesinowy 50 WP control seed infestation with fungi most effectively then both microwave seed treatments, however, negatively influenced mean germination time and uniformity of germination.

Gnyandev *et al.* (2012) tested storage potential of China aster varieties. The variety Kamini found relatively best in storability with high germination and seedling vigour with low electrical conductivity up to six months of storage. Seeds stored in aluminium foil and polythene bags maintained higher germination and seedling vigour parameters compared to seeds stored in paper and cloth bag.

Yogeesha *et al.* (2017) studied the effects of ultra-dry seed storage on longevity of China aster under ambient and controlled temperature (15°C). Seeds with 5.9 per cent moisture showed decline in seed viability and vigour under ambient temperature after 18 months of storage and rapid decline was noticed after 24 months of storage, reaching zero at 36 months but at 15°C showed no decline in viability up to 48 months of storage. Ultra dry seeds maintained higher viability and vigour at both ambient and at 15°C even after 48 months of storage.

Rao and Kumar (2019) suggested that seeds stored in polythene bags and kept in cold storage at 2°C increased viability to 24 months. Chakraborty *et al.* (2020) recommended that the freeze condition for storing China aster seeds is better for maintenance of seed quality.

8.0 Soil and Climate

China asters grow best in rich, well-drained loamy soil and prefer open sunny situation. Heavy soils with high calcium are not suitable. Although it is a winter annual in the open in tropical and subtropical climate, it is grown in summer in temperate countries, but China aster is not frost hardy. Temperature range of 20-30°C during day and 15-17°C during night with relative humidity of 50-60 per cent is the most suitable weather condition. It needs sufficient sunlight for better growth and flowering, limited rains of 500-700 mm spread over June to September, followed by provision of frequent but light irrigation. Asters can be grown year round in greenhouse by maintaining optimum temperature and light. Supplementary lighting should be provided from the seedling stage until flowering where the day length is short. Artificial lighting is not necessary from May to August (Auman, 1980). In mild subtropical climate, asters are also grown outdoors in most parts of the year.

9.0 Cultivation

9.1.0 Sowing Time

In Bangalore, the crop can be sown round-the-year. The normal sowing season is between May and June (Jayanthi and Mukhopadhyay, 1988). The seed rate is 650 g/ha. Seeds can be sown in seed pans for small quantities of seedlings. For commercial cultivation, seeds are sown in raised beds. The raised beds are prepared with fine mixture of red soil, sand and farm yard manure (1:1:1 v/v). The nursery beds are drenched with captan or dithane M-45 @ 2 g/l of water before sowing the seeds to avoid fungal diseases like damping off *etc.* Seedlings can also be raised in pro-trays with cocopeat as a substrate. The seeds are sown thinly about 0.5 cm deep in rows across the length at 10-12 cm apart and covered with a mixture of soil and farm yard manure. After sowing, beds should be covered with shade net (50 per cent) and watered gently with a rose can. After germination, shade net may be removed (Rao and Kumar, 2019).

In South Korea, Hong and Jeong (1988) sowed the seeds of *C. chinensis* at monthly intervals from February to July. Later sowing produced early flower buds but duration of flowering was shorter, whereas late sowing resulted in a shorter duration of flowering. Plant was tolerant of the high summer temperatures and RH characteristics of Korea and they flowered over a long period, even with the latest sowing.

9.2.0 Transplanting

Aster seedlings are usually transplanted when they have developed about three to four leaves and danger of frost is over. Before transplanting, they should be hardened off. Planting should preferably be done in the evenings to avoid bright sunlight and the beds be well prepared by adding sufficient organic manure. Gowda and Gowda (1990) concluded that overall, July, August and September are the best months for transplanting of *cv.* Ostrich Plume Purple seedlings in Bangalore to obtain maximum number of flowers per plant (58.82), flower yield (12128 kg/ha) and the longest peduncle (15.10 cm). After planting, the soil around the seedling is to be firmed and watered thoroughly.

Planting at 15th September and single pinching at 20 DAT recorded maximum number of primary and secondary branches, plant spread, leaf area and stem diameter. Early emergence of first flower bud, full opening of flower from bud initiation, 50 per cent flowering and first harvesting were recorded in planting at 15th September and no pinching. However, maximum duration of flowering was obtained in 15th September planting and single pinching at 20 DAT (Gaidhani *et al.*, 2020) (Table 3).

Kaushal *et al.* (2014) found that planting at 6 April recorded best results for plant height, plant spread, number of flowering stems per plant, number of flowers per plant and yield of flower cut stems per plot. However, 5 June planting gave best results for flower size and vase life.

Table 3: Effect of different Planting Dates on Growth and Flowering of China Aster*

Planting Dates	*Plant Height*	*Number of Secondary Branches*	*Plant Spread (cm)*	*Days to 50 per cent Flowering (Days)*	*Duration of Flowering (Days)*
1st September	42.09	17.11	25.59	62.51	25.27
15th September	43.87	19.61	28.05	59.76	28.87
1st October	40.26	15.77	23.60	67.62	23.44
15th October	38.82	13.50	20.95	72.03	20.56
SE(m)	0.72	0.57	0.36	1.20	0.69
CD at 5 per cent	2.11	1.68	1.05	3.52	2.04

* Gaidhani *et al.* (2020).

9.3.0 Pinching

Pinching of main shoot at one month after transplanting increases number of nodes, branches, flowers per plant and flower yield per unit area (Rao and Kumar, 2019). Pinching reduced the plant height and delayed flowering, however, pinched plant yield more than unpinched plant (Khobragade *et al.*, 2012). Single pinching at 30 days after transplanting recorded maximum primary branches, stem diameter, weight of flower, diameter of the flower, longevity of intact flower and vase life (Sailaja and Panchbhai, 2014). Significantly higher plant height was recorded in unpinched plants, while, plants pinched at 20 DAT recorded higher number of primary branches per plant, number of secondary branches per plant and plant spread when compared to other pinching treatments (Kumar *et al.*, 2014). However, double pinching recorded more number of branches/plant, flowers/plant, flower yield/ha and flowering span, whereas, no pinching recorded maximum plant height, diameter of flower and vase life and earliest first flower bud initiation (Chopde *et al.*, 2018).

9.4.0 Spacing

Yokoi and Hosaka (1962) recommended a spacing of 30 x 20 cm or 30 x 15 cm for maximum yield of good quality cut flowers. According to Auman (1980) seedlings should be spaced 30 x 30 cm in clothhouses and 20 x 20 cm in greenhouses. In an experiment under field conditions, Yadav (1983) recorded the highest yield of flowers and seeds when seedlings were planted in the first week of October (autumn) at a distance of 20 x 30 cm having density of 166666 plants per hectare. Patil *et al.* (1987) obtained highest number of cut flowers per plant (33) and seed yield (121.7 kg/ha) when *cv.* Ostrich Plume Blue were transplanted in November at 20 x 30 cm spacing. The poorest seed yields were obtained with December planting (92.1 kg/ha). Dhemre *et al.* (1997) reported that winter planting of *cv.* Powder Puff with spacing at 15 x 15 cm gave the best results in terms of number of days required for emergence of flower buds, days required for flower opening from bud emergence and days required for first flower opening. Flowering duration was greatest at the 30 x 30 cm spacing.

Closer spacing 20 cm × 10 cm produced taller plants and flower production was higher per unit area than wider spacing. It was found that closer spacing with pinching gave higher yield (56.75 kg/plot) and economically best than wider spacing with unpinched plants (Khobragade *et al.*, 2012). Further, Kumar *et al.* (2014) revealed that closer spacing of 30 x 15 cm recorded significantly higher plant height, while, wider spacing of 45 x 30 cm recorded more number of primary and secondary branches per plant and plant spread.

A spacing of 30 cm between plant to plant and row to row is recommended (Rao and Kumar, 2019). However, planting geometry 50 cm x 50 cm recorded significantly highest flower yield (109 g/plant) and seed quality characters (Pandey and Rao, 2014; Pandey *et al.*, 2017).

9.5.0 Irrigation

Asters require continuous soil moisture throughout the entire period of growth and development. Suzuki (1979) compared three soil moisture regimes (pF 2.0, 2.5 and 2.8) and observed that plants grew best with medium regime (pF 2.5). Similarly, Kobza *et al.* (1984) studied the effect of supplementary irrigation and found it highly beneficial. Of four methods compared, 1 or 2 application of 50 mm water/month was the most suitable. The next best method was growing the plants under glass and applying 20 mm water according to tensiometrically determined requirements. Potted *C. chinensis cv.* Ostrich Plume seedlings (1/pot) were watered at 3-day intervals and treated with saline water (2, 4, 6 or 8 dS/m) every 20 days. Seedlings in the 6 and 8 dS/m treatments failed to establish. Flower yield declined from 49.8 g/plant in controls (no saline water) to 26.3 g/plant in the 4 dS/m. N, P and K contents in leaves, stems and flowers decreased with increasing salinity, while Na, Ca and Mg contents increased (Mantur *et al.*, 1996).

Irrigation requirement depends upon the weather, type of soil and season of the crop grown. Since China aster is a shallow rooted crop, it requires irrigation at an interval of 3 to 7 days depending on soil moisture. Drip irrigation will help in better crop growth and also saves irrigation water (Rao and Kumar, 2019).

9.6.0 Manuring and Fertilization

Proper manuring and fertilization are very important for aster cultivation. Deficiencies of different nutrients result in poor growth and flowering. Nitrogen deficiency caused the most marked symptoms, namely stunted growth and the production of fewer flowers which were dull in colour. Phosphorus deficiency also markedly retarded vegetative growth and delayed flowering (Bose and Das, 1966).

Nitrogen (300 kg/ha) greatly influenced the vegetative characters such as plant height, plant spread, primary branches/plant, fresh weight and dry weight of plants but delayed first flower to open. Maximum number of flowers per plant and yield of flowers were recorded with 200 kg N/ha. However, highest nitrogen content of plant at harvest (1.68 per cent) was recorded at 300 kg N/ha and uptake of nitrogen at harvest (12.54 kg/ha) was highest with 200 kg N/ha (Chavan *et al.*, 2010).

Maximum plant height, plant spread, stem girth and number of secondary branches, flower initiation and flowering, flower stalk length, flower diameter, ray florets, number of flower buds/plant, duration of flower as well as field-life was recorded with N 150 kg/ha. The same dose also showed earliness in flowering, increased flower stalk length, flower diameter, ray florets, number of flower buds/plant, duration of flower as well as field-life. Thus, application of N 150 kg/ha was more beneficial to get higher yield and quality flowers (Kumar *et al.*, 2016).

N at 300 kg/ha in two splits (first half as basal and remaining half at 30 days after transplanting) and 200 kg P_2O_5/ha as basal dose has been found the best for var. Poornima (Maheta *et al.*, 2016) (Table 4).

Table 4: Effect of Nitrogen and Phosphorus on Growth and Yield of China Aster cv. Poornima*

Treatments	*Plant Height (cm)*	*No. of Branches/Plant*	*Number of Flower per Plant*	*Yield of Flower (t/ha)*
Nitrogen (kg N/ha)				
N_1 = 150	42.44	14.44	20.22	19.33
N_2 = 200	42.67	14.67	21.89	20.00
N_3 = 250	44.68	15.67	22.22	21.22
N_4 = 300	51.56	17.67	24.78	22.67
S.E. ±	1.14	0.63	0.74	0.54
C.D. (P=0.05)	3.34	1.85	2.16	1.59
Phosphorus (kg P_2O_5/ha)				
P_1 = 100	42.25	14.50	21.25	20.00
P_2 = 150	46.75	15.42	22.00	20.33
P_3 = 200	47.08	16.92	23.58	22.08
S.E. ±	0.99	0.55	0.64	0.47
C.D. (P=0.05)	2.89	1.60	1.87	1.37

* Maheta *et al.* (2016).

Application of 240: 160: 80 kg N: P_2O_5: K_2O/ha recorded higher seed germination and seedling vigour as compared to 180: 120: 60 kg N: P_2O_5: K_2O/ha (Doddagoudar *et al.*, 2004).

Merlo (1955) studied the influence of mineral fertilizers and reported that application of a complete fertilizer greatly improved growth, flowering and seed production and wilt resistance in asters. With the cultivar Hercules, best results were obtained with 34 g ammonium nitrate, 50 g superphosphate and 18 g potassium salt per sq m. According to Mantrova (1959), the best manurial treatment for asters which produced strong plants with abundant bud formation and high quality flowers, was found to be an autumn application of decayed manure as 5-6 kg per sq m, followed by a pre-planting application of 45-60 g sulphate of ammonia, 45-60 g superphosphate and 15 g potassium chloride per sq m. Spraying the plants every 10 days with 0.02 per cent $KMnO_4$ increased plant height and flower yield. Cherneva (1976) concluded that nitrogen was the most effective nutrient, significantly increasing flower number, diameter and yield of seeds. Top dressing with 120 kg N, 80 kg P_2O_5 and 120 kg K_2O/ha at the 10-12 leaf stage gave better results than at bud formation or shortly before flowering.

Investigation on the influence of major nutrients and the time of application of fertilizers on the alluvial soil revealed that a basic NPK application and a top dressing of NPK at the seedling stage (10-12 leaf stage) up to a total of 240 kg each of N, P_2O_5 and K_2O gave the best flower yield. Flowers were larger, if the top dressing was applied just before flowering (Tschernewa-Dimowa, 1976). Debnath and Maiti (1998) obtained more number of flowers/plant of *C. chinensis cv.* Ostrich ranged from 9.7 with no fertilizer to 17.1 with 10 g N + 20 g P + 20 g K/m^2.

West *et al.* (1981), however, found that with increase in the levels of NPK, there was a corresponding reduction in dry weight, leaf area and rate of leaf growth. Yadav (1983) concluded that application of 40 kg N alongwith 200 kg P_2O_5/ha produced the highest yield of flowers. For seed production, however, treatment with 300 kg N was adequate.

A fertilizer dose of 90 kg nitrogen, 60 kg phosphorus and 60 kg potash per hectare at the time of field preparation and top dressing of nitrogen at 90 kg/ha 40 days after transplanting is recommended under Bangalore condition (Rao and Kumar, 2019).

Effects of different manures [farmyard manure (FYM); mechanical plant compost (MPC), prepared from city waste (refuse compost); and phosphocompost (PC), prepared from plant waste with or without enrichment with low grade ground rock phosphate and with iron pyrites at (0, 5 or 10 per cent) and chemical fertilizers (NPK), in several combinations], on the growth and cut flower yield of *C. chinensis* was investigated by Sreenivas *et al.* (1998). Plant height and number of leaves were greatest with FYM + NPK at the recommended rate. Flowering was delayed with the appli-cation of FYM (15 t/ha) alone. The highest number of flowers/plant was recorded with the FYM + NPK at the recommended level, followed by MPC + NPK. The maximum yield of flowers/ha (846 kg/ha) was achieved with plant waste enriched with rock phosphate and 10 per cent from pyrites + NPK, followed closely by FYM + NPK (288 kg/ha). Sreenivas and Gowda (1999) concluded that application of the recommended dose of NPK and 15 t FYM/ha or 8 t phosphocompost/ha is optimal for high flower yields. Kulkarni *et al.* (1996) recorded flower yield of China aster *cv.* Ostrich Plume Mixed in excess of 11 t/ha by applying 15 t FYM/ha with the recommended rate of fertilizer (RDF), 2.5 t vermicompost/ha with 75 per cent RDF, 5 t vermicompost/ha with 50 per cent RDF or with *in situ* vermiculture with 2000 earthworms/ha. Munikrishnappa and Chanshekar (2014) reviewed role of integrated nutrient management in China aster. The recent approaches of saving fertilizers by the use of fertigation will enhance the nutrient and water use efficiency.

Das (2013) concluded that the vermicompost as an organic manure could be used as an alternative of the chemical fertilizer. Nethra *et al.* (1999) also obtained higher yield (34.33 flowers/plant; 6840 kg/ha) after application of 10 t vermicompost/ha + 100 per cent NPK. The longest vase life (11.16 days) of cut flowers was recorded with the 15 t vermicompost/ha + 50 per cent NPK treatment while in control 8.83 days.

Gigaspora margarita (VAM) inoculation under field condition gave the greatest increases in shoot DW, P uptake and flower yield of *cv.* Ostrich Plume at 75 per cent of the recommended P rate (120 kg/ha), suggesting a net saving of 25 per cent of the recommended P (Naik *et al.*, 1995, 1999).

NPK 100 per cent + Azospirillum + PSB imposed plants recorded maximum leaf area, early days to 1st flowering and 50 per cent flowering, flower stalk, flower diameter. N 75 per cent and P + 100 per cent K+ Azospirillum + PSB registered higher yield per hectare (7.87 lakhs/ha), while, 100 per cent NPK + Azospirillum + PSB recorded maximum vase life (8.67 days). Application of bio-fertilizers significantly enhanced nitrogen content in leaves. Hoverer, 100 per cent NPK + Azospirillum + PSB recorded higher N content (4.22 per cent), P content (0.48 per cent) and K content (2.47 per cent) respectively; 75 per cent N and P + 100 per cent K+ Azospirillum + PSB registered maximum benefit cost ratio of 4.94 (Sowmya and Prasad, 2017).

NPK 70 per cent + Vermicompost + Biofertilizer with April planting, recorded maximum plant height, plant spread, early blooming, number of flower per meter square, azotobacter count and Arbuscular Mycorrhizae (AM) spore count (Abrol *et al.*, 2019). Use of Azo + PSB + VC + 50 per cent RDF helped in realizing, higher quality of flower ield and C : B ratio, in the production of China aster *cv.* Prince in eastern Uttar Pradesh (Kumar *et al.*, 2019).

FYM at 18 ton/ha (50 per cent) + vermicompost at 6 ton/ha (50 per cent)+ PSB @ (50 ml/15l) + Azotobacter at 30ml/15l recorded maximum number of primary branches, plant spread, number of leaves per plant, duration of flowering, number of flowers per plant, number of flowers per m^2, vase life and early bud initiation and flowering. The quality traits *viz.*, stalk length, flower diameter and average weight of flower were found maximum from vermicompost at 9.6 ton/ha (80 per cent) + PSB at 50 ml/15l +Azotobacter at 30ml/15l in *cv.* Kamini (Bohra *et al.*, 2019).

For cut flower production, 70 per cent RDF + Biofertilizer + Vermicompost with April planting improved plant height, plant spread, flower diameter, number of days taken to bud formation, days taken to blooming, vase life, number of cut flowers per plant, azotobacter count and AM spore count (Abrol *et al.*, 2020).

Seeds of *C. sinensis* grown for cut flower production, were inoculated with *Glomus manihotis, G. versiculiferum, G. intraradix* [*G. intraradices*], *G. aggregatum, G. versiforme* or *Scutellospora pellucida* by pelleting the seed. All pelleting treatments improved plant development, with *G. mosseae, G. versiculiferum, G. versiforme* and *S. pellucida* producing the best results (Soroa *et al.*, 1998).

China asters respond well to micronutrients. Application of Zn, Cu, B and Mn improved the quality of flowers. Boron in excess of 1 g of boric acid per sq m was found phytotoxic. Asters were moderately sensitive to salinity caused by sodium salts which resulted in decreased plant height and flower size, excess B did not have these effects even when leaf necrosis was severe. Boron injury was severe when the B concentration in old leaf tissue was 1000 ppm of dry weight (Kohl *et al.*, 1957).

Foliar sprays of boric acid (0.1 per cent) twice (once at one month after planting and repeat spray after 15 days) recorded highest seed yield/plant of 4.99g and seed yield of 554 kg/ha followed by PBZ 200 ppm (4.59 g) and (509 kg/ha), respectively which was significantly higher than untreated control (1.82 g/plant and 201 kg/ha) in *cv.* Poornima (Padmini *et al.*, 2013).

Spray of ferrous sulphate (0.2 per cent) recorded maximum plant height, diameter of main stem number of leaves, plant spread, number of laterals, length and width of longest leaf, green and dry weight of plant canopy, days to visibility of flower bud, colour break, full blooming and harvesting of floral heads, fresh weight of floral head, length and width of floral head of floral head, length of floral stalk and diameter of floral stalk in *cv.* Princess (Verma *et al.*, 2018).

9.7.0 Weed Control

Control of weeds is essential in aster cultivation. Manual weeding, though easy and safe, is very expensive and may not be economic for large-scale cultivation. Certain investigations have been made to select suitable herbicides and their effective doses. Application of CDEC and chlorpropham at 6-9 lb/acre and decthal (dimethyl 2, 3, 5, 6-tetra chloroterephthalate) at 8-12 lb/acre 4-6 weeks after transplanting was found very effective in controlling the weeds (Byrne and Lert, 1962). Pre-plant treatment with liquid or granular forms of EPTC and vernolate (S-propyl dipropyl thio carbamate) at 7.5 lb/acre and trifluralin and dichlobenil at 2 lb/acre also showed excellent weed control (Haramaki, 1965). Of the herbicides tested by Baranowski (1974), the largest number of flowers was obtained after treatment with nexoval (chlorpropham) or alipur (cycluron + chlorbufam) each at 4 l/ha. Treatment one day before planting out was recommended. All herbicides gave good weed control but tenoran (chloroxuron) at 6 kg/ha applied 10 days after planting was phytotoxic. Haramaki and Kuhns (1983) reported that treatment with napropamide and oryzalin gave excellent control of grasses and broad-leaved weeds. They did not record any phytotoxic effect of these chemicals. Phetpradap and Hampton (1991) reported that a single application of oryzalin (3.75 kg/ha at 4 days after transplanting) provided excellent weed control and a 10-fold increase in seed yield to 568 kg/ha of *C. chinensis cv.* Powder puff and Kurenai. Pre-emergent diuron (0.25-1.25 kg/ha), simazine (1.0-1.5 kg), metolachlor (1.0-1.5 kg), alachlor (1.0-1.5 kg) and pendimethalin (0.75-1.25 kg) applications reduced the number of weeds in China aster field, with the exception of sedges. Higher numbers of branches/plant were observed after treatment with diuron (1.25 kg). Basavaraju *et al.* (1992) observed that China aster tolerated all treatments except pendimethalin (1.25 kg). Diuron (1.0 kg) gave higher crop LAI, maximum number of flowers/plant and highest flower yield/ha.

Phetpradap and Hampton (1991) reported that pre-sowing application of trifluralin at 2 kg/ha significantly reduce weed population but did not affect the emergence of Chinese aster seedlings. Oryzalin (applied at 4.5 kg/ha 10 days after sowing) provided the best aster seedling survival (83.2 per cent).

Herbicide Oxyfluorfen 0.10 kg *a.i.*/ha followed by earthing up at 35 DAT and Metolachlor 1.0kg *a.i.*/ha followed by earthing up at 35 DAT were effective in controlling weeds, which eventually decreased the dry matter production of weeds (Kumar and Gowda, 2010).

Pendimethalin 0.1 kg *a.i*/ha followed by hand weeding at 30 DAT recorded significantly increased number of flowers, yield and quality of flowers. Days to first flower opening, 50 per cent flowering, duration of flowering, flower diameter, fresh weight of flower, number of flowers per plant, flower yield per plant (g) and vase life exhibited maximum with pendimethalin (30EC) at 1.0 kg *a.i*/ha followed by hand weeding (Nagapushpa *et al.*, 2018).

10.0 Diseases and Pests

Asters are susceptible to a number of diseases and pests. Piwoni (1991) reported that *Alternaria alternata, Botrytis cinerea, Fusarium culmorum, F. oxysporum* and *F. solani* were the most important diseases of *C. chinensis*. Aster yellows, rust, *etc.*, can become problem in aster cultivation. Leaf-hopper, aphids, spider mite, thrips, whitefly, mealybug and caterpillar are important among the pests.

10.1.0 Diseases

Fungal Diseases

10.1.1 Wilt

This is a very serious disease generally caused by *Fusarium oxysporum* f. *callistephi*. *Acrostalagmus vilmorinii* and *Verticillium albo-atrum*, together with *F. oxysporum* f. *callistephi* also cause this disease (Hoffmann and Reber, 1963). Another species, *F. Conglutinans* var. *callistephi* was reported by Stankova-Opocenska (1957).

Phialophora fastigiata was recorded for the first time causing vascular wilt disease of aster in New Zealand. Studies on the wilting and withering of hot bed and field grown plants revealed the cause to be mainly *F. oxysporum*. It frequently occurred with soil fungi (*Botrytis cinerea, Rhizoctonia* sp.) which accelerated withering (Rosner, 1985).

Petrenko (1976) suggested certain cultural practices to reduce the infection. These included pre-soaking of seeds in micronutrient solutions, combined in unfavourable years with additional spraying during bud development and early sowing for seedling production. Autumn or early spring sowing outdoors decreased infection and increased flower production. There are some resistant cultivars which should be used for cultivation (Hoffmann and Kotter, 1963; Petrenko, 1974). Skrzypczak and Orlik-owski (1996) grew aster (*Callistephus chinensis*) in a substratum and regularly fertilized with top-dressing medium containing calcium and Fe-EDDHA, fusarium wilt was significantly reduced. Flachsbart (1972) reported that *cvs*. Bouquet Scharbach and Balls Blue were resistant to the *callistephi* race of *F. oxysporum* and suitable for growing repeatedly on same site. Kratka and Duskova (1991) also reported that cultivars Srdce Francie and Bukett Bila were resistant, and Gaja and Denisa were susceptible to *Fusarium oxysporum* f. sp. *callistephi*.

10.1.2 Collar or Root Rot

This disease is caused by *Phytophthora cryptogea* and the symptoms are similar to those caused by wilt fungus. Stems and roots of infected plants appear water soaked and black. The rot caused by the fungus is more pronounced and lacks the pink coating of spores characteristic of plants infected by wilt. The control measures include strict limitation of irrigation and avoiding planting in the vicinity of alternative hosts (Pointsi and Feldman, 1959). If possible, planting should not be done in the field which has shown the disease in previous years. If it is necessary to use the same beds, the soil should be sterilized (Pirone *et al.*, 1960).

Combined soil application of talc-based formulation of ANGRAU Pf1 at 2g/kg of soil and thiophanate methyl (0.1 per cent) was found to be superior by recording significantly high per cent germination and least root rot incidence of 16.7 per cent (Babu *et al.*, 2013).

10.1.3 Gray Mould

This disease, caused by *Botrytis cinerea*, may appear 3-4 days after sowing or 10-12 days after pricking out the seedlings, and spreads rapidly during cool and humid weather. It has been known to cause or accompany blossom blight. The disease can be reduced by using light well drained soil, thorough removal of residues and weeds, drying the seeds at 18-20°C, eliminating small and light seeds and by treating the sowing medium with NIUIF-2 at 1 g/l water (Vladimirskaja, 1959).

10.1.4 Rust

This is also one of the major diseases causing several damage to the plants and the fungus responsible for this malady is *Coleosporium solidaginis*. Bright yellowish orange spots appear on the lower surface of the leaves, particularly on those of young plants. Spraying of wettable sulphur during the growing season is effective to control the disease (Pirone *et al.*, 1960). Raabe and Pyeatt (1990) reported that triforine (Funginex) effectively control *C. asterum* of China aster followed by triadimefon (Bayleton) and mancozeb (Fore).

10.1.5 Leaf Spot

This disease is caused by several kinds of fungi such as *Ascochyta asteris*, *Septoria callistephi* and *Stemphylium callistephi*. The spots are at first yellowish, then become dark brown and black, increasing in size. The lower leaves are infected first. Weekly spray with maneb or zineb provides good control.

10.1.6 Stem Rot

Pellicularia filamentosa has been identified as the causal organism of this disease. The fungus causes rotting of stem which can be controlled by the use of clean soil and treatment with PCNB (terraclor).

Sreenivasulu *et al.* (2004) revealed that China aster cultivars Phule Ganesh Purple followed by Phule Ganesh Pink, Phule Ganesh Violet, Namadhari Pink and Phule Ganesh White were moderately resistant to Alternaria leaf spot, while, the genotype Kamini was moderately susceptible.

10.1.7 Canker

Phomopsis callistephi causes canker on the lower parts of the stem. The distal portion is completely collapsed but the roots are not affected.

10.1.8 Southern Blight

It is caused by *Sclerotium rolfsii*. Mahadevakumar *et al.* (2018) reported a severe outbreak of southern blight disease of China aster during post rainy season (September–November) in the Mysore and Mandya districts of Karnataka. The disease incidence ranged between 12 and 47 per cent. The typical disease symptoms include water-soaked lesions on leaves, stems and on the lower stem surfaces followed by quick wilting of the whole plant with abundant production of sclerotia near the stem-soil interface. Molecular identification of the pathogen by ITS-rDNA sequences of *S. rolfsii* showed 100 per cent similarity with reference sequences. Based on the cultural, morphological and molecular characteristics, the fungal pathogen was identified as *Sclerotium rolfsii* Sacc. (Sexual morph: *Athelia rolfsii* (Curzi) C.C. Tu and Kimbr).

Bacterial Diseases

10.1.9 Bacterial Wilt

In China aster, it was first reported at Hokkaido, Japan during July 2000. Bacteriological characteristics of the isolated bacteria were identical to *Erwinia chrysanthemi.* This is the first report on bacterial wilt of China aster caused by *E. chrysanthemi* (Kotsu *et al.*, 2002).

Viral Diseases

10.1.10 Yellows

A serious viral disease in aster is caused by *Chlorogenus callistephi*. The development of pale yellowish tinge on the leaves and a large number of adventitious shoots are the main symptoms. Plants become dwarf and blooms are off-colour, the rays being usually yellowish green. The virus is transmitted by a leaf hopper, *Macrosteles fascifrons*. Control of leaf hoppers with suitable insecticide and destruction of diseased plants immediately after the appearance of the symptoms are helpful for minimizing the spread of the disease. Disease development was suppressed by the use of antibiotics like chlortetracycline, tetracycline, chloramphenicol but not by penicillin (Davis *et al.*, 1968). Chiykowski (1972) reported that treatment with

Diseases and Pests in China aster

Caterpillar

Leaf Hopper

Phyllody Phytoplasma

Botrytis

Fusarium Wilt

***Source*: www.pnwhandbooks.org, www.gardeningknowhow.com, www.springer.com, www.commons.wikimedia.org, www.canolacouncil.org, www.wikiwand.com**

oxyte-tracycline, tetracycline or oxycycline, immediately after inoculation delayed the development of disease symptoms. Oxytetracycline and tetracycline were highly effective at 100 ppm regardless of the method of application but their phytotoxic effects ranged from chlorosis and leaf deformation to plant death. Oxycycline at 1000 ppm was highly effective as leaf dip or spray but as a root dip killed all the plants. Root dips, foliage dips and sprays with all three antibiotics at 100 ppm also effectively delayed the development of symptoms of the disease. Similarly, Paterson and Sinha (1977) observed that development of disease symptoms was checked by root treatment with oxytetarcycline during the incubation period of the disease agent.

Phyllody disease in China aster was first reported during 1986 from Bengaluru, India and known to be transmitted by *Orosius albicinctus*. The disease was characterized by chlorosis, upright growth, small leaf, short internode, stunting, profuse vegetative growth and phyllody (transformation of floral organs into leaf-like structures) (Bandakkanavara *et al.*, 2019).

Win *et al.* (2011) observed symptoms of flower virescence in diseased China aster plants in Yezin, Myanmar. The presence of a phytoplasma was detected and identified by applying Polymerase Chain Reaction (PCR)/Restriction Fragment Length Polymorphism (RFLP) techniques and sequencing the 16S ribosomal DNA. The phytoplasma was identified as belonging to ribosomal subgroup 16SrII-A, never reported before in China aster. It showed almost 100 per cent similarity with a 16S rDNA sequence of sunn hemp witches' broom phytoplasma (AB558143), which belongs to the peanut witches' broom phytoplasma group. This is the first record of China aster as a new host for a subgroup 16SrII-A phytoplasma.

Mahadevakumar *et al.* (2017) during field surveys in regions of Karnataka showed virescence and phyllody symptoms, suggestive of phytoplasma infections, were present on China aster plants. The disease incidence ranged from 28 to 34 per cent in 78 ha of cultivated area surveyed. Affected young plants showed crown of small leaves with reduced internodes devoid of floral organs. Plants during post flowering stage turned green and floral parts were replaced by green leafy sepaloid structures. This is the first report on association of phytoplasma (16SrII-D) with China aster. Phyllody and virescence adversely affect quality and quantity of China aster production because the affected plants remain stunted without bearing any floral organs which has the potential to cause economic losses.

10.1.11 Chrysanthemum Mosaic Virus

An extreme distortion of blooms and certain amount of dwarfing observed on California Giant asters were found to be caused by chrysanthemum mosaic virus. It was indicated that the disease was seed-borne.

Californicus strain of aster-yellows, spotted wilt, *Lethum australiense* and curly top, *Rugaverrucosans* are other viruses reported on China aster.

10.2.0 Pests

China aster is also attacked by a number of pests.

10.2.1 Flower Caterpillars

They feed and damage various floral parts. Spray fenvalerate 20EC at 0.5 ml/l or neem seed kernel extract 5 per cent. If the incidence is severe, spray indoxacarb 14.5SC at 1 ml/l.

10.2.2 Stem Borer

Larvae bore into stem and side branches making them hollow resulting in wilting. Collect and destroy infested shoots. Spray profenophos 50EC at 1 ml/l or ethofenprox 10EC at 1ml/l. If the incidence is severe, apply carbofuran 3G granules at 1 kg.a.i./ha (33 kg/ha).

10.2.3 Leaf Hopper (*Macrosteles fascifrons*)

This is the most common pest feeding on aster plants and as already mentioned, capable of transmitting virus. Treatment with parathion (Jefferson and Eads, 1952) or any other suitable insecticide will control this pest.

10.2.4 Black Blister Beetle (*Epicauta pennsylvanica*)

This pest, also known as aster beetle, feeds on foliage and flowers, often completely destroying the plants. Sprays or dust with methoxychlor at an interval of 5-7 days during the period of peak infestation was found to be effective in eliminating the pest (Pirone *et al.*, 1960).

10.2.5 Leaf Miner

This is very serious pest in Los Angles. It feeds on leaves as a result of which foliage becomes badly disfigured. It causes serpentine like leaf mines. Jefferson and Eads (1952) recorded that the use of parathion was more beneficial than chlordane, toxaphene and ethyl-p-nitrophenyl thiobenzene phosphonate. Spray triazophos 40EC at 1.5 ml/l for management. Repeated sprays are to be avoided.

10.2.6 Root Aphid (*Anuraphismaidi-radicis*)

This pest feed in large numbers on the roots causing yellowing-dwarfing and wilting of the plant. Pre-planting treatment of the soil with chlordane or post-planting treatment with lindane solution was suggested to control the insects (Pirone *et al.*, 1960).

10.2.7 Potato Aphid (*Macrosiphum salanifolii*)

It attacks the leaves, stems and flowers of asters. They mainly infest lower side of flowers and base of petals causing discoloration, deformation and withering. Spray acephate 75 SP at 1.5 g/l or dimethoate 30EC at 2 ml/l. Spray Neem or Pongamia oil 1 per cent at early incidence.

10.2.8 Spider Mite (*Tetranychus telarius*)

These are tiny pests frequently infesting asters. Mite feeding results in discoloration and drying of leaves, buds and flowers. Spray with kelthane at 1 ml/l keeps these pests under control. Spray dicofol 18.5EC at 2.5 ml/l or fenazaquin 10 EC at 1 ml/l or profenofos 50 EC at 1 ml/l.

10.2.9 Asiatic Beetle (*Autoserica castanea*)

This beetle prefers the leaves of China asters to those of many other plants. They feed on foliage at night and during the day hide themselves in the soil at the base of the plant or in the vicinity. Spraying the plants once or twice with lead arsenate will control insects. Application of chlorpyrifos to the soil destroys the pest at larval stage.

10.2.10 Tarnished Plant Bug (*Lygus lineolaris*)

These are light brown, variously spotted and very destructive insects. They puncture the terminal shoots below the flower buds which cause the flowers to drop and die.

10.2.11 Nematodes

Asters are susceptible to root-knot nematode, *Meloidogyne incognita*. The growth of infested plants is checked and the leaves are dwarfed. Another nematode, *Aphelenchoides ritzemabosi* also infects the plants. In seedlings the nematode population increased rapidly in 4 weeks at 20°C causing severe leaf malformation (Burckhardt, 1972). Soil treatment with suitable nematicide controls the infestation.

11.0 Harvesting and Postharvest Management

11.1.0 Harvesting of Flowers

Since the stage of development of flowers at harvesting determines its keeping quality, they should be harvested at the right stage. Aster flowers along with stalk are cut when they develop their original colour. For distant markets, they should be harvested earlier as compared to those for local markets.

China aster is harvested in two ways. Individual flowers are harvested for decoration and worship purpose whereas flowers along with stalk or whole plant is harvested when 60 per cent flowers are fully opened. The flowers are generally harvested in the early morning or late evening.

11.2.0 Harvesting of Seeds

Mechanical harvesting of China aster as seed crop was found viable despite 11-18 per cent seed losses; the seeds were commercially acceptable. The optimum time for mechanical harvesting occurred when inflorescences of the 1st order became completely covered by pappus and the seed moisture content was about 20 per cent (Duczmal, 1989).

11.3.0 Postharvest Management

Immediately after harvest, the flowers with peduncles are put vertically in a container having clean water. Lower leaves attached to the stalk should also be removed to minimize transpiration. Grading is done according to stalk length, size and shape of flowers, freshness and other qualities. While packing, care should be taken that there is no direct contact between the flower and packing material. Clear plastic films apart from providing adequate protection, improves the attractiveness of flower.

China aster are of diverse forms and colours used in making bouquets, buttonholes, garlands, boxes and other garden places where enhancement of colour is required. Large compact flowers with straight stalk are considered ideal for cut flower industry, while for making garlands the thickness of flower is of vital importance. Vase life of flower is also having prime importance during vase decoration.

In China aster, there are no grade standards as such, apart from the standard quality attributes of stem length, foliage quality, uniformity, and freedom from defects. Depending on the species and cultivar, bunches may be prepared by stem number or by bunch size. Forty-two genotypes of China aster evaluated for postharvest life (Bhargav *et al.*, 2019) revealed that the genotypes IIHRCC39, IIHRI69-2, IIHRG13 and IIHRJ3 was found most promising for cut flower and IIHECC42 for loose flower.

Asters are packed in traditional flower boxes, hampers, or Proconas. It is very difficult to make broad flower care recommendations for asters because of the large number of species and cultivars. Keep stems and vase solution clean. It is especially important with asters to remove leaves that might be in the water since bacteria grow quickly on leaves that are under water, contaminating the vase solution and leading to early wilting. Treat with a hydrating solution and prepare the preservative solution properly to minimize contamination by debris and microorganisms. Vijayakumar *et al.* (2017) studied the effect of plant growth regulators on flower quality, and post harvest shelf life of China aster *cv.* Local. Among the postharvest treatments, flowers maintained in the non-ventilated, 200 gauge polyethylene bags with pre-harvest sprays of 150 ppm GA_3 enhanced the shelf life of flowers (8.68 days) followed by salicylic acid at 100 ppm (8.55 days).

Motonaga and Azuma (1990) packed cut asters (*C. chinensis*) in film before shipment. Film wrapping lengthened the period over which the bunches remained fresh by up to 10 days compared with unwrapped bunches. Wrapping also increased the market value of the bunches.

11.3.1 Vase Life

The vase life of aster cut flowers, in general, is more than other annuals grown for cut flowers. Microbial population in vases determines the life of cut flowers. Marousky (1978) reported that slow release chlorine compounds, sodium dichloroisocyanurate and 1, 3, dichloro 5, 5-dimethyldydantoin reduced the population of bacteria in vase water. A preservative containing 60 g/l sucrose + 250 mg/l 8-HQS (chinosol) + 70 mg/l CCC + 50 mg/l $AgNO_3$ extended the vase life of cut flowers by 30.6 per cent (Piskornik, 1981). Another preservative, zeatin, was also found to improve the vase life of flowers of *cvs*. Schnittaster Xenia and Enzett Krallenaster (Mugge, 1983). Vase life of fully opened flowers of China aster with 10 cm long stalks was about 8 days when kept in solution containing 0.2 per cent $Al_2(SO_4)_3$, followed by those containing 2.0 per cent sucrose, which varied according to the season of harvest (Mantur and Nalawadi, 1989).

Pulsing of China aster cut flowers with solution comprising of sucrose (10 per cent) + 8-HQC (400 ppm) for 12 hours improved flower size, vase life, minimum per cent weight loss and appearance of cut bloom (Prakash *et al.*, 2018).

Chander *et al.* (2018) observed that pulsing of cut flowers of *cv.* Kamini with sucrose (10 per cent) + 8-HQC (400 ppm) for 12 hours improved flower size, amount of pulsing solution consumed (2.54 ml), vase life (12.83 days), distilled water consumed, minimum per cent weight loss and appearance of cut bloom. Poor water relations often limit the vase-life of asters. This can be demonstrated by wilting of the flower and/or buds. Their vase life has been shown to be extended by a 10-second pre-treatment with high concentrations (1000 ppm) of silver nitrate, which is a very effective germicide.

Application of 1500 ppm alar on China aster at 25th and 40th days after transplanting increased the vase life (Patil *et al.*, 2013). However, foliar spray of BA at 40 ppm at 30 days after transplanting enhanced vase life of China aster flower (Maurya *et al.*, 2017).

Sidana *et al.* (2019) studied the effect of seed priming on postharvest quality of China aster.

Seeds primed with GA_3 200 ppm showed maximum effect in postharvest quality of China aster *cv.* Kamini. Gibberellic acid promotes hydrolysis of starch and sucrose (contents of dry matter) into glucose and fructose. Hence, there is decrease in dry matter constituents in flower head and stalk. The higher contents of reducing sugar in flower head and stalk may increase the osmotic potential, thus, enhancing the water absorption and maintain their turgidity which leads to prolonged postharvest life of China aster (Vijayakumar *et al.*, 2017).

Citric acid (200 ppm), $AgNO_3$ (200 ppm), sucrose (5 per cent) and Tween-20 (0.02 per cent) was best in extending the vase-life and weight gain at third day and total solution uptake. Correlation revealed that floret diameter had a strong positive correlation with loss in weight of vase solution up to ninth day, total weight loss of vase solution, floret opening per cent and vase-life in various cultivars of both crops (Tiwari *et al.*, 2010).

Food dyes *viz.*, Apple Green, Lemon Yellow and Orange Red used for tinting in China aster *var*. Local White. Tinted flowers at 4 per cent concentration expressed full bright coloured flowers in the inflorescence with quick uptake of colours in a short period of two hours duration. But, the vase life (one day) and water uptake (0.8, 1.6 and 1 ml) was recorded minimum in all the tinted flowers at 4 per cent concentrations. By using the tinted flowers, flower arrangements can be made which will be of very novel and attractive. Value addition ensures high premium price to the China aster grower (Ranchana *et al.*, 2017).

11.4.0 Value Addition

Meman *et al.* (2006) optimized conditions for dry flower production in China aster flowers. Two drying conditions *viz.*, room drying and sun drying and four media *viz.*, Sand, Sand: Borax (1:1), Borax and Silica gel were used. Per cent weight loss and moisture loss were significantly higher under sun drying and silica gel during the entire process of drying from first day to fourth day. Moisture content was higher under room drying and borax from first day to fourth day.

11.5.0 Yield of Flowers

China aster *cv.* Ostrich Plume Blue produced 33 flowers per plant when planted in November at 20 x 30 cm spacing (Patil *et al.*, 1987). Nethra *et al.* (1999) recorded 34.33 flowers/plant (6840 kg/ha) after application of 10 t vermicompost/ha + 100 per cent NPK.

Raghupathi *et al.* (2020) studied the cost returns and input use pattern for China aster cultivation and revealed that farmers realized 12 per cent higher gross returns (Rs. 2,56,500 per acre) from cultivation of Arka Kamini crop compared the cultivating check variety (Rs. 2,25,750 per acre) which indicated cultivation of Arka Kamini was more efficient compared to check variety. The results of the descriptive statistics analysis shows that the human labor, machine labor, seedlings, chemical fertilizers and plant protection chemicals of Arka Kamini variety was Rs. 11,012, 908 5,225 3,865 and 2,790 per acre, respectively, while, in check variety it was Rs. 9,842, 571, 6,588, 2,400, and 2,756 per acre, respectively.

11.6.0 Yield of Seeds

Phetpradap and Hampton (1991) recorded 568 kg/ha seeds of *C. chinensis cv.* Powder Puff and Kurenai. Patil *et al.* (1987) obtained only 121.7 kg/ha seeds of *cv.* Ostrich Plume Blue in November planting at 20 x 30 cm spacing. One flower can produce more than 100 seeds.

12.0 Biotechnology

12.1.0 Molecular Markers

Bandakkanavara *et al.* (2019) carried out molecular detection through PCR assay by extracting the total DNA from phyllody infected aster leaf by using nested PCR phytoplasma specific universal primers R16F2n/R16FR2and results revealed that the amplification of phytoplasmal specific PCR product of 1.2 kb fragment corresponding to the 16S rDNA. The BLAST analysis revealed that aster phyllody phytoplasma had 99 per cent sequence similarity with Tomato big bud TBB1 (KX358564.1), Alfalfa phytoplasma (Sudan) AP2 (KY449416.1), *Pisum sativum* phyllody (KX358571.1) phytoplasma. Aster phyllody phytoplasma clustered with the Tomato big bud TBB1 (KX358564.1) phytoplasma.

Mallesh *et al.* (2020) conducted research aimed to understand the leaf hoppers species and defining the prospective cause of the aster-yellow disease in China aster, Marigold and Chrysanthemum. Through the use of mtCOI molecular marker eight different leafhoppers species were identified as *Sogatella furcifera, Homalodisca insolita, Amrasca biguttula, Balclutha incise* and *Balclutha abdominalis* and *Japanagallia trifurcate.* Whereas at genus level identified as *Toya, Empoasca, Perkinsiella, Hishimonus, Tambocerus, Phaconeura, Curena, Psammotettix* and *Graphocophala* species.

The genetic diversity of China aster genotypes using SRAP markers was carried out using twenty-five primer combinations out of which, eleven primer combinations were polymorphic and revealed a total of 98 scorable bands, among which 89 were polymorphic with an average of 7.27 polymorphic bands per pair of primers (Kore *et al.*, 2016).

Molecular detection was carried out through PCR assay by extracting the total DNA from phyllody infected aster leaf by using nested PCR phytoplasma specific universal primers R16F2n/R16FR2 and results revealed that the amplification of phytoplasmal specific PCR product of 1.2 kb fragment corresponding to the 16S rDNA (Bandakkanavara *et al.*, 2019). A 16S rDNA sequence comparison of aster phyllody phytoplasma with the 16S rDNA gene sequences of other phytoplasmas obtained from NCBI database. The BLAST analysis revealed that aster phyllody phytoplasma had 99 per cent sequence similarity with Tomato big bud TBB1 (KX358564.1), *Alfalfa phytoplasma* (Sudan) AP2 (KY449416.1), *Pisum sativum phyllody* (KX358571.1) phytoplasma.

The genetic diversity, phylogenetic relationships, and geographic distributions of *A. spathulifolius* populations were previously evaluated using allozymes and inter-SSR (ISSR) markers (Maki and Morita, 1998; Nguyen *et al.*, 2013). These populations were genetically differentiated into two distinct groups, mainland and island types, which were consistent with their geographic distributions (Nguyen *et al.*, 2013). The mainland populations showed higher levels of genetic diversity, while island *A. spathulifolius* populations exhibited lower diversity, probably due to their geographic isolation by hundreds of kilometers of ocean and the resulting genetic drift (Nguyen *et al.*, 2013). The complete cp genome for island *A. spathulifolius* was previously sequenced (Choi and Park, 2015; Shen *et al.*, 2018) and shows relatively high divergence from *Aster altaicus* and *Aster tataricus* (Choi and Park, 2015; Shen *et al.*, 2018). Molecular phylogeny methods were also applied to study the wider Aster genus, including the analysis of Eurasian Aster species using short DNA sequences (ITS, ETS, trnL-F), from which *A. spathulifolius* was excluded (Li *et al.*, 2012). The complete chloroplast genome of mainland *A. spathulifolius* was sequenced by Tyagi *et al.* (2020) and this study will be helpful for species identification and the analysis of the genetic diversity, evolution, and phylogenetic relationships in the Aster genus and the Asteraceae.

12.2.0 Anthocyanin Biosynthesis

Flower colour in China aster is determined by accumulation of chalcones (yellow), flavones and flavonols (cream) and anthocyanins (red to blue). Several genes (A, Ch, F, G, M and R) are involved in biosynthesis of these pigments (Forkmann, 1977). Enzymatic studies revealed that the genes Ch, F, R and M control the activities of the enzymes chalcone isomerase (CHI), dihydroflavonol 4-reductase (DFR), flavonoid 3′,5′-hydroxylase (F3′5′H) and anthocyanin 5-glucosyltransferase (A5GT), respectively (Kugn *et al.*, 1978; Ruhnau and Forkmann, 1988; Seyferrt, 1982; Wit 1937). The genes A and G could not be correlated with specific enzymes but most probably concern late steps in the anthocyanidin pathway. Mutants concerning chalcone synthase (CHS), flavanone 3-hydroxylase (FHT), flavone synthase II (FNSII) and flavonol synthase (FLS) have not yet been described.

12.3.0 Transcriptome Analysis

A flower-specific cDNA library was constructed using the wild type line 06 (Henkel *et al.*, 2000). Using heterologous probes, full-length clones of CHS, FHT, DFR and anthocyanidin synthase (ANS) as well as an almost full-length clone of CHI were isolated and characterized. The ANS cDNA clone was found to contain a 208 bp intron. They used cDNA clones for further characterization of the wild type line 06 and of the genetically stable anthocyaninless mutant lines 07, 08 and 10h. Southern blot analyses revealed that the genes encoding CHI, FHT, DFR, and ANS are single copy genes. In confirmation of the enzymatic studies, northern blot analyses revealed that the yellow-coloured mutant lines 07 and 10h are CHI-deficient, and line 10h is additionally DFR-deficient. Lines 07 and 08 with recessive alleles of gene A were found to be ANS-deficient, confirming our hypothesis that a late step of the pathway is controlled by this gene. The gene G seems to exert a regulatory effect, at least on DFR. Thus, in line 08 with dominant F-alleles but recessive alleles of gene G, DFR gene transcription was found to be extensively reduced (Henkel *et al.*, 2000)

13.0 References

Abraham, J. and Thomas, T.D. (2016) *In*: Plant tissue culture: Propagation, Conservation and Crop Improvement, (eds.) Anis, M. and Ahmad, N., Springer, Singapore, pp. 161-195.

Abrol, A., Chaudhary, S.V.S., Dhiman, S.R., Gupta, R.K. and Kaushal, R. (2019) *Int. J. Curr. Microbiol. Appl. Sci.*, **8**(12). doi: 10.20546/ijcmas.2019.812.214.

Abrol, A., Chaudhary, S.V.S., Dhiman, S.R., Gupta, R.K. and Kaushal, R. (2020) *Green Farming*, **11**(1) : 9-13.

Anonymous (2019) Proceedings of XXXXth meeting of plant germplasm registration committee (PGRC) held at ICAR-NBPGR, New Delhi, October, 21, 2019.

Aswath, S., Gowda, J.V.N. and Murthy, G.M.A. (1995a) *J. Orna. Hort.*, **2** : 9-13.

Aswath, S., Gowda, J.V.N. and Murthy, G.M.A. (1995b) *Curr. Res.*, University of Agri. Sci, Bangalore, **24** : 167-169.

Auman, C.W. (1980) In : *Introduction to Floriculture* (ed. R.A. Larson), Academic Press, New York.

Babu, T.K., Kumar, C.P.C., Reddy, D.R.R. and Singh, T.V.K. (2013) *Indian J. Plant Prot.*, **41**(1) : 50-60.

Bailey, L.H. (1963) *The Standard Cyclopedia of Horticulture*, Vol. I. The Macmillan Company, New York.

Bandakkanavara, M., Prameela, H.A., Mali, S., Hurakadli, M.S., Basavaraj, S., Achari, K.R. and Rangaswamy, K.T. (2019) *Int. J. Curr. Microbiol. Appl. Sci.*, **8**(8) : 3070-3076.

Baranowski, T. (1974) *RocznikiAkademiiRolniczej w Poznaniu*, Ogrodnictwo, **69** : 3-9.

Basavaraju, C., Gowda, J.V.N. and Muniyappa, T.V. (1992) *Curr. Res. University Agri. Sci.*, Bangalore, **21** : 50-51.

Beale, G.H. (1941) *J. Genet.*, **42** : 197-213.

Bhargav, V., Kumar, R. and Rao, T., Manjuntha, Sane, A., Venugopalan, R. and Kumari, P. (2019) *Indian J. Plant Genet. Resour.*, **32**(1): 80.

Bhargav, V., Kumar, R., Rao, T.M. and Bharathi, T.U. (2019) *Indian J. Agric. Sci.*, **89**(10) : 1643-1648.

Bhargav, V., Kumar, R., Rao, T.M., Bharathi, T.U. and Venugopalan, R. (2018) *Int. J. Curr. Microbiol. Appl. Sci.*, **7**(12) : 1225-1232.

Bhargav, V., Kumar, R., Rao, T.M., Bharathi, T.U., Dhananjaya, M.V. and Kumari, P. (2018) *Indian J. Agric. Sci.*, **88**(1) : 138-144.

Bhargav, V., Kumar, R., Rao, T.M., Bharathi, T.U., Dhananjaya, M.V. and Venugopalan, R. (2019) *Electron. J. Plant Breed.*, **10**(1) : 277-284.

Bhargav, V., Rao, T.M., Kumar, R., Bharathi, T.U., Venugopalan, R. and Dhananjaya, M.V. (2019) *J. Appl. Nat. Sci.*, **11**(1) : 1-6.

Biebel, J. (1937) *Proc. Amer. Soc. Hort. Sci.*, **34** : 635-641.

Bohra, M., Rana, A., Punetha, P., Upadhyay, S. and Nautiyal, B.P. (2019) Indian J. Hortic., **76**(2) : 329-333.

Bose, T.K. and Das, S.M. (1966) *Indian J. Hort.*, **23** : 88-97.

Burckhardt, F. (1972) *Nachrichtenblatt des DeutschenPflanzenschutzdienstes*, **24** : 132-133.

Byrne, T.G. and Lert, P.J. (1962) *Proc. 14th Calif. Weed Conf.*, pp. 5-10.

Carneiro, J.W.P. (1996) *RevistaBrasileira de Horticultura Ornamental*, **2** : 41-47.

Cathey, H.M., Halperin, J. and Piringer, A.A. (1963) *The Exchange*, **140** : 16.

Chakraborty, A., Bordolui, S.K. and Mahato, D.N.M.K. (2020) *Int. J. Curr. Microbiol. Appl. Sci.*, **9**(3). doi : 10.20546/ijcmas.2020.903.175.

Chakraborty, A., Bordolui, S.K., Mahato, M.K., Sadhukhan, R. and Sri Veda, D.J.M.S.N.K. (2019) *J. Crop Weed.*, **15**(1) : 201-204.

Chander, P., Sharma, B.P., Gupta, Y.C., Thakur, K.S. and Gupta, R.K. (2018) *J. Ornam. Hortic.*, **21**: 43- 50.

Chavan, M.D., Jadhav, P.B. and Rugge, V.C. (2010) *Indian J. Hortic.*, **67**(4) : 378-381.

Cherneva, S. (1976) *Gradinardtvo*, **57** : 36-38.

Chiykowski, L.N. (1972) *Canadian J. Plant. Sci.*, **52** : 29-33.

Choi, K.S. and Park, S. (2015) *Gene*, **572**: 214–221.

Chopde, N., Palekar, A.R. and Kuchanwar, O. (2018) *Int. J. Chem. Stud.*, **6**(6) : 1355-1357.

Das, D.C. (2013) *Int. J. Bioassays*, **2**(2) : 398-402.

Davis, R.E., Whitcomb, R.F. and Stecre, R.L. (1968) *Science*, **161** : 793-794.

Debnath, S. and Maiti, R.G. (1998) *J. Interacademicia*, **2** : 11-16.

Dhemre, J.K., Shirsath, N.S. and Naphade, A.S. (1997) *J. Soils and Crops*, **7** : 136-138.

Doddagoudar, S.R., Vyakaranahal, B.S. and Sheshargouda, M. (2004) *Karnataka J. Agric. Sci.*, **17**(4) : 701-704.

Domsch, K.H. (1955) *Gartenbauwiss.*, **2** : 109-113.

Doorenbos, J. (1959) *Euphytica*, **8** : 69-75.

Duczmal, K.W. (1989) *BiuletynInstytutuHodowliiAklimatyzacji Roslin*, **169** : 49-58.

Flachsbart, E. (1972) *Deutsche Gartnerborse*, **72** : 265-266.

Fleming, W.M. (1937) *Yearb.of Agriculture*, U.S. Dept. Agriculture, p. 985.

Forkmann, G. (1977) *Phytochem.*, **16**: 299-301.

Forkmann, G. and Seyffert, W. (1977) *Theor. Appl. Genet.*, **42** : 279-287.

Gaidhani, A., Dalal, S.R. and Nagre, P.K. (2020) *Int. J. Chem. Stud.*, doi: 10.22271/chemi.2020.v8.i2q.8918.

Gnyandev, B., Kurdikeri, S. and Vishalkumar (2012) *Indian Hortic. J.*, **2**(3-4) : 83-87.

Gnyandev, K., Patil, A.A. and Channappagoudar, B.B. (2014) *Int. J. Trop. Agric.*, **32**(3-4) : 377-380.

Górnik, K. and Grzesik, M. (2005) *Folia Hortic.*, **17**(2) : 119-127.

Gowda, J.V.N. and Narayana Gowda, J.V. (1990) *Mysore J. Agri. Sci.*, **24** : 72-78.

Graaf Van der Zande, M.T., Blacquiere, T. De, Graaf van der Zande, M.T., Zande M.T.G. van der, Van der Zande, M.T.G. and Moe, R. (1992) *Acta Hortic.*, **327** : 87-93.

Gruis, J.T. and Joiner, J.N. (1960) *Proc. Fla. St. Hort. Soc.*, **73** : 378-381.

Gupta, Y.C., Sharad, K., Sharma, B.P., Dhiman, S.R. and Sharma P. (2015) *Prog. Hortic.*, **47**(2) : 260-266.

Guzenwski, W.A. (1976) *Acta Hortic.*, **63** : 229-231.

Han, F. (2010) *J. Agric. Sci.*, **2**(4) : 201-210.

Hanzelka, P. and Kobza, F. (2001) *Zahradnictvi Horticultural Sci.*, **28** : 21-27.

Haramaki, C. (1965) *Proc. 19th N. East Weed Control Conf.*, pp. 175-179.

Haramaki, C. and Kuhns, L.J. (1983) *Proc. 37th Ann. Meet. Northeastern Weed Sci.*, pp. 357-361.

Harishkumar, K., Shiragur, M., Kulkarni, B.S. and Patil, B.C. (2017) *Agric. Res. J.*, **54**(3) : 407- 409.

Henkel, J., Forkmann, G., Min, B.W., Dedio, J. and Forkmann, G. (2000) *Acta Hortic.*, **508**: 213-214.

Hoffmann, G.M. and Kotter, C. (1963) *Gartenwelt*, **63** : 123-124.

Hoffmann, G.M. and Reber, H. (1963) *Phytopath. Z.*, **48** : 178-179.

Hong, Y.P. and Jeong, J.H. (1988) *Res. Reports of the Rural Development Administration, Horticulture*, **30** : 104-110.

Hosalli, A., Shiragur, M., Patil, B.C., Masuthi, D.K. and Tatager, M.H. (2019) *Int. J. Curr. Microbiol. Appl. Sci.*, **8**(9) : 822-828.

Houbowicz, R. and Borysiak, A. (2008) Bulletin of University of Agricultural Sciences and Veterinary Medicine Cluj-Napoca: Horticulture, **65**(1) : 184-187.

https: //www.gardeningknowhow.com/ornamental/flowers/asters/edibility-of-aster-flowers.htm

Hughes, A.P. and Cockshull, K.E. (1964) *Nature*, **201** : 413.

Hughes, A.P. and Cockshull, K.E. (1965) *Ann. Bot.*, **29** : 131-151.

Hughes, A.P. and Cockshull, K.E. (1966) *Exp. Hort.*, **16** : 44-52.

Huziwara, Y. (1954) *Kromosomo*, **21** : 773-776.

Ichihara, J. (1954) *J. Hort. Ass. Japan*, **23** : 115-122.

Janakiram, T. and Rao, T.M. (1996) *Indian Hort.*, **41** : 2-3.

Jayanthi, R. and Mukhopadhyay, A. (1988) *Indian Hort.*, **33** : 8-11.

Jayanthi, R., Gowda, J.V.N. and Pierik, R.L.M. (1990) *Curr. Plant Sci. Biotech. Agri.*, **12** : 389-392.

Jefferson, R.N. and Eads, C.O. (1952) *J. Econ. Ent.*, **45** : 476-481.

Kaushal, S., Dilta, B.S., Chaudhary, S.V.S., Sharma, B.P. and Gupta, Y.C. (2014) *Int. J. Farm Sci.*, **4**(1) : 60-71.

Khangjarakpam, G., Kumar, R., Seetharamu, G.K., Rao, T.M., Dhananjaya, M.V., Venugopalan, R., and Padmini, K. (2014) *J. Hortic. Sci.*, **9**(2) : 141-144.

Khangjarakpam, G., Kumar, R., Seetharamu, G.K., Rao, T.M., Dhananjaya, M.V., Venugopalan, R., and Padmini, K. (2015) *Curr. Hortic.*, **3**(1) : 35-40.

Khangjarakpam, G., Kumar, R., Seetharamu, G.K., Rao, T.M., Dhananjaya, M.V., Venugopalan, R., and Padmini, K. (2014) *Prog. Agric.*, **14**(1) : 189-191.

Khobragade, R.K., Bisen, S., Thakur, R.S. (2012) *Indian J. Agric. Sci.*, **82**(4) : 334-339.

Kobza, F., Pall, J. and Lernerova, J. (1984) *Acta Universitatis Agriculturae Brno*, A, **32** : 117-128.

Kofranek, A.M. (1951) *Bloom News*, **3** : 4-7.

Kohl, H.C., Kofranek, A.M. and Lunt, O.R. (1957) *Proc. Amer. Soc. Hort. Sci.*, **70** : 437-441.

Kore, R.K., Shiragur, Mukud, Pujari, Ravi, Nishani, S. and Akshata, G.H. (2016) *Indian J. Ecol.*, **43**(1): 192-196

Kotsu, T., Horita, H. and Kitaya, M. (2002) *J. Gen. Plant Pathol.*, **68**(1) : 105-107.

Kratka, J. and Duskova, E. (1991) *Sbornik UVTIZ OchransRostlin*, **27** : 127-135.

Kuhn, B. Forkmann, G. and Seyfert, W. (1978) *Planta*, **138**: 199-203.

Kulkarni, B.S., Nalawadi, U.G. and Giraddi, R.S. (1996) *South Indian Hort.*, **44** : 33-35.

Kumar, K.P., Padmalatha, T. and Pratap, M. (2014) *Plant Arch.*, **14**(2) : 961-966.

Kumar, K.P., Padmalatha, T., Pratap, M. and Reddy, S.N. (2015) *Indian J. Agric. Res.*, **49**(4) : 348-352.

Kumar, P., Raghava, S.P.S., Misra, R.L. and Singh, K.P. (2003) *J. Ornam. Hortic.*, **6**(2) : 110-112.

Kumar, R., Kumar, A., Singh, R.P. (2019) *Int. J. Chem. Stud.*, **7**(1) : 395-397.

Kumar, S., Dewan, N., Nongrum, H.B. and Marwein, B. (2016) *Curr. Hortic.*, **4**(2) : 25-26.

Kumar, V.C. and Gowda, J.V.N. (2010) *Asian J. Hort.*, **5**(2) : 472-474.

Kumar, V.C. and Gowda, J.V.N. (2010) *Asian J. Hort.*, **5**(2) : 486-490.

Kumar, R., Rao, T.M. and Janakiram, T. (2015) *Indian J. Plant Genet. Resour.*, **28**(3) : 360-361.

Kumari, P., Kumar, R., Rao, T.M., Bharathi, T.U., Dhananjaya, M.V. and Bhargav, V. (2018) *Indian J. Agric. Sci.*, **88**(3) : 453-457.

Kumari, P., Kumar, R., Rao, T.M., Dhananjaya, M.V. and Bhargav, V. (2017) *HortFlora Res. Spect.*, **6**(4) : 278-282.

Laurie, A., Keplinger, D.C. and Nelson, K.S. (1958) *Commercial Flower Forcing*, McGraw-Hill Book Co., Inc., New York.

Lee, J.S., Lee, Y.S., Choi, M.J. and Lee, K.H. (1996) *J. Korean Soc. Hort. Sci.*, **37** : 455-461.

Li, W.P., Yang, F.S., Jivkova, T. and Yin, G.S. (2012) *Ann. Bot.*, **109**: 1341–1357.

Lin, L.C. and Donald, P.W. (1950) *Proc. Amer. Soc. Hort. Sci.*, **55** : 441-446.

Lindstrom, R.S., Wittwer, S.H. and Bukovac, M.J. (1957) *Quart. Bull. Mich. Agric. Exp. Sta.*, **39** : 673-681.

Mahadevakumar, S., Chandana, C., Deepika, Y.S., Sumashri, K.S., Yadav, V., Janardhana, G.R. (2018) *Eur. J. Plant Pathol.*, **151**(4) : 1081-1087.

Mahadevakumar, S., Thorat, V., Yadav, V., Yadav, A. and Janardhana, G.R. (2017) *Plant Dis.*, **101**(1). doi: 10.1094/PDIS-06-16-0881-PDN.

Maheta, P., Polara, N.D. and Rathod, J. (2016) *Asian J. Hortic.*, **11**(1) : 132-135.

Maki, M. and Morita, H. (1998) *Int. J. Plant Sci.*, **159**: 148–152.

Mallesh, M.S.H., Asokan, R., Gadad, H., Kumar, S.D., Kumar, R. and Prakash, T. (2020) *Mitochondrial DNA A DNA Mapp. Seq. Anal.*, **31**(2) : doi: 10.1080/24701394.2020.1735378.

Mantrova, E.Z. (1959) *Cvetovodstvo*, **2** : 29.

Mantur, S.M. and Nalawadi, U.G. (1989) *South Indian Hort.*, **37** : 361-363.

Mantur, S.M., Nalawadi, U.G. and Biradar, D.P. (1996) *Karnataka J. Agri. Sci.*, **9** : 168-170.

Marousky, F.J. (1978) *Proc. Fla. St. Hort. Soc.*, **90** : 294-296.

Massante, H. (1963) *Gartenbauwiss.*, **28** : 173-197.

Mathad, R.C., Vyakaranahal, B.S. and Raikar, S.D. (2009) *Indian J. Agric. Res.*, **43**(2) : 95-100.

Maurya, R., Singh, S.P. and Singh, A.K. (2017) *Int. J. Curr. Microbiol. Appl. Sci.*, **6**(12) : 3148-3151.

Meman, M.A., Barad, A.V. and Raval, L.J. (2006) *J. Hortic. Sci.*, **1**(1) : 48-51.

Merlo, A.S. (1955) *Sad iOgorod*, **5** : 69-70.

Mishra, P.P., Pande, G. and Moharana, D.P. (2018) *Int. J. Chem. Stud.*, **6**(3) : 89-92.

Miteva, N. and Vasileva, S. (2015) Scientific Papers: Series B. Horticulture, LIX : 381-384.

Motonaga, H. and Azuma, M. (1990) *Agriculture and Horticulture*, **65** : 715-720.

Mugge, A. (1983) *Gartenbau*, **30** : 218-219.

Munikrishnappa, P.M. and Chanshekar, S.Y. (2014) *Trends Biosci.*, **7**(12) : 1217-1221.

Munikrishnappa, P.M. and Chandrashekar, S.Y. (2014) *Agric. Rev.*, **35**(1). doi: 10.5958/j.0976-0741.35.1.007

Nagapushpa, K., Vijaya, M., Devi, K.B.S., Girwani, A. and Joshi, V. (2018) *Int. J. Curr. Microbiol. App. Sci.*, **7**(5) : 3284-3292.

Naik, B.H., Nalawadi, U.G. and Sreenivasa, M.N. (1999) *Karnataka J. Agri. Sci.*, **12** : 154-158.

Naik, B.H., Nalawadi, U.G., Sreenivasan, M.N. and Patil, A.A. (1995) *Scientia Hort.*, **62** : 129-133.

Naikwad, D., Kandpal, K., Patil, M.G., Hugar, A. and Kulkarni, V. (2018) *Int. J. Curr. Microbiol. Appl. Sci.*, **7**(2) : 3353-3362.

Naikwad, D., Kandpal, K., Hugar, A., Patil, M.G. and Kulkarni, V. (2018) *Int. J. Curr. Microbiol. App. Sci.*, **7**(4) : 3329-3338.

Nataraj, K., Seetharamu, G.K., Srinivasa, V., Kulkarni, B.S., Kumar, R., Lakshmana, Venugopalan, R. and Munikrishnappa, P.M. (2020) *Int. J. Curr. Microbiol. App. Sci.*, **9**(5). doi: 10.20546/ijcmas.2020.905.072.

Nataraj, S.K., Seetharamu, G.K., Srinivasa, V., Kulkarni, B.S., Kumar, R., Lakshmana, D., Venugopalan, R. and Munikrishnappa, M. (2020) *Int. J. Curr. Microbiol. App. Sci.*, **9**(6) : 1620-1628.

Nechansky, F. and Jirasek, V. (1967) *Preslia*, **39** : 122-150.

Negi, S.S. and Raghava, S.P.S. (1990) *Euphytica*, **48** : 117-122.

Negi, S.S., Raghav, S.P.S., Sharma, T.V.R.S. and Srinivasen, V.R. (1983) *Indian J. Hort.*, **40** : 102-106.

Negi, S.S., Raghava, S.P.S., Rao, T.M., Janakiram, T. and Kumar, R. (2015) *Indian J. Plant Genet. Resour.*, **28**(3) : 360-360.

Nethra, N.N., Jayaprasad, K.V. and Kale, R.D. (1999) *Crop Res.*, **17** : 209-215.

Nguyen, H.T.T., Choi, K.S. and Park, S. (2013) *J. Korean Soc. Appl. Biol. Chem.*, **56**: 255–262.

North, C. (1979) *Plant Breeding and Genetics in Horticulture*, MacMillan Press Ltd., London.

Novotna, I. (1989) *Sbornik UVTIZ, Zahradnictvi*, **16** : 55-64.

Novotna, I. (1990) *Sbornik UVTIZ, Genetika a Slechteni*, **26** : 101-108.

Padmini, K., Janakiram, T. and Naik, L.B. (2013) *Indian J. Agric. Sci.*, **83**(3): 349-351.

Padmini, K., Janakiram, T., Naik, L.B. (2014) *Indian Hortic.*, **59**(1).

Pandey, N. and Rao, V.K. (2014) *J. Hill Agric.*, **5**(2) : 139-143.

Paterson, E.A. and Singh, R.C. (1977) *Phytopathologische Zeitschrift*, **90** : 250-256.

Patil, J.D., Bhat, N.R., Chougule, B.B. and Patil, B.A. (1987) *Curr. Res. Reporter*, **3** : 46-48.

Patil, P.V., Patil, S.D., Patil, M.R. and Kantharaju, K.T. (2013) *Trends in Life Sciences*, **2**: 55-56.

Patil, S.S.D. and Rane, D.A. (1994a) *J. Maharashtra Agri. Universities*, **19** : 372-374.

Patil, S.S.D. and Rane, D.A. (1994b) *J. Maharashtra Agri. Universities*, **19** : 418-420.

Payne, R.N. and Kays, W.R. (1954) *Bull. Okla, Agric. Exp. Stat.*, **13-439** : 15.

Petrenko, N.A. (1974) *Trudy Po PrikladnoiBotanike*, GenetikeiSelektsii, **53** : 229-235.

Petrenko, N.A. (1975) *Byulletin Vsesoyuznogo Ordena Lenina Institut Rastenievodotva N.I. Vavi-lova*, **54** : 72-75.

Petrenko, N.A. (1976) *Referativnyi Zhurnal*, **8**(55): 957.

Phetpradap, L. and Hampton, J.G. (1991) *Proc. forty-fourth New Zealand Weed and Pest Control Confer.*, pp. 300-303.

Pirone, P.P., Dodge, B.O. and Rickett, H.W. (1960) *Diseases and Pests of Ornamental Plants*, Constable and Co. Ltd., London.

Piskornik, Z. (1981) *Zeszytz Naukowe Akademii Rolniczejim. Hugona Kollataja W Krakowie*, Ogrodnictwo, **8** : 29-39.

Piwoni, A. (1991) *Prace Instytutu Sadownictwai Kwiaciarstwa w Skierniewicach. Seria B, Rosliny Ozdobne*, **16** : 139-144.

Pointsi, R.E. and Feldman, J.M. (1959) *Rev. Argent. Agron.*, **26** : 17-22.

Post, K. (1935) *Proc. Amer. Soc. Hort. Sci.*, **32** : 626-630.

Prakash, C., Sharma, B.P., Gupta, Y.C., Thakur, K.S. and Gupta, R.K. (2018) *J. Ornam. Hortic.*, **21**(1-2) : 43-50.

Raabe, R.D. and Pyeatt, L. (1990) *J. Environ. Hortic.*, **8** : 89-92.

Raghava, S.P.S. (1984) *Ph. D. Thesis*, University of Agricultural Sciences, Bangalore, India.

Raghava, S.P.S. and Negi, S.S. (1984) *Annual Report of Indian Institute of Horticultural Research*, Bangalore, 1984, pp. 136-137.

Raghava, S.P.S. and Negi, S.S. (1992) *Indian J. Hort.*, **49** : 281-283.

Raghava, S.P.S. and Negi, S.S. (2001) *J. Ornam. Hortic.*, **4**(1) : 7-12.

Raghupathi, Mahadevaiah, G.S., Kumar, M.J.A. and Gaddi, G.M. (2020) *Int. J. Curr. Microbiol. App. Sci.*, **9**(3). doi: 10.20546/ijcmas.2020.903.185.

Rai, T.S., Chaudhary, S.V.S., Dhiman, S.R., Dogra, R.K. and Gupta, R.K. (2017) *Indian J. Agric. Sci.*, **87**(4) : 540-543.

Ramya, H.M., Nataraj, S.K. and Kumar, R. (2019) *Int. J. Curr. Microbiol. App. Sci.*, **8**(4) : 1314-1318.

Ramya, H.M., Nataraj, S.K., Lakshmana, D. and Kumar, R. (2019) *Int. J. Curr. Microbiol. App. Sci.*, **8**(4) : 1230-1233.

Ranchana, P., Ganga, M., Jawaharlal, M. and Kannan, M. (2017) *Int. J. Curr. Microbiol. App. Sci.*, **6**(9) : 27-31.

Rao T.M., Negi, S.S., Raghava, S.P.S., Janakiram, T. and Kumar, R. (2015) *Indian J. Plant Genet. Resour.*, **28**(3) : 361-362.

Rao, T.M. and Kumar, R. (2019) China aster, **In:** Production Technology of Flower Crops - A Hand Book. ICAR-IIHR, Bengaluru. ISBN 978-81-942095-1-5. pp. 15-19.

Rao, T.M., Negi, S.S., Raghava, S.P.S., Janakiram, T. and Kumar, R. (2015) *Indian J. Plant Genet. Resour.*, **28**(3) : 362-363.

Rawat, A., Bohra, M. and Punetha, P. (2018) *Int. J. Agric. Sci.*, **10**(13) : 6599-6602.

Reddy, Y.T.M. (1978) *Mysore J. Agric. Sci.*, **12** : 526.

Rosinska A., Rosinski, G. and Holubowicz, R. (2009) *Acta Agrobotanica*, **62**(2) : 187-191.

Rosner, B. (1985) *GorskpstopanskaNauka*, **22** : 32-38.

Ruhnau, B. and Forkmann, G. (1988) *Phytochem.*, **27**: 1035-1039.

Runger, W. and Wehr, B. (1967) *Gartenbauwiss.*, **32** : 27-38.

Sailaja, S.M. and Panchbhai, D.M. (2014) *Asian J. Hortic.*, **9**(1) : 36-39.

Schwantiz, F. and Schewantiz, H. (1950) *Ziichter*, **20** : 336-346.

Seyffert, W. (1962) *Proc. 16th Int. Hort. Congr.*, Brussels, **1** : 295.

Sharma, M.K. and Joshi, K.I. (2015) *Int. J. Agric. Sci. Res.*, **5**(4) : 105-110.

Shedeed, M.R., Hashim, M.E. and Abu-Taleb, N.S. (1991a) *Annals of Agri. Sci., Cairo*, **36** : 229-234.

Shedeed, M.R., Hashim, M.E. and Abu-Taleb, N.S. (1991b) *Annals of Agri. Sci. Cairo*, **36** : 235-240.

Shen, X.F., Guo, S., Yin, Y., Zhang, J.J., Yin, X.M., Liang, C.L., Wang, Z.W., Huang, B.F., Liu, Y.H., Xiao, S.M. *et al.* (2018) *Molecules*, **23**: 2426.

Sidana, G., Hembrom, R., Padhi, M., Sisodia, A. and Singh, A.K. (2019) *Journal of Pharmacognosy and Phytochemistry*, **8**(4): 3127-3130.

Sidana, G., Hembrom, R., Sisodia, A., Padhi, M. and Singh, A.K. (2019) *J. Pharmacogn. Phytochem.*, **8**(5) : 372-375.

Sindhuja, M. and Prasad, V.M. (2018) *Int. J. Curr. Microbiol. App. Sci.*, **7**(12) : 1208-1212.

Skrzypczak, C. and Orlikowski, L.B. (1996) *Phytopathologia Polonica*, **11** : 31-40.

Soroa, M.R., Cortes, S.L. and Noval, B. de la (1998) *CultivosTropicales*, **19** : 29-31.

Sowmya, K.A. and Prasad, V.M. (2017) *Int. J. Curr. Microbiol. Appl. Sci.*, **6**(10) : 3204-3210.

Sreenivas, K.N. and Gowda, J.V.N. (1999) *Crop Res.*, **18** : 104-107.

Sreenivas, K.V., Narayanagowda, J.V. and Narayanaswamy, P (1998) *Karnataka J. Agri. Sci.*, **11** : 858-861.

Sreenivasulu, G.B., Kulkarni, B.S., Reddy, B.S., Adig, J.D. and Thammaiah, N. (2004) *J. Ornam. Hortic.*, **7**(3-4) : 345-346.

Stankova-Opocenska, E. (1957)*Sborn. esl. Akad. zemed. Ved, Rostl. Vyroba*, **30** : 727-740.

Strube, H. (1965) *DischeGartenb.*, **12** : 134-136.

Suzuki, S. (1979) *J. Agric. Sci. Japan*, **23** : 235-242.

Syamal, M.M., Rajput, C.B.S., Upadhyay, R.K. and Singh, J.N. (1990) *Indian J. Hort.*, **47** : 439-441.

Tamberg, T.G. (1963) *Tr. Priklad. Bot. Genet. Selekc.*, **35** : 85-93.

Thompson, P.A. and Cox, S.A. (1979) *Seed Sci. and Tech.*, **7** : 201-207.

Tiwari, A.K., Bhuj, B.D. and Mishra, S.K. (2010) *Indian J. Hortic.*, **67**(2) : 255-259.

Tschernews-Dimowa, S.S. (1976) *Arch. Gartenbau*, **24** : 335-349.

Tyagi, S., Jung, J.A., Kim, J.S. and Won, S.Y. (2020) *Plants*, **9**: 568, doi: 10.3390/ plants9050568

Varbanova, K. (1995) *Resteniev"dniNauki*, **32** : 61-62.

Verma, V.K., Verma, J.P., Verma, H.K. and Meena, R.K. (2018) *Int. J. Agric. Sci.*, **14**(1) : 160-164.

Vijayakumar, S., Rajadurai, K.R. and Pandiyaraj, P. (2017) *Int. J. Agric. Sci. & Res.*, **7**: 297-304.

Vijayakumar, S., Rajadurai, K.R., Pandiyaraj, P. and Elangaivendan, A. (2017) *Int. J. Agric.*, **9**(17) : 4148-4150.

Vladmirskaja, M.E. (1959) *Bjul. Glav. Bot. Sada*, **35** : 101-103.

Watts, L. (1980) *Flower and Vegetable Breeding*, Grower Books, London.

West, D., Merrigan, I., Taylor, J. and Collins, G. (1981) *Australian Hort.*, **79** : 7, 9, 11-12.

Wit, F. (1937) Genetica, **19**: 1-104.

Yadav, L.P. (1983) *Ph. D. Thesis,* Bidhan Chandra Krishi Viswavidyalaya, West Bengal, India.

Yasuda, I. (1955) *Okayama Univ.,* **7** : 31-35.

Yokoi, M. and Hosaka, H. (1962) *Tech. Bull. Fac. Hort. Chiba,* **10** : 143-155.

8

Zinnia

Anamika Gurung, Rajiv Kumar, B.K. Jana, M. Mitra and T.N. Saha

1.0 Introduction

Zinnia is a popular garden plant which flowers almost throughout the year in moderate climate. It is named after Johann Gottfried Zinn, a German Professor of Medicine.

2.0 Importance and Uses

2.1.0 Aesthetic Uses

Zinnia is grown in beds, borders and pots. It is also useful for rock gardens, edging and window boxes. The cut blooms remain fresh for several days in vases. It is a useful plant as it has nematicidal properties (Yassin and Ismail, 1995; McScorley, 1995) and thus is also used as organic amendments for its effectiveness in suppressing *Meloidogyne arenaria* in soil (Ritzinger and McScorley, 1998).

2.2.0 Therapeutic and other Uses

Zinnia elegans contains of a kaempferol 3-O-glycoside, and a guanidine alkaloid (Burlec *et al.*, 2019). The chemical composition of *Zinnia elegans* based on their metabolomic profiles showed that it contains coumarins, sesquiterpene lactones and phenylethanoids and the ethyl acetate fraction of *Z. elegans* showed a significant antimalarial activity (Gomaa *et al.*, 2020). Zinnia flower extract can be used as a substitute for synthetic acid base indicator in various acid-base titrations (Jagtap *et al.*, 2018).

3.0 Species and Cultivars

3.1.0 Species

Zinnia belongs to the family *Asteraceae*. A genus of 20 species of spreading to erect, annual and perennials with beautiful flowers of various colours, native of Mexico, United States and Central and South America. Much improvement has been done in the USA (Swarup, 1967). *Zinnia elegans* and *Z. haageana* with many types or groups and numerous cultivars are widely grown for the attractive flowers. The species of *Zinnia* grown in gardens are described below.

Zinnia elegans

It is a very useful species due to wide variation in plant height (25 to 120 cm), flower size (2.5 to above 12 cm across) having white, green, red, pink, yellow, orange, salmon, lavender, mauve and various intermediate shades of colour, single, semi-double and double; ray florets flat, quilled or ruffled.

- ✰ Belvedere – dwarf, with double flower-heads, weather-resistant.
- ✰ Dreamland Scarlet – dwarf and compact, ideal for containers or bedding.
- ✰ Hobgoblin Series – dwarf, bushy, with small, single, weather-resistant, good for bedding (Brickell, 1996).

Uses of Zinnia

In Pot | Hanging Basket | In Window Basket

In Garden | In Beds | Along Paths

Cut Flower | Floral Wreath | Floral Bouquet

Floral Crown | Floral Decoration | In Landscaping

Source: www.aerogarden.com, www.rareseeds.com, www.harrisseeds.com, www.i.pinimg.com, www.in.pinterest.com, www.theenglishgarden.co.uk, www.gardenerspath.com, www.flowerspicture.org, www.deborahsilver.com

- ✰ Persian Carpet –dwarf, compact and spreading, semi-double and double flowers, and weather-resistant. Star White – white flowers with golden yellow centres (Brickell, 1996).
- ✰ Peter Pan Series – extremely dwarf and compact, early flowering, 10 cm across, suitable for containers or bedding.
- ✰ Peter Pan Gold – golden yellow flowers.
- ✰ Peter Pan White – short (<38 cm), highly tolerant to powdery mildew (Gombert *et al.*, 2001).
- ✰ Ruffles Series – medium height, highly tolerant to bacterial leaf and flower spot and blight (Gombert *et al.*, 2001).
- ✰ Short Stuff Series – dwarf and compact, flowers double.
- ✰ Short Stuff Orange – short height, highly tolerant to powdery mildew (Gombert *et al.*, 2001).
- ✰ Small World Series – plant dwarf, with double flower heads.
- ✰ Sombrero – dwarf, 30-40 cm high, ligulate florets, arranged in 1-2 rows, colour red-brown and orange-yellow, suitable for beds (Kirilov, 1990).
- ✰ State Fair – plants tall, vigorous, with large, double flowers (Brickell, 1996).
- ✰ Thumbelina Series – very dwarf and spreading, flowers single or semi-double, yellow, red, magenta or pale pink.

Z. haageana (Z. angustifolia)

Several cultivars of this species are available, the height varying between 30 and 45 cm, single, semi-double or double-flowered, in orange, reddish brown, red, gold and bicolours and tricolours; excellent for bedding and ground covers.

Z. linearis

The species is perennial but usually grown as annual for bedding, border and rock garden and in pots and hanging baskets. The plant grows upto 20-25 cm high, spreading, bearing numerous flowers, orange or white in colour.

Rzedowski*et al.* (1996) identified two new species of zinnia from Central Mexico.

Z. acerosa var. *guanajuatensis*– prostrate habit, wider inflorescences, larger number of flowers, pappose achenes.

Z. zamudiana – prostrate habit, short peduncles, shorter and less intensely coloured ligules.

3.2.0 Cultivars

'Cut and Come Again' is a useful strain, about 40 cm high having yellow, dark red, rose, scarlet, violet and white blooms. Zinnia Liliput, Button's, Cupid are dwarf (30-40 cm) types.

As bedding plants, Howe and Waters (1983) identified some promising zinnia cultivars and these were Fantastic Light Pink, Dasher Scarlet, the Ruffles series and the Pulchino series. Zinnia *cvs*. Pink Buttons, Cherry Buttons, Krasnaya Shapochka,

Species of Zinnia

Zinnia elegans | *Zinnia angustifolia* | *Zinnia linearis*

Zinnia acerosa | *Zinnia zamudiana* | *Zinnia haageana*

Zinnia peruviana | *Zinnia grandiflora* | *Zinnia bicolor*

Zinnia maritime | *Zinnia anomala* | *Zinnia purpusii*

***Source*:** **www.portugal.inaturalist.org, www.plants.ces.ncsu.edu, www.conabio.gob.mx; www.en.wikipedia.org, www.commons.wikimedia.org, www.inaturalist.org, www.applewoodseed.com**

Cultivars of Zinnia

Belvedere | Dreamland Scarlet | Persian Carpet

Peter Pan Gold | Peter Pan White | Scarlet Ruffles

Yellow Ruffles | Short Stuff | Short Stuff Orange

Small World Cherry | Sombrero | State Fair

Source: www.all-americaselections.org, www.applewoodseed.com, www.thespruce.com, www.swallowtailgardenseeds.com, www.lucasgreenhouses.com, www.etsy.com, www.weseeds.com, www.shootgardening.co.uk

Cultivars of Zinnia

Cut and Come Again

Thumbelina

Profusion White

Profusion orange

Royal Purple

Holi Scarlet F_1

Magellan Coral F_1

Profusion Double Deep Salmon

Queeny Lime with Blotch

Queen Lime Green

Green Scarlet Flame

Cherry Queen Light Scarlet Flame

Source: www.swallowtailgardenseeds.com, www.images-na.ssl-images-amazon.com, www.applewoodseed.com, www.all-americaselections.org, www.americanmeadows.com

Envy and Snowman were recommended for cultivation under cover (Vashchenko, 1985). Schmidt (1979) in a glasshouse trial, noted that Early Wonder, Scabious-flowered and Liliput were suitable cultivars and types.

Kumar and Marwein (2018) found cultivars Dreamland Takii, Double Orange, Royal Purple, Orange King, Double Mixed, Tall Red and Cherry Queen showed better performance for vegetative, flowering and quality characters under West Garo Hills, Meghalaya.

4.0 Genetics and Breeding

4.1.0 Genetics

Selaru and Costea (1991) presented of linkage between the genes *L1* and *Ii* controlling, respectively, the shape of the ray florets and the degree of floral doubleness. The dominant *L* gives normal ligulate ray florets and the recessive *i* double ones. The cactus type is conferred by the homozygous recessive *iill*, while *IILL* and *IiLl* give the daisy type.

4.1.1 Stigma Receptivity and Pollen Viability

Zinnia elegans blossoms at 8:30-16:00 in day time and the viability of its pollen is 18 h. When it blossoms, the stigmas of florets stand upright and keep high turgidity for 7-10 days. The stigmatic pollen receptive on 1-3 d, 4-6 d, 7 d and 10 d are 79.9-83.1 per cent, 56.1 -63.0 per cent, 37.0 -43.3 per cent, respectively based on benzidine-H_2O_2 test. The seed set at the Y stigma phase, the tip of slightly curling stigma phase and columnar stigma phase is 76.3 per cent, 48.5 per cent and 36.8 per cent, respectively by hand-pollination. The stigma top with hair and its big surface makes it easy to accept pollen and gets a higher seed set (Ye *et al.*, 2007).

4.2.0 Breeding

4.2.1 Hybridization

A large number of cultivars have been developed through hybridization in the various parts of the world. Boyle and Stimart (1982) found that in 13 cross combinations of *Z. angustifolia* and 13 *Z. elegans* cultivars, embryos were produced in 47.2 per cent of the florets. Six cross combinations of *Z. elegans* × *Z. angustifolia* produced embryos in only 4.6 per cent of the florets. Inter-specific crosses resulted in seed propagated hybrids when *Z. angustifolia* was the maternal parent. Hybrids were obtained from reciprocal crosses when embryos were excised and cultured. Morphologically most of the hybrids resembled *Z. angustifolia* but they were sterile. Colchicine treatment of axillary buds in some of these hybrids resulted in the production of fertile inflorescence.

Terry-Lewandowski *et al.* (1984) reported that fertility of the sterile F_1 hybrids (2n = 23) between *Z. angustifolia* (2n = 22) and *Z. elegans* (2n = 24) was partially restored by inducing amphidiploidy by colchicine treatment of axillary buds. Inter-specific hybrids were developed by Boyle and Stimart (1989) from crosses between 4 *Z. angustifolia* clones (maternal parents) and an inbred line of *Z. elegans*

Orange King. The genotypes of *Z. angustifolia* significantly affected plant height, fresh weight, and days to flowering of inter-specific hybrids.

True-breeding lines of *Zinnia marylandica* (2n=46) were reciprocally back crossed with diploid and autotetraploid forms of *Z. angustifolia* (2n=22 or 44) and *Z. Violacea* (2n=24 or 48) (Boyle, 1996). In most of the cases, backcrosses were more successful with *Z. angustifolia* and *Z. violacea* as autotetraploids than as diploids. Back crossed (BC_1) hybrids of *Z. marylandica* and autotetraploid *Z. violacea* produced larger capitula and more ray florets, and exhibited novel combinations of floral pigments, and had commercial potential as seed-propagated bedding plants.

Double type as male parents produces the greater proportion of pompom type in the F_1 population through cross breeding (Gultom and Gultom, 2015). Studies of F_1 population from crossing of pompom type with single and double type of kembang kertas (*Zinnia elegans*) through cross breeding was carried out (Gultom and Gultom, 2006). Pompom type of kembang kertas is not perfect flowers (do not have male reproductive organ). To determine which type of male (single and double type) of perfect flowers (having both male and female reproductive organs) can produce greater pompom proportion, crossing between pompon type that originated of selfing results of M_6 population (M_6S_1) as female parent and single and double type of M_7 population as male parents to determine the appropriate type of male to get a descent pompom was done. The results showed that crossing between ♂ double x ♀ pompom types produced the pompom type flower to the proportion of 0.78 and crossing between ♂ single x ♀ pompom produced the pompom type flower at the proportion of 0.31.

4.2.2 Mutation Breeding

Nikolova and Vasileva (1979) noted that irradiation of dry zinnia seeds with ^{60}Co at 0.5, 1, 10 or 20 kR improved the flower size by increasing the number of ligulate florets. Irradiation of selfed seeds of *Z. elegans cv.* Crimson Red with ^{60}Co at 7.5 kR gave significant increase in plant height, branch number, flower number, more number of petals and flowers diameter. Irradiated plants showed mutation of flower structure and four new flower colour (magenta, yellow, red and red with white spots) (Venkatachalain and Jayabalan, 1997).

Kole and Meher (2005) found that plant height, number of branches, pedicel length and flower diameter increased at lower doses of 5 kR followed by decrease at higher doses of 10 kR onwards as compared to non-irradiated plants in two varieties of Zinnia in M_1 generation. Appearances of bud and consequently of flower were delayed in irradiated population while number of flowers and retention of flower were decreased. Frequencies of plants with double flower, puckered leaves and wrinkled leaves increased in radiation treatment. Survival at maturity was linearly decreased with increasing doses of gamma rays. Three plants with mutation in flower colour from orange to vermillion red and light marooned colour and from yellow to faint brown colour with florets tubular in appearance near the base were obtained at 15 kR dose. Another plant at 5 kR in Yellow variety had disc florets changed to ray florets and the flowers were compact resembled *Tagetes*. Sectorial mutation in one branch with circular leaves occurred in Yellow variety. In general, Orange variety was more sensitive to gamma irradiation as compared to Yellow variety.

Three doses of gamma radiation (75, 100 and 125Gy) were utilised to obtain new and novel varieties of *Zinnia elegans* var. Dreamland. The higher doses of gamma rays were found to be detrimental for the germination and survival of seeds and height of the seedlings. Phenotypical variations such as plant height, the number of flowers and flower diameter of the third generation mutants were highly significant as compared to the control. Eight floral variations could be obtained with novel form and colour (Kole and Meher, 2005).

4.2.3 Incompatibility

Samaha *et al.* (1989) studied the self-incompatibility (SI) in 3 *Z. angustifolia* clones. Examination with UV epifluorescence microscopy revealed that pollen load was higher and little or no callose accumulated in stigmatic papillae following compatible crosses, whereas in incompatible crosses pollen load was low and callose lenticules were deposited in stigmatic papillae; the correlation between pollen load and percen-tage embryo set was 0.89. The intensity of callose fluorescence of the pollen tube-papillae attachment sites was quantitatively measured via microspectrofluorometry. Callose fluorescence intensity ranged from 47.9 to 62.6 per cent for incompatible and from 6.4 to 9.9 per cent for compatible crosses, and was negatively correlated (r= –0.95) with percentage embryo set. A population of 14 full-sib F_1 progeny was inter crossed in a full diallel and then was reciprocally backcrossed to both parents. Two intra-incompatible and reciprocally compatible classes were obtained in the F_1 diallel (Samaha and Boyle, 1989). Results of the crosses were consistent with a one locus, multiallelic sporophytic incompatibility system. The presence of a linear dominance series of S alleles in the pollen and either a linear dominance series or a combination of dominance and independent action of S alleles in the pistil.

The three major problems suggested for the failure in seed formation in pollinated florets of zinnia included failure in fertilization, ovules lacking embryo sacs, and abortion of developing seeds. The structure of ovules that failed to form seeds after compatible pollination in double-flowered zinnias (*Zinnia violacea* Cav.) was investigated by Miyajima (2006). In normally developing seeds, the cellular endosperm became vague at 5–6 days after pollination (DAP) and the developing embryos occupied the entire portion of the seeds up to 10 DAP. A number of ovules failed in fertilization for the lack of an embryo sac. In most of the aborted seeds, the embryo sacs degenerated, whereas the aborted embryos were still alive in a small number of aborted seeds. A small percentage of the aborted seeds exhibited an aborted embryo and an unfertilized fused nucleus of two polar nuclei or 'single fertilization'.

4.2.4 Male Sterility

Shoot cultures initiated from axillary buds of identified male-sterile of zinnia plants proliferated best on a medium with 1 µM benzyl adenine. *In vitro* produced male sterile plants were more compact, more heavily branched, and earlier flowering than seed-propagated plants (Rogers *et al.*, 1992).

The differences in the structure of floral organs between male sterility line and maintainer line AB201 of *Zinnia elegans* were studied (Zhao *et al.*, 2007). The male sterility was controlled by a pair of recessive gene. The inheritance of the male sterility was correspondent to Mendel's law. The sterility was stable and could be trans selected easily.

In a study on male sterile lines of *Zinnia elegans* and their cytological observations by Ye *et al.* (2008), crimson male sterile material, obtained from selfed offsprings, backcrossed with yellow F, of dream series, got 1:1 fertile ratio of segregation. Then the male sterile lines controlled by a single recessive gene were obtained from sib-mating and selfing of fertile plants and test crossing with maintainer and restoring lines. The genotypes of the two types of male sterile materials were msms and Msms, respectively. As a result of cytological observation, the anther chamber of fertile plants was consisted of epidermis, inner wall, middle and tapetum, the normal meiosis of pollen mother cells initiated by the division of cytoplasm and mature pollens were trinucleate. On the other hand, in the male sterile plants, the petals degraded or became filamentous, and stamen degraded into filamentous without pollens and started to appear villous, which were obviously different from fertile plants. The male sterile materials, which could not develop anther chamber structure, pollen and pollen mother cells followed by formation of anther sporogenous cells, were entirely structural male sterility.

Lou *et al.* (2010) analyzed the heterosis, combining ability and heritability of eight cross combinations of *Z. elegans* using two recessive nuclear male sterile lines crossed with four restorers and also described the breeding value of parental lines with respect to their potential to produce hybrids. Six horticultural traits were improved over mid parent and best parent. The traits of plant height, number of whorls of ray florets across capitulum and pedicel length were primarily controlled by paternal additive effects, whereas crown size was mainly controlled by non-additive effects. Number of branches and length of node were affected both by paternal additive effects and non-additive effects. Among the parental lines, AH003A and restorer A3 were chosen as primary female and male combiners, respectively. AH001A and restorer S5 were chosen as secondary combiners. The cross AH003A x A3 was determined as the most promising combination for producing potted plant characteristics, and AH001A x S5 was the best hybrid obtained in this study for cut flower traits. The analysis of combining ability for the parental lines showed that there was no causal relationship between general combining ability and specific combining ability effects indicating the promising varieties of *Z. elegans* should attained by extensive test cross.

Lou *et al.* (2011) selected five male sterile two-type lines and fourteen restorer lines for incomplete diallel crossing, and analyzed the combining ability and heritability of main ornamental traits. Plant height, crown size and length of node had obvious additive genetic effects, high (>0.50) broad sense heritability and high narrow sense heritability (length of node was medium). Pedicel length showed approximately equivalent maternal and paternal additive genetic effects, high broad sense heritability and medium (0.30–0.50) narrow sense heritability. Number of whorls of ray florets across capitulum and number of branches were able to take advantage of heterosis. The relationship between general combining ability and specific combining ability in *Zinnia elegans* depended on materials and traits. Male sterile two-type line was pivotal in the hybridization breeding of *Z. elegans*. Further, ideal parents and combinations were chosen for potted flowering plants and cut flowers.

Zinnia elegans F_1 variety for landscaping was developed by cross combination which had a red filial generation F_1 derived from the cross of red male sterile line and red self-bred line. By means of character contrast test between a red filial generation F_1 and the other domestic and overseas F_1 varieties, it showed good double petal, bright flower color, no colour fading, developing healthy and strong, preferable resist-stress (Jun *et al.*, 2011).

4.2.5 Disease Resistance

True-breeding lines of *Z. marylandica* (2n=4x=46) were back crossed with autotetraploid *Z. angustifolia* (2n=4x=44) and *Z. violacea* (2n=4x=48). Seed-generated, backcross (BC_1) families were screened for resistance to *Alternaria zinniae, Xanthomonas campestris* pv. zinniae and *Erysiphe cichoracearum*. The results suggested that one *Z. angustifolia* genome in BC_1 allotetraploids is sufficient to confer resistance to *A. zinniae* and *E. cichoracearum*, but at least two *Z. angustifolia* genomes are required in BC_1 allotetraploids to provide resistance to *X. Campestris* pv. zinniae (Boyle and Wick, 1996).

Jones and Strider (1979) suggested *Z. linearis*, Classic, *Z. haageana*, Chippendale Daisy, Old Mexico and Persian Carpet, and *Z. elegans*, Cut and Come Again, Tangerine as sources of resistance to bacterial leaf spot caused by *Xanthomonas nigromaculans*. Swarup and Raghava (1974) were able to develop leaf curl resistant cultivars by X-ray irradiation of seeds and repeated selection.

Fifty-seven cultivars of *Zinnia elegans* were studied for 17 weeks to determine their resistance to Alternaria blight (*Alternaria zinniae* Pape), powdery mildew (*Erysiphe cichoracearum* DC ex Merat) and bacterial leaf and flower spot (*Xanthomonas campestris* pv. *Zinnia*). A disease severity scale was used to determine acceptability for landscape use. At week 4, all cultivars were acceptable. At week 10, eleven cultivars were acceptable. At week 17, all cultivars were unacceptable. Ten cultivars had been killed by one or more pathogens by week 17. Only two cultivars showed any tolerance to any disease (powdery mildew) at week 17 (Gombert *et al.*, 2001).

5.0 Growth and Flowering

5.1.0 Environmental Factors

5.1.1 Light

Armitage *et al.* (1981) reported that the time of flowering, plant height and weight of several *Z. elegans* cultivars were reduced by 8-hours photoperiod for 4 weeks. Reduction in flower size in *cvs*. Exquisite, Golden Dawn and Bonanza F1 hybrid was caused by short day treatment (Reimherr, 1980). While long days increased the plant height of both tall and dwarf cultivars of zinnia but the start of flowering in dwarf cultivars was not affected by day length (Han and Yeam, 1978). Daily light integral (DLI) describes the rate at which photosynthetically active radiation is delivered over a 24-hour period and is a useful measurement for describing the greenhouse light environment. Zinnia produced commercially acceptable plants at 12 mol/m^2/d, however, 19 mol/m^2/d resulted in improved growth and flowering for zinnia. Therefore, zinnia could be categorized as high to very high DLI crop (Faust *et al.*, 2005).

The vegetative growth and flowering of zinnia as affected by night interruption (NI) at different timings was studied (Park *et al.*, 2013). The NI was employed at 18:00-22:00 HR (NI18), 22:00-02:00 HR (NI22), or 02:00-06:00 HR (NI02). The height of zinnia was shorter under NI than under SD, and those grown under NI02 were shorter than those under NI18 or NI22. It flowered later under NI02 than under NI22 or NI18. So, plant height and the number of nodes or leaves were affected by prolonging photoperiod with NI regardless of NI application timings.

The light-emitting diodes could be an effective and reliable source of lighting under controlled environment for production of quality flower seedlings. Zinnia plants under blue light showed a significant increase in leaf number. Generally, the pure red light enhanced seedling emergence and during growth period, high quality seedlings could be produced when 25 per cent blue light was added to red light (Akbarian *et al.*, 2016).

Z. elegans could be grown under full sunlight conditions, however, Mahdavifard *et al.* (2020) observed its growth and flowering quality were higher under 1200 µmol/m^2/d. Therefore, it is possible to grow it under semi-shading conditions.

5.1.2 Temperature

Schmidt (1979) suggested that for growing zinnia in greenhouse, temperature should be at least 18° to 20°C for the first 2 weeks and the plants need sufficient humidity during the period of vegetative growth. Seedlings of *Z. elegans cv.* Peter Pan took a long time to flower when grown in greenhouses at day temperature of 27±3°C (9 h) and 7±3°C night temperature (15 h) (Merritt and Kohl, 1991).

Holcomb (1982) in an experiment with zinnia cultivars Dasher Scarlet and Small World Cherry noted that the plant flowered within 9 to10 weeks and the plants receiving soil heat had more flowers.

The hypocotyl elongation in zinnia decreased as the difference between day and night temperature increased. Temperature decrease during the dark period might be the primary factor causing decreased elongation of Zinnia during the dark period (Shimizu *et al.*, 2008).

Twumasi *et al.* (2009) investigated the effects of different average daily temperatures (ADT) and the difference between day-time (DT) and night-time (NT) temperature (DIF) on stem size, the length of xylem vessels within the stem, and the length of individual vessel elements within a vessel, in *Zinnia elegans* two cultivars. Increasing the ADT from 17°C to 25°C increased stem lengths by approx.

15 per cent, but more than doubled the lengths of the xylem vessels. The increase in xylem vessel lengths was only partly (< 10 per cent) due to an increase in the lengths of individual vessel members, which implies that temperature (ADT) had a greater influence on the number of fused vessel elements per xylem vessel. A negative DIF (*i.e.*, lower DT than NT temperatures) decreased stem lengths and a positive DIF increased stem lengths. DIF had no effect on xylem vessel length, probably because, other than in stem length, xylem vessel length was positively correlated with NT temperature.

Zivdar *et al.* (2011) observed that at 25°C upto 12 dS/m, 57 per cent of the *Zinnia elegans* seeds germinated which is considerably high soil salinity and temperature regimes. Riaz *et al.* (2011) compared performance of two zinnia cultivars 'Dreamland Red' and 'Super Yoga Pink' under high-temperature and limited water conditions in Pakistan where hot winds, scorching heat, and water deficiency are problems. Cultivar 'Dreamland Red' performed better than 'Super Yoga Pink' in morphological as well as physiological attributes, sustaining its acceptable beauty and quality at 70 per cent of field capacity.

5.2.0 Growth Regulators

Cathey (1965) recorded retardation of growth and improvement in flowering in zinnia by using SADH. Application of phosfon improved the growth of zinnia plants (Cathey and Stuart, 1961).

Das *et al.* (1977) recorded that *Z. elegans* cv. Double Rose flowered under short days. GA_3 failed to induce flower under long days, while IAA at 100 mg/l or kinetin at 50 mg/l enhanced flowering with additional light. Application of SADH at 5000 ppm reduced plant height and flower diameter but increased time of flowering without affecting fresh weight. Cox and Keever (1988) obtained good compactness of zinnia *cv.* Scarlet Ruffles grown in pots containing 1:1 mixture of sphagnum peat and perlite, by spraying paclobutrazol at 250, 500 or 1000 ppm and drenching at 0.5 to 1.0 mg/pot. Ancymidol at 100 ppm restricted the height and fresh weight but not the time to flowering (Armitage *et al.*, 1981). Seedlings of *Z. elegans cv.* Peter Pan Scarlet sprayed with 5000 ppm daminozide, 200 ppm ancymidol, 40 and 90 ppm of paclo-butrazol resulted in reduced leaf expansion and stem elongation but showed no effect on shoot dry weight (DW) gain. While daminozide and ancymidol delayed flowering (Latimer, 1991). Shedeed *et al.* (1986) noted that SADH at 500 and 2000 ppm increased seed yield in *Z. elegans.* Davis (1991) suggested that seedling height of *Z. elegans cv.* California Giant can be controlled by applying 5-25 mg uniconazole/l without affecting anthesis and with no adverse effect on subsequent growth after flowering.

Sperry and Chaney (1999) studied the effect of paclobutrazol and flurprimidol on phototropism of *Z. elegans*. The mode of action involves complexing of the growth retardant with cytochrome P450-dependent enzymes in the metabolic pathway for gibberellins. Because cytochrome P450 also is a part of the blue light receptor system that controls phototropism. Phototropic curvature of etiolated *Z. elegans* 'Scarlet' was reduced by both growth regulators.

Pinto *et al.* (2003) found that Daminozide, paclobutrazol and chlormequat significantly reduced plant height and side branches length in 'potted plants of zinnia *cv.* 'Persian Carpet'. Paclobutrazol (1.0 mg *a.i.*/pot) and chlormequat (1.0 g/l) increased foliage and flowers harvest index. Plant spread diameter and canopy shape were improved with paclobutrazol (0.75 mg *a.i.*/pot), while Chlormequat (2.0 and 3.0 g/l) caused phytotoxicity symptoms.

Daminozide, paclobutrazol and chlormequat supressed *Zinnia elegans* 'Lilliput' growth and did not affect flower diameter and production cycle, but plants were not short enough to meet market quality standards (Pinto *et al.*, 2005). However, cycocel at 2000 ppm caused most lateral shoot, flower number and lowest lateral shoot length, root fresh and dry weight, root number and length and shoot carbohydrate. Paclobutrazel at 30 ppm recorded in the most leaf chlorophyl content (Hojati *et al.*, 2009).

Under high temperature stress, plant height of *Zinnia elegant* with different concentrations of S3307 (uniconazole) was reduced, whereas, stem diameter was increased with increased photosynthetic rate and water use efficiency. The treatment of S3307 significantly slowed down the inhibition of high temperature stress on photosynthetic system II of seedlings, alleviated the decrease of maximum photochemical efficiency (*Fv/Fm*), actual photochemical efficiency under light (ÖPS II), photochemical quench (qP) and the increase of no photochemical quench (NPQ) and prevented the damage of high temperature on photosynthetic apparatus. Among them, the treatment effect of 30 mg/l S3307 is the best (Li *et al.*, 2012).

Iqbal *et al.* (2012) found that salicylic acid (SA) at 50 mg/l recoded maximum vase life of flower, however, maximum structural integrity with SA at 150 mg/l of zinnia cut flowers. Foliar spray of NAA at 100 mg/l resulted maximum percentage of flower color and physical appearance.

The use of 60 mg/l of salicylic acid (SA) in *Zinnia elegans* reduced the heat damage indices and increase the recovery index remarkably, increase root vitality, reduce relative conductivity and malondialdehyde (MDA) contents, and increase proline (Pro) contents. The SA treatment could alleviate the membrane lipid per oxidation and increase osmotic regulatory metabolites contents, raise heat resistance (Cao and Li, 2014).

Soil application of 10 mg/l uniconazole is effective in controlling plant height in zinnia and also increased flower diameter and chlorophyll content. However, soil application of cycocel at 1000 and 1500 mg/l recorded shortest time for flowering (Alami and Karimi, 2020).

5.3.0 Growth Regulators and Temperature

Zinnia elegans cv. Pompon were treated with GA_3 or daminozide under 3 day/night temperature difference (DIF) regimes (+5 DIF : 21/16°C day/night; 0 DIF : 18.7°C constant; –5 DIF : 16.5/21.5°C day/night), each with a daily average temperature of 18.7°C, at two developmental stages : stage 1, the period of vegetative growth before flower bud formation; and stage 3 growth just before anthesis (Table 1).

Table 1: Percentage of Diurnal Stem Growth, Occurring during the Day Period only in Zinnia Treated with Growth Regulators (GR) at Two Stages of Development and Exposed to Three DIF Regimes*

DIF	*Days Period Growth (Per cent)*							
	Stage 1: Vegetative				*State 3: Pre-anthesis*			
	GA_3	*Control*	*Daminozide*	*Mean*	*GA_3*	*Control*	*Daminozide*	*Mean*
+5	50.2	65.8	61.7	59.2	43.6	53.6	48.3	48.5
0	47.1	61.2	58.9	55.7	36.4	40.9	45.2	40.8
-5	44.7	52.6	48.6	48.6	32.1	39.1	38.4	36.5
Mean	47.3	59.9	56.4	–	37.4	44.5	44.0	–

SEM (standard error of the mean) 3.09.

* Neily *et al.* (2000).

DIF regime influenced stem elongation primarily by altering the size of an early morning peak in stem elongation rate (SER); peak height increased as DIF became more positive. GA_3 increased SER throughout the diurnal period with a proportionately larger effect on night time growth. Conversely, daminozide decreased SER more or less equally throughout the diurnal period. Neither GA_3 or daminozide, transformed growth patterns to match those of positive or negative DIF plants, but instead simply increased or decreased growth amplitude. Neither growth regulator altered the basic diurnal SER pattern at any DIF, or influenced the observed shift to greater night time growth as plants matured from stage 1 to stage 3 (Neily *et al.*, 2000).

6.0 Propagation

Zinnia is propagated by seeds. The seeds are sown in beds, boxes or seed pan. Zinnia also grows and flowers well when sown *in situ*. Mittra *et al.* (1990) observed a faster seedling emergence and development when zinnia seedlings were raised in trays under clear UV stabilized film plastics tunnels of 6 x 3 x 1 m than when grown in open. Hexacosanol, octacosanol and triacontanol when applied at 0.001, 0.01 or 0.1 mg/l to seeds of *Z. elegans cv.* Illumination, lower concentration increased seed germination and stimulated seedling growth but at higher concentration both processes were inhibited (Blamowski and Borowski, 1995).

The exposure to magnetic fields produces biochemical and physiological changes in zinnia seeds. Magnetic seed stimulation with 100 mT for 15 min showed maximum potential as a beneficial seed physical enhancement technique for better stand establishment of *Z. elegans* (Afzal *et al.*, 2006). Similarly, Zamiran *et al.* (2015) recorded enhanced seed germination and seed emergence when seeds were treated at 400 μT (micro-Tesla) electromagnetic field. The exposure durations of 240 and 30 min showed promising results over all magnetic fields in seed germination and emergence, respectively.

Propagation and Growing of Zinnia

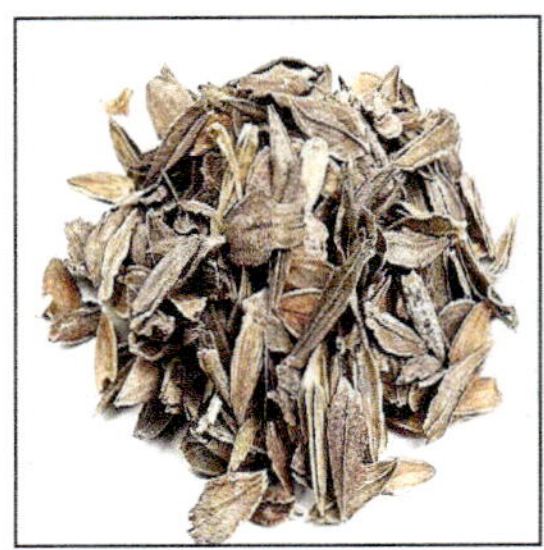
Seeds

Seedling in Trays

Seedling in Polybags

Transplanting Seedling in Field

Cuttings

Cuttings Planted in Trays

Micropropagation

Organogenesis from Callus Culture

In Polyhouse

In Greenhouse

Cultivation in Open

Source: www.gardenerspath.com, www.svaec.ifas.ufl.edu, www.thatgirlsflowers.com, www.i.pinimg.com, www.researchgate.net, www.scialert.net, www.i.ytimg.com, www.i0.wp.com, www.st3.depositphotos.com

Ascorbic and citric acids had no effect on the total seed infestation with fungi. Moreover, the acids significantly stimulated growth of *Botrytis cinerea* on the seeds. Lactic acid decreased the number of seeds infected with fungi, especially with *Alternaria zinniae* and *Fusarium* spp., however at the highest concentration negatively affected germination parameters and deteriorated seed vigour (Szopinska, 2013).

6.1.0 Seed Priming

Seed priming with salicylic acid can be used for improving zinnia seed germination under drought conditions. Under pot experiment, seed priming with 1 mM concentration of salicylic acid improved zinnia vegetative traits at -1 and -1.5 MPa drought levels. Seed priming with salicylic acid increases the relative water content of the leaves under drought stress conditions (Abdollahi *et al.*, 2016).

Szopiñska and Politycka (2016) observed that regardless of the initial quality of the treated seeds, osmopriming was more effective method of improving the germination and vigour of zinnia seeds than hydropriming. The activity of α-amylase may be potentially useful for the seed industry as physiological marker of zinnia seed vigour and the effectiveness of osmopriming. Pre-sowing seed treatment particularly halopriming with 50 mM $CaCl_2$ was the most effective for invigoration of zinnia seeds (Ahmad *et al.*, 2019).

7.0 Soil and Climate

Zinnia requires a well-drained and deep soil, fairly rich in organic matter, water-logging or moist soil condition adversely affect the growth and flowering. *Zinnia elegans* grown in 100 per cent peat produced shortest plants and dry matter accumulation was least (Yahya *et al.*, 1997). Zinnia prefers full sun and a mild climate, hot summer or cold climate retards growth. Blooms are easily damaged by heavy rains. In a two-year trial, Vashchenko (1985) reported that plants under cover had longer shoots and flowered 20-25 days later than in open.

Leaf manure mix (silt + leaf manure + coconut compost at 1:1:1) improved number of flowers per plant, while, coconut compost improved flower size in *Zinnia elegans cv.* Blue Point. However, silt alone produced flowers earliest with the longest blooming period (Riaz *et al.*, 2008).

The saw dust associated with superphosphate and ammonium sulfate provide excellent results in substrate composition for plant emergence and establishment of seedlings zinnias (de Moraes *et al.*, 2016). Application of Codaphos and Codasting (organic fertilizer containing N-Codasting and P+K-Codaphos) in floating system (alternative type of hydroponic) resulted in better development of the aboveground parts and roots of seedlings comparing with other organic and inorganic fertilization (Akoumianaki-Ioannidou *et al.*, 2019).

8.0 Cultivation

8.1.0 Transplanting and Spacing

The seedlings are transplanted at 4 leaf stage at a distance between 20 and 40 cm depending on the cultivar. If the seedlings are overgrown before planting, they may be pinched back at the time of transplanting.

Jhon *et al.* (1991) reported that spacing of 30 x 60 cm for *Z. elegans cv.* Giant Flowered Crimson, grown for 2 subsequent years, gave highest plant height, number of branches per plant, number of flowers/plant and flower diameter. The planting distance mainly depends on the cultivars, purpose and climate.

Saleem *et al.* (2003) observed that spacing of 30 cm between plants appeared better for growth rate and longer blooming period. However, Chaudhary *et al.* (2007) found that spacing 30×45 cm found best for plant height, branches per plant, length of branches, internodal length and of nodes per plant.

Zinnia elegans plants spaced at 30.0 × 30.0 cm exhibited highest plant canopy diameter, number of leaves per plant and number of flowers per plant, while, 15.0 × 15.0 cm spacing had highest plant height and leaf area. Flower diameter, stem diameter, flower quality, internodal distance, fresh weight of a stem and vase life were similar irrespective of plant spacing (Sharif *et al.*, 2019).

The yield (272 stems/plot) of zinnia planted in high tunnel on 12 March was higher than plants planted on 2 April which was 106 stem/plot in Mississippi (Zhao *et al.*, 2014).

8.2.0 Irrigation

Zinnia requires thorough irrigation instead of light sprinkling at frequent intervals. Generally, irrigation is done at an interval of 7-10 days, if the weather is dry. The plants treated (with irrigation) with a 1-2 per cent solution of photosynthetic bacterium increased resistance to drought (Xia *et al.*, 1994).

Growth parameters decreased overall as irrigation water salinity increased; however, marketable cut flowers based on stem length could still be produced (Carter and Grieve, 2010). Niu *et al.* (2012) screened different varieties of response of *Zinnia marylandica* and *Z. maritima* 'Solcito' to saline water irrigation. These varieties had both high Na+ and Cl^- accumulations in shoots, which caused salt damage and eventually death as salinity in the root zone increased.

Elhindi *et al.* (2016) found that the drip irrigation system was more efficient than surface drip irrigation system to obtain maximum yield accompanied by the highest nutrients concentration in zinnia plants and soil fertility after harvest.

The lowest cell membrane stability and highest free radicals in zinnia were observed in 40 per cent field capacity (FC) without arbuscular mycorrhizal fungi (AMF). The maximum antiradical and antioxidant properties occurred at the lowest irrigation level with the highest AMF inoculation level. Significant increase in flower diameter and longevity, in 70 per cent FC with 5 per cent AMF inoculation treatment was recorded. Water use efficiency was significantly increased at a 5 per cent AMF inoculation level (Heidari *et al.*, 2016).

Irrigating with blended water (50 per cent + 50 per cent : medium-quality water) having salinity levels as high as 2.3 dS/m would be acceptable for watering zinnia plants under circumstances where high quality water is not available (El-Nashar and Aboelsaadat, 2020).

8.3.0 Manuring and Fertilization

Growth and flowering of *Zinnia elegans* is influenced by the application of balanced fertilizer. Bose and Das (1966) reported that nitrogen deficiency caused marked reduction in growth and flower production. Phosphorus deficiency retarded vegetative growth and delayed flowering while potassium deficiency resulted in reduction in the number and quality of flowers. Tsurushima and Date (1971) observed improvement in growth and flowering by the application of nitrogen and phosphorus. Jhon *et al.* (1991) recommended application of 150 kg N/ha in 2 split doses 10 and 30 days after transplanting, in addition to a basal dose of 100 kg/ha each of P and K and 30 t/ha of well rotten weed manure to *Z. elegans cv.* Giant Flowered Crimson, best flowering in number and quality. Application of NPK increased seed yield by 24.62 per cent over that under control, the requirement of K was high (Oberthova, 1981b).

Application of N and K at 200 kg each and P at 100 kg/ha in zinnia increased number of flowers per plant, seeds per flower and per plot and flower weight. N, P and K at 200 kg/ha exhibited more number of branches and leaves and increased plant height and leaf area (Dar and Chattopadhyay, 2002). However, Khan *et al.* (2004) observed that nitrogen at 20 and 10 g/pot were most effective in improving the vegetative and floral traits in *cv.* Meteor.

Javid *et al.* (2005) observed that NPK g/m^2 at 30 + 20 + 20 performed best for plant height, number of lateral shoots, number of leaves, leaf area, number of flowers per plant and size of flower in Zinnia *cv.* 'Giant Dahlia Flowered Blue Point Series'. Ahmad *et al.* (2007) recommended that for producing excellent quality, healthy and large sized blooms and for vigorous growing zinnia plants, NPK should be applied at 67: 33: 33 kg/ha for a plant population of 10,000. Optimum phosphorus and potash dose (33 kg/ha of each) should be supplied in a single dose at the time of soil preparation before transplanting while nitrogen should be splitted in two equal doses, first half at the time of plantation followed by remaining half dose applied after one month for getting best results. Baloch *et al.* (2010) observed linear positive effect of increasing NP rates on flower life, number of branches per plant, flowers per plant and plant height of *Zinnia elegans*, while, inverse effect on days to emergence of first flower bud and opening of first flower. However, N and P at 50+10 g/1.5 m^2 proved to be an optimum level for zinnia flower production.

Aboul (1994, 1996) recorded early flowering and more number of flowers, when seeds of *Z. elegans* were sown in soil inoculated with VAM fungus *Glomus etunicatum*. Zinnia plants drenched with a 1-2 per cent solution containing photosynthetic bacterium improved soil fertility level and K availability (Xia *et al.*, 1994).

Slathia *et al.* (2009) recorded maximum plant height, number of primary branches, plant spread, minimum days taken to anthesis, maximum flowering duration, seed yield per plant in zinnia by application of NPK 21:12:7.5 g/m^2 + Azotobacter + PSB + KSB.

Long *et al.* (2010) inoculated zinnia with four arbuscular mycorrhizal fungi (AMF) *i.e. Gigaspora margarita, Gigaspora rosea, Glomus intraradices,* and *Glomus mosseae,* either singly or mixture of two species of *Gigaspora* and *Glomus*. Results indicated that Glomus significantly enhanced the leaf size and the shoot biomass. *G. mosseae* was more effective than *G. intraradices*. Only *G. mosseae* increased number and size of flowers. Mixed inoculations were not much effective in the growth-promotion than the corresponding singly inoculation with *Glomus*. Comparison of colonization per cent demonstrated that the highest colonization by *G. mosseae*, and followed by *G. intraradices* and *Gigaspora* species. In semi-quantitative PCR amplifications, *Glomus* was dominant in the roots. So, *G. mosseae* is good for inoculation to zinnia and the interaction between different AMF species should be given full consideration in the application.

Arbuscular mycorrhizal symbiosis *i.e., Rhizophagus irregularis, Rhizophagus intraradices*, and *Funneliformis mosseae* protected zinnia plants against different water regimes (WR) through improving gas-exchange capacity, higher chlorophyll and carotenoid concentrations, chlorophyll fluorescence, and water-use efficiency. Regardless of WR, comparison of colonization percentage demonstrated that the highest colonization was recorded for mixed arbuscular mycorrhizal fungi (AMF), while mixed inoculation was not so effective for the growth promotion as the corresponding single inoculation (Bagheri *et al.*, 2019).

Saini *et al.* (2020) observed that the consortium of microbial inoculants (*Glomus mosseae + Acaulospora laevis + Pseudomonas fluorescens*) fed with the low dose of superphosphate and urea resulted best growth parameters *viz.*, shoot and root length, shoot and root weight, floral traits such as floral head number and diameter and for other physiological attributes like shoot phosphorus content, root phosphorus content, acidic phosphatase, alkaline phosphatase.

Flower bud blasting in zinnia, characterized by blackening of the developed bud as a result of necrosis of the scales, unopened calyx and enclosed tissue, was caused due to boron deficiency. Plants receiving boron at 0.25 ppm were the healthiest and most vigorous (Redington and Peterson, 1983).

The application of vermicompost at 2.5 per cent is recommended to improve the qualitative properties and decrease losses due to deficient irrigation in Zinnia elegance 'Dreamland Red' (Nazarideljou and Heidari, 2014).

8.4.0 Weed Control

After zinnia has been planted, weeding is an important operation in commercial cultivation. Removal of weeds in time influences the growth and productivity of zinnia. Oryzalin was the most effective treatment for weed control, whereas mulching provided effective weed control with maximum floral display in *Z. elegans* Gold Medal Mixture (Harkess and Lyons, 1997).

9.0 Diseases and Pests

9.1.0 Diseases

Fungal Diseases

In pathogenicity trials the seed-borne fungi, *Alternaria alternata, A. zinniae, Glomerella cingulata, Cochliobolus lunatus, Phoma exigua* and *Fusarium* spp. caused seed rot and death of seedlings when seeds were inoculated. Inoculations of aerial parts showed that *A. zinniae, A. alternata, G. cingulata* and *Myrothecium verrucaria* were severe foliar pathogens while *P. exigua* caused mild infection. Dithane M-45 (mancozeb) was the most suitable fungicide for seed treatment to improve germinability (Srivastava and Gupta, 1983). Zamorshi and Milczarek (1977) recorded that zinnia seeds treated with thiram (0.3 per cent) and captan (0.5 per cent) gave the best result in controlling *Alternaria* sp. during seed germination.

Blight in zinnia caused by *Alternaria zinnae* is characterized by reddish brown spots with greyish centre. Dark brown cankers also occur on the stems, and flowers are spotted. Because it is a seed-borne disease, treatment with hot water (51.6°C) for 30 minutes has been suggested. Spraying with Bordeaux mixture also provides protection (Pirone *et al.*, 1960).

The most susceptible growth stage of zinnia for *Alternaria zinniae* causing leaf spot is at 8 weeks after planting. The infection of *A. zinniae* in zinnia seed is located in the seed coat and embryo (Seehachai and Sangchote, 2009).

The efficacy of extracts from different materials of thuja against *A. zinniae* performed (Moharam and Soliman, 2019) *in vitro*, leaf methanol and water extracts were the most effective treatments, where they completely inhibited spore germination at a concentration of (5 and 3 per cent, respectively) and mycelial growth at (7 and 7 per cent, respectively) followed by seed methanol and water extracts. In the field experiments, a highly significant reduction in the per cent of leaf area infected with *A. zinniae* of zinnia occurred after seeds were treated by soaking in methanol extracts of 3 per cent leaf and 7 per cent seed for 15 min before sowing.

Zinnia is frequently infected by *Erysiphe cichoracearum* causing powdery mildew disease. *Podosphaera fusca* (syn. *Sphaerotheca fusca, S. fuliginea*) is also known to cause powdery mildew infections on *Zinnia elegans* (Park *et al.*, 2011). Ash gray mould covers the leaves, spraying with wettable sulphur or karathane controls this disease. Root and stem-rots may cause severe losses in zinnia.

Park *et al.* (2011) observed the outbreak of powdery mildew caused by *Golovinomyces cichoracearum* on *Zinnia elegans*. Xiao *et al.* (2020) first reported *Podosphaera xanthii* causing powdery mildew on *Zinnia elegans* in Taiwan. Locke *et al.* (2006) observed delayed of expression of powdery mildew on zinnia grown hydroponically in Hoagland's solution fortified with silicon.

Kiecana and Mielniczuk (2010) observed root infection of zinnia at seedling stage to be the species of *F. equiseti* and *S. sclerotiorum*. Considerable amounts of *B. cinerea* and *F. culmorum* and *A. alternata* were also detected from diseased seedlings. Hegazi and El-Kot (2010a) found that culture filtrates of *Trichoderma harzianum*, *Epicoccum* sp., *Streptomyces endus* and an actenomycetal isolate in addition to two plant extracts *i.e.* miswak (*Salvadora persica*) and henna (*Lawsonia inermis*) significantly decrease in both disease incidence and severity. The highest powdery mildew disease incidence inhibition was obtained when *Epicoccum sp.* and *T. harzianum* were used as 50 per cent (v/v) sterilized water followed by henna and miswak extracts were used, respectively.

Diseases and Pests of Zinnia

Powdery Mildew

Bacterial Leaf Spot

Cercospora Leaf Spot

Alternaria Leaf Spot

Gray Mold

Aster Yellows Phytoplasma

Zinnia Leaf Curl Virus

Cucumber Mosaic Virus

Alternaria Blight

Four-lined Plant Bug

Botrytis Blight

Japanese Beetle

Diseases and Pests of Zinnia

Tomato Spotted Wilt Virus

Whiteflies

Spider Mites

Thrips

Source: **www.ag.umass.edu, www.apps.extension.umn.edu, www.greenhousegrower.com, www.pnwhandbooks.org, www.imagesvc.meredithcorp.io, www.bugwoodcloud.org, www.live.staticflickr.com, www.hgic.clemson.edu, www.gardenerspath.com**

Also, significant decrease in disease incidence and severity and the best results for most studied growth and flowering parameters and total green colour were recorded with essential oils obtained from ginger, cinnamon and clove oils, respectively each at 0.1 ppm (Hegazi and El-Kot., 2010b).

Cercospora leaf spot caused by *Cercospora zinniae* was controlled by trimming the diseased leaves followed by spraying of streptocycline (0.01 per cent) plus carbendazim (0.025 per cent) immediately after the symptoms appeared (Jindal *et al.*, 1991). He *et al.* (2020) reported *Z. elegans* leaf spot caused by *Stemphylium lycopersici* in China. The symptoms appeared on the edge of the leaves as small brown spots. In serious cases, the leaves were withered, which affected the ornamental value of the plants.

The soil-inhabiting fungi causing diseases in zinnia are *Fusarium* sp. *Pellicularia rolfsii, Phytophthora cryptogea.* Juber and Saeed (2015) identified 6 *Fusarium* species associated with zinnia seeds. These species were *F. acuminatum, F. culmorum, F.*

oxysporum, F. solani, F. thapsinum, and *F. verticillioides*. The existence of *F. solani* and *F. culmorum* in 27.3 per cent of seed samples with 31.16 per cent and 7.59 per cent frequency rate, respectively recorded. Other species were presented in 18.2 per cent of seed samples. *F. culmorum* isolate gave the highest rate of decrease in seed germination (0 per cent) whereas; isolates of *F. oxysporum* and *F. solani* caused decrease in seed germination to 3 per cent and 7 per cent, respectively. Isolates *F. culmorum, F. oxysporum* and *F. verticillioides* caused a significant superiority in disease incidence and severity (97.5-100 per cent and 96.5-98 per cent).

Damping off of zinnia seedlings caused by *Rhizoctonia solani* can be prevented by formulation of pregelatinized starch-flour granules containing fermenter-produced biomass of the biocontrol fungi *Gliocladium virens*, isolates GI-3, G-21 and *Trichoderma hamatum* isolates TRI-4 (Lewis *et al.*, 1995) or with a mixture of vermiculite, powdered wheat bran, and dry fermenter-produced biomass (FB) of isolates of *T. hamatum* (TRI-4) and *G. virens* (G1-3) (VBA-FB) with 0.05 M HCl (Lewis and Lumsden, 2001).

The gray mold disease caused by *Botrytis cinerea* in zinnia seedlings grown in greenhouse was observed by Kwon *et al.* (2004).The symptoms of infected plants were started with water-soaking lesions in flower bud, leaves and stems. The lesions gradually expanded and infected plants became withered and discolored to gray or dark from the tip.

Bacterial Diseases

Bacterial disease caused by *Xanthomonas campestris cv. Zinnia* was eradicated from seed with a 30 min. soak in 1 per cent sodium hypochlorite. Soaking for 30 min in water at 53°C did not eliminate the pathogen from the seed and significantly reduced germination (Strider, 1979). He (1980) also noted that the bacterial leaf spot disease was reduced from 100 per cent to 2 per cent when infected seeds were soaked 30 minutes in 360 g *a.i.* captan/l of water. Under outdoor condition, the spread of disease was reduced by weekly foliar application of captan 4.8 g *a.i.*/l of water. Streptomycin was effective but moderately phytotoxic. Removing of diseased leaves followed by a foliar spray of streptocycline (0.01 per cent) plus carbendazim (0.025 per cent) is reported to be effective in controlling *Xanthomonas campestris* pv. *zinniae* (Jindal *et al.*, 1991). Catara and Sesto (1997) reported infection by *Xanthomonas campestris* pv. *zinniae* from Italy, which caused total rot of infected tissue under high humid condition.

Zhao *et al.* (2016) observed leaf and flower spot disease caused by *Xanthomonas campestris* pv. *zinniae* in Jilin Province, China. Symptoms first appeared on leaves as diffuse, water-soaked, circular lesions 1 to 2 mm in diameter, surrounded by large chlorotic halos. The lesions gradually enlarged to 3 to 5 mm in diameter and became angular in shape with a brown to reddish brown center. Eventually, the lesions coalesced into ragged patches of decaying tissue. In wet conditions, flowers developed small brown spots then withered

Viral Diseases

Many viral diseases affect zinnia. Curly top caused by *Rugaver rucosans*, mosaic caused by several strains of *Marmor cucumeris*, spotted wilt by *Letham australiense* and aster yellows caused by *Chlorogenus callistephi* are frequently observed in zinnia. Removal and destruction of affected plants and application of dimethoate to control the vectors have been suggested (Pirone *et al.*, 1960). Leaf curl virus is transmitted from diseased to healthy plants by *Bemisia tabaci* but is not seed transmissible (Padmanabhan and Pathmanavan, 1978).

Min *et al.* (2020) first reported of cucumber mosaic virus isolated from *Zinnia elegans* in Korea. Symptoms included mosaic symptoms on upper leaves. Mottle symptoms, necrotic ring spots, dwarfism, and severe wilt were observed on *Zinnia elegans*. Tomato spotted wilt virus (TSWV) was identified as the causal agent of the disease.

Candidatus Phytoplasma Asteris associated with yellowing, witches' broom and little leaf disease of *Zinnia elegans* in India was first reported by Rao *et al.* (2012). Although Zinnia elegans phyllody (ZeP) was reported from Iran in 2017, no insect vector was identified (Hemmati and Nikooei 2017). Hemmati and Nikooei (2019) reported successful transmission of ZeP phytoplasma to *Z. elegans* by *Austroagallia sinuata*. Rihne *et al.* (2020) observed mixed infection of 'Candidatus Phytoplasma asteris' and begomovirus with zinnia leaf curl and witches' broom symptoms. Symptoms of excessive leaf curling, little leaf, stunting and witches' broom were observed on *Z. elegans*. Sequence comparison, phylogenetic and virtual RFLP analysis of 16S rRNA gene sequences of zinnia isolate confirmed the presence of 'Candidatus Phytoplasma asteris' (16SrI-B subgroup). Further coat protein sequence comparison and phylogenetic tree analysis of coat protein gene confirmed association of begomovirus with zinnia symptomatic samples and identified it as a strain of tomato leaf curl virus.

9.2.0 Pests

Zinnia is also infested by several pests, including the black bean aphid (*Aphis fabae*), the red-banded leaf hopper (*Graphocephala coccinea*), the six-spotted leaf hopper (*Macrostelesfascifrons*), the tarnished plant bug (*Lygus lineolaris*), the fourlined plant bug (*Poecilocapsus lineatus*) and long-tailed mealy bug (*Pseudococcus adonidum*). It is also severely damaged by a number of beetles. These pests are effectively controlled by spraying of Cartap 50 per cent SP Hydrochloride 0.5-1 g/l of water. The bishop bug (*Lygus rugulipennis*) damages buds, flower and foliage of zinnia (Becker, 1974). Three types of mites-broad, cyclamen and two-spotted affect growth and flowering. These pests are controlled by spraying kelthane.

Treatment of *Z. elegans* with soluble silicon provides a modest increase in resistance levels to *M. persicae*, which may be caused in part by defense-related compounds (Ranger *et al.*, 2009).

Zinnia is infested by root knot nematode (*Meloidogyne incognita*) and *Aphelenchoides ritzemabosi* causing angular spots on leaves (Pirone *et al.*, 1960). Application of carbofuron or phorate has been recommended as a control measure.

10.0 Harvesting and Postharvest Management

10.1.0 Harvesting

Zinnia flowers are generally cut with stem when they have attained the full size. Harvesting of flowers should be done in cool hours of the day (morning or evening). Flowers remain fresh for 6-7 days in vase.

10.2.0 Yield

The yield of cut flowers depends on a number of factors such as cultivars, age of seedlings, spacing, pinching, nutrition and other cultural practices adopted time to time. Jhon *et al.* (1991) obtained 32-38 flowers/plant in cultivar Giant Flowered Crimson when planted at 30 × 30 cm.

Oberthova (1981a) found that the seed quality varied with the method of sowing, date of harvesting and growing season. The highest seed yield was recorded when the seeds were sown in April.

10.3.0 Postharvest Management

Water loss from cut flower may be reduced by placing them immediately in air-free water at 10°C. Entry of air into vessels can be reduced by cutting back the stems. Control of bacteria and fungi and inhibition of ethylene evolution can be achieved by pre-treatment with solutions of chrysal, rosal or flora-2000. Chrysal universal at 12 g/l was the most effective in increasing the vase life of zinnia (Lemper, 1984). The postharvest life of zinnia flowers was prolonged by holding in deionized water containing 8-HQC and sucrose. Flower lasted longest in a solution of 200 mg/1 8-HQC and 1 per cent sucrose. The fresh weight of flower however, decreased when held in deionized water (Stimart and Brown, 1982; Stimart *et al.*, 1983). Mugge (1983) reported that vase life of cut flower was best with AKN preservative [0.08 per cent $Al_2(SO_4)_2$ + 0.3 per cent KCl + 0.02 per cent NaCl at 20 ml/l + 1.5 per cent sucrose], but cold storage did not improve the quality of flowers. Ethylene concentration in zinnia cut flower *cv.* Dasher was reduced by growth promoting rhizobacteria (*Pseudomonas putida* GR 12-2) which lowers the concentration of 1-aminocyclo-propane-1-carboxylate (ethylene precursor) thereby prolonging vase life of flowers (Seema-Nayani *et al.*, 1998).

Harvested zinnia 'Benary Giant Mixed' flowers, were either kept dry, placed in plain water or placed in a hydrating solution (Hydroflor® 100 at 30 ml/3.78l) and a floral preservative (Floralife Crystal Clear®), and stored at 2°C for 2 days. Afterwards, flowers were placed in a simulated postharvest interior environment consisting of either water or a preservative solution. Water vs. preservative solutions resulted in 8 days vs. 14 days longevity, respectively using an 80 per cent acceptance threshold. Experiment where flowers replenished with water, water and a hydrating solution, and preservative replenished with preservative. Flowers in a hydrated-preservative water treatment lasted 9 days vs. 13 days for those in the hydrated-preservative-preservative treatment (Emino and Hamilton, 2002). 'Sun Gold' and 'Benary's Giant Scarlet' zinnia flowers could be held in unamended tap or deionized (DI) water with no effect on vase life. A vase solution of 2 per cent sucrose without foam extended consumer vase life for 'Benary's Giant Scarlet' zinnia and it also had a longer vase life with a 10 per cent sucrose pulse than a water-only pulse. Both the cultivars of zinnia could not be cold stored at 2°C but unaffected by exogenous ethylene (Possiel, 2009). Stems stored dry at 5°C regained turgidity and averaged a vase life of 14 days; however petals remained slightly twisted and curled after being in the vase for several days. Floral foam reduced vase life to 9-10 days (Dole *et al.*, 2005).

Cut *Zinnia* stems were not influenced by solution pH, but as EC increased from 0.00 to 4.00 dS/m vase life decreased from 10.6 to 6.8 d. The use of commercial holding solution reduced the negative effects of high EC *Zinnia* and increased solution uptake (Carlson and Dole, 2013).

Dry flowers have great export potential as nearly 70 per cent of total export of floricultural commodities consists of dried products (Singh, 2000). Dehydration of zinnia flower upto 8 to 11.5 per cent moisture content resulted into six months longevity and good flower shape. Dehydration at low temperature (40°C) or flower drying at ambient temperature with sand or silica gel as embedding media, exhibited better quality with good petal-texture, colour and low degradation of carotenoids and anthocyanins (Singh *et al.*, 2004a). Dehydration at 40°C with sand for 30 h duration or with silica gel for 24 h was found to be effective techniques for dehydration of zinnia flowers (Singh *et al.*, 2004b).

11.0 Biotechnology

11.1.0 Biochemical and Molecular Markers

Investigations on biochemical and molecular markers have been carried out mainly to elucidate tissue differentiation.

11.1.1 Isozymes

Alam and Zarin (1998) examined the intervarietal diversification of zinnia with fluorescent bandings and isoenzyme. Application of fluorescent bandings together with isoenzyme was useful for varieties verification.

11.1.2 Molecular Markers

Milioni*et al.* (2001) used RNA fingerprinting technology to detect DNA fragments derived from RNA using cDNA synthesis and subsequent PCR-amplified fragment length polymorphisms (cDNA-AFLP), to systematically characterize hundreds of the genes involved in the process of tracheary element formation in the zinnia mesophyll cell system. Lin and Northcote (1990) reported that the absolute amounts of BA and NAA rather than the ratio of them were important for tracheary-element formation in zinnia cultures. Cells cultured in the presence of 0.001 mg BA and 0.06 mg NAA/l expanded and divided but did not differentiate. The level of phenylalanine ammonialyase (PAL) gene activity, synthesis of *PAL* protein and the level of *PAL* mRNA peaked during 72 to 96 h when lignin was actively deposited. This indicated that the *PAL* gene was temporally and preferentially expressed in association with lignification during tracheary-element differentiation and thus it can be regarded as a molecular marker for the process. Three cDNA clones, TED2, TED3 and TED4 isolated from mesophyll cells of *Z. elegans* leaves, whose corresponding mRNAs expressed in a close association with tracheary element differentiation, were isolated by differential screening of a lambdagt 11 cDNA library. Northern analysis indicated that mRNAs corresponding to the clones were expressed preferentially in cells differentiating into TEs prior to the morphological changes. Analysis of the nucleotide sequence of TED4 showed that the cDNA contained on ORF of 285 bp, encoding a polypeptide comprising 95 amino acid residues with a predicted molecular mass of 10.0 kDa (Demura and Fukuda, 1993). Igarashi *et al.* (1998) studied the regulatory mechanism of gene expression in the early stages of tracheary element (TE) differentiation and characterization of a genomic fragment of TED3, in the *in vitro Z. elegans* system.

Yoshimura *et al.* (1996) isolated complementary DNA clones for three different β-tubulin isotypes (ZeTubB1, ZeTubB2 and ZeTubB3) from a cDNA library generated from RNA of cultured mesophyll cells of *Z. elegans* (cv. Canary Bird) that were differentiating into tracheary elements and/or dividing. RNA gel blot analysis of the expression of the ZeTubB transcripts indicated that transcripts that corresponded to each clone were differently expressed during culture of mesophyll cells. Results of an *in situ* hybridization experiment with an antisense RNA probe derived from ZeTubB1 cDNA suggested the preferential expression of ZeTubB transcripts in differentiating xylem cells, as well as in the ground meristem and the procambium, of zinnia seedlings. RNA gel-blot analysis revealed that in the TE-inductive cultures the level of ZCAD1 mRNA increased from 36 h and peaked at 48 to 60 h. This result indicated that both the gene expression and the activity of cinnamyl-alcohol dehydrogenase (CAD) are strictly regulated, in association with lignification, during TE differentiation in *Z. elegans* (Sato *et al.*, 1997).

Pesquet *et al.* (2005) constructed and characterized a subtractive library from zinnia cultures enriched in genes that are specifically expressed at the onset of secondary wall deposition and tracheary element (TE) programmed cell death. This Late Xylogenesis Library (LXL) consisted of 236 nonredundant cDNAs, 77 per cent of which encoded novel sequences in comparison with the 17,622 expressed sequence tag sequences publicly available. cDNA arrays were constructed to examine dynamic global gene expression during the course of TE formation. As a first step in dissecting auxin and cytokinin signaling during TE differentiation, macroarrays were probed with cDNAs from cells cultured in different hormonal conditions. Fiftyone per cent of the LXL genes were induced by either auxin or cytokinin individually, the large majority by auxin during TE differentiation. To determine the potential involvement of these categories of genes in TE differentiation, multiplex *in situ*-reverse transcription-PCR was performed on cells for two genes encoding putative cell wall proteins: Gibberellin stimulated transcript-1, induced by auxin alone, and expansin 5, induced by cytokinin alone. All transcriptionally active TEs expressed both genes, indicating that, although these genes may not be considered as specific markers for TE differentiation per se, they are nevertheless an integral part of TE differentiation program. Among the non-TE population, four different gene expression-based cell types could be distinguished.

Pesquet *et al.* (2013) provided evidence in *Zinnia elegans* tracheary element (TE) cell cultures that TE lignification occurs postmortem (*i.e.*, after TE programmed cell death). *In situ* RT-PCR verified expression of the lignin monomer biosynthetic cinnamoyl CoA reductase and cinnamyl alcohol dehydrogenase in not only the lignifying TEs but also in the unlignified non-TE cells of *Z. elegans* TE cell cultures and in living, parenchymatic xylem cells that surround TEs in stems. These cells were also shown to have the capacity to synthesize and transport lignin monomers and reactive oxygen species to the cell walls of dead TEs. Differential gene expression analysis in *Z. elegans* TE cell cultures and concomitant functional analysis in Arabidopsis thaliana resulted in identification of several genes that were expressed in the non-TE cells and that affected lignin chemistry on the basis of pyrolysis–gas chromatography/mass spectrometry analysis. These data suggest that living, parenchymatic xylem cells contribute to TE lignification in a non-cellautonomous manner, thus enabling the postmortem lignification of TEs.

In order to isolate genes related to anther development and understand molecular basis of male sterility, cDNA library of *Zinnia elegans* was constructed using suppression subtractive hybridization (SSH) approach (Pang *et al.*, 2012). PCR results showed that cDNA inserts were ranged from 100 to 750 bp. The homology alignment showed that 76 expressed sequence tags (ESTs) had functional annotations in GenBank, the other 11 ESTs without any homology to the known gene. In addition, 87 ESTs were divided into 17 groups according to MIPS of Arabidopsis thaliana database. 7 genes related to anther development were isolated. Results from semi-quantitative RT-PCR showed 6 genes were expressed only in disk florets of fertile plants compared with that of male sterile plants. These ESTs obtained provide important clues for further isolation and identification of fertility-related genes in *Z. elegans*.

The EST sequences of *Zinnia* from NCBI were analyzed, 20 667 unigenes were obtained to develop new EST-derived SSR markers. Then with 20 μL red enzyme PCR reaction, taking the male sterile two-type lines MJ16AB of *Zinnia elegans* as templates, primers were screened 622 SSRs, whose repeated motifs were not less than 5 times and repeated length greater than 12 bp, were obtained. Among them, the dominant types of the dinucleotide and trinucleotide repeats were AT/TA and CTA/GAT, accounting for 9.5 per cent and 10.6 per cent respectively. 5 among 136 pairs of primers could produce stable and expected polymorphism PCR products between fertile pool and sterile pool of *Zinnia elegans*, they were primers 28, 30, 33, 69 and ZeM63 respectively. The results indicated that it is a feasible way to develop EST-SSR markers from *Zinnia elegans* EST, 5 polymorphic primers could be used for the establishment of molecular markers of *Zinnia elegans* male fertility (Hu *et al.*, 2017).

11.1.3 Genetic Transformation

Transgenic *Arabidopsis* plants with a chimaeric gene comprising the 537 bp TED3 promoter and the β-glucuronidase (*GUS*) reporter gene indicated strong expression of the *GUS* gene by the TED3 promoter in TEs, in particular in immature TEs, as well as stipules and trichomes. GUS expression drived by the promoter was also induced in callus, in which GUS activity was localized in immature TEs and slender cells around TEs that may be TE precursor cells. The TED3 promoter was not significantly activated by wounding. This pattern of expression differed significantly from that of other vascular tissue-related genes such as *PAL, 4CL* and *GRP1.8.* It is suggested that the nature of the TED3 promoter will enable it to be used to monitor TE differentiation in tissue and to introduce foreign genes preferentially into immature TE.

Six cultivars of *Z. elegans* 'Dreamland' series petals were compared for anthocyanin content (Qian *et al.,* 2021). No anthocyanin was detected in ivory and yellow cultivars, while four acetylated anthocyanins, cyanidin 3-O-(6′-acetyl) glucoside-5-O-glucoside, pelargonidin 3-O-(6′-acetyl)glucoside-5-O-glucoside, cyanidin 3-O-(6′-acetyl) glucoside, and pelargonidin 3-O-(6′-acetyl) glucoside, have been detected in the petals of other four cultivars. 5-bp insertion causes a premature termination codon in ZeF3H transcripts of ivory and yellow cultivars. The main anthocyanin type in coral and pink cultivars is pelargonidin. Main anthocyanin type in coral and pink cultivars is pelargonidin, because of lower expression of ZeF32 H. Cyanidin is the main anthocyanin type in rose and red cultivars.

12.0 References

Abdollahi, F., Sari, E.R.K. and Jafari, L. (2016) *Iran. J. Seed Res.,* **3**(4) : 105-116.

Aboul-Nasr, A. (1994) *Alexandria J. Agril. Res.,* **39** : 453-470.

Aboul-Nasr, A. (1996) *Mycorrhiza,* **6** : 61-64.

Afzal, I., Abbasi, K.Y., Iqbal, A., Younis, A., Bakhtavar, M.A. and Rehman, H.U. (2016) *Acta Sci. Pol. Hortorum Cultus,* **15**(5) : 173.

Ahmad, I., Ahmad, T., Zafar, M.S. and Nadeem, A. (2007) *Sarhad J. Agric.,* **23**(2) : 309.

Ahmad, I., Saleem, A.M., Mustafa, G., Ziaf, K., Afzal, I. and Qasim, M. (2017) *Sarhad J. Agric.,* **33**(2) : 199-205.

Akbarian, B., Matloobi, M. and Mahna, N. (2016) *J. Ornam. Hortic.,* **6**(2) : 115-123.

Akoumianaki-Ioannidou, A., Spentza, R.P., Roussis, I. and Bilalis, D. (2019) *Bull. Univ. Agric. Sci. Vet. Medicine Cluj-Napoca. Hortic.,* **76**(1) : 120-123.

Alam, S. and Zarin, N. (1998) *Cytologia,* **63** : 387-393.

Alami, Ù. and Karimi, M. (2020) *Int. J. Hort. Sci. Tech.,* **7**(1) : 81-91.

Armitage, A.M., Bass, R.E., Carlson, W.H. and Ewart, L.C. (1981) *HortScience,* **16** : 218-220.

Bagheri, V., Shamshiri, M.H., Alaei, H. and Salehi, H. (2019) *Photosynthetica,* **57**(2) : 409-419.

Baloch, Q.B., Chachar, Q.I. and Panhwar, U.I. (2010) *J. Agric. Sci. Technol.,* **6**(1) : 193-200.

Becker, P. (1974) *Pests of Ornamental Plants,* Her Majesty's Stationery Office, London, No. 46, p. 136.

Blamowski, Z.K. and Borowski, E. (1995) *Materialyog Olnopolskiej Konferencjinaukowe Nauka Praktyce Ogrodniczej z okazji XXV-lecia Wydzialu Ogrodniczego Akademil Rolniczej w Lublinie*, pp. 721-724.

Bose, T.K. and Das, S.M. (1966) *Indian J. Hort.*, **23** : 88-97.

Boyle, T.H. (1996) *J. Amer. Soc. Hort. Sci.*, **121** : 27-32.

Boyle, T.H. and Stimart, D.P. (1982) *Euphytica*, **31** : 857-867.

Boyle, T.H. and Stimart, D.P. (1989) *Euphytica*, **44** : 73-79.

Boyle, T.H. and Wick, R.L. (1996) *HortScience*, **31** : 851-854.

Brickell, C. (1996) *A-Z Encyclopedia of Garden Plants*, Dorling Kindersley, London, p. 1071.

Burlec, A.F., Pecio, £., Mircea, C., Cioancã, O., Corciovã, A., Nicolescu, A., Oleszek, W. and Hãncianu, M. (2019) *Molecules*, **24**(16) : 2934.

Cao, S. and Li, N. (2014) *J. Shenyang Agri. Uni.*, **45**(1) : 91-94.

Carlson, A.S. and Dole, J.M. (2013) *Sci. Hortic.*, **164** : 277-286.

Carter, C.T. and Grieve, C.M. (2010) *HortScience*, **45**(7) : 1058-1063.

Catara, V. and Sesto, F. (1997) *InformatoreFitopatologico*, **47** : 62-64.

Cathey, H.M. (1965) *Flor. Rev.*, **136** : 17-18.

Chaudhary, V.R., Jitendra, K., Yeshpal, S., Singh, R.K. and Ram, P. (2007) *Asian J. Hort.*, **2**(1) : 242-243.

Cox, D.A. and Keever, G.J. (1988) *Res. Report Series alabama Agricultural Experiment Station*, Auburn University, 1987, publ. 1988, No. 5, pp. 18-19.

Dar, Q.A.H. and Chattopadhyay, T.K. (2002) *Proceedings of The National Symposium on Indian Floriculture in The New Millennium, Lal-Bagh, Bangalore*, 203-205.

Das, P., Bose, T.K. and Roy, P.K. (1977) *Orissa Hort. J.*, **5** : 1-6.

Davis, T.D. (1991) *HortTechnology*, **1** : 49-52.

de Moraes, E.R., Santana, A.S., Peixoto, J.V.O.M. and da Silva, R.V. (2016) *Afr. J. Agric. Res.*, **11**(33) : 3090-3096.

Demura, T. and Fukuda, H. (1993) *Plant Physiol.*, **103** : 815-821.

Dole, J.M., Fanelli, F.L., Fonteno, W.C., Harden, B. and Blankenship, S.M. (2005) *HortScience*, **40**(4) : 1123B-1123.

Elhindi, K., El-Hendawy, S., Abdel-Salam, E., Elgorban, A. and Ahmed, M. (2016) *Bragantia*, **75**(1) : 96-107.

El-Nashar, Y. and Aboelsaadat, E.M. (2020) *Sci. J. Flowers Ornam. Plants*, **7**(4) : 409-423.

Emino, E.R. and Hamilton, B. (2002) In: *Proceedings of the Florida State Horticultural Society*, **115** : 111-114.

Faust, J.E., Holcombe, V., Rajapakse, N.C. and Layne, D.R. (2005) *HortScience*, **40**(3) : 645-649.

Gomaa, A.A.R., Samy, M.N., Abdelmohsen, U.R., Krischke, M., Mueller, M.J., Wanas, A.S., Desoukey, S.Y. and Kamel, M.S. (2020) *Nat. Prod. Res.*, **34**(18) : 2612-2615.

Gombert, L., Windham, M. and Hamilton, S. (2001) *HortTechnology*, **11**(1) : 71-74.

Gultom, T. and Gultom, E.S. (2015) *ARPN J. Agric. Biol. Sci.*, **10**(1) : 21-23.

Han, L.S. and Yeam, D.Y. (1978) *J. Korean Soc. Hort. Sci.*, **19** : 117-128.

Harkess, R.L. and Lyons, R.E. (1997) *HortScience*, **7** : 39-41.

He, L., Liu, H.F., Liu, F.Y., Ni, Y., Lu, B.B. and Deng, J.X. (2020) *Plant Dis.*, **104**(2) : 578-578.

Hegazi, M.A. and El-Kot, G.A. (2010a) *J. Agric. Sci.*, **2**(4) : 221.

Hegazi, M.A. and El-Kot, G.A.N. (2010b) *J. Agric. Sci.*, **2**(4) : 63.

Heidari, Z., Nazarideljou, M.J., Rezaie Danesh, Y. and Khezrinejad, N. (2016) *Int. J. Hortic. Sci.*, **3**(1) : 19-32.

Hemmati, C. and Nikooei, M. (2017) *Aust. Plant. Dis. Notes*, **12**(11): 1-4.

Hemmati, C. and Nikooei, M. (2019) *J. Pl. Plant Pathol.*, **101**(4) : 1223-1223

Hojati, M., Etemadi, N. and Bani Nasab, B. (2009) *JWSS-Isfahan Univ. Technol.*, **13**(47) : 649-656.

Holcomb, E.J. (1982) *Bull. Pennsylvania Flower Growers*, No. 339, pp. 1-5.

Howe, T.K. and Waters, W.E. (1983) *Proc. Fla. Sta. Hort. Soc.*, **96** : 131-135.

Hu, M., Qin, M., Li, J. and Ye, Y. (2017) *Acta Agric. Univ. Jiangxi.*, **39**(5) : 976-982.

Igarashi, M., Demura, T. and Fukuda, H. (1998) *Plant Mol. Biol.*, **36** : 917-927.

Iqbal, D., Habib, U., Abbasi, N.A. and Chaudhry, A.N. (2012) *Pak. J. Bot*, **44**(3) : 1091-1094.

Jagtap, A.M., Vambhurkar, G.B., Gavade, A.S., Kengar, M.D., Jagtap, N.M. and Patil, O.A. (2018) *Asian J. Pharm. Anal.*, **8**(3).

Javid, Q.A., Abbasi, N.A., Saleem, N., Hafiz, I.A. and Mughal, A.L. (2005) *Int. J. Agric. Biol.*, **7**(3) : 471-473.

Jhon, A.Q., Paul, T.M. and Tanki, M.I. (1991) *Adv. Plant Sci.*, **4** : 1-7.

Jindal, K.K., Madhu-Meeta and Meeta, M. (1991) *Indian J. Agri. Sci.*, **61** : 156-158.

Jones, J. J. and Strider, D.L. (1979) *Plant Dis. Reptr.*, **63** : 449-453.

Juber, K.S. and Saeed, R.I. (2015) *Iraqi J. Agric. Sci.*, **46**(2) : 220-227.

Jun, L., Tao, W. and Xixi, Z. (2011) *Short communication about male sterile hybridization breeding in Zinnia elegans.* Chinese Agricultural Science Bulletin.

Khan, M.A., Ziaf, K. and Ahmad, I. (2004) *Asian J. Plant Sci.*, **3** : 571-573.

Kiecana, I. and Mielniczuk, E. (2010) *Acta Sci. Pol.-Hortorum Cultus*, **9**(3) : 147-160.

Kirilov, D. (1990) *Rasteniev' dniNauki*, **27** : 43-46.

Kole, P.C. and Meher, S.K. (2005) *J. Ornam. Hortic.*, **8**(4) : 303-305.

Kumar, S. and Marwein, B. (2018) *Int. J. Curr. Microbiol. App. Sci.*, **7**(5) : 2202-2212.

Kwon, J.H., Son, K., Jeong, S.G. and Park, C.S. (2004) *Pl. Dis. Res.*, **10**(4) : 337-340.

Latimer, J.G. (1991) *HortScience*, **26** : 557-560.

Lemper, L. (1984) *Gb + Gw*, **84** : 158-162.

Lewis, J.A. and Lumsden, R.D. (2001) *Crop Protection*, **20** : 49-56.

Lewis, J.A., Fravel, D.R., Lumsden, R.D. and Shasha, B.S. (1995) *Biological Control*, **5** : 397-404.

Li, N., Han, X. and Song, Y. (2012) *J. Jilin Univ.*, **34**(1) : 5-10.

Lin, Q. and Northcote, D.H. (1990) *Planta*, **182** : 591-598.

Locke, J.C., Omer, M.A., Widrig, A.K. and Krause, C.R. (2006) *Phytopathology*, **96**(6) : 70.

Long, L.K., Yao, Q., Huang, Y.H., Yang, R.H., Guo, J. and Zhu, H.H. (2010) *World J. Microbio. Biotechnol.* **26**(8) : 1527-1531.

Lou, X., Lu, T., Li, M., Pang, R., Ye, Y. and Bao, M. (2011) *Sci. Hortic.*, **129**(4) : 862-868.

Lou, X.Y., Hu, Q.S., Bao, M.Z. and Ye, Y.M. (2010) *Euphytica*, **174**(1) : 91-103.

Mahdavifard, M., Rezaeinejad, A. and Mosavifard, S. (2020) *J. Plant Res.*, **32**(4) : 784-791.

McScorley, R. (1995) *Proc. Florida State Hort. Soc.*, **107** : 430-432.

Merritt, R.H. and Kohl, H.C. Jr. (1991) *Scientia Hort.*, **45** : 295-302.

Milioni, D., Sado, P.E., Stacey, N.J., Domingo, C., Roberts, K. and McCann, M.C. (2001) *Plant Mol. Biol.*, **47** : 221-238.

Min, D.J., Park, J.S. and Hong, J.S. (2020) *J. Plant Pathol.*, **102**: 963.

Mittra, B.N., Ghosh, B.C., Behera, M.C., Rao, K.A., Tiwari, K.N. and Rao, Y.P. (1990) In : *Proc. 11th International Congress on the use of plastics in agriculture*, New Delhi, India, 26th February to 2nd March, pp. 135-140.

Miyajima, D. (2006) *Sci. Hortic.*, **107**(2) : 176-182.

Moharam, M.H. and Soliman, Y.M. (2019) *Australas. Plant Pathol.*, **48**(6) : 583-594.

Morales-Díaz, M., Alcacio-Rangel, S. and La Torre-Almaraz, D. (2008) *Agrociencia*, **42**(3) : 335-347.

Mugge, A. (1983) *Gartenbau*, **30** : 218-219.

Nazarideljou, M.J. and Heidari, Z. (2014) *Int. J. Hort. Sci. Tech.*,**1**(2) : 141-150.

Neily, W.G., Hicklenton, P.R. and Kristie, D.N. (2000) *HortScience*, **35** : 39-42.

Nikolova, N. and Vasileva, R. (1979) *NauchniTrudove, VisshSelskosgopanski Institute 'Vasilkoharov'*, **24** : 117-120.

Niu, G., Wang, M., Rodriguez, D. and Zhang, D. (2012) *HortScience*, **47**(6) : 793-797.

Oberthova, K. (1981a) *Acta Fytotechnica*, **37** : 55-70.

Oberthova, K. (1981b) *Acta Fytotechnica*, **37** : 71-80.

Padmanabhan, C. and Pathmanavam, G. (1978) *Curr. Res.*, **7** : 100-101.

Pang, R.H., Lou, X.Y., Bao, M.Z. and Ye, Y.M. (2012) *J. Integr. Agric.*, **11**(4) : 567-575.

Park, M.J., Park, J.H., Kim, H.G., Lee, S.G., Koh, Y.J., Kim, B.S., Cha, B., Lee, H.B. and Shin, H.D. (2011) *Plant Pathol. J.*, **27**(1) : 85-88.

Park, Y.J., Kim, Y.J. and Kim, K.S. (2013) *Hortic. Environ. Biotechnol.*, **54**(3) : 236-242.

Pesquet, E., Ranocha, P., Legay, S., Digonnet, C., Barbier, O., Pichon, M. and Goffner, D. (2005) *Plant Physiology*, **139**(4) : 1821-1839.

Pesquet, E., Zhang, B., Gorzsás, A., Puhakainen, T., Serk, H., Escamez, S., Barbier, O., Gerber, L., Courtois-Moreau, C., Alatalo, E. and Paulin, L. (2013) *The Plant Cell*, **25**(4) : 1314-1328.

Pinto, A.C.R., Rodrigues, T.D.J.D., Leite, I.C. and Barbosa, J.C. (2005) *Sci Agric.*, **62**(4) : 337-345.

Pinto, A.C.R., Rodrigues, T.J.D., Leite, I.C. and Barbosa, J.C. (2003) In: *V International Symposium on New Floricultural Crops*, **683** : 399-406.

Pirone, P.D., Dodge, B.O. and Rickett, H.W. (1960) *Diseases and Pests of Ornamental Plant*, Constable and Co. Ltd., London.

Possiel, E.Y. (2008) *Horticultural Science*. North Carolina State University, Raleigh, North Carolina, 149.

Qian, J., Lai, W., Jiang, L., Zhan, H., Zhai, M., Fu, J. and Zhang, C. (2021) *Sci. Hortic.*, **277** : 109809.

Ranger, C.M., Singh, A.P., Frantz, J.M., Cañas, L., Locke, J.C., Reding, M.E. and Vorsa, N. (2009) *Environ. Entomol.*, **38**(1) : 129-136.

Rao, G.P., Tiwari, A.K., Singh, M., Chaturvedi, Y. and Srivastava, R. (2012) *Phytopathogenic Mollicutes*, **2**(1) : 33-36.

Redington, C.B. and Peterson, J.L. (1983) *Bull. Torrey Bot. Club*, **110** : 77-79.

Reimherr, P. (1980) *Gb + Gw*, **37** : 828-830.

Riaz, A., Arshad, M., Younis, A., Raza, A. and Hameed, M.A.N.S.O.O.R. (2008) *Pak. J. Bot.*, **40**(4) : 1579-1585.

Riaz, A., Martin, P., Riaz, S., Younis, A. and Hameed, M. (2011) In *I International Trials Conference: Assessment of Ornamental Plants*, **980** : 69-78.

Rihne, T., Mitra, S., Bahadur, A., Panda, P., Banyal, N. and Rao, G.P. (2020) *Indian Phytopathol.*, **73**: 527-532.

Ritzinger, C.H.S.P. and McSorley, R. (1998) *Nematropica*, **28** : 173-185.

Rogers, R.B., Smith, M.A.L. and Cowen, R.K.D. (1992) *Euphytica*, **61** : 217-223.

Rzedowski, G.C. de, Rzedowski, J. and De Rzedowski, G.C. (1996) *Acta Bot. Mexicana*, **36** : 77-83.

Saini, I., Yadav, V.K., Aggarwal, A. and Kaushik, P. (2020) *Indian J. Exp. Biol.*, **58**(10) : 730-737.

Saleem, B.A., Nafees, M., Farooq, M. and Sadaqat, H.A. (2003) *Int. J. Agric. Biol.*, **5**(4) : 496-498.

Samaha, R.R. and Boyle, T.H. (1989) *J. Heredity*, **80** : 368-372.

Samaha, R.R., Boyle, T.H. and Mulcahy, D.L. (1989) *Sexual Plant Rep*roduction, **2** : 18-26.

Sato, Y., Watanabe, T., Komamine, A., Hibino, T., Shibata, D., Sugiyama, M. and Fukuda, H. (1997) *Plant Physiol.*, **113** : 425-430.

Schmidt, E. (1979) *Gartenbau*, **33** : 1937.

Seehachai, W. and Sangchote, S. (2009) *In Proceedings of the 47th Kasetsart University Annual Conference, Kasetsart*, 498-505.

Seema-Nayani, Shimon-Mayak, Glick, B.R., Navani, S. and Mayak, S. (1998) *Indian J. Exp. Biol.*, **36** : 836-839.

Selaru, E. and Costea, M. (1991) *Lucrari Stiintifice, Institutul Agronomic Nicolae Balcescu, Bucuresti, Seria B, Horticultura*, **34**: 103-108.

Sharif, M.M., Ali, I. and Ahmad, I. (2019) *J. Hort. Sci. Tech.*, **2**(1) : 1-4

Shedeed, M.R., El-Gamassy, K.M., Hashim, M.E. and Almulla, A.M.N. (1986) *Annals Agri. Sci.*, Ain Shams University, Egypt, **31** : 667-689.

Shimizu, H., Tsushima, Y., Komatsu, K., Shiigi, T., Nishizu, T., Kiong, C.V. and Kondo, N. (2008) *Journal of Science and High Technology in Agriculture (Japan)*, **20**(4): 253-256.

Singh, A., Dhaduk, B.K. and Shah, R.R. (2004b) *Indian J. Hortic.*, **61**(3) : 249-252.

Singh, A., Shah, R.R. and Dhaduk, B.K. (2004a) *Indian J. Plant Physiol.*, **9**(4) : 383-387.

Singh, H.P. (2000) *Floriculture Today*, **5** : 6-11.

Slathia, D., Khan, F.U., Masoodi, N.H., Khan, F.A., Wani, J.A. and Iqbal, U. (2019) *Curr. J. Appl. Sci. Technol.*, 1-7.

Sperry, C.E. and Chaney, W.R. (1999) *J. Arboriculture*, **25** : 43-47.

Srivastava, R.N. and Gupta, J.S. (1983) *Indian Phytopath.*, **36** : 14-16.

Stimart, D.P. and Brown, D.J. (1982) *Scientia Hort.*, **17** : 391-396.

Stimart, D.P., Brown, D.J. and Solomos, T. (1983) *Proc. Amer. Soc. Hort. Sci.*, **108** : 651-655.

Strider, D.L. (1979) *Plant Dis. Reptr.*, **63** : 873-916.

Strider, D.L. (1980) *Plant Dis. Reptr.*, **64** : 920-922.

Swarup, V. (1967) *Garden flowers*, pp. 145.

Swarup, V. and Raghava, S.P.S. (1974) *Indian J. Genet.*, **34** : 17-21.

Szopinska, D. (2013) *Acta Sci. Pol.-Hortorum Cultus*, **12**(5).

Szopiñska, D. and Politycka, B. (2016) *Folia Hortic.*, **28**(1) : 3-11.

Terry-Lewandowski, V.M., Bauchan,G.R. and Stimart, D.P. (1984) *Canada J. Genet. Cytol.*, **26**: 40-45.

Tsurushima, H. and Date, N. (1971) *J. Japanese Soc. Hort. Sci.*, **40** : 407-415.

Twumasi, P., Schel, J.H.N. and Van Ieperen, W. (2009) *J. Hortic. Sci. Biotech.*, **84**(5) : 531-535.

Vashchenko, M.A. (1985) *Referativnyi Zhurnal*, **55** : 847.

Venkatachalam, P. and Jayabalan, N. (1997) *Indian J. Genet. Plant Breed.*, **57** : 255-261.

Xia, Y.P., Chen, S.M., Qian, Z.S., Xia, Y.P., Chen, S.M. and Qian, Z.S. (1994) *Adv. Hort.*, **1** : 660-663.

Xiao, Y.T., Shen, Y.M., Wang, C.J. and Huang, T.C. (2020) *Plant Dis.*, 1-6.

Yahya, A., Mohd, R. I. and Roeber, R.U. (1997) *Acta Hortic.*, **450** : 32-38.

Yassin, M.Y. and Ismail, A.E. (1995) *Tropenlandwirt*, **96** : 221-225.

Ye, Y.M., Hu, Q.S., Chen, T.H. and Bao, M.Z. (2008) *Agric. Sci. China.*, **7**(4) : 423-431.

Ye, Y.M., Zhang, J.W., Qi, Y.C., Chen, T.H. and Bao, M.Z. (2007) *Scientia Agricultura Sinica*, 10.

Yoshimura, T., Demura, T., Igarashi, M. and Fukuda, H. (1996) *Plant Cell Physiol.*, **37** : 1167-1176.

Zamiran, A., Saffari, V.R. and Maleki, M.R. (2015) *J. Ornam. Plants*, **3**(3) : 203-214.

Zamorshi, C. and Milczarek, K. (1977) *Acta Agrobotanica*, **33** : 335-340.

Zhao, S.Y., Shen, X.Q., Wang, P., Wang, Z.G. and Zang, S.Z. (2007) *Northern Hortic.*, 3.

Zhao, Y., Gu, M., Bi, G., Evans, B. and Harkess, R. (2014) *J. Crop Improv.*, **28**(1) : 27-37.

Zhao, Y.T., Lu, B.H., Bai, Q.R. and Gao, J. (2016) *Plant Dis.*, **100**(1) : 208.

Zivdar, S., Khaleghi, E. and Dehkordi, F.S. (2011) *J. Applied Hortic.*, **13**(1) : 48-51.

Lupine

Namita, T. Rihne and Shisa Ullas P.

1.0 Introduction

Lupinus, commonly known as lupine, is a genus of flowering plants. It mostly includes herbaceous perennial plants, but some are annual plants and a few are shrubs. Among commercial cut flowers, lupine has shown potential as cut flower in the floriculture industry. The genus name *'Lupinus'* was derived from Latin word *'lupus'* meaning wolfish, related to the erroneous belief of consuming nutrients from the soil by the plant. However, lupine enhances soil fertility with the aid of nitrogen-fixing bacteria present in its root nodules.

2.0 Importance and Uses

2.1.0 Aesthetic Uses

The popular garden *Lupinus* (*Lupinus* × *hybrida*) are commonly grown in gardens for its attractiveness and ornamental values. They are ideal for growing as bedding plant, pot plant and as herbaceous border. Hybrid lupines are the most ornamental among the garden lupines and offer fuller flower spikes and arrays of flower color options. The lupine spikes are great as a cut flower, alone or in an arrangement. It attracts many beneficial pollinators to the garden including bumblebees, honeybees, butterflies, humming birds, *etc.* It has the ability to improve poor sandy soils with their nitrogen fixing ability like the other members of the Fabaceae family. There are numerous cultivars of lupine having flower colour of red, pink, white, blue, and multicoloured with different colours on different petals. It can be used for planting in flower beds, border and pots. Lupines are also used as specialty cut flower and can be stored at 2°C wet or dry for 2 weeks (Dole *et al.*, 2009). Lupine flowers and leaves are also used for making dry flower value added products.

2.2.0 Therapeutic and other Uses

Lupine is a high-value protein source and used as an alternative to soya bean in plant protein requirement (Lucas *et al.*, 2015). In addition to protein, lupine is also a source of non-starch polysaccharides (NSP), fats and several alkaloids (Van Barneveld, 1999; Stanek *et al.*, 2015). Low alkaloid containing lupines are also used as a supplementary diet in animal feed (Bryszak *et al.*, 2020). Few lupine species such as *Lupinus albus*, *Lupinus luteus* and *Lupinus angustifolius* were commonly used in the animal diets. Recently, narrow leaved blue coloured lupine (*L. angustifolius*) has been introduced as a vital feed source for livestock animals.

3.0 Origin and History

The genus *Lupinus* is one of the most diverse and widespread taxonomic groups of flowering plants. Its wild species are found distributed in the Mediterranean region including areas of northern Africa and in North and South America. The center of diversity of wild white lupine (sub-species *graecus*, *termis* and *albus*), and the primary center of diversity of its initial cultivated forms is the Balkans area of south-eastern Europe. *Lupinus albus* L., was found to be cultivated around the Nile valley and the Mediterranean, however, the primary centre of diversity

Uses of Lupine

For Outdoor Landscaping

As Bedding Plant

As Pot Plant

As Herbaceous Border

In Cottage Gardens

As Border Plant

As Cut Flower

In Floral Baskets

In Flower Vase

In Bouquet

In Flower Arrangement

Dried Lupine Flowers

Source: https://in.pinterest.com; https://www.etsy.com

was reported from the distribution of the wild *L. albus* var. *graecus* in the southern Balkans and into North-Eastern Greece, Southern Italy, and Western Turkey. The natural distribution of cultivated landraces, *L. albus* var. *albus* is around northern Mediterranean basin, the mid-Atlantic islands, the Canary Islands, North Africa and the Nile valley, Kenya, and Ethiopia (Gladstones, 1998). Generally, it is distributed on mildly acidic or neutral soils of light to medium texture. *Lupinus luteus* L. is distributed on sandy, acidic soils and natural distribution is much less than *L. albus* L. and *L. angustifolius* L. The proposed place of origin is the Iberian Peninsula (Gladstones, 1974) and the primary centre of distribution is in the western parts of this region. *L. luteus* landraces are distributed along with the wild forms which are used as garden ornamental.

The center of diversity of *L. angustifolius* was found to be the Aegean region where small seeded and finer-leafed genotypes occur at higher elevations of northern Greece and some islands (Cowling, 2001), whereas, wild types were distributed in North Africa and Iberia (Gladstones, 1998). *L. pilosus* has distributed from the mountainous regions of Crete and Turkey to the Greek Islands to the coastal areas of Israel and Syria. This species grows on sandy soils of neutral to alkaline pH. *L. palaestinus* is naturally distributed from the central and southern plains of Sharon and Philistia, down the coast of Israel, and on the Sinai Peninsula in the Jebel El Tih. *L. digitatus* is distributed in number of regions in southern Sahara which are mostly associated with mountainous areas. Lupines of the 'New World' group are distributed across a wide range of climates involving alpine, temperate, and subtropical. The centers of diversity for new world lupine are considered primarily North and Central America, the Andean South American region and the Atlantic South American region including most of the primitive simple-leafed species. 'New World' group include ornamental and forage species *L. polyphyllus* Lindl., soil conservation species *L. arboreus* Sims and *L. nootkatensis* Donn ex Sims, and ornamental *L. mexicanus* Cerv. ex Lag. (Synonyms *L. hartwegii*, *L. ehrenbergii*).

4.0 Taxonomy

Lupinus, belongs to the family Fabaceae, comprises of 200 to 500 species (Aïnouche and Bayer, 1999), with its origin in North and South America (Drummond *et al.,* 2012). The genus *Lupinus* has been classified into two subgenera namely *Platycarpos* and *Lupinus.* The subgenus *Platycarpos* comprises of perennial and annual species having minimum two or more ovules or seed buds from the Western Hemisphere, however the subgenus *Lupinus* consists of 12 species having minimum of four ovules or seed buds from Africa and the Mediterranean, (Kurlovich and Stankevich, 2002).

The genus *Lupinus* is large and has simple leaf and palmate compound leaf species. It has abundant flowers on terminal racemes. Flowers have a deep cleft calyx and erect standard petals. Wings of lupine connate at the apex and keel incurved, beaked and enclosed within the wings. Flowers have 10 alternately long and basifixed, short, and versatile stamens, sessile ovary, incurved and glabrous style, terminal stigma, oblong pod, more or less compressed, septate between the seeds, thick and leathery, dehiscent valves, thick and fleshy cotyledons (Allen and Allen, 1981).

Lupinus is a comparatively large and one of the most geographically widespread genera with a rich diversity of species. It is divided into two major groups *viz.*, 'Old World' group (the Mediterranean/North and East African species) and 'New World' group (the North and South American species) (Wolko *et al.*, 2011). Wild and landrace lupins are distributed across climatic ranges from subarctic regions, Mediterranean, and semi-desert climates; high land and mountain regions of East Africa, Mexico, the Andes; the warm temperate climates of South-Eastern USA and subtropical regions of eastern South America (Gladstones 1998). Lupine that belongs to 'Old World' group are mostly annual species and some of the species are *Lupinus albus* L., *L. hispanicus* ssp. *hispanicus*, *L. luteus* L., *L. micranthus* Douglas, *L. angustifolius*, *L. pilosus* Murr., *L. digitatus* Forsk, *L. palaestinus* Boiss., and *L. atlanticus* Gladst. On the basis of assessed genetic similarities by number of researchers, the centers of diversity of *Lupinus* are North and Central America and Andean South America, Atlantic South America, and the Mediterranean and northern and eastern African region.

5.0 Species and Cultivars

5.1.0 Species

Species of this genus vary from annual to perennial having growth habits *viz.*, acaulescent or small prostrate to tree-like shrubs of 4.5 m (Dunn 1984) or even to 8 m high with woody trunks of 30 cm across such as in *L. jaimehintoniana*. The species of *Lupinus* grown in gardens are described hereunder.

Lupinus polyphyllus Lindl. (Large-leaved lupine)

It is also known as big-leaved lupine, many-leaved lupine (Dickinson *et al.*, 2004), blue-pod lupine (Whitney, 1985) or garden lupine. The name of species was derived from Greek world *polyphyllus* which means "many-leaved". It is a tall perennial herbaceous plant with stout stems. The leaves are palmately compound and divided into 9-17 leaflets. Flowers produced on a tall spike are blue to purple colour in wild plants. Flower looks like a pea arranged compactly on stems of 60 cm length. The fruit of the lupine is a brown pod of 4 cm length, covered with entwined woolly hairs. The pod has 5-9 spotted seeds. The *polyphyllus* spp. is a progenitor of numerous hybrids and cultivars which are generally grown as garden lupines including the 'Russell hybrids'- *Lupinus* × *russellii* hort. and 'Woodfield' lupine in which flower spikes are dense, large, and more colourful or bicoloured. There are five varieties under this species *L. polyphyllus var. burkei*, *L. polyphyllus var. humicola*, *L. polyphyllus var. pallidipes*, *L. polyphyllus var. polyphyllus*, *L. polyphyllus var. ammophilus*, and *L. polyphyllus var. prunophilus*. Many hybrids were developed by using this species in crossing with *L. nootkatensis* (Nootka lupine).

L. arboreus Sims

This is known as tree lupine or yellow bush lupine. It is a tall evergreen shrub that grows up to 2 m and lives for up to seven years. It is hardy and thrives even at low temperatures down to –12 °C. It bears many fragrant 30 cm long racemes having yellow, lilac or purple-coloured flowers (Kindersley, 2008). It has green to

Species of Lupine

Lupinus polyphyllus Lindl. ***Lupinus arboreus*** Sims ***Lupinus angustifolius*** L.

Lupinus albus L. *Lupinus perennis* *Lupinus arcticus*

Lupinus texensis *Lupinus albifrons* *Lupinus cosentinii*

Lupinus mutabilis *Lupinus pilosus* *Lupinus hirsutum*

Source: www.commons.wikimedia.org, www.en.wikipedia.org, www.sbs.utexas.edu, www.gardenia.net

gray-green palmate leaves and each leaf has 5–12 leaflets. The leaflets are 2–6 cm in length, often lightly covered with fine silky hairs. Like other species of lupine, it is an effective nitrogen fixer in the soil. It is also used to bind drifting sand dunes. It can also hybridize with other lupine species including *L. littoralis* (seashore lupine) and *L. rivularis* (riverbank lupine). The yellow-flowering forms of the species are used widely as ornamental plants and can be find growing in cracks in vertical stone walls because of sharp drainage. Few species such as *L. chamissonis* (Chamisso bush lupine or silver bush lupine), *L. albifrons* (Bush lupine), *L. excubitus* (Grape soda lupine) and *L. longifolius* (Longleaf bush lupine) are closely related to *L. arboreus* Sims and used as shrubs in the gardens.

L. angustifolius L.

It is also known as narrow leaf lupine and blue lupine. This species is used as a food crop for its edible legume seeds, as a fodder for livestock and for green manuring. This lupine species fixes nitrogen in a symbiotic interaction with various bacteria *Bradyrhizobium lupinii* and *Kribbella lupini*. It has palmate leaf which is divided into 5- 9 leaflets. Its inflorescence bears numerous flowers in shades of blue, violet, pink, or white. This species was found to be grown on acid, well-drained, non-calcareous, light to medium textured soils.

L. albus L.

It is commonly known as white lupine or field lupine. It is annual and has more or less pubescence. The height is 30-120 cm tall. It has a broad distribution in the Mediterranean region. This species occurs in grassy slopes, meadows, and pastures, primarily on sandy and acid soils.

L. perennis (Wild Lupine)

It is commonly known as blue lupine, wild lupine, wild perennial lupine, sundial lupine, old maid's bonnets or Indian beet. This is 30-75 cm tall and compactly branched. The stems are light green to reddish green, angular or terete and canescent-hairy. It has alternate compound leaves which occur along the stems. Individual leafis palmate having 7-11 leaflets. Single leaflet is 2-6 cm long, oblanceolate in shape and have smooth and ciliate margins. The upper surface of the leaflets is medium green and hairless, however lower surface is pale green. Each compound leaf has light green to reddish green coloured, 2- 10 cm long ascending petiole. The stems, petioles and leaf undersides are sparsely to moderately canescent-hairy. The upper stems of the plants terminate into 10-25 cm long narrow racemes of flowers. The other species are *L. hartwegii* Lindl. (Synonym of *L. mexicanus*), *L. palaestinus* Boiss., *L. texensis* (Texas blue bonnet), *L. densiflorus* var. *aureus* (golden lupine), *etc.* They can be grown as ornamental plants in any garden.

5.2.0 Cultivars

The varieties/hybrids of garden lupine are given hereunder.

Lupinus Russell Hybrids - Mixed: These are mixed seed strain of lupine including various colours such as blue, purple, yellow, red, pink flowers, *etc. Lupinus polyphyllus* 'The Chatelaine' is a member of this Russell Hybrids family and have

Cultivars/Hybrids of Lupine

Lupinus
Russell Hybrids - Mixed

Lupinus
'Manhattan Lights'

Lupinus polyphyllus
'The Chatelaine'

Lupinus polyphyllus
'Gallery Yellow'

Lupinus polyphyllus
'Tutti Frutti'

Lupinus texensis
'Alamo Fire'

Lupinus densiflorus
var. *aureus* (Golden lupine)

Lupinus hartwegii nanus
'Pixie Delight'

***Lupinus* Popsicle**
Series

***Source*: www.gardendesign.com**

sweet scented, rosy pink flowers with accented white banners. Other varieties in the series are 'The pages' (Brick Red), 'The Governor' (Violet blue and white bicolour), 'Chandelier (Buttery Yellow), ' My Castie' (Scarlet red), 'The Maiden' (Creamy White), *etc.*

Gallery series: Dense flower spikes produced on the compact plant grow not more than 60 cm tall. This series have many hues in colours such as pink, red, indigo, white, *etc. Lupinus polyphyllus* 'Gallery Yellow' (Yellow colour flower spike) is an example of the Gallery series.

'Dwarf Lulu': These plants grow about 60 cm tall in a rainbow of hues and feature unusually dense racemes.

'Minarette': This 45 cm tall variety looks stunning in drifts along a border's edge or in containers.

Westcountry™ series: It is a collection of hybrids developed in UK. It include varieties such as *Lupinus* 'Manhattan Lights' (having a contrast of deep purple and neon yellow blooms); 'Blacksmith' (Royal purple and white coloured flowers); 'Desert Sun' (Golden yellow); 'Gladiator' (Yellow and Orange); 'Persian Slipper' (Periwinkle blue and white); 'Masterpiece' (Purple and Orange); 'Reed Rum' (Magenta and Red colour), *etc.*

***L. polyphyllus* 'Tutti Frutti':** It involves various colours such as magenta, dark purple, persimmon and gold, butter and cream, pink and primrose, purple with coral, *etc.*

***L. densiflorus* var. *aureus* (Golden lupine):** It is used as pollinator habitat restoration in California. It can tolerate drought and grow on poor soils.

***Lupinus* Popsicle Series:** The flowers in the series are of vanilla yellow, grape purple, cherry red, bubble-gum pink or bicolour.

Texas Bluebonnet Series: *Lupinus texensis* 'Alamo Fire' is an annual variety of this series. It has reddish maroon petals with white tips and bears more numerous flower clusters. It is tolerant to drought and thrives in sandy soil.

***L. hartwegii nanus* 'Pixie Delight':** It is ideal for small gardens and have light blue, pink, white and violet colours.

6.0 Genetics and Breeding

Lupinus is a self-pollinated crop. Its chromosome number ranges from n = 12 to n = 26 or 2n = 32 to 52 (Wolko *et al.*, 2011). Breeding of lupine was first started by George Russell during 1930s in England with the aim of selecting brighter flower colours. Crosses were made between North, Central, and Southern American species. Cross combinations that were apparently successful include *L. elegans* × *L. mutabilis* viable seeds; *L. pubescens* Benth. × *L. mutablis*, F_1 seeds; *L. nanus* Benth. × *L. mutablis* F_1 seeds; *L. polyphyllus* × *L. mutabilis* viable plants, used for backcrossing with *L. mutabilis*; *L. hartwegii* (synonymous with *L. mexicanus*) × *L. mutabilis*, viable seeds; *L. mutabilis* × *L. hartwegii*, embryo rescue hybrid F_1 plants, F_2 seed (Clements *et al.*, 2005). Variation for cold and frost tolerance has been used for breeding and sources of cold resistance are likely to be found in germplasm of *L. albus var. albus* and var. graecus from Italy, Spain, Turkey, and the Greece (Huyghe, 1997).

Pollination decreases floral longevity in several *Lupinus* spp. (Halevy, 1986). Pollination type mediated floral longevity and floral attraction in arctic lupine (*Lupinus arcticus* S. Wats.), a common herbaceous perennial in north-western North America which further may be due to influence of short growing seasons and low annual temperatures during the study (Reid, 2019).

7.0 Growth and Flowering

Lupines are very hardy flowering plants which can survive under extreme temperatures withstanding frost to at least -25°C. Normal plant growth occurs in a fairly narrow range of temperatures 0-35°C.

Inflorescence and flower development begins with floral initiation, when the apex of the main stem changes from creating leaves to flowers. The apex initiates many flowers over a number of days to form a raceme of 40 or more flowers. Individual flowers on the main stem inflorescence open from the bottom upward, starting about 75-100 days after sowing and continuing for about 20 days. Non branching lupine has shorter growing period (90-150 days) than branching type (150-180 days) (Gresta *et al.*, 2017). Anthesis occurs one to two days before flowering so that cross-pollination with other plants is rare (White *et al.*, 2008).

8.0 Propagation

Lupines can be propagated from seed, cuttings, or divisions.

8.1.0 Seeds

Lupine is propagated by seeds. Lupine seeds have very tough outer shells that need to be softened up before sowing. The seed can be soaked overnight in lukewarm water to allow the seed coat to be easily penetrated. Alternatively, seed scarification using sandpaper or a small file can help them to absorb water. Seeds of the lupine plant may also be chilled for a week in the refrigerator prior to planting. For cold stratification, the seeds are placed on slightly damp paper towels and then stored in a Ziploc bag in a refrigerator for seven days. The seeds should be sown at a shallow depth of about ¼ inch in loose topsoil. They should be kept evenly moist until they germinate. Seedlings generally emerge 15 to 25 days after planting. The seed rate for sowing of lupines varies depending on the size and germination percentage of the seeds. Seed rate varies from 70 - 130 kg/hectare land and sown at a depth of 3 -5 cm below the soil surface. Early sowing brings earlier flowering and a longer effective growing season. The best planting time is September-November.

8.2.0 Cuttings

Lupine can also be propagated through basal cuttings. Unlike a stem cutting, basal cutting is obtained from the base of the plant where the stem meets the crown. The advantage of cuttings (or division) over seed is that the resulting plants will be genetically identical. March or April is a good time for propagation through basal cuttings as the plant is initiating new growth suitable for this method.

Propagation of Lupine

Seeds

Seedling in Protray

Transplanting in Pot

Planting in Soil

Plantlets

Separation of Plantlets

Basal Cuttings

Propagation of Lupine

Planting in Pot

Rooted Cuttings

Micropropagation

(a) Half Cotyledon Node inoculated on media (b) After 7 days cotyledon develop new bud (c) After 14 days small plantlet appears (d) After 21 days more shoots generate (e) After 28 days shoot length increase (f) after 28 days plant bud turns into dark brown color (g) Plant inoculated on medium containing (KT 4.0 mg/l + NAA 0.1 mg/l) (h) Individual shoot cultured on medium with (IAA 1.0 mg/l + KT 0.1 mg/l).

***Source*: www.en.wikipedia.org, www.gardenofeaden.blogspot.com, www.fivecedars.com, www.thepropagatorblog.wordpress.com, www.gardeningelsa.com, Aslam *et al.* (2020).**

Pniweski *et al.* (2002) elaborated the complete protocols for long-term micropropagation of some cultivars of four lupine species *viz. Lupinus luteus, L. albus, L. angustifolius* and *L. Mutabilis*. The shoots were regenerated *in vitro* via induction of axillary buds development. Plantlets were multiplicated on lowered salts MS-derived media containing BAP in diverse and generally low concentrations. Significant differences in regeneration capacity between species and cultivars were observed. The highest multiplication ratio revealed in *L. mutabilis* and *L. luteus*. Regenerated shoots were rooted *in vitro* on low-salts MS-derived media with B5 vitamins. Media were supplemented with different auxins that affected roots formation of particular species and cultivars. However, rooting ability of regenerated shoots decreased rapidly through *in vitro* culture. Grafting method was applied as an alternative method of transfer of shoots to *in vivo* conditions.

Plant regeneration from different cotyledonary nodes of white lupine were established by Aslam *et al.* (2020). Five days old aseptically cultured seedlings were used to prepare three type of explants (half cotyledonary node HCN; whole cotyledonary node WCN; and traditional cotyledonary node TCN). These were cultured on four concentrations of M519 medium (M519, 1/2 M519, 1/3 M519, and 1/4 M519), containing four carbohydrate sources (sucrose, fructose, maltose, and glucose), and stimulated with combinations of KT (kinetin), and NAA (naphthalene acetic acid) for direct shoot regeneration. High frequency of 80 per cent shoot regeneration was obtained on 1/2 M519 medium (KT 4.0 mg/l + NAA 0.1 mg/l) by using HCN as an explant. Augmentation of 0.25 g/l activated charcoal (AC) not only reduced browning effect but also improved shoot elongation. Among the all carbohydrate sources, sucrose showed the highest regeneration frequency with HCN. Good rooting frequency (80 per cent) was recorded on 1/2 M519 containing IAA 1.0 mg/l + KT 0.1 mg/l (indole acetic acid) after 28 days of culturing.

9.0 Soil and Climate

9.1.0 Soil

Lupine grows best in light soils, rather than heavy and clay types. It prefers well-drained, coarsely textured, neutral to acidic soils. The organically rich, moderately fertile, slightly acidic, evenly moist, well-drained soils are ideal for lupine cultivation. Poorly drained soil with higher pH often results in Iron chlorosis and disease problems. *L. luteus* is the most drought resistant and can be grown in soil with low pH of 4.0 while *L. angustifolius* require a minimum of pH 5.0 and a maximum of 6.5-6.8.

9.2.0 Climate

Lupine is a cool-season crop, and is relatively tolerant to spring frosts. The flowering process is affected by high temperatures which may cause blasting of flowers and a subsequent yield reduction. They require full sun to survive. The garden lupine can survive in partial shade, but will produce far less blossoms in shady locations than they would in full sun. Lupines grown under deep shady locations will not produce flowers at all, but will focus entirely on growing foliage instead. For optimum growth of *Lupinus,* temperature should be maintained at 15 °C (night) and 24°C (day).

10.0 Cultivation

10.1.0 Transplanting

Lupine seedlings should be transplanted at the 2–3 leaf stage to obtain the longest and thickest stems (Dole and Greer, 2009). The recommended plant density for narrow-leafed lupine crops is 40 - 45 plants/m^2.

10.2.0 Irrigation

After planting, keep the soil moist to ensure good root growth and development. The deep rooted plants can tolerate dry conditions and will only need water during drought periods. Mulching application in the crop helps in keeping the soil moist. Irrigation by sprinkler method successfully increased the seed germination percentage and yield in *Lupinus* (Winnicki *et al.,* 2019).

10.3.0 Fertilization

Because of lupine's nature of N absorption from atmosphere, it does not require much fertilizer in the soil. Inoculation of *Lupinus* with its symbiont *Bradyrhizobium* sp. with $(NH_4)_2SO_4$ at 30 ppm N enhanced N_2-fixation and N absorption. Soil application of 3 ppm Mo, 2 ppm Co^{2+}, 1 ppm B with 30 ppm N increased nodule weight, plant dry weight, seed yield, seed size and the N and P contents of seed (Yanni, 1992). Nitrogen application at flowering markedly increased total dry matter, seed yield and total N in white lupines but remained without effect in blue lupines simply replacing nitrogen fixation (Schulze *et al.,* 1999).

10.4.0 Pruning and Deheading

It often encourages a second flush of blooms in early fall, especially in cool summers areas. The perennial species can cut back to the ground once the foliage turns yellow at the end of the season. However, for encouraging self-sowing of the lupine, avoid pruning and deheading and let the flowers to form seedpods.

11.0 Diseases and Pests

11.1.0 Diseases

11.1.1 Root Rot

Fusarium solani, Rhizoctonia solani, and *Macrophomina phaseolina* is one of the most destructive fungi of lupine. Bio-agents (Rhizo-N, Bio-Arc and Plant Guard), chemical inducers (salicylic acid, K_2HPO_4 and hydrogen peroxide), fungicides (Rhizolext-T, Vitavax-200 and Topsin-M70) reduced root-rot significantly and promoted growth parameters of lupine (El-Sayed and Abdel-Monaim, 2017).

Diseases and Pests of Lupine

Anthracnose, Powdery Mildew, Brown Leaf Spot

Collar Rot, Root Rot, Phomopsis

Phomopsis on Seeds, Cucumber Mosaic Virus, Bean Yellow Mosaic Virus

Aphids, Thrips, Budworm

Source: www.plantpath.ifas.ufl.edu, www.s1te.net, www.agriculture.vic.gov.au, www.agric.wa.gov.au, www.grdc.com.au, www.ipmguidelinesforgrains.com.au

11.1.2 Collar Rot

It is caused by the fungus *Sclerotinia minor*. Infected plants have a characteristic white fungal growth that attacks the stem at ground level. The fungal growth contains small black fruiting bodies called sclerotia.

11.1.3 Brown Spot

Brown spot is caused by the fungus *Pleiochaeta setose.* Leaves develop dark brown spots, often net-like in appearance and can be distorted and reduced in size before prematurely dropping off. Dressing seed with iprodione or procymidone fungicides protects the cotyledons and leaves of seedlings and should be used in conjunction with stubble retention for prolonged protection.

11.1.4 Anthracnose

Anthracnose is caused by the fungus *Colletotrichum lupini*. The most distinctive symptom is the bending and twisting of stems with a lesion in the crook of the bend on leaf petioles. Sowing of uninfected seeds or resistant varieties and application of thiram fungicide as seed dressing are the best management options.

11.2.0 Pests

11.2.1 Cut Worms (*Agrotis* spp.)

Cut worms chew through the stems at ground level. They feed on plants at night and remain hidden in the soil during the day. 2 to 3 cutworm/m^2 is the threshold level.

11.2.2 Aphids

Aphids usually are grey or green small bugs that grows up to 3 mm long, with or without wings. Leaves may be distorted and the growth of plants stunted. They cluster together on flower heads and flowers may be damaged and lost. If 30 per cent or more of plants are infested, control measures should be taken.

11.2.3 Thrips

Thrips may be found in the growing points and flowers. Sometimes leaves are distorted.

11.2.4 Native Budworm (*Helicoverpa punctigera*)

They are dark brown to light green caterpillars (larvae) with longitudinal stripes up to 40 mm in size. They feed on seeds, flowers and leaves.

11.2.5 Lucerne Seed Web Moth (*Etiella behrii*)

Caterpillars (larvae) 20 mm long with reddish-brown head and pinkish-green body and several light brown stripes running along. Penetrated pods and eaten seeds. Webbing is usually found in pods.

12.0 Harvesting and Postharvest Management

Lupine flowers were harvested when 1/4 to 1/3 of florets had reflexed petals (Dole and Greer, 2009). Citric acid and aluminium sulphate considerably enhanced the number of fresh flowers of lupine (*Lupinus hartwegii* Lindl.) and a maximum vase-life of 5.3 days was noted at 500 ppm of citric acid as compared to 2.9 days in the controls was observed by Mohan Ram and Ramanuja Rao (1977).

The use of Chrysal Professional 2 solution (CP2) or Flora life Professional (FLP) improved consumer vase life (10-12 days) in Sunrise' lupine (*Lupinus hartwegii* ssp. *cruickshankii*) compared with deionized water (8 days). Lupine can be stored at 2°C wet or dry for 2 weeks but it is susceptible to 1 µL/l exogenous ethylene, which can be effectively suppressed by methylcyclopropene and silver thiosulfate (Dole *et al.*, 2009).

Lupine leaves can be dried using different drying methods and the best drying method with suitable drying time in lupine leaf is press drying at room temperature for 5-6 days (summer) and 9-10 days (Winter) or microwave drying at 50°C for 120 seconds. For maintaining flower shape and colour, embedded drying with silica gel for 14 hrs at 50°C in Hot air drying was found suitable (Deepti, 2018).

13.0 Biotechnology

Based on internal transcribed spacer (ITS) sequences of nuclear ribosomal DNA using 44 *Lupinus* taxa and five outgroup taxa, Ainouche and Bayer (1999) clearly supported previous classifications that the *Lupinus* genus is a strongly monophyletic genus and should be included in tribe Genisteae but as a distinct lineage from the Genistinae ("Cytisus-Genista complex") as subtribe Lupininae (Hutch.) Bisby. The results were consistent with serological data (Cristofolini 1989) and molecular-based phylogenies of the Papilionoideae (Doyle *et al.*, 1997; K€ass and Wink, 1997).

14.0 References

Aïnouche, A.K. and Bayer, R.J. (1999) *American Journal of Botany*, **86**(4): 590-607.

Allen, O.N. and Allen, E.K. (1981) *The Leguminosae: A Source Book of Characteristics, Uses and Nodulation.* MacMillan, London.

Aslam, M.M., Karanja, J.K., Zhang, Q., Lin, H., Xia, T., Akhtar, K., Liu, J., Miao, R., Xu, F. and Xu, W. (2020) *Plants*, **9**: 318. doi: 10.3390/plants9030318

Bourne, V. (2002) Falling in love again. The Daily Telegraph. (Retrieved in 2017).

Bryszak, M., Szumacher-Strabel, M., Huang, H., Pawlak, P., Lechniak, D., Ko³odziejski, P., Yanza, Y.R., Patra, A.K., Váradyová, Z. and Cieslak, A. (2020) *Animal Feed Science and Technology*, **267**: 114590.

Clements, J.C., Buirchell, B.J., Yang, H., Smith, P.M.C., Sweetingham, M.W. and Smith, C.G. (2005a) Chapter 9: Lupine. In: Singh R, Jauhar P (eds) *Genetic resources, chromosome engineering, and crop improvement*, series-II grain legumes. CRC, Boca Raton, FL, pp. 231–323.

Cowling, W. (2001) Current Plant Science and Biotechnology in Agriculture, Springer, Dordrecht, **39**: 191-206.

Deepti, S. (2018) *International Journal of Agriculture Sciences, ISSN*, pp. 0975-3710.

Dickinson, T., Metsger, D., Bull, J. and Dickinson, R. (2004) The ROM Field Guide to Wild flowers of Ontario. Toronto: Royal Ontario Museum, pp.270.

Dole, J.M. and Greer, L. (2009) *Scientia horticulturae*, **122**(2): 233-237.

Dole, J.M., Viloria, Z., Fanelli, F.L. and Fonteno, W. (2009) *Hort Technology*, *19*(3), pp.593-600.

Doyle, J.J., Doyle, J.L., Ballenger, J.A., Dickson, E.E., Kajita, T. and Ohashi, H. (1997) *American Journal of Botany*, **84**: 541-554.

Drummond, C.S., Eastwood, R.J., Miotto, S.T. and Hughes, C.E. (2012) *Systematic biology*, **61**(3): 443-460.

Dunn, D.B. (1984) *Proceedings of the 3rd International Lupine Conference* (Ed. Anon.), pp. 65-85, La Rochelle, France.

El-Sayed, S.A. and Abdel-Monaim, M.F. (2017) *J. of Plant Sci. and Agric. Research*, **1**: 1-8.

Gladstones, J.S. (1974) *Lupins of the Mediterranean Region and Africa*. Technical Bulletin No. 26. Department of Agriculture of Western Australia, Perth.

In: Gladstones JS, Atkins C, Hamblin J(eds) *Lupins as crop plants*. Biology: production and utilization. CABI, Oxon, pp. 1–39.

Gladstones, J.S. (1998) In: Gladstones JS, Atkins C, Hamblin J(eds) *Lupins as crop plants*. Biology: production and utilization. CABI, Oxon, pp. 1–39.

Gresta, F., Wink, M., Prins, U., Abberton, M., Capraro, J., Scarafoni, A. and Hill, G. (2017) *Legumes in cropping systems*, pp. 88-108.

Halevy, A. H. (1986) *International Society for Horticultural Science*, **181**(2): 25–32.

Huyghe, C. (1997) *Field Crops Research*, **53**(1-3): 147-160.

Kindersley, D. (2008) *RHS A-Z encyclopedia of garden plants*. United Kingdom. pp.1136.

Kass, E. and Wink, M. (1997) *Plant Systems and Evolution*, **208**: 139-167.

Kurlovich, B.S. and Stankevich, A.K. (2002) Eds. *Classification of Lupins*. In: *Lupins: Geography, Classification, Genetic Resources and Breeding*, St. Petersburg: Intan. pp. 39-87.

Lucas, M.M., Stoddard, F.L., Annicchiarico, P., Frias, J., Martinez-Villaluenga, C., Sussmann, D., Duranti, M., Seger, A., Zander, P.M. and Pueyo, J.J. (2015) *Frontiers in plant science*, **6**: 705.

Mohan Ram, H.Y. and Ramanuja Rao, I.V. (1977) *Scientia Horticulturae*, **7**: 377-382

Pniewski, Kapusta, T.J and Legocki, A.B. (2002) *Acta Physiologiae Plantarum*, **24**(4): 417-424.

Reid, C. (2019) *The Arbutus Review*, **1**(1): 83-99.

Russell, G. (2017) stillingtonvillage.org.

Schulze, J., Beschow, H. and Merbach, W. (1999) *Isotopes in Environmental and Health Studies*, **35**(1-2): 85-95.

Stanek, M., Rotkiewicz, T., Sobotka, W., Bogusz, J.U., Otrocka-Domaga³a, I. and Rotkiewicz, A. (2015) *Acta Vet. Brno.*, **84**: 55–62.

Van Barneveld, R.J. (1999) *Nutr. Res. Rev.*, **12**: 203–230.

White, P., French, B. and McLarty, A. (2008) In: *Producing lupins*, Bulletin 4720, Department of Agriculture and Food, Western Australia, Perth.

Whitney, S. (1985) *Western Forests* (*The Audubon Society Nature Guides*). New York: Knopf. pp. 601.

Winnicki, K., Ciereszko, I., Leœniewska, J., Dubis, A.T., Basa, A., ¯abka, A., Ho³ota, M., Sobiech, £., Faligowska, A., Skrzypczak, G. and Maszewski, J. (2019) *Planta*, **249**(6): 1731-1746.

Wolko, B., Clements, J.C., Naganowska, B. and Nelson, M.N. (2011) In : *Wild crop relatives: Genomic and breeding resources* (pp. 153-206). Springer, Berlin, Heidelberg.

Wolko, B., Clements, Jon C., Naganowska, B., Nelson, M. N. and Yang, H. (2011): In: *Wild Crop Relatives: Genomic and Breeding Resources, Legume Crops and Forages*, C. Kole (ed.). 153-206.

Yanni, Y.G. (1992) *World Journal of Microbiology and Biotechnology*, **8**(6): 607-613.

10

Salvia

M. Visalakshi, M. Ganga and T.N. Saha

1.0 Introduction

Salvia commonly called as 'Sage' is the largest genus in mint family with nearly 1000 species including shrubs, herbaceous perennials and annuals. The most commonly cultivated species in the genus *Salvia* is *Salvia officinalis*, also known as common sage and mainly utilised for its culinary use. *Salvia splendens*, commonly called as 'Scarlet Sage' is widely used as an ornamental plant owing to its attractive scarlet red flowers (Griffiths, 1994). The word '*Salvia*' is derived from the Latin word '*Salvere*' which means 'to feel well and healthy, referring to the herb's healing properties of this plant.

2.0 Importance and Uses

2.1.0 Aesthetic Uses

Horticulturally, Salvia species can be grouped as tender or hardy plants. Tender types are mostly herbaceous perennials or shrubs in tropical and sub-tropical areas. Among the tender species, *S. splendens*, are used as ornamental plants highly valued for their flowers and for their aromatic foliage. Scarlet sage is among the 25 leading bedding plants of the world and widely used as summer and autumn flowering bedding plants. Although, Scarlet sage is best known for its scarlet variety, other colours including white, salmon, pink, purple, lavender, burgundy and orange are also available. Hardy species are used as border plants for spring and summer flowering shrubs. Salvias are versatile, colourful plants and are widely grown in mixed borders, formal mass plantings, wild flower gardens, cottage gardens and containers. The stunning red salvia flowers form a striking accent when massed together as bedding plants or lined up in a row as edging plants. They are also popular in container gardens, where they can serve as a vertical accent. Wild species grow tall, while smaller cultivars are used as bedding plants and borders. Salvia also make great cut flowers. The blooms of most salvia plants are long-lasting and attract hummingbirds, butterflies and other pollinators. Mealy cup sage (*Salvia farinacea*), a native of the South Western United States, is used for bedding or in a mixed border.

2.2.0 Therapeutic and other Uses

Plants from the genus Salvia have been important ingredients in traditional medicine all over the world since earliest times. Many species of this genus are extensively used for culinary purposes, essential oil production and as Chinese herbal remedies. Common sage is extensively grown for its culinary use as commercial herb as well as for decorative foliage in the garden. *Salvia miltiorrhiza* (Danshen or Red sage) is a popular traditional herb in Chinese, Korean, and Japanese medicine. It has been widely used to treat diabetes and its related chronic complications. There is an extensive literature mostly from Europe, on the chemical composition, production, and extraction of essential oils from common sage and related species. The whole plant, especially the inflorescence of Clary sage (*S. sclarea*) L., possesses a strong aromatic scent and the essential oil, is used in high-grade perfumes and the

Uses of Salvia

Source: www.almanac.com, www.wikiwand.com, www.in.pinterest.com, www.gardenia.net, www.naturehills.com, www.dreamstime.com, www.funflowerfacts.wordpress.com

essential oil constituent sclareol is used as base material for the industrial synthesis of the highly valued ambergris odorants (Singh *et al.*, 2008).

3.0 Origin and History

Pliny the Elder was the first to describe *Salvia officinalis.* The genus *Salvia* occurs worldwide in temperate, tropical and subtropical regions. Its three centres of diversity are South-West Asia, Sino- Himalaya and the New World species range continuously from the Great Lakes in North America to the Argentine Plains in South America.

S. splendens is native to Brazil and probably Columbia. The species is an introduced species to the West Indies, Central Africa and most tropical and temperate climates. It is cultivated in the Pacific region. Mealy cup sage (*S. farinacea*) is a native of the South Western United States.

4.0 Taxonomy

Salvia is the largest genus in the family Labiatae (Lamiaceae), the mint family. Within the Lamiaceae, *Salvia* is a part of the tribe Mentheae within the subfamily Nepetoideae. *Salvia splendens* was first described and named in 1822. Some of species, such as scarlet sage (*S. splendens* Sell ex Roem. and Schult) are perennial herbs in their native habitat but are used as annuals in temperate climates. Additionally, mealy cup sage (*S. farinacea* Benth) is a perennial herb in its native Mexico and Southern Texas, but is grown as an annual throughout the temperate regions of the United States. Such plants are sometimes referred to as tender perennials or as half hardy perennials. Although cosmopolitan in distribution, salvias are generally characterized as growing in dry or stony soils (Liberty Hyde Bailey Hortorium, 1976). One notable exception is bog sage (*S. uliginosa* Benth.), which is native to wet soils in Brazil, Uruguay, and Argentina.

S. splendens is mostly annual/biennial/herbaceous perennial growing upto 90 cm tall. Petioles are 3.0-4.5 cm and glabrous. Foliage is attractive and remains fragrant. The fragrance varies with species. *S. splendens* give spicy flavour while *Salvia elegans* give a pineapple fragrance. Leaf blades are ovate to triangular-ovate, glabrous, abaxially glandular with truncate or rounded base, serrated margin and acuminate apex. Inflorescence is racemes with 2-6-flower to length of 20 cm. The bracts are ovate, red, enveloping flowers in bud, apex caudate-acuminate. Pedicel 4-7 mm, red glandular villous. Calyx red, campanulate, 1.6 cm in flowering, dilated to 2 cm after anthesis, red glandular, veins villous, 2-lipped, upper lip triangular-ovate, 5-6 × 10 mm, apex mucronate, lower lip slightly longer than upper, deeply 2-toothed, teeth triangular. Corolla scarlet, 4-4.2 cm, pubescent, tube slightly dilated at throat, upper lip straight, somewhat concave, oblong, lower lip shorter than upper. Nutlets dark brown, ellipsoid, apex irregularly pleated, margin (or mid-vein) narrowly winged. Flowering occurs during March-October. Hummingbirds, bees, bumblebees, and bee-flies harvest nectar from salvias and thereby act as pollinators (Boring *et al.*, 1995).

5.0 Species and Cultivars

5.1.0 Species

Although, about 140 species are under cultivation, only a few salvias are grown commercially grown as medicinal herbs and as ornamental plants in gardens. Most species under the genus *Salvia* are generally designated as sage and is usually preceded by at least one modifier, such as common sage (*S. officinalis* L.), Scarlet sage (*S. spendens* L.) or Mexican bush sage (*S. leucantha* Cav.). A few salvia species are known as clary (*S. sclarea* L.), meadow clary (*S. pratensis* L.) and wild clary (*S. verbenaca* L.) and at least two have neither clary nor sage in the common name: Jupiter's-distaff (*S. glutinosa* L.) and cancer weed (*S. lyrata* L.) (Liberty Hyde Bailey Hortorium, 1976). Clary also has its linguistic roots in the plant's usage; *S. sclarea* was used for afflictions of the eye and therefore called clear eye or clary (Stearn, 1992).

Three species of Salvia namely, *S. splendens, S. coccinea* and *S. farinacea.* are grown as annuals in India.

Salvia splendens

It is commonly known as scarlet sage, bonfire sage or coral sage. Height of the plant may vary from 20-70 cm depending on the variety. Bright scarlet flowers appear on long spike at the end of the branches, other colours include salmon, pink and violet.

S. coccinea

It is commonly called as tropical sage. Several varieties of this species bearing red, scarlet, pink or white flowers on terminal spikes are grown in Southern India.

S. farinacea

It is also known as mealy cup sage. The plants are 50-70 cm high, produces spikes of small blue or white flower. The foliage part of this plant is relatively low and compact, but from this rises slender spike of light purple flowers on darker purple stems. The name mealy cup refers to the fine white hairs around the calyx or casing that carries each flower. The foliage is grey-green that sets off the flowers well. The greyness is due to fine hairs that cover the leaves and stem.

5.2.0 Cultivars

Among the ornamental salvias, cultivar development has been most extensive for scarlet sage. Some of the popular cultivars of *S. splendens* includes 'Alba' (white),

'Atro purpurea' (violet to purple), 'Atrosanguinea' (dark red), 'Bicolor' (white and red), 'Bruantii' (small and red), Compacta' (small, dense racemes, white or red) 'Grandiflora', (large, red) 'Issanchon' (small, white striped pink to red), 'Nana' (early-flowering, red), 'Scarlet Pygmy' (very dwarf, early flowering, intense scarlet), 'Semperflorens', (continuous flowering), 'Souchetii', (small, white or red), 'St. John's Fire', (dwarf, dense, abundant, scarlet, early-flowering, long-lasting) 'Violacea', (dark violet to purple).

Species of Salvia

Salvia greggii

Salvia clevelandii

Salvia regla

Salvia leucantha

Salvia splendens

Salvia coccinea

Salvia farinacea

Salvia officinalis

Salvia sclarea

Salvia pratensis

Salvia verbenaca

Salvia lyrata

Source: www.repository.arizona.edu, www.hgtv.com; www.en.wikipedia.org, www.in.pinterest.com, www.ballseed.com, www.davesgarden.com

Cultivars of Salvia

Salvatore Blue | Pink Sunday | Strata

Blaze of Fire | Purple and Bloom | Dyson's Joy

Lips Collection | Amistad | Hot Lips

Schneehugel | Alba | Purpurascens

Source: www.in.pinterest.com, www.prairieblossomnursery.com, www.thompson-morgan.com, www.gardeningexpress.co.uk

Table 1: Salvia Species with Cultivars and their Landscape Characteristics*

Common Name	*Botanical Name*	*Propagation*	*Height and Flower Colour*	*Landscape Uses*	*Cultivars*
Texas salvia/ Tropical sage	*Salvia coccinea*	Seed or cuttings	0.6 – 0.9 m Red, pink, or rarely white	Annual, Used for mass effect or as a tall ground cover	Bi colour, Nana, Nana carmine, Nana compacta, Lactea, lacteal var major, Punicea, Rosea Nana, Punicea nana
Dorri sage/ Purple sage/ Desert sage	*Salvia dorrii*	Seed or cuttings	0.6–0.9 m Tubular blue flowers nestled in pink to violet bracts.	Semi evergreen shrub, Attracts hummingbirds, Small hedge in Landscape	
Mealycup sage	*Salvia farinacea*	Seed or cuttings	0.3–0.6 m Lavender-blue or white flowers	Perennial herb used as an annual, Common bedding plant Good container plant and cut flower	Alba, Blue bedder, Victoria, Victoria Blue, Victoria white, Rhea, Strata
Gregg sage/ Autumn sage/ Cherry sage	*Salvia gregii*	Cuttings	0.6–0.9 m Red, pink, white or coral	Perennial herb or shrubBlooms from spring to frost	Alba, Cherry Red, Rosea, Furmans Red
Mexican bush sage	*Salvia leucanth*	Cuttings or division	1.2 m White or purple flowers surrounded by purple to lavender-blue calyces	Good cut flower	
Common sage/Garden sage	*Salvia officinalis*	Cuttings or division	0.3-0.6 m Lilac-blue	Bedding plant,Herbal garden	Hardy, Tricolor, purple, and variegated cultivars
Scarlet sage/ Red salvia	*Salvia splendens*	Seed	0.2-0.8 m Scarlet, purple, lavender, red, cream, orange	Common bedding plant	Blaze of Fire, Fire ball, Grandiflora, Calypso, Gypsy Rose, Bruantii, Arto sanguinea, Lavender Love and Harbinger, Red Blazer, salmon beauty
Perennial salvia	*Salvia x superba*	Cuttings or division	0.6-0.9 m Violet or purple	Hardy	

* Boring *et al.* (1995), Clausen and Ekstrom (1989), Clebsch (1997), Griffiths (1994), Still (1994), Wasowski and Wasowski (1997)

6.0 Genetics and Breeding

6.1.0 Genetics

The chromosome numbers of Salvia species are unusual and shows extreme variability. The chromosome counts in genus *Salvia* ranges from 2n = 12 in *S. hispanica* (Harley and Heywood, 1992) to as high as 2n = 88 in the octoploid *S. guaranitica* (Alberto *et al.*, 2003; Gedic *et al.*, 2016). In addition, the genus *Salvia* exhibits wide variation in ploidy level and the basic chromosomes number. The basic chromosome number (x) has been reported as x = 6, 7, 8, 9, 10, 11, and 15 for different species within the genus *Salvia* (Goldblatt and Johnson, 1996; Harley and Heywood, 1992). Seven of 13 species studied by Alberto *et al.* (2003) had at least one known supernumerary chromosome (non-essential chromosomes absent in some individuals). It is possible that closely related species that appear to have different basic chromosome numbers might simply have supernumerary chromosomes that have not been properly identified as such. The chromosome number of common sage is 2n=14, while that of scarlet sage is 2n = 44 (Goldblatt and Johnson, 1996). The type B chromosome is most prominent in *Salvia* taxa (Dogan *et al.*, 2019).

A recent work was carried out on 15 *Salvia* species for identification of genetic similarity. The data obtained from ISSR and RAPD profiles depicted that variation in similarity index varied in between 0.54 (*S. rosifolia, S. sclarea, S. rosifolia, S. limbata,* and *S. staminea, S. verticillata*) and 0.93 (*S. sclera-S. divaricata*) (Sunar *et al.*, 2020).

6.2.0 Breeding

Traditional breeding activities using phenotypic selection as well as targeted hybridizations between elite cultivars have resulted in a large number of new cultivars with different performances regarding flowering characteristics (*e.g.*, related to color, flowering time and flowering period), individual growth performance, height and tolerance to moisture and temperature extremes (Dong *et al.*, 2018). For genetic improvement of Salvia species intraspecific hybridization is most suitable (Bernath and Nemeth, 2000). With around 900 species, native to every continent except Antarctica and Australia (Mabberly, 2008) and possessing a myriad of growth habits and flower colors, the genus Salvia represents an extraordinary range of diversity. The diversity of the genus represents a largely untapped resource for crop improvement, which might be accessed by means of interspecific hybridization. There are a number of known interspecific hybrids in the genus. However, with the exception of intentional crosses between culinary sage (*S. officinalis*) and *S. lavandulifolia* (Gomez *et al.*, 1995), *S. fruticosa* and *S. tomentosa* (Putievsky *et al.*, 1990), all reported hybrids have occurred spontaneously either in the wild or in cultivation. There is a paucity of information concerning the interspecific fertility of Salvia species.

The crossability of two species is often related to phylogenetic relationship and chromosome number. Crosses between closely related species with the same chromosome number are more likely to hybridize and produce fertile progeny, whereas crosses between distantly related species with different chromosome numbers are more likely to either not hybridize or to produce sterile offspring resulting from genetic incongruity or meiotic errors (Harlan and DeWet, 1971; Sybenga, 1992). Tychonievich and Warner (2011) evaluated the crossability of nine *Salvia* species selected to cover a wide range of the diversity in European and American species under full diallel mating scheme. It was revealed that crossability of the selected species was low with only five out of 72 interspecific cross combinations producing viable seed, whereas all nine species were self-fertile. Successful crosses were mostly within close phylogenetic groupings. The majority of successful crosses were between species with different chromosome numbers, suggesting that chromosome number differences alone are not a major barrier to hybridization in this genus. A *Salvia nemorosa* × *Salvia transslyvanica* F_2 population exhibited transgressive segregation for several horticulturally important traits, including flower size, plant height and time to flower. Plant height was correlated positively with flower length, inflorescence branch number and time to flower. Time to flower was correlated positively with flower length. Individuals with desirable trait combinations were identified within the population.

It has only been proven in the last ten years that the induction of mutant and polyploid variants have lead to the improvement of *S. sclarea's* genetic history. The construction of these mutant lines is being carried out with the use of colchicine (Bernath and Nemeth, 2000).There is an increase in the essential oil accumulation.

7.0 Growth and Flowering

7.1.0 Temperature

Scarlet sage responded strongly to a difference between day and night temperatures (DIF), a response that can be used to regulate plant height (Barrett and Erwin, 1994). Internode length was greatest for scarlet sage with a cooler night than day temperature (DIF = +3.3°C), was intermediate with a constant temperature and shortest with a higher night than day temperature (DIF = –2.8°C). Using temperature to reduce internode length is most effective if the temperature drop is rapid, occurs at sunrise and during high light intensity.

7.2.0 Light

Production of scarlet sage requires high light. Supplemental photosynthetic lighting of seedlings accelerates development and limits excessively leggy growth but photoperiod requirements for flowering are cultivar dependent. Carlson (1978) suggested three categories for the photoperiod response of scarlet sage cultivars: 1) cultivars that flower faster under short days (facultative short day plants), 2) cultivars that flower in the same length of time under either long or short days (day neutral plants) and 3) cultivars that flower faster under long days (facultative long day plants). Of the 44 cultivars studied, Carlson found 18 to be quantitative short day, 11 to be day neutral, and 14 to be quantitative long day plants (one cultivar did not flower at all).

7.3.0 Growth Regulators

Plant size can be controlled during production with the use of growth regulators. The growth retardants like Uniconazole, Paclobutrazol and Daminozide are commercially used for controlling the plant height. Among the retardants, cuttings when soaked in Uniconazole for 24 h and foliar spray at 50 ppm are most effective in reducing the plant height. The foliar spray of 0.5 mg/l of flurprimidol (Topflor®) to 25 days old seedlings was found effective in height control in *S. splendens, cv.* Flamex 2000 (Vernieri, 2003). Singh *et al.* (2008) assessed the effects of growth regulator namely kinetin, IAA and paclobutrazol on field grown *S. sclarea* L. The maximum height and leaf numbers were observed after application of kinetin 10 µl/l and IAA 50 µl/l respectively. The application with 40 µl/l paclobutrazol resulted in maximum flower and oil yield. Further, it was revealed that the application of 80 µl/l paclobutrazol significantly increased the linalool-linalyl acetate content of the plant.

Cytokinins can also be used to induce branching, control plant growth and influence flowering. Carey *et al.* (2013) studied the effects of foliar sprays of benzyladenine (BA) at concentrations of 100 to 1600 mg/l applied 14 days after planting (DAP) onto *Salvia nemorosa* (L.). BA foliar sprays ≥ 400 mg/l controlled plant height and ≥ 200 mg/l resulted in smaller plant diameter. Poyh and Ono (2006) observed higher essential oil content in sage (*S. officinalis*) treated with 100 mg/l of gibberellic acid (GA) compared to control plants. The chemical composition changed with gibberellic acid (100 mg/l) application, with a significant reduction of β-tujone and α-humulene (Povh and Ono 2007).

8.0 Propagation

Salvia can be propagated by seeds, cuttings and divisions.

8.1.0 Seed Propagation

Scarlet sage is mainly propagated through seeds while perennial species may be produced from either seed or cuttings. Sage does not remain true to type when grown from seed, thus named cultivars should be propagated from cuttings. Certain cultivars of Mealy cup sage come true from seed and therefore may be propagated by either method. Salvia seeds should not be stored for more than one year since they lose not only viability but seedlings may have reduced vigour (Dole and Wilkins, 1999). Optimal germination occurs when seeds were stored at a Relative humidity between 11 and 32 per cent regardless of temperature (Carpenter *et al.*, 1995). In areas with severe winters, seeds can be sown under protected structure and transplanted outside in main field after frost. The seeds can be sown indoor in propagation chambers 10-12 weeks before the last frost. Under outdoor conditions, salvia seeds can be sown once the danger of frost has passed in the spring. Salvia seeds requires a temperature of 22-24°C for optimum germination. The seeds takes about 12-15 days to germinate. Scarlet sage and Mealy cup sage require light for germination while Common sage germinates best when covered and Perennial salvia may be grown under either condition (Nau, 1993).

8.2.0 Vegetative Propagation

Salvia can be vegetatively propagated through soft wood or tip cuttings. The cuttings (10-12 cm) should be taken from tip portion of vegetative (non-flowering)

Propagation and Growing of Salvia

Seeds

Sowing in Tray

Germination

Seedlings in Protray

Twig Selection for Cuttings

Cutting

For Rooting

Rooted Cuttings

Planting in Pot

Micropropagation

Micropropagation of ***Salvia miltiorrhiza*** (Liu *et al.*, 2018)

(A) Green-colored protuberances from leaf explants; (B) Adventitious shoots induced after 3 weeks; (C) Shoot proliferation and elongation; (D–G) The plantlets rooted from the shoot base; (H) Acclimatized plants in the greenhouse.

Open Field Cultivation In Field

In Greenhouse

Source: www.lcsd.gov.hk, www.alamy.com, www.gardentags.com, www.flower-db.com

branches before the development of flower buds. The cuttings of small leaved Salvias are best taken in spring before flowering or in late summer when flowering is over. It is recommended to take cuttings from the young new growth where stems have not yet hardened much or when the sap is actively flowing. The cuttings are taken from the stock plant by trimming just below the nodes. While preparing the cuttings, the lower leaves should be trimmed off. The growing tips/flower buds should be removed to encourage a bushier plant. Salvia cuttings can be taken in the spring or early fall. The lower leaves are removed and cuttings are planted either in the plug trays or pots into the pre-watered potting media consisting of peat: perlite and vermiculite (1:1:1). The cuttings should be placed in a cool greenhouse under shade to prevent scorching from strong sunlight. After three weeks, the cuttings become ready for transplanting.

8.2.1 Division

The herbaceous species of salvia producing basal clumps can be propagated through division. The plants can be divided after winter when the frosts are over and the new shoots start to appear. The whole clump is dug out and is divided/ sliced into either half or quarters using a sharp. Enough shoots and roots must be retained in each divided clump to ensure the new growth of the shoots after division.

8.3.0 Micropropagation

Sharma *et al.* (2014) developed protocol for high-frequency in vitro propagation and short-term conservation using encapsulation technology in *Salvia splendens*. Among the different explants, nodal segments exhibited higher multiplication potentials than shoot tips. Murashige and Skoog's (MS) basal medium augmented with 5.0 µM 6-benzyl adenine (BA) was optimum for shoot induction through both explants. For synseed preparation, nodal segments were encapsulated in 4 per cent Na-alginate with different gel matrices and 100 mM $CaCl_2{\cdot}2H_2O$ (complexing agent). Maximum conversion (63.6 per cent) of synseeds into plantlets was achieved when they were prepared with the gel matrix of MS basal medium supplemented with 5.0 µM BA and 2.5 µM IAA. Pretreatment of synseeds with 100 mM KNO_3 for 30 min, prior to incubation on the planting media significantly improved conversion frequency (94.6 per cent). These synseeds could be stored up to 10 weeks at low temperature (4°C).

9.0 Soil and Climate

9.1.0 Soil

Salvias are resilient and come up well in any type of soil. It prefers average to slightly dry, well-drained soil with a pH of 7.0. Soil pH have significant influence on germination of seeds in salvia. Germination of scarlet sage 'Red Hot Sally' seed was 0 per cent at pH 4.5-5.0, 44 per cent at 5.5 and 70 per cent at 7.0 and 7.5 (Shoemaker and Carlson, 1990). If the soil becomes too dry, *S. splendens* may show wilting. In dry conditions, the blue bedding *S. farinacea* also suffers.

9.2.0 Climate

Salvias grow well at 2000-3000 m from the mean sea level, with warm climate and high humidity. *S. splendens* is a subtropical species that does not survive freezing temperatures, but can grow in cold climates as an annual plant. It thrives in full sun in cooler climates but tolerates shade, especially in areas with very hot summers. The optimum growth temperature for salvia ranges from 15°C to 30°C. It requires a day temperature of 21-24°C and night temperature of 13-15°C for optimum growth. Temperatures lower than 5°C can lead to cold damage, while temperatures above 25°C hinder the growth of flowers and leaves, making them weak and small. *S. coccinea* (Tropical Sage) and *S. farinacea* (Mealy cup sage) require a day temperature of climate of 14-15°C. Salvias can grow in full sun, filtered light, or partial shade. *S. koyamae* (Japanese Yellow Sage) loves to come in shade condition with moist soil (Armitage 1994).

10.0 Cultivation

10.1.0 Seed Sowing and Spacing

Seeds are sown thinly in seed boxes or raised beds. Seedling emerges within 15-21 days after sowing. Hardened seedlings are transplanted at two-four pairs of true leaved stage. The seed rate requirement is 400-500 g/ha. The seedlings or cuttings are planted with a spacing of 20-60 cm apart depending upon the variety. For cut flower production, plants should be spaced at least 40 - 90 cm.

10.2.0 Irrigation and Weeding

Salvias are dry crops and require less water. Irrigate the plants regularly once in a week. Watering regularly will help the plant in rapid regeneration of plants after pruning in first season. Hand weeding and mulching are practised. Mulching help in reducing weed growth, retain soil moisture and maintain even soil temperatures.

10.3.0 Manuring and Fertilization

A light dose of organic manure for rich garden soil and a medium dose of complete fertilizer are required for poorer soils. A balanced fertilizer like N:P:K at 20:20:20 can be applied in split doses. Good nutrition results in more number of spikes. Over watering, low calcium/high calcium-inhibiting magnesium, low phosphorus, or low magnesium can all lead to poor root and root hair development in Scarlet sage (Koranski *et al.*, 1996). For high-quality salvia plants, the optimum leaf tissue nutrient levels are (in percent) N 3.0 to 4.5; P 0.3 to 0.7; K 3.5 to 5.0; Ca 1.5 to 2.5 and Mg 0.3 to 0.6 (values based on most recent fully developed leaf) (Nelson, 1994).

10.4.0 Aftercare and Management Practices

In rich soils, staking is necessary to prevent the blooms from lodging. Removal of completely bloomed and senescent spikes encourage additional and prolonged flowering. After one season the plants are pruned to one third height. Withered spikes should be removed in time. Liquid feeding or top dressing with FYM is necessary when plants are seen exhausted.

11.0 Diseases and Pests

Salvia species are generally free of pest problems in the landscape. In greenhouse cultivation as cut flower Salvias are susceptible to many common disease and insect problems including aphids, whiteflies, spider mites, common grey mould, root rot and damping off. Some cultivars of scarlet sage were slightly susceptible to root-knot nematodes. Downy mildew and rust have reported on some species.

11.1.0 Diseases

11.1.1 Leaf Spot

Many fungi cause leaf spots on different hosts. Spots may vary from small discrete dots and raised areas to irregular yellow or brownish patches that cover much of the leaf surface. Leaves may fall off the plant if the problem is severe, but these pathogens rarely cause long-term damage to plants. Remove fallen leaves and debris promptly. Many of the pathogens are favored by moisture, so avoid overhead sprinklers and irrigate early in the day so that the foliage dries more quickly.

11.1.2 Powdery Mildew

This disease is more prevalent in humid weather conditions and infects the top of the leaves. The leaves appear to have a whitish or greyish surface and may curl. Most powdery mildew fungi grow as thin layers of mycelium on the surface of the affected plant parts. Spores, which you can see with a hand lens, are part of the white, powdery appearance of this fungus and are produced in chains on upper or lower leaf surfaces or on flowers, fruits or herbaceous stems. The best method of control is prevention. Biological fungicides (*e.g.*, Serenade) are commercially available beneficial microorganisms formulated into a product that, when sprayed on the plant, destroys fungal pathogens. The active ingredient in Serenade is a bacterium, *Bacillus subtilis*, that helps prevent the powdery mildew. Sulphur based fungicides are effective against this disease. Myclobutanil (Immunox) is also effective as an eradicant and protectant against both powdery mildew and rust.

11.1.3 Rust

Infected foliage may become spotted, turn yellow or brown overall and drop prematurely. Infected plants may develop bushy growth (*e.g.*, witches' brooms), cankered, galled and oozing bark and limbs and spotted needles. Avoid overhead watering. Removing nearby alternate hosts of the fungus may help to reduce new infections in certain situations.

11.2.0 Pests

11.2.1 Aphids

Aphids (*Aphis passeriniana*) pierce stems, leaves and other tender plant parts and suck out fluids. Large populations can turn leaves yellow and stunt shoots; aphids also produce large quantities of a sticky exudate known as honeydew, which often turns black with the growth of a sooty mold fungus. Some aphid species inject a

Diseases and Pests of Salvia

Downy Mildew | Leaf Spot | Stem Rot

Foliar Nematode | *Aphis passeriniana* | Chrysolina Larvae

Dysmicoccus angustifrons | *Eucarazzia elegans* | *Eupteryx gyaurdagicus*

Thrips tabaci | Leaf Hopper | Mites

Source: www.pnwhandbooks.org, www.apsnet.org, www.mrec.ifas.ufl.edu, www.apps.extension.umn.edu, Zarkani and Turanli (2019).

toxin into plants, which causes leaves to curl and further distorts growth. Natural enemies can be very important for controlling aphids. Among the most important natural enemies are various species of parasitic wasps that lay their eggs inside aphids. Many predators also feed on aphids. The most well-known are lady beetle adults and larvae, lacewing larvae, soldier beetles and syrphid fly larvae. The tips of young plants where aphids are concentrated should be pinched off to get rid of the aphids. The affected plants can be sprayed with a pyrethrin based insecticides.

11.2.2 Spider Mites

Red spider mite infestation is seen on the salvia plants. The tops of lower leaves on affected plants become stippled with tiny yellow dots or red spots. Leaves, stalks and adjacent stems may become distorted or swathed in fine webbing, while foliage becomes discolored and deformed as these pests suck juices from plant cells. Mites typically attack plants that are under stress. Sometimes their populations explode when frequent use of broad spectrum insecticides in the area have killed the resident beneficial insects that normally.

11.2.3 Thrips

Myoporum laetum and *M. pacificum* feed on soft parts of plants, stunt plant growth and cause damaged leaves to become papery and distorted, develop tiny pale spots (stippling) and drop prematurely. Infested terminals may discolour and become rolled. Petals may exhibit "color break" which is pale or dark discolouring of petal tissue that are killed by thrips feeding before buds opened. Predatory thrips, green lacewings, minute pirate bugs, mites and certain parasitic wasps help to control plant-feeding thrips. Silver or gray synthetic reflective mulch/mesh are very effective in controlling thrips. Contact insecticides that do not leave persistent residues can be effective for greenhouse thrips and other species that feed openly on plants. Neonicotinoids vary in effectiveness for thrips control.

11.2.4 Whiteflies

These are small white flying insects clearly visible on the undersides of salvia leaves. Nymphs and adults of this insect suck juices from plant leaves, buds and stems weakening the plant. As the infestation worsens, the plants themselves weaken and the leaves can begin to turn yellow and die. The undersides of the leaves may be covered with shiny, sticky honeydew secreted by the whiteflies, and a black fungus that is encouraged by the honeydew may appear. Affected plants will become stunted and pale. Spray on visible whiteflies with insecticidal soap as directed on the label.

11.2.5 Slugs or Snails

Slugs and snails emerge at night, attracted to moist, well-mulched gardens and acidic soil. Although these pests are always most destructive in shaded gardens and during rainy spells, they may attack red salvia in a sunny bed. They make holes with their file-like tongues in a wide variety of plants and disappear in daylight, often leaving behind a trail of silvery mucous. They hide under boards or leaf litter during the day. The slugs can be controlled by trapping them early in the season when they first appear. The commercial slug traps can be set up near the salvia beds. If slugs persist, sprinkle Diatomaceous earth (DE) or put slug barrier on the soil around each salvia plant to discourage them.

12.0 Harvesting and Postharvest Management

12.1.0 Harvesting

Scarlet sage requires 5 to 6 months from sowing to flowering. Mealy cup sage reaches to marketable size in 5 months from seed. Salvia blooms continuously for a long duration after about 2 to 2½ months of planting (Armitage, 1994). Mexican bush sage and Mealy cup sage can be grown in the greenhouse or in the field for the specialty cut flower market. Inflorescences should be harvested when white flower petals appear on the lower three to four individual flowers.

12.2.0 Herbage Yield

In the growing season, salvia herbage can be harvested twice at two cutting heights at various stages of development, the first harvest on nine dates at one- or two-week intervals in the spring and the second on a single date in the summer. The yield of crop depended on plant growing phase and was 3.0–10.0 t/ha of fresh and 0.5–2.7 t/ha of dried herb.

12.3.0 Postharvest Management

The cut stems last about 7 days when floral preservatives are used. A Mexican bush sage spike also makes good dried flowers. Individual flowers will shatter if allowed to wilt (Armitage, 1993).

Although scarlet sage cannot withstand freezing temperatures, optimum temperature for storage and transport of flowers ranges from 7 to 13°C (Heins *et al.*, 1994). However, reddening of foliage may occur below 10°C (Armitage, 1994). Perennial salvia (*Salvia* × *superba*), an herbaceous perennial of garden origin, is a cold hardy plant able to withstand temperatures upto (–10 °C).

13.0 Biotechnology

With the advancement of plant transformation technology, the genetic modification of crop plants has been used to improve many varieties, but only a few examples of Salvia species could be found. Experiments with activation tagging mutagenesis (ATM) showed that transgenic *Salvia miltiorrhiza* Bunge plants, obtained after transformation with pTAG8 binary vector, caring a viral CaMV35S enhancer and hptII (hygromycin phosphotransferase) as selectable marker, had increased leaves size with bigger guard cells, stomata and trichomes (Ho *et al.*, 2013). These transgenic plants accumulated 3.7 times more tanshinone I and 2 times more tanshinone IIA than the control plants. However, the biggest disadvantage of ATM remains the unpredictable result after transformation, since the Agrobacterium tumefaciens introduce its T-DNA randomly into the plant genome. This obstacle could be overcome by cloning and overexpression of entire key gene into the transgenic plant. Wu *et al.* (2014) demonstrated, that cloning of the gene, responsible for biosynthesis of late embryogenesis abundant (LEA) proteins in *Salvia miltiorrhiza* Bunge and its overexpression under the control of CaMV35S promoter in transgenic *S. miltiorrhiza* plants could significantly increase the plant tolerance to abiotic stress as increased salinity and drought.

The advanced methods of genetic and metabolic engineering, synthetic biology and combinatorial biosynthesis can be used as powerful tools to overcome any obstacles limiting the accumulation of biologically active phytochemicals and thus, to improve significantly the crops yields. The transcriptome profiling of different species of *Salvia*, has helped in the identification of functionally important genes, their role in regulating a specific pathway. In *Salvia miltiorrhiza* Bunge, the genes regulating the expression of tanshinones (have medicinal importance) were identified through transcriptome profiling (Chang *et al.*, 2019).Through illumina sequencing, genes involved in volatile tarpenoid biosynthesis is being identified in *S. officinalis* (Ali *et al.*, 2017). A transcriptomics study carried out by Wang *et al.* (2019) in *Salvia miltiorrhiza* demonstrated that phenylalanine ammonialyase (PAL) is involved in rosmarinic acid and salvianolic acid B biosynthesis and flavone synthase (FLS) is an essential enzyme catalysing flavonoid biosynthesis measures.

14.0 References

Alberto, C.M., Sanso, A.M. and Xifreda, C.C. (2003) *Bot. J. Linn. Soc.*, **141**: 483 490.

Ali, M., Li, P., She, G., Chen, D., Wan, X. and Zhao, J. (2017) *Sci. rep.*, **7**(1): 1-21.

Armitage, A.M. (1994) *Ornamental bedding plants*. CAB International, Wallingford, United Kingdom

Barrett, J.E. and Erwin, J.E. (1994) *Height control. In*: E.J. Holcomb (ed.). Bedding plants IV. Ball Publ., Batavia, Ill.

Bernath, J.E.N.O. and Nemeth, E. (2000) *Kintzios, SE, Sage-The genus Salvia*. Harwood Academic Publisher, pp.109-124.

Boring, J.K., Glasener, E., Keator, G., Knopf, J., Scott, J. and Wasowski, S. (1995) *Natural gardening*. Time-Life Books, New York.

Carey, D. J., Fair, B. A., Buhler, W. *et al.* (2013) *J. Appl. Hort.*, **15**: 87-89.

Carlson, W. (1978) *Amer. Veg. Grower*, **26**: 20–21.

Carpenter, W.J., Ostmark, E.R. and Cornell, J.A. (1995) *Hort Science*, **30**: 1003–1006.

Chang, Y., Wang, M., Li, J. and Lu, S. (2019) *Sci. Rep.*, **9**(1): 1-12.

Clausen, R.R. and Ekstrom, N.H. (1989) *Perennials for American gardens*. Random House, New York.

Clebsch, B. (1997) *A book of salvias, sages for every garden*. Timber Press, Portland, Ore.

Dogan, H., Eroglu, H.E., Senkal, B.C., Uskutoðlu, C.C.T., Özkan, U. and Altay, D. (2019) *Acta Biologica Turcica*, **32**(2): 74-77.

Dole, J.M. and Wilkins, H.F. (1999) *Floriculture: Principles and species*. Prentice Hall, Inc., Upper Saddle River, N.J.

Gedik, O., Kýran, Y., Emre, I. and Kursat, M. (2016) *Cytologia*, **81**(4): 395-402.

Goldblatt, P. and Johnson, D.E. (1996) *Monogr. Syst. Bot. Mo. Bot. Garden*, **58**: 134.

Gomez, P. S., Cano, M. S., Sánchez, J. S. *et al.* (1995) *J. Essent. Oil Res.*, **7**(3): 317-318.

Griffiths, M. (1994) *Index of garden plants*. Timber Press, Portland, Ore

Harlan, J.R. and DeWet, J.M.J. (1971) *Taxon*, **20**: 509 517.

Harley, R.M. and Heywood, C.A. (1992) *In*: Harley, R.M. and Reynolds, T. (Eds.). *Advances in labiate science*, Royal Botanic Gardens Kew, UK. pp. 211-246.

Heins, R., Lange, N., Wallace, Jr.T.F. and Carlson, W. (1994) *Plug storage. Greenhouse grower*. Meister Publishing, Willoughby, Ohio.

Ho, H.S., Vishwakarma, R.K., Chen, E.C-F *et al.* (2013) *Botanical Studies*, **54**: 1-8.

http: //ipm.ucanr.edu/PMG/GARDEN/PLANTS/DISEASES/rusts.html

http: //ipm.ucanr.edu/PMG/PESTNOTES/pn7404.html

https: //www.burpee.com/gardenadvicecenter/encyclopedia/annual-flowers/learn-about-salvias/encyclopedia_Salvia-article.html#howToCommonPests

Koranski, D., Kessler, R., Khademi, M. and McNertney, D. (1996) *Troubleshooting: Diagnose your plug problems*. In: D. Hamrick (Ed.). Grower talks on plugs II. Ball Publ., Batavia, Ill

Liberty Hyde Bailey Hortorium. (1976) *Hortus third: A concise dictionary of plants cultivated in the United States and Canada*. 3rd ed. Macmillan, New York

Liv, B.L., Fan, Z.B., Liu, Z.Q., Qiu, X.H. and Jiang, Y.H. (2018) *J. Amer. Soc. Hort. Sci.*, **53**(7): 1038-1044.

Louis, B.A. (2000) *Hort Technology*, **10**(4) : 800-805

Ma, X. H., Ma, Y., Tang, J. F., *et al.* (2015) *Molecules*, **20**(9): 16235-16254.

Mabberly, D.J. (2008) *Mabberly's plant book* 3rd Ed., Cambridge University Press Cambridge, UK.

Nau, J. (1993) *Ball culture guide*. 2nd Ed. Ball Publ. Batavia, Ill.

Nelson, P.V. (1994) *Fertilization*. In: E.J. Holcomb (ed.) Bedding plants IV. Ball Publ., Batavia, Ill, pp. 151– 175.

Povh, J.A. and Ono, E.O. (2006) *Acta Sci. Biol. Sci.* **28**(3): 189-193.

Povh, J.Á. and Ono, E.O. (2007) *Publ. UEPG Biol. Health. Sci.*, **13**(1/2): 7-10.

Putievsky, E., Ravid, U., Diwan-Rinzler, N. and Zohary D. (1990) *Flavour Fragrance J.* **5:** 121-123.

Sharma, S., Shahzad, A., Kumar, J. and Anis, M. (2014) *Rend. Fis. Acc. Lincei*, **25**: 359–368.

Shoemaker, C.A. and Carlson, W.H. (1990) *Hort Science*, **25**: 762– 764.

Singh, V., Sood, R., Ramesh, K., and Singh, B. (2008) *J. Herbs Spices Med. Plants*, **14**(1-2): 29-36.

Stearn, W.T. (1992) *Stearn's dictionary of plant names for gardeners*. Cassell Publ. Ltd., London.

Sunar, S., Korkmaz, M., Sigmaz, B. and Agar, G. (2020) *Turk J Pharm Sci*, **17**(5): 480-485.

Sybenga, J. (1992) *Cytogenetics in plant breeding*, Springer-Verlag Berlin, Germany.

Tychonievich, J. and Warner, R. M. (2011) *J. Amer. Soc. Hort. Sci.*, **136**(1): 41-47.

Vernieri, P., Mugnai, S. and Tognoni, F. (2002) *In*: *VI International Symposium on Protected Cultivation in Mild Winter Climate: Product and Process Innovation*, **614:** 307-312.

Wang, R., Jiang, H., Zhou, Z., Guo, H. and Dong, J. (2019) *BMC genomics*, **20**(1): 1-14.

Wang. (2010) *Genetica*, **138**(2): 241-249.

Wasowski, S. and Wasowski, A. (1997) *Native Texas plants: Landscaping region by region*. Gulf Pub. Co., Houston, Texas.

Wu, Y., Liu, C., Kuang, J., Ge, Q., Zhang, Y. and Wang, Z. (2014) *Protoplasma*, **251**(5): 1191–1199.

Xu, Z., Luo, H., Ji, A., Zhang, X., Song, J. and Chen, S. (2016) *Front. Plant Sci.* **7**: 100.

Zarkani, A. and Turanli, F. (2019) *Yyu. J. Agr. Sci.*, **29**(1): 34-42.

Stock

Anamika Gurung, Rajiv Kumar, Parvathi Bennurmath, K. Dutta, J.S. Arora and B. Biswas

1.0 Introduction

Stock (*Matthiola incana*) is an annual or biennial plant producing flowers in terminal clusters. Both single and double forms exist. In Europe and USA, it is cultivated for the production of cut flowers. In India, its cultivation is very limited and grown in open field or pots. The earliest record of *M. incana* describes it as a hoary biennial with purple and white flowers with blackish green seeds (Saunders, 1928).

2.0 Importance and Uses

2.1.0 Aesthetic Uses

Stock is a very showy and popular garden flower. In European countries and America it is extensively grown for cut flowers. It is also grown in herbaceous borders and pots. Flowers of stock are well arranged on a long column and the colour is most variable running from white through rose, crimson, purple, yellow, mauve, pink, *etc.* It is marketed with 14 commercial flower colors ranged from pure white to lavender, purple, blue, red, and pale yellow (Dole and Wilkins, 2005). With all these, stock proves to be a valuable plant material in any garden planning. It is regarded as an important ornamental plant due to its pleasant aroma and beautiful form (Sanchez *et al.*, 2005).

2.2.0 Therapeutic and other Uses

In China, its flowers are edible, sold in health food stores and consumed in the form of tea (Jin *et al.*, 2016). Hammami *et al.* (2006) isolated and identified three phenolic glycosides from the flowers of *M. longipetala*, these are 4-O-β-D-glycopyranosyl zingerone, 4-O-β-D-glycopyranosylhomovanillyl alcohol and eugenol glycoside. The chemical composition of the volatile parts of *M. longipetala* flowers were analyzed by Hammami *et al.* (2007), most representative compounds found were eugenol (19.93 per cent), bicyclogermacrene (13.60 per cent), heptacosane (7.85 per cent) and tetradecanoic acid (5.57 per cent).

Taviano *et al.* (2020) detected potential pharmacological applications by evaluating the hydroalcoholic extract from aerial parts of *M. incana*. The extract was neuroprotective inhibitor of central nervous system (CNS) enzymes such as monoamine oxidase A, tyrosinase, acetylcholinesterase, and as a natural enzyme inhibitor of α-glucosidase and lipase involved in some metabolic disorders such as obesity or type 2 diabetes. The extract also displayed moderate to low cytotoxicity against the human colorectal adenocarcinoma (CaCo-2) cells. The extract acted as a superoxide radical scavenger and enzymatic inhibitor of monoamine oxidase A, tyrosinase, α-glucosidase, and lipase.

3.0 Origin and History

In the wild state, stocks are found in the Mediterranean, Egypt, South Europe, and in South Africa and two species, *Matthiola incana* and *M. sinuata*, are among

Uses of Stock

As Cut Flower | In Pot | In Vase

Flower Arrangement | In Bouquet | In Container

Window Box | In Flower Bed | Garden Basket

Flower Show | Border Plant | Rock Garden

Source: www.plantsofdistinction.co.uk, www.in.pinterest.com, www.blog.flowersexpress.ph, www.ballseed.com, www.gardensonline.com.au, www.maltawildplants.com

the rare British natives. Presently, this has been naturalised in many parts of the world, particularly the southern parts of California. Earlier, stock was called as *Leucoium album* or *Viola alba* but later on the genus was renamed as *Matthiola* after P.A. Matthioli, an Italian physician and writer on plants (Coats, 1968).

This was being grown by the ancient Greeks and was liked for its fragrance. In Europe, it started growing in the 16th century and both single and double stocks in many colours were common in gardens as described by Lyte in 1578. John Tradescent purchased many plants in France for Lord Salisbury in 1611 and double white stock was one of them. These all show the popularity of stock amongst European gardeners. Sowing of stock seed mixed with radish seed in order to distract flea beetle was a common practice. It was also a standard practice that only double stocks were allowed in the garden and singles were used to grow in nursery for seed production from which doubles were raised. The identification of double and single flowered plants was done on the basis of root system, and it is presumed that single flowered plant has tap root system whereas double flowered plant has fibrous root system. Hansen, a Danish seedsman, developed a strain in which all the double flowered plants have pale green leaves and single flower plants have dark green leaves so that double plants are easily selected in nursery stage (Coats, 1968). An algorithm based on the fuzzy evaluation theory was developed to discriminate between double and single flowering stock *cv.* Sourei in the seedling stage. The accuracy of discrimination was 90.8 per cent. Three basic characteristics (area, degree of complexity and saturation of the cotyledon) of the seedling image were measured by a colour image processing system to fix the membership function of fuzzy sets (Koku *et al.*, 1999). From Europe, stock spread to different parts of the world by British Settlers.

4.0 Species and Cultivars

4.1.0 Species

Stock belongs to family *Brassicaceae.* There are about 55 species of the genus *Matthiola* (Brickell, 1996). Important species are given here.

Matthiola fruticulosa

Dwarf, tufted, base woody, up to 60 cm height; leaves simple or pinnatifid, linear to oblong, woolly; racemes upright, long, flowers 1.5-3.0 cm across, yellow to purplish violet. *M. fruticulosa* subsp. *valesiaca*: tufted, produce runners; racemes dense, flowers mauve-purple to red-purple. It is found in calcareous rocky habitats (screes, rocky pastures and shrublands) (Šegota *et al.*, 2016).

M. incana (Gilly flower)

Annual (10-14 weeks) as well as biennial, up to 80 cm height; leaves entire, or lobed, 5-10 cm long, inversely lance-shaped to linear lance-shaped, hairy; racemes upright, flowers 2.5 cm across, mauve, purple, violet, pink and white; fragrant. Large number of cultivars of *M. incana* developed from crosses between *M. Incana* × *M. sinuata.* The important cultivars are: Cinderella series– plants 20-25 cm tall; racemes 15 cm long, flowers double, dark, blue-purple, red, rose-pink, silvery blue

Species of Stock

Matthiola fruticulosa | *Matthiola incana* | *Matthiola sinuata*

Matthiola longipetala | *Matthiola odoratissima* | *Matthiola fruticulosa*

Matthiola tricuspidata | *Matthiola vallesiaca* | *Matthiola fragrans*

Matthiola parviflora | *Matthiola crassifolia* | *Matthiola maroccana*

Source: www.species.wikimedia.org, www.alanbuckingham.photoshelter.com, www.first-nature.com, www.plantarium.ru, www.plantsoftheworldonline.org, www.flickr.com

Cultivars of Stock

Ruby Punch, Apricot Punch, Lavender Blue Punch

Turkish Delight, White Punch, Lemon Punch

Cherry Blossom Punch, Rainbow Quartet, Tosca Blue Sea

Aida Apricot F_1, Aida Purple F_1, Tosca White F_1

Source: www.plantsofdistinction.co.uk

and white. Excelsior Mammoth Column Series – plants 75 cm height; racemes 30 cm long, flowers double, pink, red, pale blue and white. Legacy Series – plants up to 30-45 cm long; panicles 30 cm; flowers double, scarlet-red, rose-pink, lavender-blue, white and creamy yellow. Midget Series – plant up to 75 cm tall; racemes 30 cm long, flowers double, white, pink, caramine-red and light to dark blue. Ten Week Mixed – dwarf, 30 cm height, bushy; racemes 15 cm long, flowers double, crimson, pink, lavender-pink, purple and white.

M. incana cultivars are divided by horticulturist into 4 groups : Brompton stocks–biennial; panicle tall, flowers single or double; East Lothian – annual or biennial; panicle compact, flowers single or double, flowers smaller than Brompton group; Ten Week Stocks – annual, dwarf or tall; bear single or double flowers, dwarf cultivar suitable for bredding; tall *cvs.* racemes dense, spike-like, unbranched, flowers double. Column stocks – racemes long, dense, upright; flowers double; suitable for cut flower.

M. longipetala subsp. *Bicornis* syn. *M. bicornis*

Annual, erect to spreading, single stem or branched; leaves narrowly linear, 8 cm long; racemes open, flowers 2 cm across, pink, mauve and purple; strong fragrant at night. It has very high γ-linolenic acids (GLA) content (Karaman *et al.*, 2011).

M. odoratissima

Plant evergreen, sub-shrubby, up to 30-60 cm; leaves hairy, toothed; flowers creamy white, fading to purplish brown; sweet scented in the evening.

M. sinuata

Plant erect, branched; leaves oblong, hairy, sinuate; stigma compressed, velvety. It is a short-lived (biennial) species typical of the sand beach drift lines (Magrini *et al.*, 2019).

M. tristis

Plant perennial, woody base; leaves linear, entire or toothed, hairy; flowers creamy white.

Other species are *M. tricuspidata*, *M. vallesiaca*, *M. pedemontana* and *M. fenestralis*. The characteristics of 64 cultivars belonging to 8 races of *Matthiola* were studied by Loeser (1986). The dark blue form of Buchholzer Riesen Treib had the shortest panicles while the pink form of Super Giant had the longest panicles; cultivars Royal Treib and Treib Wunder flowered earliest (20 May) with Buchholzer Riesen Treib (12 June); Excelsior Mammut was the tallest (81 cm) and Royal Trieb the shortest (46 cm).

5.0 Genetics and Breeding

5.1.0 Genetics

The genetics of stock has been reviewed by Crane and Lawrence (1956). Stock is an old and popular flower in European gardens. Earlier, stock (*M. incana*) was a hairy, bushy plant with purple flowers and blackish green seeds and these characters have been changed due to gene mutation. The change in various forms was recorded by Saunders (1928) as follows:

Year of Record	*Original*	*Change in Character*	*Mutant Gene*
1536	*CR*-purple	Only two forms purple or white	*cR* or *Cr* (white)
1542	*B*-blue	Red flowers	*b*-red
1563-68	*S*-single	Double flowers	*s*-double
1588		Shades of purple	
1629	*W*-white	Yellow plastids	*w*-cream
1660		Branching - Unbranched single spike	
1762		Hoary - half hoary	
1861		Glabrous	

Inheritance and linkage relationships of a leaf morphology gene of *M. incana* was investigated by Ecker *et al.* (1994a). The allele for sinuate leaf shape, c, was found to be recessive to the allele for normal entire leaf, C. The c allele was tightly linked to the recessive allele for double flowering, s. The recombination frequency between the two loci was close to zero. The mode of inheritance of the C gene was in accordance with the hypothesis that a pollen lethal gene is responsible for the constant 1:1 segregation ratio of double-flowered (= male sterile) to single-flowered (= fertile) plants in most *M. incana* breeding lines. The sinuate leaf allele seemed to reduce the frequency and delayed the flowering of double-flowered plants. Ecker *et al.* (1994b) also studied the inheritance of flowering time (FT) in a cross between 70 early flowering (P1) and 120 late-flowering (P2) genotypes. The distribution of FT in F_1 and BC_1 generations indicated additive genetic control with earliness alleles partially dominant, particularly in double-flowered plants. Single-flowered plants flowered earlier than double-flowered when averaged within both parental lines. However, the mean difference between singles and doubles was 16 days for P2. This flower-doubleness related delay in FT was found to be heritable.

Ardelean *et al.* (2009) studied the genetic effects involved in the heritability of several quantitative characters of ornamental value in six cultivars of *Mathiola incana* belonging to series Giant Imperial (White, Yellow, Lavender, Violet, Crimson and Rose). These cultivars were used in a cyclic crossing system with Giant Imperial White as tester parent and F_1 and F_2 hybrids as well as BCP_1 and BCP_2 backcross generations were produced and grown on individual plots with 92-166 plants/plot. 50 plants with double flowers were randomly taken from each plot and were analyzed for plant height, number of inflorescences/plant and length of inflorescence. It is noticed that the additive effects were the largest component of the total genetic effects involved in the inheritance characters like number of inflorescences/plant and length of inflorescence while for plant height the dominance effects were slightly higher than the additive ones but with very small differences.

5.1.1 Flower Colour

It has been reported by Saunders (1928) that gene *C* and *R* are both necessary for the production of anthocyanin in the flowers. Other than these *A* and *A′* are also essential for the production of anthocyanin even though *C* and *R* are present. *B* gene is responsible for the conversion of red anthocyanin into blue whereas *V* gene governs purity and *v* impurity of colours. Thus, the various colours are obtained by combinations of four genes. Important colours are as follows:

CRBV	Purple
CRBv	Plum
CRbV	Red
CRbv	Copper
All other genotypes	White

It has also been reported that dilute colours like rose, pale mauve, azure and flesh are dominant, respectively, to the deep colours like carmine, purple and terracota. There are also white, sulphur-white and cream colours and these colours are due to incomplete dominance of single gene effect, *i.e.*, white (*W*) which is dominant over cream (*w*) and intermediate is sulphur-white.

An experimental system has been developed of all possible genotype combinations of 4 loci involved in the modification of the anthocyanin molecule structure by oxidation, glycosidation and acylation in *M. incana*. These are varied against a background which is nearly isogenic with respect to the flower colour (Seyffert, 1971). Jana and Seyffert (1971) found dominance to the most important type of gene action in the inheritance of anthocyanin content in the flower tissue of *M. incana*. There was considerable epistasis, but effect was very unstable over the years and genetic background. The relative magnitude of the additive effect was the most stable. Heterosis was observed and was found to be largely due to dominance and additive × dominance interaction.

Petals of genetically defined lines having different coloured petals of *M. incana* were studied by Teusch (1986). All genotypes contained a xylosyl transferase which catalyzed the transfer of the xyloslymoity of uridine 5-diphosphate-xylose to the glucose of cyanidin-3-glucoside. No significant difference in enzyme activity was detected amongst the lines that were either dominant or recessive for gene *u*, indicating that this gene is not responsible for xylosyl transferase activity. Accumulation of cyanidin-3-glucosides during flower development was correlated with xylosyl transferase activity. In an investigation, the expression of the key enzyme of flavonoid biosynthesis, chalcone synthase (CHS) was followed in different genotypes which are generally defined with respect to anthocyanin production. During flower development of a line with coloured petals (line 07) a clear pattern of CHS enzyme production could be observed, with a large increase in the amount of CHS enzyme protein in the petals occurring at the stage of bud opening. These results differed from results obtained with petals of the white flowering mutant line 18 in which a reduced and nearly constant level of CHS enzyme protein was observed during flower development (Rall and Hembleben,

1984). Saito *et al.* (1996) observed that the acylated pelargonidin 3-sambubioside-5-glucosides were present as the dominant pigments in red-purple, salmon-pink and apricot cultivars. Pelargonidin 3-glucoside was the dominant anthocyanin in copper-coloured flowers. The simple recessive factor for male sterility in stocks showed linkage with that for opaque flower colours. The other simple recessive factor investigated with reduced vitality and a linkage was suspected between it and red colour (Schnack and Fehleisen, 1954).

Anthocyanins, particularly pelargonidin and cyanidin derivatives, are responsible for the flower color, but delphinidin derivatives are not detected in commercial cultivars of *M. incana*. The flower colour diversity in cultivars is thought to be caused by a slightly different modification of the major anthocyanins derived from pelargonidin or cyanidin and differences in anthocyanin levels (Tatsuzawa *et al.*, 2012).

Hemleben *et al.* (2004) reported three white-flowered mutant lines, which were influenced by *e*, *f*, and *g* genes. *F* encodes a chalcone synthase (CHS). The white-flowered and glabrous *e* mutant exhibits one nucleotide change, leading to a direct amino acid substitution in the WD motif (W158R) of TRANSPARENT TESTA GLABRA1 TTG1 (Dressel and Hemleben 2009). Transient expression of two basic helix–loop–helix (bHLH) genes, *i.e.* MYC146 and GLABRA 3 (GL3), derived from *Arabidopsis thaliana*, can activate anthocyanin biosynthesis in white-flowered g mutant (Ramsay *et al.*, 2003).

Chen *et al.* (2018) found differentially expressed genes (DEGs) among three flower colour varieties *viz.*, white, light purple, rose bengal. Successfully identified CHS, chalcone isomerase (CHI), favanone 3-hydroxylase (F3H), anthocyanidin synthase (ANS), UGT78D2, and favonol synthase 1 (FLS1) as DEGs of flavonoid biosynthetic genes.

Nuraini *et al.* (2020) expressed that *MiMYB1* and *MibHLH2* genes play key roles in anthocyanin biosynthesis in flower.

5.1.2 Single *vs* Double Flowers

Stock plants are of following types according to production of single and double flowers. Single flowers have 4 petals, 4 stamens and a pistil having 30-60 ovules whereas double flowers have 40-70 petals and male and female sterile.

1. Singles – which produce only single flowers.
2. Singles – which produce 75 per cent single and 25 per cent double flowers.
3. Eversporting singles – which produce 54-56 per cent double flowers and 44-46 per cent singles.
4. Doubles – flowers are double and do not have ability to produce seeds due to complete sterility, *i.e.*, without male and female reproductive organs.

There are two types of single races which have 7 basic number of chromosomes. Homozygous singles breed true and give rise to only singles whereas heterozygous follows the Mendelian behaviour of complete dominance and on selfing gives rise to 3 singles and 1 double. The gene which is responsible for single flower has been designated as *S* and its recessive allele, *i.e.*, *s* for double. Eversporting singles which give rise to 54-56 per cent double flowers is due to the fact that eversporting transmits double character only due to occurrence of pollen lethal of singleness whereas female germ cell carries normal gene for single and doubleness. This process is similar to back cross a normal heterozygote to a pure recessive, *i.e.*, *IS* : *Is* × *s* which gives rise to *ISs* and *Iss* in ratio of 50 : 50 approximately. This phenomenon was explained by Philip and Huskins (1931) who made cytological studies and observed that in eversporting singles, one of the chromosomes carrying the genes for singleness lacked a trabant (*i.e.*, small terminal portion of chromatio) which was present in pure singles. The recessive *s* is borne on the normal chromosome. The pollen without trabant is lethal and incapable of fertilization. Therefore, all the viable pollens carry gene *s* for doubleness. On the female side, however, gametes without the trabant are practically viable and only 3-4 per cent fails to take part in fertilization and, hence, percentage, instead of 50, increases to 53-54 for double flowers.

From the reciprocal crosses between eversporting and true breeding single flowered stocks, Johnson (1953) observed that *S* allele is transmitted through eggs but not through the pollen. The altered ratio is due to non-functioning of the pollen carrying *S* allele; however, *S* allele in eversporting single is apparently linked with pollen lethal. This linkage is not complete and is broken by unequal crossing over. Lesley (1973) gave the concept of unstable *S* chromosome and observed crossing over between a very small fragment and a whole chromosome in stock in a progeny of 41 double flower. Fehleisen and Schnack (1972) described a method based on the crossing of line A which carries a dominant gene for single flowers and B which carries paired recessive genes for double flowers. Marker genes influencing plant pigmentation are linked to these genes. Progeny from crosses include yellowish and green seedlings and 93 per cent of the latter yielded double flowers. In the original method double flowers were produced on somewhat inferior yellowish plants.

Irani and Arab (2017) selected eight stock cultivars namely, Nobel, Cinderella, Pacific, Avalanche, Midblue, Lavender, Goddess and Esfahan, with different percentage of double-flowered plants to examine the relationship with three morphological types of cotyledons; heart-shaped cotyledon (HC), cup-shaped cotyledon (CC) and Dumbbell-shaped cotyledon (DC). The seedlings with heart-shaped and cup-shaped cotyledons produced plants with double-flowered in heart shaped (HC) and cup-shaped cotyledon (CC) populations, the number of plants with double flowers was much more than that of single flowers and CC seedlings rarely produced single flowers. Therefore, increasing the number of CC seedlings can improve the percentage of double flowers. Aneuploid cells were found in cultivars Nobel and Goddess, which showed the highest percentage of CC seedling. Based on morphological measurements, the highest value of inflorescence size was observed in the seedlings with cup-shaped cotyledons.

Although the double-flower trait of *M. incana* corresponds to the *s* locus, the gene responsible for double flowers was not been identified. Therefore, to obtain double-flowered plants, breeders have relied on the correlation between double flowers and several morphological traits, such as cotyledon shape, serrate leaf, seed color, and leaf color when selecting *M. incana* plants (Ecker *et al.*, 1993). However, this form of selection is extremely labour intensive. Nakatsuka and Koishi (2018) identified the deletion mutation alleles of *M. incana* AGAMOUS (MiAG), which have been known as doubled flower s locus in genetic analysis.

Twenty cultivars of *Matthiola incana* growing in various areas with different percentages of double flowers were chromosomally investigated by Irani *et al.* (2016). Distinct somatic chromosome numbers, karyotype parameters and asymmetry indices, idiograms were observed. The results collectively showed great genetic variability between cultivars in cytological features and chromosome number that can be considered as a significant gene pool for stock breeding programs. The cultivars with additional chromosomes can lead to genetic diversity in *M. incana* as they can create new chromosome numbers and differentiation of cultivars.

5.1.3 Trisomic

It is a desire of floriculturists to produce 100 per cent double flowers and the method which was suggested by Frost (1928) solved this problem. He observed that a certain form of eversporting single 'Snow Flake' characterized by slender leaves had an extra segment of chromosome, *i.e.*, 14+1 which carried the *Ss* genes. On selfing, 37 per cent population, being weak and lacking decorative value, was of no commercial use whereas 90 per cent of remaining plants were double. It was suggested that to get the high percentage of double flowering plants, seeds must be collected from slender types and not from normal singles. Irani *et al.* (2016) observed aneuploidy in two cultivars with high percentage of double flowers. An extra chromosome was observed as a rod univalent or chain trivalent in PMCs at diakinesis stage.

5.1.4 Hoariness *vs* Glabrous

Hoariness character of stock leaves is polygenic in nature and is governed by three pairs of genes, *i.e., Cc, Rr* and *Kk* and three multiple allelomorphs, *i.e., H, H1* and *h*. The degree of hoariness is regulated by the presence of dominant genes and any combination of *C, R, K* and *H* (or *H1*) is hoary, *H1H1KK* is half-hoary, *H1H1Kk* is mid-hoary and *H1h1KK* is quarter-hoary. All others are glabrous. Other characters which have been found dominant are tallness, branching, glandless, blue embryo colour and brown seed coat colour over dwarf habit, unbranched, glands on leaves, yellow embryo colour and light yellow seed coat, respectively.

5.2.0 Breeding

5.2.1 Disease Resistance

Interspecific F_1 hybrid plants resulting from the cross *M. incana* × *M. tricuspidata* exhibited high resistance to the bacterial blight (*Xanthomonas campestris* pv. *incanae*) (Ecker *et al.*, 1995).

6.0 Growth and Flowering

6.1.0 Environmental Factors

Growth and flowering in stock are influenced by photoperiod as well as by temperature. Genetic variation is observed at flowering time, which could be attributed to differential sensitivity to low temperature and day length in addition to varying lengths of the juvenile phase (Hisamatsu *et al.*, 2000).

6.1.1 Light

Klein (1990) suggested that supplementary UV radiation (in UV-A and UV-B wave length bands) did not enhance floral anthocyanin content in *M. incana* grown in greenhouse. The flowering of stock was promoted by the irradiation of compact self-ballasted fluorescent lamps of far-red light, the effects of incandescent lamps were also similar (Yoshimura *et al.*, 2006).

The effects of red light and far-red light on the main shoot elongation and flowering of long-day plant stock were compared by Yoshimura *et al.* (2002). During the day, plants were covered with the films which convert light quality of sunlight into different red/far-red photon flux ratio, whereas at night the plants were irradiated with compact self-ballasted fluorescent lamps of red or far-red light. Plants exposed to high red/far-red photon flux ratio during the day and red light at night, flowered late and developed short internodes, whereas plants grown under low red/far-red photon flux film during the day and irradiated with far-red light at night advanced flowering and internodes were elongated.

6.1.2 Temperature

Seventeen cultivars of stock were grown under glass in constant temperatures of 12°, 17° or 22°C. Late flowering cultivars initiated flowers only at 12°C, mid-season at 12° and 17°C and early cultivars formed flowers at all temperatures. High temperature encouraged the formation of more leaves before flower initiation (Sakanishi and Fukuzumi, 1968). In an investigation on the effect of temperature on growth and flowering, Fujita (1979) grew branching and non-branching cultivars of stocks with a daily minimum temperature of 0°C from the 5th day after sowing until flowering. The number of nodes to the inflorescence was lower in early flowering cultivars than in late ones of branching stocks. With non-branching stocks, however, there was little difference in the number of nodes to the inflorescence among extremely early, early and medium cultivars. When stocks of both types, grown initially at temperatures higher than 18°C, were transferred at various stages to a low temperature of 1° or 4°C the earlier the start of low temperature treatment the lower was the number of nodes to the inflorescence, except with the extremely early cultivars.

Fujita and Nishitani (1978) carried out a series of experiments and reported that seeds of 25 cultivars were sown on 25 August and the plants, were in an unheated greenhouse. Flower initiation in extremely early cultivars occurred by mid-October (mean temperature 21.5°C) whereas in early, mid-season and late cultivars it occurred from mid-November to early December at temperatures of 14.4°, 13.4° and 12.4°C, respectively. In another experiment, plants of 7 cultivars from sowing on 17 February and 21 December were transferred to houses with minimum temperatures above 0-5°C, 8-10°C and 13-15°C. All plants of extremely early cultivars and Seikai (a mid-season cultivar) flowered at all temperatures but in the remaining two cultivars the flowering percentage fell at 15°C. In the third experiment, 100 per cent of the plants of 3 extremely early cultivars flowered at 8°, 13° and 18°C but the highest temperature which could induce 100 per cent flowering in early and mid-season cultivars was 13°C and for a late cultivar 8°C. Flowering in extremely early, early and mid-season *cvs.* was hastened by increasing temperatures up to 15°C but higher temperatures delayed it.

In a 2-year trial, stock was grown in double layer polyethylene-covered heated or unheated greenhouses. The results revealed that the warmer greenhouses (minimum night temperature 13°C) reduced production time while colder greenhouses minimum night temperature 2 or 10°C gave a higher profit (Caviews *et al.,* 2000).

The classification of *cvs.* of stock on temperature requirement has been done by Fujita and Nishitani (1979), as follows :

Very Early	*Early*	*Mid-season*	*Late*
23°C	18°C	13°-18°C	13°C
Iwaiaka No.1, Seven Week	Iwaiaka No. 2,	Kanchidori	Matsudoaka
Trisomic Double	Wakazakura		
Dwarf	Dwarf Ten Week		

Flowering is also influenced by duration of low temperature and it varied with cultivars. Early cultivars require less days than late ones. It is quite clear from the study made by Fujita and Nishitani (1979) who observed that shortest duration of 10 days was required for 100 per cent flowering by early cultivars, 20 days were required by medium and 40 days required by late flowering ones. Good flowering was observed, the flower buds had been well differentiated.

Ito *et al.* (1997) studied the effect of altering diurnal fluctuation of day and night temperature (DIF) at the seedling stage and reported that under positive DIF seedling height, internode length and 1st -leaf petiole length of stock plant were greater than in plants grown under negative DIF. When seedlings were switched from the negative DIF regime to the positive DIF regime (or vice versa) on day 13, significant differences in final (*i.e.,* 10 days later) plant height, 1st internode length and 1st -leaf petiole length between the corresponding treatments. It was concluded that DIF during the early seedling state had a small effect on subsequent growth.

6.1.3 Light and Temperature

Effects of light and temperature on stem extension and flowering were investigated by Hisamatsu *et al.* (2002). Stem extension was markedly suppressed by red (R)-rich spectrum treatment until the flower buds became visible, whereas it was slightly promoted or was not affected by the far-red (FR)-rich spectrum treatment. The FR-rich spectrum treatment accelerated flowering and the R-rich spectrum treatment delayed flowering under weakly flower-inducing temperatures. However, these phenomena were eliminated under strong flower-inducing temperatures. Dark treated plants did not initiate flower buds in spite of cold treatment. These results indicate that light quality is involved in regulation of stem extension and flowering but temperature is more responsible than light quality in the flowering of stock.

6.2.0 Growth Regulators

Growth and flowering of stock can also be regulated by the application of growth substances. Grzesik (1989) reported that seeds of *M. incana* soaked in 100 mg GA_3 or 100 mg GA_{4+7}/l resulted in higher germination percentage compared with the control, whereas triapenthenol 25-400 mg/l RSW 0411 and diniconazole 100-400 mg/l (sumi-7) markedly reduced it. Soaking of seeds with GA for 24 h improved seed germination. The seedlings obtained from the seeds treated with GA_3, GA_{4+7} and NAA were taller and flowered earlier than plants from the other treatments (Grzesik, 1995).

A series of investigations on the role of gibberellins on growth and flowering and GA biosynthesis in stock have been carried out by Hisamatsu and co-workers. Hisamatsu *et al.* (1998) reported that *M. incana* treated with acylcyclohexanediones promoted stem elongation and flowering, seemed to be caused by an inhibition of beta-hydro-xylation of endogenous GA_4 and a consequent accumulation of endogenous GA_4. Hisamatsu *et al.* (2000) also recorded that stem elongation and flowering were most effectively promoted by GA_4 when combined with GA biosynthesis inhibitor prohexadione-calcium (PCa) and, followed by 2,2-dimethyl-GA_4, PCa, GA_4 + trinexapac-ethyl (TNE), GA_9 + PCa and by GA_4. The results indicated that : (i) GAs with 3beta-OH and without 13-OH groups (*e.g.*, GA_4) are the most important for stem elongation and flowering of stock. Hisamatsu (2001) concluded that GA_4 was the most active component for stem elongation and flowering and GA biosynthesis inhibitors of the acyclohexanedione type could promote stem elongation and flowering.

The influence of temperature on the effectiveness of GA_4 and GA biosynthesis inhibitors was reported by Hisamatsu *et al.* (1998). In early flowering Sourei, uniconazole inhibited stem elongation and flowering. Compared with controls, GA_4, PCa and TNE promoted flowering at 20°/15°C and 25°/20°C (day/night) under 12 h photoperiods, though only PCa and TNE induced flowering of late flowering Banrei at 20°/15°C. Flowering was not induced in at 25°/20°C or 30°/25°C, but stem elongation of both cultivars was promoted by GA4 and PCa at all temperatures. It is suggested that biosynthesis and deactivation of GA_4 are important factors in the control of stem elongation and flowering. It was further observed that cold treatments enhanced responsiveness to GA_4 of stock in both the stem elongation and flower bud initiation process and that the development of responsiveness to GA_4 is temperature and duration of cold treatment dependant (Hisamatsu and Koshioka, 2000).

Treatment with GA_3 at 100 ppm increased overall growth, improved plant appea-rance, the florets being larger, and flower stem thicker and more compact than in control (Parups and Helson, 1970). Sakanishi and Fukuzumi (1968) observed that treatment of GA_3 every 2 weeks from planting generally initiated flowers more rapidly than the untreated plants, this effect was marked in May-sown plants. In the cultivar Snow White, flowering was induced by GA application in the absence of chilling.

Weekly foliar application of GA_{4+7} at 25 ppm during seedling growth promoted reproductive development of *M. incana* whether grown in short day (8 h) or long days effected by night breaks. GA application substituted long day requirement for flowering (Peat and Summerfield, 1977). Treatment with GA_3 at 0.02 per cent advanced flowering by 5-7 days, increased the length of the flower stems and improved flower quality. A combined treatment of CCC and GA_3 limited the excessive elongation without reducing the time of flowering. CCC alone inhibited growth more markedly in dwarf than tall cultivars (Verzilov and Runkova, 1972). Spraying of *cv.* 'Midget Red' with a solution of chlormequat/daminozide with concentrations of 1000/800, 1250/1250 or 1500/5000 mg/l when new growth was nearly 5 cm in height significantly reduced plant size (Kuehny *et al.*, 2001). Treatment with ancymidol suppressed growth of main shoot and induced lateral branches in cultivars of columner type. Ethephon and dikegulac (atrinal) inhibited flowering. Dikegulac also inhibited growth of main shoots but failed to induce lateral growth (Roh, 1978).

Hamza and Helaly (1983) reported decrease in plant, inflorescence and peduncle length and increase in the number of lateral shoots and flowers by using CCC. GA_3 counteracted the effect of CCC on shoot growth and time of flowering. Foliar spray of SADH markedly reduced plant height and also affected the response to N (Thomas and Leong, 1980). Loeser and Essig (1968) obtained excellent pot plants of stock by 3 sprays of 5 per cent SADH or 3 drenches of 2 per cent CCC.

Drenching of the substrate in pot grown stock of *cv.* Midget Red and Lavender with paclobutrazol (0.25, 0.5 or 2.0 mg/pot in 50 ml solution per pot) delayed the time to flowering of Lavender with increasing paclobutrazol concentration. The inflorescence length of both cultivars, and only the stem length of Lavender, decreased with increa-sing paclobutrazol concentration. Increasing paclobutrazol concentration resulted in a decrease in the percentage of senesced flowers of Midget Red, recorded 30 days after the onset of flowering. Paclobutrazol at 0.5 mg/pot produced marketable pot plants (Ecker *et al.*, 1992).

7.0 Propagation

7.1.0 Seed Propagation

Stocks are propagated from seeds (16,000-20,000/28 g) which required 18-24°C temperature for their normal germination (Auman, 1980). Seeds are sown in well prepared nursery beds, boxes, pans or pots. In India, the time of sowing is from middle of September to middle of October in plains and in February-March in the hills. In temperate countries, Ten week mixed stock seeds are sown under glass in March. Seeds are scattered thinly on the surface and are covered with a layer of sand and/or sieved manure. Additional covering of seeds by paper or cloth sheet is recommended till seeds start germinating. The double flowered stock is completely sterile, lacking reproductive organs (Philp and Huskins, 1931) and thus, double-flowered stock has to be produced from seeds of single-flowered plants. Dressing of

Propagation of Stock

Seeds

Seedlings in Pan

Pricking of Seedlings

Planting in Bed

Planting in Pot

Production in Polyhouse

Micropropagation

a) Seeds cultured in MS medium without plant growth regulators, b) Germination of cultured seeds after a week, c) Produced seedling 4 weeks after cultivation, d) Excised buds as explants and their cultivation, e) Regeneration of explants 4 weeks after culturing on medium containing NAA and KIN, and f) Produced plantlets. (*Source*: Kaviani (2014b)

Harvesting and Postharvest Management

Harvesting **Packaging**

Retail Outlet

***Source*: www.bluecloudfarm.com, www.flickr.com, www.forum.gardenersworld.com, www.123rf.com, www.in.pinterest.com, www.originallaflowermarket.com, www.coloriginz.com, www.semanticscholar.org**

seeds with captan preparation to control damping off and foot-rot diseases, resulted in higher percentage of seedling emergence (Niebisch, 1985).

7.2.0 Micropropagation

Shoot formation in leaf explant culture of *M. incana cv.* Ginsho was enhanced when the mother plants received 75 per cent shading and half-strength Ootsuka liquid fertilizer. However, in *cv.* Yukimatsuri shoot formation was inhibited when mother plants were shaded and supplied with half- or quarter-strength liquid fertilizer. Reducing the $NH_4^+ : NO_3^-$ ratio in the culture medium increased shoot formation in both cultivars (Mizozoe *et al.*, 1992). *In vitro* regeneration of *M. incana cv.* White King was achieved from apical microcuttings and leaves. Only adventitious shoots formed, either indirectly, from compact nodular callus at the basal end of the microcuttings, or directly, from nodular structures on leaf explants. One and 10 mg/l BA (for young and old explants, respectively) in the culture medium was critical for shoot regeneration (Mensuali *et al.*, 1994). The best callus growth of *M. Incana in vitro* was promoted with 10 µM NAA and 1 µM 2,4-D. Callus growth differed between leaf explants taken from single and double flowering plants. Cytokinins promoted regeneration of shoot and 2iP was most effective. Adventitious roots formed on medium without any growth regulators and cytokinins inhibited root formation, especially kinetin (Okuse *et al.*, 1996).

Abdoosi *et al.* (2014) reported that callus of cotyledons explants with 2 mg/l BAP+ 0.5 mg/l NAA) recorded 98.66 per cent of regeneration. Hesar *et al.* (2011) developed simple and reliable strategy for micro-propagation of *Matthiola incana* in the presence of the single growth regulator, kinetin. *In vitro* single nodes of *Matthiola incana* were cultured in MS basal medium with different concentrations of kinetin to produce shoots and roots. The best shoot length and most number of nodes were obtained in 2 mg/l of kinetin. The largest number and longest length of roots were achieved using 1 mg/l kinetin.

Kaviani (2014a) expressed that micropropagation by organogenesis is an efficient *in vitro* propagation method for *Matthiola incana*. MS medium supplemented with 5 mg/l BA + 1 mg/l IBA resulted in the highest shoot length. Largest number of node was obtained in MS medium containing 1 mg/l BA + 1 mg/l IBA. Shoot tips inoculated in medium containing 0.5 mg/l IBA without BA recorded best for root number. Media containing 0.5 mg/l IBA without BA showed maximum root length. About 85 per cent of the micropropagated plantlets were established successfully in acclimatization medium. This protocol has proven useful for tissue culture propagation of *Matthiola incana*.

Kaviani (2014b) found that four-week-old *in vitro* plants obtained from micro-cuttings showed successful shooting and rooting. MS medium supplemented with 2 mg/l kinetin without NAA resulted in the best shoot length and largest number of node, while, shoot tips inoculated with 2 mg/l NAA without kinetin and 1 mg/l NAA + 2 mg/l kinetin recorded best for root number and root length.

8.0 Soil and Climate

Stock is successfully grown in moderately fertile, moist but well-drained soil. It prefers soil having pH between 6.5 and 7.0 and does not like acidic soil.

Stock prefers full sun in a sheltered position. Temperature plays an important role on growth and flowering of stock. The crop should be grown at the night temperature of 16°C until ten fully developed leaves are produced and then be maintained at 10°C for atleast 3 weeks. The temperature can then be raised again to 16°C. Stock will not flower if it is exposed more than 6 h per day to 18°C. For obtaining better quality flowers the night temperature should be 2-4°C (Auman, 1980). Plants are frost hardy.

9.0 Cultivation

9.1.0 Planting

When the seedlings are 2.5-3 cm tall or have 3-4 true leaves, they should be transplanted. The tall cultivars should be transplanted at a distance of 20-25 cm between rows and plants, whereas, for dwarf ones a distance of 15 cm is sufficient. Auman (1980) suggested that seedlings should be planted 7 × 15 cm apart. Transplanting should be done in cool hours of the day, especially in the afternoon and watered immediately. The cultivars should be grown using staggered sowing times for extending harvest season. The cultivars may be sown on 15th September to 1st October and transplanted at 22.5 × 22.5/ cm spacing for best cut stem quality (Abdullah *et al.,* 2020).

9.2.0 Irrigation

Stock plant needs adequate moisture throughout its growth period. It has been experienced that stock need frequent irrigation at the interval of 6-7 days. Frequency of irrigation will also depend upon other factors like soil type and prevailing weather conditions. In lighter soil, frequent irrigation is required.

Demiral (2003) suggested that irrigation with saline water is feasible *M. tricuspidata*. Therefore, it can be grown in arid and semi-arid regions by using non-conventional water resources such as saline water and treated sewage effluent, which is also quite saline. However, Heuer and Ravina (2004) reported that irrigation with saline water of <6 dS/m in *Matthiola incana* resulted in no reduction in yield.

Heuer *et al.* (2005) reported that omega-3 content in stock under greenhouse and field conditions was increased by use saline irrigation water, however, seed yield, oil content, and fatty acid composition was not affected (Table 1).

Grieve *et al.* (2006) suggested that stock is a relatively a salt tolerant plant and may be produced under saline irrigation without loss of neither quality nor reduction of marketability. Cultivars 'Cheerful White' and 'Frolic Carmine' grown in greenhouse sand culture irrigated with SJV and CCR waters at salinity levels 2.5, 5, 8, 11, and 14 dS/m produced high quality, marketable flower stem length and mass were significantly higher in plants irrigated with CCRW where external-Cl concentrations was 107 mM. Further, Grieve *et al.* (2008) reported that saline solution (2 to 11 dS/m) amended with substrate N concentrations (35 to 100 ppm) produced marketable stems. Neither growth parameters nor yield components were affected by increases in substrate N. Therefore, in closed-loop irrigation systems, the N requirements for stock are low and growers could minimize costs and limit offsite pollution potential by reducing N inputs.

Table 1: Effect of Irrigation Water Salinity on Yield Parameters of Stock in the Greenhouse*

Line Parameter	*Salinity Levels*				
	Control	*2 dS/m*	*4 dS/m*	*6 dS/m*	*8 dS/m*
Total fruit yield (g/plant)	43.49	42.35	38.32	34.26	26.24*
Seed yield (g/plant)	12.00	10.88	9.83	8.22*	5.99*
Seed Number	13680	13428	13277	11392	8566*
Seed yield (per cent of total yield)	27.59	25.69	25.76	23.99*	22.74*

$P < 0.05$ for difference from control.

* Heuer *et al.* (2005).

However, Jafari and Garmdareh (2019) suggested high quality stock plants can be achieved through proving irrigation with salinity up to 4 dS/m during growing period. Jafari *et al.* (2019) found that drought stress has undesirable effects on morphological, physiological and biochemical characteristics of stock plants, therefore, irrigation should be carefully controlled.

9.3.0 Manuring and Fertilization

Stocks need fertile soil for its healthy growth and normal flowering. Nitrogen promoted growth which was further enhanced by high K (Thomas and Leong, 1980). Application of 30 tonnes of farmyard manure at the time of land preparation and 400 kg nitrogen and 150 kg each of phosphorus and potassium per hectare will be sufficient. Removal of N from the nutrient solution supplied to stocks 20 days before opening of first flower resulted in a reduction in plant height, flower number and seed yield. A reduction of N level from 84 ppm to 42 ppm at 20 days before flowering or its removal or reduction of N, 7 days before flowering had no effect on growth, flowering and seed production. Moderate changes in P and K also showed no effect on seed yield (Semenuik and Stewart, 1965). Stocks are subject to potassium deficiency which is seen as burning on the older leaf margins. Therefore, care should be taken to avoid its deficiency. A level of 100 ppm potassium in the soil has been recommended as optimum (Holms *et al.*, 1981). Effect of 3 levels of N, P and K on growth and seed production was studied by Semenuik (1964). Medium level of nitrogen (48 ppm) resulted in the tallest plants. Plants deficient in K or in a high (N+P)/K ratio showed the typical symptoms at the time of flowering. The highest number of seeds per plant was obtained from a high level of K and low levels of N and P. Medium and high levels of N resulted in more vigorous growth with lower seed production than low level of nitrogen. In sand culture, Yamanouchi and Yang (2000) studied relationship between minimum levels of nitrogen concentration in leaves and quality of cut flower stock *cv.* Asanami. The level of nitrogen concentration that resulted in good cut flower quality was around 30-32 g/kg in leaves at the upper 11th position. This level was sufficient for nitrogen absorption to continue until visible flower bud or the beginning of flowering. There was a high positive correlation between N concentration and K (stem only), Ca (except 180 N), P and Mg concentrations in the leaves and stem, respectively, but K concentration in the leaves decreased when N increased above 40 g/kg.

Penningsfeld and Heussler (1965) reported beneficial effect of 1 mg borax and 2 mg sodium molybdate to each litre of basic fertilizers when stocks were grown in North German peat. Handreck (1986) suggested that stock needed 25-27 mg/l sulphur in liquid feed. Drainage water containing less than 12-16 mg/l was associated with deficiency.

In an experiment on different potting media with fertilizer mixture, foliage growth and dry weight of plants were best in peat + sand (equal parts by volume) and osmocote (14:6:12) (Thomas *et al.,* 1980).

Waseem *et al.* (2013) observed that leaf mould (100 per cent) as potting media encouraged the growth and flowering of Stock plants as it produced maximum plant height, flowering cluster, flowers per cluster and flowers per plant.

9.4.0 Greenhouse

Vita and Agnello (2000) reported that the availability of cut flowers (cv. Sentinel White) could be extended by phasing the planting in cool greenhouse. The effect of EVA 'foamed' film (210 µm) as a covering material on tunnels for *M. incana* cultivation on growth and flowering were recorded by Magnani *et al.* (1997). Harvesting of flowers was earlier, quality of flowers (stem height, flower length) and stem FW and DW were also higher under EVA 'foamed' film than under standard film (200 µm). Dansereau *et al.* (1998) studied the effects of single-layer glass and double-layer antifog polyethylene films on growth and flowering of stock. Plants showed higher photosynthetic capacity (PC) under single-layer glass than under other covers regardless of light regimes. Supplementary lighting reduced PC of leaves, especially under single-layer glass, and diminished differences in PC among covering treatments. Dry mass was influenced more by larger leaf area caused by higher leaf temperature than by net photosynthetic rate. Generally, anti-fog 3 year polyethylene was a good covering material when plant quality and energy savings were considered.

10.0 Diseases and Pests

10.1.0 Diseases

Fungal Diseases

10.1.1 Damping Off

This is a serious disease in seedlings and older plants, caused by *Pythium debaryanum* and *Pellicuraria filamentosa*. Basal parts including roots of the infected plants turn black and die, stem girdling at the soil line also results in wilting. Excess moisture and high temperature favour occurrence and spread of the disease (Sohi, 1984). *Phytophthora megasperma* and *Sclerotinia sclerotiorum* also attack seedlings and cause foot or root-rot of older plants. The control measures include disinfection of seeds and compost with heat or chemical.

10.1.2 Gray Mould

Botrytis cinerea also causes damping off of seedlings in seed bed and they die or decay. The fungus also produces mass of gray mould on the flowers and causes rotting. Surface sterilization of seeds and spraying of seedlings with zineb have been suggested as control measures (Pirone *et al.*, 1960).

10.1.3 Wilt

Fusarium oxysporum f. sp. *matthioli* and *Verticillium albo-atrum* cause wilting of seedlings and older plant which is characterized by yellowing, premature leaf drop, stunting, vascular discolouration and wilting. It is serious in cool moist weather.

Planting should be avoided in wilt-infested soil. Fumigation of infested soil with chlorpicrin or dazomat is suggested to eradicate the soil borne fungus.

10.1.4 Leaf Spot

Leaf spots of stocks is caused by *Alternaria raphani*. The disease is characterized by formation of round spots on leaves, stem and flowers. The spot on close examination reveals the formation of concentric rings and exhibits profuse blackish sporulation under moist conditions. This disease can be controlled by spraying with fungicides like dithane M-45, dithane Z-78 or blitox at 0.3 per cent at regular intervals.

10.1.5 Downy Mildew

Downy mildew is caused by *Peronospora parasitica*, occasionally referred to as *P. matthiolae*. The infected leaves develop pale green, depressed spots on the upper side. The lower side of the spot shows downy mould growth at the point of infection. The disease can be controlled by copper oxychloride or dithane M-45 sprays at 0.2 per cent at regular intervals. Dirkse (1981) reported effective control of *P. matthiolae* with 2 application of 0.1 per cent or 0.2 per cent fongaride WP (furalaxyl) or 0.15 per cent previcur, at weekly interval. A downy mildew of stocks caused by *P. parasitica* was reported by Koike (2000) from California and the disease symptoms developed after 10 days.

10.1.6 White Rust

The disease is caused by a fungus *Albugo candida*. It is characterized by the formation of white raised pustules on the under surfaces of lower leaves (Sohi, 1984). The disease spreads in the field by white powdery spores released from these pustules. Removal of diseased plant and destruction of the plant debris and spraying of copper oxychloride (blitox) and follow a 2-year rotation are suggested to control the disease.

Diseases and Pests of Stock

Bacterial Blight

Downy Mildew

Blossom Blight

Diamondback Moth

Aphids

Springtails

Source: www.pnwhandbooks.org, www.semanticscholar.org

Bacterial Diseases

Lintzel (1964) reported a bacterial disease caused by *Pseudomonas syringae* on stocks and the bacteria were found within the seeds of infected plants. The spread of the disease was restricted by spraying with copper fungicide. Minardi *et al.* (1988) reported bacterial blight on greenhouse grown stocks, causing losses of 40-90 per cent. The disease appeared at different stages, from emergence to flowering, with acropetal discolouration and withering of the leaves associated with longitudinal slits or cracks in stems as the most common symptoms. Seed treatment at 53-55°C for 10 minutes followed by rapid cooling is suggested as a preventive measure.

Viral Diseases

Stocks are infected by several viruses, mosaic (*Marmor brassicae*), curly top (*Ruga verrucosans*), spotted wilt (*Lethum australiense*) and flower breaking (*Marmor cruciferarum*). Mixed infection of cauliflower mosaic caulimovirus and turnip mosaic potyvirus caused stunting, mosaic, severe leaf malformation, breaking and greening of flowers, and abortion of the inflorescence of greenhouse grown stock plants (Alioto *et al.*, 1994). The methods of eliminating these diseases include roguing out and destruction of plants showing symptoms and control of aphids and leaf hoppers by using malathion (Pirone *et al.*, 1960).

10.2.0 Pests

10.2.1 Aphids (*Aphis mathiolae*)

This is major pest on stocks, when they feed on inflorescence stalk, colour of the flower fades. Rogor and malathion are effective to control the pest.

10.2.2 Diamond Back Moth (*Plutella maculipennis*)

The greenish caterpillars feed on the lower surface of leaves and holes develop due to the decay of the upper epidermis. These caterpillars also feed on leaves and shoots by forming webs as leaf-miners. Several control measures have been suggested including hand picking of cocoons and spraying of rogor or malathion.

10.2.3 Flea Beetles

Three types of flea-beetles – striped, western black and western striped infest stocks and feed on the leaves forming holes. In severe infestation, the plants may die. Spraying of cartap hydrochloride 50 per cent sp. and malathion has proved effective.

10.2.4 Springtails (*Bourletiella hortensis*)

Small dead spots appear on the leaves and the lower leaves wither due to the infestation of these small insects. Dusting the soil surface with lindane and malathion powder in early morning or evening controls the pest (Pirone *et al.*, 1960).

11.0 Harvesting and Postharvest Management

11.1.0 Harvesting

Spikes should be cut when one-half to two-third of the florets have opened

(Auman, 1980) or with a minimum of 6 open flowers (Barendse, 1980). Harvesting should be done early in the morning and the stalk placed in cool water.

11.2.0 Postharvest Management

11.2.1 Vase Life

Cut flowers of stock stored at 4°C for up to 3 days. The vase life is about 3-4 days (Auman, 1980). In order to improve the vase life of stocks, Barendse (1980) suggested the following practices : (i) harvesting with a minimum of 6 open flowers, (ii) transporting the flowering stems with roots still attached or pre-treating the bunches in a solution of commercial preservatives, (iii) transporting bunches upright to prevent the stems growing crooked, (iv) cutting the stems above the root collar in the retail store, and (v) ensuring that the buyer uses a solution of a cut flower preservative in the vase water.

Stock flowers rapidly desiccate and shatter when exposed to ethylene, (Celikel and Reid, 2002). Silver thiosulfate (STS) and 1-methylcyclopropene (1-MCP) are two common anti-ethylene agents effective in increasing the vase life of stock (Regan and Dole, 2010).

In an experiment on vase life of stock *cv.* Rubin, proflovit 72 (0.3 g 8-hydroxyquinoline + 0.05 g CCC + 50 g sucrose/l) was found to be best preservative (Nowak and Rudnicki, 1975). The effects of pretreatments [water, sucrose (7 per cent) + silver thiosulfate (STS; 0.5 or 1.0 mM) or sucrose + STS (0.5 mM) + GA_3 (20 mg/l) for 2 h], cold storage in distilled water or newspaper at 4°C for 1 or 2 weeks, and various preservative solutions [water, carbonated water (30 per cent) or 8-hydroxyquinoline sulphate (HQS; 50-2000 mg/l) alone or in combination with sucrose (1 or 7 per cent), GA_3 (5 mg/l) or $CaNO_3$ (0.5 or 1 mM)] on the vase life of cut flowers of *M. incana* (cv. Chohong) were investigated by Song *et al.* (1996). Pre-treatments containing STS + sucrose or STS + sucrose + GA_3 extended vase life by about 2 days compared with the control regardless of storage durations and methods. Preservative solutions containing HQS at 100-200 mg/l + sucrose at 1 per cent increased vase life by about 4 days compared with the control and improved the quality of cut flowers. Preservative solutions containing $CaNO_3$ at 0.5 or 1 mM decreased tip bending of flower stalks and increased vase life. The vase life of flowers held in HQS + sucrose after pretreatment with STS + sucrose was increased by about 5 days; the quality of flowers following 1 or 2 weeks dry or wet storage was good following these treatments.

The use of 200 mg/l 8-HQS, 15 per cent sucrose (in holding solution) and 500 mg/l 8-HQS (in pulsing solution) treatments improved stock cut flower quality by increasing vase life as measured by improving water balance and longevity of cut stock *cv.* 'Gold Cut Series' (Khan *et al.*, 2015).

Combination 1 per cent sucrose + 200 mg/l 8-HQS + 25 mg/l $AgNO_3$ + 50 mg/l $(Al_2SO_4)_3$ extended the vase life of *M. incana* cut flowers as well as maintain the shape and colour of cut flowers and leaves (Huang, 2014).

Pre-treatment for 6 h with low concentrations of 1-MCP (1-methylcyclopropene) prolonged vase life of stock in an ethylene-free environment. Treatment with 1-MCP was as effective in inhibiting ethylene effects as treatment with silver thiosulphate. In the presence of ethylene at 1 µl/l, the vase life of 1-MCP-treated flowers was up to 4 times that of the controls (Serek *et al.*, 1995).

Çelikel and Reid (2002) observed increase in flower opening, water uptake, and vase life in a vase solution containing 50 ppm NaOCl. Exposure to exogenous ethylene resulted in rapid desiccation and abscission of the petals, these effects were prevented by pretreatment with 1-methylcyclopropene (1-MCP). Even in the absence of exogenous ethylene, vas life was significantly increased by inhibiting ethylene action using pretreatment with silver thiosulfate (STS) or 1-MCP.

Leaf yellowing of cut stock flowers during vase life can be best avoided by using pulse treatments with thidiazuron (TDZ) 5 µM (Ferrante *et al.*, 2009) but combination of TDZ with 1-MCP is not advisable because the 1-MCP reduce the efficiency of TDZ (Ferrante *et al.*, 2012). Abscisic acid (ABA) is not the main factor involved in leaf yellowing, but its presence and accumulation may have a synergic effect with other senescence inducers (ethylene, hormone imbalance, *etc.*) and it dramatically accelerates leaf senescence in stock (Ferrante *et al.*, 2004).

Regan and Dole (2010) recommended different post handling practices for stock *cv.* 'Vivas Blue'. Cut stock 'Vivas Blue' stems recorded longer vase life when stored dry, as compared to stored in water. Although sucrose pulsing is not necessary as it did not increase vase life, but use of sucrose in the final vase solution increased the vase life as well as increased bud opening and developed a deep purple colour.

The mild vacuum packaging can be used for storing cut stock flowers if transportation or storage is limited to three days. In this period of time, Pacifici *et al.* (2013) observed no differences of vase life and quality cut flowers stored in conventional cold wet storage or mild vacuum packaging. Therefore, mild vacuum packaging can be used for lowering the transportation costs by increasing the loading capacity per unit of volume of cut flowers. Moreover, the absence of water facilitates all handling operations, contributing to a reducing in the high costs.

12.0 Biotechnology

12.1.0 Protoplast Culture

Protoplasts were isolated from leaves of 21-28 days old plants of wild-type line MSC and mutant lines MCh1 (chlorophyll deficient) and MutF (petioleless leaves and stalkless flowers). High division frequencies and plating efficiencies (0.9-2.2 per cent) were obtained from protoplasts of MSC and MChl embedded in sodium alginate and cultured in MI medium. The highest shoot regeneration rates of 41, 37 and 27 per cent were obtained for MChl, MSC and MutF, respectively, on AIIR medium (Siemens and Sacristan, 1995). The cotyledons from two-week-old seedling (at 25°C, 2000 lx, 14h light/d) are appropriate for isolating protoplasts for plant regeneration (Ma *et al.*, 2005).

12.2.0 Molecular Markers

A cDNA clone (pcM12) of the chalcone [naringenin-chalcone] synthase (CHS) of *M. incana* was isolated from a cDNA library, sequenced and analysed. It comprises the whole coding sequence for the CHS and 5′ and 3′ untranslated regions. The deduced amino acid sequence shows that matthiola CHS consists of 394 amino acid residues. Comparison with CHS amino acid sequences of other plants indicated >82 per cent homology (Epping *et al.*, 1990).

13.0 References

Abdoosi, V., Mousavi, E.S., Abdali, A. (2014) *Bull. Env. Pharmacol. Life Sci.*, **3**(10): 175-181.

Abdullah, B., Ahmad, I. and Dole, J.M. (2020) *Sci. Hortic.*, **261**: 108909.

Alioto, D., Stavolone, L. and Aloj, B. (1994) *InformatoreFitopatologico*, **44** : 43-46.

Ardelean, M., PUI, D.A., Cordea, M. and Mihalescu, L. (2009) *Bull. Univ. Agric. Sci. Vet. Med. Cluj-Napoca, Hortic.*, **66**(1) : 13-16.

Auman, C.W. (1980) in: *Introduction to Floriculture* (Ed. R.A. Larson), Academic Press, New York.

Barendse, L.V.J. (1980) *Vakbladvoor de Bloemisterij*, **35** : 38-39.

Brickell, C. (1996) *A-Z Encyclopedia of Garden Plants*, Dorling Kindersley, London, pp. 660-661.

Caviews, T.J., Dole, J.M. and Stamback, V. (2000) *HortTechnol.*, **10** : 793-799.

Çelikel, F.G. and Reid, M.S. (2002) *HortScience*, **37**(1) : 144-147.

Chen, D., Liu, Y., Pan, Q., Li, F., Zhang, Q., Ge, X. and Li, Z. (2018) *Sci. Hortic.*, **240** : 592-602.

Coats, A.M. (1968) *Flowers and Their Histories*, Billing and Sons. Ltd., London.

Crane, M.M. and Lawerence, W.J.S. (1956) *The Genetics of Garden Plants*, MaCmillan and Co., Ltd., New York, pp. 50-54.

Dansereau, B., Zhang, Y., Gagnon, S. and Xu, H.L. (1998) *HortScience*, **33** : 668-671.

Demiral, M.A. (2003) *Turk. J. Agric. For.*, **27**(4) : 229-235.

Dirkse, F.B. (1981) *Vakbladvoor de Bloemisterij*, **36** : 35.

Dole, J.M. and Wilkins, H.F. (2005) *Floriculture: Principles and Species, 2nd ed.* Prentice Hall, Upper Saddle River, NJ.

Dressel, A. and Hemleben, V. (2009) *Plant Biol.*, **11** : 204-212.

Ecker, R., Barzilay, A. and Osherenko, E. (1993) *Euphytica.*, **74**(1-2) : 133-136.

Ecker, R., Barzilay, A. and Osherenko, E. (1994a) *Euphytica*, **74** : 133-136.

Ecker, R., Barzilay, A. and Osherenko, E. (1994b) *Euphytica*, **72** : 153-156.

Ecker, R., Barzilay, A., Afgin, L. and Watad, A.A. (1992) *HortScience*, **27** : 1330.

Ecker, R., Zutra, D., Barzilay, A., Osherenko, E. and Rav-David, D. (1995) *Genet. Resour. Crop Evol.*, **42** : 371-372.

Epping, B., Kittel, M., Ruhnau, B. and Hemleben, V. (1990) *Plant Mol. Biol.*, **14** : 1061-1063.

Fehleisen, S. and Schnack, B. (1972) *Revista de la Facultad de Agronomica*, Universidad Nacional de la Plata, **68** : 125-130.

Ferrante, A., Mensuali-Sodi, A. and Serra, G. (2009) *Cent. Eur. J. Biol.*, **4**(4) : 461-468.

Ferrante, A., Trivellini, A. and Mensuali Sodi, A. (2012) *The Open Hortic. J.*, **5** : 1-5.

Ferrante, A., Vernieri, P.A.O.L.O., Serra, G. and Tognoni, F. (2004) *Plant Growth Regul.*, **43**(2) : 127-134.

Frost, H.B. (1928) *J. Hered.*, **19** : 105-111.

Fujita, M. (1979) *J. Japanese Hort. Sci.*, **48** : 327-335.

Fujita, M. and Nishitani, T. (1978) *J. Japanese Soc. Hort. Sci.*, **47** : 217-226.

Fujita, M. and Nishitani, T. (1979) *J. Japanese Soc. Hort. Sci.*, **48** : 213-223.

Grieve, C.M., Poss, J.A. and Amrhein, C. (2006) *HortScience*, **41**(1) : 119-123.

Grieve, C.M., Poss, J.A., Shouse, P.J. and Carter, C.T. (2008) *HortScience*, **43**(6) : 1787-1793.

Grzesik, M. (1989) *Acta Hortic.*, **251** : 71-74.

Grzesik, M. (1995) *Seed Sci. and Technol.*, **23** : 801-806.

Hammami, S., Ciavatta, M.L., Ben Jannet, H., Cimino, G. and Mighria, Z. (2006) *Croat. Chem. Acta.*, **79**(2): 215-218.

Hammami, S., Khoja, I., Khoja, I., Jannet, H.B., Halima, M.B. and Mighri, Z. (2007) *J. Essent. Oil Bear. Pl.*, **10**(2): 162-167.

Hamza, A.M. and Helaly, M.N.M. (1983) *Acta Hortic.*, **137** : 197-209.

Handreck, K.A. (1986) *Scientia Hort.*, **30** : 1-17.

Hemleben, V., Dressel, A., Epping, B., Lukacin, R., Martens, S. and Austin, M. (2004) *Plant Mol. Bio.*, **55** : 455-465.

Hesar, A.A., Kaviani, B., Tarang, A. and Zanjani, S.B. (2011) *Plant Omics J.*, **4**(5) : 236-238.

Heuer, B. and Ravina, I. (2004) *Aust. J. Agric. Res.*, **55**(8) : 907-910.

Heuer, B., Ravina, I. and Davidov, S. (2005) *Aust. J. Agric. Res.*, **56**(1) : 45-47.

Hisamatsu, T. (2001) *Bull. Nat. Res. Inst. Veg. Orna. Plants Tea.*, **16** : 79-133.

Hisamatsu, T. and Koshioka, M. (2000) *J. Hort. Sci. Biotech.*, **75** : 672-678.

Hisamatsu, T., Koshioka, M., Kubota, S. and King, R.W. (1998) *J. Japanese Soc. Hort. Sci.*, **67** : 537-543.

Hisamatsu, T., Koshioka, M., Kubota, S., Fujime, Y., King, R.W. and Mander, L.N. (2000) *Physiologia Plantarum*, **109** : 97-105.

Hisamatsu, T., Oyama-Okubo, N., Ichimura, K., Esaki, S., Oi, R. and Koshioka, M. (2002) *J. Hortic. Sci. Biotech.*, **77**(1) : 1-8.

Holms, R.L., Coorts, G.D. and Rosenfeld, T.L. (1981) *HortScience*, **16** : 560-561.

Huang, H., Jiang, T., Wang, W., Zheng, J. and Huang, M. (2014) *J. Yunnan Agric. Univ.*, **29**(4) : 528-532.

Irani, S.F. and Arab, M. (2017) *Hortic. Sci. and Tech. J.*, **35**(2) : 265-275.

Irani, S.F., Arab, M., Norouzi, M. and Lotfi, M. (2016) *Asian J. Adv. Basic Sci*, **4**(2) : 65-73.

Ito, A., Hisamatsu, T., Soichi, N., Nonaka, M., Amano, M. and Koshioka, M. (1997) *J. Japanese Soc. Hort. Sci.*, **65** : 817-823.

Jafari, S. and Garmdareh, S.E.H. (2019) *Sci. Hortic.*, **257** : 108731.

Jafari, S., Garmdareh, S.E.H. and Azadegan, B. (2019) *Sci. Hortic.*, **253** : 128-133.

Jana, S. and Seyffert, W. (1971) *Theor. Appl. Genet.*, **41** : 329-337.

Jin, L., Li, X.B., Tian, D.Q., Fang, X.P., Yu, Y.M., Zhu, H.Q., Ge, Y.Y., Ma, G.Y., Wang, W.Y., Xiao, W.F. and Li, M. (2016) *Ind. Crops Prod.*, **87** : 198-209.

Johnson, B.L. (1953) *Genetics*, **38** : 229-243.

Karaman, S., Gulseven, M., Comlekcioglu, N. and Ilcim, A. (2011) *Int. J. Agric. Biol.*, **13**(4) : 581-585.

Kaviani, B. (2014a) *Iranian J. Plant Physiol.*, **4**(3): 1071-1078. 10.22034/ijpp.2014.540651.

Kaviani, B. (2014b) *Acta Sci. Pol., Hortorum Cultus*, **13**(1) : 141-154.

Khan, M.A., Younis, A.D.N.A.N. and Riaz, A.T.I.F. (2015) *Agricul. Commun*, **3** : 37-42.

Klein, R.M. (1990) *HortScience*, **25** : 307-308.

Koike, S.T. (2000) *Plant Dis.*, **84** : 103.

Koku, K., Aoki, N., Dohi, M., Fujiura, T. and Takeyama, K. (1999) *J. Japanese Soc. Hort. Sci.*, **68** : 49-54.

Kuehny, J.S., Painter, A. and Branch, P.C. (2001) *HortScience*, **36** : 321-323.

Lesley, M.M. (1973) *J. Hered.*, **64** : 359-362.

Lintzel, B. (1964) *Arch. Gartenb.*, **12** : 3-11.

Loeser, H. (1986) *Dutsche Gartenb.*, **40** : 1474-1475.

Loeser, H. and Essig, W. (1968) *Dtsche Gartnenborse*, **68** : 51-59.

Ma, Z.Q., Lin, L.B., Dong, N. and Qi, Y.Q. (2005) *J. Yunnan Agric. Univ.*, **2**.

Magnani, G., Falleri, F. and Vedrani, G. (1997) *Plasticulture*, **113** : 2-10.

Magrini, S., Barreca, D., Neri, L., Iacomino, A., Barbarese, F., Superchi, S. and Zucconi, L. (2019) *Flora*, **29** : 293.

Mensuali-Sodi, A., Brea, M., Panizza, M., Serra, G. and Tognoni, F. (1994) *Gartenbauwissenschaft*, **59** : 77-80.

Minardi, P., Mazzucchi, U. and Parrini, C. (1988) *Informatore Fitopatologico*, **38** : 43-46.

Mizozoe, M., Inagaki, N., Okano, M., Kanechi, M. and Maekawa, S. (1992) *J. Japanese Soc. Hort. Sci.*, **61** : 625-633.

Nakatsuka, T. and Koishi, K. (2018) *Plant Sci.*, **268** : 39-46.

Niebisch, R.M. (1985) *Machrichtenblatt fur den Pflanzenschutzinder DDR*, **39** : 190-191.

Nowak, J. and Rudnicki, R.M. (1975) *PraceInstytutuSadownictwa w Skierniewicach. B*, pp. 173-179.

Nuraini, L., Ando, Y., Kawai, K., Tatsuzawa, F., Tanaka, K., Ochiai, M., Suzuki, K., Aragonés, V., Daròs, J.A. and Nakatsuka, T. (2020) *Planta*, **251**(3) : 61.

Okuse, I., Satoh, Y. and Saga, K. (1996) *Bull. Faculty Agri.*, Hirosaki University, **59** : 71-78.

Pacifici, S., Ferrante, A. and Mensuali Sodi, A. (2013) *Aust. J. Crop Sci.*, **7**(6) : 861-869.

Parups, E.V. and Helson, V.A. (1970) *Greenhouse, Garden, Grass*, **9** : 1-3.

Peat, J.R. and Summerfield, R.J. (1977) *Scientia Hort.*, **7** : 81-89.

Penningsfeld, F. and Huessler, A. (1965) *Iber. st. Lehr-U.*

Philip, J. and Huskins, C.L. (1931) *J. Genet.*, **24** : 359-404.

Pirone, P.P., Dodge, B.O. and Riekett, N.W. (1960) *Diseases and Pests of Ornamental Plants*, Constable Co. Ltd., London.

Rall, S. and Hembleben, V. (1984) *Plant Mol. Biol.*, **3** : 137-145.

Ramsay, N.A, Walker, A.R, Mooney, M. and Gray, J.C. (2003) *Plant Mol. Biol*, **52** : 679–688.

Regan, E.M. and Dole, J.M. (2010) *Postharvest Biol. Technol.*, **58**(3): 268-273.

Roh, S.M. (1978) *Korean Soc. Hort. Sci.*, **19** : 154-166.

Saito, N., Tatsuzawa, F., Hongo, A., Khin, W.W., Yokoi, M., Shigihara, A. and Honda, T. (1996) *Phytochem.*, **41** : 1613-1620.

Sakanishi, Y. and Fukuzumi, H. (1968) *J. Japanese Soc. Hort. Sci.*, **37** : 357-367.

Sanchez, J.L., Domina, G. and Caujape-Castells, J. (2005) *Plant Syst. Evol.*, **253** : 81-93.

Saunders, E.R. (1928) *Genetica.*, **4** : 141-170.

Saunders, E.R. (1928) *Bibliogr. Genet.*, **4** : 141-170.

Schnack, B. and Fehleisen, S. (1954) *Rev. Fac. Agron. B. Aires*, **35** : 25-30.

Šegota, V., Jasprica, N., Bogdanoviæ, S., Pandža, M., Miloviæ, M. and Alegro, A. (2016) *Glasnik Hrvatskog Botanièkog Društva*, **4**(2) : 20-25.

Semenuik, P. (1964) *Bot. Gaz.*, **125** : 62-65.

Semenuik, P. and Stewart, R.N. (1965) *Proc. Amer. Soc. Hort. Sci.*, **86** : 695-700.

Serek, M., Sisler, E.C. and Reid, M.S. (1995) *Plant Growth Regul.*, **16** : 93-97.

Seyffert, W. (1971) *Theor. Appl. Genet.*, **41** : 285-291.

Siemens, J. and Sacristan, M.D. (1995) *Plant Cell Rep.*, **14** : 466-449.

Sohi, H.S. (1984) *Diseases of Ornamental Plants*, ICAR, New Delhi.

Song, C.Y., Bang, C.S., Huh, K.Y. and Song, J.S. (1996) *RDA J. Agri. Sci. Hort.*, **38** : 598-603.

Tatsuzawa, F., Saito, N., Toki, K., Shinoda, K. and Honda T. (2012) *J. Japan. Soc. Hort. Sci.*, **81**(1) : 91–100.

Taviano, M.F., Miceli, N., Acquaviva, R., Malfa, G.A., Ragusa, S., Giordano, D., Cásedas, G., Les, F. and López, V. (2020) *Biology*, **9**(7) : 163.

Teusch, M. (1986) *Planta*, **169** : 559-563.

Thomas, M.B. and Leong, A. (1980) *Comb. Proc. Intern. Pl. Prop. Soc.*, **30** : 156-164.

Thomas, M.B., Oates, M.R. and Spurway, M.I. (1980) *N.Z.J. Expt. Agric.*, **8** : 281-286.

Verzilov, V.F. and Runkova, L.V. (1972) *Izvestiya Akademii Nauk*, USSR Biologisches Kaya, **5** : 761-766.

Vita, M. de and Agnello, S. (2000) *Fiori. ColtureProtette* (special issue), **29** : 113-117.

Waseem, K., Hameed, A., Jilani, M.S., Kiran, M., Rasheed, M., Javeria, S. and Jilani, T.A. (2013) *Pak. J. Agri. Sci*, **50**(3) : 523-527.

Yamanouchi, M. and Yang, X.Z. (2000) *Japanese J. Soil Sci. Plant Nutrition*, **71** : 204-209.

Yoshimura, T., Nishiyama, M. and Kanahama, K. (2002) *J. Jpn. Soc. Hortic. Sci.*, **71**(4) : 575-582.

Yoshimura, T., Sasaki, A., Moriyama, T., Shibahara, Y., Katsuta, K. and Kanahama, K. (2006) *Hortic. Res.(Jpn)*, **5**(3) : 297-301.

Celosia

Parvathi Bennurmath, Rajiv Kumar, C. Aswath and Swati Saha

1.0 Introduction

Celosia is an herbaceous summer annual plant. The generic name *Celosia* is derived from the ancient Greek word *keleos*, meaning 'burning' and refers to the flame-like flower heads. It is highly heliotrope and drought resistant (Dole and Wilkins, 1999). It is grown as an ornamental plant in most of the parts of the world because of the attractive and vibrant colored inflorescences.

The stem is erect, glabrous. Leaf blade is unlobed, variable, mostly lanceolate or ovate. The leaves are either green or bronze/maroon, depending upon the cultivar. The inflorescence is terminal with broad, showy fan-shaped heads.

In some parts of the World like Africa, China, Indonesia, India, and other parts of Asia, its leaves and inflorescences are eaten as vegetables (National Research Council, 2006). The plants are well known in East Africa's highlands with an interesting and increasingly popular crop in specialty cut flower production (Zuck, 2015).

2.0 Importance and Uses

2.1.0 Aesthetic Uses

Celosia is used as ornamental plants, as cut flowers, or dry arrangements. The plant has eye-catching flowers that come in many different colors and forms. It is commonly known as Red Cockscomb, Chinese wool flower, feathery amaranth because of its resemblance to rooster head. *Celosia argentea* is most popular for growing in containers as house plants.

2.2.0 Therapeutic and other Uses

As ethno-medicine, different parts of *C. cristata* are used by traditional healers. The dried red flowers are effective against hematuria, abdominal pains, leucorrhea, heavy menstrual bleeding, hemoptysis, hemorrhoid bleeding and osteoporosis (Motaleb *et al.*, 2015). Traditional healers (*Hakims*) practicing in the Kashmir Valley mostly follow the Yunani system of medicine. Besides being used in traditional medicine, the inhabitants of the Kashmir valley have found a novel way of using this edible flower. It is used as a natural food colorant in local dishes to impart a fiery hot color to gravies, without affecting taste and aroma. The coloring property of *C. cristata* flower is due to the presence of betalains, which are water soluble pigments found in plants belonging to the family Amaranthaceae (Cai *et al.*, 2003).

C. cristata is a potential source of betalain pigment which is responsible for its brilliantly colored inflorescences. The pigment demonstrates high antioxidant and anti- inflammatory properties, suggesting that it may be adopted at commercial scale for the production of an important food additive with promising health benefits (Sayeed *et al.*, 2020). The Ethno-medicinal uses of Celosia is given in Table 1.

Its dried, ripe seed is reported to have medicinal properties because of its β-carotene, vitamin E, folic acid, ascorbic acid, phosphorus, calcium, iron, and

Uses of Celosia

In Pots In Hanging Basket In Vase

Floral Bouquet Dry Flowers in Pot Cut Flowers

Window Box In Terrace Gardening Flower Wreath

Along Paths In Border In Landscape Garden

Source: www.in.pinterest.com; www.gardenerspath.com; www.guzmansgreenhouse.com; www.bloomr.com; www.curiouscountrycreations.com; www.junketjuice.blogspot.com; www.floretflowers.com; www.newplantsandflowers.com

protein contents. The leaves contain amaranthine (betacyanin), oxalic acid, and phytic acid (Tang *et al.*, 2016).

Celosia argentea have reported to have higher API (Air pollution tolerance index) grades as tested for industrial plant species and are recommended for green belt development in the industrial area to mitigate air pollution (Singare and More, 2020). *C. cristata* is highly valued for its use as a natural food colorant, as well as a medicinal and ornamental plant.

Table 1: Ethno-medicinal Uses of Celosia*

Sl.No.	*Ethnomedicinal Uses*	*Part of the Plant Used*	I_P	I_u	*Fidelity Level (per cent)*
1	Diarrhoea	Flower	07	10	70
2	Dysentry	Seed, leaves	06	10	60
3	Heavy menstrual bleeding	Flower, seed	10	10	100
4	Leucorrhoea	Flower	10	10	100
5	Bloodshot eyes	Seed	05	10	50
6	Blurry vision	Seed	05	10	50
7	Hypertension	Flower	10	10	100
8	Cardiovascular	Flowers	10	10	100
9	Internal Hemorrhage	Flower	10	10	100
10	Liver Heat	Seed	04	10	40
11	Treatment of Helminths	Flower, leaves	08	10	80
12	Hemorrhoid Bleeding	Flowers	10	10	100

* Sayeed and Thakur (2020).

Fidelity Level (FL) is a quantitative method that measures the importance of a species for a given purpose. It's a helping tool that identifies the key informants' most preferred method of using the plant species for food or medicinal purposes. The fidelity level was calculated by using the formula proposed by Friedman and co-workers 10.

$$FL\ (\text{per cent}) = I_p/I_u \times 100$$

Where, FL = Fidelity level, Ip = the number of informants who cited the species for a particular use, and Iu= total number of informants that mentioned the plant for any use.

3.0 Origin and History

It is native to tropical American and Africa, naturalized pan-tropically. *Kashmiris* have found a unique way of using *C. cristata* as a natural food colorant in local dishes. This invention dates back to the 14th century when wazwan as a cuisine was introduced in Kashmir under Mongolian rule (Ahmad *et al.*, 2012). Celosia is now distributed in warm temperate climates in Asia, Africa and the America. It is thought to be first cultivated in East Asia from *C. argentea*. Some sources say that the *Celosia* was brought to Japan from China in the Nara period (A.D. 710-794) where it was established as an ornamental crop and various cultivars were developed (Foster and Chongxi, 1992).

4.0 Species and Cultivars

4.1.0 Species

Genus *Celosia* consists of about 60 species in the family Amaranthaceae. Celosia species ranged from *Celosia argentea, C. cristata, C. leptostachya, C. nitida, C. palmeri,* to *C. spicata*. Seed production is very high in these species, ranged from 200-700 kg/ha. One ounce of seed may be containing up to 43,000 seeds. One thousand seeds can be weight up to 1.0-1.2 grams. *Celosia cristata* L. in some cases is considered a species separate from *Celosia argentea* L., but can also be treated as varieties of the species, denoted as *Celosia argentea* L. var. *cristata* Kuntze (Zuck, 2015).

Celosias can be divided into three groups.

I. A crested type, *Celosia cristata*, also known as cockscomb celosia. The flower heads resemble the structures of a brain and in some varieties a rooster's comb. It blooms in red, yellow, pink, white, and orange.

II. A spike type, *Celosia spicata*, also known as wheat celosia. The flower heads resemble a head of wheat or a bottle brush. The flower colors are not as bright as with other types, but they come in rose-pink, white and dark red varieties.

III. A plume type, *Celosia argentea*, also known as plumed celosia. The flowers appear as feathery plumes and come in the shape of an arrowhead. They come in green-yellow and apricot-orange plumes.

Celosia argentea L. is a leafy vegetable crop commonly known as Lagos spinach. It is one of the most versatile herbaceous plants characterized with fast growth. It has erect stem with alternate leaves and few branches. The inflorescence is found at the tip of stems. The larger pink appendages of the central open flower comprise the calyx. The pink rimmed, white crown-like structure bearing anthers is the androecium. The pistil with its green ovary and pink style can be seen in the center of the flower. The bracts and bractlets, smaller than the flowers in this case, are best seen in association with the unopened flower. It is propagated by seed. The seedlings emerge 5-7 days after sowing and flowering 6-7 weeks after sowing. Pollination is by wind and insects especially bees and flies which visit the flowers regularly. Seed maturity starts from the basal part of the inflorescence and gradually moves up to the tip. Seeds get matured in 10-16 weeks after sowing (Okeke *et al.*, 2020).

4.2.0 Cultivars

In Kashmir, *C. cristata* is widely planted as an ornamental plant. With its striking appearance and magnificent hues of red, magenta, yellow, pink and orange, this flower adds beauty and life. The red and magenta varieties are mainly cultivated for culinary and medicinal purposes. All the other varieties are strictly ornamental. These are ideal choices as border plants, fresh cut flowers, for pots, garden edges, flower beds and can be incorporated into dry bouquets (Sayeed and Thakur, 2020).

Species of Celosia

Celosia plumosa *Celosia cristata* *Celosia spicata*

Celosia trigyna *Celosia caracas* *Celosia nitida*

Celosia virgata *Celosia caracas* *Celosia floribunda*

Celosia isertii *Celosia argentea* *Celosia palmeri*

***Source*: www.backyardnature.net, www.mygarden.org, www.google.com, www.swbiodiversity.org, www.jircas.affrc.go.jp, www.upload.wikimedia.org, www.i.pinimg.com**

Cultivars of Celosia

Bombay Fire | Jewel Box | Chief Red Flame

Celway Mix | Chief Dark Rose | Chief Orange Peach

Celosia Sunday Mix | Supercrest Mix | Tall Mix

Forest Fire | Apricot Brandy | Toreador

Source: www.silvertreeforest.com, www.all-americaselections.org, www.in.pinterest.com, www.google.com, www.mullerseeds.com

Cultivars of Celosia

Chief, Dragons Breath, Kimono, Yukata, Amigo Mahogany Red, Kurume, Chief Gold, Castle, Bright Sparks, Spring Green, Century, Prestige Scarlet

Source: www.in.pinterest.com, www.sakataornamentals.com, www.sakataornamentals.eu, www.floranova.com, www.takii.com, www.whiteseed.com

There are three main series or cultivars of *C. cristata* on the market for cut flower production: the 'Bombay' series, which are bred specifically for greenhouse production, the 'Chief' series, and 'Spring Green'. Other notable cultivars are 'Kurume' and 'Supercrest' (Kieft-Pro Seeds, 2010; Sakata 2015). Since *C. cristata*is also produced as a potted flowering plant, other cultivars are available such as 'Amigo' and 'Jewel Box', 'Sparkler' Series, 'Prestige Scarlet', 'Toreador'. The 'Chief' series was developed by the Sakata Seed Co. and is currently produced as seed. It has a height and width of 30 cm x 20 cm and is notable for produces large, uniform globe-shaped flower heads which is good for drying and cutting purposes. They are also highly suitable for production during periods of high temperature and can be used in both field production and in greenhouses (Sakata, 2015).

The 'Bombay Fiora' cultivar is also notable with a bicolor inflorescence. The Century cultivars are usually taller (1-2 feet), and are bright red, yellow, orange, or pink. The Kimono cultivars are generally smaller (4 inches - 1 foot), and have more muted colors, though similar to the Century cultivars. Other colors, such as white, burgundy, orange-red, *etc.*, can be found. Certain varieties will grow to 3–4 feet in height (Barash, 2006).

5.0 Genetics and Breeding

Celosia has chromosome number of 2n = 36. *Celosia cristata* is a branched, erect annual, 15 to 90 cm in height and 15 to 30 cm in spread. It has alternate, oval to lance-shaped leaves, 5 to 12 cm long that are medium green in color. The inflorescence is terminal with broad, showy fan-shaped heads. Colors for inflorescences are white, pink, red, violet, orange and yellow.

The complete chloroplast genome of *C. cristata* 'Xiaguang' was sequenced and analyzed using Illumina and PacBio sequencing. The chloroplast genome is 153,472 bp in length with a quadripartite structure containing a pair of inverted repeat regions of 25,408 bp (IRa and IRb) separated by a large single-copy region (LSC) of 84,975 bp and a small single-copy region (SSC) of 17,681 bp. The *C. cristata* 'Xiaguang' chloroplast genome encodes 117 genes, comprising 73 protein-coding genes, 36 tRNA genes, and eight rRNA genes. The A þ T content of the whole chloroplast genome is 63.3 per cent, whereas, the LSC, SSC, and IR region are 65.5 per cent, 69.4 per cent, and 57.4 per cent, respectively (Liu *et al.*, 2020).

One influential breeding company for *Celosia* is Sakata Seed Corporation. This company developed the 'Flame of Fire' cultivar that was awarded the All American Selections (AAS) bronze medal in 1935 (Sakata, 2005) (Table 2).

Breeding of *C. cristata* through hybridization is constrained in the small size of the flower which ranged from 6-8 mm gathered to form a set of inflorescences (Van Steenis *et al.*, 1978), and is self pollinated plant (UPOV 2001). This fact makes castration and isolation of reproductive organs more complicated.

Induce mutation using physical or chemical mutagens combined with *in vitro* culture technique are an effective method for plant improvement in several vegetative plants. Radio-sensitivity of *C. cristata* planlets LD_{50}, LD_{30} and LD_{20} were 68.73, 46.68 and 35.65 Gy, respectively. The mutant response of plantlets produced in MV_3 generation is abnormal growth such as shortening of internode and curly leaves (Hayati *et al.*, 2016).

Table 2: Celosia as Cut Flower (*Celosia cristata*) in the Market and their Growth Size (height x width), Flower Type, the Usual Production Type (Kieft-Pro Seeds, 2010; Sakata, 2015)*

Cultivars on the Market	*Plant Height x Width (cm)*	*Flower Type*	*Production Type*
'Chief' (Sakata)	30cm x 20cm	Globe -shaped	Field/Greenhouse
'Bombay' (Kieft)	70cm-1m x 13cm	Comb-shaped	Greenhouse
'Spring Green' (Kieft)	1-1.2m x 13cm	Comb-shaped	Greenhouse
'Kurume' (Takii)	70cm-1.2m x 13cm	Globe-shaped	Field
'Super Crest' Mix	70cm-1m x 13cm	Mix of comb and globe shaped	Field

* Kieft-Pro Seeds (2010); Sakata (2015).

Irradiation 0, 25, 50 and 75 Gy on *C. cristata* plantlets on MV_6 generation showed changes in morphological characters in the stem, leaves, and flowers. These changes occurred in 4 plants. Changes in chemical content occurred in clone C1U3 2.3.1 that irradiated at 25 Gy. Change in chemical content was indicated by the triterpenoid content (Muhallilin *et al.*, 2019).

6.0 Growth and Flowering

The models developed to predict flowering time and plant quality could be used by commercial growers to determine the impacts of changing growing temperature, growing plants at different times of the year and providing supplemental lighting. For example, celosia plants grown under an average DLI if 8 mol/m^2/d are predicted to flower from transplant in 45 d at 18°C and in 34 d at 23°C. Growing plants at these temperatures but with a DLI of 16 mol/m^2/d would accelerate flowering by 8 and 4 d, respectively. In addition, plant dry mass at first flowering is predicted to increase by 41 per cent and 31 per cent when the average daily temperature from 23 to 18°C would delay flowering from transplant by 6 or 5 d under an average DLI of 8 or 16 mol/m^2/d, respectively (Pramuk and Runkle, 2005).

7.0 Propagation

7.1.0 Seed Propagation

The plants can be propagated easily from seeds. The best time for sowing the seed is from the months of May to July. The seeds are normally broadcast on top of the soil. Germination takes place within 2 weeks after which the plantlets are plotted into prepared beds (Sayeed and Thakur, 2020).

Seeds are also sown in a plug tray one per cell with media that has a pH of 5.8-6.2. The plug stage, from sow to transplant, is 2-3 weeks long. During this time *Celosia* require a temperature of 20-22°C. The light requirement for plug production is light up to 26,900 Lux in the first three stages and 54,000 Lux in the final stage of growth. It is important to maintain the soil moisture and not let the media dry out. Fertilizers in plug production can start after germination (3-4 days) at less than 100 ppm N and later in stage three this can be increased to 100-175 ppm N. *Celosia* is also sensitive to root damage, which can result in early bud formation or deformed flowers. Plantings should be done before plugs get root bound, this usually being at the 2-3 week point or when the first pair of true leaves unfolds (Sakata, 2015).

The effect of sodium chloride (NaCl) concentration on seed germination and vegetative growth of Cockscomb varieties 'Amigo Red' and 'Super Master Yellow' was investigated. It is concluded that induced salinity up to 7 dS/m improved seedling and vegetative growth of cockscomb (Khan *et al.*, 2019).

It was observed that the temperature is a limiting factor for the germination of cockscomb, having its optimal thermal range between 20 and 30°C (Menegaes *et al.*, 2018).

It was found that the seed germination of *Celosia argentea* is solely dependent on appropriate temperature (25°C), alternating light and dark and a sowing depth of 1 cm, therefore, this species has potential of thriving under varying environmental conditions (Adegbaju *et al.*, 2018).

PGPR bacteria isolated from the rhizosphere of *Celosia cristata* L. The bacterium was identified as *Pseudomonas aeruginosa* (MCC 3198). Seed germination, seedling vigor, root length, shoot length, leaf count, dry and fresh weight, chlorophyll content and some defense enzymes of the plant *C. cristata* have significantly been improved as a result of application of this bacteria (Dutta *et al.*, 2019).

7.2.0 Micropropagation

Warhade and Badere (2015) observed that shoot tip derived from 12 days old seedlings were the best explants for micropropagation through recurrent regeneration when they were cultured over MS medium containing 0.5 µM NAA and 6.6 M µBAP and also found that 6 benzyladenine (BAP) was more effective than kinetin for shoot generation.

The effectiveness of BAP and NAA on establishment of plant regeneration for *C. cristata* was studied. Various explants (leaf, stem, shoot tip and bulb) derived from one-month-old aseptic seedlings were utilized to achieve complete plant regeneration. MS medium supplemented with various hormones, with an emphasis on BAP and NAA was tested to obtain direct and indirect regeneration. *C. cristata* (shoots) formed complete plantlets on MS added with 0.5-2.0 mg/l BAP and NAA (Yaacob *et al.*, 2014).

8.0 Soil and Climate

C. cristata is a tropical climate crop. It prefers fertile, moist soil, but will grow in most soils including acidic, sand, loam andclay. The preferred soil temperature is 16°C.The plant prefers a pH of 6.0–7.0. The plants do best in full sun but will tolerate partial shade.

Phytoextraction using *Celosia argentea* Linn. can potentially decontaminate Cd-contaminated soils. Cd in soils was efficiently removed by *C. argentea* combined

Propagation and Growing of Celosia

Collection of Seed | Celosia Seed | Sowing of Seed

Germination | Thinning | Pinching

Micropropagation of Celosia

In vitro Regeneration | Rooting of Explants | Hardening of Celosia

In Field | In Polytunnel | In Greenhoues

Source: www.scialert.net, www.scienceasia.org, www.gardengatemagazine.com, www.gardenoid.com, www.asergeev.com, www.in.pinterest.com, www.shop.floretflowers.com

with soil amendments such as ammonium chloride, *Bacillus megaterium* and citric acid increased the overall Cd reduction rate in soil by 40.5 per cent, 46.1 per cent, and 105 per cent, respectively and it was greater in the rhizoshere. Phytoextraction increased soil enzyme activities and thus improved soil functions (Yu *et al.*, 2020).

9.0 Cultivation

Celosia is a quantitative short day plant, meaning that flowers will initiate under short days. The growth of celosia is affected by water logging. High tunnel production compared to field production of *C. cristata* 'Chief Red' in the Midwest region was carried out. High tunnel production yielded 14 more stems/m^2 and TTH (Time to harvest) was reduced by 13 days compared with field production. Therefore, high tunnels can potentially offer new possibilities in production of *Celosia* (Ortiz *et al.*, 2012).

9.1.0 Planting

The seedlings are sown into the plots with 6-12 inches spacing. *Celosia* should be transplanted at 64 plants/m^2 and netting can be used for support. The temperature for planting through start of flower development should be 17-18°C during nights and 18-24°C during days. The total time from transplant to flowering is 10-14 weeks (Sakata, 2015). Exposure to cold temperatures causes of premature blooming.

A pot experiment was conducted in greenhouse study the effect of putrescine and alpha-tocopherol on the vegetative growth and flowering of *Celosia argentea* var. *Cristata* L. plants. Foliar application of putrescine at 200 ppm recorded highest increase values in the vegetative growth and inflorescence characters, and spraying with alpha-tocopherol at 400 ppm recorded highest values in the vegetative growth and flowering except for fresh and dry weight of the inflorescence, which gave the highest increase values when the plants were treated with 200 ppm alpha-tocopherol (Badawy *et al.*, 2014).

9.2.0 Irrigation

Effects of irrigation with municipal secondary-treated effluents on growth of celosia (*Celosia argenteavarcristata* L.) plants for cut flower production were examined. The effect of the two types of irrigation water on the chemical composition of soil and leaves, flower yield and postharvest performance was also examined. In celosia, boron accumulated to higher levels in leaves of plants irrigated with effluents. In addition, under irrigation with effluents, celosia accumulated higher levels of N and Mn. The secondary-treated municipal effluents can be used for production of celosia for cut flowers, in the northern-eastern part of the 'Negev' desert in Israel (Friedman *et al.*, 2007).

9.3.0 Manuring and Fertilization

Cockscomb requires nutrients, especially nitrogen, fertilizers with nitrogen, phosphorus-potassium ratio of 3:1:2, 20:10:20, or 15:5:15 serve the cockscomb best. Regular fertilization is needed for celosia to reach its full vigor of 10-30 inches.

It was found that the soil + Pitmose + Worms residues (1:1:1) in combination of GA_3 improved most growth characteristics such as number of leaves, flower diameter, wet weight and dry weight (AL-Chalabi, 2019).

The growth and flowering of *Celosia cristata* were the greatest when grown in a mixture of 70 per cent cocopeat : 30 per cent burnt rice hull and perhaps linked with a good balance in the aeration and moisture relationship of the media (Awang *et al.*, 2009).

Nitrogen at 45 kg/ha and phosphorus at 50 kg/ha is recommended for better growth of cockscomb flower (Muhammd *et al.*, 2018). Application of vermicompost at 5 t/ha along with Panchagavya at 100 ppm significantly improved flowering parameters *viz.*, days taken for first flowering, duration of flowering, number of flower per plant, single flower weight and flower head width and flower yield traits *viz.*, flower yield per plant, flower yield per plot and flower yield per hectare was achieved in Celosia (Sendhilnathan *et al.*, 2019).

9.4.0 Plant Growth Regulators

The Azospirillum (5 x 10 cfu/ml) inoculation or gibberellic acid (50 ppm) significantly improved floral characters of *Celosia cristata cv.* Red Velvet Moreover, Azospirillum application increased the contents of N and P in shoot, total carbohydrates, total chlorophyll and some endogenous hormones-like GA and ABA (Shanan *et al.*, 2014).

A plant spacing of 37.5 cm in combination with GA_3 at 50 mg/l significantly increased chlorophyll contents, leaf area and number of leaves, fresh and dry weight of stem, flower diameter and fresh and dry weight of flower of *Celosia cristata* (Arslan *et al.*, 2013).

10.0 Diseases and Pests

10.1.0 Diseases

Common celosia diseases include powdery mildew and fungal leaf spot (www.gardeningchannel.com). Ornamental cultivars are subject to few diseases, although fungal and bacterial leaf spots, Botrytis blight, Rhizoctonia root and stem rot, damping-off, and root-knot nematode may occur where humidity is high or where good growing practices, sanitation, and other cultural disease management practices that reduce inoculum and spread of propagules are not followed (Gould, 2017).

Disease	*Pathogen*	*Stage of Crop*	*Best Cultural Controls*
Alternaria leaf spot	*Alternaria alternate*	Can occur in plug trays but usually in landscape	Use pathogen free seeds and avoid over-head irrigation
Botrytis stem rot	*Botrytis cinerea*	Following transplant	Avoid rough handling at transplant control fungus and do not over water
Cold damage	Low temperatures that don't cause necrosis	Any stage	Try to keep production temperatures adequate and protect from frost in the landscape

Disease	Pathogen	Stage of Crop	Best Cultural Controls
High soluble salts injury	Harmful salts concentration in media	Mid-production	Use appropriate fertilizer concentration and check salts level periodically
Pythium root rot	*Pythium* spp.	Anytime in production or the landscape	Avoid using potting medium with poor drainage and don't overwater or over fertilize
Rhizoctonia damping-off	*Rhizoctonia solani*	Plug production mainly	Practice careful sanitation, use new potting media and pots
Xanthomonas leaf spot	*Xanthomonas* sp.	All stages of production	Use pathogen free seeds and avoid over-head irrigation

* Chase and Daughtrey (2002).

10.2.0 Pests

There are several pests, aphids and mites that are reported to feed on *Celosia.* Spider mites cause yellow spots on the leaves which results in decreased growth and under production. Aphids feed on the underside of the leaves which causes yellow spots to form. Nymph and adult spider produce webbing which also causes reduction in production. Aphids feed on the plant sap, which in turn causes the leaves to become deformed (Khan *et al.,* 2019).

The production of *C. argentea* is challenged by the sedentary endoparasite root-knot nematode (*Meloidogyne incognita*) which causes 60–80 per cent yield loss due in part to stunted growth, rotted, and galled roots. *Trichoderma asperellum* improved the growth and development of *C. argentea* infected with *M. incognita,* indicating that *T. asperellum* has the potential to be used as a biocontrol agent in *C. argentea* production (Temitope *et al.,* 2020).

Survey was conducted in different flower gardens of Vellayani area of Thiruvanthapuram district of Kerala to identify the *Meloidogyne* spp. associated with *Celosia plumosa* and the effect of different inoculum level on the growth parameters. The root knot nematode species was identified as *Meloidogyne incognita* based on the morphological and morphometric data. The plants affected with the nematode showed delay in flowering, heavy galling in roots and incipient wilting. On comparing the effect of different inoculum levels of root knot nematode on Celosia plants, maximum reduction in shoot and root length was observed in plants with 1,00,000 J2/pot and the reduction in growth parameters and nematode infestation were found to be directly proportional to the inoculum level (Narayana *et al.,* 2017).

11.0 Harvesting and Postharvest Management

11.1.0 Harvesting

The flowers could be expected around 3-4 weeks after transplanting. Flowers are harvested when they reach full bloom around September and are harvested in early October. These may be dried for culinary, medicinal or ornamental purposes.

Diseases and Pest of Celosia

Botrytis Blight | Root Rot | Bacterial Leaf Spot

Alternaria Leaf Spot | Alternanthera Mosaic Virus (AltMV) | Leaf Spot

Wilting | Celosia Bug | Root Galling Symptom

Phytoplasma | Mosaic and Mottling | Broad Mite

Source: www.ncbi.nlm.nih.gov; www.researchgate.net; www.canr.msu.edu; www.ask2.extension.org; www.commons.wikimedia.org; www.scialert.net; www.i1.wp.com; www.e-gro.org; www.insectimages.org; www.pnwhandbooks.org; www.invasive.org; www.content.ces.ncsu.edu

Within-row spacing of 15 or 20 cm combined with two-row or three-row per bed plantings resulted in field planting densities ranging from 100,005 to 200,010 plants/ha of cockscomb celosia (*Celosia argentea* var. *cristata* 'Chief Mix') and wheat celosia (*Celosia spicata* 'Pink Candle'). Cockscomb celosia produced four harvests. Both celosia species had a 4.5-month harvest season, one early summer peak harvest, and fewer than 300 harvested stems totaling 0.6 to 0.7 kg dry weight per plot for the year. For that species, three-row planting (high density) increased the total number of spray flower (multiple head) stems, provided longer stems later into the season and wider heads midway into the season, and prolonged the production of spray stems (15-cm spacing only) and reported as excellent candidates as new specialty crops in semi-arid conditions (Green *et al.*, 2010).

11.2.0 Postharvest Management

At harvesting, the stems are often dried to prevent *Botrytis*. *C. cristata* can be stored and transported cool with a minimum temperature of 5-6°C. A constant temperature during transportation can prevent condensation and, therefore, *Botrytis* infection. By packing the plants securely this can help prevent movement and breaking of the stems. *C. cristata* have a vase life that lasts 2 weeks (Sakata, 2015).

The flowers with red inflorescences are usually chosen to be dried for medicinal and ornamental purposes but 60 per cent of the flowers are dried for its use as an active food coloring agent. The drying technique requires the cut flowers to be hung upside down in bunches of 8-10 flowers. The flowers are then shade dried under the roof in order to retain the deep red hue and all the bioactive compounds. It may take several weeks for the flowers to completely dry. They are stored in open, dark and dry places until use. National Research Council, Washington D.C, documented similar methods for drying *C. cristata* (Gilman and Howe, 1999).

11.2.1 Marketing

The flowers are largely sold in dried forms by local spice vendors known as "*Bohris*". They are sold either in powdered form or as whole flowers. The seeds are also available for sale in the market. The dried flower can be bought for USD 20-25 per kilogram and powdered form is available for USD 32-35 per kilogram. Fresh flowers are also available for a limited period of time in the months of May to September. The plantlets can cost around USD 2-3 for a dozen (Sayeed and Thakur, 2020).

Celosia argentea can be dried with press drying method (5-7 days drying time) for ornamental value as there is no shattering and have good shelf life after drying (more than 170 days) and can be well utilized for making of dry flower products (Singh *et al.*, 2017).

After harvesting each batch of seeds, in their respective years, they were stored with an average moisture content of 11 and average germination of 98 per cent in cold chamber in packages of Kraft paper (15°C and 40 per cent RH). It was observed that the average vigor (first count of germination) of the seeds of these species of cockscomb was 88 per cent, and the average germination of normal seedlings was 89 and 96 per cent for the *C. argentea* and *C. cristata* species, respectively, with mean germination of 2.9 days for all the storage periods. The relative frequency demonstrated homogeneity in the germination of the two species of cockscomb maintaining germinative viability in all periods of storage. It is concluded that the physiological potential and sanitary quality of the seeds of the *C. argentea* and *C. cristata* were preserved for periods ranging from zero to sixteen years of storage (Menegaes *et al.*, 2019).

It was observed that the temperature is a limiting factor for germination of cockscomb, having its optimal thermal range between 20° and 30°C (Menegaes *et al.*, 2018).

12.0 Biotechnology

12.1.0 Tissue Culture

The shoot-tip explant harvested from ethyl methane sulphonate (EMS) and gamma ray (GR) mutagenized seedling was cultured over MS medium fortified with NAA and BAP for five generations to amplify the mutated sector. Mutagens reduced the regeneration efficiency of the explant and affected its plant growth regulator-dependence for multiple shoot induction. While the 12d-old shoot-tip from GR-treated seedling induced shoots with 0.5μM NAA+6.6μM BAP; that from EMS-treated seedling induced shoots with 8.8μM BAP. It establishes that mutagens affect the regeneration process in the explant in *Celosia cristata* (Rinkey and Badere, 2020).

The combination of silicon and agar at 1.0 and 8.0 g/l, respectively, improved *C. cristata* anatomical attributes, such as leaf blade width and thickness of the palisade and spongy parenchyma (Assis *et al.*, 2018).

Marcelina *et al.* (2016) described the influence of sugar, ammonium nitrate and plant growth regulators on *in vitro* flowering from nodal explants of *Celosia argentea* var. *cristata* cultured for 6 weeks on basal MS medium under aseptic and light conditions. Supplementation with 1.0 mg/l KIN induced maximum number of new shoots (2.52). Flowering occurred on 100 per cent of *in vitro* shoots cultured on modify MS medium supplemented with 16.50 and 8.25 g/l of NH_4NO_3 after 40 days.

Ning *et al.* (2011) expressed that, in *Celosia cristata*, IAA and NAA could promote the rooting and flowering *in vitro*, the flowering rate after 30 d of culture was 100 per cent in medium of MS+IAA 0.5 mg/l or NAA 0.1 mg/l.KT 0.5~1.5 mg/l promoted differentiation and blooming, but 6-BA had inhibition for flower bud of plantlet. GA_3 profited for root inducing and flower formation, all the plantlets flowered in MS medium added GA_3. Rooting rate was positively correlated with flowering rate (R=0.902, P0.01) and flowering number (R=0.729, P0.01).

12.2.0 Cryopreservation

Oxidative stress is a potential cause of damage in plant cells during cryopreservation, and hydrogen peroxide (H_2O_2) may be the leading reactive oxygen species (ROS) components. In an investigation it was found that adding low concentration of H_2O_2 (4 μmol/l H_2O_2) at early cryopreservation steps improves cockscomb seedlings cryo-tolerance by enhancing the antioxidant capacity, which prevents ROS accumulation (Li *et al.*, 2018).

13.0 References

Adegbaju, O.D., Otunola, G.A. and Afolayan, A. J. (2018) *Asian J. Plant Sci.*, **17** : 47-55.

Ahmad, M., Akhter, S., Masoodi, S.R. and Wazwan. (2012) *J. Med. Sci.*, **15** : 174-75.

AL-Chalabi, A.T.M. (2019) *Int. J. Agric. and Stat. Sci.*, **15** : 611-615.

Anonymous (1996) *National Research Council, Lost crops of Africa*: volume I : grains. National Academies Press.

Arslan, A., Muhammad, M., Amir, M.S., Seema, M., Ahsan, I. and Ali, A. (2013) *Russian J. Agric. and Socio-Economic Sci.*, **4**(16) : 3-10.

Assis, F.A.D., Souza, G.A.D., Dias, G.D.M.G., Assis, G. A. D., Rodrigues, F. A., Pasqual, M. and Carvalho, F. J. (2018) *Pesquisa Agropecuária Brasileira*, **53** : 30-41.

Awang, Y., Shaharom, A.S., Mohamad, R. B. and Selamat, A. (2009) *Amer. J. Agric. Biol. Sci*, **4** : 63-71.

Badawy, E.M., Shanan, N.T., Mahgoub, M. H. and Hegazi, N.A. (2014) *Bulletin of Faculty of Agriculture, Cairo University*, **65** : 78-83.

Barash, C.W. (2006) *Horticulture update, Celosia cristata*. Texas Cooperative Extension. Texas.

Cai, Y., Sun, M. and Corke, H. (2003) *J. Agric. Food Chem.*, **51** : 2288-2294.

Chase, A.R. and Daughtrey, M. (2002) Pest and Diseases: This month in diseases celosia, www.gardeningchannel.com.

Dole, J.M. and Wilkins, H.F. (1999) Floriculture: Principle and Species. Viacom Press, New Jersey, pp. 613.

Dutta, S., Sarkar, A. and Dutta, S. (2019) *Int. J. Curr. Microbiol. App. Sci.*, **8** : 985-997.

Foster, S. and Chongxi, Y. (1992) Inner Traditions/Bear and Co.

Friedman, H., Bernstein, N., Bruner, M., Rot, I., Ben-Noon, Z., Zuriel, A. and Hagiladi, A. (2007) *Sci. Hortic.*, **115** : 62-69.

Gilman, E. F. and Howe, T. (1999) *Fact sheet FPS113*, Institute of Food and Agricultural Sciences, University of Florida.

Gould, A. B. (2017) *Diseases of celosia*, Handbook of Florists' Crops Diseases. pp.1-40.

Green, S.R., Picchioni, G.A., Murray, L.W. and Wall, M.M. (2010) *HortTechnology*, **20** : 612-619.

Hayati, D., Syarifah, I.A. and Krisatini. (2016) *J. Trop. Crop. Sci.*, **3** : 61-65.

Khan, Z., Gola, A.Q., Abro, J.A., Badini, M.A., Aziz, T., Mahesar, M.A. and Kasi, M.U. (2019) *J. Entomol. Zool. Stud.*, **7** : 495-499.

Kieft-Pro Seeds. (2010) Grower Facts: Bombay Celosia. Ball Horticultural Company.

Li, B.L., Zhang, M., Wang, X.Z., Jiang, X.R. and Liu, Y. (2018) In: *III International Symposium on Plant Cryopreservation* 1234. pp.73-78.

Liu, X.F., Ye, Y.J., Liu, J.M., Yu, B. and Xu, Y.C. (2020) *Mitochondrial DNA Part B*, **5** : 1338-1339.

Marcelina, K.M. and Oktawia, M. (2016) *J. Basic Appl. Sci.*, **12** : 135-139.

Menegaes, J.F., Barbieri, G.F., Bellé, R.A. and Nunes, U.R. (2018) *Ornam. Hortic.*, **24** : 408-414.

Menegaes, J.F., Barbieri, G.F., Bellé, R.A. and Nunes, U.R. (2019) *Ornam. Hortic.*, **25** : 34-41.

Motaleb, M.A., Abdullah-Al-Mamun, M.M., Hossain, M.K., Alam, M.K. and Sultana, M. (2015) *Eur. J. Med. Plants.*, 23-52.

Muhallilin, I., Aisyah, S.I. and Sukma, D. (2019) *Biodiversitas J. Biol. Diversity.*, **20** : 862-866.

Muhammd, W., Amin, J., Imtiaz, M., Shah, A., Khan, M.M., Ghani, F. and Anjum, M.M. (2018) *Int. J. Env. Sci. and Nat. Resour.*, **8** : 79-83.

Narayana, R., Babu, C.P. and Thomas, S. (2017) *Pest Manag. Hortic. Ecosystems*, **23** : 160-164.

Ning, S., Xiaoqiang, C., Xinhai, Z. and Lei, Z. (2011) *Chinese Agric. Sci. Bull.*, **13**.

Okeke, N.F., Ilodibia, C.V. and Okoli, B.E. (2020) *Univ. J. Plant Sci.*, **8** : 1-10.

Ortiz, M.A., Hyrczyk, K. and Lopez, R.G. (2012) *HortScience*, **47** : 1265-1269.

Pramuk, L.A. and Runkle, E.S. (2005) *J. Ame. Soc. for Hortic. Sci.*, **130** : 813-818.

Rinkey, P.K. and Badere, R.S. (2020) *Indian J. Genet.*, **80** : 351-353.

Sakata (2005) Traditional Japanese flowers expanding the international market for the celosia.

Sayeed, R. and Thakur, M. (2020) *Ind. J. Tradit. Knowl.*, **19** : 788-794.

Sayeed, R., Thakur, M. and Gani, A. (2020) *Heliyon*, **6** : 05792.

Sendhilnathan, R., Balaraman, E., Rajkumar, M. and Sureshkumar, R. (2019) *Plant Arch.*, **19** : 938-940.

Shanan, N.T., Medhat, Y., Abou, Z. and Zaenab, E.H.S. (2014) *World J. Agric. Sci.*, **10** : 146-153.

Singare, P.U. and More, S.N. (2020) *SN Applied Sciences*, **2** : 1-13.

Singh, A., Bhandari, A., Chhatrola, H.N. and Patel, B.N. (2017) *Current Agri. Res. J.*, **5** : 292.

Tang, Y., Xin, H.L. and Guo, M.L. (2016) *Revistabrasileira de farmacognosia*, **26**: 787-796.

Temitope, A.E., Patrick, A.A., Abiodun, J., Olasekan, A.A., Onye, A.C., Vincent, A.O.T. and Elliseus, R.J. (2020) *Open Agric.*, **5** : 778-784.

UPOV. 2001. Draft guideline for the conduct of test for distinctness, uniformity, and stability; Celosia: Celosia L. Intl. Union for The Protection of New Varieties of Plants, Geneva.

Van Steenis, C.G.G.J., Bloembergen, S. and Eyma, P.J. (1978) Flora. PT. Pradnya Paramita, Jakarta

Warhade, M.I. and Badere, R.S. (2015) *Ind. J.Biotech.*, **14** : 402-410.

Yaacob, J.S., Saleh, A., Elias, H., Abdullah, S., Mahmad, N. and Mohamed, N. (2014) *Sains Malaysiana*, **43** : 715-722.

Yu, G., Jiang, P., Fu, X., Liu, J., Sunahara, G.I., Chen, Z. and Wang, X. (2020) *Environ. Pollut.*, **266** : 115-408.

Zuck, C. (2015) *Reformation of specialty cut flower production for Celosia cristata.*

13

Gypsophila

Safeena, P. Pal, T.K. Bose and R. Jyothi

1.0 Introduction

Gypsophila is one of the most important and popular annual and perennial flowering plants. It is useful as cut flowers. Plants are hardy and excellent for mixed borders, dry-stone wall or rock gardens and prefer alkaline soil. The leading *Gypsophila* producing countries are Holland and Israel. Cultivation of *Gypsophila* has spread to countries like Columbia, Ecuador, and Peru also (Kwadwo *et al.*, 2009). It is considered as one of the ten best-selling cut flower species worldwide (Zvi *et al.*, 2008). It is commonly known as Baby's breath.

2.0 Importance and Uses

2.1.0 Aesthetic Uses

The species *Gypsophila paniculata* is grown worldwide as a commercial filler cut flower, bedding plant or as potted plant. *Gypsophila paniculata* is used not only as cut flowers, but also as a "filler" in flower arrangements. Because of its ornamental value, *Gypsophila paniculata* has been grown not only for flower arrangements but also as a perennial garden plant (Wang *et al.*, 2013). It is used in all types of banquets and floral designs. It can be used for fresh or dried flower arrangements. Gypsophila consists of number of annual and perennial flowering plants grown almost throughout the world. They are useful for mist-like effects in mixed borders, beds and edges and are also suitable for rock garden. They make an excellent effect as filling amongst shrubbery and good for covering unkept places with a mass of delicate blooms. Flowers of *G. elegans* and *G. paniculata* are popular as cut flower and used in flower arrangement and bouquets (Plomacher and Hagemann, 1976; Bremer, 1982). Gypsophila is a major cut flower in Korea (Lee *et al.*, 1993). Cut flower of gypsophila is exported from Zimbabwe to the Netherlands (Riddell, 1990). Results of survey in Japan revealed that *G. paniculata* and chrysanthemum were the most popular species for interior decoration and gypsophila, roses and carnation were popular flowers as gifts (Matsuo and Nonaka, 1991).

2.2.0 Therapeutic and other Uses

The roots of *G. paniculata* and *G. arrostii* (soap root) are known as saponin drugs and have been used as detergents and expectorants (Frechet *et al.*, 1991). The genus is a source of saponins that can be used for many purposes, including the production of photographic film and hemolytic laboratory reagents. Their detergent qualities make them useful in soap and shampoo. *G. rokejeka* is used to make the dessert halva (Henry, 1993). Species are also ingredients in liqueur, cheese, and ice cream, providing flavour, aroma, and crispness to foods.

Several species are hyper-accumulators of boron, and may be planted to absorb the element from polluted soils (Korkmaz *et al.*, 2010).

3.0 Origin and Distribution

Gypsophila paniculata L. originates from many regions including Europe, Central Asia and North America. It belongs to the family family Caryophyllaceae.

Uses of Gypsophila

Flower Bouquet

In Pot

In Vase

In Wreath

In Floral Crown

In Garland

Table Decoration

Cake Decoration

As Centerpiece

As Filler

Uses of Gypsophila

Near Water Pool

Flower Mix Border

Hanging Basket

Edge Plant

Flower Bed

In Rockery

Source: www.farmergracy.co.uk, www.caroledrake.com, www.norseco.com, www.flowersacrossindia.com, www.pinterest.co.uk, www.woodflowers.com, www.in.pinterest.com, www.emmaloves weddings.com, www.pinterest.ca, www.welwynflorist.co.uk, www.amazon.com, www.gpnmag.com

Gypsophila is native to Central and Eastern Europe, Eurasia, Africa, Australia, and the Pacific Islands. Its natural habitat is on the Steppes in dry, sandy and stony places, often on calcareous soils (*gypsophila* = chalk-loving). The genus name is from the Greek gypsos ("gypsum") and philios ("loving"), a reference to the gypsum-rich substrates on which some species grow (James 1975). Turkey has a particularly high diversity of Gypsophila taxa, with about 35 endemic species. Major Gypsophila growing countries in the world are Asia, North America, South America, Europe and Australia. The main areas of gypsophila production are Belgium, Germany, Israel, Italy, Netherlands *etc.* Major Gypsophila growing states in India are Maharashtra, Karnataka, Andhra Pradesh, Haryana, Tamil Nadu, Rajasthan and West Bengal.

4.0 Morphology

These are branched or spreading slender herbs, with scanty foliage when in bloom. Leaves are small, entire and opposite. Flowers are borne in panicles; sepals 5, united below, petals 5, clawed, very small, usually white, stamens 10; styles 2 (Bailey, 1963). *Gypsophila* species showed major variation in inflorescence type ranging from many-flowered lax thyrses or panicles *e.g., G. elegans* M. Bieb., *G. paniculata* L., *G. pilosa* Huds. to compact head-like cymes *e.g., G. capitata* M. Bieb., *G. capituliflora* Rupr. and *G. caricifolia Boiss.*, and few uni-flowered raceme-like monochasia *e.g., G. bazorganica* Rech.f., *G. saponarioides* Bornm. and Gauba). Each small flower has a cup-like calyx of white-edged green sepals containing five petals in shades of white or pink. The fruit is a rounded or oval capsule opening at valves, contains several brown or black seeds which are often shaped like a kidney or a snail shell.

5.0 Species and Cultivars

Gypsophila is an annual or perennial herbaceous flowering plant belonging to the family Caryophyllaceae. The genus comprises approximately 150 species of annual or perennial herbaceous nature, creeping or cushion-forming plants, inhabiting primarily the mountainous steppes in the north temperate part of the Old World with a diversification hotspot in the Irano-Turanian region.

5.1.0 Species

Some of the important species are described below:

Gypsophila acutifolia

It is native to North Caucasus and South Russia. Plants are deep rooted herbaceous perennial with many branched stems; height 1.0 m to 1.5 m; leaves greyish green, lance-shaped and pointed. Panicles spreading, flowers 2-5 cm across. It produces cymes of light purplish pink five-stellate flowers from July to September.

G. aretioides

It is native to the Caucasus and the mountains of North Iran. Plants are very dense, hardy, cushion-forming, evergreen; height is 5-15 cm; leaves tiny, oblong and fleshy, flowers small, usually solitary, star-shaped white. Tiny, fleshy, oblong, grey-green leaves form a dense foliage. It is extremely dense, grown more for its form than flowers. Flowering is rather scarce with solitary blooms being very small

Species of Gypsophila

Gypsophila acutifolia — *Gypsophila aretioides* — *Gypsophila cerastioides*

Gypsophila elegans — *Gypsophila paniculata* — *Gypsophila repens*

Gypsophila muralis — *Gypsophila pacifica* — *Gypsophila perfoliata*

Gypsophila rosenschleier — *Gypsophila fastigiata* — *Gypsophila tenuifolia*

Source: www.commons.wikimedia.org

and stemless. Small, star-shaped, 5-petaled, white flowers bloom infrequently in June-July.

G. cerastioides

It is native to the Himalayas. Plants are mat-forming, perennial; leaves are 6-10 mm long, spoon-shaped or broadly obovate and hairy; flowers on loose panicles, 1.5 cm across, trumpet-shaped, white in colour, veined and faintly tinged pink.

G. elegans

It is native to South Ukraine and Turkey. Plants are erect, branching, annual; leaves 2-4 cm long, oblong-lance-shaped to linear-lance-shaped; flowers 4-petalled, star-shaped, 1 cm across, produced on long and slender stalks. It is commonly known as the annual baby's-breath or showy baby's-breath.

G. paniculata

Tap rooted herbaceous perennial, native to central and eastern Europe. Plants are branched, leaves glaucous and hairless; flowers are loose and borne on panicles. It grows 1.2 m tall and 1 m wide (Vettori *et al.*, 2015; Toaima *et al.*, 2016). It is has rosette of greyish green basal leaves and linear lanceolate from which the flower stems propagate and end in mounds of branching stems covered in clouds of tiny white or pink flowers (Cronquist, 1981; Atasxlar *et al.*, 2009). It is widely known as baby's breath.

G. repens

This species is a trailing, perennial, native to the mountain of Central and South Europe. Plant height is variable ranging from 20 cm to 50 cm; leaves linear, often sickle shaped; loose corymb-like panicles bear star-shaped flowers. Useful in rockeries. It is commonly called as the alpine gypsophila or creeping baby's breath. It is a prostrate, mat-forming herbaceous perennial, growing around 20 cm tall by 30–50 cm wide. During summer, it bears masses of star-shaped flowers which may be white, lilac or light purple in colour, in loose panicles. Plants are often grown in rock gardens or against dry stone walls. It is used as a cut flower.

G. petraea

It is endemic to the Eastern Carpathians and Southern Carpathians, Romania.

G. fastigiata

It is a montane, mat-like species with wiry stems of 30 to 45 cm long and open sprays of white flowers.

G. tenuifolia

It forms a tight, bright-green dome with wiry stems to 20 cm topped with a loose cluster of relatively large, white flowers.

G. perfoliata

It is a maritime sand-dune species naturally occurring in Asia and south eastern Europe (Waclaw and Alina, 2016).

Other species of ornamental importance are *G. muralis, G. pacifica, G. perfoliata, G. rosenschleier, G. stevenii* and *G. fastigiata* (Mlynarcyk, 1993).

5.2.0 Cultivars

G. elegans

Bright Rose (rose-pink), Carminea (caramine-pink), Covent Garden (white), Giant White (white), Red Cloud (caramine-pink) and Rosea (rose-pink).

Table 1: Characteristic Features of some Gypsophila Cultivars

Cultivars	*Species*	*Flowers*	*Height*	*Other Features*
var. *alba*	*repens*	Clear white	4-6"	
'Bristol Fairy'	*paniculata*	White, double	24"	Repeat bloom
'Caucasica'	*aretoides*	White	18"	Dense, blue-green foliage
'Compacta Plena'	*paniculata*	White, semi-double	18"	
'Dorothy Teacher'	*repens*	Soft pink	4-6"	Blue-green leaves
'Double Snowflake': 'Schneeflocke'				
'Dubia'	*repens*	White flushed pink	4-6"	Bluish leaves, red stems
'Festival'	*paniculata*	Pink, white mix	15"	From seed, single, double
'Flamingo'	*paniculata*	Pink, double	24"	
'Fratensis'	*repens*	Rich pink	4-6"	
'Perfecta'	*paniculata*	White, double, large	30"	
'Pink Fairy'	*paniculata*	Pink, double	18"	Blooms until frost
'Pink Star'	*paniculata*	Pink	18"	
'Red Sea'	*paniculata*	Rose, double	3-4'	
'Rosea'	*repens*	Pink	4-6"	
'Rosenschlier'	hybrid	Pink, semi-double	15"	*Paniculata x repens*
'Rosy Veil':'Rosenschlier'				
'Snowflake':'Schneeflocke'				
'Schneeflocke'	*paniculata*	White, double	36"	
'Viette's Dwarf'	*paniculata*	Pinkish, double	18"	

Source: Leonard Perry, University of Vermont.

G. paniculata

Bristol Fairy (white), Compacta Plena (pink to white) and Flamingo (pale pink). There are many named forms of *G. paniculata*, which may have single or double flowers in white or pink shades. Most popular and commonly grown cultivars of *Gypsophila paniculata* are Bristol Fairy (white), Bridal Veil, Compacta Plena (pink to white), Covent Garden, Early Snowball, Danfestar, Double Snowflake, Fairy Perfect, Festival White, Festival Pink, Festival Star, Flamingo (pale pink), Happy

Cultivars of Gypsophila

Bright Rose — Carmenia — Covent Garden

Rosenschleier — Bristol Fairy — Compacta Plena

Flamingo — Dorothy Teacher — Huixing (Mutant)

Gypsy White — Gypsy White Improved — Rosea

Source: www.stokeseeds.com, www.all-americaselections.org, www. id.pinterest.com

Festival, Perfecta, Pink Fairy, Pink Star, Red Sea, Schneeflocke, Snowflake, Virgo *etc.* All *Gypsophila* varieties have white flowers except Flamengo which has pale-pink flowers. Amongst these 'Festival Star', 'Snowflake' and 'Compacta' are single-flowered; 'Bristol Fairy', 'Double Snowflake', 'Early Snowball', 'Perfecta', 'Virgo', Festival White' and 'Happy Festival' are double-flowered; 'Red Sea' is single pink-flowered and 'Flamingo', 'Pink Fairy' and 'Festival Pink are double pink-flowered '.

G. repens

Dorothy Teacher (pale pink but darken with age), and Fratensis (pale pink). The cultivars, Floriana Mist and Floriana Cascade, have been developed from clones of *G. paniculate* cv. Bristol Fairy. Both the cultivars produce double flowers of 0.7-0.9 cm diameter on 1 and 1.2 m long inflorescences, respectively. Floriana Mist flowers earlier than Floriana Cascade (Wilfret *et al.*, 1986). In Australia, several cultivars of gypsophila were registered. Happy Festival is compact (height 37.3 cm), produces semi-double, pinkish-white (RHS 155C) flowers, 7.99 mm in diameter (Anon., 1997b). Another cultivar Magic Arbel is semi-compact (height 67.1 cm), produces double, shiny white (RHS 155D) flowers (aging to very pale pink), 11 mm in diameter (Anon., 1997a). Festival, Magic Tavor and White Festival are also some cultivars from Australia (Anon., 1997c).

G. elegans

The white-flowered cultivars of the annual *G. elegans* include 'Covent Garden', 'Grandiflora Alba', 'Giant White', 'White Elephant', 'Lady Lace', 'White Monarch' and 'Snow Fountain'. Some of the pink-flowered cultivars of *G. elegans* are 'Red Cloud', 'Rosea' and 'Carminea'. Fratensis (pale pink) and Dorothy teacher (Pale pink but darken with age) are some of the cultivars.

6.0 Genetics and Breeding

6.1.0 Genetics

The basic chromosome number of *Gypsophila* is n = 17. Most of the species are diploid (2n=34) as reported by Blackburn (1928), Favarger (1946) and Skalinska (1950). In case of *G. elegans*, however, Furusato (1940) recorded 2n=40. *G. pacifica* was reported to be tetraploid (Favarger, 1946). The chromosome number of 2n=34 is most commonly reported although there are isolated reports of 2n=26 and 2n=68 (IPCN Chromosome Reports, 2014).

6.2.0 Breeding

Kishi *et al.* (1994) reported successful interspecific hybrid between *G. paniculata cv.* Red Sea (2n = 34) and *G. manginii* (2n = 66), produced by ovule-embryo culture, the combination is incompatible using ordinary cross breeding methods. Ovules were cultured on MS medium supplemented with 2.9 µM GA_3, 3 per cent sucrose and 0.8 per cent agar. On this medium, they grew into torpedo-shaped embryos which regenerated into plantlets. The hybrid plant (2n=50) had double, pale purplish-pink flowers. Hybridity was confirmed by morphological intermediacy, chromosome number and peroxidase isozyme patterns.

Classical hybridization is seriously hampered by the sterility of *G. paniculata* (Wang *et al.*, 2013). As a consequence, very few variations exist among commercial cultivars used as fresh and dried cut flowers for fillers in flower arrangements (Zvi *et al.*, 2008). Flower type is the key ornamental trait and the top interest in *G. paniculata* breeding. Several commercial varieties of *G. paniculata* have been released by breeders, such as 'Million Stars', 'Bristol Fairy', 'Snowball', and 'Cloudstar 5' (Li *et al.*, 2019; Shibuya *et al.*, 2017; Zvi *et al.*, 2008).

The study of basic chromosome number and the reproductive system (potential barriers to intraspecific and interspecific incompatibilities) of the wild species as well as the commercial hybrids forms the basis of cross breeding and selection of variety. A significant difference in ploidy was found between the species *paniculata*, its cultivar Flocon de Neige doubles, the species *libanotica* Boiss. and Pacifica Kom. A significant difference in the chromosome number was found among the commercial hybrids of Gypsophila and this difference was observed among them as well as when compared with the wild type species. The flower structure of commercial hybrids can be remarkably complex. Some hybrids such as Biancanievesw and MeryGw showed empty pollen sacs, whereas others such as "Mirabella" and New Lovew showed pollen grains trapped into anthers. However, the pollen when present, exhibits immature or degenerated appearance. The examined commercial hybrids showed a gynoecium morphologically and anatomically conforming to that of wild species and it was potentially fertile. Therefore, the fertile pollen collected from the wild species could be used to fertilize the compatible eggs of commercial hybrids, thereby speeding the process of breeding varieties (Vettori *et al.*, 2015).

To develop new cultivars with a novel double-flower type, gamma radiation mutagenesis was conducted using a variety of 'Cloudstar 4', which has a semi-double flower (Fan *et al.*, 2020). In 2018, a new *G. paniculata* cultivar was selected from the mutagenized population and was named 'Huixing 1'. Compared with its progenitor ('Cloudstar 4'), 'Huixing 1' has bigger flowers and more flower petals. In April 2020, the new cultivar was officially authorized to be released as 'Huixing 1' by the Yunnan Forestry and Grassland Bureau with accession no. 20200013.

7.0 Growth and Flowering

The growth and flowering behaviour of *Gypsophila paniculata* is strongly influenced by several factors in growth conditions, mainly daylength and temperature (Thakur *et al.*, 2013). Doi *et al.* (1990) studied flower development in *cv.* Bristol Fairy using a scanning electron microscope. In plants which had been sufficiently chilled in the previous winter, flower bud initiation occurred when the stem reached approximately 18 cm in length. Sepals and petals subsequently differentiated. Stamen primordia in the terminal floret were initiated around 30 April. Between 5 and 10 May, when the buds were visible, the pistil primordium appeared in the terminal floret, subsequently, the style elongated and the petaloid stamens appeared. The effect of environmental factors, growth substances, nutrition and soil moisture are discussed separately.

7.1.0 Environmental Factors

Gypsophila paniculata is an obligate long day plant. Successful commercial production of *Gypsophila* depends upon knowledge of the environmental conditions necessary to produce flowers at specific times of the year in selected cultivars (Hicklenton *et al.*, 1993).

7.1.1 Temperature

Doi *et al.* (1984) exposed the plants of *cv.* Bristol Fairy up to 61 days to natural low temperature in winters (to break dormancy) and then transferred to a plastic house maintained at 15-27°C with an extended day length of 16 hours. Flowering response varied widely, few plants produced some flowers even after the longest low temperature exposure. They also recorded that the pink-flowered cultivars Flamingo and Red Sea had the least low temperature requirements, followed by the white-flowered *cv.* Diamond. The cultivar Perfecta did not flower even after the longest low temperature exposure. Doi *et al.* (1991a) also reported that after winter chilling, plants were exposed to 30°C from 06.00 to 18.00 h and 25°C from 18.00 to 06.00 h, rosette formation occurred in cultivars Perfecta and Bristol Fairy line 20 which require long chilling. About 50 per cent of the shoots of Diamond and Bristol Fairy line 03, which require less chilling, produced rosettes while Flamingo, Red Sea and Bristol Fairy line 08, which require little chilling, flowered normally. In an experiment on the effect of temperature, Doi *et al.* (1991b) found that shoot elongation was delayed and percentage of flowering was reduced in 13/20°, 20/13° and 20/20°C compared with 13/13°C and ungrafted line 13 plants. Plants of line 04 grew rapidly and all flowered when their roots and shoots were chilled at 2°C for 55 days. All the plants with the roots chilled, flowered during chilling; removal of the flowers after chilling resulted in rapid shoot growth and a high level of flowering (80 per cent) in lateral shoots which had not been chilled. It is suggested that chilling effects may be translocated from root to shoot.

The optimum temperature for the growth of Gypsophila is 15°C. Higher temperature of more than 30°C will reduce flowering. Gypsophila plantlets are grown at higher relative humidity. Nishidate *et al.* (2012) reported that plants of gypsophila will stay in the vegetative growth stage and form a rosette under short daylighting and lower temperatures. The formation of the rosette significantly affects the flower production of *Gypsophila paniculata* because these rosette plants cannot bloom.

7.1.2 Light

Gypsophila is in general Long Day Plant (LDP) and flower inducement is more effective after 45 days of vernalization and 16 hours of photoperiod. Hicklenton (1986) reported that supplemental lighting (93 μmol/s/m^2) supplied from high pressure sodium lamps from 20.00 to 07.00 h each day applied for 42 or 63 days during September-February, or January-June enhanced vegetative growth and flowering. In another experiment, vegetative growth and yield of flowers increased with cumulative PAR up to about 745 mol/m received over a 115-day period. Higher irradiance during early development was most effective for enhancing yield (Hicklenton, 1987). To evaluate the efficacy of different lamps for photoperiod lighting of gypsophila *cv.* Perfecta, a light gradient method was used. During day, extensions up to daylight 18 h and dusk-to-dawn lighting treatments, plants received lighting levels (PPF) ranging from 0.1 at the edges of the greenhouse to 2.0 μmol/m^2/s underneath the lamps (from 8 August to the end of November, 1990). Only plants in the incandescent treatment were induced to flower; plants from the other treatments remained in a rosette stage. The rate of elongation (bolting) correlated positively with the lighting level in the incandescent treatment. In the second experiment (spring flowering) in all treatments some of the shoots started to elongate. Only in the incandescent and red+far-red treatments did the majority of the shoots bolt and flower; in the other treatments the majority remained at the rosette stage. In the third experiment (from 31st July to the end of November, 1991) plants in all treatments readily bolted and flowered; the rate was a little lower with blue fluorescent or red light emitting diodes (Table 2). Bolting rate correlated positively with the lighting level in all treatments (Graaf-Van der Zande *et al.*, 1997). Dijk and Wolter (1990) found that with a day length of 24 h, the yield of primary stems was slightly higher in gypsophila and that of lateral stems slightly lower than with a day length of 16 h. Total yield was highest with a combination of 16 h day length and an early starting date of supplementary lighting. Karaguzel and Altan (1999) reported that growth, yield and quality were best from the plants grown under 16 and 15 h photoperiods. A 14 h photoperiod was not conducive to obtaining desirable yields and flower quality from *G. paniculata cv.* Perfecta during winter and early spring. The results concerning the time from planting to harvest also indicated that flowering in the plants grown under the natural short photoperiod depended upon the beginning of natural long days. The rate at which the changes in plant development were induced suggested that they were primarily due to a morphogenetic response of the plants to the increased red/far-red ratio of the light transmitted through the fluorescent sheets of the greenhouse (Novoplansky *et al.*, 1990). Fluorescent dyes efficiently convert light from the green part of the spectrum into red light. When under field conditions, the use of the fluorescent sheets increased the agricultural yields of *G. paniculata*. The rate at which the changes in plant development were induced suggest that they were primarily due to a morphogenetic response of the plants to the increased red/far-red ratio of the light transmitted through the fluorescent sheets greenhouse (Novoplansky *et al.*, 1990).

Table 2: Percentage of Bolted and Rosette Shoots of *Gypsophila paniculata* under different Light Sources*

Light Source	*Far-red*	*CD*	*Inc.*	*R+FR*	*CWF*	*LED*	*Blue*	*LPS*
Rosette shoots	81	69	19	36	70	59	84	71
Bolted shoots	19	31	81	64	30	41	16	29

* Graaf-Van der Zande *et al.* (1997).

Light plays a significant role in the plant growth and flowering of *Gypsophila*. Therefore, for year-round production of *Gypsophila paniculata*, specific techniques such as artificial lighting and heating are used to provide optimal light and temperature conditions. This is a species native to Mediterranean, hence it is suited to intermediate environmental conditions for its natural growth, and although, being a long day plant, it needs a great number of light hours to bloom. These plants show excellent vegetative and floral development in non-natural growth condition cycles as long as their thermic and lighting needs are covered (Alberto *et al.*, 2006). It can be grown continuously with 2 to 3 cycles of flowering per year and can last 2 to 3 years.

The effects of light quality on the flowering of *Gypsophila paniculata* were investigated using light-emitting diodes emitting the following lights: FR, R, and blue (B). Flowering and flower budding were observed under long-day conditions with FR, while no flowering was observed under short-day conditions. Promotion of flowering and flower budding was increased under FR supplemented with R compared to FR alone. Although generally R inhibits flowering in long-day plants, a certain intensity of R, for example, R:FR ratio between 0.23 and 0.71, may be necessary for effective promotion of flowering. B supplementation of FR was not effective at ratio in this study in inducing flowering and flower budding. Quality of cut flowers produced under long-day conditions with LEDs that promoted flowering was not lower than that under incandescent lamps. It provide basic knowledge for the development of LED bulbs as a replacement for incandescent bulbs (Nishidate *et al.*, 2012).

Shibuya *et al.* (2017) reported that flowering was promoted in *G./ paniculata* 'Million Star', under long-day conditions with blue light, suggesting that there is variation in *G./paniculata's* flowering response to blue light. Expression of GpFT and GpSOC1 in 'Million Star' was analyzed. The expression of GpFT and GpSOC1 was promoted with flowering in 'Million Star' under long-day conditions with blue light in contrast to 'Bristol Fairy'. *G./paniculata* homologs (GpFKF1, GpGI) of FLAVIN-BINDING KELCH REPEAT F-BOX 1 and GIGANTEA genes were analyzed, which participate in the flowering response to blue light. GpFKF1 and GpGI amino acid sequences were well conserved; gene expression showed a diurnal rhythm with different peaks under short-day and long-day conditions. GpFKF1 interacted with GpGI. There were no important differences in GpFKF1 or GpGI amino acid sequences between the two cultivars. The results suggested that variation in the flowering response to blue light is associated with GpFT and GpSOC1, rather than GpFKF1 and GpGI.

7.1.3 Temperature and Light

In an experiment, Shillo and Halevy (1982) observed that *G. paniculate* was very responsive to photoperiod and temperature. Long day (16 hours) promoted flowering in *cv.* Bristol Fairy. Plants treated with GA_3 in short day (8 hours) did not promote flowering. The long photoperiod was effective only at relatively high temperatures (27/22°C or 22/17°C day/night) (Table 3). At night temperature below 12°C, the plants remained vegetative even in long days, artificial lighting (60 to 100 lux) for 4 h as a night break or at the end of the night was equally effective. Continuous lighting promoted flowering more than any of the 4 h lighting regimes. High light intensity during the day had a beneficial effect on flower production compared with 40 per cent shade (Table 3 and 4). In an experiment on *G. elegans cv.* Covent Garden Market, Takeda (1996) observed that the threshold day length that induced vegetative or reproductive growth was about 10 h. Intermediate photoperiods of 12-14 h led to bolting plants with very short internodes. Photoperiods >10 h had a quantitative effect; the longer the day length, the greater was the degree of flower promotion. High temperature in summer inhibited or delayed bolting but promoted vegetative growth. Supplemental lighting promoted bolting of the rosetted plants induced under natural short days at the end of summer. The effects of night temperature (NT) and photo-synthetic photon flux (PPF) on time to flowering and flower yield in cultivars Bristol Fairy and Bridal Veil were studied in controlled environment (Hicklenton *et al.*, 1993a). Plants were grown with NT of 8, 12, 16 or 20°C, and a day temperature of 20°C and PPF of 450 or 710 mol/m^2/s. Failure of Bristol Fairy plants to reach anthesis was common at 8°C NT and either 450 or 710 mol/m^2/s, whereas in Bridal Veil nearly all plants flowered, regardless of environmental conditions. Flower yield (measured as FW of inflorescences) decreased with NT in Bristol Fairy but was highest at 8 or 12°C in Bridal Veil. In a 2nd experiment using the same cultivars, Bristol Fairy required more LD cycles (>56) than Bridal Veil for maximum stem elongation and flower yield. Terminating LD conditions before the start of inflorescence expansion resulted in lower yields and shorter plants in both cultivars. Hicklenton *et al.* (1993) also studied the growth and flowering responses of *G. paniculata* cultivars Bristol Fairy and Bridal Veil to temperature and photosynthetic photon flux (PPF) by growing plants in controlled environment rooms under PPF values of 210, 450 and 710 µmol/s/m^2 and temperatures of 12, 20 and 28°C in all combinations. Bristol Fairy had a higher relative growth rate and root:whole plant weight ratio, but lower leaf area ratio than Bridal Veil over the first 30 days of treatment. Temperature and PPF responses of net photosynthesis were similar in the 2 cultivars and were characterized by a temperature optimum which shifted from 20°C at 450 µmol/s/m^2 to between 12 and 20°C at 710 µmol/s/m^2. Stem elongation occurred earliest at 28°C and 710 µmol/s/m^2 and was delayed progressively at 20°C and 12°C, and at lower PPF. Plant height at harvest was inversely related to temperature, but increased with increasing PPF. Generally, bolting and flower bud formation occurred earlier in Bridal Veil than in Bristol Fairy; flower buds developed to anthesis more rapidly in Bridal Veil at 12°, but not at 20° or 28°C. Flower yield was inversely related to temperature and was higher in Bridal Veil than in Bristol Fairy under all except 28°C growth conditions. Flower yield decreased with decreasing PPF in both cultivars, but Bristol Fairy displayed more blindness and flower bud abortion at 210 µmol/s/m^2.

Table 3: Interaction of Day Length and Temperature on Percentage of Blooming Plants*

Temperature (°C)	*Flowering Plants (Per cent)*	
Day/night	*SD*	*LD*
27/22	0	83
22/17	0	12
17/12	0	4

* Shillo and Halevy (1982).

The reproductive and vegetative growth of *G. paniculata* was interpreted using photoperiod and vernalization under greenhouse conditions (Amezquita *et al.*, 1999). The flower inducement was more effective after 45 days of vernalization and 16 h light/day over 70 days.

Table 4: Effect of Shading on Flowering of *Gypsophila* Plants. Four beds of 150 plants each were used for each treatment. Differences are significant at *P* = 0.01*

Treatments	*No. of Flowers per Plant*	*Blind Plants (per cent)*
Full light	26.4	2.3
40 per cent shade	15.1	13.2

* Shillo and Halevy (1982).

7.1.4 Carbon Dioxide (CO_2)

Zebian and Reekie (1998) reported CO_2 concentration was positively correlated with the height of gypsophila in both the low and high light regimes. The greatest increase in stem height was between the 0 and 350 µl/l CO_2 treatments, and relative stem elongation decreased as CO_2 concentration increased. CO_2 and affected rate of stem elongation in darkness in a more complex manner. Elevated CO_2 concentration increased plant weight in the high light treatment.

7.1.5 Temperature, Light and Chemicals

Doi (1992) found that natural chilling requirement of *G. paniculata cv.* Bristol Fairy was in the range of 0-10°C and application of 300 ppm BA promoted flower initiation in a number of lines without chilling. Initial exposure to high temperature (day/night 30/25°C) resulted in a vegetative rosette in cultivars such as Perfecta and Bristol Fairy, which require prolonged chilling. The cross grafting of selections with long and short chilling requirements resulted in lack of flower initiation after a short chilling period in all combinations in which shoot or root had a long chilling requirement. Plants of *G. paniculata* cultivars Bristol Fairy and Bridal Veil were grown in controlled environments at 20°C day and either 20° (WN) or 7°C (CN) night temperatures with the day length being either 18 h (LD) or 9 h (SD). This provided 4 day length/night temperature combinations : LD/WN, LD/CN, SD/WN and SD/CN. One quarter of the plants were subjected to a vernalization pretreatment (8 weeks at 11/5°C, day/night; 9 h day length), while unvernalized plants were sprayed with either GA_3 or BA at 300 mg at the start of the environmental regimes. Both cultivars bolted and formed flower buds very rapidly in LD/WN irrespective of pretreatment, but both final stem height and yield were low. Vernalized plants flowered in the other regimes (LD/CN, SD/CN, SD/WN) and best yields and stem height were obtained under LD/CN conditions. CN prevented flowering in unvernalized Bristol Fairy plants, but not in unvernalized Bridal Veil plants, SD prevented flowering in unvernalized plants of both cultivars, GA_3 or BA promoted flowering and improved yield in all plants grown in LD/CN, but only Bridal Veil responded weakly to these treatments in SD (Davies *et al.*, 1996).

7.2.0 Growth Regulators

The plants treated with Phosphon-D, CCC and SADH showed slight retardation of growth (Kaukovirta, 1966). Treatment with MH at 1000 ppm inhibited both growth and flowering, TIBA at 100 ppm accelerated the height of the plants but did not affect flowering (Bose, 1965).

Application of GA_3 (250 mg/l) on 80 days old plants of *cv.* Perfecta at 3 and 5 times a week significantly increased plant height, number of flowering shoots/plant and the length of flowering shoots at harvest (Karaguzel, 1996). In Japan, Doi *et al.* (1992) reported that a foliar spray of 100-300 ppm BA before 30 September advanced flowering time for the first flush in November. However, advancing flowering time increased flower malformation and reduced flower quality. Exposure of rooted cuttings to 2°C for 40 or 55 days before planting on 15 September and spraying with 100 ppm BA on 30 September advanced flowering and improved cut flower quality in the first flush during mid-November. Exposure to natural chilling for more than 50 days after the first flush was required for vigorous shoot regrowth in line 09. Application of 300 ppm BA in combination with natural chilling promoted flowering for the second flush in mid-April. The third flush was harvested from late June to early July. Doi *et al.* (1989) reported that application of 300 ppm BA to plants grown in a heated plastic house under long day conditions promoted flower initiation and advanced flowering by about 2 weeks in the vegetative lines BF04, BF08 and BF13, but not in BF20. BA also promoted flowering in BF08 and BF 13 grown in an unheated plastic house under a natural day length. BA treatment appeared to substitute for chilling in the induction of flowering, and so enabled plants to flower in normally adverse conditions. Treatment with GA_3, however, did not promote flower initiation in any of the lines grown in an unheated plastic house under natural daylight; it advanced flowering only in BF08 in the heated plastic house. Song *et al.* (1991) reported that 50 or 100 ppm uniconazole reduced plant height of *Gypsophila oldhamiana*.

Vegetative development out of the season is improved by the application of growth regulators like gibberellins. Effects of Gibberellic acid treatment on the growth and development of four cultivars of *Gypsophila paniculata viz.*, 'My Pink'; 'Million Stars'; 'Cloudstar 8'; and 'Cloudstar 5'was investigated (Fan *et al.*, 2019). GA_3 treatment not only significantly promoted vegetative growth but also stimulated early flowering and suppressed the occurrence of rosettes during winter. This was more specific to cultivar 'My Pink', which showed 40 per cent and 80 per cent reductions in rosette rates with four and eight GA_3 treatment applications, respectively (Figure 1).

8.0 Propagation

Gypsophila can be propagated through both sexual and asexual methods.

8.1.0 Seed Propagation

Gypsophila elegans takes about 11 weeks to flower after germination, and flowering continues for nearly 20 days. Seeds can be sown at intervals to maintain a continuous blooming period in winter. Propagation of *G. elegans* is exclusively

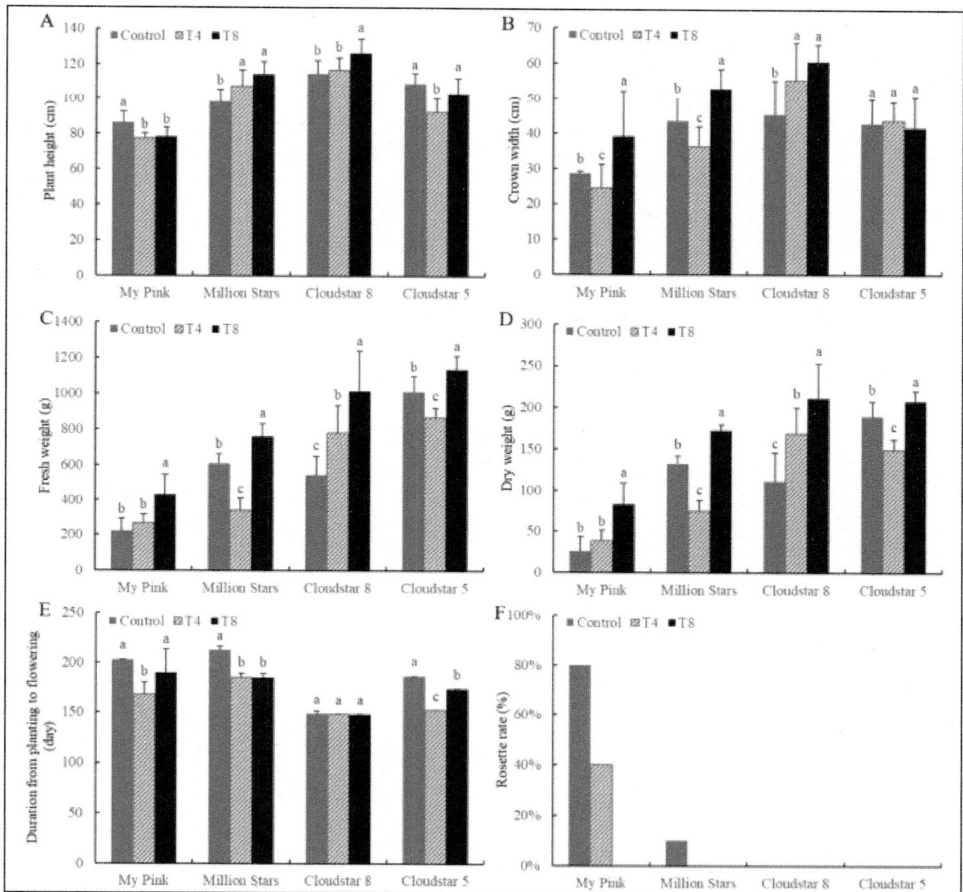

Figure 1: Effects of GA_3 Spray Times on Growth of Four Cultivars of *Gypsophila paniculate* (Fan *et al.*, 2019).

done through seeds which did not impose any dormancy (Ichihara, 1954). The seeds are sown both in beds or pans. Direct sowing in permanent beds also gives good crop. They are sown thinly, and after germination, the seedlings are thinned out 20 to 25 cm apart. Kantartzis (1964a) recorded 98.9 per cent germination within 2 days of sowing. Nordestgaard (1988) recommended that seeds of *G. elegans cv.* Covent Garden should be sown at 6 kg/ha at a depth of 1-1.5 cm. The seeds of gypsophila are generally sown during spring season in India as it allows the fast growing of the crop (Subhankar *et al.*, 2020). The seeds of gypsophila are reported to germinate in two to three weeks at 21 to 26°C (70 to 79°F). The levels of N had far more effect on yield of seeds than those of K or P and 40 kg N/ha was most satisfactory. *G. paniculata* also blooms in about 3 months after germination, produces loose panicles of numerous single or double white flowers. All except the double flowered forms are readily raised from seed. De Groote (1957) observed that the seeds of this species germinated and developed rapidly. Seeds treated with GA_3 at 0.001 per cent caused marked stem elongation of seedlings.

Askim *et al.* (2012) determined the responses of *Gypsophila oblanceolata* to salt stress during germination and vegetative growth. The effects of salinity (0, 50, 100, 150, 300 mM NaCl) on germination and changes in the activities of antioxidant enzymes/isoenzymes (SOD, CAT and POX) during germination under stress and recovery after stress were determined. During vegetative growth (60-day old plants), changes in physiological parameters, ion concentrations, proline (Pro) content, lipid peroxidation (MDA), H_2O_2 content, NADPH oxidase activity and, antioxidant enzyme/isoenzyme system (superoxide dismutase (SOD), catalase (CAT), peroxidase (POX), ascorbate peroxidase (APX), glutathione reductase (GR)) were also investigated. Salt stress decreased both the germination percentage and rate. Few seeds germinated at 100 mM NaCl. Exposure to high concentrations of NaCl did not permanently inhibit germination. Salinity (50 mM NaCl) caused a decrease in activities of SOD, CAT and POX during germination. However, after stress, the activities of all enzymes were increased in recovered plants. During vegetative growth, increased activities of SOD, CAT and APX in 50 and 100 mM NaCl treated plants may help to avoid oxidative damage in *G. oblanceolata*. However, at higher doses, induced stimulation of the CAT, POX and GR was not sufficient to cope with the enhanced ROS production and MDA level. The study suggested that *G. oblanceolata* is a moderately salt-tolerant species.

Double-flowered cultivars of *Gypsophila paniculata* can be difficult to root and can be propagated by seeds that yield approximately 60 per cent double plants (Dole *et al.*, 2006).

8.2.0 Vegetative Propagation

Perennial gypsophila can be propagated through both sexual and asexual methods.

8.2.1 Stem Cutting

Dorsman and Ravensberg (1957) noted that *G. paniculata* can be propagated through stem cuttings under mist, 8 cm long basal shoots being ideal for propagation.

The leaves at the base are removed before inserting the cuttings into the media. Diaz and Orozco (1990a) reported that cuttings taken from upper, middle and lower sections of gypsophila mother plants and dipped in 8000 ppm IBA and placed in scoria, produced roots in 98-100 per cent of cuttings taken from the middle and lower sections of the mother plant after 21 days compared with 85-90 per cent in cuttings from the upper section. After 45 days, root biomass was significantly greater in lower section cuttings than in cuttings taken from the middle and upper sections. By using cuttings from the middle and lower sections of the mother plant it was possible to reduce the time to commercial rooting from 45 days to 37 days. Diaz and Orozco (1990b) also found that flowering stem production was greatest in plants from middle section cuttings which had been stored at 0°C for one month. Cold treatment advanced flowering by 15 days. Rooting of *G. paniculata cv.* Bristol Fairy herbaceous cuttings was significantly improved by the use of IBA. Dipping the basal part of the cuttings, for 10 s in 6000 mg IBA/l, gave 66 per cent rooting during April while the same concentration in July gave no rooting at all. It is recommended

Propagation and Growing of Gypsophila

Seeds

Seedlings in Pan

Ready for Transplanting

Bare Rooted Stalk

Planting in Pot

Planting in Polyhouse

Micropropagation

(A) Explants inoculated in MS medium. (B) Clustered shoots formed after 30 d. (C) The clustered shoots separated and cultured in rooting medium. (D) The seedlings were grown in a 10-h photoperiod (light intensity, 5000 Lx) at 22 °C. (E) The seedlings were transplanted into perlite for root formation. (F) The seedlings were planted in plant tray (*Source*: Li *et al.,* 2020.)

Field Preparation   Planting

Open Field Cultivation   In Polyhouse

Source: www.floraldaily.com, www.hartsnursery.co.uk, www.projectpurityseeds.com, www.gardentags.com, www.web.fleurametz.com, www.oregonwholesaleseed.com, www.violetaflowerssa.net

that cuttings be taken in early spring, late summer or early autumn and the cutting base be dipped in IBA (6000 mg/l) (Sawwan 1993). In experiment on rooting in cuttings Han *et al.* (1992) dipped in a powder containing 0.1, 0.2, 0.3, 0.5 or 1.0 per cent of NAA or IBA or soaked for 3 h in 1, 5, 10, 20, 30 or 50 per cent solution of the 2 growth regulators. Rooting was best with 0.3 per cent IBA in powder form (86 per cent) or soaking in 30-50 ppm NAA (86 per cent). Rooting was >90 per cent when cuttings were dipped in solutions of 2000-3000 ppm IBA or 500-1000 ppm NAA for 1 min. Millions of plants of gypsophila are produced in South Korea from cuttings of tissue cultured elite plants (Paek *et al.*, 1998). Terminal stem cuttings of *Gypsophila paniculata* treated with 3,000 ppm IBA can be rooted under mist (Dole *et al.*, 2006).

The effects of root zone volume were studied in *G. paniculata*, using trays with 4 cell sizes (cell volume 6, 12, 21 and 63 cm^3) (Goto *et al.*, 1997). One cutting was placed in each cell and experiments were initiated after rooting. Stem length increased significantly with increasing root zone volume, and this effect was observed after only 10 days. Growing in small cells significantly reduced the shoot and root dry weight. The total number of lateral shoots, node number and total FW increased significantly with increases in root zone volume. Goto *et al.* (2000) conducted experiments to evaluate the effects of cell (root zone) volume during rooting and transplanting age. Cuttings of *cv.* Bristol Fairy were grown in trays, comprised of 406 and 128 cells pre-filled with 6 and 20 ml potting medium/cell, respectively and they were transplanted when 22-52 days old. Older plants with root balls were easier to transplant than the younger ones. However, the older ones grew more slowly after transplanting than the younger ones. Younger transplants had more succulent roots per plant than the older ones. Older transplants of Bristol Fairy not only had longer harvesting periods, but they also had weaker shoot growth following pinching which resulted in low flower production. It was suggested that transplanting of cuttings should be done before root ball formation occurs on account of their slow growth rate.

Goto *et al.* (1998) reported that 20-day-old cutting of gypsophila *cv.* Bristol Fairy grown in cell tray (12 ml) could be stored for 20 days at 1 and 10°C in the dark. The greatest rooting of *Gypsophila paniculata* occurred in various propagation media with a pH near, or slightly above, neutrality and poorest rooting in acidic media. Addition of ground limestone to acidic media significantly improved rooting. Basal dipping of cuttings in sulfuric acid, sodium hydroxide, and potassium hydroxide resulted in improved rooting over nontreated cuttings. Lignin specific phloroglucinol staining of freehand sections revealed the presence of a band of sclerenchyma tissue, external to the vascular region of the stem, that may impede the emergence of newly formed roots (James *et al.*, 1983).

8.2.2 Grafting

Double-flowering types of *Gypsophila paniculata* have been wedge grafted on seedling *G. paniculata* (single-flowering) roots or crowns. Grafting can be done in summer and fall, using outdoor-grown plants for rootstocks and placing them in a cold frame for healing of the graft; grafting is also done in winter and early spring, using greenhouse-grown stock plants.

8.2.3 Micropropagation

Gypsophila. paniculata plants are sterile and do not produce seeds, therefore, breeding programs are severely restricted (Shillo and Abraham, 1982). The major issue is phytoplasma disease, which has been identified in Gypsophila (Gera *et al.*, 2005). Another main problem is low rooting frequency of vegetatively propagated cuttings which hinders propagation so, the application of tissue culture techniques for Gypsophila is very vital to overcome problems for breeding and propagation.

An efficient adventitious shoot regeneration procedure was developed for *G. paniculata* by Ahroni *et al.* (1997). Using *cv.* Arbel, shoot regeneration from the three upper internodes of the stem was monitored on MS media supplemented with different cytokinins (thidiazuron, benzyladenine, kinetin or zeatin) and auxin (NAA). Thidiazuron was the most efficient cytokinin, with up to 100 per cent of the explants forming shoots, at an average of up to 19 shoots per explant regenerated. *G. paniculata cv.* Perfecta shoot tip and leaf segments were cultured on MS medium supplemented with various combinations and concentrations of 2,4-D, NAA, IBA, IAA and BA. Shoots were regenerated with all combinations of IAA and BA, but the highest number of shoots (8/shoot tip explant, within 5 weeks) was obtained with 0.5 mg IAA + 3.0 mg BA. Root formation occurred on all media, even in controls with no growth regulators (Zamorano-Mendoza and Mejia-Munoz, 1994). Shoots were cultured on MS medium supplemented with 1.0 mg BA and 0.3 mg IAA/l at 5 different agar concentrations, *viz.*, 0.7, 1.0, 1.3, 1.7, 2.1 or 2.5 per cent for 6 weeks. The effect of the growth inhibitors CCC [chlormequat], ancymidol, TIBA and ABA (abscisic acid) added to the medium at 0.1, 0.5, 1.0 or 1.5 mg/l and culture temperatures of 22, 25 or 28°C were also studied. As the agar concentration increased, FW, shoot number and the rate of vitrification decreased. Water potentials lower than –6.0 bars (1.3 per cent agar) reduced the FW:DW ratio significantly. Growth inhibitors had no effect on vitrification. Shoots cultured at 22°C had significantly less vitrification than those cultured at 25° or 28°C (Han *et al.*, 1991b). MS medium containing kinetin and BA resulted in the best differentiation of buds from callus (Han *et al.*, 1996). Shoots showing root initiation were replanted; the survival rate was high.

In cultures of shoot tips of *G. paniculata*, BA at 0.5-2 mg/l was the most effective cytokinin for shoot proliferation, but it resulted in a higher percentage of vitrified plantlets than kinetin or 2ip. Increasing the nitrate : ammonium ratio of the medium decreased the number of vitrified plants and the FW:DW ratio. Addition of 1 mg IBA/l induced root formation (Han *et al.*, 1991a). Node cultures of *G. paniculata* showed vitrified growth when cultured under standard conditions in bottles closed with a plastic lid, resulting in poor survival of tissue culture derived plants. The deviant growth pattern appeared to be determined in the lag phase of the cultures. A limited evaporation period in this stage induced normal growth (Dillen and Buysens, 1989). Meristem tips, 2 mm long, taken from cuttings of *G. paniculata* were cultured on half-strength medium supplemented with sucrose, myo-inositol, agar, kinetin and IAA at various concentrations. Rooting percentage was similar (approximately 50 per cent) following transfer to sand or a 3:1 mixture of sand :

peatmoss, but the number of roots was higher in the sand : peatmoss substrate. From a single explant an average of 7 plantlets were produced during 5 months of culture (Castro *et al.*, 1996). Adventitious shoots were successfully regenerated from leaf explants of cuttings of *G. paniculata* (cultivars Arbel, Golan, Gilboa, Tavor, Perfecta and Flamingo), obtained from a commercial source in Israel (Zuker *et al.*, 1997). The efficiency of shoot regeneration for Arbel was tested on 18 media based on MS basal medium containing different concentrations of TDZ and BA (1, 3 or 6 mg/1) in combination with NAA (0, 0.3 or 1 mg/1). The highest efficiency of adventitious shoot regeneration was obtained with the oldest leaves originating from the youngest cutting analyzed. Experiments were conducted by Lee and Bae (1999a) to find out the best cultural conditions for adventitious shoot regeneration from leaf segments of *G. paniculata*. Thidiazuron (TDZ) was remarkably effective for the regeneration of leaf segments compared with BA and kinetin. TDZ showed the highest rate of regeneration at 3 mg/l, while kinetin did not affect regeneration. BA in the medium increased vitrification. Shoot formation efficiency was higher in the presence of IAA (0.3 mg/l) than NAA. Lee and Bae (1999b) also determined the best conditions of *in vitro* proliferation using shoot tip of *G. paniculata* cultivars Bristol Fairy and Red Sea. Formation and fresh weight of adventitious shoots were promoted by culture with 0.2 mg/l BA and 0.1 mg/l NAA (Red sea). Vitrification was suppressed by using 7 × 13 cm (diameter × height) vessel. Large explants (shoot tips of about 1 cm) showed the best survival and regeneration. The optimum culture medium was MS supplemented with 0.5 mg BA + 0.5 mg NAA + 1.0 mg kinetin which resulted a high multiplication rate (10.4). Plantlets rooted easily on MS medium supplemented with 5.0 mg IAA + 0.1 mg NAA + 0.2 mg IBA/l (95 per cent rooting after 14 days) (Song *et al.*, 1996). When plantlets were transferred to soil, covering the transplants with glass during the first week reduced transpirational water loss and increased survival rate; fungicides effectively controlled plantlet decay and death. Explants (shoot-tip) in the MS medium containing $CaCl_2$ at 1760 mg/l showed the most vigorous growth and the leaves were deep green (Takane *et al.*, 1994).

Results of experiment indicated that PP_{333} [paclobutrazol] had a dwarfing effect on tube-shoots (*in vitro*) of *G. elegans*. Combination of PP_{333} with 6-BA [benzyl adenine] promoted the multiplication of shoots and produced strong shoots with dark green leaves. On transplanting these shoots exhibited a good survival rate (Jiang *et al.*, 2000). Gribble (1999) found that small decreases in relative humidity *in vitro* were sufficient to increase plantlet transpiration to the extent that vitrification was greatly reduced and *ex vitro* survival increased. Small differences in relative humidity also influenced plantlet morphology; plantlets grown at a higher relative humidity were larger and consisted of more shoots than plantlets grown at a lower relative humidity. It is suggested that *in vitro* relative humidity may be manipulated to produce plantlets of a particular morphology.

The effect of different kinds of sealing materials on vitrification was reported by Han *et al.* (1995). Extended time of culture in vessels sealed with aluminium foil resulted in increased vitrification, causing decreased carbohydrate and increased P and K concentrations of plantlets, and increasing the ratio of FW : DW.

Ahroni *et al.* (1997) and Zuker *et al.* (1997) reported that adventitious shoots can be regenerated from internodes and leaf segments of Gypsophila.

Asaph *et al.* (1997) developed an efficient adventitious shoot regeneration procedure for *Gypsophila paniculata* L. Using cultivar Arbel, shoot regeneration from the three upper internodes of the stem was monitored on MS media supplemented with different cytokinins (thidiazuron, benzyl adenine, kinetin or zeatin) and an auxin (naphthaleneacetic acid). Thidiazuron was found to be the most efficient cytokinin, with up to 100 per cent of the explants forming shoots, at an average of up to 19 shoots per explant being regenerated. The highest percentage of shoot formation was observed in the stem explants originating from the first internode, with all cytokinins tested. The adventitious origin of shoots regenerated from stem explants was confirmed by scanning electron microscopy. The regeneration procedure was found to be applicable to five other gypsophila cultivars (Perfecta, Golan, Gilboa, Flamingo and Tavor). Regenerating plants were successfully transferred to soil, and did not differ in flower color, size or shape from standard vegetatively propagated plants derived from cuttings.

Salman (2002) reported that leaf segments of *Gypsophila paniculata* are amenable to callus production and cell suspension cultures. The use of cytokinins and anticytokinins have been used in *in vitro* propagation of Gypsophila as a means of studying action mechanism of cytokinins (Kapchina-Toteva and Stoyanova, 2003).

Rady (2005) established protocol for regeneration of the economically *Gypsophila paniculata* L. using shoot tip explants. Multiple shoots were obtained on MS-medium fortified with 0.5 mg/l each of NAA and BA. Addition of 10 g/l agar promoted shoot proliferation and reduced the degree of shoot vitrification. Transfer to 3 mg/l IBA containing medium produced optimum root initiation and development.

Micropropagation of *Gypsophila* has been reported using shoot tips as explants which produced adventitious shoots with NAA and BA as growth regulators (Rady, 2006).

Kwadwo *et al.* (2009) investigated *in vitro* culture techniques to efficiently micro propagate *Gypsophila paniculata* L., using stem segments as explants. Three different media designated as Alt1, RS1 and RS2 were used which were modified form of the Murashige and Skoog medium were used. The media differed from each other in terms of the macro elements, micro elements, and growth regulators. RS2 gave the average maximum biomass of 13.0 g. *In vitro* plants from RS2 medium gave the highest average percentage survival of 89.8 per cent when micro shoots were transferred to pots and kept in the greenhouse. Plants from RS2 medium also flowered a week earlier than Alt1 and RS1 media used. The results also showed that a 3-week in vitro transfer interval coupled with a lower reduction in cytokinin 0.5 mg/l reduced vitrification.

Apical meristem as explant was used to micro propagate *Gypsophila paniculata* L. by using *in vitro* techniques (Rashid *et al.*, 2012). The highest number of shoot emergence, multiple shoot formation, and maximum shoot length was obtained in medium supplemented with BAP 1mg/l after 4 days of inoculation, which shows the maximum shoot length of 6.53 cm with 86 percentage response observed in same medium after 21 days. For root initiation the best medium was NAA 0.5 mg/l which showed 85 per cent of root initiation with length of 2.66 cm after 11 days of inoculation.

Studies on the mass production of high-quality plantlets in *Gypsophila paniculata* L. using a bioreactor and microponic system (a hydroponic system in which micropropagation shoots are planted) indicated that both aeration treatments, in which bioreactors were aerated from the top of explants by sparger (AS) and by tub (AT), were more effective than unaerated treatment for shoot proliferation and growth, and the maximum shoots (15.7 shoots per explant) with low hyper-hydricity rate (2.9 per cent) were found in the AS group. The *ex vitro* culture was more efficient for rooting when compared to the *in vitro* culture; the better shoot and root growth was obtained in the *ex vitro* culture, with rooting rate reaching 100 per cent after 20 d of culture, but only 65 per cent of *in vitro* shoots rooted; all stomata of *ex vitro* shoots closed, and their length was more than their width, but the stomata in *in vitro* shoots were all opened, the length close to the width. Furthermore, the stomata numbers were less in *ex vitro* (67.8) than *in vitro* (267.2). The survival rate of *ex vitro* plants reached 83.3 per cent when plantlets derived *in vitro* and *ex vitro* were transferred to pots, while only 23.3 per cent of *in vitro* plantlets survived. During *ex vitro* rooting with the microponic system, foam as the supporter material, 90 µmol/m/s of light, and 80 shoots of planting density were favorable for shoot and root growth. The combination of bioreactor and microponic systems is an efficient way to produce high-quality plantlets of *G. paniculata*. Their application can reduce costs during large-scale industrial production (Wang *et al.*, 2013).

Highest multiplication coefficient, which reached 6.5, and best growth parameters were obtained on modified MS medium supplemented with 1.0 mg/l 2iP and 0.2 mg/l IAA. Obtained cultures were treated with different concentrations of lead nitrate (0.1, 0.5, and 1.0 mM $Pb(NO_3)_2$ or cadmium chloride (0.5, 2.5, and 5.0 µM $CdCl_2$). It was ascertained that Pb ions induced growth disturbances and contributed to shoot wither. On the contrary, the proliferative shoot cultures were established on media containing Cd ions and the multiplication coefficients and shoot length increased on all media enriched with $CdCl_2$. Chlorophylls and carotenoid contents were negatively affected by application of 5.0 µM of cadmium; nevertheless, in shoots treated with 2.5 µM $CdCl_2$, increased accumulation of photosynthetic pigments occurred and their amount was similar to untreated culture. Adaptation to Cd was associated with stimulation of phenolic compound synthesis (Ewa *et al.*, 2018).

9.0 Soil and Climate

9.1.0 Soil

Gypsophila thrives in any well drained garden soil. Light soil containing lime more than normal is best suited for its growth and development. A soil pH of 6.5 to 7.5 and Electrical conductivity (EC) below 0.7ms/cm is essential. In an experiment on substrates and nutrient solution, Zhou *et al.* (1994) reported best result on growth and flowering using a vermiculite : perlite mixture (1:1) and with a nutrient solution containing (meq/l) 17.0 N, 4.0 P, 8.0 K and 4.0 Mg and trace elements. *G. paniculate* cvs. Super Pink, Super Bristol, Super White and Bristol Fairy have been found to grow well in sandy soil in polyethylene clad greenhouse (Vergara Gongora, 1984).

Fertigation in Gypsophila depends upon the soil fertility. The soil required for growing of Gypsophila should be well fertilized and mixed with organic matter so that the plants will get adequate amount of nutrients. The soil should not be saline and it should be well drained and medium to light textured. It needs sandy or porous soil in order to regulate its moisture uptake.

Different hydroponics systems for the commercial production of gypsophila was attempted (Paul *et al.*, 2011). Three hydroponics systems (elevated tray, ground lay bed and bag culture systems) were used as main plots. The sub-plots were allocated to three different aggregate/medium component (saw dust, river sand and vermiculite). Highest vegetative growth of gypsophila in terms of plant height was obtained from plants grown using vermiculite in ground lay bed and bag culture hydroponics systems. However, highest number of shoots/plant was observed in plants grown using sawdust. Lowest plant height and number of shoots/plant were obtained from plants grown using sand under all hydroponics systems. The highest plant height and number of shoots/plant were observed in plants grown under bag culture when compared to the elevated tray and ground laid bed hydroponics systems. The highest quality cut flowers in terms of stem length and number of branches/flower were obtained from plants grown using sawdust and under bag culture hydroponics system. For the hydroponics culture of gypsophila, sawdust should be used as growing medium. To induce highest vegetative growth, flower yield and quality, plants should be grown using bag culture hydroponics system.

9.2.0 Climate

It is a cool season crop, but susceptible to very severe winter. Kantartzis (1964b) reported that gypsophila required protection against frost. It can also grow well in places with light rainfall during the monsoon on medium elevation. An open and sunny situation is necessary for good growth and flowering. Gypsophila is a long day plant, requiring a day length between 12 to 18 hours to initiate flowering, depending on the variety. Crop requires light intensity of 55000 to 70000 lux at plant level, Gypsophila responds to long days. Flowering is induced under 13-14 hours of light. Optimum Humidity requirement of gypsophila is 70 to 75 per cent. Long days with warm temperatures promote good flowering. Low night temperatures will cause the plant to remain in its vegetative stage. If minimum temperatures average 10°C, it will take 120 days to flower (Tregea, 2003). High temperatures cause the plant to bloom quickly, flowers to dry out, and stem lengths to be short. Shade cloth (50 per cent) can be used to cover the plants during summer to improve the quality.

10.0 Cultivation

10.1.0 Selection of Cultivars

In Italy, outdoor trials on *G. paniculate* cultivars Dana and Perfecta demonstrated that Dana was the better cultivar for autumn cut flower production with its cultural cycle of 6-7 months compared with 11 months for Perfecta. Planting should be carried out on 1-10 July (Farina and Paterniani, 1989). Trials were carried out in a commercial greenhouse in Madrid producing cut *Gypsophila* sprays for export. Highly homogenous cuttings, originating from 3 sources of propagation material *viz.*, Balborts (T1), Raham Meristem (T2) and Cor van Duyn (T3), were treated with NAA and rooted under mist. Rooted cuttings were planted in beds 45 days later and received identical cultural treatment. Of the 3 clones T1 was the most long-flowering and productive, followed by T2 and T3. In all cases flower quality was very good (Arens-Blanco and Chaparro, 1992).

In Indian condition, many varieties of *Gypsophila* are available for cultivation. Some varieties show different characteristics *e.g.* flower size, light needs, yield, duration of cultivation cycle and flowering phase, among others. Few of the species which are suitable for cultivation under Indian cultivation are as follows: *Gypsophila paniculata, Gypsophila elegans, Gypsophila repens, Gypsophila muralis, Gypsophila oldhamiana* and *Gypsophila perfoliata.*

10.1.1 Life Cycle

Gypsophila paniculata is reported to have a life span of two years and it yield three flushes in one year if planted in greenhouse (total six flushes are harvested, when once planted), whereas, if it is planted on open field condition it yields two flushes in one year total four flushes are harvested (https://www.plantopedia.com/gypsophila/).

10.1.2 Greenhouse Cultivation

Cultivation of *Gypsophila paniculata* under greenhouse condition requires a day temperature of 26-27° C and night temperature of 18-20°C, humidity of 60-70 per cent and a light intensity of 55000-70000 lux (https://en.Wikipedia.org/wiki/Gypsophila). After the crop is planted, it is left up to fourth week for bushing of the plants. Pinching is done in the fifth week after planting and 15 ppm GA treatment should done. From 6th to 9th week the crop is artificially lighted for elongation then from 10th to 12th week induction is done. Flowering starts from 13th week of planting and from 14th to 17th week harvesting can be done. The procedure for second flush starts from 18th week of planting. The 18th week is kept as a dry period. Subsequently, on 19th week pruning should be done and Benlate at 2 g/l should be sprayed. Then the same process from making bush to harvesting is repeated for the second flush of the crop and it is completed on 34th week and the third flush starts from 35th week and completed on 51st week. Similarly, in the second year, another three flushes of *Gypsophila paniculata* can be harvested.

For greenhouse cultivation of *Gypsophila*, few vital measures should be taken care such as the proper land levelling, availability of good quality water, good connectivity with market and proper communication and proper electricity facility. There should not be any wind breaks or big structures near the greenhouse. The greenhouse should have a height of 5.0 - 6.5 meter and the orientation should be north-south direction facing, the vent should be opening along the wind direction. There should be gap of at least 4 meter between two adjoining polyhouses. Adequate space should be there at the top and side for ventilation and provision for adequate light should be there inside the polyhouse. Optimum humidity for plant growth should be maintained inside the polyhouse. Cleanliness and hygiene should be maintained inside the polyhouse. The rapid growth of gypsophila with resulting numerous, small secondary stems mean that it needs a guide, which is usually provided by several meshes of support nets at different heights. Support netting is essential for successful growing of *Gypsophila*. Support netting should be fixed within 3 weeks of planting. First net of size 20x20 cm should be provided at 30 cm from the base. Second net of same sizes should be installed at 45 cm if required. Side string at 60 cm should be tied at both sides of the bed to avoid crop lodging in beds. If support netting is not done properly then the gypsophila plant will fall, which leads to crop failure. Therefore, inside the greenhouse system of support netting should be ensured (KF Bioplants, 2018).

10.1.3 Open Field Cultivation

For cultivation of *Gypsophila paniculata* on open field condition, it can be cultivated in both winter and summer season. Support netting is necessary for cultivation of gypsophila in open field also. For winter season it takes 19 weeks for one flush, whereas it takes about 15 weeks for summer crop.

For winter crop, the crop is planted in the first week and left up to fourth week for bushing of the plants. Pinching is done in the fifth week and 15 ppm GA treatment should be done. From 6th to 10th week the crop is artificially lighted for elongation then from 11th to 14th week induction is done, then from 15th week flowering starts and from 16th to 19th week harvesting can be done.

For summer crop, in the first week the crop is planted and left up to fourth week for bushing of the plants, in the fifth week pinching is done and 15 ppm GA treatment is done, from 6th to 8th week the crop is artificially lighted for elongation then from 9th to 10th week induction is done, then from 11th week flowering starts and from 12th to 15th week harvesting is done.

10.1.4. Land Preparation

Soil solarization improved the growth and yield of *G. paniculata* plants growing in a commercial greenhouse in soil with an 8-year history of monoculture (Gamliel *et al.*, 1989). Solarization caused an increase in the population of fluorescent pseudomonas (FP) and a decrease in the deleterious fungi causing stunting. The intensified root colonization by FP in these plots was noted within 2 weeks after planting. The roots were free of *Pythium* spp. throughout the 10-week test period (Tables 5 and 6).

In a plot with a 3-year history of monoculture, yield increases with solarization and fumigation with methyl bromide were 15 per cent and 27 per cent, respectively, during the first growth cycle. The effect of disinfestation by solarization, fumigation with methyl bromide, and treatment with metam sodium was studied in commercial greenhouses growing *G. paniculata* and experiencing yield decline under continuous cropping (Gamliel *et al.*, 1993). Disinfestation of soil or tuff container medium (volcanic ash), resulted in improved growth and increased flower yield. Flower weight was increased by 17-65 per cent following solarization, 26-97 per cent by methyl bromide, and 51 per cent by metam sodium. A long-term effect on yield increase over 2 to 5 successive crop cycles was apparent with solarization and methyl bromide fumigation, but not with metam sodium. The quality of flowers was also improved by solarization of a tuff container medium. Disinfestation increased the percentage of early yield.

Table 5: Effect of Soil Solarization and Methyl Bromide (MB) Fumigation on Root Colonization of *Gypsophila* by Several Micro-organisms (Qidron)*

Treatments	*Colonization (Per cent)*			
	FPz	*Minorz*	*Pythium*	*Fusarium*
Non-treated	24	45	20	74
Solarized	61	2	0	54
MB	0	4	0	55

z FP = Fluorescent pseudomonas; Minor = deleterious fungi.

* Gamliel *et al.* (1993).

Yield was increased due to solarization by 15, 46 and 48 per cent at the first, second and third cycle of flower picking, respectively (Table 5).

Table 6: Effect of Soil Solarization and Fumigation with Methyl Bromide (MB) on Number and Weight of *Gypsophila* Flowers during Two Crops Cycles (Qidron)*

Treatments	*Crop Cycle*			
	1		*2*	
	Number kg/m²	*Weight*	*Number kg/m²*	*Weight*
Non-treated	95 az	1.73 b	90 c	1.6 c
Solarized	97 a	2.0 a	97 b	1.74 b
MB	108 a	2.2 a	104 a	2.09 a

z Within columns, different letters indicate significant differences between treatments (p=0.05).

* Gamliel *et al.* (1993).

Witomska and Her-Zachariasiewicz (1997) reported increased shoot length in field grown plants propagated *in vitro*. Plants are spaced at 0.7 × 1.3 or 1 × 1.7 m in the field (Auman, 1980).

Hayashi *et al.* (1992) found change in inflorescence composition with planting date. From spring to summer, the number of days to flowering, inflorescence FW, stem length, number of nodes, number of primary and secondary branches and number of florets tended to decrease but in units, FW, diameter of the unit axis, highest order of visible florets and number of florets tended to increase.

The land preparation for Gypsophila should be done with a deep ploughing up to 20 cm depth because the planting of seedlings is done by digging 20 cm depth. Raised beds with dimensions of one meter width and 30 cm height should be prepared and the length of beds is dependent on the length of field. There should be a pathway of 50 cm width in between two beds.

10.2.0 Sowing and Planting

Sowing is generally done in spring and the flowering occurs in summer. For Indian condition, sowing of the seeds should be done in early spring because early growth stage of the crop thrives well in cold condition. The row to row spacing is 50 cm and plant to plant spacing is 30 cm in between two plants of Gypsophila.

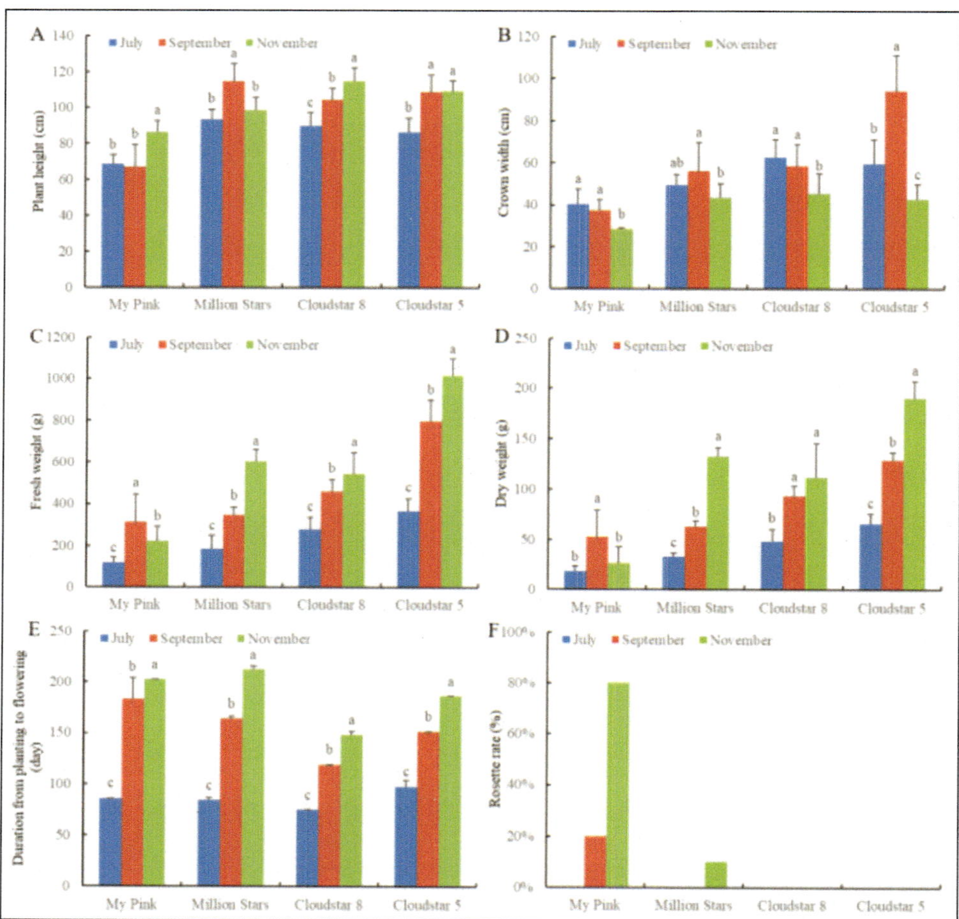

Figure 2: Effects of different Planting Seasons on the Growth and Development of Four Cultivars of *Gypsophila paniculata* (Fan *et al.,* 2019).

The effects of different planting seasons on the growth and development of *Gypsophila paniculata* were investigated to explore new approaches to controlling the flowering period. Four different cultivars of *Gypsophila* were selected and continually planted in July, September, and November in the low-latitude and high-altitude region of Kunming, China (25° N, 102° E). Results indicated that the vegetative growth and flowering time of *Gypsophila paniculata* were prolonged and postponed when the planting time was delayed. Specifically, cultivar 'My Pink' showed 20 per cent and 80 per cent rosette rates when grown in autumn and winter, respectively (Fan *et al.*, 2019) (Figure 2).

The effect of plant density and variety on allometry in *Gypsophila paniculata* L. was investigated (Takashi *et al.*, 2007). Increasing plant density significantly reduced fresh weight, total branch length, number of dichasia, number of florets per dichasium, and total number of florets in cut flowers in 'New Face'. Relationship between fresh weight and the total branch length, number of dichasia, and total number of florets were negatively allometric, isometric, and positively allometric, respectively. There were allometric relationships between fresh weight and the number of dichasia, number of florets per dichasium, and total number of florets on each branch order. The allometric exponents of regression relating the number of dichasia and the total number of florets to fresh weight increased from lower- to higher-order branches. On the other hand, the allometric exponents of regression relating the number of florets per dichasium to fresh weight were similar in almost all orders. Significant differences were found in fresh weight and parameters of inflorescence architecture among varieties grown under identical plant density. Allometric exponents relating the total branch length and number of dichasia to fresh weight, varied, and those relating the total number of florets to fresh weight were stable among varieties grown under identical plant density.

10.3.0 Watering

Watering is an important aspect of growing gypsophila. Plants need adequate watering during active vegetative growth. Most of the species dislike winter wet, although *G.* 'Rosenschleier' tolerates moist soil (Brickell, 1996). Harbough *et al.* (1989) observed that the flowering status of plants in 2 commercial fields was related to soil moisture levels. In wet (14-21 per cent moisture) and dry areas (3-4 per cent moisture), 43-62 per cent of the plants did not flower compared with only 14-16 per cent in areas with intermediate soil moisture levels.

The crop cannot tolerate any types of water stress, whether physical or physiological, means it is susceptible to drought and water-logging both. Excess water can seriously harm the crop. Water stress can result in poor flowering. Therefore, there must be limited, localised and controlled watering as and when required along with proper drainage facility (Subhankar *et al.*, 2020). Once established, gypsophila is relatively drought resistant due to its deep root system. Drip irrigation with emitter spacing of 30 cm is recommended. Each row of plants will require an irrigation line. Use overhead irrigation for the first week after planting. Keep the soil at field capacity level. As the plants start to flower the amount of irrigation can be reduced substantially. Reduce water to 50 per cent at floral bud formation. Reduce water to 30 per cent one week prior to the harvesting of gypsophila flowers. The irrigation should be stopped during pruning. Restart irrigation 30 per cent after pruning and increase it gradually to 100 per cent as leaves appear. Tensiometers are useful instruments to monitor and assess water requirements in gypsophila.

10.4.0 Manuring and Fertilization

As it is a high value crop, organic manures are preferred in cultivation of gypsophila. Generally, the organic manures are incorporated in the soil during land preparation. Different types of organic manures like farmyard manure, compost, oil cakes and different animal manures like bone meal, meat meal, horn meal, fish meal, blood meal *etc.* can be applied. Need based application of chemical fertilizers are done as fertigation. The influence of cultural practices on yield and incidence of non-flowering gypsophila was investigated by Harbaugh *et al.* (1989). They found that fertilizer application rate from 200 lb N to 400 or 800 lb N/acre delayed but did not prevent flowering. Mechanical damage 3 and 6 weeks after planting to simulate root injury by hoeing or insects reduced flower yield by about 50 per cent compared with undamaged plants or those damaged before planting; 13 per cent of the damaged plants did not flower while all control plants flowered. In greenhouse trials during 2 harvest periods, plants grown for cut flowers received K_2O at 0, 125, 250, 375 or 500 Kg/ha applied in 6 splits every 15 days starting from the planting of rooted cuttings at 5 m^2 (Molina *et al.*,1991). It was recorded that the highest yield of cut stems was obtained with 250 kg/ha and correlated well with the concentration of exchangeable K in the soil with values ranging between 2.00 and 2.35 cmol (+)/ kg. At higher K concentrations in the soil no fertilizer effect on yield was observed. Inoculation of seedlings with the vesicular arbuscular mycorrhizal fungus (VAM), *Glomus* spp. increased root length and dry weight of the whole plant. Field tests indicated that inoculation could decrease the application of phosphate fertilizers used in the production of *G. paniculata* (Chen *et al.,* 1998). A general fertigation formula for Gypsophila could be N-P-K = 100:50:150 ppm; Ca-Mg = 120-50 ppm. Delayed flowering can be caused by over fertilising during the growing period.

Fertigation of *G. paniculata* mother plants with a 20:20:20 compound fertilizer (N, P, K, plus microelements) at an N concentration of 720 mg/l reduced the expression of deformation disease compared with the conventional N concentration of 360 mg/l (Ben *et al.,* 2006). Increasing the 20:20:20 fertilizer concentration reduced the disease level from 82 to 96 per cent with N at 180 mg/l to 6 to 10 per cent with N at 720 mg/l. Among tested N sources, NH_4 was the most effective in reducing the disease. Low disease incidence (0 to 10 per cent) was always associated with effluent pH of 6 or lower. Irrigating with acidified water (pH 5.5) in the presence of N, as NH_4 at 180 mg/l, additionally reduced disease from 56 per cent under tap water (pH 7.8) irrigation to 11 per cent. Similar acidification in the presence of NO^{-3} N at 180 mg/l was ineffective in reducing the disease, but water basification in the presence of NO^{-3} N reduced disease incidence from 93 to 38 per cent 90 days after planting.

10.5.0 Weed Control

A study was conducted to evaluate weed control and phytotoxicity of pre-emergence herbicides. Pennant EC (metolachlor), Pennant G (metolachlor) and Surflan A.S. (oryzalin) andpost-emergence herbicides Fusilade (fluazifop–butyl) and Poast(sethoxydim) gave effective control of weeds (Smith and Treaster, 1992). Oxadiazon at 4 l/ha applied in early spring to gypsophila infested with *Digitaria sanguinalis, Mercurialis annua, Portulaca oleracea,* and *Setaria media* gave good to excellent control of weeds without injuring the crops (Katalinic, 1989). *G. paniculate* was found to tolerate herbicide Oxadiazon at 1-2 kg/ha (Cox *et al.*, 1990). Evaluations indicated that herbicides containing oryzalin resulted in phytotoxicity or stunting *G. pacifica* (Staats and Klett, 1993).

The effects of mulching materials (control, black polyethylene film or reflective film) and planting dates (6 May, 6 June or 5 July) on growth and yield of cut flowers of *G. paniculata cv.* Bristol Fairy in Korea Republic, were investigated by Nam *et al.* (1997). Mulching did not influence minimum soil or air temperature, but did influence maximum soil and air temperature (increased in the order of black PE film, control and reflective film). Photon flux density was highest in the reflective film treatment and lowest in the control. Flowering time was promoted by early planting (in May and June) and by mulching with black film. In the July planting, flowering was observed only in the reflective film treatment. Cut flower yield was increased by early planting and the use of reflective film mulching.

Gypsophila paniculata can be planted through a plastic mulch if weeds are a severe problem. Pre-emergent herbicides can be used prior to planting to stop many weed seeds from germinating. Grass weeds in the crop can be controlled using a selective herbicide.

10.6.0 Pruning and Pinching

One-year-old gypsophila plants were pruned to their base on 3 dates. Two weeks after pruning, light was given as a 4-h night break of 100 lux from incandescent lamps. Early pruning advanced flowering and shortened the period of harvesting flush (Table 7) (Shillo and Halevy, 1982).

Table 7: Effect of Pruning-Date on Flowering of 1-year-old Gypsophila Plants. All differences between pruning-dates and significant at *P* = 0.05*

Pruning-Date	*No. of Days to First Flower Cut*	*No. of Days in Harvesting Flush*	*No. of Flowering Stems per Plant*
25/8	62	34	13
5/9	97	50	18
20/9	118	75	40

* Shillo and Halevy (1982).

Vergara Gongora (1984) pruned 4 cultivars, *viz.*, Super Pink, Super Bristol, Bristol Fairy and Super White on 1st February, 1984. He observed that pruned plant came to flowering stage in the middle of April whereas non-pruned plant produced flower in May.

The effects of pruning time (26 June, 11 or 24 July, and 7 August), and pinching and chilling treatments (2, 4 or 6 weeks at 2°C) on the incidence of rosette formation and the rate of flower malformation induced by high temperatures were investigated in *G. paniculata cv.* Bristol Fairy (Cheong *et al.*, 1999). Flowering was delayed by later pruning dates and by pinching. The length of flower stalks, number of nodes, 1st branch number, internodal length and cut flower length were also increased in these treatments, while the rates of abnormal inflorescences and flower malformation decreased. Commercial yield was 4209 bundles/10 a in the former treatment and 4170 bundles/10 a in the latter. Chilling at 2°C, both before and after emergence of shoots, reduced the severity and rate of flower malformation, and increased the length of flower stalks, number of nodes, 1st branch number and cut flower length. In chilling treatments, rosette formation rate was lowest (12.5 per cent) following 6 weeks of chilling before emergence. Chilling for 6 weeks before and after emergence promoted commercial yield (3473 and 3434 bundles/10 a, respectively, compared with 2400 bundles/10 a in the control).

Blooming of the gypsophila flowers occur simultaneously with the crop growth, so the practice of pruning in between is necessary in gypsophila. The dead and diseased parts of the gypsophila plants should be pruned immediately when located, otherwise it will damage the whole plant. As the crop gives 2-3 harvesting with a single planting, it must be taken care during the crop growth period. Therefore, pruning is necessary for this crop.

Pinching is done when the gypsophila plants are 15 cm height. After 4 days of pinching, GA_3 at 250 ppm should be sprayed. Second GA_3 spray can be done selectively 8 days after the first spray to plants with delayed growth.

11.0 Diseases and Pests

11.1.0 Diseases

Fungal Diseases

11.1.1 Stem Rot

Pythium aphanidermatum is the main cause of damping off in *G. paniculata*. The disease affects the cuttings, when rooted at a high temperature and relative humidity. Previeur (prothiocarb) sprayed weekly at 0.3 per cent is most effective in controlling air and mist sprays infection. Soil treatment, followed by weekly sprays gives good control of soil-borne infection (Vigodsky-Hass *et al.*, 1979). Infestation of *Fusarium, Rhizoctonia, Pythium* and *Phytophthora* species has been reported from Italy (Belmonte, 1986). *Phytophthora nicotianae* var. *parasitica* was isolated by Sekiyama and Uematsu (1992) from diseased *G. paniculata* and pathogenicity was confirmed. Soil fumigation with chloropicrin, chloropicrin methyl bromide or methyl isothiocyanate D-D, or soil application of metalaxyl or a soil drench with fosetyl, effectively controlled the disease. The pathogen in naturally infected gypsophila was identified as *P. nicotianae* var. *parasitica* (Kimishima and Kobayashi, 1991). A total of 34 isolates of *P. nicotiana* var. *parasitica* and 19 isolates (all A1 mating type) of *P. capsici* were isolated from diseased tissues of basal stems or roots of *G. paniculata* collected from 18 fields in central and southern Taiwan. In most cases, *Phytophthora* attacked from basal stems. Infected tissues became bleached, water soaked and collapsed. The affected plants drooped and wilted and eventually died due to severe stem and root rot. *Phytophthora* killed plants at every growing stage from young seedlings to flowering plants in the fields (Ann and Liu, 1993). Crown rot and wilt of *G. paniculata* is reported for the first time from South Africa

(Thompson and Naude, 1992). The disease occurs during hot, moist periods with wilting of plants and internal browning of tissues and *P. nicotianae* was consistently isolated from diseased crowns and pathogenicity was confirmed in greenhouse trials. Mycoflora of *G. paniculata* plants from greenhouses and plastic tunnels was analyzed by Orlikowski *et al.* (1991). *F. culmorum* was isolated from 60 per cent and *P. nicotanae* var. *nicotianae* from 35 per cent of the investigated plants. Other, less important fungi included *F. solani, Penicillium, Mucor* and *Trichoderma*. The disease development was stimulated at temperature of 25-35°C. Fosetyl applied as spray 24 h after inoculation or soil drench gave the best disease control. Other effective fungicides included mancozeb + metalaxyl and oxadixyl + mancozeb (Orlikowski and Wojdyla, 1991).

Basal rot of *G. paniculata* caused by *P. nicotianae* var. *parasitica* was reported in Italy by Garibaldi *et al.* (1990). A sudden wilt was detected in the Perfecta variety of *G. paniculata* in Mexico with an incidence of 25 per cent (Ojeda and Zak, 1993). The disease was associated with high humidity in soil after irrigation or heavy rain and with *Phytophthora cryptogea, Rhizoctonia* sp., *Fusarium* sp. and 8 bacterial strains were also isolated from the diseased plants, but pathogenicity was confirmed only for *P. cryptogea*. This is thought to be the first report of *P. cryptogea* on *G. paniculata*. In greenhouse conditions, Werner and Szpanbriker (1997) and Werner (1997) observed the most diseased gypsophila plants in peat infested with *Fusarium oxysporum* f.sp. *dianthi* (isolate No. 1), *F. culmorum* and *Rhizoctonia solani*. Among the plants originating from cuttings, the percentage of infected plants was the highest (96-100) in peat infected with *F. oxysporum.* A high percentage of diseased plants was also observed in peat infested with *P. nicotianae* var. *nicotianae* while the fewest diseased plants were observed in peat infested with *R. solani*.

11.1.2 Powdery Mildew

Matsuura *et al.* (2000) suggested that the flower bud emergence stage is favourable for the inception of powdery mildew (*Oidium* sp.) and that the optimum temperature for the development of the fungus is about 20°C. The infection cycle of the pathogen was found to be highly dependent on the existence of *G. paniculata* and the disease can be eliminated by the removal of this infection source. Triflumizole, bitertanol, fenarimol, kresoxim-methyl and diflumetorim were highly effective in control of this disease in laboratory tests.

Bacterial Diseases

11.1.3 Crown Gall

Crown gall (*Agrobacterium gypsophilae*) is a major disease of *G. paniculata*. Soft nodullar galls develop in the collar region of the grafted plants. Propagation from plants that bear galls should be avoided. Disinfection of plants and soil was recommended (Verona and Loprieno, 1956); streptomycin dip may also help (Pirone *et al.*, 1960). Crown and root galls on nursery rooted cuttings of *G. paniculata* were first observed in Taiwan in 1991. Infected plants showed leaf yellowing, and eventually wilted and died. The bacterium was identified as *Erwinia herbicola* pv. *gypso-philae* based on the physiological characteristics, the biologGn Micro plate system, the

MIDI system for cellular fatty acid analysis, and pathogenicity. Most strains of the bacterium were resistant to copper hydroxide and copper oxychloride, but were sensitive to streptomycin, kasugamycin, tetracycline and a copper oxychloride – mancozeb combination (Hseu and Hsu, 1998). Three PCR primer pairs, based on the cytokinins (*etz*) or IAA biosynthetic genes, were used for detecting *Erwinia herbicola* pv. *gypsophilae* in *paniculata* plants. The primers were specific to all gall forming *E. herbicola* strains and distinguished them from saprophytic strains associated with gypsophila plants or from other gall forming bacteria (Manulis *et al.*, 1998).

11.1.4 Bacterial Wilt

Bacterial wilt of *G. paniculata* caused by *Pseudomonas caryophylli* commonly occurred in Taiwan during the summer season, and caused serious losses. Several chemicals including streptomycin, atakin, cuprosan 311 SD [pyrifenox], kinset, dithane M-45 [mancozeb] and dikar [mancozeb + dinocap] inhibited growth of *P. caryophylli in vitro* (Liu, 1990).

11.1.5 Botrytis

Botrytis cinerea causes blight in *G. paniculata* which can be controlled with spray of captan, ferbam or maneb.

11.2.0 Pests

The major pests in *Gypsophila* include grubs and caterpillars, aphids, grasshoppers, leaf miner, thrips and mites.

11.2.1 Beet Armyworm

Armyworms are troublesome pests. *G. paniculata* plants damaged by *Spodoptera exigua* in the early growing stage did not grow and flower normally (Park *et al.*, 1991). Laboratory tests were conducted to determine whether 18 candidate adjuvants would increase *S. exigua* on *G. paniculata* leaves (Kao *et al.*, 1991). Bioassays showed that 15 adjuvants increased effectiveness; Bivert, Agral 90 and Nu-film 17 were the most effective. Mixing various insecticides with potato dextrose agar (PDA) medium or spore suspension of *Metarhizium anisopliae* var. *anisopliae* isolate Ma_2 was found to inhibit spore germination, mycelial growth and sporulation of the entomogenous fungus (Tsai *et al.*, 1992). Among the insecticides tested, chlorpyrifos had the most inhibitory effect on the fungus. The leaf area of *G. paniculata* consumed by *S. exigua* larvae after 24 or 48 h infection with *M. anisopliae* var. *anisopliae* was reduced by over 10 times after treatment of the leaves with EPN, bifenthrin or chlorpyrifos. Mixing methomyl with spore suspensions of *M. anisopliae* var. *anisopliae* had a synergistic effect, reducing the rate of damage to *G. paniculata* leaves and the LD_{50} to *S. exigua* larvae. Beet armyworm (Spodoptera) may be controlled by combined application of insecticides and miticides (Price *et al.*, 1981).

11.2.2 Leaf Miner

The infestation of leaf miner (*Liriomyza* sp.) may be controlled by combined application of insecticides, miticides and fungicides (Price *et al.*, 1981). Glasshouse experiments showed that *Liriomyza huidobrensis* affected the growth of *G. paniculata*, especially during the vegetative phase. Most of the agromyzids were located in the lower half of the plants. The eulophids *Diglyphus begini* and *Chrysocharis* sp., and the braconid *Opius* sp. reduced the levels of *L. huidobrensis*, provided that no chemical control methods were used (Saray *et al.*, 1988). *Liriomyza trifolii* is an economically important pest of gypsophila in glasshouses. It could be controlled by imidacloprid (Paradikovic, 1998).

11.2.3 Spider Mites

Attack of spider mite on *G. elegans* and *G. paniculata* has been reported by Belmonte (1986).

11.2.4 Other Pests

Red spider mite and leaf hopper are serious pests of *G. paniculata*. The infestation of two-spotted mite (*Tetranychus urticae*), flower thrips (*Frankliniella* sp.) and other lepidopterous larvae may be controlled by combined application of insecticides, miticides and fungicides (Price *et al.*, 1981).

11.2.5 Nematode

Gypsophila is a good host of *Meloidogyne incognita* (Wilcken and Ferraz, 1998).

12.0 Harvesting and Postharvest Management

12.1.0 Harvesting

Generally, the gypsophila crop is harvested by cutting from the stem, when 30-40 per cent of the flowers have started to open up. Similar good quality results can also be obtained by applying artificial opening techniques when only 5 per cent of the flowers are open. Harvesting should be done in the cool hours of the day. The flowered stem should be harvested leaving one internode. Flowers should be placed in water and refrigerated immediately after the harvest. The harvested stems of gypsophila should be immediately bunched and dipped in the pot containing postharvest solution; otherwise, the cut flowers won't get nutrients and water and immediately get damaged. The harvesting of gypsophila should not be delayed, because there will be browning of flowers if not harvested at the optimum or proper stage.

Harvest period lasted 82 days in 3 clones of *cv.* Perfecta, Balborts being the earliest clone. There was considerable variation in the number of bunches produced by each clone. The most productive clone was Balborts and the least productive was Cor Van Duyn. Vase life and quality were greatest for the first harvest of Raham Meristem and Cor Van Duyn clones and for the second harvest in Balborts. Quality and vase life were similar in all clones when the flowers were cut on day 30, 45 or 55. Senescence began slowly 7 days after placing the flowers in vases in all clones. Fading increased with time (Chaparro and Arenas *et al.*, 1999). In South Korea flowering characteristics, cut flower quality, percentage rosette formation and the degree of flower malformation were investigated by Cheong *et al.* (1999) in *cv.* Bristol Fairy in June and September. Flower stalk length, node number and 1st branch number of the June harvest (JH) were much better than those of the

September harvest (SH). Cut flower length was 82.5 cm in JH, and 52.9 cm in SH. Rosette formation rate of SH (24.4 per cent) was much higher than that of JH (1.3 per cent). The flower malformation rate in SH was 92.5 per cent, with 37.5 per cent of cut flowers showing abnormal inflorescences; no flower malformation occurred in JH. SH florets were larger and had more petals than JH florets. Flowers from the 2 harvest times differed in their contents of pigments.

12.2.0 Yield

Farina *et al.* (1991) reported that plants propagated *in vitro* tended to give higher cut flower yields (1716.1 g/plant from 2 flower flushes) than those propagated *in vivo* (1552.0 g/plant), but these differences were not statistically significant. Among the 4 clones, cut flower yields ranged from 1524.0 to 1761.2 g/plant. It is concluded that plants propagated by direct rooting of *in vitro*-raised plants can be used for cut flower production but care must be taken during the early field establishment phase because of the tenderness of the plant tissues.

In Japan, Kono (1997) surveyed the wholesale market in urban areas and concluded that the value of gypsophila depended on the external factors, especially selection of varieties and stability of shipment. Internal factors emphasized the whiteness of flower and prolonged vase life.

For normal varieties, the yield is 8-10 stems per plant per flush and for improved varieties it is 10-12 stems per plant per flush. Plants of *Gypsophila paniculata* L. 'Bristol Fairy' were grown in boxes at nine different shoot densities to know how the number of florets per shoot, a basis of flower quality, was determined during plant growth and development. Four components were defined to analyze the number of florets per shoot (FN): number of nodes (NN), primary branches (PB), secondary branches (SB), and units (= dichasia: UN). NN, PB, SB, UN, and FN were determined in this order during plant growth and development. Fresh weight, stem diameter (SD), PB, SB, UN, and FN of cut flowers decreased as shoot density increased. Total weight of cut flowers and total number of florets per box were almost equal, except in three low-density plots. Days to visible bud and days to flowering tended to decrease as the number of shoots per plant decreased. Path analysis was used to estimate the interrelationship among components. Stem diameter was defined as an outer variable. Path coefficients of stem diameter for NN, PB, SB, UN, and FN were –0.520, 0.686, 0.725, 0.589, and 0.408, respectively. Path analysis indicated that stem diameter had strong direct effects on each component. The correlation coefficient between SD and FN was 0.985. Stem diameter has a strong indirect influence on the number of florets per shoot and can be used as an index to predict FN (Takahiro *et al.*, 2009).

12.3.0 Postharvest Management

12.3.1 Handling of Flowers

Shinbori (1992) considered cooling as an essential process in the preservation of cut flowers of gypsophila and vacuum cooling gave a more rapid decrease in temperature than other methods. Gao *et al.* (1996) in an experiment concluded that water absorption through the stem of cut flowers of gypsophila could be used to prevent water loss during vacuum cooling. Precooling to 1°C followed by

Harvesting and Postharvest Management

Harvesting of Gypsophila

Grading

Packing

Storing in Bucket Solution

Packaging

CFB Packing

Auction Centre

Retail Store

Dried Gypsophila

***Source*: www.shutterstock.com, www.dreamstime.com, www.etsy.com, www.coloriginz.com, www.eagle-linkflowers.com, www.violetaflowerssa.net, www.blogdelflorista.es, www.highcontrolgroup.com**

transportation at 5°C, improved the longevity of cut flowers which were hydrated with the preservative, but precooling alone had no effect, especially on those hydrated with deionized water. Subjecting cut flowers to 18-25°C or sudden rise in temperature from 1°C to 18°C increased ethylene and carbon dioxide concentrations within carton boxes during the transport period. This resulted in an increasing number of florets exhibiting petal browning.

Gypsophila flowers are sensitive to ethylene and also susceptible to drying after harvest. Hence, flowers should be placed into a high humidity (98 per cent RH) cool room at 2°C as soon as possible. Forced air cooling is the best for rapid cooling of the flowers. Strip the basal leaves and grade and bunch to market specifications.

12.3.2 Pulsing

In an experiment on ethylene inhibitors for postharvest treatment, Newman *et al.* (1998) reported that overnight treatments with STS were as effective as short pulse treatments. 1-methyl-cyclopropene, pre-treatment helped to prevent the effects of ethylene on flowers that were open at the time of pre-treatment, it provided no protection for flowers that opened subsequently. No marked differences in ethylene sensitivity was observed between cultivars.

The pulsing solutions containing 5 per cent sucrose + 510 ppm benzalkonium chloride + 300 ppm GA; or 5 per cent sucrose + 300 ppm 8-hydroxyquinoline sulfate + 300 ppm GA, in cut *G. paniculata cv.* Bristol Fairy flowers proved very effective in increasing flower opening and retarding wilting. After pulsing for 20 h at 3 or 21°C, flowers were stored for 12 h at 3°C (simulation of transportation). Benzalkonium chloride showed a synergistic effect of flower opening when combined with GA. Treatment with 5 per cent sucrose + 510 ppm benzalkonium chloride + 300 ppm GA, at 21°C, resulted in the longest vase life, with 99 per cent flower opening (Son *et al.*, 1995). Pulsing of cut *G. paniculata cv.* Bristol Fairy stems with 30 per cent of buds open for 3 h in a preservative that contained 0.2 mM silver thiosulfate and 4 per cent sucrose promoted bud opening and prolonged vase life (Doi *et al.*, 1999a).

12.3.3 Storage of Flowers

Flowers can be stored for 1 to 2 days at 4°C. Doi *et al.* (1999b) reported that inflorescences of gypsophila *cv.* Bristol Fairy cut at the bud stage and treated with STS could be stored with their stem base in a solution containing sucrose and 8-HQS at 1°C or 4 weeks without lowering cut flower quality and longevity. A mixture prepared from 40 mg silver nitrate, 50 mg sodium benzoate, 30 mg sugar, 7.5 mg aluminium sulfate and 2 mg kinetin per litre of water gave the longest vase life of gypsophila (Yildirim *et al.*, 1995). Dried flowers will last at least for 1 year.

Store the gypsophila flowers with stems in water containing 5 ml/l bleach (household bleach - sodium hypochlorite) and 20 g/l sugar. Place in a cool room. This solution should be replaced after every 3 days, or earlier if the solution becomes cloudy. Try not to store the gypsophila flowers for more than 2 weeks. Always scrub and clean buckets in bleach (Tregea, 2003).

12.3.4 Transporting and Marketing

Gypsophila flowers are best transported in tissue paper as plastic sleeves may cause flower browning due to the heat build-up. Prices are generally higher during the winter months. Growers should continually monitor and compare prices in the various markets during the harvesting season.

12.3.5 Value Addition

When markets are flooded, and prices are low gypsophila can be dried. Bunches should be sleeved before stems are put into the drying solution. The drying solution consists of 1 part glycerine to 2 parts water and 1 cap of bleach. Fill a bucket to a depth of 50 mm with this solution. Leave the stems to soak for 48 hours at room temperature. After drying, pack 70-80 gypsophila bunches to a box of dimension: 900 mm x 300 mm x 300 mm. Dried flowers of gypsophila will last for 12 months.

12.3.6 Vase Life

The vase life of *G. elegans* is short and the longevity of cut flowers can be prolonged more than 7 days by dipping in a solution containing 400 mg DICA (sodium dichloroisocyanurate) and 20 mg sucrose per litre of water (Marousky, 1978). Farnham *et al.* (1978) reported that the solution containing 25 ppm $AgNO_3$, 5 to 10 per cent sugar produced the largest blooms with longest life of cut gypsophila. Physan-20 at 200 ppm when added in the vase water together with 10 per cent sucrose also produced the high quality blooms. The flower stalks of *G. paniculata* were held immediately after cutting in various conditioning solutions or water before marketing. Best results were obtained with chrysal-avb, which increased their vase life, compared with water-held controls. The addition of sugar to the chrysal-avb solution at 50 g/l lengthened vase life by a further 2.1 days (Kalkman, 1983). Treatment with a sugar solution + a bactericide, or with silver thiosulphate improved the flower quality and vase life; the flowers opened better, the colour was whiter and the keeping quality doubled to about 10 days. It was suggested that postharvest treatment should continue for at least 4 hours and in some cases, depending on solution, for up to 24 hours (Barendse, 1986). Quaternary ammonium compound delayed flower wilting and prolonged vase life of gypsophila flowers (Tandler *et al.*, 1986). Treatments in which these compounds were combined with sucrose had the added effect of enhancing flower opening. Silver thiosulphate suppressed the initial opening of flowers but after a few days there were more open flowers on the stem than in other treatments. Van Doorn and Reid (1992) in an experiment on flower senescence reported that in harvested panicles of gypsophila kept in water, the flowers lasted for 4 days prior to petal wilting, for 6 days when physan (a disinfectant; 200 mg/l) was included in the vase solution, and for 10 days when both physan and sucrose (15 g/l) were included. Petal wilting was preceded by a sharp rise in ethylene production. Treatment directly after harvest with silver thiosulphate (4 mM) for 30 min. increased flower longevity in each of the foregoing solutions.

Flowering stems of gypsophila were treated with a commercial formula before shipment to study its effect on floret water uptake and to see if it can ameliorate damage caused by exposure to air that is encountered during harvest (Ilona and Haya., 2010). By employing this dye, it was determined that desiccation of flowering stems of more than 15 per cent of their fresh weight, followed by rehydration with water, decreased floret water uptake. Reduction in water uptake was not observed in whole stems; nevertheless, floret opening was reduced. To improve floret water uptake following desiccation, the effects of rehydration with the detergents polyoxyethylene (10) lauryl ether, mixture of quaternary ammoniums, sodium dodecyl sulphate and Triton X-100 were examined. Triton X-100 was the most effective detergent in enhancing floret water uptake and it was better than TOG3 (included in the commercial formula), which was also shown to improve floret water uptake. Since flower opening was correlated with higher floret water uptake, the TOG3 in the commercial formula was replaced with Triton X-100. These solutions were equally effective in improving flower opening; however, neither completely reversed the desiccation-enhancement of flower wilting. These results support the concept that water absorption by florets is not the sole contributor to flower opening. Thus, it is possible to replace TOG3 with Triton X-100 in the commercial formula and rehydration with this solution following desiccation can overcome the reduction in floret water uptake.

13.0 Biotechnology

13.1.0 Protoplast Culture and Somatic Hybrid

Protoplasts isolated from cell suspension cultures of *Dianthus caryophyllus* and *Gypsophila paniculata* were fused using polyethylene glycol and cultured in MS medium containing 1 mg picloram/l and 0.7 M sorbitol. Among the calluses produced, one showing green coloration and vigorous growth was identified as an intergeneric somatic hybrid based on isoenzyme analyses for esterase and glutamate dehydro-genase. Hybridity of this callus was also confirmed by Southern hybridization analysis using a rice rDNA probe. The callus regenerated roots after 2 months of culture on MS medium containing 1 mg picloram/l, but shoot regeneration was not observed on any of the media tested (Nakano and Mii, 1993). Hypocotyl-derived protoplasts of *Dianthus barbatus* that had been pre-treated with iodoacetamide were fused electrically with cell suspension culture derived protoplasts of *G. paniculata* that could divide to form callus but could not regenerate shoots under the culture conditions used in this study.

13.2.0 Molecular Markers

Electrofusion derived calli which produced shoots were selected as putative somatic hybrids, and plantlets were subsequently regenerated from 2 of these selected calli. These plantlets, which *in vitro* produced flowers precociously, were identified as intergeneric somatic hybrids by nuclear ribosomal DNA analysis. Normal plants have not been established up to the present (Nakano *et al.*, 1996).

Although Gypsophila is one of the most economically important ornamental crops, little is known about its genetic variability and the relationships among the different wild species, cultivars, and commercial hybrids. Calistri *et al.* (2014) analysed genetic distances among 5 wild species and 13 commercial hybrids of *Gypsophila* with similar phenotypes but unknown origin. They have used amplified fragment length polymorphism, target region amplification polymorphism, and inter simple sequence repeat whole-genome markers and chloroplast simple sequence repeat (cpSSR), targeting chloroplast DNA. Nuclear markers were found to distinguish all the analyzed samples while cpSSR markers were found to discriminate the different wild species, but could not sufficiently separate the commercial hybrids. The data obtained allowed to cluster the commercial hybrids into different sub-groups and to determine the relationships with the putative species of origin.

The genetic diversity of 14 Gypsophila species from Turkey was analyzed using random amplification of polymorphic DNA (RAPD) and inter-simple sequence repeat (ISSR) markers. Sixteen RAPD and 6 ISSR primers produced 132 polymorphic bands. RAPD, ISSR, and RAPD + ISSR primers for the 14 species showed 92.7, 93.8, and 92.9 per cent polymorphism, respectively. The results indicate that RAPD and ISSR markers are reliable and effective for assessing the genetic diversity of *Gypsophila species*. Gypsophila species studied were separated into 2 clusters and the clusters correlated with geographic and phytogeographic regions (Korkmaz and Dogan, 2015).

13.3.0 Genetic Transformation

Gypsophila paniculata was genetically transformed by co-culturing cut stems with *Agrobacterium tumefaciens*. Different types of gall were formed after infection, although occurrence of a crown gall was essential for successful genetic transformation. After selection at 50 mg kanamycin/l, a transformation rate of 89.8 per cent was attained (Yu and Yang, 1997).

The establishment of a transformation procedure for gypsophila (*Gypsophila paniculata* L.) based on Agrobacterium inoculation of highly regenerative stem segments was described by Moyal *et al.* (2008). The transformation procedure employs stem explants derived from GA_3-pretreated mother plants and a two-step selection scheme. The GA_3 treatment was crucial for obtaining high gene-transfer frequencies (75–90 per cent GUS-expressing explants out of total inoculated explants), using three different gypsophila varieties. An overall transformation efficiency of five GUS-expressing shoots per 100 stem explants was demonstrated for *cv.* Arbel. The applicability of the transformation system to gypsophila was further reinforced by the generation of transgenic plants expressing Agrobacterium rhizogenes rolC driven by a CaMV 35S promoter. Transgenic gypsophila plantlets exhibited extensive rooting and branching, traits that could be beneficial to the ornamental industry.

13.4.0 Cryopreservation

Shoot tips from plants of 3 species of *Gypsophila*, were cryopreserved (Fukai *et al.*, 1991). Excised shoot tips were cooled at a rate of 0.5°C/min down to -40°C in the presence of 10 per cent dimethyl sulfoxide and 3 per cent glucose prior to immersion in liquid nitrogen. The shoot tips of all species and cultivars were viable after thawing. High viability of the cryopreserved shoot tips was maintained for up to 4 years. The plants derived from the cryopreserved shoot tips grew and flowered normally.

14.0 References

Ahroni, A., Zuker, A., Rozen. Y., Shejtman. H. and Vainstein, A. (1997) *Plant Cell Tissue Organ Cult.*, **49** : 101-106.

Alberto, G.B.G., López-Marin, J. and Gonzalez, A. (2006) In book: Floriculture, Ornamental and Plant Biotechnology, **1** : 276-281.

Amezquita, M.O. de, Contor, F., Bernard, J. and Cure, R. (1999) *Acta Hortic.*, **482** : 33-37.

Ann, P.J. and Liu, S.S. (1993) *Plant Pathology Bull.*, **2** : 106-110.

Anonymous (1997a) *Plant Var. J.*,**10** : 25-26.

Anonymous (1997b) *Plant Var. J.*,**10** : 25-29.

Anonymous (1997c) *Plant Var. J.*, **10** : 26-27.

Arens- Blanco, G. and Chaparro, de Barrera, A. (1992) *Agronomia Colombiana*, **9** : 119-130.

Asaph A., Amir Z., Yael R., Hadas S. and Alexander V. (1997) *Plant Cell Tiss. Org.*, **49** : 101–106.

Askim, H.S., Turkan, I., Tanyolac, Z.O., Ozfidan, C. and Dinc, A. (2012) *Environ. Exp. Bot.*, **77** : 63-76.

Atasxlar, E., Erkara, I.P. and Tokur, S. (2009) *Turk. J. Bot.*, **33** : 335–351.

Auman, (1980) In : *Introduction to Floriculture* (*ed. R. A. Larson*), Academic Press. Inc., New York.

Bailey, L.H. (1963) *The Standard Cyclopedia of Horticulture*, Macmillan and Co., New York.

Barendse, L.V.J. (1986) *Acta Hortic.*, **181** : 338.

Belmonte, A. (1986) *ColtureProtette*, **15** : 35-39.

Ben-Yephet, Y., Reuven, M., Zviebil, A., Szmulewich, Y., Lavkovits, I., Markovits, T., Soriano, S. and Bar-Yosef, B. (2006) *Phytopathol.*, **96**(7) : 771-776.

Blackburn, K.B. (1928) *Z.I.A.V. Suppl.*, **1** : 439.

Bose, T. K. (1965) *Sci. and Cult.*, **31** : 34-36.

Bremer, G. (1982) *Zierpflanzenbau*, **22** : 280-283.

Brickell, C. (1996) *A-Z Encyclopedia of Garden Plants*. Dorling Kindersely, pp. 486.

Calistri, E., Buiatti, M., and Bogani, P. (2014) *Plant Biosys.*, http: //dx.doi.org/10.1080/11263504.2014.944609

Castro, V.M., Dardel, H. C. and Verdugo, R. G. (1996) *Agricultura Tecnica Santiago*, **56** : 224-228.

Chaparro de Barrera, A. and Arenas, G. (1999) *Acta Hortic.*, **482** : 71-76.

Chen, C.W., Tsai Y. F. and Young, C.C. (1998) *Bull. Taichung District Agricultural Improvement Station*, **58** : 1-10.

Cheong, D.C., Cheong, J.S., Chang, C.J. and Park, H. B. (1999) *J. Korean Soc. Hort. Sci.* **40** : 376-380.

Chromosome Reports IPCN (2014) Tropicos website. St. Louis, Missouri, USA: Missouri Botanical Garden. http: //tropicos.org/Project/IPCN.

Cox, T. I., Kerr, R. M. and Ingle, A. (1990) *Proc. 43rd New Zealand Weed and Pest Control Conf.*, pp. 43-47.

Cronquist, A. (1981) An integrated system of classification of flowering plants. New York.

Davies L.J., Hicklenton, P.R. and Catley, J. L. (1996) *J. Hort. Sci.*, **71** : 1-9.

De Groote, R. (1957) *Bull. Hort. Liege*, **12** : 279-280.

Diaz- Arteaga, A. and Orozco-de-Amezquita, M. (1990 a) *Agronomia Colombiana*, **7** : 47-53.

Diaz–Arteaga, A. and Orozco-de-Amezquita, M. (1990b) *Agronomia Colombiana*, **7** : 54-60.

Dijk, J. V. and Wolters, G. (1990) *Vakblad. voor-de-Bloemisterij*, **45** : 40-42

Dillen, W. and Buysens, S. (1989) *Plant Cell Tissue Organ Cult.*, **19** : 181-188.

Doi, M. (1992) *Buss Univer. Osaka Prefecture Series B, Agric. and Biology*, **45** : 107-154.

Doi, M., Kobata, T., Ohizumi, K. and Imanishi, H. (1992) *J. Japanese Soc. Hort. Sci.*, **61** : 421-429.

Doi, M., Morita, E., Ogasawara, N., Takeda, Y. and Asahira, T. (1991b) *J. Japanese Soc. Hort. Sci.*, **60**: 119-124.

Doi, M., Morita,T., Takeda,Y. and Asahira, T. (1991a) *J.Japanese Soc. Hort. Sci.*, **59** : 795-801.

Doi, M., Nagai, N. and Imanishi, H. (1999a) *J. Japanese Soc. Hort. Sci.*, **68** : 635-639.

Doi, M., Saito, T., Nagai, N. and Imanishi, H. (1999b) *J. Japanese Soc. Hort. Sci.*, **68** : 854-860.

Doi, M., Takeda, Y. and Asahira, T. (1984) *Memoirs of the College of Agriculture*, Kyoto University, No. 124, pp. 27-34.

Doi, M., Takeda, Y. and Asahira, T. (1989) *J. Japanese Soci. Hort. Sci.*, **58** : 221-226.

Doi, M., Takeda, Y., and Asahira, T. (1990) *J. Japanese Soc. Hort. Sci.*, **59** : 621-626.

Dole, J. M. and Gibson, J. L. (2006) Batavia, IL: Ball Pub.

Ewa, M., Hanus-Fajerska, E. and Kozminska, A. (2018) *Water Air Soil Pollut.*, **229** : 42.

Fan, L., Wang, G., Yu, R., Wu, M., Shan, Q., Wu, L., Ruan, J. and Yang, C. (2019) *HortScience*, **54**(8) : 1370–1374.

Farina, E., Damiano, C., Cervelli, C. and Costantino, C. (1991) *ColtureProtette*, **20** : 115-117.

Farnham, D. S., Kofranek, A. M. and Kubota, J. (1978) *Amer. Soc. Hort. Sci.*, **103** : 382-384.

Favarger, C. (1946) *Ber. Schweiz. Bot. Ges.*, **56** : 365.

Frechet, D., Christ, B., Monegier-du-Sorbier, B., Fischer, H., Vuilhorgne, M., D -Sorbier, B.M. and Sorbier, B.M-du (1991) *Phytochemistry*, **30** : 927-931.

Fukai, S., Goi, M. and Tanaka, M. ((1991) *Euphytica*, **56** : 149-153.

Furusato, K. (1940) *Bot. and Zool. Tokyo*, **8** : 1303.

Gamliel, A., Hadar, E. and Katan, J. (1993) *Plant Disease*, **77** : 933-938.

Gamliel, A., Hadar, E., and Katan, J. (1989) *Acta Hortic.*, **255** : 131-138.

Gao, J.P., Sun, Z. R. and Zhou, S. T. (1996) *Scientia Agricultura Sinica*, **29** : 66-72.

Garibaldi, A., Rapetti, S., Aloi, C., Gullino, M.L. and Bozzano, G. (1990) *InformatoreFitopatologico*, **40** : 5, 9-15.

Gera, A., Phyllus, G., Ladmila, M., Sara, S. and Mohammad. Z. (2005) A new disease Causing stuting and shoot proliferation in Gypsophila is associated with phytoplasma. *Arab. J. Biotec.*, **8**(1): 155- 168.

Goto, T., Kageyama, Y. and Konishi, K. (1997) *Scientific Reports*, Faculty Agri. Okayama University, **86** : 43-49.

Goto, T., Kageyama, Y. and Konishi., K. (2000) *J. Japanese Soc. Hort. Sci.* **69** : 749-757.

Goto, T., Masaoka, K., Kageyama, Y. and Konishi. K. (1998) *Scientific Reports*,Faculty Agri. Okayama University, **87** : 141-148.

Graaf. van der zande M T. de, Blacquiere, T. and Gude, H. (1997) *Acta Hortic.*, **418** : 119-125.

Gribble, K. (1999) *Plant Growth Regulation*, **27** : 179-188.

Han, B.H., Joung, H.Y., Choi, J. K. and Peak, K.Y. (1995) *J. Korean Soc. Hort. Sci.*, **36** : 886-892.

Han, B.H., Pack, K. Y. and Choi, J. K. (1992) *J. Korean Soc. Hort. Sci.*, **33** : 73-78.

Han, B.H., Paek, K.Y. and Choi, J.K. (1991a) *J. Korean Soc. Hort. Sci.*, **32** : 518-524.

Han, B.H., Paek, K.Y. and Choi, J.K. (1991b) *J. Korean Soc. Hort. Sci.*, **32** : 394-400.

Han, S. Y., Qi, L.W., Wei, Y.T., Yang, Y. L. and Zhang, Y.W. (1996) *Acta HorticulturaeSinica*, **23**: 175-178.

Harbaugh, B.K., Cooper, V.F. and Nell, T.A. (1989) *Fla. State Hort. Soc.*, **102** : 284-286.

Hayashi, T., Miyata, H. and Konishi, K. (1992) *J. Japanese Soc. Hort. Sci.*, **61** : 135-141.

Henry, M. (1993) In : *Medicinal and Aromatic Plants*, **IV** : 187-206.

Hicklenton, P.R. (1986) *Canad. J. Plant Sci.*, **66** : 653-658.

Hicklenton, P.R. (1987) *Acta Hortic.*, **205** : 103-111.

Hicklenton, P.R., Newman, S.M. and Davies, L.J. (1993) *HortScience*, **28** : 888-890.

Hicklenton, P.R., Newman, S.M. and Davies, L.J. (1993a) *Scientia Hort.*, **53** : 319-331.

Hseu, S. H. and Hsu, S. T. (1998) *Plant Protection Bull. Taipei*, **40** : 329-338.

https: //en.wikipedia.org/wiki/Gypsophila

https: //www.plantopedia.com/gypsophila/.

Ichihara, J. (1954) *J. Hort. Ass. Japan*, **23** : 115-122.

Ilona, R. and Friedman, H. (2010) *Postharvest Biol. Technol.*, **57** : 189-195.

James H. L., Preece, J.E. and Coorts, G.D. (1983) *J. Environ. Hortic.*, **1**(4) : 83–86.

Jiang, Z.P., Liu, G.L. and Ni, J.D. (2000) *J. Jiangsu. Forestry Sci. Technol.*, **27** : 29-31.

Kalkman, E. C. (1983) *VakbladVoor de Bloemisterij*, **38** : 50.

Kantartzis, N. A. (1964a) Reprint from Geoponika No. 116/117, pp. 79-88.

Kantartzis, N.A. (1964b) *Reprint from Geoponika*, No. 116/117, pp. 79-88.

Kao, S.S., Hsia, W.T. and Huang, L.H. (1991) *Chinese J. Ento.*,**11** : 330-334.

Kapchina-Toteva, V. and Stoyanova, D. (2003) *Biol. Plant.*, **46** : 337-341.

Karaguzel, O. (1996) *Bahce*, **25** : 55-59.

Karaguzel, O. and Altan, S. (1999) *Turkish J. Agric. and Forestry*, **23** : 275-280.

Katalinic, M. (1989) *ZastilaBilja.*, **40** : 217-222.

Kaukovirta, E. (1966) *Puutarha*, **69** : 16-18.

KF Bioplants. (2018) Gypsophila Manual, KF Bioplants, Pune, India. pp. 1-17.

Kimishima, E. and Kobayashi, Y. (1991) *Res. Bull. Plant Protetion Service, Japan*, **27** : 47-54.

Kishi, F., Kagami, Y., Shinohara, M., Hatano, S. and Tsurushima, H. (1994) *Euphytica*,**74** : 85-90.

Kono, Y. (1997) *Bull Kyushu National Agricultural Experiment Station*, **32** : 51-73.

Korkmaz, M. and Dogan, N.Y. (2015) *Genet. Mol. Res.*, **14**(3) : 8829-8838.

Korkmaz, M., *et al.* (2012) *Biyoloji Bilimleri Araþtýrma Dergisi* (*BÝBAD*) **5**(2) : 111-125.

Korkmaz, M., *et al.* (2010) Archived 2016-03-04 at the Wayback Machine In: 2nd International Symposium on Sustainable Development, June 8–9, 2010, Sarajevo.

Kwadwo Owusu Ayeh, Ravichandran Selliah and Anne Kathrine Hvoslef-Eide. (2009) *Acad. J. Plant Sci.*, **2**(4) : 260-266.

Lee, I. K., Lee, B. N. and Kang, K. H.(1993) *J. Agric. Sci.*, **35** : 2, 661-671.

Lee, S. W. and Bae J. J. (1999a) *Korean J. Hort. Sci. Technol.*,**17** : 765-767.

Lee, S. W. and Bae, J. J. (1999b) *Korean J. Hot. Sci. Technol.*, **17** : 768-769.

Leonard Perry. *http: //pss.uvm.edu/pss123/pergypso.html*

Li, F., Mo, X., Wu, L. and Yang, C. (2020) *Hort. Sci.*, **55**(9): 1531-1532.

Liu, H.L. (1990) *Bull. Taichung District Agricultural Improvement Station*, **28** : 33-42.

Madhani, Hossein; Rabeler, Richard; Pirani, Atefeh; Oxelman, Bengt; Heubl, Guenther; Zarre, Shahin (2018) Taxon., **67**(1) : 83–112.

Manulis, S., Kogan., N., Valinsky. L., Dror, O. and Kleitman, F. (1998) *European J. Plant Pathol.*, **104** : 85-91.

Marousky, F. J. (1978) *Proc. Fla. St. Hort. Soc.*, **90** : 297-298.

Matsuo, E. and Nonaka, A. (1991) *Agriculture and Horticulture.*, **66** : 1181-1186.

Matsuura, S., Kajihara, S., Iwamoto, T., Kakou, T. and Kohguchi, T. (2000) *Bull. Hiorshima Prefectural Agril. Res. Center*, **68** : 21-29.

Mlynarcyk, M. P. (1993) *Vestsi Akademii Navuk BSSR. Seryya Biyalagichnykh Navuk.*,**L** : 33-37.

Molina, E., Gamboa, J. and Gonzalez, P. (1991) *Agronomia Costarricense*, **15** : 185-187.

Moyal, M., Amir, Z., Marianna, O., Elena, S., Hagit, B.M., Shamir, Z. and Alexander, V. (2008) *Mol. Breed.*, **22** : 543-553.

Nakano, M. and Mii, M. (1993) *Scientia Hort.*, **53** : 13-19.

Nakano, M., Hoshino, Y. and Mii, Y. (1996) *Theor. Appl. Genet.*, **92** : 107-172.

Nam, C. W., Ryu, S.Y., Yoo, D. L., Kim, W.B. and Paek, K.Y. (1997) *RDA. J. Hort. Sci.*, **39** : 75-79.

Newman, J. P., Dodge, L. L. and Reid, M.S. (1998) *Hort Techno.*, **8** : 58-63.

Nishidate, K., Kanayama, Y., Nishiyama, M., Yamamoto, T., Hamaguchi, Y., and Kanahama, K. (2012) *Jpn. Soc. Hort. Sci.*, **81** : 198–203.

Nordestgaard, A. (1988) *TidsskriftPlanteavl*, **92** : 217-219.

Novoplansky, A., Shoshany, H., Assenhaim, D., Schnizer, M., Eyal, M., Chernyak, V., Reisfeld., R. and Segal, I. (1990) *Proc. Inter. Seminar British-Israel Workshop on Greenhouse Technology, Bet Dagan, Israel*, pp. 119-129.

Ojeda, M.S. and Zak, L.F. (1993)*Revista Mexicana de Micoloia*, **9** : 169-171.

Orlikowski, L. B. and Wojdyla, A. (1991) *Prace Instytutu Sadownictwai Kwiaciarstwa w Skiernie-wicach. Seria B, Rosliny Ozdobne*, **16** : 163-169.

Orlikowski, L.B., Wojdyla, A. and Skrzypezak, C.S. (1991) *Prace Instytutu Sadownictwai Kwiaciarstwa w Skiernie-wicach. Seria B, Rosliny Ozdobne*, **16** : 157-162.

Paek, K.Y., Seon, J.H., Son, S. H., Han, B. H. and Yae B. W. (1998) *Korean J. Hort. Sci. Technol.*, **16** : 272-278.

Paradikovic, N. (1998) *Poljoprivreda*, **4** : 87-90.

Park, J.D., Goh, H.G., Lee, J.H., Lee, W.J. and Kim, K.J. (1991) *Korean J. Appl. Ento.*, **30** : 124-129.

Paul K. Wahome, Tajudeen O. Oseni, Michael T. Masarirambi and Victor D. Shongwe. (2011) *World J. Agric. Sci.*, **7**(6) : 692-698.

Pirone, pp., Dodage, B.O. and Rickett, H. W. (1960) *Diseases and Pests of Ornamental Plants*, Cons-table and Co. Ltd., London.

Plomacher, H. and Hagemann, H. (1976) *Zirepflanzenbau*, **16** : 246-247.

Price, J. F., Englehard, A. W., Overman, A. J., Yingst, V. W. and Iverson, M. K. (1981) *Proc. Fla. St. Hort. Sci.*, **93** : 187-190.

Rady, M.R. (2005) *Arab J. Biotech.*, **8**(1) : 155-168.

Rady, M.R. (2006) *Biol. Plant.*, **50** : 507-513.

Riddell, R.C. (1990) *ODI working paper*, pp. 54.

Rise 'N' Shine. Gypsophila growing manual.

Rashid, S., Ilyas, S., Naz, S., Aslam, F. and Ali, A. (2012) *Pak. J. Sci.*, **64**(1) : 6-10.

Salman, M.N. (2002) *Plant Cell Tiss.Org.*, **69** : 189-196.

Saray, M.P., Sarmiento, C.J. and Acosta, G.A. (1988) *RevistaColombiana de, Entomologia*,**14** : 34-42.

Sawwan, J.S. (1993) *Adv. Hort. Sci.*, **7** : 103 -104.

Sekiyama, K. and Uematsu, S. (1992) *Proc. Kanto Tosan Plant Protection Soc.*, **39** : 157-160.

Shibuya, T., Murakawa Y., Nishidate, K., Nishiyama, M., and Kanayama, Y. (2017) *Hort. J.*, **86** : 94–104.

Shillo, R. and Halevy, A. H. (1982) *Scientia Hort.*, **16** : 385-393.

Shinbori, F. (1992) *Agriculture and Horticulture.*, **67** : 97-104.

Skalinska, M. (1950) *Acta Soc. Lot. Polon.*, **20** : 45.

Smith, E.M. and Treaster, S.A. (1992) *Special Circular Ohio Agri. Res. and Development Centre*, **140** : 21-23.

Son, K. C., Gu, E. G. and Chae, S. C. (1995) *J. Korean Soc. Hort. Sci.*, **36** : 273-280.

Song, J.S., Lee, M.S. and Han, I.S. (1991) *Res. Rep. Rural Dev. Administration Hort.*, **33** : 48-54.

Song, R.H., Lu, J.P., Zhou, X.G. and Zhu, Z.Y. (1996) *Acta Agriculturae Shanghai*, **12** : 71-73.

Staats, D. and Klett, J.E. (1993) *Enviro. Hort.*, **11** : 78-80.

Subhankar, P., Palai, J.B., Dash, L. and Tripathy, B (2020) *In*: Protected Cultivation and Smart Agriculture edited by Sagar Maitra, Dinkar J Gaikwad and Tanmoy Shankar. New Delhi Publishers, New Delhi, pp. 227-231. Doi: 10.30954/NDP-PCSA.2020.24.

Sunset Western Garden Book (1995) pp. 606–607.

Takahiro, H., Kakeya, H. and Konishi, K. (2009) *J. Japan. Soc. Hort. Sci.*, **78**(3) : 330–334.

Takane, R.J., Minami, K., Lucchesi, A.A., Almeida M. de. and De, Almeida,M. (1994) *Scientia Agricola.*, **51** : 235-239.

Takashi, F., Hayashi, T. and Yazawa, S. (2007) *J. Japan. Soc. Hort. Sci.*, **76**(4) : 327–332.

Takeda, T. (1996) *J. Japanese Soc. Hort. Sci.*, **64** : 875-882.

Tandler, J., Mor, Y., Spiegelstein, H. and Mayak, S. (1986) *Acta Hortic.*, **181** : . 443-449.

Thakur, S., Dhiman, S., Sharma, Y., and Singh, A. (2013) *J. Hill Agr.*, **4** : 74–77.

Thompson, A.H. and Naude S.P. (1992) *Phytophylactica*, **24** : 349-350.

Toaima, N., Bosila, H., and El-Ateeq, A.A.E.-F. (2016) *Middle East J.*, **5** : 313–323.

Todd, B. (2008) *https: //davesgarden. com/guides/articles/view/1734.*

Tregea (2003) *Gypsophila Growing in Central Australia*. (2003) **615**(A28) Agdex No: 282/20 ISSN No: 0157-8243.

Tsai, Y.S., Kao, S.S., Kao, C.W. and Hou, R.F. (1992) *Plant Protection Bull.*,Taipei., **34** : 216-226.

Van, D., W.G. and Reid, M.S. (1992) *Postharvest Biol. Technol.*, **1** : 265-272.

Vergara Gongora, J. (1984) *BoletinInformativo Centro de Investigacion Y Desarrollo Hoifola*, No. 8, pp. 79-84.

Verona, O. and Loprieno, N. (1956) *Ann. Sper. Agrar.*, **10** : 133-143.

Vettori, L., Schiff, S., Tani, C., Pasqualetto, P., and Bennici, A. (2015) *Plant Biosyst.- An Int. J. Dealing with all Aspects Plant Biol.*, **149** : 322–328.

Vigodsky-Haas, H., Farkash, E. and Sharoni, M. (1979) *Hassadeh.*, **59** : 1941-1946.

Wac³aw, B. and Alina, S.S. (2016) *Polish Bot. J.*, **61**(2) : 257–262.

Wang, S.M., Piao, X.C., Park, S.Y. and Lian, M.L. (2013) *In Vitro Cell Dev*-PL, **49** : 70–78.

Werner, M. (1997) *Roczniki Akademii Rolniczej-W- Poznaniu Ogrodnictwo*, **25** : 137-143.

Werner, M. and Szpanbriker, A. (1997) *RocznikiAkademiiRolniczej-w-poznaniu Ogrodnictwo*. **25** : 127-136.

Wilcken, S.R.S. and Ferraz, L.C.C.B (1998) *Summa Phytopathologica*, **24** : 171-176.

Wilfret, G.J., Weiler, T. C., Harbaugh, B.K. and Hammer, P. A. (1986) *HortScience*, **21** : 190-191.

Witomska, M. and Her- Zachariasiewicz A. (1997) *Zeszyty-Problemowe Postepow NaukRolniczych*, **449** : 229-236.

Yildirim, T. B., Hatipoglu, A., Zager, B. and Gulgun, B. (1995) *Ege. Universitesi Ziraat Fakultesi Dergisi*, **32** : 175-182.

Yu, P. T and Yang, Y. P. (1997) *Acta Agriculturae Shanghai*, **13** : 17-20.

Yücekutlu, A. Nihal (2000) *Gazi Üniversitesi Fen Bilimleri Enstitüsü, Kimya, Yüksek Lisans Tezi.*, **64s.** Ankara.

Yücekutlu, A.N. and Bildacý, I. (2008) *J. Biol. Chem.*, **36**(2) : 129-135.

Zamorano–Mendoza, J.J. and Mejia-Munoz, J.M. (1994) *RevistaChapingo., Serie Horticulturea*, **1** : 67-71

Zebian, K.J. and Reekie, E.G. (1998) *Anna. Bot.*, **81** : 185-193.

Zhou, X.Q., Lian, B.Z., Zhang, H.L., Cai, Y.M. and Lin, D.W. (1994) *J. Shanghai Agri. College*, **12** : 107-112.

Zuker, A., Ahroni, A., Shejtman, H. and Vainstein, A. (1997) *Plant Cell Rep.*, **16** : 775-778.

Zyi, M.M.B., Zuker, A., Ovadis, M., Shklarman, E., Ben-Meir, H., Zenvirt, S., and Vainstein, A. (2008) *Mol. Breed.*, **22** : 543–553.

Rudbeckia

Solanki Bal and Swati Saha

1.0 Introduction

Rudbeckia is a plant genus in the Asteraceae or composite family (Linnaeus, 1753). Rudbeckia flowers feature a prominent, raised central disc in black, brown shades of green and in-between tones, giving rise to their familiar common names of coneflowers and black-eyed-susans. Many species are cultivated in gardens for their showy yellow or gold flower heads that bloom in mid to late summer. All are native to North America. The species are herbaceous, mostly perennial plants growing up to 0.5–3.0 m tall with simple or branched stems. They can be even grown as biennials or half-hardy annuals and may be sown indoors in warmth. The annual varieties may survive from year to year if left in the garden, but flowering may be reduced in subsequent years. The leaves are spirally arranged, entire to deeply lobed, and 5–25 cm long. The flowers are produced in daisy-like inflorescences with yellow or orange florets surrounding a prominent, cone-shaped (as the flower head opens, the ray florets tend to point out and are down as decumbent) head or a conical disk. Several currently accepted species have many accepted varieties. Some of them (for example the black-eyed susan, *R. hirta*), are popular garden flowers distinguished for their long flowering times. Rudbeckia is one of at least four genera within the flowering plant family Asteraceae whose members are commonly known as coneflowers; the others are Echinacea, Dracopis and Ratibida.

2.0 Importance and Uses

2.1.0 Aesthetic Use

The colorful yellow/gold, and/or orange/gold, daisy-like ray petals serve to attract butterflies, bees and other pollinators to the slightly raised, black/brown, central cone, or disk. This flower disk/cone is comprised of scores of florets arranged in a spiral formation. Each floret contains its own stamens and pistils. During autumn, numerous species of birds that feed on the tiny seeds produced within these multiple disk florets which remain erect even after all the ray petals have dried and fallen away, eventually providing support during winter even in a snow-covered landscape. Considering its aesthetinc values, *Rudbeckia fulgida, var.* sullivantii 'Goldsturm', orange coneflower was awarded by the Royal Horticultural Society's Award of Garden Merit in 1993 and Perennial Plant of the Year in 1999 by the Perennial Plant Association in America.

2.2.0 Therapeutic and other Uses

Black-eyed Susan has been used for centuries as a medicinal herb by various native North American tribes to treat a wide range of ailments and inflammatory conditions such as colds, earache, sores and snakebites (Barker, 2004; Michael *et al.*, 2014) even if it is not one of the most recognized medicinal plants it still has its uses in modern herbal medicine. Recent studies indicate that extracts made from the black-eyed Susan root can be beneficial in stimulating the immune system and in this regard being even more effective than the better known medicinal plant Echinacea. It should be noted that they are in the same family and thus have similar immune

Uses of Rudbeckia

In Pot — In Flower Vase — Cut Flower

In Flower Arrangement — Garden Path — Side of Water Pool

In Flower Bed — In Home Garden — As Border in Garden

Home Entrance — As Specimen Plant — In Landscape

Source: www.atozflowers.com; www.gardenclubofcamaswashougal.org; www.stonehousenursery.com; www.pinterest.com.au; www.courier-journal.com; www.thepaintboxgarden.com

boosting properties. This herb has been used to treat some of the symptoms of the common cold, usually in the form of tea or infusion made from the root. The root has been used traditionally as an herbal remedy to rid the body of parasitic worms. An herbal remedy in the form of an infusion from the roots is used traditionally to treat dropsy. A yellow dye is obtained from the flowers. Previous research has also demonstrated that herbs belonging to the genus *Rudbeckia* possess antimicrobial and antileukaemic (Michael *et al.*, 2014) properties.

3.0 Origin and History

Rudbeckia hirta, commonly called black-eyed Susan, is a North American flowering plant in the sunflower family, native to Eastern and Central North America and naturalized in the Western part of the continent as well as in China. Botanist Carolus Linnaeus gave the genus name Rudbeckia as a nod to the renowned Swedish father and son duo, both named Olof Rudbeck. The older Rudbeck was a professor of medicine at Uppsala University and a scientist recognized for his works in the field of human anatomy and linguistics. Olof Rudbeck the Younger, his son, is a distinguished botanist like Linnaeus. The species name *hirta*, on the other hand, comes from the Latin word "hirtus." It translates to "hairy," describing the plant's leaves, stems and phyllaries.

These are indigenous to many sections of the United States. They were introduced into Europe soon after Columbus's visits and were named by Linnaeus in 1753. Linnaeus gave them the Latin name *Rudbeckia hirta* after his mentor Olaf Rudbeck and hirta meaning rough, hairy. The plant was known and used by many Native American tribes. The Native Americans used the plant to wash sores and swellings, as a poultice for snake bites and for treating worms and colds in children. In 1918, the Black Eyed Susan, Rudbeckia Hirta was designated as the Floral Emblem of the State of Maryland. Each year this flower has its moment of glory when a huge drape made of hundreds of black eyed susans is lifted onto the back of the horse that wins the Preakness – the second jewel of the triple crown of horse racing.

Petals of *Rudbeckia hirta* L. contain flavonols patulitrin, quercetagetin and 6,7-dimethoxy-3′,4′,5-trihydroxyflavon-3-O-glucoside. The flavonoid composition of flowers and leaves of *Rudbeckia hirta* L. is identical. The content of patulitrin in flowers is 1.90 per cent, in leaves 0.82 per cent and in stems 0.12 per cent. Only patulitrin was found in stem culture. The total content of flavonoids in the petals of *Rudbeckia hirta* L. is 4.72 per cent (Raman and Natalia, 2019).

4.0 Morphology

Rudbeckia hirta is an upright annual (sometimes biennial or perennial). Black-eyed susan is a relatively large wildflower, ranging from 30-90 cm. in height and 30–45 cm (12–18 in) wide. As indicated by its name, the flower head has a prominent black or dark-brown central cone that is surrounded by rich, yellow, petal-like rays. The leaves are long, lanceolate, and rough to the touch. The stalk is robust and also coarsely textured. Each black-eyed susan has a singular flower-head positioned atop the terminal shoot. The flower-heads are quite large, measuring 5-7.5 cm. in width, and are radially symmetrical. The bloom has a prominent black or dark-brown hairy central cone that is surrounded by rich-yellow, long, ray-like, prominently linearly-veined, bristle-tipped petal-like rays. The rays grow form the under portion of the central cone and droop slightly downward. The seed-like fruit is small and brown. The leaves of the black-eyed susan are quite long, measuring approximately 5-17.5 cm. They are thin and lanceolate-to-ovate in shape. The leaves have winged petioles, are prominently veined, rough to the touch, and sometimes sparsely toothed. Leaves are several and grow irregularly along the length of the stalk.

5.0 Species and Cultivars

5.1.0 Species

A large number of species have been proposed within *Rudbeckia*, but most are now regarded as synonyms of the limited list given below. Several currently accepted species have several accepted varieties. Some of them (for example the black-eyed susan, *R. hirta*), are popular garden flowers distinguished for their long flowering times.

Rudbeckia hirta

It is a short-lived perennial that should be treated as an annual. It may likely endure few winters, but will often self-seed prolifically. The flowers are showier than other *Rudbeckia* species and have been widely hybridized.

R. triloba

This species is native to all over the eastern North America. The species is a short-lived perennial that will self-seed and works well in naturalized areas. The bright yellow flowers have a dark black center and are a great addition to compliment native plantings.

R. laciniata

It is affectionately known as the "Out House Plant". The large yellow flowers have a whimsical look and make excellent cut flowers or even a good choice for a children's garden or fairy garden.

R. maxima

This species is native to the Southern United States. It blooms in mid-summer through mid-fall with large, 3 inch yellow flowers and black centers. The leaves grow more upright, which gives the plant a different look than other *Rudbeckia* species.

R. grandiflora

It is a rhizomatous perennial species native to the Eastern United States. The plant is huge, growing to 39–47 inches in height and spreading 23–29 inches. Colonies of this species can be quite large, covering massive areas. The flowers are a bright and showy yellow and make perfect cut flowers and are attractants for pollinators and beneficial insects.

List of other accepted species are given below.

- ☆ *Rudbeckia alpicola* Piper – showy coneflower - Cascades in Washington

Species of Rudbeckia

Rudbeckia hirta

Rudbeckia trilobata

Rudbeckia laciniata

Rudbeckia maxima

Rudbeckia grandiflora

Rudbeckia fulgida

Rudbeckia alpicola

Rudbeckia auriculata

Rudbeckia californica

Rudbeckia mollis

Rudbeckia occidentalis

Rudbeckia graminifolia

Source: www.worldoffloweringplants.com; www.georgiabiodiversity.org; www.botanyphoto.botanicalgarden.ubc.ca; www.coastalwildscapes.org; www.fnps.org

- *Rudbeckia auriculata* (Perdue) Kral – eared coneflower - Alabama, Georgia, Florida Panhandle
- *Rudbeckia californica* A.Gray – California coneflower - California
- *Rudbeckia flava* T.V.Moore - Colorado, Wyoming
- *Rudbeckia fulgida* Aiton – orange coneflower - eastern USA + Canada, Texas to Connecticut + Quebec
- *Rudbeckia glaucescens* Eastw. – waxy coneflower - northwestern California, southwestern Oregon
- *Rudbeckia graminifolia* (Torr. and A.Gray) C.L.Boynton and Beadle – grassleaf coneflower - Florida Panhandle
- *Rudbeckia heliopsidis* Torr. and A.Gray – sunfacing coneflower - Mississippi to Virginia
- *Rudbeckia klamathensis* – Klamath coneflower - northwestern California
- *Rudbeckia laciniata* L. – cutleaf coneflower, Green-head Coneflower - widespread in USA + Canada
- *Rudbeckia maxima* Nutt. – great coneflower - mostly east Texas to Missouri
- *Rudbeckia missouriensis* Engelm. ex C.L.Boynton and Beadle – Missouri coneflower - Texas to Illinois; mostly in Ozarks
- *Rudbeckia mohrii* A.Gray – Mohr's coneflower - Florida Panhandle, southern Georgia
- *Rudbeckia mollis* Elliott – softhair coneflower - Florida Panhandle, southern Georgia, Alabama, South Carolina
- *Rudbeckia montana* A.Gray – montane coneflower - Colorado, Utah, Oregon
- *Rudbeckia newmannii* Loudon
- *Rudbeckia nitida* Nutt. – shiny coneflower - Florida, Georgia, Alabama, Louisiana
- *Rudbeckia occidentalis* Nutt. – western coneflower - from Colorado to Washington + northern California
- *Rudbeckia scabrifolia* L.E.Br. – roughleaf coneflower - Louisiana, eastern Texas
- *Rudbeckia speciosa* – showy coneflower - eastern USA
- *Rudbeckia subtomentosa* Pursh – sweet coneflower - mostly Mississippi Valley
- *Rudbeckia texana* (Perdue) P.B.Cox and Urbatsch – Texas coneflower - Louisiana, eastern Texas

5.2.0 Cultivars

The *Rudbeckia* genus has a rich history as a native wildflower, popular first as a medicinal herb used by pre-colonial Native Americans and then finding its way into 19th century cultivated flowerbeds. Two of the most common species in the genus are *Rudbeckia hirta,* often known as black-eyed Susan, and *R. triloba*, commonly called brown-eyed Susan. These two species are often confused with one another. The "brown-eyed" and "black-eyed" labels are sometimes used interchangeably to refer to any of the commonly grown *Rudbeckia* species and their cultivars – even those that have been bred to eliminate the dark flower centers. However, black-eyed Susans are larger and have more petals than brown-eyed Susans. There are some additional rudbeckia varieties that are also common garden plants. Some of the popular cultivars are listed below.

Indian Summer (*Rudbeckia hirta* 'Indian Summer')

It is a drought-tolerant plant as a short-lived perennial, but it blooms from seed its first year and volunteers freely, making it suitable as an annual as well. It is native to North America and grows at a height of 2–3 feet.

Prairie Sun (*Rudbeckia hirta* 'Prairie Sun')

It is native to North America and grows at a height of 30–36 inches. The green centers of this orange and yellow flower give 'Prairie Sun' versatility in cut flower arrangements, popping nicely against blue flowers like delphiniums, bachelor's buttons, and asters.

Moreno (*Rudbeckia hirta* 'Moreno')

It is native to North America and grows at a height of 12-18 inches. *Rudbeckia hirta* 'Moreno' is a shorter cultivar that deserves a special place at the front of a sunny flower border. Burgundy and orange flower petals provide the rich jewel tones that accent classic fall flower arrangements. The 'Moreno' variety blooms from June to the first frost. It tolerates heat, drought, and nearly all conditions except poorly drained, wet soil. The flowers self-seed easily.

Irish Eyes (*Rudbeckia hirta* 'Irish Eyes')

It is native to North America and grows at a height of 27-31 inches. This is a black-eyed Susan without black eyes: The centers of the blooms are greenish-yellow.

Cherokee Sunset' (*Rudbeckia hirta* 'Cherokee Sunset')

It is native to North America and grows at a height of 24-30 inches. *Rudbeckia hirta* 'Cherokee Sunset' looks more like a dahlia than a humble black-eyed Susan flower. The semi-double to fully double blooms are deer-resistant and attract pollinating insects, butterflies, and birds. Like other rudbeckia varieties, it self-seeds. It thrives in full sun and is drought-tolerant once established. The plant blooms profusely from early summer to the first frost.

Maya (*Rudbeckia hirta* 'Maya')

It is native to North America and grows at a height of 1-2 feet. *Rudbeckia hirta* 'Maya' is reminiscent of fall mums with its layers of frilly petals. 'Maya' thrives in full sun and well-drained, medium-moisture soil. Once established, the plant is drought-tolerant. Overall, this is an easy plant to grow, especially because it's deer-resistant and attractive to pollinating insects.

Cultivars of Rudbeckia

Toto | Amarillo Gold | Cappuccino

Little Gold Star | Cherry Brandy | Enchanted Embers

Toto Rustic | Rustic Dwarf | Storm of Gold

Black Nipple | Goldilocks | Toto Lemon

Source: www.dobies.co.uk, www.ipm.missouri.edu, www.thompson-morgan.com, www.lucangardencentre.ie, www.dgiplants.com, www.thursd.com

Prairie Glow (*Rudbeckia triloba* 'Prairie Glow')

It is native to North America and grows at a height of 3-4 feet. The petals of *Rudbeckia triloba* 'Prairie Glow' look like they've been dipped in bright yellow paint. The casual form of these tall plants looks great swaying in the breeze alongside ornamental grasses or *Verbena bonariensis*. It's an incredibly easy flower to grow and blooms from mid-summer to the first frost.

Cherry Brandy (*Rudbeckia hirta* 'Cherry Brandy')

It is native to North America and grows at a height of 1-2 feet. It has numerous flowers with a long bloom time and is attractive to pollinators while being deer-resistant and drought-tolerant. The blooms of this rudbeckia type are a vibrant reddish-maroon. 'Cherry Brandy' will bloom continually throughout the summer. This plant needs weekly watering and is not quite as tolerant of drought as other varieties.

Autumn Colors: It grows to 20–23 inches in height and spreads 12–18 inches. The flowers are a dark mix of bronze orange and red. The variety makes an excellent addition to any garden.

Denver Daisy: It grows to 18 – 20 inches in height and spreads 12 – 18 inches. The flower is unique, having yellow petals with a dark brown center and a mahogany eye.

Goldilocks: It grows to 20–23 inches in height and spreads 12–18 inches. The yellow flowers are large and showy. They make excellent cut flowers.

Goldrush: It grows to 18–23 inches in height and spreads 18–23 inches. This showy variety is ideal for use as cut flowers.

Prairie Sun: It grows to 27– 31 inches in height and spreads 12–16 inches. The flowers are bright yellow with greenish yellow centers.

Rustic Colors: It grows to 18–23 inches in height and spreads 12–18 inches. The flowers are a mix of orange and bronze with yellow edges.

Sonora: It grows to 12–16 inches in height and spreads 12–18 inches. The flowers have large dark brown centers with yellow edges.

Toto Gold: It grows to 12–16 inches in height and spreads 10–12 inches. With its classic yellow showy flowers with black centers, it is perfect miniature *Rudbeckia* species for small garden areas or containers.

Toto Lemon: It grows to 12–16 inches tall and spreads 10–12 inches. This variety is ideal for small gardens, containers, and as cut flowers.

6.0 Genetics and Breeding

6.1.0 Genetics

Species in Rudbeckia subgenus Rudbeckia have a base chromosome number of x = 19 and polyploids have been reported to exist in *R. hirta*, *R. fulgida var. speciosa*, and *R. triloba* L. (McCrea, 1981; Urbatsch *et al.*, 2000). Chromosome counts documented by Palmar *et al.* (2009) that *R. hirta* 'Toto Gold' was a diploid

(2n = 2x = 38) and *R. hirta* 'Indian Summer' was a tetraploid (2n = 4x = 76) with an average 1Cx value of 3.8 rg. Seedlings of *R. missouriensis* and *R. subtomentosa* were confirmed to be diploids with 1Cx values of 8.8 rg and 11.0 rg, respectively. *Rudbeckia fulgida* var. *sullivantii* 'Goldsturm' was found to be tetraploid with a 1Cx value of 8.9 rg. Based on these standards, ploidy levels of the remaining cultivars were then estimated for each species.

The tetraploid cultivar of var. *pulcherrima,* commonly known as the *Gloriosa* daisy, has a drastically shortened lifespan as a tradeoff for its almost double-sized flower heads. (Blakeslee, 1948). Pseudogamy, a form of apomixis, has also been implicated as the sole means of reproduction in triploid *R. triloba* (McCrea, 1981). A greater understanding of reproductive pathways and occurrence of apomixis would provide valuable information for both breeding and commercial propagation of Rudbeckia.

6.2.0 Breeding

6.2.1 Hybridization

Most of the cultivars of the perennial forms have occurred as spontaneous mutations or seedling variations (Barnes, 2007). Perhaps the most well known is *R. fulgida* var. *sullvantii cv.* Goldsturm. This plant was found in Germany and after some effort made its way back home to North America. It is a tetraploid and is self sterile and produces seeds apomictically (Palmer *et al.*, 2007.) It has many characteristics which make it very desirable. It blooms heavily, is naturally short in stature, and is generally immune to most pest problems. Because it gives rise to parthenocarpic seeds its propagation is straight forward because all of the seeds give rise to daughter plants that are identical to the mother. Hundreds of thousands of 'Goldsturm' are raised every year from seed. Other perennial species, such as *R. missouriensi,* and *R. subtomentosa* are also good garden plants and exhibit good disease and pest resistance. *Rudbeckia fulgida* var. fulgida, although considered by some to be a worthy plant is not as versatile as 'Goldsturm'. Few of the natural forms of *R. missouriensis* or *R. subtomentosa* lend themselves to the massive production capabilities of 'Goldsturm'. *Rudbeckia missouriensis* is desirable however its stature is tall and rangy where as 'Goldsturm' and its even shorter cousin, *R. fulgida* var. *sullivantii* 'Pot of Gold' are much more compact and more applicable to usual garden conditions. All rudbeckia have interesting flowers with 'Goldsturm', 'Pot of Gold', and *R. missouriensis* being some of the best for the perennial species. Breeding efforts to combine the positive flowering characteristics of any of the more wild species with that of 'Goldsturm' should give rise to some superior plants both in terms of disease resistance and to enhanced flowering capability.

6.2.1.1 The Male Parent

'Goldsturm' is self-sterile and under normal circumstances in cultivation the flowers rarely if ever produce pollen. Flowering in 'Goldsturm' occurs in two cycles. The initial one produces large showy flowers and the cycle will last for several weeks from the onset. None of these flowers will produce pollen. Being self sterile it seems that theoretically there might not be a need for pollen as it serves no purpose. However, in the second cycle of flowering much smaller flowers are produced and they originate from small spurs emanating from the main stem that produced the original terminal flowering flush (Barnes, 2007). If the whole stem is removed and the terminal flower is discarded, what remains is the much smaller and less showy round two flowers. If the stem is then wounded much as though it were being prepared as a cutting and then cut under water to eliminate air bubbles in the vascular tissue similar to that of a cut flower, the second flush of flowers can be made to produce pollen. Once the spur flowers are prepared accordingly they can be inserted in a vessel or vase filled with a 50 per cent solution of water and either ginger ale or lemon-lime soda. After about 3 to 5 days, approximately 50 per cent of the spur flowers will show the emergence of stamens and copious amounts of pollen. Caution should be exercised to be on the lookout for an unknown species of inchworm that devours the newly formed stamens. If a good pollen set is at hand the stage is ready for moving this pollen to its new home.

6.2.1.2 The Female Parent in Waiting

Rudbeckia missouriensis flowers almost at the same time as 'Goldsturm' albeit about 5 to 10 days after the initial flush of 'Goldsturm'. This is actually good, because by the time the flowers of *R. missouriensis* are receptive, the spur flowers of 'Goldsturm' if forced are ready. The flowers of *R. missouriensis* are quite showy and attract a good many pollinating insects such as bees, flies, and butterflies. If, however, the ray flowers are removed, leaving nothing but the central cluster of fertile female flowers, the visitation of pollinating insects is greatly reduced. During the course of development the fertile female flowers will swell and change color from a pink lavender to a much darker red giving rise to a red-brown. At this point pollination can be achieved by directly applying the pollen from the 'Goldsturm' spur flowers to the female portions of the *R. missouriensis*. *Rudbeckia missouriensis* is a diploid and should yield triploid seedlings if fertilization with 'Goldsturm' pollen is effective. The appearance of the flowers that have been pollinated will change again to a much darker color and after a few days, the flowers of *R. missouriensis* will produce their own pollen. But because of the absence of the rays, interloper insects fail to visit the flowers. While not full proof, it is thought that self pollination is limited by this move although this does not preclude the formation of parthenogenic seed (Barnes, 2007).

6.2.1.3 Gestation and New Forms

Once the fertilized flowers are starting to undergo senescence by turning brown throughout it is prudent to bag them so that any loose seed will not fall out by the actions of wind or other physical movement. The dried seed heads can be cut and stored for several months and allowed to dry out completely. At this stage the resultant seed can be easily removed, subjected to 60 to 90 days cold-moist stratification and then after that period, sown and placed in a warm environment to germinate. Palmer *et al.* (2007) reports though their breeding produced only one hybrid seedling from a multitude of interspecific crosses, although many apomictic seedlings did develop. Apomictic seed commonly develops as a result of pollination even though the pollination is ineffectual. Obviously, the pairing of rudbeckia for hybridization requires more understanding.

6.2.2 Cross-Compatibility

Rudbeckia is valuable nursery crop that offers broad adaptability and exceptional ornamental merit. However, there is little information on interspecific and interploid crossability and ploidy levels of specific cultivars. Reciprocal interspecific crosses were performed by Palmer *et al.* (2009) between *R. hirta* cultivars and *R. fulgida* Ait., *R. missouriensis* Engelm. ex C.L. Boynton and Beadle, and *R. subtomentosa* Pursh. as well as reciprocal interploid crosses among four *R. hirta* cultivars. A combination of relative DNA content analysis and chromosome counts was used to test for hybridity and to determine ploidy levels for selected species, cultivars, and interploid *R. hirta* F_1 hybrids. Of the specific clones tested, *R. subtomentosa* and *R. missouriensis* were diploid, *R. fuligida* varieties were tetraploid, and *R. hirta* include both diploid and tetraploid cultivars. Mean 1Cx DNA content varied over 32.0 per cent among species. The interploid *R. hirta* crosses produced triploids as well as pentaploids and hexaploids. Seedlings from open-pollinated triploid R. hirta appeared, based on diverse phenotypes and DNA contents, to be aneuploids resulting from sexual fertilization, not apomixis. Of the 844 seedlings from interspecific F_1 crosses, only one individual, *R. subtomentosa* × *R. hirta* had a DNA content intermediate between its parents and was confirmed as the only interspecific hybrid. Overall, interspecific crossability among these species was found to be extremely low with production of only one successful hybrid: *R. subtomentosa* (2C =21.9 rg) × R. hirta 'Toto Gold' (2C = 7.2 rg), which yielded a seedling with an intermediate 2C DNA content of 14.7 rg supporting hybridity. Cytology also documented two sets of 19 chromosomes of disparate size, consistent with both parents. McCrea (1981) also found poor crossability among species in this subgenus and reported that among interspecific crosses attempted with six species, the only successful hybrids were between *R. missouriensis* and *R. fulgida*. Although most taxa had low self-fertility, seedlings (with genomic sizes similar to their maternal parent) resulted after interspecific crosspollination, indicating that pseudogamy is one reproductive pathway in *Rudbeckia* species. Breeding and crossability components determined a high level of self-incompatibility among these taxa and found that apomixis (pseudogamy) appears to be more prevalent after interspecific pollination in some taxa. Interploid crosses among R. hirta cultivars resulted in 98 viable seedlings, 53 of which were triploids. *Rudbeckia hirta* 'Cherokee Sunset' (4x) × *R. hirta* 'Toto Rustic' (2x) produced two triploid, 39 tetraploid (most likely from pseudogamy), two pentaploid, and one hexaploid offspring. *Rudbeckia hirta* 'Toto Rustic' (2x) · R. hirta 'Cherokee Sunset' (4x) produced 18 triploids and three tetraploids. The larger than expected ploidy in some individuals may be the result of unreduced gametes in one or both parents *Rudbeckia hirta* 'Prairie Sun' (4x) × *R. hirta* 'Goldilocks' (2x) resulted in 17 triploids, and the reciprocal cross generated 16 triploids whereas, interploid cross progeny exhibited intermediate traits (Barnes, 2007).

6.2.3 Mutation Breeding

Induced mutations are commonly used in plant breeding to create genetic variability. Mutation treatment of vegetative tissue such as buds or cuttings often results in chimeras as a result of the complexity and number of initial/stem cells in the apical meristem and the potential for each cell to be affected differently. *In vitro* treatment of organogenic callus can result in fewer chimeric mutations because plants are being regenerated from one or a few individual cells. Gamma irradiation of callus tissue has been used to induce mutations in several crop species. To date, there have been few reports of the use of gamma irradiation to induce mutations in embryogenic callus of *Rudbeckia* spp.

Rudbeckia subtomentosa 'Henry Eilers' is an adaptable and popular garden plant; however, reduction in height and increased novelty in flower color would be desirable. The effect of gamma radiation dose on *in vitro* survival and development, and *ex vitro* fertility, phenology, and morphology of tetraploid *Rudbeckia subtomentosa* 'Henry Eilers' was investigated (Oates *et al.*, 2013). *In vitro* embryogenic callus was treated with gamma radiation at different doses 0, 5, 10, 20, or 40 Gy. Treating callus with low levels of gamma radiation (5–10 Gy) resulted in relatively high *in vitro* and field survival while inducing a range of other mutations that could be selected. Several accessions with a reduced height and a desirable flower form were selected from the 5- and 10-Gy treatments for further assessment of commercial traits. Higher doses resulted in increased and undesirable irregularity of the ray floret morphology. Development of larger populations of plants derived from 5- to 10-Gy treatments would be a promising approach to recover combinations of desirable mutations.

7.0 Growth and Flowering

7.1.0 Day Length

Rudbeckia hirta is considered an obligate long-day plant (LD) (Garner and Allard, 1925). Beckwith (1991) observed that a Texas ecotype had a weak facultative LD flowering response, while a northern ecotype was an obligate LD plant. Celik (1996) also noted a difference in flowering among *Rudbeckia hirta* ecotypes. In a study conducted in north Florida (Norcini *et al.*, 1998), a northern Florida ecotype of *Rudbeckia hirta* flowered until the end of the study (mid-October) whereas a Texas selection senesced and/or succumbed to disease by mid-August. There were no differences in the level of insect damage (which was minimal) and neither ecotype exhibited any signs of drought stress. A naturally-occurring selection from central Florida also outperformed two *Rudbeckia hirta* cultivars ('Sonora' and 'Indian Summer') in a formal bedding plant evaluation conducted in central Florida (Howe, 1998).

Plants grown in a 20°C greenhouse under long days generally flower in 12 weeks while those grown under short days do not flower. Rudbeckia 'Indian Summer' has a lengthy juvenile period (about 4 weeks) before plants can be fully induced to flower in the plug tray. It is found that 'Indian Summer' seedlings with two leaves had to first be grown under long days for at least 25 to 30 days for complete flowering to occur under naturally short day lengths. Increasing the number of long days plants were exposed to also increased flower number. It is also noted that inducing Rudbeckia plants under long days and transferring them to short days resulted in extremely compact plants.

7.2.0 Plant Growth Regulators

Gibberellins applied under short day (SD) simultaneously stimulate cell proliferation in the subapical tissues and the central zone of the apical meristem (Bernier and Kinet, 1986). In *Rudbeckia,* stem elongation and flowering are not inextricably linked (Mumeek, 1940). Therefore, enhanced flowering in the terminal inflorescence is likely an early promotion of floral initiation. Yet, as is often observed following exogenously applied growth regulators, the response was temporary, and GA4+7 did not enhance axillary inflorescence initiation in *R. hirta.* In this case, apical floral initiation alone, and not subsequent development, was affected by GA_{4+7}; nor were axillary meristems influenced by the single GA_{4+7}, treatment at the start of Long days (LD). Even under SD with GA applied daily for 5 days, Kochankov *et al.* (1987) observed only apical floral initiation. These results indicate that BA did not affect flowering in *R. hirta* when applied at the onset of LD. In *Rudbeckia,* GA_{4+7} applied at the onset of LD may enhance the effects of LD in the apical meristem. When applied during floral initiation of the apical meristem, GA_{4+7}, did not enhance floral initiation in the axillary meristems. The net result was no increase in either inflorescence size or count.

8.0 Propagation

Rudbeckia can be propagated by seed, but the best way to propagate them is by division.

8.1.0 Seed Propagation

If propagating from seed, sow seeds in early to mid-fall, or early to mid-spring. Seeds sown in the spring will need to be stratified (subjected to a cold treatment to mimic winter conditions) for 3 months at 4.5°C. Cover the seeds lightly with ¼ to ½ inch of soil and keep moist. Germination should take about 2 weeks. Seeds can be sown indoors up to 2 months before frost; however, it may take 2 years or more to develop a sizeable plant. However, in tropical and subtropical region, seeds are sown in August-September for flowering in winter months and thus protected from moisture stress condition.

8.2.0 Vegetative Propagation

Rudbeckia can be divided every 3–4 years in early spring or fall to prevent overcrowding. *Rudbeckia* can be divided in early spring or in the fall, usually every 4-5 years. It is best to divide when the plants have become too large or exhibit diminished or small flowers. Dig them up, divide using a spade or garden fork and replant in newly spaded soil, improved with compost or other organic matter. Black-eyed Susans (and coneflowers) are gaining popularity outside of their natural habitat in moist, sun-soaked, well-drained meadows.

8.3.0 Micropropagation

Tissue culture is an efficient tool for mass propagation of Rudbeckia species. Protoplasts isolated from leaves, cotyledons, ray florets, and callus culture of seedlings of *R. hirta, R. laciniata* L., and *R. purpurea* L. regenerated shoot buds on regeneration medium (Al-Atabee and Power, 1987). Somatic hybrids of *R. hirta*

Propagation and Growing of Rudbeckia

Seeds

Seedlings in Plug Tray

Planting of Seedling

Planting in Pot

Division of Clump

Separation

Replanting in Pots

Micropropagation

Open Field Cultivation

In Greenhouse

Mass Production in Pot

Source: www.thompson-morgan.com, www.marksvegplot.blogspot.com, www.rhs.org.uk, www.brooksidenursery.co.uk, www.svenmagnussen.com, www.plantcelltechnology.com, www.sargentsgardens.com, www.pinterest.it

'Marmalade' and *R. laciniata* 'Irish Eyes' were regenerated via rhizogenesis on Murashige and Skoog (MS) medium (Murashige and Skoog, 1962) supplemented with 0.1 mg/l 3-indoleacetic acid (IAA), and 0.4 mg/l 6-furfurylaminopurine (kinetin, KIN) (Al-Atabee *et al.*, 1990).

Cotyledon sections of *R. hirta* were initiated to induce callus on Schenk and Hildebrandt basal medium (Schenk and Hildebrandt, 1972) supplemented with 5 mg/l 2.4-dichlorophenoxyacetic acid (2,4-D), and 5 mg/l KIN (Luczkiewicz and Cisowski, 2001). Daimon and Mii (1995) reported that adventitious shoots were induced on the surface of seedling-derived hairy roots of *R. hirta* after 30-50 days of culture on half strength MS agar medium supplemented with 0.5 or 1.0 or mg/l 6-benzylaminopurine (BA) and 0.1 mg/l α-naphthaleneacetic acid (NAA). Plants regenerated from hairy roots had morphological alterations such as wrinkled leaves, small size in flowers, and abundant lateral branching of roots. Szarvas *et al.* (2006) developed a micropropagation protocol using seedlings' meristems of *R.hirta*. Shoot-buds developed from seedling axillary meristem on half-strength MS medium containing 10mg/l KIN, or 2 mg/l KIN plus 0.1 mg/l N6-(2-isopentenyl)-adenine (2iP) and elongated roots formed on MS medium with 2 mg/l IAA.

9.0 Soil and Climate

Rudbeckia needs a fertile soil enriched with lots of organic matter which holds plenty of moisture in spring and summer and doesn't dry out or become waterlogged. They mostly tolerate well-drained soils under full sun. Provide excellent drainage for best growth. Seeds can be sown in loosely covered soil. It's best if soil is fertile (not poor) though they can tolerate tough conditions. Black-eyed Susans generally grow between 1 and 3 feet tall (though they can grow taller) and can spread between 12 to 18 inches, so plant seeds closer to prevent lots of spreading or plant further apart to make a nice border. High salt tolerance makes Rudbeckia well-suited for planting them close to the beach.

Both Black-eyed Susan (*Rudbeckia hirta)* and Orange Coneflower (*Rudbeckia fulgida)* are some of the more recognizable and eye-catching flowers which stand erect from 2 to 3 to 4-feet-high and usually remain in bloom from mid-summer right to early autumn. These two, along with about two-dozen other different species of *Rudbeckia* are all wild flowers of North American native. Annuals required a temperature for Rudbeckia to grow perfectly in positions in either full sun or light shade.

10.0 Cultivation

10.1.0 Planting Time

When the soil temperature has reached 21.1°C it is for best for Rudbeckia seed for germination. In many parts of North America, the planting period is March to May to obtain flower during June to September. Germination takes 7 to 30 days. These hearty flowers really enjoy the Sun. They prefer full sun, though they may also grow in partial sun.

10.2.0 Manuring and Fertilization

Black-eyed Susans are to be planted in full sun in spring or early fall. Plants are to be watered thoroughly at the time of planting and as needed throughout the season. Feed plants once at the beginning of the growing season with Miracle-Gro® Shake 'n Feed® Rose and Bloom Plant Food. Deadhead to keep plants tidy and encourage more blooms. Rudbeckia does not need liquid feeding or even granular fertilizer applications. For perennial rudbeckia it is recommended to plant them in spring in a blend of soil mix and garden soil with a distance of about 16 inches (40cm) between plants. For cut rudbeckia stems, plants needed with 100 mg/l N from 20N–4.4P–16.6K fertilizer (Peter's Professional 20–10–20) (Dole *et al.*, 2009)

10.3.0 Pruning

On Rudbeckia that grow a single flower on each stem, cut the stem back to the base of the plant. For Rudbeckias with multiple flowers on a stem, just snip off the spent blooms. In autumn, cut Black Eyed Susan back to about 4" tall (10 cm) or, to grow a few more Black Eyed Susan plants, let the last blooms go to seed for the birds. The seed heads can also be cut and dried to propagate new plants. To boost flower-bearing, remove wilted flowers regularly (deadheading). Deadheading Black Eyed Susan flowers is not necessary but can prolong the blooming period and prevent the plants from seeding all over the landscape.

10.4.0 Watering

Rudbeckia has huge needs in water, especially in summer. Plants are to be watered whenever necessary to keep the soil or compost moist during spring, summer and early autumn, as this will prolong flowering. Proper mulch will help avoid a good deal of watering chores since it helps retain moisture in the ground. It is better to water the soil or compost, rather than over the foliage.

11.0 Diseases and Pests

11.1.0 Diseases

Fungal Diseases

11.1.1 Downy Mildew (*Plasmopora* sp.)

This pathogen typically causes leaf spots with downy white or gray patches under the leaves. The downy growth results from the production of spores called sporangia which are wind-dispersed between plants. Disease is usually favored by cool wet weather. Control may include cultural means of reducing humidity and leaf wetness. Control may also be achieved with the use of fungicides applied as soon as symptoms are visible. Among the compounds registered for use is mancozeb at recommended dose level.

11.1.2 Leaf Spots (*Alternaria, Cercospora, Colletotrichum, Phyllosticta* or *Septoria* spp.)

Leaf spots are very common, typically sharply delimited necrotic areas on plant leaves caused by a wide variety of pathogenic species. Leaf spots usually are favored by wet conditions and may become important if a large number of lesions are present or if they start to coalesce. Under those conditions, control may also be achieved with the use of fungicides applied as soon as symptoms are visible. Among the compounds recommended for use are thiophanate-methyl and sulfur.

11.1.3 Powdery Mildew (*Erysiphe* sp.)

These fungi are obligate plant parasites which grow vegetatively on the plant leaf surface, sending haustoria, structures which absorb food from the host, into epidermal cells. The white mildew seen on the leaf is a combination of vegetative mycelium and spores borne in chains on upright conidiophores. Wind-dispersed mildew spores can germinate without free water under high humidity conditions, and disease is often severe when conditions are humid but dry. Small black over-wintering structures called perithecia are often found in powdery mildew affected areas. Control may also be achieved with the use of fungicides applied as soon as symptoms are visible. Among the compounds registered are potassium bicarbonate, ultra fine oil, sulfur, triadimefon, or thiophanate-methyl fungicides.

11.1.4 Rust (*Puccinia, Uromyces,* or *Aecidium* spp.)

The term rust refers to both the disease and pathogen causing the disease. Rusts are specialized obligate parasites which can cause disease on one (monoecious) or two (heteroecious) host species. Symptoms of rust infection include rust-colored spores or gelatinous horns in powdery pustules on leaves or stems. Surrounding tissue is discolored and yellowed, and plants are often stunted. Control of heteroecious rusts may be aided by removal of the alternate host, but for most perennials, control may also be achieved with the use of fungicides applied as soon as symptoms are visible. Among the compounds registered for use are sulfur and mancozeb.

11.1.5 Stem Rot (*Sclerotium rolfsii,* or *Sclerotinia sclerotiorum*)

Symptoms include yellowing of lower leaves followed by wilting and death of the rest of the plant. A white cottony mass of mycelium growing around the crown or on the soil near the crown distinguishes this crown rot from others. In this fungus web may be found the whitish to cream color or buff or reddish-brown seed-like sclerotia the size of a pinhead. Control may include removing and destroying all infected plant parts and removal of top soil around the plant. New soil may replace sclerotia-filled soil.

11.1.6 Wilt (*Verticillium dahliae*)

This pathogen infects the vascular or water-conducting tissues of plants, and cause wilt symptoms by impairing water flow. As a result, symptomatic plants may flag or wilt on one side of the plant, leaves may be twisted and yellow on one side, turn brown and hang down prior to drying up. Overall, the plant exhibits drought symptoms despite adequate soil moisture. If the stem is cut open near the base, the vascular tissues are typically brown or discolored. Wilt control involves removal of infected plants and associated roots and soil. The pathogens survive in soil and plant debris for long times, so disposal of plants and soil without spread is important.

Diseases and Pests of Rudbeckia

Downy Mildew **Powdery Mildew** **Leaf Spot**

Rust **Fourlined Plant Bug** **Fuller Rose Beetle**

Stalk Borer **Asiatic Garden Beetle** **Aster Yellow**

Aphids **Gall** **Flies**

***Source*: www.dengarden.com, www.houzz.com, www.bugguide.net, www.bugguide.net, www.my.chicagobotanic.org**

Viral Diseases

11.1.7 Aster Yellows

The pathogen is a prokaryotic organism without cell walls. It infects the phloem of susceptible plants and causes a general yellowing and dwarfing symptom. The phytoplasma is spread by a leafhopper vector. Infected plants should be removed and destroyed. Early season control of the leafhopper vector and removal of weed hosts may help prevent re-infection.

11.2.0 Pests

11.2.1 Asiatic Garden Beetle (*Maladera castanea*)

Adults are 3/8″ long, dull cinnamon brown, with a faint iridescent sheen. They are active night fliers, attracted to lights, and are injurious as adults when they feed on many kinds of plants. During the day, the beetles hide in the soil around the plants and are seldom seen unless one knows where to find them. If needed, azadirachtin or carbaryl, which are among the compounds registered for control of this pest can be applied to foliage when adults are present. Otherwise, treating with imidacloprid as a systemic may kill adults feeding on the foliage. Consult the label for dosage rates and safety precautions. Grubs of this and other beetle species can eat *Rudbeckia* roots as well as those of turf. Treating nearby lawns for grubs will reduce the population.

11.2.2 Flea Hopper (*Halticus bractatus*)

Both adults and nymphs of this plant bug damage plants by sucking plant sap. They overwinter as adults in protected areas. Females puncture leaves and stems with their mouthparts and then lays eggs in the damaged tissue. Nymphs are green but get darker as they mature. There are multiple generations each year. Foliar symptoms include bleached or dark spots. Some plants show red spotting. Leaves may be distorted along with flowers. If control is needed, malathion or methoxychlor, both of which are among the products registered for use against this pest, can be applied as foliar sprays. Imidacloprid can be used as a systemic treatment.

11.2.3 Fourlined Plant Bug (*Poecilocapsus lineatus*)

This bug lays eggs in the soft stems. They hatch about the middle of May, and the young bugs suck the sap from the tender leaves. They molt five times and when mature, about the middle of June, they have wings and are nearly 0.8 cm long. The insect body is yellow, marked lengthwise on the wings with four black stripes alternating with three green stripes. The injury to the leaves consists of sunken areas around the punctures. These areas later appear as circular transparent spots and finally as circular holes. This insect injures the new leaves of many different kinds of annual and perennial plants and shrubs. There is only one generation each year. Among the products registered for use against this pest are azadirachtin, ultrafine horticultural oil, insecticidal soap, or malathion applied as sprays, or imidacloprid applied as a systemic to be taken up by the roots.

11.2.4 Fuller Rose Beetle (*Asynonychus godmani*)

Many species of plants are sometimes injured by Fuller rose beetle, which feeds on the leaves at night and rests in the leaves or in some protected place during the day. This is a brown weevil marked with patches of gray scales on wing covers that are fused. Only females exist. They vary in length from 7-9 mm. The eggs are laid under the edges of bark near the ground or directly on the ground and can be found throughout the season. The white legless grubs live in the soil and chew upon feeder roots, as well as the larger roots of various plants. At maturity, they measure about 10-12 mm. This weevil overwinters as a pupa near the soil surface or as an adult in a protected spot. Adults do the most damage. Control treatments are rarely needed. When the infestation is light, handpicking the insects is a possibility.

11.2.5 Sawflies (*Macrophllya intermedia*)

The adult of this damaging caterpillar is wasp-like, but has a fat waist. The caterpillar is gray with dark stripes and is a nocturnal feeder doing most of its damage in May and June. Spraying young caterpillars with spinosad or malathion which are among the compounds registered for control of this pest will control them.

11.2.6 Stalk Borer (*Papaipema nebris*)

This borer infests an occasional stalk of many kinds of herbaceous plants. As a rule, its presence escapes notice until the plant begins to wilt. Then it is too late for the plant to recover. The larva tunnels up and down inside the stem and the top portion usually wilts and later dies. There is one generation each year. The moths emerge in September and October and lay eggs on the stalks of their food plants, in which stage the insect passes the winter. The eggs hatch in May or early June. The young larva begins to feed upon the leaves of the nearest food plant, and later tunnels in the stem. The mature larva is nearly 3.80 cm long, grayish brown with one white stripe on top and two white stripes on each side. On the front half of the body, the lateral stripes are interrupted, and the lower brown stripe extends forward onto the side of the head. Disposing of all the old stalks and destroying weeds at the edges of the garden helps reduce the population. If needed, methoxychlor dust, which is among the compounds registered for control of this pest applied in June should control this pest.

11.2.7 Two Spotted Spider Mite (*Tetranychus urticae*)

This pest infests the undersides of the leaves, which become light yellow in color, and the plants have a generally unhealthy appearance. Sometimes the mites form webs, which more or less enclose the upper as well the lower leaf surface. Among the compounds registered for control of this pest are insecticidal soap and ultrafine horticultural oil. Spraying with insecticidal soap will give sufficient control if applied at least twice at 7-10 day intervals. The predatory mite, *Neoseiulus fallacis* is most commonly found feeding where there are mite infestations. A single application of ultrafine horticultural oil (0.5-1 per cent dilution) can be effective if predatory mites are present. Special care should be taken with soap or oil to obtain thorough spray coverage, because they only work on contact. Additional materials appropriate for commercial growers include hexythiazox and abamectin (a restricted use product). Consult the label for dosage rates and safety precautions. Avoid applying carbaryl or pyrethroids, which tend to be much more toxic to the predators than to the pest spider mites.

11.2.8 White Fly

The greenhouse white fly, *Trialeurodes vaporariorum*, the sweetpotato white fly, *Bemisia tabaci* and silverleaf white fly, *Bemisia argentifolii* commonly infest plants grown under glass and are often carried into the field where they may persist. The life cycles of these species are similar. The tiny, white moth-like adult has a mealy appearance due to the small particles of wax that it secretes. It lays groups of eggs on the underside of leaves. The eggs hatch into small oval crawlers, which then settle down on the undersides of leaves and become scale-like. These insects spend about 4 days in an immobile pupal stage before becoming adults. About 5 weeks are required to complete the life cycle in the greenhouse. Yellow sticky traps are an effective way to monitor populations of whiteflies, and may even be attractive enough to reduce minor infestations. Biological controls can be effective against white flies, especially in a greenhouse environment. The predatory ladybeetle *Delphastus pusillus* specializes in whiteflies and feeds on all three whitefly species. The parasitoid, *Encarsia formosa,* can control the greenhouse whitefly in the greenhouse. Another parasitoid, *Eretmocerus californicus* attacks all three species and can assist in controlling minor infestations in the greenhouse. Insecticidal soap or ultra-fine horticultural oil, which is among the compounds registered for control of this pest is sprayed on the undersides of leaves, can be used against whiteflies in the greenhouse or the field. Azadirachtin (neem) or fenoxycarb directed to the undersides of the leaves, can also be used. Repeat applications of sprays will probably be needed because some stages in the life cycle may not be affected by insecticides. Chemical control using conventional insecticides is difficult because of widespread insecticide resistance. Season-long control can be achieved with imidacloprid applied as a systemic taken up by the roots.

12.0 Harvesting and Postharvest Management

Rhudbeckia stems are harvested according to criteria set for each species. Usually when the outer petals are fully opened, harvest and place into buckets of tap water (pH 6.3–7.1; EC 0.18–0.23 dS/m), sorted into similar groups based on stem length and diameter, flower bud number and flower size, as appropriate for the species, re-cut and subjected to appropriate pulsing or any other post harvest treatments. Stem length after recutting was 30 cm for most species (Dole *et al.*, 2009). However, consumer vase life is considered up to the stage before the stem collapses or the petals begin to turn brown.

Dole *et al.* (2009) determined the vase life of 'Indian Summer' rudbeckia (*Rudbeckia hybrida*) after being subjected to postharvest handling procedures. Cut rudbeckia flowers could be held in unamended tap or deionized (DI) water with no effect on vase life. Floral foam or 2 per cent or 4 per cent sucrose had no effect on the consumer vase life of Rudbeckia, longer vase life was observed with 10 per cent sucrose pulse than a water-only pulse but it could not tolerate a 20 per cent sucrose treatment for 24 hour. Rudbeckia could be cold stored (2°C to 90 per cent relative humidity in the dark) for 3 weeks.

13.0 Biotechnology

13.1.0 Molecular Markers

Rudbeckia fulgida var. *sullivantii* (Sullivant's Coneflower, Asteraceae) is a target species for conservation and restoration at the Midewin National Tallgrass Prairie in Will County, Illinois because it is classified as a sensitive species by the USDA Forest Service. This species also has an extensively used cultivar named 'Goldsturm.' To aid in the restoration efforts of this species, the genetic diversity and cross pollination potential of *R. fulgida* var. *sullivantii* populations in Illinois were investigated (Scott *et al.*, 2007). Random amplified polymorphic DNA results show high levels of genetic diversity in most of the populations surveyed. Little difference in genetic variation was observed between wild and cultivar populations. Cross pollination potential was confirmed as reciprocal crosses between wild and cultivar plants set seeds.

Few studies on other *Rudbeckia* species have been done to compare levels of genetic diversity with *R. fulgida* var. *sullivantii.* However, one study that used Restriction Fragment Length Polymorphisms (RFLPs) to look at isolated *Rudbeckia missouriensis* populations in Missouri glades, found very low levels (5 per cent) of overall diversity (King and Schaal, 1989). While higher diversity levels are expected when using RAPDs, the discrepancy between 5 per cent and an average of about 50 per cent in *R. fulgida* var. *sullivantii* is substantially different.

The oomycete *Plasmopara halstedii* emerged at the onset of the 21st century as a destructive new pathogen causing downy mildew disease of ornamental *Rudbeckia fulgida* in the United States. To determine whether Rudbeckia and sunflower downy mildew are caused by the same genotypes, population genetic and phylogenetic analyses were performed (Rivera *et al.,* 2016). A draft genome assembly of a *P. halstedii* isolate from sunflower was generated and used to design 15 polymorphic simple sequence repeat (SSR) markers. SSRs and two sequenced phylogenetic markers measured differentiation between 232 *P. halstedii* samples collected from 1883 to 2014. Samples clustered into two main groups, corresponding to host origin. Sunflower-derived samples separated into eight admixed subclusters, and Rudbeckia-derived samples further separated into three subclusters. Pre-epidemic Rudbeckia samples clustered separately from modern strains. Despite the observed genetic distinction based on host origin, *P. halstedii* from rudbeckia could infect sunflower, and exhibited the virulence phenotype of race 734. These data indicate that the newly emergent pathogen populations infecting commercial rudbeckia are a different species from sunflower-infecting strains, notwithstanding cross-infectivity, and genetically distinct from pre-epidemic populations infecting native rudbeckia hosts.

14.0 References

Al-Atabee, J.S. and Power, J.B. (1987) *Plant Cell Reports*, **6**(6): 414-416.

Al-Atabee, J.S., Mulligan, B.J. and Power, J.B. (1990) *Plant Cell Reports*, **8**(9): 517-520.

Barker, J. (2004) In: *The encyclopedia of North American wild flowers.* Parragon. ISBN-13 : 978-1405430357.

Barnes, H.W. (2007) *Combined Proceedings International Plant Propagators' Society,* **57**: 409-411.

Beckwith, D.D. (1991) *M.Sc. Thesis,* Dept. of Horticulture, Va. Polytech. Inst. and State Univ., Blacksburg.

Bernier, G., Bodson, M., Kinet, J.M., Jacqmard, A. and Havelange, A. (1974) In: *Plant growth substances,* Hirokawa, Tokyo, pp. 980.

Blakeslee, A.F. and Stephens, S.G. (1948) *Science,* **107**(2784): 458. PMID: 18938464.

Celik, M. (1996) *M.Sc. Thesis,* Dept. of Plant and Soil Sci., Miss. State Univ., Miss. State.

Dole, J. M., Viloria, Z., Fanelli, F. L. and Fonteno, W. (2009) *HortTechnology,* **19**(3) : 593-600.

Garner, W.W. and Allard, H.A. (1925) *J. Agric. Res.,* **31**: 555-566.

Harkess, R.L. and Lyons, R.E. (1994) *HortScience,* **29**: 134,227.

Howe, T. (1998) Gulf Coast Res. Educ. Ctr.-Bradenton Res. Rpt. BRA 1998-11. Univ. of Fla./Inst. of Food and Agr. Sci., Gulf Coast Res. Educ. Ctr., Bradenton, Fla.

King, L.M. and Schaal, B.A. (1989) *Evolution,* **43**: 1117-1119.

Kochankov, V.G. and Chailakhyan, M.K. (1985) In: *CRC handbook of flowering* (Ed. A.H. Halevy) vol. 5. CRC Press, Boca Raton, Fla, pp. 295–320.

Linnaeus, Carl von. (1753) *Species Plantarum,* **2**: 906-907.

Luczkiewicz, M. and Cisowski, W. (2001) *Plant cell, Tissue and Organ Culture* **65**(1): 57-68.

McCrea, K.D. (1981) *Ph.D. Thesis,* Purdue Univ., West Lafayette, IN.

Michael, B. R., Gedara, S.R., Mohamed Amer, M., Stevenson, L. and Ahmed, A. F. (2014) *Nat. Prod. Res.,* **28** (12): 909–913.

Murashige, T. and Skoog. F. (1962) *Physiol. Plant.,* **15**: 473–497.

Murneek, A.E. (1940) *Bot. Gaz.,* **102**: 269-279.

Norcini, J.G., Aldrich, J.H., Halsey, L.A. and Lilly, J.G. (1998) *Proc. Fla. State Hort. Soc.,* **111**: 4-9.

Oates, K.M., Touchell, D.H. and Ranney, T.G. (2013) *HortScience,* **48**(7): 831–834.

Palmer I.E., Ranney, T.G., Lynch, N.P., and Bir, R.E. (2009) *HortScience,* **44**(1): 44–48.

Palmer, I.E., Ranney, T.G., Lynch, N.P. and Bir, R.E. (2007a) *Proc. SNA Res. Conf.,* 52nd Ann. Rpt. Southern Nurseryman's Association, Athens, GA.

Palmer, I.E., Thomas G., Nathan, P.L. and Richard, E.B. (2007b) North Carolina State University, Dept. Of Horticultural Science, Mountain Horticultural Crops Research and Extension Center, 455 Research Drive Fletcher, NC 28732.

Raman, L. and Gurina, N. (2019) *Acta Scientific Med. Sci.,* **3** (10): 65-70.

Rivera, Y., Salgado-Salazar, C., Gulya, T. J. and Crouch, J.A. (2016) *Phytopathol.,* **106**: 752-761.

Schenk, R.U. and Hildebrandt, A.C. (1972) *Canadian J. Bot.,* **50**(1): 199-204.

Scott, L., Molano-Flores, B. and Koontz, J.A. (2007) *Transactions of the Illinois State Academy of Science,* **100**: 129-144.

Szarvas, P., Zsila-André, A., Fari, M.G. and Kovats, Z. (2006) *Acta Hortic.,* **725** : 527-533.

Szarvas, P., Zsila-André, A., Kováts, Z. and Fári, M.G. (2005) *International Journal of Horticultural Science,* **11**(2), 105–108. https: / /doi.org/10.31421/IJHS/11/2/588

Urbatsch, L.E., Baldwin, B.G. and Donoghue, M.J. (2000) *Syst. Bot.,* **25**: 539–565.

Calendula

Namita, Reeta Bhatia, Sapna Panwar and Bibin Poulose

1.0 Introduction

Calendula (*Calendula officinalis* Linn.), belongs to the family Asteraceae, is an herbaceous annual or short-lived perennial plant, widespread across temperate climates (Mohammad and Kashani, 2012). The common names for calendula are pot marigold, field marigold, gold blooms, garden marigold, poet's marigold, holligold, marigold, maravilla, marybud, Oculus Christi, *etc.* It is a very popular annual from North America to Asia. Since, 12th century, Egyptians, Greeks and Arabs have been cultivating calendula for medicinal purpose as well as for their ornamental use (Khalid and Teixeira da Silva, 2012). Calendula plants are tolerant to cold weather, have a long flowering period, and bloom from spring to winter. In ancient English culture, Calendula is often called as Mary's gold due to its association with 'Virgin Mary' and 'Queen Mary'. The genus name is derived from a Latin word '*calends*' which means the first day of every month, due to its long flowering period. It has also been called the 'herb of the sun,' due to blooming of flowers in the morning and closing in the evening. According to Greek mythology, it was named after four wood nymphs that fell in love with Apollo. As the flowers follow the sun, it has become an astrological sign of summer, Leo (Dinda and Craker, 1998) and also linked to treating the heart and conditions caused by heat.

2.0 Importance and Uses

2.1.0 Aesthetic Uses

Calendula officinalis L. is one of the widely grown annuals in India, China, Europe and the United States (Muley *et al.*, 2009). Due to its attractive flowers, calendula is widely used as cut flower, potted plant, in flower beds and border plant. It is ideal for planting in patio and mixed border. It has thick, fibrous roots and grows in thick patches, hence, it can be used as a cover crop or as a living mulch to protect the soil. In gardens, it can benefit soil and helps to repel pests, and aid healing (Clark *et al.*, 2010).

2.2.0 Therapeutic and other Uses

Besides ornamental use, it is also cultivated for medicinal/herbal or culinary purpose (Bcerentrup and Robbelen, 1987; Cromack and Smith, 1988; Mohammad and Kashani, 2012) and used in phytotherapy (Von and Die, 1992). It is used to treat fever, cancer and promote menstruation. In traditional system of medicine, decoctions, infusions, tinctures, balms and salves prepared from flowers were applied directly to skin to heal wounds and to soothe inflamed and damaged skin, psoriasis, leprosy, *etc.* (Krag, 1976; Horvath, 1992; Zahra *et al.*, 2000; Cordova *et al.*, 2002; Harrison, 2003). Calendula is used as an antipyretic and anti-inflammatory in Italian folk medicines. Calendula tea is used as eye wash, gargles or compresses to treat conjunctivitis, pharyngitis, aphthous stomatitis and gingovostomatitis, diaper rashes and other inflammatory conditions of the skin and mucus membranes (Mozherenkov and Shubina, 1976; Nedelka *et al.*, 1988; Fleming, 1998). It is used to treat haemorrhoids in India. In homeopathy, calendula cream alone or in combination with other remedies is used to treat abrasions and minor burns.

Uses of Calendula

In Pot | As Cut Flower | In Vase

In Bouquet | Floral Crown | Floral Wreath

In Hanging Basket | In Garden Bed | In Garden Border

Calendula Oil | Edible Flower Dishes | Herbal Ice Cubes

***Source*:** **www.chestnutherbs.com, www.shop.floretflowers.com, www.thompson-morgan.com, www.dtbrownseeds.co.uk, www.in.pinterest.com, www.jparkers.co.uk**

The flowers are used in ointments for treating wounds, ulcers, skin damage, herpes, frostbite, blood purification and scars whereas, leaves of calendula are used in infusion for treating varicose veins externally (Kirtikar and Basu, 1993; Khare, 2004). In Europe, leaves of calendula are used as a resolvent and diaphoretic. However, flowers are utilized as a stimulant, emmenagogue and antispasmodic (Kirtikar and Basu, 1993). Khare (2004) explained that flower decoction can be used as a potent drink for the treatment smallpox and measles, while fresh juice as a remedy for constipation, jaundice and suppression of menstrual flow. European folk medicine healers used calendula to produce sweat during fever and cure jaundice. Calendula preparations were also used to treat liver complaints, conjunctivitis, ulcers and wounds in USA during 19th century. According to Grieve (1931), calendula flowers are used in the form of fluid extract, tincture, infusions, cold infused oil and ointments for healing of skin inflammations, wounds, bruises, burns and cuts, *etc.*

Different essential/volatile oils, carotenoids, flavonoids, terpenoids, coumarins, quinines, sugars, oils, amino acids, lipids, hormones, tocopherols and other small compounds with varying health benefits are present in calendula (Muley *et al.*, 2009) flowers and leaves, hence it has wide application in folk medicine (Yoshikawa *et al.*, 2001). The detail of the volatile oil constituents are given the Table 1.

Table 1: Volatile Oil Constituents Extracted from Dried Flowers of *Calendula officinalis* Linn.*

Sl. No.	*Component*	*Whole Plant*	*Dried Flowers*	*Sl. No.*	*Component*	*Whole Plant*	*Dried Flowers*
1	Butanone	Traces	0.02	39	β-Cubebene	1.48	2.53
2	Pentan-2-one	Traces	0.04	40	γ-Cadinene	0.21	0.31
3	α-Pinene	3.46	0.20	41	γ-Patchoulene	0.07	-
4	β-Fenchene	8.52	0.65	42	α-Muurolene	2.87	1.54
5	Camphene	0.07	0.02	43	γ_2-Cadinene	-	0.50
6	β-Pinene	0.33	0.24	44	6-Cadinene	17.43	5.30
7	Sabinene	0.63	0.11	45	Cubebenene	0.34	0.17
8	$C_{10}H_{24}$	0.05	-	46	α-Cadinene	1.03	0.46
9	Δ-Carene	0.09	0.21	47	Calamenene	0.10	0.10
10	α-Phellandrene	0.16	0.02	48	$C_{13}H_{26}O$	0.74	0.28
11	Myrcene	0.54	0.17	49	Calacorene	0.15	-
12	α-Terpinene	0.20	0.12	50	Unknown	-	1.00
13	Limonene	0.48	1.17	51	Unknown	0.05	-
14	1,8-Cineole	0.11	-	52	Unknown	0.69	0.78
15	β-Phellandrene	0.09	0.25	53	Unknown	0.15	0.13
16	*cis*-β-Ocimene	Traces	-	54	$C_{20}H_{38}$	-	0.65
17	γ-Terpinene	0.65	0.37	55	Unknown	-	0.111
18	*trans*-β-Ocimene	0.26	0.40	56	Unknown	0.63	-
19	*p*-Cymene	0.20	0.59	57	Unknown	0.86	2.76
20	Terpinolene	0.11	0.09	58	Nerolidol	3.04	8.14

Sl. No.	Component	Whole Plant	Dried Flowers	Sl. No.	Component	Whole Plant	Dried Flowers
21	$C_{10}H_{16}O$	Traces	-	59	Cubenol	0.90	0.59
22	Menthone	Traces	0.05	60	Epicubenol	0.88	7.26
23	α-Copaene	0.12	-	61	Unknown	0.07	0.04
24	Unknown	0.13	0.12	62	Unknown	Traces	0.06
25	β-Ylangene	Traces	0.13	63	Bisabolol oxide	Traces	0.12
26	Camphor	Traces	-	64	$C_{21}H_{44}$	0.12	-
27	Unknown	0.43	0.32	65	*T*-Cadinol	5.22	3.21
28	α-Gurjunene	-	0.58	66	α-Muurlol	6.97	5.67
29	β-Cubebene	0.12	0.11	67	*T*-Muurlol	1.69	1.06
30	$C_{10}H_{18}O$	0.24	0.06	68	Eudesmol	0.30	7.83
31	$C_{10}H_{18}O$	0.34	0.15	69	α-Cadinol	24.76	20.09
32	Bornyl acetate	Traces	1.37	70	$C_{23}H_{48}$	0.18	0.14
33	Aromadendrene	0.28	2.25	71	$nC_{25}H_{52}$	0.63	0.10
34	*trans*-Caryophyllene	0.49	0.82	72	$nC_{27}H_{56}$	0.88	0.16
35	Terpinene-4-ol	0.65	0.04	73	γ-Amrophene	0.11	0.10
36	ε-Cadinene isomers	0.21	0.07	74	β-Farnesene	0.05	0.22
37	γ-Cadinene isomers	0.31	0.13	75	γ-Muurolene	1.42	0.45
38	α-Humulene	0.56	0.94	76	α-Patchoulene	1.35	2.93

* Khalid and Teixeira da Silva (2012).

In an ethno-botanical study performed at Usak University, tea made from dried calendula flowers, used for treatment of liver and gallbladder disease, stomach, gastritis and ulcer and cream made from flowers were used against psoriasis, eczema and fungal diseases such as dermatological problems (Deniz *et al.*, 2010). Plant floral extracts have an anti- HIV effect (Kalvatchev *et al.*, 1997). Plant extracts of *Calendula officinalis in-vitro* are anti-cancerous in different tumour cell lines derived from leukaemia, fibrosarcoma, melanomas, breast, cervix, uterus, pancreas and lung (Medina *et al.*, 2006). It is used as an internal medicine for gastritis, colitis and duodenal ulcers (Bone, 2003).

Calendula is used in cosmetics as both pre- and post-sun moisturising creams. Petals, stems and leaves have culinary applications. Petals can serve as a colouring agent in dessert and salads, and their flavour is mildly salty. Flowers are used to flavour soups and stews, golden yellow blossoms for adding colour to butter and cheese. The flower buds contain 0.009 per cent pyrethrin, so an extract is also effective as an insecticide. Petals can be used in the preparation of dye to colour clothes or hair. The flowers are used as an alternative to saffron and in various ointments to increase their appearance by adding a gold colour (Khalid and Teixeira da Silva, 2012).

3.0 Origin and History

Calendula has been a part of human civilization for perhaps thousands of years. The ancient Romans and other civilizations around the Mediterranean were well acquainted with this plant. It was a symbol of joy and happiness for Romans. Greeks and Romans used the golden calendula in many rituals and ceremonies, sometimes wearing crowns or garlands made from the flowers. In India, its flowers are sacred flowers and have been used to beautify the statues of Hindu idols since early times. Aztecs and Mayans also used the calendula flowers in their ancient ceremonies. In Mexico and Central America, flowers are still used on home altars on the Day of the Dead.

Calendula is believed to be originated in southern Europe, Egypt, parts of Asia and Mediterranean area, but is now widely spread worldwide. Since 12th century, calendula has been cultivated in European gardens. It is also grown in Mediterranean countries such as Balkan states, Eastern Europe, Northern America and Germany (Samsam, 2003). Romans has been cultivating calendula for its medicinal properties (Macht, 1955). It was used as an ornamental plant until its beneficial properties were recognized and afterwards started to be used as a medicinal plant. Its flowers have no calycles and have been introduced in some pharmacopoeia and are used to treat diseases of the stomach and intestine (Stary, 1991). It is bitter and has long been used as an effective pesticide against insects (Martin, 2005). On the other hand, Egyptians, Arabs, Indians, Europeans and Greeks have grown it for medicinal purposes (Zaman, 2003). Calendula is widely cultivated in Germany, Australia, Czech Republic, Austria, Switzerland, Hungary and recently in Egypt and Syria as a medicinal crop.

4.0 Taxonomy

Calendula is a taxonomically complex genus (Norlindh, 1977; Heyn and Joel, 1983), mainly because of the high levels of hybridization between different taxa (Lanza, 1919; Heyn and Joel, 1983), and because of their large morphological variability, leading to the appearance of a wide range of intermediate forms, recognized by various names under different taxonomic categories (Heyn and Joel, 1983). The genus *Calendula* L. is placed in the Class Calenduleae, which is one of the smallest tribes of the family Asteraceae (Bremer, 1994). Several authors have tried to study the taxonomy of the genus *Calendula* using different approaches (Norlindh 1946, 1977; Meusel and Ohle 1966; Heyn *et al.*, 1974; Ohle 1974; Heyn and Joel 1983). The works of Heyn *et al.* (1974) in annual taxa and Ohle (1974) in perennials are complete taxonomic reviews of the genus. Heyn *et al.* (1974) has recognised five annual species: *C. tripterocarpa* Rupr. (2n = 30), *C. arvensis* L. (2n = 44), *C. stellata* Cav. (2n = 14), *C. palaestina* Boiss. (2n = ±85) and *C. pachysperma* Zoh. (2n = ±85; also reported to have 88 by Pazy, 2000). *Calendula arvensis* is a widely distributed synanthropic species, while the remaining taxa are more geographically restricted. Basically, these taxa were categorized into three groups: two groups from the western Mediterranean region, *C. suffruticosa* and *C. incana*, (both with 2n = 32), and a third group consisting of four different species with the same chromosome number (2n = 18); *C. maroccana* (Ball.), *C. meuselii* Ohle, *C. eckerleinii* Ohle and *C. lanzae* Maire (the first three species are perennial and the last one is annual) (Ohle, 1975). The numbers of chromosomes have acted as the foundation for numerous theories on the evolution of the genus *Calendula* (Ehrendorfer, 1970) and have been considered to be a significant method in the delimitation of many organisms, which in several instances have not been distinguished so easily by means of morpho-anatomic studies.

The plants under genus *Calendula* are annual or herbaceous perennial in nature. The plant grows upto 20 to 50 cm height, while cut flower varieties can grow upto 120 cm height. This annual has simple and oblong leaves having 7-14 cm leaf length and 1-4 cm leaf width. Leaves are lance-shaped with narrow bases, narrowly parallel-sided or spoon-shaped, blunt, glandular-hairy to sparsely spidery, cottony hairy, typically with a smooth edge. Leaf arrangement is alternate. the edge of the leaf blade has teeth or lobes and sometimes no teeth. Flower heads are solitary 4-10 cm in diameter and have pale yellow, orange or brown colours. There are disc and ray florets in the inflorescences (Bailey and Bailey, 1976). Fruits (achenes) are incurved or sometimes flat of 2-2.5 cm size and narrowly beaked. Achenes are naviculate and muricate on dorsal surface. The achenes have no pappus and are heteromorphic (Nordenstam, 1994). There is no plume, or the plume is made up of scales, awns, a crown, or a rim.

5.0 Species and Cultivars

5.1.0 Species

Around 25 species are included in the genus *Calendula*, among them *Calendula officinalis, C. arvensis, C. maritima, C. tripterocarpa, etc.* are more common. The most widespread and cultivated plant is *C. officinalis* L. The other important species under genus *Calendula* include *C. denticulata* Schousb. ex Willd., *C. eckerleinii* Ohle, *C. incana* Willd., *C. incana* subsp. *algarbiensis* (Boiss.) Ohle, *C. incana* subsp. *maderensis* (DC.) Ohle (Madeiran marigold), *C. incana* subsp. *microphylla* (Lange) Ohle, *C. lanzae* Maire, *C. maroccana* (Ball) Ball, *C. maroccana* subsp. *maroccana*, *C. maroccana* subsp. *murbeckii* (Lanza) Ohle, *C. meuselii* Ohle, *C. palaestina* Boiss., *Calendula stellata* Cav., *C. suffruticosa* Vahl, *C. suffruticosa* subsp. *balansae* (Boiss. and Reut.) Ohle, *C. suffruticosa* subsp. *boissieri* Lanza, *C. suffruticosa* subsp. *fulgida* (Raf.) Guadagno, *C. suffruticosa* subsp. *lusitanica* (Boiss.) Ohle, *C. suffruticosa* subsp. *maritima* (Guss.) Meikle, *C. suffruticosa* subsp. *monardii* (Boiss. and Reut.) Ohle, *C. suffruticosa* subsp. *tomentosa* Murb *etc.*

Calendula arvensis

It is also known as marigold and is native to southern and central Europe. *C. arvensis* is a 10-50 cm-tall annual or biennial plant. The leaves are lance-shaped and hairy stem. The inflorescence is a single flower head up to four centimetres across with bright yellow to yellow-orange ray florets circling a core of yellow disc (Clavijo, 2005).

C. maritima

It is commonly known as the sea marigold or trailing calendula and is a very rare in the genus. Some scientists named it as *Calendula suffruticosa* subspecies *maritima*. This halophyte plant is limited in coastal ecosystems and is native to western Sicily. It is an essentially endangered species. This perennial herb reaches 20-40 cm

Species of Calendula

Calendula arvensis *Calendula maritima* *Calendula officinalis*

Calendula tripterocarpa *Calendula maroccana* *Calendula palaestina*

Calendula suffruticosa subsp. *lusitanica* *Calendula incana* subsp. *algarbiensis* *Calendula incana* subsp. *maderensis*

Source: www.commons.wikimedia.org, www.en.wikipedia.org, www.teline.fr, www.floressilvestresdelmediterraneo.blogspot.com, www.flickr.com, www.florealpes.com

Cultivars of Calendula

Greenheart Orange | Porcupine | Radio Extra Selected

Nana Citrus Cocktail | Pink Surprise | Indian Prince

Fiesta Gitana | Fruit Twist | Bronzed Beauty

Calypso Orange | Neon | Triangle Flashback

***Source*: gardenersworld.com and Calendula guide by Herb Society of America**

height. The stems can be quickly lignified and the leaves are covered with small sticky hairs. At first the young stems are upright, but later they hang and scatter on the soil. Unlike, *C. officinalis*, the leaves are fleshy with a heavy scent. The leaves, form differs from egg-shaped to linear according on their placement on stems. The main flowering period is May-June. Sea marigold is grown as an ornamental plant and used as ground cover as well as pot plant. Cultivars such as yellow flowering 'Skyfire' and 'Summer Lovers', 'Skyfire Yellow' are grown in parks, along street and highways, and as flower pots on patios and balconies.

C. *Officinalis* L.

It is among the most widely grown species of *Calendula*, commonly known as marigold or Scotch marigold or pot-marigold or ruddles. It is most probably native to southern Europe, but its long cultivation background renders its exact origin uncertain and likely to be garden origin.

5.2.0 Cultivars

Calendula cultivars are primarily ornamental selections with single or double flowers, which differ in colour from deep orange to pale yellow with a distinctively coloured disc. There are several varieties and cultivars in calendula and some of the popular ones include: 'Radio Extra' (Orange cactus like flowers), 'Pink Surprise' (Ruffled gold and yellow flowers with pink borders and dark apricot centres), 'Touch of Colour' (Flowers with a blend of red and orange petals), 'Neon' (Bold double-petalled blooms), 'Greenheart Orange' (flowers with orange petals surrounding lime green disc), 'Tangerine Cream' (Double-petalled flowers with vivid bi-coloured orange and cream flowers), 'Bronzed Beauty' (cream and peach colour flowers), 'Citrus Cocktail' (Orange and yellow flowers and best suitable for planting in containers), 'Sherbet Fizz' (Buff flowers with deep red petal ends), 'Dwarf Gem' (Double flowered, compact variety suitable for containers with orange, yellow and apricot colour), 'Variegata' (Variegated leaves with yellow colour), 'Party Fiesta Gitana' (Won the Garden Merit Award from the Royal Horticultural Society). Other cultivars are 'Jane Harmony' (bright and yellow), 'Alpha' (deep orange), 'Lemon' (pale yellow) 'Orange Prince' (orange), 'Sun Glow' (bright yellow), 'Indian Prince' (dark orange), 'Pink Surprise' (double with inner florets darker than outer florets), 'Chrysantha' (yellow, double), 'Green-heart gold' (double, bright yellow) and 'Apricot Pygmy' (double, light peach), 'Nova', 'Orange Beauty' (Orange) 'Orange Porcupine', 'Yellow Gem'.

6.0 Genetics and Breeding

6.1.0 Genetics

In order to circumcise and assume their relationships between the taxa, large number of cytological characters, such as the number, morphology and meiotic behaviours of chromosomes as well as their nuclear DNA have been used especially for the resolution of the taxonomy of complex groups and are essential for the systemic community (Kron *et al.*, 2007). *Calendula* species show a large variation in chromosome numbers. The basic chromosome numbers are 7, 8, 9, 11, and 15

(Rice *et al.*, 2015). The basic chromosome number of this genus identified so far are n=7 (*C. stellata*), n=8 (most perennial taxa of northern Africa and south-western Europe), n=9 (some perennial taxa of North Africa), n=11 (*C. arvensis*) and n=15 (*C. tripterocarpa*) (Darlington and Wylie 1955; Norlindh 1977, Heyn and Joel 1983; Nordenstam 1994).

Table 2: Ploidy Level and Genome Size in *Calendula* spp. from Iberian Peninsula (including two *taxa* from Morocco) Taxon*

Sl.No.	Species	Habit	Ploidy Level	Genome Size [(1Cx) M bp]
1.	*C. arvensis*	Annual	$2n = 4x = 44$	1323
2.	*C. incana* ssp. *incana*	Perennial	$2n = 4x = 32$	792
3.	*C. incana* ssp. *algarbiensis* var. *algarbiensis*	Perennial	$2n = 4x = 32$	746
4.	*C. incana* ssp. *algarbiensis* var. *cinerea*	Perennial	$2n = 4x = 32$	778
5.	*C. incana* ssp. *maderensis*	Perennial	$2n = 4x = 32$	785
6.	*C. incana* ssp. *microphylla*	Perennial	$2n = 4x = 32$	817
7.	*C. maroccana* ssp. *maroccana*	Perennial	$2n = 2x = 18$	856
8.	*C. Officinalis* L.	Annual/Perennial	$2n = 4x = 32$	726
9.	*C. suffruticosa* ssp. *carbonellii*	Perennial	$2n = 4x = 32$	785
10.	*C. suffruticosa* ssp. *greuteri*	Annual/Perennial	$2n = 4x = 32$	836
11.	*C. suffruticosa* ssp. *lusitanica*	Perennial	$2n = 4x = 32$	804
12.	*C. tripterocarpa*	Annual	$2n = 2x = 30$	1696
13.	*C. stellate*	Annual	$2n = 2x = 14$	1002

* Nora *et al.* (2013).

The evolution of *Calendula* species with varied basic chromosome numbers is owing to hybridization and back-crossing in the genus. *C. persica* with a higher chromosome number has smaller flowers than the cultivated tetraploid species. In calendula, tetraploids are ideal for cultivation as a garden plant. *Calendula officinalis* is available in two colours: orange and lemon yellow; where the hue in the former is attributed to plastids including carotene and xanthin, whereas in the latter, xanthin only. *Calendula tripterocarpa* is believed to have originated through hybridization between *C. stellata* and: (1) *C. maroccana*, *C. stellata* hybrid, or (2) *C. maroccana* after the loss of two chromosomes.

The proposed scheme of speciation in Calendula based on chromosome numbers (entire numbers) as adapted from Heyn and Joel (1983) is shown in Figure 1.

6.2.0 Breeding

The major focus in calendula breeding is to develop genotypes of specific decorative value, to obtain large number of flowers, to ensure the development of large quantities of seeds, high oil yield (Diaconu, 1992), to improve the quality of plant-derived medicinal products (Zitterl-Eglseer *et al.*, 2001) and to obtain cultivars

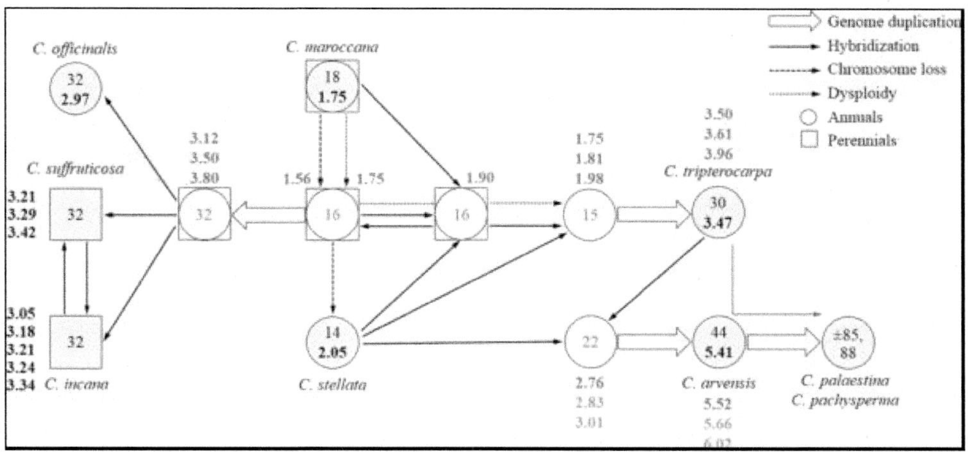

Figure 1: Proposed Scheme of Speciation in Calendula Based in Chromosome Numbers (Entire numbers) Adapted from Heyn and Joel (1983): Grey values represent hypothetical genome sizes of ancestral species, while black values represent genome size estimations of current species (Source: Nora *et al.*, 2013).

that are tolerant to major insects and pests (Baciu *et al.*, 2010). Massoud *et al.* (2020) evaluated 12 genotypes of *Calendula officinalis* for flowering, vegetative and chemical characteristics in order to study their genetic parameters and reported highest heritability values for chlorophyll A, chlorophyll B and carotenoids. Further, they reported maximum genetic advance which was confined to plant height, number of inflorescences and number of branches, as well as chlorophyll A, chlorophyll B, and carotenoids. As a consequence of the additive gene action, data are strong in genetic advance as a percentage of mean and heritability.

6.2.1 Hybridization

Hybridization has undoubtedly played a role in the radiation and diversification of flowering plants (Arnold 2006; Mallet 2007). Baciu *et al.* (2012) estimated the heterosis in the F_1 and F_2 calendula hybrids for their response to the *Aphis fabae* attack. A Negative heterosis was measured as the mean Attack Degree (AD) per cent per hybrid combination which is needed for the phenotype. The idea that partial dominance induces heterosis has been clarified by the assumption that the inbreds of these cultivars are fixed for recessive or partly recessive deleterious alleles. The maximum desirable negative value of heterosis was the result of self-pollination of the susceptible cultivar 'Rech.f.'. Baciu *et al.* (2012) evaluated the length, width and thickness of the seed of the F_3 hybrids of *C. arvensis*. F_3 Hybrid *C. arvensis* '121GE' × *C. arvensis* '121GE' had the maximum seed length, whereas the hybrid family 'Prycosnovjenie' × 'Bon-Bon Orange' had high seed width and seed thickness values. Zito (2018) reported that *Calendula maritima* and *C. fulgida* emitted varying absolute quantities of fragrance while the hybrid produced moderate amounts of fragrance. Parental and hybrid taxa scents were both regulated by monoterpenes, with some compounds in common among the taxa. According to Plume *et al.* (2013) rare, endangered Sicilian endemic, *Calendula maritima*, may be hybridized with other congeners in the wild. Hybridization may be placing *C. maritima* at risk of extinction via. introgression. Anthropogenic habitat disturbance and expansion of *C. suffruticosa* ssp. *fulgida* into remaining populations could pose a serious threat to its successful conservation.

6.2.2 Mutation Breeding

In recent times, mutagenesis is used in association with transgenic approach and recombinant DNA technology to obtain new plant cultivars with useful economic characteristics. In particular, chemical mutagenesis is an essential and simple way to obtain useful base material for plant breeding. The chemical mutagen (*e.g.*, azide, diethyl sulphate (DES), dimethyl sulphate (DMS), ethyl methane sulphonate (EMS) and N-nitroso compounds) generates a higher rate of non-lethal point DNA mutations and induces novel genomic variability in various crops (Rapoport, 1993). In certain instances, mutagen-induced damage may be resolved at the early stages of ontogenesis of plant cells. Environmental factors during the growing period should also be considered in the determination of these mutagenic impacts.

DMS and DES mutagens induced variability in morphological traits of *C. officinalis* cultivar 'Zolotoe more' (Samatadze *et al.*, 2019). Since an optimum plant height is a favourable agronomic trait that leads to a higher yield but the mean plant height was found lower in the M_1 and M_2 plants of cultivar 'Zolotoe more' than the original cultivar. El-Nashar and Ammar (2016) treated calendula seeds with different concentrations of Sodium Azide (SA) and DES (0, 1000, 2000, 3000, 4000 and 5000 ppm) and observed that calendula plants appeared to bloom faster at low mutagen concentrations (1000 ppm), whereas higher concentrations delayed flowering considerably. Lower levels of mutagens may be suggested for improved floral and physiochemical efficiency. The largest overall soluble protein was detected at 1000 ppm SA and 2000 ppm DES and it was found that lower doses of SA and DES had beneficial benefits on the percentage of seed germination, plant height and leaf area.

6.2.3 Ploidy Breeding

Occurrence of polyploidy evidently played a significant role in the growth and speciation of plants. Polyploidization incidents also tend to be correlated with a rise in vigour and adaptation of freshly developed polyploids to novel circumstances. Polyploids may have larger and thicker leaves and stems, higher biomass, bigger inflorescences, larger pollen, better tolerance to environmental stress and higher concentration of secondary metabolites (Mori *et al.*, 2016; Li *et al.*, 2018; Luo *et al.*, 2018; Eng and Ho, 2019). Polyploidy has an influence on biosynthesis pathways and gene expression (Dhooghe *et al.*, 2010; Sattler *et al.*, 2016; Salma *et al.*, 2017; Zhang *et al.*, 2018). Polyploidy has led to an increase in secondary metabolites in medicinal plants mainly due to enhancement of biomass in chromosome doubled plants and the increase in number of gene copies related to the pathways of the secondary metabolites. Esmaeili *et al.* (2020) carried out chromosome doubling of three allotetraploid *Calendula officinalis* cultivars 'Nova,' 'WUR 1553-7' and 'Orange Beauty' using three antimitotic agents: colchicine, oryzaline and trifluralin at varying concentrations to find the optimal combination of the best agent and the best dosage to produce octoploids. A few octoploids were obtained in all three cultivars, they displayed thicker leaves and larger cells. Higher dosages of antimitotic agents had toxic effect on seed germination and survival of seedlings of calendula. They observed that application of colchicine at low concentrations (about 200–400 ppm) is the most efficient antimitotic agent for chromosome doubling. The number of octoploids obtained were low (maximum 2.2 per cent), however the plants obtained are valuable for further breeding as well as the study of the agricultural value and metabolite profiles among ploidy levels. Such insights can lead to development of better performing cultivars.

7.0 Growth and Flowering

7.1.0 Environmental Factors

7.1.1 Light

Calendula officinalis L. has been referred to as a day neutral plant (Van De Pol, 1971; Vince-Prue, 1975). Sawhney *et al.* (1981) reported calendula as qualitative long day plant with an exceptionally low photoperiodic requirement of 6 h 30 min in a 24 h cycle. However, inflorescence initiation period did not differ quantitatively with an increase in the length of inductive photoperiod from 6 h 30 min to 24 h. The increase in the length of photoperiod beyond 12 h promoted extension growth but did not affect leaf differentiation. The plant is insensitive to photoperiod till 4-leaf stage (Kamlesh, 1977). Post (1949) stated that calendula is a profitable summer, field grown crop, but should be grown under shade which will increase quality, stem length, and flower size. Armitage *et al.* (1985) found that 50 per cent shade did not delay flowering. Aliniaeifard (2018) reported that red light increases growth and induces early flowering in *Calendula officinalis*, but full spectrum of light was required to avoid deformation of leaves and interruption in electron transport under red light to prevent photosynthesis.

Chlorophyll and carotenoid content, chlorophyll fluorescence and electron transport levels were examined in pot marigolds grown under low and high light conditions (35 per cent and 100 per cent of maximum daylight) by Posada *et al.* (2015). Shaded plants reported to have a greater proportion of chlorophyll a to chlorophyll b and a lower ratio of carotenoids to chlorophyll and hence, it was observed that calendula plants are highly susceptible to low-light conditions.

7.1.2 Temperature

Calendula prefers cooler temperatures (< 30°F) and often stops blooming in extreme heat. In frost-free areas, calendula can be grown from February to May for winter field production and from June to September for summer production in the North is from (Post, 1949). Greenhouse production is mainly from October to May. Armitage *et al.* (1987) grew 10 cm dwarf potted cultivars at 21°C day and 16-17°C night temperature and found that 13°C nights could be used if 2 weeks were added to the time schedule of production. Calendula prefers cold temperatures, and its flowering and marketable quality is lost at high temperatures (Warner and Erwin, 2005). High germination percentage of calendula seeds can be achieved at temperatures of 16.5–16.8°C. As the temperature rose by more than 17°C, germination decreased. Heat shock temperatures (35–40°C) of less than 50 h duration reduced seed germination below 50 per cent. While, at 45°C, 100 per cent seed lethality was induced within 24 h of heat treatment (Eberle *et al.*, 2014).

7.1.3 Growth Regulators

Kumar *et al.* (2017) studied the impact of plant growth regulators on growth and flower yield of calendula (*Calendula officinalis*) and reported that GA_3 at 300 ppm was successful in improving the pre-and postharvest parameters *viz.*, flower diameter, Flower weight and flower yield/plant of calendula.

8.0 Propagation

8.1.0 Seed Propagation

Calendula is commonly propagated by using seeds. Seeds are sown in plugs or seed beds 4 to 6 weeks before planting. The average seed rate at sowing is within the range 7–12 kg/ha. Calendula seeds require light to germinate. Due to the small seed size, calendula seeds should be sown at a depth of 1-2 cm. The seedlings are transplanted into the field in rows spacing 70 cm apart with plants spacing 15 to 20 cm apart in rows.

8.2.0 Micropropagation

Cocu (2004) studied the *in vitro* regeneration response of different explants *viz.* hypocotyl, cotyledon and cotyledonary nodes of *Calendula officinalis* L. The explants were cultured on Murashige and Skoog (MS) media supplemented with various concentrations of thidiazuron (TDZ), kinetin (KIN), α-naphthaleneacetic acid (NAA) and indole-3-butyric acid (IBA) to induce adventitious shoot regeneration. The highest frequency of regeneration was achieved from hypocotyl and cotyledon explants cultured on MS media supplemented with 0.75 mg/dm^3 TDZ and 0.25 or 0.50 mg/dm^3 IBA. Regenerated shoots were successfully rooted on MS medium supplemented with 1 mg/dm^3 NAA.

9.0 Soil and Climate

9.1.0 Soil

Calendula may thrive in a wide variety of soil. It requires well-drained, fairly nutrient-rich soils with a pH of 6-7. Dense, wet soils can cause the roots to rot.

9.2.0 Climate

Cold temperate regions with hot summers are ideally adapted to calendula. When summers are too humid, yields are depressed. Calendula likes maximum light, and can accommodate a wide variety of soils if the drainage is strong. Calendula is less sensitive to frost.

10.0 Cultivation

10.1.0 Transplanting

The ground should be finely worked to provide a smooth surface for sowing.

Propagation and Growing of Calendula

Seeds

Sowing in Plug

Sowing in Soil

Seedling

Bed Preparation

Pot Planting

Open Field Cultivation

Polyhouse Cultivation

Mass Production in Pot

Harvesting and Postharvest Management

Harvested Flower

Drying

Tea Infusion

Dried Petals

Calendula Salve

Seed Harvesting

Source: www.commons.wikimedia.org, www.gapphotos.com, www.alamy.com, www.dreamstime.com, www.gardenersworld.com, www.hudsonvalleyseed.com, www.homesteadandchill.com

Calendula is a hardy annual. Planting is done in early spring or, where winters are mild, in the fall. The plants are grown in rows and in the main field at suitable season. In order to speed up and organise the development of crops, the field must be watered after planting. Transplant seedlings into the field in rows that are 70 cm apart, with plants that are 15 to 20 cm apart in rows.

10.2.0 Direct Seeding

For rows 70 cm apart with a final plant density of 5 to 7 per m^2, the generally prescribed seed rate is 2 to 3 kg/hectare. For dry flower production 24 kg/hectare is required under broadcast, but the decision to sow or broadcast depends on weed control, harvesting methods and seed costs. Broadcast sowing is only practicable if the usage of herbicide is expected. In the case of oil extraction, broadcast sowing is favoured. For oilseed production, a seeding rate of 24 kg/hectare is advised, resulting in a plant density of approximately $60/m^2$.

10.3.0 Nutrition

Higher dose of nitrogen lowers flower yields and increase vegetative growth. Nitrogen dose of 50 kg/hectare is appropriate for its growth. Phosphate (P_2O_5) and potassium (K_2O) can be used at rates of 100 kg/ha and 50-100 kg/ha respectively. Organic farmers can use compost or composted manure at a rate of 10 tonnes per acre whenever soil fertility is low. The micronutrient Boron, Molybdenum, Zinc, Manganese and Cobalt are known to promote the development of flowers and the carotenoid content. Maleic hydrazide (50 mg/l) causes branching and flowering. Armitage *et al.* (1987) used a constant 200 ppm N feed, but leached every fourth irrigation to reduce soluble salts. If irradiance is low, nitrate nitrogen is preferred and should be comprise atleast 50 per cent of the source. When flowering initiates, nutrition levels should be lowered to 100 ppm N. Reducing nutrition at flowering increases postharvest longevity.

10.4.0 Irrigation

Calendula requires 2.5 to 4 cm water per week during dry periods. The media should be kept moist. Cut flowers grown in benches are shorter with smaller flowers when compared with plants grown in ground beds due to more constant moisture and cooler temperature in ground beds (Post, 1949). Flowering is delayed if plants grown dry.

10.5.0 Pinching, Disbudding and Support

Depending on the cultivar, terminal bud can be removed to enhance axillary shoot elongation and to increase the number of flowers harvested during a limited time span. For potted plants, two or more flowers will usually open simultaneously; otherwise the terminal bud will flower, followed by three to five axillary flowers (Armitage, 1985). Removal of the old flowers from the plant will encourage re-bloom. Two layers of wire or plastic mesh can be used for cut flowers and layers then raised as plants grow.

10.6.0 Weed Control

When grown in rows, regular weeding and hoeing is adequate to hold the weeds under control. The plastic weed barrier works for transplanted areas, but may not be cost-effective. Efficient chemical controls include pre-plant herbicide ethalfluralin (Edge) and trifluralin (Bonanza), pre- or post-emergent herbicide propyzamide (Kerb).

11.0 Diseases, Pests and Physiological Disorder

Major diseases that affect calendula are listed as under:

Powdery mildew: (*Sphaerotheca fuliginea* and *Erysiphe cichoracearum*): It damages flowers during cool seasons. Preventive applications of sulphur sprays may protect from mildew. Picking flowers early and often prevents late cool season powdery mildew.

Aster yellows: The disease results in stunting and deformation of flowers and is transmitted by six spotted leaf hoppers. Elimination of contaminated plants is important to protect plants. Control steps involve tracking leafhopper with sweeping nets and treatment with neem oil, soap and pyrethrum sprays or with chemicals such as carbaryl (Sevin). Other diseases that may affect plant growth and flowering are Botrytis blight (*Botrytis cinerea*), Southern blight (*Sclerotium rolfsii*) and stem rot (Sclerotinia sclerotiorum) (Horst, 1990).

The major pests are aphids, leaf minors, blister beetles, white flies, cabbage loopers, mealy bugs and caterpillars.

Physiological disorder such as weak stems and small flowers may develop when greenhouse light intensities are too low and temperatures are too high.

12.0 Harvesting and Postharvest Management

Plants bloom from the beginning of July to the winter (70-120 days). Flowers may be collected by hand-picking or using a comb. The best time to harvest calendula flowers is in the morning after the dew has dried. Hand-picking results in a better quality as the comb process gathers seeds, ripe flowers and seed heads. The combed flowers must either be picked up to extract the undesirable materials or sold at a cheaper price. Flower yield vary from 2.5 to 4 tonnes per acre. A single labour can harvest 12 to 20 kg of fresh flower heads per hour. Harvesting labour cost account for up to 80 per cent of the total labour needed to produce a flower crop. For oilseed development, the crop is desiccated with diquat (Reglon Desiccant) and then mixed.

Spring produced flowers have better keeping quality than those from winter production. Flowers should not be harvest prematurely, but rather when the flowers are fully developed. Stem should be placed in warm water with a floral preservative. Transport the plants or flowers at 4-5°C for 3-6 days (Nowak and Rudnicki, 1990). The flower does or does not require more processing, such as cleaning, depending on the market. No further processing is normally needed for whole flower heads. Only product winnowing may be done for petals to clean the product.

Diseases and Pests of Calendula

Bacterial Leaf Spot

Powdery Mildew

Virescence

Phyllody

Aster Yellow

Smut

Blister Beetle

Aphids

Source: www.omafra.gov.on.ca, www.pnwhandbooks.org, www.thestarphoenix.com, www.twitter.com, www.susansinthegarden.com, greenfusestock.photoshelter.com, www.omafra.gov.on.ca, Hosseini *et al.* (2011).

Drying calendula flower heads need shade, airflow and external heat for drying. Since there are several sites on flower heads where moisture may accumulate, careful drying must be undertaken. The flower heads should be dispersed across the screens with no more than one dense layer. A brief period of high heat 60°C to expel surface moisture is accompanied by a prolonged drying period of not more than 35°C.

13.0 Biotechnology

13.1.0 Molecular Markers

Many molecular techniques are currently available for the analysis of genetic variation including, RAPD (Smolik *et al.*, 2011), RFLP (Botstein *et al.*, 1980), AFLP (Vos *et al.*, 1995) and SSR (Király *et al.*, 2012). Molecular methods have significant application in crop improvement, genetic taxonomy, mapping and phylogenetic studies (Vos *et al.*, 1995). Among these molecular tools, RAPD (Randomly Amplified Polymorphic DNA) markers have many advantages and have been often used in modern research. The technique is easy, fast and needs only a small amount of DNA and, most importantly, no prior knowledge of DNA sequences is needed (Hadrys *et al.*, 1992). In addition, the relatedness and distinctiveness of various genotypes can be clearly calculated by RAPD fingerprinting (Thomas *et al.*, 2006). Baciu *et al.* (2013) revealed that genotypic and phenotypic variation and relationship among the 34 varieties of pot marigold using RAPD markers and all the accessions tested tend to be monophyletic (Cluster I includes Groups II and III and Group II also included Group IV).

14.0 References

Aliniaeifard, S., Seif, M., Arab, M., ZareMehrjerdi, M., Li, T. and Lastochkina, O. (2018) *Int. J. Hort. Sci. Tech.*, **5**(1): 123-132.

Armitage, A. (1985) *Greenhouse Grower,* **3**(9): 68-69.

Armitage, A.M., Bergmann, B. and Bell, E.L. (1987) *HortScience*, **22**: 611-612.

Arnold, M.L. (2006) Oxford: Oxford University Press.

Baciu, A.D., Mihalte, L., Sestras A.F. and Sestras, R.E. (2010) *Not Bot HortiAgrobo.,* **38** (3): 265-270.

Baciu, A.D., Pamfil, D., Sestras, A.F., Mihalte, L. and Sestras, R.E. (2012) *Bulletin UASVM Horticulture*, **69**(1).

Bcerentrup, H. and Robbelen, G. (1987) *Fat Sci. Tech.*, **89**: 227-230.

Bone, K. (2003) *A clinical Guide to blending liquid herbs.* St Louis Missoun, Churchill livinigstone, pp. 120-123.

Botstein, D., White, R.L., Skolnick M. and Davis, R.W. (1980) *Am. J. Hum. Genet.*, **32**: 314-331.

Bremer, K. (1994) *Asteraceae: cladistics and classification.* Timber Press, Portland

Casierra-Posada, F., and Ávila-León, O.F. (2015) *Revista UDCA.*

Clark, E.M.R., Dole, J.M., Carlson, A.S., Moody, E.P., McCall, I.F., Fanelli, F.L. and Fonteno, W.C. (2010) *Horttech*, **20**(6): 1016-1025.

Clavijo, E. (2005) *Acta Oecol.*, **28**(2): 119-126.

Çöçü, S., Uranbey, S., Ýpek, A., Khawar, K.M., Sarihan, E.O., Kaya, M.D. and Özcan, S. (2004) *Biol. Plant.*, **48**(3): 449-451.

Colonel Kirtikar K.R. and Major Basu, B.D. (1993) *Indian Medicinal Plants* (Vol II), Deharadun, India, pp. 1413-1414.

Cordova, C.A., Siqueira, I.R., Netto, C.A., Yunes, R.A., Volpato, A.M., Cechinel, S., Filho, V., Curi-Pedrosa, R. and Creszynski-Pasa, T. (2002) *Redox Rep.*, **7**: 95-102

Cromack, H.T.H. and Smith, J.M. (1998) *Ind. Crop. Prod.*, **7**: 223-229.

Darlington, C. D, Wylie, A. P. (1955) 2nd edn. George Allen and Unwind Ltd., London

De Clavijo, R. (2005) *Acta Oecol.*, **28**: 119-126.

Deniz, L., Serteser, A., Kargioglu, M. (2010) *Afyon Kocatepe Univ. J. Sci. Eng.*, **01**: 57-72.

Dhooghe, E., Van Laere, K., Eeckhaut, T., Leus, L., and Van Huylenbroeck, J. (2010) *Plant Cell Tiss. Org.*, **104**: 359–373.

Diaconu, P. (1992) *Agronomie: Lucraristiintifice*, **34**: 17-21.

Dinda, K. and Craker, L.E. (1998) HSMP Press, Amherst, pp. 35-37.

Eberle, C.A., Forcella, F., Gesch, R., Peterson, D., and Eklund, J. (2014) *Ind. Crop Prod.*, **52**: 199-204.

Ehrendorfer, F. (1970) *Feddes Repert*, **81**: 3–32.

El-Nashar, Y.I. and Ammar, M.H. (2016) *Genet. Mol. Res.*, **15**(2): 1-15.

Eng, W.H. and Ho, W.S. (2019) *Sci. Hortic.*, **246**: 604–617.

Esmaeili, G., Van Laere, K., Muylle, H. and Leus, L. (2020) *Front*. Plant *Sci.*, **11**: 622.

Fleming, T. (1998) *PDR for Herbal Medicines Montvale*, Medical Economics Co., Inc., NJ, pp. 567.

Grieve, M. (1931) Jonathan Cape Ltd, London, pp. 456.

Hadrys, H., Balick M. and Schierwater, B. (1992) *Mol. Ecol.*, **1**: 55-63.

Harrison, D. (Inventor) (2003) *Canadian Application Publication*, pp. 675.

Harrison, D. (2003) Canadian Application Publication, pp. 675.

Heyn, C.C., Dagan, O., Nachman, B. (1974) *Israel J. Bot.*, **23**: 169–201.

Heyn, C.C., Joel, A. (1983) *Plant Syst. Evol.*, **143**: 311–329.

Horst, R.K. (1990) In: *Westcott's Plant Disease handbook*, 5th edition. Van Nostrand Reinhold, Newyork. pp. 572-572.

Horvath, F.I. (1992) *Therapeutical compositions against psoriasis*. US Patent 5165932, Nov 24, pp. 454.

Jan, N. and John, R. (2017) *Proc. Indian National Sci. Acad.*, **83**(4): 769-787.

Kalvatchev, Z., Walder, R. and Garzaro, D. (1997) *Biochem. Pharmacol.*, **51** : 176–80.

Kamlesh. (1977) Effect of some regulatory substances on growth and development of *Calendula officinalis* in relation to photoperiod. M.Sc. thesis submitted to G.N.D. University, Amritsar.

Khalid, K.A. and Teixeira da Silva, J.A. (2012) *Med. Aromat. Plant. Sci. Biotechnol.*, **6**: 21-27.

Khare, C.P. (2004) *Encyclopedia of Indian Medicinal Plants*, Springer-Verlag, Germany, pp. 116-117.

Kiraly, I., Redeczki, R., Erdelyi, E., and Toth, M. (2012) *Not Bot HortiAgrobo.*, **40** (1): 269-275.

Krag, K. (1976) *Plants used as contraceptives by the North American Indians: an ethnobotanical study*. Botanical Museum, Cambridge, Thesis BS, Harvard University, pp. 196.

Kron, P., Suda, J., Husband, B.C. (2007) *Annu. Rev. Ecol.Evol. S*, **38**: 847–876.

Leach, M.J. (2008) *Wounds*, **20**: 1-7.

Li, M., Ding, B., Huang, W., Pan, J., Ding, Z. and Jiang, F. (2018) *Biomed. Res. Int.* **5**: 1–8.

Luo, Z., Iaffaldano, B. J. and Cornish, K. (2018) *Ind. Crop. Prod.*, **112**: 75–81.

Macht, D.I. (1955) *Bull. Hist. Med.*, **29**(6): 491-502.

Mallet, J. (2007) *Nature*, **446**: 279–283.

Martin, F. (2005) ADAS Bridget Research Centre.

Massoud, H.Y., El-Kafie, A., Omaima, M., Helaly, A.A., and Ghanem, M.E. (2020) *J. Plant Prod.*, **11**(5): 515-518.

Medina, E.J., Angel, L.G., Paco, L., Algarra, I., Collado, A. and Garrido, F. (2006) *BMC Cancer*, **6**: 1-4.

Meusel, H., Ohle, H. (1966) *Oesterreichische botanische Zeitschrift.*, **113**: 191–210.

Mohammad, S., Kashani, H.H. (2012) *Sci. Res. Essays*, **7**(14): 1468-1472.

Mori, S., Yamane, T., Yahata, M., Shinoda, K. and Murata, N. (2016) *Hort. J.*, **4**: 366–371.

Mozherenkov, V.P. and Shubina, L.F. (1976) *Medica Sestra*, **35**: 33-34.

Muley, B.P., Khadabadi, S.S. and Banarase, N.B. (2009) *Trop. J. Pharm. Res.*, **8**(5): 455-465.

Nedelka, A., Golovkin, V., Solovieva, V. and Nedelka, V. (1988) *Oftalmologicheskii Zhurnal.* **43**: 429-431.

Nora, S., Castro, S., Loureiro, J., Gonçalves, A.C., Oliveira, H., Castro, M., Santos, C. and Silveira, P. (2013) *Plant Systematics and Evolution*, **299**: 853–864.

Nordenstam, B. (1994) In: Bermer K (ed) *Asteraceae*. Cladistics and Classification. Timber Press, Portland, pp. 365–376.

Norlindh, T. (1946) *BotaniskaNotiser*, **4**: 471–506.

Norlindh, T. (1977) In: Heywood VH, Harborne JB, Turner BL (eds) *The biology and chemistry of the Compositae*. Academic Press, London, pp. 961–987.

Nowak, J. and Rudniki, R.M. (1990) *Post harvest handling and storage of cut flowers, florist greens and potted plants*. Timber press, Portland, Oregon.

Ohle, H. (1974) *Feddes Repert*, **85**: 245–283.

Ohle, H. (1975) *Feddes Repert*, **86**: 1–17.

Post, K. (1949) *Calendula*, **172**: 358-360.

Rice, A., Glick, L., Abadi, S., Einhorn, M., Kopelman, N. M., Salman-Minkov, A., Mayzel, J., Chay, O. and Mayrose, I. (2015) *New Phytol.*, **206**(1): 19–26.

Salma, U., Kundu, S., and Mandal, N. (2017) *J. Crop Sci. Biotechnol.*, **20**: 9–19.

Sattler, M.C., Carvalho, C.R., and Clarindo, W.R. (2016) *Planta*, **243**: 281–296.

Sawhney, S., Kamlesh and Kohli, R.K. (1981) *Indian J. Plant Physio.*, **24**(4): 299-303.

Van De Pol, P.A. (1971) *Proc. Kon. Ned. Akad. Wet.*, **74**: 449-454.

Vince-Prue, D. (1975) Photoperiodism in Plants. McGraw-Hili., London.

Von, O. and Die, O. (1992) Wissenschftliche Ver- lagsgellschaft mbH, Stuttgart, Bundesrepublik Deutschland, pp. 786.

Yoshikawa, M., Murakami, T., Kishi, A., Kageura, T., Matsuda, H. (2001) *Chemical and Pharmaceutical Bulletin*, **49**: 863-870.

Zahra, A., Mohammed, A. and Mohammed, H.K. (2000) *J. Ethnopharmacol*, **72**: 167-172.

16

Gaillardia

Chetna Jyoti, Rajiv Kumar, P. Pal, M. Mitra and K. Dutta

1.0 Introduction

Gaillardia is popularly known as blanket flower. Plants are hardy and grown in a wide range of climatic conditions. In tropical climate gaillardia can be grown throughout the year, but it is a major annual in summer. It is also frost hardy. Gaillardia flourishes well in any garden soil and can withstand high light intensities, high temperature and drought better than most of the flowering plants. It is also tolerant to salinity (Tija and Rose, 1988).

2.0 Importance and Uses

2.1.0 Aesthetic Uses

The plants are grown as annuals, biennials and perennials. They are grown in herbaceous borders and bed and are also suitable for cut flowers. Gaillardia flowers are extensively used in bouquets, garlands and floral decoration (Gawade *et al.*, 2017). Besides, its utility in landscape *Gaillardia pulchella* is useful in reducing erosion in coastal dune areas (Carig, 1977). *G. aristata* is suitable for dryland and requires low maintenance (Cox and Klett, 1984; Robinson and Schultz, 1995).

2.2.0 Therapeutic and other Uses

Panchaude (1990) reported the nematicidal property of gaillardia, when grown as catch crop and green manure. The essential oil from *Gaillardia pulchella* Foug. flowers was obtained by hydro-distillation and its chemical composition was analyzed by gas chromatography-mass spectrometry. Twenty-eight compounds representing 92.63 per cent of the essential oil were identified, of which the most prominent were *n*-Hexadecanoic acid (26.90 per cent), Phytol (7.58 per cent) and Cyclopropaneoctanoic acid, 2-[[2-[(2-ethylcyclopropyl) methyl] cyclopropyl] methyl]-, methyl ester (6.73 per cent). The essential oil showed certain antioxidant activity in 1,1-diphenyl-2-picrylhydrazyl (DPPH) with an EC_{50} of 70.95 µg/ml (Yao *et al.*, 2013).

3.0 Species and Cultivars

It is a native of North America and South America (Brickell, 1996). The generic name *Gaillardia* was proposed in honour of M. Gaillard, a French patron of botany. It belongs to the family *Asteraceae* and there are about 30 species, out of which *G. pulchella*, *G. aristata* and *G.* × *grandiflora* are of horticultural importance, being annual and perennial forms, respectively. *Gaillardia pulchella* is diploid (2n = 36) as reported by Morinaga *et al.* (1929) and *G. aristata* is both diploid (Cooper and Mahony, 1936) and tetraploid, 2n = 78 (Atwood, 1937). Plants are bushy with spreading branches. The leaves are alternate, more or less toothed and rough. Flowers are borne in solitary, usually showy heads, involucre broad, bracts in 2 or 3 series, hairy, ligules 3-toothed, giving a fringed appearance to the flowers.

The cultivars of *G. pulchella* have single or double flowers with different colours, red, orange yellow, yellow-tipped red, red-tipped yellow, bronzed-scarlet and others.

Uses of Gaillardia

In Bed

As Border in Garden

In Pot

As Specimen in Garden

In Flower Vase

In Rock Garden

Display in Pots

Near Water Body

In Garden Decoration

Species of Gaillardia

Gaillardia pulchella ***Gaillardia aristata*** ***Gaillardia grandiflora***

Gaillardia aestivalis ***Gaillardia suavis*** ***Gaillardia amblyodon***

Cultivars of Gaillardia

Maxima Aurea **Burgundy** **Dazzler**

Goblin **Red Plume** **Sunset**

G. aristata has more number of strains and cultivars than in *G. pulchella*, the height varying between 35-37 cm and the range of colour includes yellow, coppery-scarlet, orange-brown, golden-yellow, orange, crimson-banded yellow.

G. grandiflora, a hybrid between *G. aristata* and *G. pulchella*, a short lived perennial, about 30 cm high, bushy, bears large, attractive flower heads (8-14 cm across) and flower colour varies in many cultivars, wine red, orange red, yellow-tipped red ray floret, yellow-tipped orange-red and many others.

Different phylogeny relationship studies carried out by Marlowe and Hufford in 2007. Phylogenetic analyses based on maximum parsimony, maximum likelihood and Bayesian inference criteria consistently recovered several well-supported clades. *Gaillardia comosa*, *G. gypsophila*, and *G. suavis*, forming sect. Agassizia, were placed sister to the rest of Gaillardia, in which the two major clades correspond to sects. Gaillardia and Hollandia. Section Hollandia consists of the Gulf Coast *G. aestivalis* complex. Three major, well supported clades were consistently recovered in sect. Gaillardia: (1) *G. multiceps* and *G. turneri* were sister to an Intermountain clade of *G. parry*, *G. spathulata*, and *G. flava*; (2) a paraphyletic *G. pinnatifida* and *G. arizonica*; and (3) *G. pulchella*, *G. aristata*, *G. amblyodon*, *G. coahuilensis*, *G. mexicana*, *G. henricksonii*, and *G. powellii*. Dispersal Vicariance Analysis reconstructed a Chihuahuan Desert origin for Gaillardia, an inference consistent with earlier hypotheses. Major range expansions in Gaillardia included the following: (1) eastward expansion into central Texas and along the Gulf Coast in sect. Hollandia; (2) north western expansion, resulting ultimately in the Intermountain clade, but including also *G. multiceps and G. turneri*; (3) northward expansion of the *G. pinnatifida* complex; (4) easterly and especially northward expansion of *G. amblyodon*, *G. pulchella*, and *G. aristata*; and, possibly, (5) northward expansion onto the Great Plains by *G. suavis*.

Chief *et al.* (2018) reported different cultivars as follows *Gaillardia pulchella* - annual type are Fieata, Red giant, Fireball, Sunshine, Gaiety, Tera Fiesta, Indian Chief, Tera Giant, Lorenziana.

Gillardia grandiflora – hybrid perennialare Aurea, Aurea Pura, Baby Cole, Bremen, Burgundy, Compacta, Dazzler, General patton, Goblin, Golden goblin, Golden Goddess, Grandiflora, Kobold, Maxima, Monarch, Mr. Sherbrooke, Nana Nieske, Portola, Ruby, Sun, Sun God, Sunset, Superba, Tangerine, The King, The Prince, The Warrior, Tokaier, Tokaj, Tokay, Yellow Queen.

4.0 Growth and Flowering

4.1.0 Environmental Factors

4.1.1 Temperature and Light

The effect of temperature and light on flowering in *G. × grandiflora cv.* Goblin was studied by Yuan *et al.* (1998). The plants were exposed to 5°C for 10 weeks, and then grown in greenhouse sections set at 15, 18, 21, 24 or 27°C under 4h night interruption lighting until anthesis. Days to visible bud (VB), days to anthesis (FLW) and days from VB to FLW decreased as temperature increased. The rate of progress towards FLW increased linearly with temperature. Trippi (1965) noted that continuous light stimulated flowering and accelerated the process of ageing.

4.2.0 Growth Regulators

Bankar (1980) observed that gaillardia seeds treated with GA_3 at 60 ppm enhanced seed germination.

Plants sprayed with CCC and daminozide produced more number of flowers than control and MH. Daminozide 500 ppm produced highest number of flowers/ yields (160.60 flowers/plant, 165.20 g/plant and 27.43 t/ha) (Khimani *et al.*, 1994). TIBA at 100, 200 and 300 ppm, CCC at 750 ppm and kinetin at 50 and 75 ppm produced significantly higher flower yields in *G. pulchella cv.* Picta (Hugar and Nalawadi, 1999a). Uniconazole applied as a spray, had no effect on torch blanket flower, but when applied as a drench, growth indices were reduced by 12 per cent to 30 per cent without delayed flowering. When treated with ethephon spray, there was 15 per cent to 25 per cent more compact than untreated controls, but flowering was delayed. (Hammond *et al.*, 2007).

The field experiment was conducted to study the effect of growth retardants and pinching on growth and yield of gaillardia (*Gaillardia pulchella* var. Lorenziana) *cv.* Local Double. The results revealed that the

Foliar spray of MH 400 ppm at 45 DAT significantly retarded plant the height at 60, 90 and 120 DAT. The plants treated with CCC at 1000 ppm at 45 DAT recorded the highest numbers of branches per plant (56.53) at 120 DAT and shelf life (3.83 days). The maximum number of flowers per plant (123.67), weight of flower/ plant (350 g) and significantly the highest yield (4.20 Kg/net plot and 172.98 q/ha, respectively) was also recorded with CCC at 1000 ppm as compared to pinching and control in *cv.* Local Double (Nishith *et al.*, 2015).

5.0 Propagation

5.1.0 Seed Propagation

The annual gaillardia is propagated through seeds. The perennial forms can be propagated by division, cutting or seeds. The seeds may be sown either in pot or in raised nursery beds from January to October to obtain flowers almost throughout the year. Ichihara (1954) reported that the seeds of gaillardia have no dormancy. Seeds of *G. pulchella* can be sown all the year round (De Souza, 1958). Seed-propagated plants of *G. aristata* produced economic flowers up to 3 years (Ignateva, 1962).

The seeds of *Gaillardia aristata* packed in five different packaging materials for a period of 18 months resulted that the seeds stored in cold storage recorded maximum germination (39.51 per cent) followed by ambient conditions (31.62 per cent) and incubator (22.98 per cent). Among the packaging material, the cloth bag maintained the maximum germination (34.29 per cent) at the end of storage period. After one year of seed storage, highest germination (45.67 per cent) was obtained in desiccators kept in cold store. The seed viability was significantly better in cold store (46.05 per cent) than ambient (35.88 per cent) and incubator (34.57 per cent). The cloth and aluminium foil maintained 41.48 and 39.93 per cent seed viability that is considerably higher than all other packaging. The maximum viability after one year was also recorded in aluminium foil (51.33 per cent) followed by cloth bag under cold storage (Dhatt *et al.*, 2010).

Propagation of Gaillardia

Seeds

Planting of Seeds

Seedlings

Cultivation of Gaillardia

Soil Preparation

Cultivation in Greenhouse

Cultivation in Open

Adegbola *et al.* (2016) investigated the response of *Gaillardia pulchella* seeds to desiccation and aging stress to gain some perspective on the germplasm storage potential and seed vigour. Seed–water relations of mature, freshly harvested *G. pulchella* seeds were characteristic of desiccation-tolerant species. For example, initial seed water potential (–53 MPa) was well below the lethal water potential limit (–15 MPa) for desiccation-sensitive seeds. Desiccation tolerance was confirmed by high (>70 per cent), rapid (t50 range 4–7 days), and uniform germination following equilibration drying. Likewise, post–saturated salt accelerated aging (SSAA) germination tests indicated a high degree of vigor of fresh seeds. The substantial level of desiccation and aging-related stress tolerance in *G. pulchella* seeds suggests that these organs potentially display orthodox storage physiology and an ability to endure variable seed bed conditions.

5.2.0 Micropropagation

Young leaf and intermodal stem segments of Gaillardia pulchella, collected from wild species re-established in the greenhouse, were used to initiate callus on MS medium supplemented with NAA (2.0 mg/l) and BA (0.4 mg/l). Callus formed after 10 to 14 days in the dark. Culture were transferred to fresh medium and placed under lighted conditions were shoot formation occurred approximately 14 to 30 days after initiation. Callus sub-cultured at 14 to 21 day intervals continued to produce primordial for several weeks. Flowers were produced by regenerated shoots maintained on MS medium, but roots did not develop until the plantlets were transferred to soil conditions (Bourque *et al.*, 1989).

6.0 Soil and Climate

Gaillardia prefers well-drained sandy loam soil and sunny situation. Medium fertility status of the soil is ideal for successful cultivation of this crop. Sharma *et al.* (1989) reported an increase in plant N and P contents and decrease in K content with increasing level of salinity. Khimani and Patil (1994) observed a significant decrease in total plant, stem, flower and root DWs, flower number and yield per plant with increase in soil salinity level from 1.20 to 2.80 ds/m. Leaf proline content increased with increasing salinity (21.92 mg/g DW at 1.20 ds/m to 128.42 mg/g DW at 8.70 ds/m).

Rajvanshi *et al.* (2011) found that potting mixture ratios of Farmyad manure + coarse sand + soil (2 : 2 : 1) in gaillardia was effective. Plants grown in sand + garden soil + organic compost (1:1:1) media showed maximum plant height, plant spread, number of leaves per plant, number of branches per plant, flower size, disc size and number of rays per flower. In conclusion, combination of compost with garden soil (1:1) showed promising potential for optimal growth of Gaillardia cultivar Arizona Sun under local conditions (Alhajhoj, 2015).

Water demand and growth traits of gaillardia in peat-based substrates with different volume based ratios of field soil (clay loam) and peat moss (100 per cent peat, 80 per cent peat + 20 per cent soil, 60 per cent peat + 40 per cent soil, 20 per cent peat + 80 per cent soil and 100 per cent soil) were studied by Khandan *et al.* (2018). Results showed the root system of Gaillardia was more developed so it was a more efficient in water uptake. Combining 40-60 per cent (v/v) of peat with field soil resulted in progress of the root system and significantly increased the efficiency of water absorption. Therefore, the combination of at least 20 per cent of peat with field soil is recommended for production of gaillardia.

7.0 Cultivation

Plants should be spaced 30-40 cm apart. Mishra (1998) reported that flower size, weight and stalk length of *G. pulchella* was highest at 40 x 30 cm spacing. The plants need watering once a week or fortnight, depending on the weather.

Vaagdevi *et al.* (2020) revealed that planting during 1st week of October recorded maximum number of seeds/inflorescence and germination per cent; 1st week of January planting recorded maximum seed yield/hectare, seed yield/plot and seed yield/plant. Among cultivars, *cv.* Local Yellow showed higher germination per cent than *cv.* Local Red with yellow tip.

October planting significantly increased plant height, number of branches, number of leaves per plant, induced early emergence of flower bud to full blooming of flowers, longer duration of flowering, improved flower fresh and dry weight and enhanced flower yield in terms of number of flowers per plant and flower weight per plant (Sharma *et al.*, 2015).

In order to improve flowering, liquid feeds should be given when the plants have started bearing flowers. Optimum fertilizer rate for growth and flower yield of *G. pulchella cv.* Yellow Double was 173 kg N + 103 kg P + 63 kg K/ha (Singatkar *et al.*, 1995). Mishra (1998) recorded maximum plant height, spread of the plant, early flowering, flower size and flower weight at 30 g N/m^2, where 1/3 N were applied at transplanting and rest in 2 equal split doses at monthly interval and 5 kg FYM, 15 g of P_2O_5 and 10 g K_2O/m^2.

Nitrogen at 150 kg/ha of two splits, *i.e.* 50 per cent as basal dose and 50 per cent at initiation of flowering recorded the highest flower yield (39.04 and 33.16 t/ha) on red and black soils, respectively. The increase in flower yield was mainly due to improved yield attributes like number of flowers per plant, flower size and yield per planting (Patil *et al.*, 2004).

Nitrogen at 300 and 200 kg/ha, phosphorus at 100 and 75 kg/ha and potash at 75 kg K_2O/ha improved plant height, number of branches/plant and leaf area and yield attributes *viz.*, weight of 10 flowers and diameter of flower (Karetha *et al.*, 2011).

Early bud emergence to full blooming of flowers and flower longevity on plant and fresh and dry weight of ten flowers was observed with 130 kg N/ha. The October planting with 130 kg nitrogen per hectare proved superior for growth, flowering and yield attributes *viz.*, plant height, number of branches and leaves per plant, duration of flowering, number of flowers per plant and weight of flowers per plant (Sharma *et al.*, 2015).

Azospirillum strain OAD-2 inoculation significantly increased plant height, number of leaves per plant, branches per plant and total dry mass accumulation in *G. pulchella*. Nitrogen at 150 kg/ha further enhanced the above plant growth parameters significantly over N at 112 kg/ha. The N uptake of *G. pulchella* was enhanced due to *Azospirillum* strain OAD-2 inoculation, which was further augmented in the presence of nitrogen. The combination of *Azospirillum* strain OAD-2 and N at150 kg/ha showed the highest N uptake at 120 DAT (Gadagi, 2004).

Application of mulches is one of the suggested strategies for maintaining bed moisture and weed control. The polyethylene mulch inhibited the weed growth up to 100 per cent. Application of pine leaves delayed the flowering for seven days, while, polyethylene caused an early flowering to six days compared to the control (Kazemi and Jozay, 2020).

Seeds were sown directly into field plots of sandy soil and irrigated with saline water at electrical conductivity (ECW) of 2 (control), 5, 10 and 15 dS/m. Increase in salinity reduced the plant stand by 38 per cent at 5 dS/m, 50 per cent at 10 dS/m and 67 per cent at 15 dS/m. The number of flowers per plant declined by 57 per cent at 5 dS/m, 61 per cent at 10 dS/m and 67 per cent at 15 dS/m.The flower diameter was significantly reduced at 15 dS/m, but was not affected at the lower salinities. The results showed that *G. aristata* could be successfully cultivated up to irrigation water salinity level 10 dS/m in sandy soils (Rao and Shahid, 2011).

Hoeing and weeding is done, 3-4 times, during the crop period. Beran *et al.* (1999) reported that application of imidazolinone (imazapic) at 70 g *a.i.*/ha reduced the weed growth.

8.0 Diseases and Pests

Gaillardia is less susceptible to diseases and pests than many other garden plants.

8.1.0 Diseases

Fungal Diseases

8.1.1 Foot Rot

Sclerotinia sclerotiorum was found to cause foot rot in *G. pulchella* (Rai and Agnihotri, 1970) and *G.* × *grandiflora* (Koike, 1997). Overcrowding and over-watering enhance damping off disease caused by *Pythium* sp. Sterilization of nursery beds before seed sowing is effective.

8.1.2 Leaf Spot

Leaf spot diseases caused by *Cercospora gaillardiae* is controlled by spraying copper or dithiocarbamate fungicides. Gaillardia is also infected by powdery mildew (*Erysiphe cichorearum*) which is controlled by using karathane.

A leaf spot disease occurred on 100 per cent of *Gaillardia* × *grandiflora cv.* Goblin in coastal Monterey Co., CA. Nearly all of the affected plants were unsalable. The causal agent was determined to be *Entyloma gaillardianum* based on morphological features, host, and ITS region. This species has not been reported previously from this host in North America (Glawe *et al.*, 2010).

G. aristata plants were affected, exhibited circular spots on leaves, from 0.8 to 1.3 cm in diameter, sometimes coalescent. Amphigenous spots were light green and 6 to 9 days later turned brown and necrotic. It was reported to caused by *Entyloma gaillardianum* as leaf smut reported (Garibaldi *et al.*, 2018).

8.1.3 Powdery Mildew

The disease was reported in field grown *Gaillardia × grandiflora* (Blanket flower) in Yakima Co., WA, on the *cv.* 'Baby Cole'. It is caused by *Leveillula taurica* (Lév.) G. Arnaud (Glawe *et al.*, 2006).

Viral Diseases

Spotted wilt virus and California type of aster yellows occur in gaillardia. In October 2009 leaf yellowing symptoms of Gaillardia, were observed in more than 50 per cent of plants growing in fields of Fatehnagar village in the Udaipur district of Rajasthan, India. It was shown that the causal agent of the disease was transmitted by whiteflies. The virus infecting Gaillardia is considered to be the variant of Tomato leaf curl Bangladesh virus (Mahatma and Mahatma, 2012).

Gaillardia plants exhibiting symptoms characteristic of viral and phytoplasmal diseases were reported at botanical gardens and floriculture farms in Lithuania. Cucumber mosaic cucumovirus was isolated from diseased plants exhibiting symptoms characterized by stunting, flower breaking and malformation of petals. The virus was characterized based on electron microscopy, serology, reactions of inoculated test-plants, and reverse transcription polymerase chain reaction (RT-PCR). Symptoms of leaf yellowing and general stunting of plants, and virescence and phyllody of flowers were associated with infection by a phytoplasma. Phytoplasmal 16S rRNA gene sequences were amplified in polymerase chain reactions (PCRs) primed by phytoplasma universal primer pairs P1/P7 and R16F2n/R16R2. RFLP analysis of the 1.2 kbp rDNA product, subjected to single enzyme digestions with nine restriction endonucleases, revealed that the Gaillardia plants were infected by a phytoplasma (gaillardia yellows, GaiY) belonging to group 16SrI (aster yellows phytoplasma group), subgroup I–C (clover phyllody, CPh, phytoplasma subgroup). 16S rDNA sequence alignments and phylogenetic analysis confirmed that the GaiY phytoplasma was closely related to CPh phytoplasma. Gaillardia can be naturally infected by a broad biodiversity of phytoplasma strains belonging to subgroups 16SrIII–B (clover yellow edge subgroup), 16SrI–A (aster yellows subgroup), and 16SrI–C in Lithuania (Valiûnas *et al.*, 2008).

8.2.0 Pests

Occasionally, plants are attacked by aphids, leaf hoppers, thrips and spider mites. These can be controlled by spraying of malathion. Spraying of Kelthane at 2 per cent is effective to control the spider mites (Smith, 1967).

Investigations on bio-efficacy of botanicals and synthetic insecticides against *Uroleucon compositae* (Theobald) (Hemiptera: Aphididae) a pest of ornamental crop, *Gaillardia pulehella* Foug was carried out. Among the botanicals evaluated, neem seed kernel extract (3 per cent), neem leaf extract (10 per cent) and arduso leaf extract

Diseases and Pests of Gaillardia

Leaf Spot | Powdery Mildew | Aster Yellow

Aphids | Leaf Hopper | Spider Mites

Harvesting and Marketing of Gaillardia

Stage of Harvesting | Harvesting of Flowers | Harvesting of Seeds

Marketed as Cut Flowers | Flowers in Market | Flower Pot for Sale

(10 per cent) were found effective in reducing the pest which resulted in higher (25.39 to 24.21 kg/ha). The Incremental cost benefit ratio (ICBR) value of effective botanicals was 1:53.71, 1:47.87 and 1:41.74, respectively. Of the nine synthetic insecticides tested, dimethoate (0.03 per cent), methyl-o-demeton (0.025 per cent) and monocrotophos (0.04 per cent) provided maximum protection of gaillardia crop against *U. compositae*. Dimethoate (ICBR 1:77) yielded maximum gaillardia flowers followed by methyl-o-demeton (ICBR 1:60.91) and monocrotophos (ICBR 1:63.90) (Bhatt *et al.*, 2007).

9.0 Harvesting and Postharvest Management

9.1.0 Harvesting

9.1.1 Flowers

Fully open flowers with long stem are harvested for cut flowers which remain fresh for 5-6 days in vase. Removal of wilted flowers prolongs the flowering period of gaillardia. Hugar and Nalawadi (1999b) recorded highest flower yield of 19.6 and 18.0 t/ha when grown in winter and summer respectively and seed yield of 7.0 and 6.8 q/ha in the two seasons in *G. pulchella cv.* Picta Fouger. Bhaskarwar *et al.* (2016) reported that var. Local 2 recorded flower yield of 65.80 q/ha and weight of flower (4.8 g).

9.1.2 Seeds

The seeds are collected from the plant when they are fully ripe. Nikolova-Khristeva (1973) suggested that during storage the optimum moisture content of the seeds of *Gaillardia* should be 9 per cent for maintaining the viability for a long period. The genotypes MG-9-1, MG-2-2 and MG-6-2 showed better storage performance and quantitative characters (Salve *et al.*, 2019).

9.2.0 Postharvest Management

9.2.1 Vase Life

The effects of fluoride and other chemicals on vase life of *G. pulchella* cut flowers were studied by Rajitha *et al.* (1999), and use of silver nitrate gave best results followed by sucrose and BA. BA also reduced the adverse effect of flouride. The post harvest parameters of gaillardia in terms of vase life (days), shelf life (days) and water uptake (ml) increased considerably with 100 per cent recommended dose of fertilizers with pressmud at 10 t/ha. It is therefore, recommended that, to obtain maximum post harvest life, crop should be supplied with 100 per cent recommended dose of fertilizers with pressmud at 10 t/ha (Sowmyamala *et al.*, 2012).

Using sucrose, methylchlorflurenol (a morphactin, MOR) and 8-hydroxyquinoline (8-HQ) individually as vase solutions and also in combination of two and three of these chemicals, regulation of petal senescence was investigated in scapes of *Gaillardia pulchella*. It was interesting to note that individually sucrose and 8-HQ behaved differently in comparison to the situation when both were present in the same holding solution. A combination of three (sucrose + MOR + 8-HQ) could effectively reduce the decline in flower diameter and membrane stability index (MSI) of petals. It could also minimize protein degradation and lipid peroxidation. Higher fresh weight and cumulative uptake of solution in scapes could be maintained by the combination of sucrose + 8-HQ in vase solution (Mukherjee *et al.*, 2019).

10.0 Biotechnology

Young leaf and internodal stem segments of *Gaillardia pulchella*, collected from wild species re-established in the greenhouse, were used to initiate callus on Murashige and Skoog medium supplemented with NAA (2.0 mg/l) and BA (0.4 mg/l). Callus formed after 10 to 14 days in the dark. Cultures were transferred to fresh medium and placed under lighted conditions where shoot formation occurred approximately 14 to 30 days after initiation. Callus sub-cultured at 14 to 21 day intervals continued to produce primordia for several weeks. Flowers were produced by regenerated shoots maintained on MS medium, but roots did not develop until the plantlets were transferred to soil conditions (Bourque *et al.*, 1989).

In another study by Kim *et al.* (2015) to determine the anti-melanogenic and anti-oxidant properties of *Gaillardia aristata* flower extract (GAE). Melanogenesis inhibition by GAE was investigated in cultivated cells and in a human skin model. In cultivated cells, the melanogenesis regulatory effect of GAE was evaluated using melanin content, intracellular tyrosinase activity and anti-oxidant characteristics. In addition, the expression of melanogenesis-related proteins was determined by western blot assay and real-time PCR. Result showed that GAE reduced the amount of melanin in B16F10 and normal human epidermal melanocyte cells and suppressed intracellular tyrosinase activity in a dose-dependent pattern. Also, GAE significantly decreased the expression of melanogenesis-related proteins (microphthalmia associated transcription factor, tyrosinase, tyrosinase-related protein-1, and dopachrome tautomerase). Real-time PCR results revealed a down-regulation of the mRNAs of these proteins. GAE possessed anti-oxidant characteristics as free radical-scavenging capacity and reducing power. In the three-dimensional human skin model, GAE applied to hyperpigmented skin significantly increased the degree of skin lightening within 2 weeks of treatment. The safety of GAE on human skin was confirmed and indicated the potential of GAE for use in suppressing skin pigmentation.

11.0 References

Adegbola, Y.U. and Pérez, H.E. (2016) *HortScience*, **51** : 159-163.

Alhajhoj, M.R. (2015) *Asian J. Plant Sci.*, **14** : 45-49.

Atwood, S. (1937) *Cellule*, **46** : 389.

Bankar, G.J. (1980) *South Indian Hort.*, **28** : 60-61.

Beran, D.D., Gaussoin, R.E. and Masters, R.A. (1999) *HortScience*, **34** : 283-287.

Bhaskarwar, A.C., Jadhav, G.N., Ghube, N.N., Tayade, M.M. and Patekar, T.R. (2016) *Int. J Res. Biosci. Agri. Tech*, **4**: 7-10.

Bhatt, N.A., Borad, P.K. and Darji, V.B. (2007) *Pest Manag. Hortic. Ecosyst.*, **13** : 153-158.

Bourque, J.E., Tanner, S. and Mabry, T.J. (1989) *Plant Cell, Tissue Organ Cult.*, **16**: 67-72.

Brickell, C. (1996) *Encyclopedia of Gardening*. Dorling Kindersely, London, pp. 180.

Carig, R.M. (1977) *Proc. Fla. Sta. Hort. Sci.*, **90** : 108-110.

Chief, I., Giant, R., Fiesta, T., Giant, T., Pura, A., Cole, B., Goblin, G., Goddess, G., Sherbrooke, M., Nieske, N. and God, S. (2018) *Handbook of Flowering*. pp.118.

Cooper, D.C. and Mahony, K.L. (1936) *Amer. J. Bot.*, **22** : 843.

Cox, R.A. and Klett, J.E. (1984) *HortScinece*, **19** : 856-858.

De Souza, H.M. (1958) *Bragantia*, **17** : 15-18.

Dhatt, K.K. and Kumar, R. (2010) *Indian. J. Hortic.*, **67** : 362-371.

Gadagi, R.S., Krishnaraj, P.U., Kulkarni, J.H. and Sa, T. (2004) *Sci. Hortic*, **100** : 323-332.

Garibaldi, A., Gilardi, G., Matic, S. and Gullino, M.L. (2018) *Plant. Dis.*, **102** : 678-678.

Gawade, N.V., Bhalekar, S.G., Wadekar, V.D. and Marbhal, S.K. (2017) *J. Pharmacogn. Phytochem.*, **6** : 1946-1948.

Glawe, D.A., Barlow, T. and Koike, S.T. (2010) *Plant Health Prog.*, **11** : 51.

Glawe, D.A., G.G. Grove and Nelson, M. (2006) *Plant Health Prog.*, **7** : 60.

Hammond, H.E., Schoellhorn, R.K., Wilson, S.B. and Norcini, J.G. (2007) *HortTechnology*, **17** : 552-556.

Hugar, A.H. and Nalawadi, U.G. (1999a) *Karnataka J. Agric. Sci.*, **12** : 137-141.

Hugar, A.H. and Nalawadi, U.G. (1999b) *Karnataka J. Agric. Sci.*, **12** : 226-229.

Ichihara, J. (1954) *J. Hort. Ass. Japan*, **23** : 115-122.

Ignateva, I.P. (1962) *Bot. Zurnal*, **47** : 1258-1272.

Karetha, K.M., Giriraj, J., Virendra, S. and Gajipara, N.N. (2011) *Asian J. Hort.*, **6** : 344-347

Kazemi, F. and Jozay, M. (2020) *Desert.*, **25** : 155-164.

Khandan-Mirkohi, A., Fayyaz, H. and Eraqi, A.M. (2018) *Iranian J. Hortic. Sci.*, **48** : 459-468.

Khimani, R.A. and Patil, A.A. (1994) *Recent Horticulture*, **1** : 97-101.

Khimani, R.A., Patil, A.A. and Chetty, M.B. (1994) *Recent Horticulture*, **1** : 91-96.

Kim, M., Shin, S., Lee, J.A., Park, D., Lee, J. and Jung. (2015) *BMC Complement Altern Med.*, **15** : 449.

Koike, S.T. (1997) *Plant Disease*, **81** : 1334.

Mahatma, L. and Mahatma, M.K. (2012) *New. Dis. Rep.*, **26** : 2044-0588.

Marlowe, K. and Hufford, L. (2007) *Syst. Bot.*, **32** : 208-226

Mishra, H.P. (1998) *J. Orna. Hort.*, **1** : 41-47.

Morinaga, Y., Fukushima, E., Kano, T. and Yamasaki, Y. (1929) *Bot. Mag. Tokyo*, **43** : 589.

Mukherjee, S., Singh, S., Jakhar, S. and Mukherjee, D. (2019) *J. Indian Bot.* **98**: 36-43.

Nikolova-Khristeva, N. (1973) *Nauchni Trudove Vissh Selskostopanski Institut Vasil Kolarvo,* **22** : 43-48.

Nishith, V., Ramesh, V., Nitesh, P. and Dilip, K. (2015) *Biosci. Trends,* **8** : 4090-4095.

Panchaude-Mattei, E. (1990) *Revue Horticole,* **309** : 29-31.

Patil, B.C., Hemla, N.B., Balaji, S.K., Jagadeesh, S.L. and Nalawadi, U.G. (2004) *J. Ornam. Hortic.,* **7** : 138-143.

Rai, R.N. and Agnihotri, J.P. (1970) *Sci. and Cult.,* **36** : 284-286.

Rajitha, B.S., Rao, S.G., Rao, Y.B.N., Deena, Singh and Singh, D. (1999) *J. Phytolo. Res.,* **12** : 107-108.

Rajvanshi, S.K., Dwivedi, D.H., Meena, M.L. and Ram, R.B. (2011) *Asian J. Hortic.,* **6** : 141-144.

Rao, N.K. and Shahid, M. (2011) *J. Applied Hort,* **13** : 65-67.

Robinson, M.L. and Schultz, U.E. (1995) *Proc. Florida State Hort. Soc.,* **107** : 194-196.

Salve, S., Bhalekar, S.G., Gawade, N. and Nilwarn, Y. (2019) *J. Pharmacogn. Phytochem.,* **8** : 1032-1034.

Sharma, M.K., Parmar, P.B. and Singh, A. (2015) *Int. J. Botany Stud.,* **5** : 1-6.

Sharma, S.S., Yamdagni, R. and Arora, R.K. (1989) *South Indian Hort.,* **37** : 291-294.

Singatkar, S.S., Sawant, R.B., Ranpise, S.A. and Wavhal, K.N. (1995) *J. Maharashtra Agri. Univ.,* **20** : 392-394.

Smith, F.F. (1967) *Agriculture Information Bulletin* No. 237, U.S. Department of Agriculture, p. 34.

Sowmyamala, B.V., Kumar, R.V., Nagaraj, H.T., Kulkarni, B.S., Hiremath, J.B. and Turamari, R.M. (2012) *Int. J. Applied Biol. Pharma. Technol.,* **3** : 53-57.

Tija, B. And Rose, S.A. (1988) *Proc. Flor. St. Hort. Soc.,* **100** : 181-182.

Trippi, V.S. (1965) *Phyton, Argentina,* **22** : 113-117.

Vaagdevi, A. N., Kumar, T. S., Prasanth, P., Kumar, A. K., Kumar, A. K., and Sunil, N. (2020) *Curr. J. Appl. Sci. Technol.,* **39** : 45-52.

Valiûnas, D., Samuitienë, M., Navalinskienë, M. and Davis, R.E. (2008) *Agron. Res.,* **6** : 109-118.

Yao, X.T., Ling, P.X., Jiang, S., Lai, P.X. and Zhu, C.G. (2013) *J. Oleo Sci.,* **62** : 329-333.

Yuan, M., Carlson, W.H., Heins, R.D. and Cameron, A.C. (1998) *HortScience,* **33** : 663-667.

V

W

Y

Z